Ultra-High Voltage AC/DC Grids

Ultra-High Voltage AC/DC Grids

Zhenya Liu

ELSEVIER

AMSTERDAM • BOSTON • HEIDELBERG • LONDON
NEW YORK • OXFORD • PARIS • SAN DIEGO
SAN FRANCISCO • SINGAPORE • SYDNEY • TOKYO
Academic Press is an imprint of Elsevier

Academic Press is an imprint of Elsevier
225 Wyman Street, Waltham, MA 02451, USA
The Boulevard, Langford Lane, Kidlington, Oxford OX5 1GB, UK

ISBN: 978-0-12-802161-3

British Library Cataloguing-in-Publication Data
A catalogue record for this book is available from the British Library.

Library of Congress Cataloging-in-Publication Data
A catalog record for this book is available from the Library of Congress.

For Information on all Academic Press publications
visit our website at http://store.elsevier.com

Typeset by MPS Limited, Chennai, India
www.adi-mps.com

Contents

Preface

More than a century since its inception, the world's power grid technology has seen rapid development, featuring higher voltage levels, more expansive interconnections, and stronger resource allocation capabilities. From the beginning of the 21st century, building a strong and smart grid—a modern grid system capable of allocating electricity across nations or even continents and flexible enough to adapt to renewable energy development and diverse needs—has become the direction and strategic choice for power grid development around the world. The construction of a strong and smart grid plays an essential role in promoting coordinated development of energy, economy, and environment.

1. **Security, efficiency, and cleanliness are important goals for energy development**

   Energy is a basic need to sustain economic and social development. The increasing global resource shortage and worsening climate change have imposed mounting constraints on energy development. How to take advantage of the new round of energy revolution to accelerate the strategic transformation of energy and maintain its secure, efficient, and clean supply is a common challenge faced by all.

   Energy is a multidimensional issue that involves policy, technology, market, and environment. To address it properly, energy has to be looked at with a "Grand Energy Vision." The development mode of energy needs to be transformed with a global vision, sustainable concept, strategic initiatives, and innovative technologies. Its development should be coordinated with that of the economy, society, and environment. Additionally, efforts are needed to promote the transitions of the following dimensions:

   - *Energy mix*: from high carbon to low carbon
   - *Energy utilization*: from extensive to intensive
   - *Energy allocation*: from local to global
   - *Energy service*: from unidirectional to intelligently interactive

   Eventually, a secure, efficient, and clean modern energy supply system should be achieved.

   Since the twenty-first century, the worldwide development and use of energy has been expanding, and renewable energies have been experiencing a continuous and rapid boom, presenting a significant trend of diversified energy mix. Electricity is a secure, quality, efficient, and clean secondary energy. Using electricity to replace the share of fossil fuels in end-use consumption of energy is already an obvious trend. The power grid is a basic means to transfer electricity, allocate resources, perform market transactions, and serve customers. To realize secure, efficient, and clean energy development, we must fully exploit the functions of the grid in transferring electricity and allocating resources, highlight the central role of electricity, and diversify the mix of primary energy sources. This is the only way to enable sustainable energy development. According to Jeremy Rifkin, author of *The Third Industrial Revolution*, Internet technology and renewable energy are booming to create a powerful "Third Industrial Revolution," which would have a profound influence on global development pattern. A strong and smart grid is a prerequisite for the third industrial revolution to be made possible. In recent years, the power grid has been recognized as a worldwide strategic focus for renewable energy development.

2. **Ultra-high-voltage grids are a well-justified means to realize a secure, efficient, and clean supply of energy**

Against this background, State Grid Corporation of China (SGCC), a backbone player in the energy sector, is facing strategic options and serious challenges regarding how to ensure electricity supply and how to maintain sound development of grids.

After carefully studying the nationwide demand on electricity and the geographical mismatch between resources and demand, SGCC proposed a "Grand Energy Vision" and a global perspective to promote technological innovations and concentrate efforts regarding transforming the development mode of energy and electricity. SGCC has launched a "One Ultra Four Large (1U4L)" strategy, which involves accelerating the construction of ultra-high-voltage (UHV) grids and promoting the intensive development of large coal power, large hydropower, large nuclear power, and large renewable power bases. The strategy focuses on "replacing coal and oil with electricity generated from remote sites" using electricity as a replacement to enable sustainability.

The construction of a strong and smart grid with the UHV network as its backbone is a fundamental solution to the underlying conflicts in developing energy and electricity, as well as a pressing task to meet the requirements of extensively developing large energy bases and renewables. The UHV grids will serve to deliver electricity from northwest China, northeast China, west Inner Mongolia, west Sichuan, Tibet, and some other countries to the load centers in east and middle China. As much as 76% of China's coal resources are located in the north and northwest regions, and 80% of water resources are located in southwest. The onshore wind energies are concentrated in northwest and northeast China, and in the north part of north China. However, more than 70% of China's energy demands come from the east and middle regions. With major coal explorations being shifted to the west and north, and with large-scale intensive exploitation of hydropower in the west, the development mode of electricity is being quickly transformed from local generation–demand balance to electricity supply by interconnected large grids. The increasing environmental pressure, high transportation costs, and land shortage have determined that east China is no longer an option for extensive deployment of coal-fired power plants, and that China has to find a strategy for its energy and electricity development, which features the transfer of massive electricity over long distances and optimization of allocating resources on a nationwide level. The transmission distance between large energy bases and load centers is usually 1000–3000 km, which is beyond the cost-effective transfer range of a traditional extra-high voltage system. Therefore, the electricity has to be transmitted in large capacity over long distances and consumed in a widespread region, and in an economic and efficient way. By connecting hydropower, wind power, and solar power to large grids featured by UHV grids, we can build a complementary energy allocation platform, boost the use of green and clean energies, and reduce carbon emissions. This is a practical and inevitable choice to build a beautiful China.

3. **Innovative practices and prospect of UHV grids**

The development of a UHV transmission system has been incorporated into the outlines of the 11th and 12th Five-Year Plans, and the *Outline of the National Program for Long- and Medium-Term Scientific and Technological Development (2006–2020)*, making it an important component of the national energy development strategy.

In January 2009, the Jindongnan–Nanyang–Jingmen 1000-kV UHV AC Pilot Project, which was independently developed, designed, and built by China, was completed and put into commercial operation. This UHV AC system has the world's highest voltage, largest capacity, and most advanced technologies. In July 2010, the Xiangjiaba–Shanghai ±800-kV UHV DC Pilot Project was completed and put into commercial operation. The commissioning of and stable operation of these UHV AC/UHV DC systems demonstrate the feasibility, safety, economy, and superiority of developing UHV transmission systems. Three years later, SGCC built two more UHV AC systems and two more UHV DC systems, which have been operating stably since they were commissioned.

In April 2011, the UHV AC pilot project won the China Industry Award and was recognized by CIGRE as "a great technical accomplishment." In February 2013, the research program of "UHV AC Transmission Key Technology, Equipment and Engineering Application" won the Grand National Award for S&T Progress. China owns proprietary intellectual property rights of this technology, and it is the only country that has mastered it. According to the IEC, China's success in building the UHV AC system with the highest voltage level and largest transfer capacity in the world is "a major milestone in the history of the power industry," which establishes China's leading position in the world's UHV power transmission field.

Accomplishments made in developing UHV grids are a combined result of the central government's foresight, support from various sectors of the society, and SGCC's efforts in independent innovation and hard work. China has achieved fruitful results in UHV grid development, including:

- Four test bases (UHV AC, UHV DC, high-altitude, engineering mechanics) and two R&D centers (bulk power system simulation, DC system design) that form a full-fledged research and testing system for UHV and bulk grid, and master core technologies of UHV DC transmission and manufacturing capability of set equipment

- Accomplishing a multitude of world-leading innovations in UHV AC/DC transmission and transformation, control and protection of bulk power system, smart grid, and clean energy integration
- Formulating a series of technical standards for UHV and smart grid, involving 363 corporate standards, 145 industry standards, 66 national standards, and 19 international standards. China's UHV AC voltage has been accepted as the international standard voltage
- Acting as secretariats for four sub committees of the IEC, including High-Voltage Direct Current (HVDC) transmission for DC voltages higher than 100 kV, and Grid Integration of Large-Capacity Renewable Energy (RE) Generation

China's success in the UHV transmission field has presented new milestones in the world's grid technology developments and has initiated a new era of grids featuring UHV as the highest voltage. In the power grid community, China has become a creative leader after its historic transition from being a follower, realizing qualifying terms such as "Led by China" and "Created by China."

By 2020, the 12 east and middle provinces (municipalities) in the service area of SGCC will need 350 GW of electricity from other regions. The installed capacities of wind power and solar power will reach 200 and 50 GW, respectively. These increase the requirements of the capabilities of grids to transfer massive energy over long distances and to accommodate intermittent clean energy. As planned, SGCC will build five horizontal and five vertical UHV AC lines and 27 UHV DC projects, which will allow the allocation of 450 GW of electricity over a wide region and the delivery of 550 GW of clean energies. These systems will annually accommodate 1700 TWh of clean energies, which will replace 700 million tons of raw coal, reduce emissions of carbon dioxide by 1400 million tons, and reduce emissions of sulfur dioxide by 3.9 million tons.

With the wider use of UHV transmission and other advanced technologies, the grid is not only a carrier of electricity in the traditional sense but also a powerful platform for conversion, efficient allocation, and interaction of energies. Through this platform, primary energies such as coal, hydro, wind, solar, nuclear, biomass, and tidal are converted to electrical energy for complementary, coordinated, and rational use. The platform may connect large energy bases to load centers and realize long-distance, massive, and efficient transport of electricity, thereby optimizing energy allocation in a larger scope. It may also be integrated with Internet, the Internet of Things, and smart mobile terminals to meet the diverse needs of customers and serve the development of smart home, smart community, smart transportation, and smart city. The power grid will be an energy-based Internet platform in future China.

A comprehensive review of and wider use of our accomplishments in the UHV grid are significant. From the perspectives of power grid history, status quo, and trends, this book analyzes the driving forces behind the development of UHV grids, discusses the features of AC/DC transmission and the strengths of UHV systems, and proposes schemes for building UHV grids in China as well as the rationale for these schemes. This book describes the technical innovations and engineering practices in UHV transmission. It also summarizes experience with and technical standards for UHV practices. It is my intent, through the systemic descriptions given in this book, for the readers to get a general idea about China's UHV grid development. Comments and suggestions in this regard from more authorities and experts are welcome. This way, we will be better equipped to advance the power industry in a comprehensive, coordinated, and sustainable manner, to contribute to socioeconomic development, and to realize the Chinese dream of rejuvenating the nation.

Zhenya Liu

September 2014

CHAPTER 1

Grid Development and Voltage Upgrade

Chapter Outline

1.1 Grid Development and Interconnection

1.1.1 Basic Concepts of Grid

An electric power system is a system that involves generation, transmission, distribution, and consumption of electricity, as shown in Figure 1.1. An electrical grid is used to connect power plants and users [1].

The grid mainly consists of transmission and distribution networks. The transmission network sends the electricity produced by power plants to centers/regions of load or

Ultra-high Voltage AC/DC Grids.
DOI: http://dx.doi.org/10.1016/B978-0-12-802161-3.00001-9

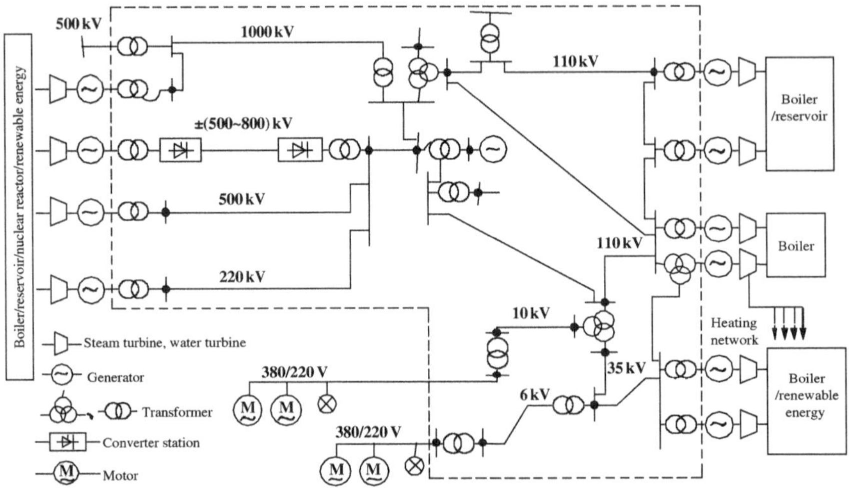

Figure 1.1
Schematic diagram of electric power system (the dotted box represents a grid).

exchanges electricity between neighboring grids through interconnection. The distribution network receives electricity from the transmission network and distributes it to urban or rural areas, including industry, agriculture, business, residence, or other special consumers.

Electricity can be transmitted and supplied in two forms: alternating current (AC) transmission and distribution, and DC transmission. The AC transmission and distribution is completed by the step-up substation, step-down substation (including primary and secondary equipment), and connected transmission lines. Transmission equipment is connected to form a transmission network, and distribution equipment is connected to form a distribution network. The DC transmission is achieved using DC transmission lines and equipment in converter stations (including primary and secondary equipment). Equipment used in substations/converter stations includes primary equipment such as transformers, reactors, capacitors, circuit breakers, grounding switch, disconnectors, surge arresters, voltage transformers, current transformers, bus, and secondary equipment such as relay protection, surveillance, monitoring and control, and power communication systems. Transmission equipment mainly includes conductors, towers, insulator strings, and grounding wires (including optical fibers). DC equipment includes converter valves, converter transformers, smoothing reactors, DC filters, DC disconnectors, grounding switches, bypass switches, DC circuit breakers, DC-measuring devices, and DC surge arresters.

Voltage classes of transmission networks are usually classified into high voltage (HV), extra-high voltage (EHV), and ultrahigh voltage (UHV). As internationally recognized for AC transmission networks, HV usually refers to voltage from 35 to 220 kV, EHV refers to voltage from 330 to less than 1000 kV, and UHV refers to voltage 1000 kV or higher.

For DC transmission, HVDC usually refers to ±660 kV or lower; UHV DC refers to ±800 kV or higher [2,3].

China has established two AC voltage series, namely, 1000/500/220/110 (66)/35/10/0.4 kV and 750/330 (220)/110/35/10/0.4 kV, and a DC voltage series of ±500 (±400), ±660, and ±800 kV. In China, HV networks refer to 110 and 220 kV, EHV 330, 500, and 750 kV, and UHV 1000 kV, which act as the backbone of the transmission network, and UHV DC systems are directly or hierarchically connected to the 1000/500 kV network. China has already constructed 1000 kV UHV AC and ±800 kV UHV DC projects, and it is currently working on ±1100 kV DC transmission technologies. The 1000-kV AC voltage proposed by China has been accepted as an international nominal voltage for UHV AC.

An electrical grid is a network to transfer electricity, like a type of logistics. The basic function of a grid is to transfer electricity from the generation side to the demand side.

Electricity is a secondary energy converted from primary energy sources, such as hydro energy (hydropower generation), thermal energy (thermal power generation), nuclear energy (nuclear power generation), wind energy (wind power generation), chemical energy (battery), and optical energy (photocell, solar cell). Electricity has several features: (i) it cannot be stored in large quantities, because generation, transmission, distribution, and consumption occur instantly; and (ii) generation and consumption must maintain a real-time balance; otherwise, the security, continuity, and quality of electricity will be undermined. Given the high synchronism of electricity, the power source, grid, and consumer must be coordinated to ensure security and quality. Notably, the grid, as the network to transmit and distribute electricity, is a hub that connects dispersed power sources and consumers. The security and quality of electricity must be ensured by properly dispatching, operating, and controlling the grid. The electrical grid plays an irreplaceable role in maintaining the power system security; therefore, it is an essential part of the energy supply system and an infrastructure indispensible to the modern economic development and social progress.

The development and wide use of smart grid technologies are bringing fundamental changes to the configuration and function of modern grids. Originally from a single-purpose means for transferring electricity, the grid is evolving into an intelligent infrastructure platform that is a powerful conduit for allocating energy resources. It is also a significant contributor to energy security, energy conservation and emission reduction, and economic and social development. In the future, the grid will become a market platform that optimizes energy allocation, a service platform that meets diverse needs of customers, and an infrastructure platform that maintains the national energy security.

The grid is a key component of a power system. The power system consists of generating equipment such as boiler, reactor, steam turbine, hydro-turbine, and generator; transforming, transmitting, and distributing equipment such as transformer and power line; consuming equipment such as motor, electric stove, and lamp; and measurement, protection, control, and energy management systems. Combined, these make the electrical power system a unified whole and a huge and complex subject to study [4].

For an electrical power system, the security and stability are the two fundamental conditions needed to operate smoothly. Security means that all the equipment in the system must operate within their permitted amplitude and limit for current, voltage, and frequency. Insecurity may lead to damage of the equipment. Stability means a state in which the power system continuously delivers electricity to the consumers [5].

In electrical power system operation, three stability requirements must be met at all times: power angle stability, frequency stability, and voltage stability. Losing power angle stability will cause system oscillation, drastic and periodic fluctuation of voltage at pilot nodes, and drastic and periodic fluctuation of current and voltage in transmission equipment. The system will no longer be able to feed the load. If poorly handled, the loss of power angle stability may cause a widespread outage. Losing frequency stability will lead to system frequency collapse, and ultimately a complete system outage. Losing voltage stability will result in system voltage collapse, causing blackouts in affected areas. Therefore, security and stability must be considered throughout the engineering, construction, dispatching, and operation of the power system.

1.1.2 History of Grid Development

In 1875, the world's first fossil-fired power plant was built in Paris, which marked the advent of the age of electricity. In 1882, Edison built the world's first commercial power station (660 kW, transmitted power over a 1.6-km-long direct current cable at 110 V) in New York, turning electricity into a commodity. According to Ohm's law, to transfer the same power, the electricity be sent over a long distance with a low loss only by increasing the voltage and reducing the current. Considering safety, the user-side voltage should not be too high, so the voltage has to be decreased after the electricity is transmitted to the users so they could consume electricity safely. If the transmission voltage is not upgraded, then the transmission loss would increase linearly when the distance further expands. With the technologies available at that time, it was difficult to increase or decrease the voltage for DC transmission; therefore, the line loss was huge for long-distance transmission. Nikola Tesla invented the AC technology, which uses voltage that can be easily and economically increased or decreased. This technology makes it possible to increase the transmission voltage and transmit electricity over longer distances through transmission lines. After arriving at the destination, the electrical voltage can be decreased to ensure its safe use [6]. In 1885–1886, Westinghouse constructed the first AC transmission system. In 1895, it built a 35-km transmission line from Niagara Falls Generating Station (with 3×3675-kW hydroelectric generating units) to Buffalo. The electricity was used for lighting and traction, and the technical barriers encountered by the power industry were overcome.

Grid development in different periods shows different technical and economic features, of which voltage class, grid size, generator unit capacity, and operating technology are the most salient. Based on these features, the development history of grid can be divided into several stages [1].

In the decades from the late nineteenth century to the mid-twentieth century, the electrical grid was featured by the use of AC in generation, transmission, and distribution. This was the initial stage that lasted until the end of World War II. At this stage, the generator unit capacity was not more than 200 MW; AC transmission at the voltage up to 220 kV was dominant. The grid was mainly small, urban, and isolated. Operation technologies were still in their infancy, with grid failure and resultant outages regularly seen. Following are the landmark events of the worldwide grid development at this stage:

1. A 35-km transmission line from Niagara Falls Generation Station (with 3×3675-kW hydroelectric generating unit) to Buffalo was built in 1895, which established the dominant role of AC transmission.
2. The 132-, 230-, and 287-kV lines were first built in the United States in 1916, 1923, and 1937, respectively.
3. The first 60-MW turbine generator was built in 1918, and the first 200-MW turbine generator was built in 1929, both in the United States.
4. The Dnieper Hydroelectric Station was put into operation in 1932 in the Soviet Union, with a unit capacity of 62 MW; in 1934, the Grand Coulee Dam Hydroelectric Power Plant was put into operation in the United States, with a unit capacity of 108 MW.

The development of electricity in China almost kept up with the world [7]. In 1879, the first electrical lamp was used for lighting in the Shanghai International Settlement. Then, in 1882, the first electric utility—Shanghai Electric Company—was set up in Shanghai. Since then, China opened the first page of its electric power industry. But the industry was sluggish before 1949. Compared with the rest of the world, it was far more backward. By the time the PRC was founded, the national generation capacity was only 1850 MW, with an annual generation of 43×10^8 kWh. Electricity was available in only a few large cities, and the grids were small but at various voltage classes. In 1908, the 22-kV Shilong Dam Hydropower Station–Kunming line was built. In 1921, the 33-kV Shijingshan Power Plant–Beijing line was built. In 1933, the 44-kV outgoing line of Fushun Power Plant was built. In 1934, the 66-kV Yanbian–Laotougou line was built. In 1935, the 154-kV Fushun Power Plant–Anshan line was built. In 1943, the 110-kV Jingpohu Hydropower Plant–Yanbian line was built. At that time, except the few 154-kV lines in the northeast, the highest voltage was only 77 kV in Beijing, Tianjin, and Tangshan, and only 33 kV in Shanghai.

After World War II, the global economy developed rapidly. Huge electricity needed for industrial production and cheap fossil energies contributed to the boom of the power industry, with unprecedented progress and innovations in grid technologies. The electrical industry was going toward higher voltage, larger generator units, and larger grids. The boom created vast benefits from economies of scale and met the increasing needs of social and economic development.

From the middle to late twentieth century, grids were expanded to form large interconnected systems. The generator unit capacity was increased to 300–1000 MW, and 330-kV (and higher) EHVAC and HVDC transmission systems emerged.

Grid interconnection in Europe and North America began to flourish in the 1950s. By the 1980s and 1990s, large cross-border and cross-region interconnections covering a wide area with large amounts of the electricity exchange took shape. In the 1950s and 1960s, low-cost, highly efficient, large-capacity generation units were introduced, which required large-capacity power grids to accommodate their operation. In the 1970s, a global oil crisis occurred, resulting in greatly increased cost of power generation. In such case, the interconnection of power grids was required to reduce the proportion of oil-based generation. In addition, more coal-based and hydro-based power sources were developed, nuclear power was introduced, the distance between the new power sources and existing load centers was increased, and the various power sources needed coordination. These also required expanding the interconnected capacity and size. In the 1980s, the world economy kept growing for many years; however, the development varied among regions and countries, and so did energy supply and demand. As environmental constraints became stringent, the risk of investing in power source construction increased. As a result, the commissioned generation capacity failed to match the demand. Many countries promoted grid interconnection to achieve higher economic efficiency. In the 1990s, the development of new remote energy bases led to long-distance, large-capacity transmission lines and interconnected lines. Landmark events at this stage include:

1. In 1952, the first 380-kV EHV transmission line was built in Sweden. It is 620 km long and has a capacity of 450 MW, sending electricity produced from Harsprånget Hydroelectric Power Station in north Sweden to the south.
2. In 1954, a 345-kV line was built in the United States, which sent out electricity from Columbia River's hydropower.
3. In 1954, an HVDC transmission line with voltage of ± 100 kV, rated power of 20 MW, and 96 km of submarine cables was constructed between native Sweden and Gotland Island. It uses mercury arc valve commutation technology and is the world's first industrial HVDC project.
4. In 1956, the 400-kV Kuybyshev–Moscow transmission line was put into operation in the Soviet Union, which is 1000 km long. The voltage of the project was upgraded to 500 kV in 1959, and it was the world's first use of 500 kV in transmission.

5. In 1965, the world's first 735-kV transmission line was completed in Canada, which transmitted power from the Manicouagan-Outardes hydroelectric generating complex to Quebec City and Montreal.
6. In 1967, a pilot 750-kV transmission line was built in the Soviet Union; in 1984, a 750-kV line from the Soviet Union to Poland was constructed.
7. In 1969, a 765-kV transmission line was built in the United States.
8. In 1972, the Eel River back-to-back DC project was completed in Canada, which was the world's first DC project to completely use thyristor valves. The DC voltage was 80 kV, and the rated capacity was 320 MW.
9. In 1985, the Soviet Union built an 1150-kV power transmission system to send electricity from the thermal power base in Kazakhstan to Europe, which, because of the collapse of the Soviet Union, was forced to enter segmented operation at a reduced voltage.

After the founding of the PRC in 1949, the voltage grade was unified to form a series. In 1952, a 110-kV transmission line was built with China's own technologies. After this, a 110-kV Beijing–Tianjin–Tangshan transmission network began to emerge. In 1954, the 220-kV Fengman–Lishizhai line was built; afterward, a number of other 220-kV lines, like Liaoning Power Plant–Lishizhai and Fuxin Power Plant–Qingduizi, were built. These lines soon formed the 220-kV backbone network in northeast China. In 1972, the 330-kV Liujiaxia–Tianshui–Guanzhong line was completed, following which the 330-kV backbone network was gradually set up in northwest China. In 1981, the 500-kV Yaomeng–Shuanghe–Fenghuangshan (Wuchang) line was built. To send the electricity produced by Gezhouba Hydropower Station, in 1983, two 500-kV circuits, Gezhouba–Wuchang and Gezhouba–Shuanghe lines, were built, which formed the 500-kV backbone network of central China. Also, 500-kV backbone networks were gradually constructed in north, east, northeast, and south China. In 1989, the ±500-kV Gezhouba–Shanghai DC transmission line was completed, which connected central China and east China network. In September 2005, China's first 750-kV pilot transmission project (Lanzhou East–Guanting (Qinghai)) was completed. After that, the 750-kV Lanzhou East–Baiyin–Yinchuan East transmission project was completed in late 2007, and the Laxiwa Hydropower Station transmission project was completed in early 2008. Based on these, the 750-kV backbone network in northwest China gradually formed. In December 2008, the 1000-kV Jindongnan–Nanyang–Jingmen UHV AC pilot project began a test run. The project connected the north China and central China network with the UHV. In 2010, Xinjiang was connected to the northwest China main grid via a 750-kV line. At the end of 2011, the Qinghai–Tibet grid interconnection project began commercial operation. In 2010, the ±800-kV Xiangjiaba–Shanghai UHV DC demonstration project and ±800-kV Yunnan–Guangdong UHV DC project were completed. In 2012, the ±800-kV Jinping–Sunan UHV DC transmission project was completed.

Along with social progress and technological improvement, the distance between the power source and the load center keeps increasing. To achieve electricity transmission over a longer distance at larger capacity and higher efficiency, the grid's voltage class should become higher, and the grid should become larger. As we look back at the history of grid development (more than a century), this is indeed the history of voltage becoming higher and grid size becoming larger.

1.1.3 Status of Grid Interconnection [8]

The grid interconnection in North America and Europe began in the 1920s, and had its prime time in the 1950s. By the 1980s–1990s, widespread cross-border and cross-region interconnected power systems took shape.

The North American interconnection is composed of the US grid and part of the grids in Canada and Mexico, including four interconnections, namely, the Eastern Interconnection, the Western Interconnection, the ERCOT Interconnection, and the Quebec Interconnection. The Eastern Interconnection is the largest of the four, with an installed capacity of approximately 600 GW and a maximum load of approximately 500 GW. The interconnection covers approximately 5.2×10^6 km^2, serving a region from Nova Scotia in Canada to Florida in the United States. The Western Interconnection is the second largest and is connected to the Eastern Interconnection via several DC back-to-back links. The ERCOT Interconnection is the only independent AC interconnection that does not cross states on the North American Continent. It provides services to most parts of Texas and is connected to the Eastern Interconnection via several DC back-to-back links. The Quebec Interconnection is located in Canada and is connected to the Eastern Interconnection via the DC link. According to statistics in 1998, there were 79 power transmission lines between the United States and seven provinces of Canada. The AC interconnection lines were operated mainly at 500, 230, and 115 kV. In addition, there was also a multiterminal EHVDC interconnection. The United States and Mexico had 27 transmission lines, most of which were AC.

The main European grids include five cross-border synchronous grids, namely, the European continent grid, the Nordic grid, the Baltic Grid, the British grid, the Irish Grid, as well as two independent grids, the Iceland grid and Cyprus grid. Among them, the five synchronous grids form one of the world's largest interconnections—the synchronous grid of Continental Europe. It supplies 730 million people in 24 countries, has a production capacity of 670 GW, and covers an area of 4.5×10^6 km^2. The major impetus to further promote interconnection is that the countries concerned wish to achieve a higher degree of interconnection to take full advantage of a larger grid. After establishing connection to the Romania and Bulgaria grids in Western Balkans, the European interconnection shows a trend to include Russia, Ukraine, Belarus, and Moldova. On October 10, 2004, the Western Europe and Southeastern Europe grids were connected via Croatia using HV transmission lines.

This connection helped improve the stability of electricity supply in Southeastern Europe and provided a channel for grid expansion to Asia and trade of energy [4].

In addition to North America and Europe, Russia, Brazil, and India also set up nationwide interconnections.

The Russian unified power system is composed of 69 regional power grids that form seven cross-region grids, including the eastern, the Siberia, the Urals, the Middle Volga River, the southern, the central, and the northwestern grids, covering Russia's 79 states by 10,700 transmission lines with 110–1150 kV (of which the 1150-kV lines are currently operated at 500 kV). Except the eastern grid, the other six cross-region grids are already synchronized. Grids that are in synchronous operation with the Russian unified power system also include the grids in Azerbaijan, Belarus, Georgia, Kazakhstan, Moldova, Mongolia, Ukraine, Latvia, Lithuania, and Estonia. The grids of Kyrgyzstan and Uzbekistan in Central Asia are synchronized with the Russian system through the Kazakhstan grid. The synchronous grid formed by these 13 interconnected grids is known as the CIS–Baltic interconnection. The Finnish grid is asynchronously connected to the Russian system through the Vyborg DC converter station.

Brazil's power system consists of the southern grid (including the middle and western), the southeastern grid, the northern grid, and the northeastern grid. These regional grids are interconnected through 500- and 345-kV tie lines to form a national grid, serving 170 million people with a production capacity of more than 100 GW and covering an area of 5×10^6 km^2. With abundant hydro resources, Brazil's hydropower accounts for 71.7% of its total installed capacity. Although the hydropower resources are geographically uneven and the loads are concentrated in the south and southeast, Brazil manages to optimize the allocation of these resources through nationwide interconnection, even with a weak main network. The southern and southeastern grids are synchronously connected through the Itaipu 750 kV AC lines, the northern and northeastern grids by a single-circuit 500-kV AC line and the northern and southern grids through a single-circuit 500-kV AC line. Through these interconnections, cross-basin balance is achieved in Brazil.

India has five regional grids, namely, northern, western, southern, eastern, and northeastern grids. All the grids, except for the southern gird, which is asynchronously connected with the eastern and western grids, are synchronized. The northern grid is the largest of the five, covering approximately 31% of India's territory. Generators are concentrated in the eastern part of the northern grid, whereas the loads are mainly located in the middle and the west part. As a result, long-distance transmission is required to transport electricity. To meet the demand for high-capacity transmission, India built a $\pm$500-kV, 1500-MW HVDC transmission line that runs in parallel with the 400-kV AC transmission grid to supply the 220-kV grid.

Since 1978, the rapid development of China's economy drove the increasing demand for electricity. To meet the large-scale production and load growth, the Chinese grids have been constantly expanded, from city grid, provincial grid, regional grid, and gradually to a national interconnection (excluding Taiwan). In the 1980s, Hong Kong and Macao were interconnected with the China Southern Power Grid. In the late 1990s to the early 2000s, the Three Gorges Project was taken as an opportunity to accelerate construction of a nationwide network. This contributed to build the interconnection between central China and east China grids and central China and southern China grids. In addition, northeastern and north China grids, central China and the north China grids, and central China and northwestern China grids were also connected. Isolated provincial grids such as Sichuan, Chongqing, Shandong, Fujian, and Hainan have also been incorporated into regional networks. At the end of September 2013, China had set up six synchronous networks, namely, the north-central China, the east China, the northeast China, the northwest China, the south China, and the Tibet grids, forming a nationwide interconnection, except in Taiwan. The cross-province and cross-region transmission increased from 1326×10^8 kWh in 2003 to 6940×10^8 kWh in 2012. In addition to optimization of resource allocation, a large nationwide interconnection also helps significantly improve the mutual support capability of the grids.

1.1.4 Grid Development Trend

1.1.4.1 Continually enhancing capabilities of the grid for optimal allocation of energy resources

The development of cross-border and cross-region interconnection provides the physical conditions for power exchange, promotes the growth of an integrated electricity market, and enables the optimized allocation and use of energy resources. Considering that availability of energy resources varies among countries and regions, the worldwide need for optimized allocation is even stronger. For this reason, cross-border and cross-region interconnections in the global context are booming to cover more areas.

The European Network of Transmission System Operators for Electricity (ENTSO-E) consists of 34 European countries and 42 transmission system operators (TSO). With the European interconnection becoming more extensive, the capability of optimizing resource allocation is enhanced and the power exchange among the countries gradually increases. In 2009, the power exchange within ENTSO-E was 3757×10^8 kWh. By 2012, the figure increased to 4265×10^8 kWh. To further expand the exchange, the Western Europe grid incorporated the Central Europe grid and part of the Eastern Europe grid. More countries have been incorporated in the interconnection in the recent decade. The Nordic grid has planned five major projects to strengthen grid configuration and to eliminate transmission congestion. It also worked out an ambitious plan to connect the continental European grid and the CIS–Baltic grid via an HVDC interconnection. The Mediterranean interconnection is expanding, with the voltage increasing from 220 to 400 kV, and the grid configuration being strengthened.

Europe depends highly on fossil fuels, which offer 80% of its need for primary energy. To ensure energy security and to address climate change, many scholars, research institutions, and government agencies have proposed to establish a European "Super Grid." As early as 2003, some scholars proposed the idea of pan-Mediterranean cooperation on renewable energies [9], pointing out that the connection of Europe's offshore wind power, North Africa's and eastern Mediterranean coast's wind power, and solar power by an HVDC system to the European grid would reduce CO_2 emissions in Europe and would help solve the problems of socioeconomic development and employment in North Africa and the eastern Mediterranean coast. Moreover, the interconnection and power trade would contribute to the development and relationship of the Mediterranean countries. In 2006, Airtricity proposed the European Offshore Super Grid program [10], which was an HV submarine transmission network and could eventually cover the Baltic Sea, the North Sea, the Irish Sea, the English Channel, the Persian Gulf, and the Mediterranean. In 2007, the Trans-Mediterranean Renewable Energy Cooperation presented the "DESERTEC" concept [11], which involved establishing a centralized solar power plant, a PV power plant, and a wind farm, covering 1.7×10^4 km^2 in the Sahara desert. The electricity produced from this complex would be sent through HVDC cables to Africa and Europe. This would provide large amounts of electricity to North Africa and the Middle East countries, along with desalination. The project would further supply 15% of Europe's electricity need.

The interconnection of Southern African countries is expanding the number of connected countries and increasing the voltage class. The six countries of the Gulf Cooperation Council (Saudi Arabia, Qatar, Kuwait, Bahrain, the United Arab Emirates, and Oman) are also performing a three-stage interconnection plan. China and Russia have built a DC interconnection project. On this basis, China is actively studying and planning further cross-border interconnection and electricity exchange with Russia, Kazakhstan, Mongolia, Vietnam, Myanmar, and North Korea.

Central Asia and the Far East possess rich energy resources, whereas in the future, Europe would see an energy gap that would give rise to the need for cross-continent electricity transmission. The cross-continent interconnection of Asia and Europe will be a trend for future grid development. In 2011, SGCC proposed the concept of a large Asia–Europe intercontinental transmission channel that would use UHV transmission as a safe, economical, efficient, and sophisticated technology to optimize the allocation of global resources in a wider scope. This concept would offer a feasible solution to the electricity shortage in Europe because of low-carbon development and abandonment of nuclear power.

Optimizing the allocation of energy resources is the prime mover for China to further expand its interconnection. By 2020, China's total installed power capacity and electricity consumption are expected to reach 2000 GW and 8.4×10^{12} kWh, respectively, doubling the figures of 2010. This represents a huge increase. However, the new large coal, hydro, and wind power bases to be developed are mostly located in the western and northern regions, approximately 1000–3000 km away from the load centers. As a result, robust UHV grids are needed for long-distance, large-capacity transmission. The Three-Hua UHV synchronous grid (the large synchronous grid connecting North China, Central China, and East China Grids) will play a central role in this, with the northwest and northeast grids being the sending end. In this process, AC and DC systems should be coordinated so that a robust, safe, efficient, economic, and eco-friendly modern grid could be created for optimized allocation of energy across the country.

1.1.4.2 Continuous improvement in system security and reliability

Large generating units and large power stations have to be connected to large grids for secure and safe operation and full exploitation of their benefits. Interconnection enlarges the grids, making the conditions available for use of large generating units and power stations. For instance, the interconnections between Fujian and east China grids, Sichuan (Chongqing) and central China grids, and east China and central China grids enabled the Houshi, Ertan, the Three Gorges, and other large power plants to make full use of their capacity and operate securely and reliably.

Interconnected grids provide the capability of emergency support when an incident occurs, and, with increased system capacity, the impact of removing generators on the grid after a failure is reduced, which enhances the security of the power system. For example, the interconnection of the southern and southeastern grids in Brazil narrowed the supply gap and created huge benefits. Another example is the interconnection of the five southern provincial grids in China. When Guangdong grid at the receiving end suffers a fault such as tripping of a large unit, the western sending end will be able to provide emergency support in time.

On July 30, 2012, and July 31, 2012, India had two successive blackouts. Studies [12,13] show that the root cause is the weak network as a result of inadequate investment in generation and grid to match the demand increase. There are a lot of 400/220-kV electromagnetic loop networks in India's regional grids, which are prone to instability incidents; in the cross-region main grids, interconnection of the five regions consists of just 16 400-kV AC circuits and five DC circuits (totaling 5000 MW), and the highest operating voltage is only 400 kV. These lead to the low capacity and poor security. Insufficient capacity and old equipment are common problems for the distribution networks. Therefore, a local fault may very likely cause cascading trips and a widespread blackout. Also, there are no robust backbone grids that link the regional networks, and this results in poor support among the regional networks, extended blackout, and slow recovery. In fact, as early as around 2007, India already planned to use UHV AC and UHV DC technologies to solve these problems and increase the security and stability of the grids [14]. However, the plan was not implemented, and the grid's "bottleneck" remains unresolved, which is an important contributor to the two major blackouts.

Because the voltage class increases and the grids become stronger, China's instability incidents decreased from an annual average of 19 times in the 1970s to two times in 1997; since then, no grid-wide collapse has occurred. To accommodate longer transmission distance and larger capacity, China will build 1000-kV transmission grids to ensure secure and reliable grid operation. The north-central-east China 1000-kV transmission grids are more closely connected, having an equivalent electrical distance equal to only one-quarter of a 500-kV system. The configuration of the grids has been improved from the "chain" structure (northeast-north-central China grids) to a stronger "mesh" one, which also has established DC links to the northwest, northeast, and south China grids. As expected, by 2020, the north-central-east China grids will have an installed capacity of 1000 GW and supply an area of 3.2×10^6 km^2, approximately the same size as the world's existing large synchronous grids like the Eastern Interconnection in North America, but with a shorter electrical distance, stronger configuration, and much higher security.

1.1.4.3 Future grid development

Since the end of the twentieth century, a new energy revolution has begun worldwide. All countries are facing major challenges in transforming and upgrading their energy and electricity development. Accommodating large-scale renewable generation and promoting smart grid operation are now the trends of grid development [1].

The main features of future power grids are: (i) in power source mix, nonfossil clean energies will account for a large share (e.g., China is supposed to reach 40%−50%) and large, backbone power sources will be coordinated with distributed generation; (ii) in grid configuration, nationwide (or larger) backbone transmission grids, regional grids, and distribution networks will be coordinated; (iii) large-capacity, low-loss, eco-friendly transmission methods (such as UHV overhead transmission, superconducting cable transmission, gas-insulated pipeline transmission) will be used; (iv) intelligent grid scheduling, control, and protection; and (v) intelligent, bidirectionally interactive electricity distribution and consumption system.

Future grids will inherit some of the features of existing grids and will have further innovations in the construction of large-scale power sources and national backbone grids, as well as digital, information-based, intelligent control, and dispatch of grids. The two major features of future grids are centralized and distributed integration of renewable power and intelligent operation control and use of electricity. Realizing these features poses unprecedented challenges for development of models of power source and grid, innovation in power equipment, grid operation control, simulation analysis, intelligent use of electricity, and two-way interaction between the user and grid. These challenges can be categorized into two aspects, namely, equipment and system integration.

1. *Equipment technologies*: Efficient, energy-saving, and eco-friendly equipment are the basis for the future grids. Such equipment mainly includes cost-efficient generation equipment for renewable energy sources (wind, solar, biomass, etc.), new and efficient transmission and distribution technologies and equipment (UHV transmission, superconducting transmission, underground transmission, and intelligent green appliances), new power electronic equipment and technologies, large-capacity and distributed energy storage technologies and equipment, and various sensors and information networks.
2. *System integration technologies*: Systematic integration of advanced information and communication technologies, optimization and control system theories and technologies, and new electricity market theories and technologies are the basis for the construction and safe and economic operation of the future grid. Specifically, these technologies include integration of large-capacity centralized and distributed renewable energy; intelligent energy management and control based on advanced sensing, communications, control, computing, and simulation technologies covering all kinds of power sources and loads; modeling and analysis techniques for the next-generation grid; reliability assessment and security protection for energy flow and information flow related to grid operation; electricity market theory, pattern and operation mode that support extensive interaction of various power sources with users; asset management and integrated service system; intelligent electricity distribution and consumption system, demand-side response, and flexible integration of distributed generation, electric vehicles, and energy storage devices; and integrated energy, electricity, and information service system covering urban and rural areas.

Compared with the existing grids, the future grids will face very different missions:

1. Transmission network that is capable of accommodating and transferring bulk electricity from renewable sources
2. Flexible, efficient energy allocation and supply systems and established user demand–response mechanisms. Distributed generation and energy storage will change the end-use model, and electricity will flow bidirectionally between the grid and the user. These will greatly improve the end-use efficiency of energy.
3. Safe and reliable smart energy networks that offer a very high reliability and basically mitigate the risk of widespread blackout
4. Internet of Things and integrated service system for energy, power, and information covering the urban and rural areas, where multiple networks are combined to act as a carrier for both energy and information

These four missions are also the major requirements for constructing the future grids. Key technologies required are analyzed from six aspects, namely, generation, transmission, distribution, storage, consumption, and support, as given in Table 1.1 [6].

Table 1.1: Key Technologies Required for Development of Future Grids.

Generation technologies	Wind, solar, biomass, new nuclear technologies, and others
Transmission technologies	Smart transmission grid, large-capacity DC circuit breaker and DC network, high-temperature superconducting transmission, large-capacity cable transmission, gas-insulated pipeline transmission
Distribution technologies	Smart distribution grid, microgrid
Storage technologies	Superconducting magnetic energy storage, flywheel energy storage, pumped storage, compressed air energy storage, electric vehicle technologies
Consumption technologies	The Internet of Things, intelligent use of electricity
Supporting technologies	Electricity market, grid operation control, large-scale simulation

Based on these basic concepts and forecast of new grid technologies, some research institutions in Europe, America, and other developed countries proposed future grid development goals that are of strategic significance, of which the "Grid 2030" plan put forward by the United States Department of Energy [15] is typical. In the plan, the United States proposes to build a national electricity backbone network to connect the east coast, west coast, Canada, and Mexico. The backbone can serve to balance power supply and demand at the national level, expand the scope of electricity service, optimize allocation of resources, and enable efficient transmission, reducing line loss through the use of seasonal, regional, and climatic diversity across the country and demand-side management. In addition, the United States also had a proposal for a Unified National Smart Grid [16], which involves the construction of a smart grid that links all regions within the United States and realizes long-distance, bulk, and economic delivery of wind, solar, and other types of power.

China's alternative energies such as wind and solar power are developing rapidly, mainly in north China, northeast China, and northwest China. Their integration, delivery, and accommodation are all facing some urgent problems. The outbreak of serious haze in the east and central China is, in part, a manifestation of the unreasonable energy development pattern and structural contradictions over a long period of time. There is an urgent case to accelerate the development of electricity as an alternative by increasing its proportion in the end-use energy mix and enhancing the overall electrification level. As the UHV transmission technologies become mature, there is an urgent need to accelerate the construction of a flexible, efficient, and robust smart grid in which UHV grids act as the backbone, grids at all levels are well coordinated, and advanced technologies are fully utilized. In addition, the market platform for optimized allocation of energy resources is required to meet the need of economic and social development. In the next 20 years, China is expected to continue its rapid development; the tasks for developing grids will be to extend and expand the existing grids, and the function of grids as a resource allocation platform will be greatly enhanced. From the user's perspective, the power quality and service levels will be further improved.

1.2 Driver for UHV Transmission Development and Its History

1.2.1 Drivers for Developing UHV Transmission [17]

1.2.1.1 Meeting the requirement for bulk, long-distance, and efficient delivery of power

Many countries and regions face mismatches between energy resources and load centers, for example, the load centers have fast economic development and, hence, high and rapidly increasing load, but they often lack primary energy. However, the major primary energy bases tend to have relatively low demand growth or low consumption. This imbalance is a result of the geographical distribution of energy resources and historical socioeconomic development. Naturally, this situation calls for bulk, long-distance, and efficient delivery of electricity.

The Soviet Union was the first country in the world to construct a UHV AC transmission project. Siberia in the Soviet Union is very rich in water and coal resources, and Kazakhstan is also endowed with a lot of coal. As such, approximately 80% of the Soviet Union's primary energies for generation are located in the east, whereas its load centers are located in the European area of its territory. From the 1980s, the construction of large energy bases, coupled with the country's vast territory and extensive grid coverage, and the imbalance between energy and loads, gave rise to the need for UHV transmission. From 1980, the Soviet Union began to build a 1150-kV UHV AC transmission project to connect Siberia, Kazakhstan, and Ural for sending electricity from the east to the load centers in Ural and the European territory. The project began official operation at rated voltage in 1985, but it now operates at a reduced voltage.

More energy in the west and north and less energy in the east and south are the features of China. The mismatch between energy bases and load centers is significant. China is now at a critical period of rapid economic growth; its electricity demand will continue to increase quickly. The load centers will remain in the east and central regions for a long time in the future, but development of coal resources is being gradually shifted westward and northward, and hydropower resources are being shifted toward the southwest. Additionally, wind, solar, and other emerging energy resources are also mainly located in the west and north. Given these conditions, the energy flow in the future will see further increases in scale and distance, leading to challenges for large-scale, long-distance, high-efficiency power delivery. The distance between the large energy bases and load centers reaches 1000–3000 km, which is beyond the economic distance of traditional EHV transmission lines. The regional imbalance in electricity production and consumption will be more prominent, the pressure on electricity transmission will be higher, and the need for cost-effective bulk delivery and consumption in a wider scope will be more urgent.

Compared with EHV, UHV transmission features larger capacity, longer distance, and higher efficiency, and is capable of meeting the requirements for bulk, long-distance, cross-region transmission. Therefore, it has to be developed at a faster pace to enable the intensive development of large energy bases and reliable delivery of electricity, and to adapt to the needs of securely integrating and efficiently accommodating bulk clean energies.

UHV transmission consists of UHV AC and UHV DC, which vary in functions and features. AC transmission is mainly used to build strong networks at various levels and interconnection routes for which intermediate terminals can be provided along the transmission line, and the integration, transmission, and accommodation of electricity are highly flexible. It is a basis for the secure operation of the grid. A higher AC voltage offers a stronger network, a larger capacity, and a stronger capability for the system to withstand disturbance. For a two-terminal DC link, there can be no intermediate terminals; therefore, it is difficult to form a network. This makes the DC system a more suitable option for large-capacity, long-distance, point-to-point transmission. A multi-infeed, high-capacity HVDC system requires a reliable AC voltage and network as its support to operate properly and ensure the grid's operational security and stability.

In view of the geographical characteristics of China's energy and load, UHV AC transmission is targeted to provide a backbone and cross-region interconnection, and to provide the essential support for DC transmission. UHV DC is set to deliver massive electricity from remote large energy bases to load centers. Strong UHV-based hybrid transmission grids with robust sending and receiving ends will be used to offer the fundamental infrastructures for the cross-region, long-distance, large-capacity, and efficient delivery and allocation of electricity produced from large hydropower, coal power, nuclear power, and renewable power bases.

1.2.1.2 Protecting environment

China is in the middle of an industrialization process characterized by rapid economic growth. The exploitation and consumption of traditional fossil fuels have caused serious environmental problems in China, especially in the more developed eastern and central regions, where all of the nation's 104 cities suffering from heavy acid rain are located. Sulfur dioxide emission density from the east is 5.2 times that in the west of the country. The annual sulfur dioxide emissions per square kilometer in the Yangtze River Delta region are 45 tonnes, 20 times the national average. In recent years, the eastern and central regions have been experiencing recurring, widespread haze, and, most notably, excessive PM2.5. Pollution in Beijing, Tianjin, Hebei, the Yangtze River Delta, and central China is particularly serious, where hazy days in some areas are more than 50% of the year. Protecting the environment has become a focus of social concern.

Ecological and environmental protection is closely related to how energy is produced and consumed. The generation industry and other industries, which burn a lot of coal, are a major source of sulfur dioxide, nitrogen oxides, smoke dusts, and other pollutants. To safeguard the sustainable development of economy and steady improvement of people's quality of life and to build an ecological and beautiful China, it becomes a realistic and urgent need to accelerate the transformation of energy and power development pattern to consider the environment-carrying capacity of east and west China in a coordinated way and to optimize the allocation of environmental resources in a nationwide context.

In the medium and long term, China's energy mix will still be coal-based, with coal maintaining a high proportion of the nation's power production. Development of UHV transmission, clean energy, and improved layout of coal resources will optimize the allocation of energy, environment, and other resources in the country and bring multiple environmental benefits:

1. The development of UHV grids will drive the accomplishment of the nation's clean energy development goals and efficient use of clean energy. By 2020, the cross-region wind power transmission will exceed 100 GW, accounting for more than half of the nation's installed wind power capacity. UHV grids may create significant environmental benefits, such as promoting the cross-region delivery of hydropower, wind power, and other clean energies, reducing wind curtailment in northeast, north, and northwest China, reducing surplus water in southwest China, decreasing fossil fuel consumption and emissions, and significantly reducing the yield of PM2.5.
2. The development of UHV grids will contribute to the use of electricity as an alternative to coal and improved layout of coal bases, significantly reduce the coal consumption for the Three-Hua grid area, promote the efficient use of cleaned coal and foreign electricity, enhance the fuel quality and efficiency of existing power plants, and improve the local environment. The eastern and central regions are running out of environmental capacity, whereas the western region is different from the two regions in environmental capacity, population density, density of power production capacity, and other aspects. Developing UHV grids and more pithead power plants in north and west China, where coal bases are located, will turn a higher portion of coal into electricity and send the electricity to the eastern and central regions over long-distance channels. This may increase the share of electricity in energy consumption of these regions, mitigate the environmental pressure, make full use of the environmental capacity in the west and north, and reduce the nation's overall environmental damage.

1.2.1.3 Improving operational security of grids and their overall social benefits

The 500-kV grid could not accommodate the infeed of multiple UHV DC systems. Particularly, if a fault occurs on a UHV DC system, then the 500-kV grid would experience problems such as large transfer of power flows and insufficient support for reactive power and voltage. The transfer of power flow will cause overload on the 500-kV lines, the possibility of simultaneous blocking of multiple DC systems is much higher, and there is a risk of widespread blackout. By building UHV AC transmission grids and thus forming a hybrid network of strong UHV AC/DC systems, the reliability of the UHV DC systems could be ensured.

Developing UHV grids will prepare for the gradual layering and zoning of lower voltage grids, solve the problem of out-of-limit short-circuit current, and increase the flexibility and reliability of whole grids. The out-of-limit short-circuit current of 500 kV grids has become a prominent problem and a major barrier to grid development. In the dense grids in the Yangtze River Delta of east China, nearly 30% of the power plants and substations are experiencing out-of-limit 500-kV short-circuit current, which greatly restricts the development of grids. Constructing a grid of higher voltage class provides a solution to the excessive short-circuit current and could increase the operational flexibility and reliability of the grid.

Land resources are a material basis for humankind's survival and development. In the process of economic development and industrialization, available land resources has declined sharply in China, where the per capita land area is less than one-third of the world average and the reserves for development are insufficient. With the growing population and the increasing use of land by industry, mining, transportation, and urban construction, the shortage is becoming worse. In this situation, reserving land resources and increasing the efficiency of land use are becoming increasingly important.

The development of UHV grids would contribute to the optimal and efficient use of land at the national level. In contrast to EHV transmission, UHV transmission requires much less right of way, greatly increases the capacity carried by per-unit width of right of way, and thus reduces the requirement for precious land. With vast territory, sparse population, and relatively abundant land resources, China's western region offers more favorable conditions for building coal-fired power plants. Eastern and central regions have developed economy, dense population, expensive land, and scarce resources. Power plants can be built on the cheaper land in west and north, and the electricity generated can be delivered to eastern and central regions, where land can be saved, by expanding the cross-province and cross-region transmission. Through such nationwide optimization of industrial layouts, the overall land use efficiency can be further increased.

1.2.1.4 Enhancing capabilities for energy delivery

The coal-based energy mix in China is not expected to change significantly in the foreseeable future, which indicates that the transport of energy mainly involves coal. Transporting coal and transmitting electricity are two important means to deliver energy, both of which can play a major role for China to build a resource-efficient and eco-friendly society. In 2012, the ratio of coal to electricity exported from Shanxi, Shaanxi, and west Inner Mongolia was 15:1, and that of coal-to-electricity received in eastern China was 48:1.

Transporting coal, as a conventional way of energy delivery, has accounted for a high proportion in the energy transmission system for quite some time. In 2012, the national consumption of coal in China was 37.4×10^8 tonnes, of which 20×10^8 tonnes was used for power generation, accounting for 53% of the total consumption; the amounts of coal transported by railway and highway were, respectively, 22.6×10^8 and more than 4×10^8 tonnes.

For a long time, the development mode of regional/provincial self-sufficiency in electricity supply, coupled with high dependence on coal supply from western and northern regions (such as Shanxi, Shaanxi, Inner Mongolia, Ningxia, and Xinjiang), have resulted in the bulk, long-distance transport of coal to eastern and central regions. Pressure on production and transportation of coal has caused recurring difficulties for electricity generation in recent years, which is an important factor that affects the stable electricity supply in eastern and central regions.

In the future, with further depletion of coal resources in eastern and central regions, coal production centers will be shifted to the west and north, resulting in longer transportation distance and larger volume. The great need for transportation of energy in future China produces a vast potential for delivering coal and electricity. UHV transmission is not intended to replace transportation by railway. Transporting coal and electricity should be differently focused and should be coordinated to exploit their complementary advantages. Promoting transportation of both and promoting accelerated development of electricity are key initiatives for maintaining the production–transportation–generation balance. This is of great significance for improving the energy efficiency in production, conversion, transportation, and use, improving the use of the nationwide environmental resources, and enhancing the security of energy supply.

1.2.2 History of UHV Development Worldwide

With the rapid increase in electrical load and demand for long-distance, large-capacity transmission, the construction of high-capacity power plants, and the increasing number of HV and EHV transmission lines and substations, environmental problems have become increasingly prominent. To achieve economies of scale, reduce line loss, avoid duplicated capacity of transmission equipment, ensure power system reliability, and reduce the impact of transmission lines on environment, the United States, Japan, Russia, Italy, and other countries performed feasibility studies of UHV transmission in the late 1960s or early 1970s based on their need for electricity development [18,19].

The Soviet Union was one of the first countries to conduct research on UHV transmission technologies in the world; so far, it is the only one with experience in operating a UHV transmission system (apart from China). Since 1960, the Soviet Union organized the Technology Board of Ministry of Power and Electrification, All-Union Electrical Engineering Institute, Leningrad Direct Current Research Institute, All-Union Line Design Institute, and other organizations to conduct fundamental studies of UHV transmission. In 1973, the Soviet Union built a UHV three-phase test circuit of 1.17 km at the Beily Rast Substation, powered through a group of 1150/500/10-kV, 3×417-MVA autotransformers. Extensive tests were conducted including equipment insulation, operating overvoltage, audible noise, radio interference, electric fields in substation, installation, transportation, and maintenance of equipment. In 1978, the construction of a 270-km industrial test line from Itatsk to Novokuznetsk began. In addition, various on-site tests were performed on UHV equipment. UHV test sites with 3×1200 kV, 10–12A cascade test UHV transformers base, and 1000-kV impulse generators were also constructed. After 1985, the test line became part of the 1150-kV UHV transmission lines from Ekibastuz to Siberia.

American Electric Power (AEP) and General Electric Company first began studies of audible noise, radio interference, corona loss, and other environmental effects in 1974 regarding the UHV transmission technology research and test station in Pittsfield. Bonneville Power Administration began to conduct studies of the mechanical structure of UHV transmission lines in 1976 in Lyons Proving Ground and Morrow line segments for mechanical test, and they performed studies of corona and electric field, ecological and environmental effects, and operating and lightning impulse insulation. Electric Power Research Institute (EPRI) built a 1000–1500-kV three-phase test line in 1974. By running this line, EPRI acquired electromagnetic environmental indicators and performed tower installation tests, experimental studies of design, and checks of large transformers, with fruitful research results.

Japan launched an R&D program in 1972. Central Research Institute of Electric Power Industry (CRIEPI), Tokyo Electric Power Company (TEPCO), and NGK Insulators Company were the major players. CRIEPI led the construction of several UHV testing centers in Shiobara and Akagi. On the UHV test line in Akagi, CRIEPI measured the corona noise, radio and television interference, wind noise, corona loss of $8 \times 810\ mm^2$ and other bundle conductors, and the impact on the ecological environment, and they evaluated the measuring methods and results under various weather conditions. In the outdoor testing facilities and in the UHV fog room in Shiobara, CRIEPI conducted tests on tower air gap and insulator string, which produced useful data for the design of transmission lines. In 1988, Japan began to construct 1000-kV UHV transmission lines to supply Tokyo, which are 426 km long and are currently running at 500 kV. In 1995, packaged UHV equipment was tested energized in the UHV test field at Sin-Haruna substation.

After establishing the 1000-kV program, Enel conducted research of UHV transmission technology in different experimental stations and laboratories since 1971. The 1000-kV test facilities at the Suvereto testing site included a 1-km-long experimental line, corona and electromagnetic environment testing equipment consisting of a 40-m-long cage. Other tests conducted included operating and lightning overvoltage tests, including operating impulse characteristics of air gap, surface insulating properties under contaminated atmosphere conditions, insulating properties of SF_6, and development tests of unconventional insulators. In 1984, Italy began to build a UHV test project. In October 1995, the project was completed and used in a test run at the rated voltage (nominal voltage) of 1050 kV until December 1997; some operation experience was obtained.

Hydro Quebec's HV test chamber conducted tests of transmission equipment at a rated voltage of 1500 kV. The outdoor test field used to study conductor corona was composed of a test line and a corona cage. The test line and corona cage were both used for corona test on bundle conductors up to 1500-kV AC systems and ±1800 kV DC systems with a span of 300 m. Air insulation tests for lines and substations up to 1500 kV were conducted in an HV test chamber in Quebec. Studies were conducted on four types of bundle conductors in the outdoor test field. Hydro Quebec also studied the corona, electric field, and ion flow characteristics of ±600- to ±1200-kV DC transmission lines and conducted wind tunnel measurements for air dynamics (such as drag, lift, offset) on four-, six-, and eight-bundle conductors. Under different wind speeds, dynamic characteristics of six-bundle conductors and air dynamics of 12-bundle conductors were studied on Madeleine Island's test line.

Ukraine is also one of the few countries in the world with the capability of developing EHV/UHV equipment. Its Zaporozhie Transformer Research Institute was responsible for scientific research, equipment design, development of new products, design of tooling, research on production technology, manufacturing of prototypes and a few products, electrical equipment testing, research on and formulation of national standards, product certification, and consulting services. Major products and tests developed and conducted by the institute include 750-kV, 320-MVA converter transformers for UHV DC; smoothing reactors; disconnectors; 750-kV shunt reactors; 667-MVA, 1150/500-kV autotransformer model; 1500-kV/500-kV autotransformer model; 750- to 1800-kV bushings; DC 600-kV pulse generators; 220- to 500-kV neutral bushings; 750-kV or lower electromagnetic current transformers; 500-kV or lower electromagnetic voltage transformers; and 750-kV capacitor voltage transformers.

The efforts made by various countries on key UHV transmission technologies and equipment manufacturing over the years laid the foundation for the subsequent development and application of the technologies. Technology is no longer a constraint for the development of UHV transmission. Large-scale, long-distance power transmission is the major driver that promotes the application of UHV transmission technologies, but it also has to adapt to the country-specific conditions. Due to the slow growth in electricity load and consequent low demand for high-capacity, long-distance transmission in the Soviet Union, Japan, and other countries, their UHV transmission projects were suspended or postponed, or operated at reduced voltage. However, the United States and Italy performed research and tests to make technical preparations, not because of their actual load requirements.

1.2.3 Innovations and Practices in China's UHV Transmission

Imbalanced locations of energy resources and load centers indicate that the UHV transmission technologies have a huge potential in China.

UHV transmission technologies include UHV AC and UHV DC.

Since 2004, SGCC has, based on independent innovations and in joint efforts with others, organized studies and demonstrations of UHV transmission technical research, planning and design, equipment development and production, and construction and operation of UHV projects. It made breakthroughs in both theory and practice for UHV AC/DC systems, having demonstrated the security, economy, and eco-friendliness of UHV transmission. The commissioning and continuous secure and stable operation of UHV AC and DC pilot and demonstration projects indicate that the Chinese UHV transmission technologies are proven and ready for future large-scale application.

1.2.3.1 Development of UHV AC transmission

As early as in 1986, a "preliminary study program of UHV AC transmission" was launched, beginning studies of UHV power transmission. In 1990–1995, demonstrations of long-distance transmission mode and voltage class were conducted. In 1990–1999, studies were performed regarding subjects such as "early demonstration for UHV transmission" and "feasibility of 1000-kV class UHV AC transmission," from which a preliminary understanding of UHV transmission was established.

In 2004, SGCC launched studies of key technologies and feasibility of UHV transmission engineering by research institutions and manufacturers. According to the research framework for key UHV AC transmission technologies, studies of 46 subjects were completed. In addition, SGCC conducted frequent technical exchanges and conferences with international organizations and research institutions as well as manufacturers, including EPRI, CRIEPI, TEPCO, High Voltage Direct Current Power Transmission Research Institute, ABB, Siemens, Areva, Toshiba, Mitsubishi, Japan AE Power Systems Corporation, NGK, and AXICOM.

By 2006, China had achieved several firsthand research results of UHV AC transmission, solved all the key problems for building a UHV demonstration project, and basically grasped the basic characteristics of UHV AC transmission and grids. Preliminary results were obtained for key technologies such as UHV electromagnetic environment limit, overvoltage level, reactive configuration, insulation coordination, and lightning protection, which were the basis for the feasibility study of the UHV transmission project to pass the examination and provided a large amount of reliable and accurate data for preliminary design.

UHV transmission is a complicated and systematic program that must be preceded by a pilot and demonstration project. The purpose of such a project is to assess the performance of UHV transmission and equipment used for it, gather experience in research, manufacturing, and grid operation and control, and improve capabilities in equipment manufacturing and transmission technology. In 2005, China determined to use the Jindongnan–Nanyang–Jingmen project as a demonstration project and completed the feasibility study. In this project, the transmission system and equipment were extensively checked and tested, and the results obtained could be directly used for China's future UHV grids.

The 1000-kV Jindongnan–Nanyang–Jingmen UHV AC pilot project has 640-km-long transmission lines equipped with intermediate switching substations. Large-capacity, HV shunt reactors, high-performance arresters, and circuit breakers with closing resistor were used for the lines, consistent with China's UHV technology roadmap. Taking advantage of the pilot project, the performance of equipment required for a UHV system, such as lines, transformers, HV reactors, circuit breakers, GIS, surge arresters, voltage transformers, current transformers, and insulators, was fully examined under various conditions such as temporary overvoltage, switching overvoltage, resonance overvoltage, lightning overvoltage, load-shedding overvoltage, short-circuit current, switching in/out low-voltage capacitors, switching in/out low-voltage reactors, and switching in/out no-load lines.

The project began operation in early 2009 and was extended in December 2011. The goal of reliably transmitting 5000 MW by a single circuit was achieved.

1.2.3.2 Development of UHV DC transmission

Compared with AC transmission, although DC transmission accounts for a smaller proportion by capacity, its unique advantages (which are discussed in Chapter 3) give rise to the complementary joint use of the two technologies. Given China's geographically imbalanced energy and load distribution, hybrid grids consisting of UHV AC and UHV DC components are a strategic choice to transmit more electricity over longer distances.

Since the late 1950s, China launched studies of DC transmission following the world's trend. In the 1970s, a test HVDC transmission project was built in Shanghai. In the late 1980s, an industrial test project was constructed in Zhoushan. In 1990, the Gezhouba–Shanghai HVDC project (±500 kV, 1200 MW), China's first large-capacity, long-distance HVDC project, was constructed. After more than a decade, China's HVDC technologies have developed quickly and have been used in a number of engineering applications. China is now capable of producing conventional equipment used for UHV transmission. At the end of 2009, there were 18 EHVDC and below projects in operation with a capacity of 33,570 MW and line length of 11,841 km. China has now become a major player in the world's DC transmission and has laid a solid foundation for the use of UHV DC in their country.

1. **Development and use of DC transmission technologies**
   China began to study and investigate the feasibility of Zhoushan HVDC project in the late 1970s. As the first HVDC project in the country, the project was a national major research program. The entire project, from feasibility study to design, to manufacture, to construction, to commissioning, and, finally, to operation, was fully completed with China's own efforts. The project sends alternating electricity from the east China grid to the Daqi Town Rectifier Station in Ningbo, Zhejiang, where the electricity is rectified and sent to the Aotoupu Inverter Station on Zhoulan Island, and it is then distributed for use in Zhoushan. The transmission distance is 54.1 km, containing three sections of overhead lines totaling 42.1 km and two sections of submarine cable totaling 12 km. The first phase of the project was monopolar DC 100 kV, 500 A, 50 MW, with six-pulse converters and space for second-phase expansion. Ultimately, the project was bipolar DC ±100 kV, 500 A, 100 MW, and it was put into operation in September 1989.

   The Shengsi project is another DC transmission project that fully relies on domestic technologies and equipment after Zhoushan project, and it is China's first cross-sea DC system to an island through bipolar, two-circuit submarine cables. The lines of this project are 66.2 km, of which 59.7 km is submarine cables. The project is rated ±50 kV, 600 A, bipolar 60 MW, and monopolar 30 MW. The project was put into operation in September 2002.
2. **Application of ±400-, ±500-, and ±660-kV HVDC technologies**
   Since the 1970s, the principles of thyristor converters had been used in all HVDC projects. One 12-pulse converter at each pole of each end forms a two-end, bipolar DC transmission system. In the early 1990s, the Gezhouba–Shanghai HVDC project, China's first ±500-kV HVDC system, began bipolar operation. The project is an associated work for delivering electricity from Gezhouba Hydropower Station and China's first large-capacity, long-distance HVDC transmission system. The commissioning of the project for the first time realized the asynchronous interconnection of two cross-province (cross-region) grids, the eastern and central China grids.

   The Three Gorges projects introduced the world's leading HVDC technologies and played a very important role in driving the progress and innovations in China's HVDC technologies, as well as in increasing the proportion of domestically supplied equipment. The Three Gorges projects include three ± 500-kV, 3000-MW projects, namely the Three Gorges–Changzhou, the Three Gorges–Guangdong, and the Three Gorges–Shanghai projects, which were put into operation in 2003, 2004, and 2007. Relying on these projects and suiting local characteristics, China's power engineering and design organizations performed in-depth research of packaged converter station design, overvoltage suppression and insulation coordination, DC system control and protection strategies, and electromagnetic environment, and established comprehensive technical systems and norms. Chinese manufacturers mastered design and manufacturing technologies of converter transformers, smoothing reactors, and 5-in converter valves, and manufacturing technologies of control and protection devices for DC systems through technology transfer and absorption, technical cooperation, and independent efforts. These are the technical preparations. The Three Gorges–Shanghai II project had all of its equipment manufactured locally, and the owner fully implemented the project management on its own.

The ±660-kV Ningdong–Shandong HVDC demonstration project is the first HVDC project at this voltage in the world and was independently designed, developed, and constructed by China. The project began bipolar operation in February 2011, gathering electricity from the east and northwest Ningxia and sending it to Shandong. It is important for the construction of a modern integrated transport system of energy, optimized allocation of energy resources in a wider region, and promotion of coordinated development of regional economy. It is also a comprehensive test of China's capability of independent innovation in HVDC.

The ±400-kV Qaidam–Lhasa HVDC project (i.e., the Qinghai–Tibet project) is a large-scale HVDC transmission project implemented at the world's highest altitudes. The commissioning of this project ended the isolated operation of the Tibet grid and it was an important milestone in China's history of transmission engineering.

3. **Application of ±800-kV or higher UHV DC technologies**

The development of UHV DC technologies is accompanied by the emergence and innovation of power technologies and equipment. In 2004, China began extensive and in-depth studies of ±800-kV UHV DC transmission technologies and successfully used the results in the construction of a ±800-kV project.

From 2004 to July 2010, after more than 5 years of independent research and construction, China built the Xiangjiaba–Shanghai ±800-kV UHV DC transmission demonstration project, which had the world's highest voltage class, the largest transmission capacity, the longest transmission distance, and the most advanced technologies. The project began monopolar system operation in 2009 and bipolar system operation in 2010, with a rated power of 6400 MW and 1891-km-long DC lines.

In 2012, the ±800-kV Jinping–Sunan UHV DC project was completed and put into operation. This project was a result of effective studies of increasing capacity based on the successful experience of the Xiangjiaba–Shanghai project. The project ratings are ±800 kV, 7200 MW, 4500 A, with a DC line length of 2059 km. The ±800-kV

Haminan–Zhengzhou and Xiluodu Left Bank–Zhejiang Jinhua UHV DC transmission projects under construction would further increase the transmission capacity to 8000 MW.

When the distance and power are increased to a certain degree, the voltage class has to be increased to improve the cost efficiency and reduce operation loss. As planned, China is conducting studies of ±1100-kV UHV DC transmission technologies. The major technical difficulties include insulation coordination of equipment and increase in capacity of converter transformers and their connection to the grid. By reviewing the experience of the operating projects, combined with the studies of key technologies of major equipment and transport schemes of large equipment, an approximate conclusion can be reached that there are no insurmountable technical obstacles for ±1100-kV UHV DC systems. In addition, three options are considered for ±1100-kV UHV DC projects, namely, direct connection to 1000 kV, layered connection (high-end to 500 kV, low-end to 1000 kV), and direct connection to 500 kV. The layered connection could meet the local demand and accommodate the development of the UHV AC grid; therefore it has widened applications.

1.3 Hybrid UHV AC and UHV DC Grid

1.3.1 Features of AC and DC Transmission Technologies

An AC transmission system can have intermediate terminals and form a network based on actual needs such as locations of power sources and loads, the power transmitted, and power exchange. UHV AC transmission has a large capacity and extensive coverage. It provides a platform for a national electricity market and could flexibly adapt to the requirements of market operations. It could also significantly reduce the required right of way, with a small active power loss on lines and transformers for the unit power transmitted.

The equivalent circuit of an AC transmission line is composed of resistance, reactance, conductance, and susceptance, which are detailed in Chapter 2. The inductance is resistant to the sinusoidal AC; therefore, it is known as inductive reactance; the impedance of an inductive component is inductive reactance ($X_L = \omega L = 2\pi fL$), which is proportional to the angular velocity (frequency) and line length. The transmission capacity of an AC transmission line is related to the grid configuration, location of the AC lines in the grid, operation mode, and other factors, and is dependent on the ratio of short-circuit capacity at both ends of the lines and the line length.

DC transmission systems are mostly two-terminal systems without intermediate terminals. It is suitable for point-to-point, large-capacity, long-distance transmission directly to the load center. DC transmission can reduce or avoid heavy cross-network power flow.

It could change the flow according to the operation modes of the sending and receiving ends, with the direction and magnitude being easily controlled. Studies show that from an economic and environmental point of view, a UHV DC system higher than ±600 kV is a preferred means for large-capacity transmission over long distances. However, an HVDC system must rely on a strong AC grid to play its role.

AC and DC are both part of the grid, and they have different characteristics when used in the grid; the two complement each other. Development of the grid cannot simply rely on direct current transmission; rather, a strong grid with mutual support among AC and DC components is required. DC transmission applies to long-distance, point-to-point, large-capacity transmission that exceeds the distance when the cost efficiency of AC and DC transmissions is equal, whereas the "back-to-back" DC technology is mainly suitable for networking between systems of different/same frequencies. AC transmission is mainly intended to build strong transmission networks at all levels and a channel for interconnection between grids while satisfying the need for delivery of power sources when the distance is less than the AC−DC equal point and providing an essential support for DC transmission.

1.3.2 Features of Hybrid UHV AC and UHV DC Grids

A hybrid UHV grid is defined as a grid in which 1000-kV AC and ±800-kV or higher DC systems are synchronously or asynchronously connected to an EHVAC grid. From the worldwide voltage development history, we can see that the new higher voltage class generally should not be lower than twice the existing voltage class. The introduction of 1000 kV over 500 kV, which offers a four- to five-times higher transmission capability, is technically and economically viable. Based on China's actual conditions, it is a strategic choice to introduce 1000-kV UHV AC transmission at a proper time to strengthen the receiving-end grids. It will provide a strong support for multi-infeed connections of DC systems, effectively reduce the risk of large-scale blackouts, save right of way, and completely solve the problem of excessive short-circuit current on 500-kV grids. Choosing 1000 kV as the one-level-higher voltage above 500 kV is reasonable; 1000-kV UHV AC and ± 800-kV UHV DC systems, both as part of a UHV grid, have different characteristics and complement each other.

The stability of DC transmission is related to the electrical strength of the grid at the receiving end. The strength can be expressed by the short-current ratio of the converter station, defined as the ratio of the short-circuit capacity of the AC bus in the converter station to the rated DC power. The control of the DC system plays an important role in most of the interactions between the AC and DC systems; therefore, it must be considered when evaluating the strength of an AC system.

In cases when multiple DC inverter stations are built at the receiving-end grid and the rectifier side is under power control, when a three-phase short-circuit fault occurs on the AC lines near the inverter station, the entire AC system at the receiving end will suffer two impacts: reduced active power from the DC system and instant increase of reactive power absorption. For a large receiving-end grid, there is usually a sufficient margin to withstand the impact of instantaneously reduced active power from a single DC circuit; in this case, whether the receiving-end AC system could withstand the impact of instantaneously increased reactive power absorption is a key for it to maintain voltage stability. This capability is related to the effective short-circuit ratio (ESCR) of the inverter station, the recovery speed of DC power, the dynamic reactive power reserves of the receiving-end AC system, the synchronization capabilities of other AC systems running in parallel to the AC system, and the load characteristics. With larger number and higher capacity of UHV DC transmission projects, 1000-kV UHV AC grids must be built at the receiving end to solve the security problems caused by the DC systems. So, after seeing the interaction between AC and DC systems, UHV grids could provide strong voltage and reactive power support for a multi-DC inverting station receiving-end grid, offer a fundamental solution to the weak supporting capabilities of 500-kV grids, and provide sustainability. In the case of a hybrid AC–DC transmission system, using control functions such as DC power modulation can effectively suppress power oscillations on AC lines connected to the DC system, including regional low-frequency oscillations, thus significantly improving the transient and dynamic stability performance of the AC system.

A hybrid, robust, smart "Three-Hua" synchronous network with UHV grids as the backbone, with strong AC and DC systems, will produce a more reasonable structure. UHV grids have a stronger capability to accomplish large-capacity, long-distance transmission and accommodation of electricity, to ensure the security operation of a whole system, and to withstand various serious incidents. Strong and smart hybrid UHV AC and UHV DC grids are not only a carrier of electricity but also an important component of a modern integrated energy transport system and a strong and powerful platform for optimal allocation of resources. Its security and reliability performance and capabilities of withstanding severe incidents are much higher.

1.3.3 Basic Principles for Selecting UHV Voltage Classes

Determination of UHV voltage classes should follow the following basic principles taking into account the existing EHVAC and HVDC transmission system and future requirements:

1. Be commensurate with the size of geographic area covered and scale of the power system required
2. Have appropriate coordination with the existing EHV voltage systems in terms of technical requirements and economy

3. Be consistent with the average capacity (capability) and distance of the grid
4. Have due regard for the timescale required to take equipment from development to industrial use
5. Take into account the availability and actual requirements of UHV transmission technologies
6. Promote development in new generation technologies

With the wide use of DC transmission, selecting the voltage should be primarily based on the consideration of universal design and standardized equipment to realize economies of scale.

Voltage classes to be determined include nominal voltage for a UHV AC system, maximum operating voltage for a UHV AC system, and rated voltage for a UHV DC system. A UHV AC system's nominal voltage refers to a given value that marks or identifies the system voltage. The maximum operating voltage refers to the maximum value of voltage that appears at any time and at any point under the system's normal operating conditions. A UHV DC system's rated voltage refers to the voltage required to deliver the rated power under the rated current.

1. **Investigation of nominal and maximum UHV AC operating voltages**
   There are two main ways to determine voltage class for a UHV AC system: the empirical data and formula method and the economic comparison method. In the former, an economically viable range of UHV voltage classes is derived following the rules according to which the existing voltage classes are determined by using induction and interpolation methods after sufficient investigation and analysis of empirical data and formulas used in the determination process. In the latter method, various models are established for comparison using the comparative annual cost approach.
   Worldwide, 1000 and 1150 kV are two major nominal voltages for pilot UHV AC systems. In China, regarding nominal voltage for the UHV AC system, choosing from these two voltages is preferred.
   In determining the nominal UHV AC voltage, studies were conducted based on transmission needs, altitude and corona loss limitation, reactive power compensation capacity, and impact on parameters of main equipment.
   a. The need for large-capacity power transmission and exchange was analyzed considering the characteristics of China's imbalanced distribution of energy and load. The normal power transmitted by a UHV AC line usually ranges from 3000 to 5000 MW. Using 1000 kV as the nominal voltage can meet the need for long-distance, high-capacity transmission and power exchange; whereas using 1150 kV can increase the maximum power transmitted by the line, but it also increases the cost (15–20% higher than 1000 kV). From the perspective of meeting the maximum transmission demand, 1000 kV will suffice.

b. Altitude and corona loss limitation. China's southwestern hydroelectric bases are located at 1000–2000 m or even higher altitudes. When the altitude exceeds 1000 m, the requirements of external insulation characteristics of equipment are higher. To reduce the external insulation level, and thus the cost of equipment, choosing a lower nominal voltage is appropriate. The corona loss of a UHV AC transmission line is proportional to the operating voltage. For the same critical corona voltage, a higher operating voltage would result in greater corona loss. The corona loss under severe weather conditions can be reduced by properly reducing the operating voltage. Therefore, to reduce the external insulation level of equipment and corona loss at high altitudes, choosing 1000 kV as the nominal voltage is advantageous.
c. Impact of nominal voltage on reactive power compensation capacity. A proper number of HV reactors are required on the UHV transmission lines to limit the temporary overvoltage. The cost of HV reactors is approximately 19% higher for 1150 kV than for 1000 kV. As such, a 1000-kV nominal voltage requires less cost in reactive power compensation equipment.
d. Impact of nominal voltage on parameters of main equipment. Main equipment used for UHV transmission includes towers, transformers, and switchgears. Considering insulation, electrostatic induction, and noise, a nominal voltage of 1150 kV (corresponding to a maximum operating voltage of 1200 kV) requires more subconductors, higher towers, and more corona protection measures compared with 1000 kV (corresponding to a maximum operating voltage of 1100 kV). The tower weight would increase by approximately 30% and the construction cost by approximately 19%.

Taking all these factors into account, choosing 1000 kV as the nominal voltage for China's UHV AC systems is justified. After determining the nominal voltage, the maximum operating voltage can be finalized, constituting the basis for developing specifications for equipment manufacturing and system operation.

Internationally, maximum operating voltages for pilot UHV AC systems rated 1000 kV are 1050 kV and 1100 kV. In China, the maximum operating voltage should also be chosen from 1050 and 1100 kV.

Studies were conducted on grid dispatching and operation and manufacturing of equipment to determine the maximum AC operating voltage.

a. Impact of the maximum operating voltage on grid dispatch and operation. The maximum operating voltage of 1100 kV differs from 1050 kV in the following aspects: (i) it requires more HV reactors but far fewer LV reactors; therefore, the requirements of voltage regulation are lower; (ii) it offers a larger voltage difference margin for system operation; therefore, it increases the operational flexibility of the grid; (iii) it could reduce the line loss and increase the operational economy of the grid; (iv) it could increase the transmitted power under the normal operation of the lines and mitigate the pressure imposed on the 500-kV grid; and (v) it could increase the transient stability power limit for 1000-kV transmission and the stability of the UHV grid. However, considering the corona loss and equipment insulation at high altitudes, the maximum operating voltage of the UHV grid should not be set too high.

b. Impact on manufacturing of UHV equipment: (i) the cost of UHV equipment is basically determined by the insulation level required for lightning overvoltage and switching overvoltage. For maximum operating voltages of 1100 and 1050 kV, the required insulation level is the same, with approximately the same manufacturing cost; international manufacturing practices of circuit breakers show that when the maximum operating voltage is higher than 1100 kV, the circuit breakers will require more breaks, too many of which will affect the reliability and cost; from the line cost point of view, the average reference cost per kilometer is only approximately 0.5% higher for 1100 kV than for 1050 kV; as we look at the R&D activities of the manufacturers, especially those for UHV circuit breakers and bushings, the development and testing of the products are based on a maximum operating voltage of 1100 kV. This is also the case with the development of transformers.

Considering these two aspects, the maximum operating voltage for 1000 kV is determined as 1100 kV. On October 7, 2008, at the third meeting of the Joint Working Group on UHV of the IEC/CIGRE held in Geneva, the maximum operating voltage of 1100 kV established in Chinese standards is recognized as the maximum operating voltage for IEC-compliant equipment, indicating that China's UHV AC voltage class standard was accepted as an international standard.

2. **Investigation of UHV DC-rated voltage**

The UHV DC voltage class is usually determined using the economic comparison method, that is, the scheme with a lower annual cost of unit capacity (the ratio of annual cost to the DC transmission capacity) is deemed more cost-effective after comparing the annual costs of unit capacity for DC transmission schemes with different voltage classes.

On the basis of conventional ±500-kV DC transmission, combined with the hydropower transmission projects in southwest China, a variety of factors such as the transmission distance, economic transmission capacity, annual cost of unit capacity, equipment manufacturing and transportation, and right of way are comprehensively evaluated. Based on these, ±800 kV is proposed as the rated UHV DC voltage, with a transmission capacity of 6400–8000 MW. The voltage and transmission capacity could form a reasonable differential and create comprehensive technical and economic advantages. However, this voltage could bring the economies of scale, and it meets the transmission needs of China's large energy bases.

Given the fact that the transmission distance from energy bases in Xinjiang, Tibet, and other areas to major load centers is more than 3000 km, ±1100 kV is considered as an extension of ±800 kV, offering a transmission capacity of more than 10,000 MW. The voltage and capacity differentials are appropriate and will help improve the economies of scale for UHV DC transmission.

At the end of 2012, China had built a number of ±500- (±400-), ±660-, and ±800-kV HVDC projects. The ±1100-kV UHV DC transmission technologies are being developed to meet the diverse needs of DC transmission in China.

References

[1] Z. Xiaoxin, C. Shuyong, Lu Zongxiang, Retrospect and prospect of the grid and grid technology development—discussions on the third-generation grid, Proc. CSEE 33(22) (2013) 1–11.
[2] L. Zhenya, UHV Power Grid, China National Economic Press, Beijing, 2005.
[3] L. Zhenya, Collection of UHV DC Transmission Technologies, Theory of UHV DC Transmission, China Electric Power Press, Beijing, 2009.
[4] C. Yan, Power System Steady-State Analysis, China Electric Power Press, Beijing, 2007.
[5] W. Meiyi, Wu Jingchang, M. Dingzhong, Large Power System Technology, China Electric Power Press, Beijing, 1995.
[6] D. Yergin, The Quest: Energy, Security, and the Remaking of the Modern World, Petroleum Industry Press, Beijing, 2012.
[7] Drafting Team of China Electric Power Planning, China Electric Power Planning (The Grid Volume), China Water and Power Press, Beijing, 2007.
[8] Yi. Haiqiong, S. Fulong, Gaoyi, et al., Foreign Large Power System Development and Structure Study, 2011.4.
[9] Dr. –Ing. Dipl. –Phys. G. Czisch, Low cost but totally renewable electricity supply for a huge supply area—a European/trans-European example, Proceedings of the Third World Wind Energy Conference and Renewable Energy Exhibition the Second Wind Power Asia, 2004.
[10] Airtricity, European offshore supergrid proposal—creating a more powerful Europ-vision and executive summary, 2006.
[11] McKie, R. How Africa's desert sun can bring Europe power, The Observer (London). solarpower, 2007. 12.
[12] Z. Ming, Li Honglin, X. Song, et al., Key technology direction for construction of smart grid in the background of system security—analysis of the root cause of blackouts in India and Implications for the Chinese power sector, Proc. CSEE 32(25) (2012) 175–181.
[13] T. Yong, Bu Guangquan, Yijun, Analysis of India's "7 · 30" and "7 · 31" blackouts and their implications, Proc. CSEE 32(25) (2012) 167–174.
[14] H. Dayu, India's power construction and UHV AC/DC transmission planning, China Electric Power 41(2) (2008) 65–68.
[15] United States Department of Energy Office of Electric Transmission and Distribution, "Grid 2030" A national vision for electricity's second 100 years, July 2003.
[16] Repower America, Unified national smart grid [EB/OL], 2009.3.
[17] L. Zhenya, Electric Power and Energy in China, China Electric Power Press, Beijing, 2012.
[18] L. Zhenya, UHV AC Transmission Technology Study Collection (2005–2011), China Electric Power Press, Beijing, 2006–2012.
[19] L. Zhenya, UHV DC Transmission Technology Study Collection (2006–2011), China Electric Power Press, Beijing, 2007–2012.
[20] UHV AC Transmission Technology, Science and Technology Department of Ministry of Electric Power Industry, Beijing, 1994.

CHAPTER 2

Characteristics of UHV AC Transmission System

Chapter Outline

Ultra-high Voltage AC/DC Grids.
DOI: http://dx.doi.org/10.1016/B978-0-12-802161-3.00002-0

2.1 Parameters of UHV AC Transmission Lines

2.1.1 Unit Length Parameters of Transmission Line

The basic electrical parameters of transmission lines include resistance (R), conductance (G), inductance (L), and capacitance (C), which dictate the characteristics of transmission lines and power grids. For EHV/UHV AC transmission lines, the resistance mainly affects the power loss of transmission lines. The conductance represents the leakage resistance and corona loss of insulators and may affect the power loss (which may usually be negligible in steady-state analysis). The inductance is the dominant factor that dictates the distribution of power flow on a power grid and affects the voltage decline across transmission lines and the stability performance of the power system. Under the effects of AC voltage, the line-to-line capacitance and line-to-ground capacitance give rise to charging and discharge currents on lines that affect not only the voltage decline across transmission line but also the power transmission efficiency and the distribution of active and reactive power in the power system.

The UHV transmission lines may be simulated using series impedance $Z = R + \mathrm{j}\omega L$ and shunt admittance $Y = G + \mathrm{j}\omega C$. In the normal operating mode, the transmission lines can be represented by a Π-type or T-type equivalent network irrespective of whether they are analyzed based on distributed or lumped parameters. Generally, Π-type network is used for power system analysis, representing the impedance and admittance per unit length of a transmission line or the entire transmission line as a function of voltage and current, as shown in Figure 2.1.

As an AC current flows along a conductor, the current density increases gradually from the center to the surface of the conductor section. This phenomenon is known as skin effect. To allow uniform distribution of current in the conductor as practical as possible, full use of cross-section of the conductor, decreased resistance of the line, and reduced environmental impacts of corona, an EHV/UHV AC transmission line tends to adopt a number of small-section subconductors (instead of a single large-section solid conductor), which are separated from each other using spacers while bundled together into a conductor strand to form a phase conductor, known as a conductor bundle.

2.1.1.1 Reactance of unit length symmetrically arranged conductor bundle

A current-carrying conductor produces a magnetic field around it. According to Faraday's law of induction, the relationship between the magnetic field of a closed conductor loop and

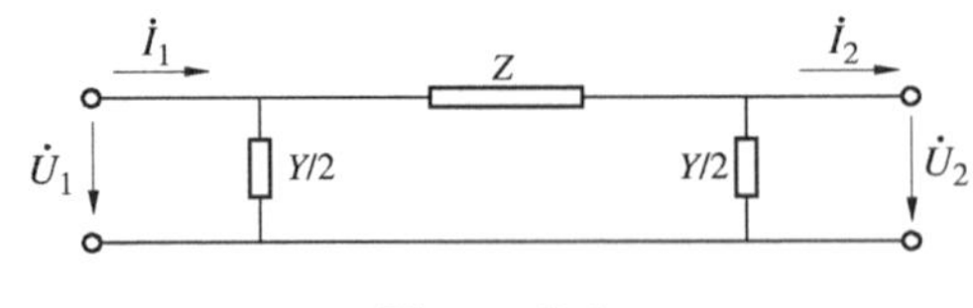

Figure 2.1
Π-type equivalent circuit for a transmission line.

the current and voltage of the conductor is known as inductance or reactance. The magnitude of inductance depends on the geometry and structure of the conductor, the spacing of conductors, and the permeability of free space.

1. **Inductance of two parallel conductors**

   For two parallel conductors that are d apart, let $l \gg d$, so the self-inductance L_S and mutual inductance M per unit length are, respectively:

$$L_S = 2 \times 10^{-7}\left[\ln\left(\frac{2l}{r}\right) + \frac{\mu}{4} - 1\right] \tag{2.1}$$

$$M = 2 \times 10^{-7}\left[\ln\left(\frac{2l}{d}\right) - 1\right] \tag{2.2}$$

   where

   l—length of the conductor, m

   r—radius of the conductor, m

   μ—relative permeability.

   If the geometrical mean distance for a conductor is defined as $D_S = re^{-1/4}$ for conductors made of a nonmagnetic material (such as aluminum conductors and copper conductors, $\mu = 1$), then

$$L_S = 2 \times 10^{-7}\left[\ln\left(\frac{2l}{D_S}\right) - 1\right] \tag{2.3}$$

2. **Reactance of two parallel conductor bundles with symmetrically arranged subconductors**

   Assuming that the current through each individual subconductor of the conductor bundle is the same, the self-inductance L_S per unit length will be:

$$L_S = 2 \times 10^{-7}\left\{\ln\left[\frac{2l}{(ND_S A^{N-1})^{1/N}}\right] - 1\right\} \tag{2.4}$$

   where

   N—number of subconductors in a conductor bundle

   A—radius of conductor bundle.

   For two parallel conductor bundles a and b, the self-reactance per unit length is:

$$X_{aa} = X_{bb} = \omega L_S = 4\pi f \times 10^{-7}\left\{\ln\left[\frac{2l}{(ND_S A^{N-1})^{1/N}}\right] - 1\right\} \tag{2.5}$$

The equation for the mutual inductance M_{ab} per unit length is the same as Eq. (2.2). Thus, the mutual reactance per unit length of the two parallel conductors a and b will be:

$$X_{ab} = \omega M_{ab} = 4\pi f \times 10^{-7}\left(\ln\frac{2l}{d} - 1\right) \tag{2.6}$$

3. **Reactance of symmetrically arranged conductor bundles in a three-phase transmission line**

   If the geometrical mean distance of conductors is defined as $D_{eq} = \sqrt[3]{d_{ab}d_{bc}d_{ca}}$ for a uniformly transposed three-phase transmission line, then the inductance L per unit length of conductor bundles will be:

$$L = L_S - \frac{1}{3}(M_{ab} + M_{bc} + M_{ca}) = 2 \times 10^{-7}\ln\left[D_{eq}/(ND_SA^{N-1})^{1/N}\right] \tag{2.7}$$

   where

   d_{ab}, d_{bc}, d_{ca}—spacing of three-phase conductors of a transmission line.

   The reactance X_0 per unit length of symmetrically arranged conductor bundles in a three-phase transmission line is:

$$X_0 = \omega L = 6.28 \times 10^{-2}\ln[D_{eq}/(ND_SA^{N-1})^{1/N}] \tag{2.8}$$

2.1.1.2 Susceptance per unit length of symmetrically arranged conductor bundles

During normal operation of a three-phase transmission line, the phase capacitance or phase equivalent capacitance comprises the ground capacitance of individual phase conductors and the capacitance between any two of the three-phase conductors.

1. **Ground capacitance of symmetrically arranged conductor bundles**

   A phase conductor bundle and its mirror image are shown in Figure 2.2. The relationship between the voltage and charge on the symmetrically arranged conductor bundles is given by:

$$\begin{bmatrix} \dot{U}_1 \\ \dot{U}_2 \\ \vdots \\ \dot{U}_j \\ \vdots \\ \dot{U}_N \end{bmatrix} = \begin{bmatrix} P_{11} & P_{12} & \cdots & P_{1j} & \cdots & P_{1N} \\ P_{21} & P_{22} & \cdots & P_{2j} & \cdots & P_{2N} \\ & & \vdots & & & \\ P_{j1} & P_{j2} & \cdots & P_{jj} & \cdots & P_{jN} \\ & & \vdots & & & \\ P_{N1} & P_{N2} & \cdots & P_{Nj} & \cdots & P_{NN} \end{bmatrix} \begin{bmatrix} \dot{Q}_1 \\ \dot{Q}_2 \\ \vdots \\ \dot{Q}_j \\ \vdots \\ \dot{Q}_N \end{bmatrix} \tag{2.9}$$

   where

   $\dot{U}_j$—voltage to ground of subconductor j, $j = 1, 2, \ldots, N$

   $\dot{Q}_j$—charge on subconductor j, $j = 1, 2, \ldots, N$

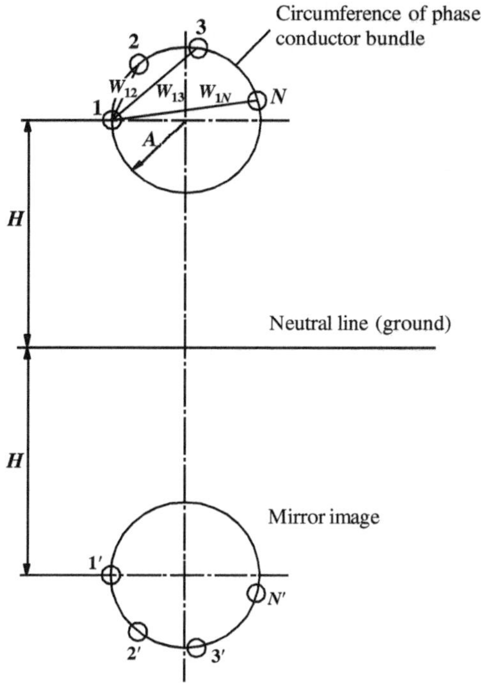

Figure 2.2
Phase conductor bundle and its mirror image.

N—number of subconductors
P—electric potential coefficient.
The self-electric potential coefficient is:

$$P_{jj} = \frac{1}{2\pi\varepsilon}\ln\left(\frac{2H}{r}\right) \tag{2.10}$$

The mutual electric potential coefficient is:

$$P_{jk} = \frac{1}{2\pi\varepsilon}\ln\left(\frac{2H}{W_{jk}}\right) \tag{2.11}$$

where
W_{jk}—spacing of subconductors j and k; $W_{jk} \ll 2H$; $j, k = 1, 2, \ldots, N$; $j \neq k$
H—conductor bundle height above ground
r—radius of subconductor
ε—dielectric constant, $\varepsilon = 1/36\pi \times 10^{-9}$ F/m.
Assuming that the charge is uniformly distributed on each individual subconductor and the subconductors are at the same voltage, namely, $\dot{Q}_1 = \dot{Q}_2 = \cdots = \dot{Q}_N = \dot{Q}$,

$\dot{U}_1 = \dot{U}_2 = \cdots = \dot{U}_N = \dot{U}$, then the total charge on the conductor bundle is $\dot{Q}_\Sigma = N\dot{Q}$. Equation (2.9) may be transformed to:

$$\dot{U} = (P_{11} + P_{12} + \cdots + P_{1N})\dot{Q}$$

$$\dot{Q}_\Sigma = \frac{N\dot{U}}{P_{11} + P_{12} + \cdots + P_{1N}}$$

The ground capacitance of the conductor bundle is:

$$C_{aa} = \frac{\dot{Q}_\Sigma}{\dot{U}} = \frac{N}{P_{11} + P_{12} + \cdots + P_{1N}} \tag{2.12}$$

Substituting Eqs. (2.10) and (2.11) into Eq. (2.12) gives:

$$C_{aa} = \frac{N}{\frac{1}{2\pi\varepsilon}\left(\ln\frac{2H}{r} + \ln\frac{2H}{W_{12}} + \ln\frac{2H}{W_{13}} + \cdots + \ln\frac{2H}{W_{1N}}\right)}$$

Through algebraic reduction, the equation becomes:

$$C_{aa} = 1\bigg/\left\{\frac{1}{2\pi\varepsilon}\ln\left[\frac{2H}{(rNA^{N-1})^{1/N}}\right]\right\} \tag{2.13}$$

$$P_{aa} = \frac{1}{2\pi\varepsilon}\ln\left[\frac{2H}{(rNA^{N-1})^{1/N}}\right] \tag{2.14}$$

2. **Susceptance of symmetrically arranged conductor bundles in a three-phase transmission line**

   The relationship between voltage and charge of a three-phase transmission line is given by:

$$\begin{bmatrix} \dot{U}_a \\ \dot{U}_b \\ \dot{U}_c \end{bmatrix} = \begin{bmatrix} P_{aa} & P_{ab} & P_{ac} \\ P_{ba} & P_{bb} & P_{bc} \\ P_{ca} & P_{cb} & P_{cc} \end{bmatrix} \begin{bmatrix} \dot{Q}_a \\ \dot{Q}_b \\ \dot{Q}_c \end{bmatrix} \tag{2.15}$$

$$\dot{Q}_a + \dot{Q}_b + \dot{Q}_c = 0$$

$$P_{ij} = \frac{1}{2\pi\varepsilon}\ln\frac{S_{ij}}{d_{ij}} \tag{2.16}$$

where

P_{ij}—mutual electric potential coefficient between three phase conductors, $i, j =$ a, b, c

d_{ij}—spacing of centers of two phase conductors i and j, $i, j =$ a, b, c

S_{ij}—distance from one phase conductor to the mirror image point of another phase conductor, $i, j =$ a, b, c.

For a uniformly transposed three-phase transmission line, assuming that $P_{ab} \approx P_{ac}$, $P_{bc} \approx P_{ba}$, $P_{ca} \approx P_{cb}$, then:

$$\dot{U}_a = (P_{aa} - P_{ab})\dot{Q}_a$$
$$\dot{U}_b = (P_{bb} - P_{bc})\dot{Q}_b$$
$$\dot{U}_c = (P_{cc} - P_{ca})\dot{Q}_c$$

Let equivalent electric potential coefficients for three phases $P_a = \dot{U}_a/\dot{Q}_a$, $P_b = \dot{U}_b/\dot{Q}_b$, $P_c = \dot{U}_c/\dot{Q}_c$, then:

$$\left.\begin{aligned} P_a &= P_{aa} - P_{ab} \\ P_b &= P_{bb} - P_{bc} \\ P_{cc} &- P_{cc} - P_{ca} \end{aligned}\right\} \tag{2.17}$$

Assuming that $S_{ij} \approx 2H$, substituting it into Eq. (2.16) gives:

$$P_{ij} = \frac{1}{2\pi\varepsilon}\ln\frac{2H}{d_{ij}}$$

Substituting P_{aa}, P_{bb}, P_{cc}, and P_{ij} into Eq. (2.17) gives:

$$\begin{aligned} P_a &= \frac{1}{2\pi\varepsilon}\ln\left[\frac{d_{ab}}{(rNA^{N-1})^{1/N}}\right] \\ P_b &= \frac{1}{2\pi\varepsilon}\ln\left[\frac{d_{bc}}{(rNA^{N-1})^{1/N}}\right] \\ P_c &= \frac{1}{2\pi\varepsilon}\ln\left[\frac{d_{ca}}{(rNA^{N-1})^{1/N}}\right] \end{aligned} \tag{2.18}$$

The equivalent electric potential coefficient for a three-phase transmission line is:

$$P = \frac{1}{3}(P_a + P_b + P_c) = \frac{1}{2\pi\varepsilon}\ln\left[\frac{D_{eq}}{(rNA^{N-1})^{1/N}}\right]$$

The equivalent capacitance for a three-phase transmission line is:

$$C = \frac{1}{P} = 2\pi\varepsilon/\ln\left[\frac{D_{eq}}{(rNA^{N-1})^{1/N}}\right] \tag{2.19}$$

The susceptance B_0 per unit length of symmetrically arranged conductor bundles in a three-phase transmission line is:

$$B_0 = \omega C = 2\pi f \times 2\pi\varepsilon \times 1000/\ln\left[\frac{D_{\mathrm{eq}}}{(rNA^{N-1})^{1/N}}\right] \tag{2.20}$$

Through algebraic reduction, the equation becomes:

$$B_0 = 1.744 \times 10^{-5}/\ln\left[\frac{D_{\mathrm{eq}}}{(rNA^{N-1})^{1/N}}\right]$$

The capacitive reactance per unit length of symmetrically arranged conductor bundles in a three-phase transmission line is:

$$X_{C0} = 0.573 \times 10^5 \ln\left[\frac{D_{\mathrm{eq}}}{(rNA^{N-1})^{1/N}}\right] \tag{2.21}$$

2.1.1.3 Resistance per unit length of a conductor bundle

Because of skin effect and proximity effect, the AC resistance of a conductor is different from its DC resistance. Depending on the application, the DC or AC resistance of subconductors is calculated first, and then the resistance of a conductor bundle, which is the resultant resistance of subconductors in parallel connection is calculated.

$$\left.\begin{aligned} R_0 &= \frac{r_1}{N} \\ r_1 &= \frac{\rho}{S} \end{aligned}\right\} \tag{2.22}$$

where

R_0—resistance per unit length of conductor bundle, Ω/km

r_1—resistance per unit length of subconductor, Ω/km

N—number of subconductors

S—rated cross-sectional area of subconductor, mm^2

ρ—resistivity of conductor materials, (Ω mm)/km.

The resistance consulted from product catalogs or relevant manuals is generally the resistance at 20°C. For other temperatures, the resistance is corrected according to the following equation:

$$r_t = r_{20}[1 + \alpha(t - 20)]$$

where

r_t, r_{20}—resistance per unit length of subconductor at temperature t and 20°C, respectively, Ω/km

α—resistance temperature coefficient, 1/°C.

The actual resistance of the conductor is somewhat different from the calculated value because the actual length of ACSR is increased as a result of the twisting operation, and the cross-sectional area used for calculation is slightly greater than the actual area.

2.1.2 Impacts of Bundle Configuration of Conductors on Inductive and Capacitive Reactance of Lines

For phase conductors with approximately the same total cross-sectional area, different bundle configurations (including the spacing of subconductors or the diameter of conductor bundle) would cause varying inductive and capacitive reactance of the conductor, as shown in Table 2.1.

2.1.3 Comparison of Parameters Between EHV/UHV AC Transmission Lines

As shown in Eqs. (2.8), (2.20), and (2.22), the reactance, susceptance, and resistance of an EHV/UHV AC transmission line are dictated by the number and radius of subconductors, the diameter of conductor bundles, and the spacing of phase conductors, which, in turn, have a bearing on the corona characteristics requirements, power frequency electric and magnetic fields limits, insulation level of power transmission lines, and power transmission costs. Because the requirements of transmission capacity, corona-induced audible noise, limits imposed on radio interference, and power frequency electromagnetic field are not exactly the same, varieties of EHV/UHV AC transmission lines, even at the same voltage

Table 2.1: Impacts of Bundle Configuration of Conductor on Inductive and Capacitive Reactance.

Number of Subconductors	Total Cross-Sectional Area (mm^2)	Spacing of Subconductors (cm)	Diameter of Conductor Bundle (cm)	X_L (Ω/km)	X_C (Ω/km)
1	2515	–	–	0.556	0.189
2	2544	45	45	0.433	0.150
3	2625	45	52	0.390	0.136
4	2544	46	65	0.357	0.125
6	2392	46	92	0.319	0.111
8	2400	39	102	0.258	0.106
12	2539	33	127	0.215	0.096

Note: Mutual geometrical mean distance $D_{eq} = 14$ m.

Table 2.2: Typical Parameters of EHV/UHV AC Transmission Line.

Parameter	Voltage Class (kV)		
	500	750	1000
Conductor bundle	4×300	6×400	8×500
Spacing of subconductors (cm)	42.0	40.0	40.0
Diameter of conductor bundle (cm)	59.4	80.0	104.5
Phase-to-phase spacing (m)	13.0	21.7	26.3
R_0 (Ω/km)	0.02625	0.01261	0.00782
X_0 (Ω/km)	0.284	0.269	0.259
B_0 (S/km)	3.910×10^{-6}	4.306×10^{-6}	4.392×10^{-6}
R_0^* (1/km)	1.05×10^{-5}	0.224×10^{-5}	0.078×10^{-5}
X_0^* (1/km)	1.136×10^{-4}	0.478×10^{-4}	0.259×10^{-4}
B_0^* (1/km)	9.775×10^{-3}	24.221×10^{-3}	43.920×10^{-3}

Note: R_0^*, X_0^*, and B_0^* are per-unit values based on reference capacity of 100 MVA.

class, would have somewhat different reactance, susceptance, and resistance per unit length. The typical parameters of EHV/UHV AC transmission lines are given in Table 2.2.

2.1.4 Equivalent Circuit of UHV AC Transmission Line

The resistance, reactance, conductance, and susceptance of the UHV AC transmission line are uniformly distributed along the line length. In the calculation and analysis of power flow and electromechanical transient of power system, lumped equivalent parameters instead of distributed parameters are generally used. In contrast, in the calculation and analysis of electromagnetic transient, distributed parameters are used.

1. **Voltage and current equations for Π-type equivalent circuit**

   The Π-type equivalent circuit is shown in Figure 2.1, from which the equations for voltage and current at the sending and receiving ends of a transmission line can be yielded:

$$\dot{U}_1 = Z\left(\dot{I}_2 + \frac{Y}{2}\dot{U}_2\right) + \dot{U}_2 = \left(\frac{ZY}{2} + 1\right)\dot{U}_2 + Z\dot{I}_2 \tag{2.23}$$

$$\dot{I}_1 = \frac{Y}{2}\dot{U}_1 + \frac{Y}{2}\dot{U}_2 + \dot{I}_2 = Y\left(\frac{ZY}{4} + 1\right)\dot{U}_2 + \left(\frac{YZ}{2} + 1\right)\dot{I}_2 \tag{2.24}$$

Equations (2.23) and (2.24) can be expressed in matrix form:

$$\begin{bmatrix} \dot{U}_2 \\ \dot{I}_1 \end{bmatrix} = \begin{bmatrix} \dfrac{ZY}{2} + 1 & Z \\ Y\left(\dfrac{ZY}{4} + 1\right) & \dfrac{YZ}{2} + 1 \end{bmatrix} \begin{bmatrix} \dot{U}_2 \\ \dot{I}_2 \end{bmatrix} \tag{2.25}$$

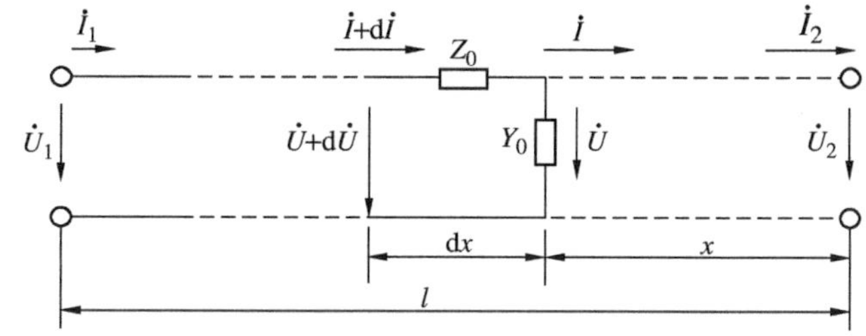

Figure 2.3
Schematic diagram of an equivalent circuit with distributed parameters.

2. **Calculation of equivalent circuit with distributed parameters**

An equivalent circuit with distributed parameters consists of a number of Π-type equivalent networks each having unit length line. Figure 2.3 depicts its schematic diagram. In this figure, the series impedance per unit length of the line $Z_0 = R_0 + \mathrm{j}X_0$ and the shunt admittance per unit length of the line $Y_0 = G_0 + \mathrm{j}B_0$. The differential equation for voltage is:

$$\frac{\mathrm{d}\dot{U}}{\mathrm{d}x} = \dot{I}Z_0 \tag{2.26}$$

The differential equation for current is:

$$\frac{\mathrm{d}\dot{I}}{\mathrm{d}x} = \dot{U}Y_0 \tag{2.27}$$

Equations (2.26) and (2.27) may be transformed and reduced to the following equation:

$$\frac{\mathrm{d}^2\dot{U}}{\mathrm{d}x^2} - Z_0Y_0\dot{U} = 0 \tag{2.28}$$

Let $P = \mathrm{d}/\mathrm{d}x$ be the differential operator; then, Eq. (2.28) is transformed to $(P^2 - Z_0Y_0)\dot{U} = 0$. The root of its characteristic equation $P^2 - Z_0Y_0 = 0$ is:

$$P_{1,2} = \pm\sqrt{Z_0Y_0} = \pm\gamma$$

The solution of the differential equation is:

$$\left.\begin{aligned} \dot{U} &= C_1\mathrm{e}^{\gamma x} + C_2\mathrm{e}^{-\gamma x} \\ \frac{\mathrm{d}\dot{U}}{\mathrm{d}x} &= C_1\gamma\mathrm{e}^{\gamma x} - C_2\gamma\mathrm{e}^{-\gamma x} \end{aligned}\right\} \tag{2.29}$$

From Eq. (2.26), we can obtain:

$$\dot{I} = \frac{1}{Z_0}\left(\frac{d\dot{U}}{dx}\right) = \frac{C_1}{\sqrt{Z_0/Y_0}}e^{\gamma x} - \frac{C_2}{\sqrt{Z_0/Y_0}}e^{-\gamma x} \tag{2.30}$$

where $\gamma = \sqrt{Z_0 Y_0} = \alpha + j\beta$ is the propagation constant of transmission line, with the real part being the attenuation coefficient and the imaginary part being the phase coefficient.

Thus, the exponential terms $e^{\gamma x}$ and $e^{-\gamma x}$ in Eq. (2.30) may be transformed to:

$$e^{\gamma x} = e^{(\alpha+j\beta)x} = e^{\alpha x}(\cos\beta x + j\sin\beta x)$$

$$e^{-\gamma x} = e^{-(\alpha+j\beta x)} = e^{-\alpha x}(\cos\beta x - j\sin\beta x)$$

$Z_C = \sqrt{Z_0/Y_0} = R_C + jX_C$ is referred to as the wave impedance of transmission line, also known as the characteristic impedance.

For EHV/UHV AC transmission lines, $G_0 \ll B_0$, $R_0 \ll X_0$, G_0 and R_0 are generally negligible, then:

$$\gamma = \sqrt{j\omega L_0 \bullet j\omega C_0} = j\omega\sqrt{L_0 C_0} = j\beta \tag{2.31}$$

$$Z_C = \sqrt{\frac{j\omega L_0}{j\omega C_0}} = \sqrt{\frac{L_0}{C_0}} \tag{2.32}$$

The boundary condition is: at $x = 0$, $\dot{U} = \dot{U}_2$, $\dot{I} = \dot{I}_2$, which are substituted into Eqs. (2.29) and (2.30) to derive two constants C_1 and C_2:

$$C_1 = \frac{1}{2}(\dot{U}_2 + Z_C\dot{I}_2)$$

$$C_2 = \frac{1}{2}(\dot{U}_2 - Z_C\dot{I}_2)$$

Substituting the constants C_1 and C_2 into Eqs. (2.29) and (2.30) gives:

$$\begin{aligned}\dot{U} &= \frac{1}{2}(e^{\gamma x} + e^{-\gamma x})\,\dot{U}_2 + \frac{1}{2}(e^{\gamma x} - e^{-\gamma x})Z_C\dot{I}_2 \\ &= \dot{U}_2\cosh\gamma x + Z_C\dot{I}_2\sinh\gamma x\end{aligned} \tag{2.33}$$

$$\begin{aligned}\dot{I} &= \frac{1}{2Z_C}(e^{\gamma x} - e^{-\gamma x})\dot{U}_2 + \frac{1}{2}(e^{\gamma x} + e^{-\gamma x})\dot{I}_2 \\ &= \frac{1}{Z_C}\dot{U}_2\sinh\gamma x + \dot{I}_2\cosh\gamma x\end{aligned} \tag{2.34}$$

If the resistance and conductance are neglected, the transmission line would be lossless, then:

$$\dot{U} = \dot{U}_2\cos\beta x + jZ_C\dot{I}_2\sin\beta x \tag{2.35}$$

$$\dot{I} = j\frac{1}{Z_C}\dot{U}_2\sin\beta x + \dot{I}_2\cos\beta x \tag{2.36}$$

In Eqs. (2.33) and (2.34), let $x = l$; one can derive the equations for voltage and current at the sending and receiving ends of AC transmission line with a length of l, which are expressed in the matrix form:

$$\begin{bmatrix} \dot{U}_1 \\ \dot{I}_1 \end{bmatrix} = \begin{bmatrix} \cosh\gamma l & Z_C\sinh\gamma l \\ \dfrac{\sinh\gamma l}{Z_C} & \cosh\gamma l \end{bmatrix} \begin{bmatrix} \dot{U}_2 \\ \dot{I}_2 \end{bmatrix} \tag{2.37}$$

By comparing Eqs. (2.25) and (2.37), one can derive the impedance and admittance of the Π-type equivalent circuit for the AC transmission line with a length of l:

$$Z = Z_C\sinh\gamma l \tag{2.38}$$

$$Y = \frac{2(\cosh\gamma l - 1)}{Z_C\sinh\gamma l} \tag{2.39}$$

3. **Wave impedance and propagation constant of EHV/UHV AC transmission lines**
   The wave impedance Z_C and propagation constant γ of an EHV/UHV AC transmission line are related to the configuration and phase-to-phase spacing of conductor bundles, regardless of the line length. In other words, the wave impedance and propagation constant are a function of the configuration and phase-to-phase spacing of conductor bundles. However, the wave impedance and propagation constant vary slightly for different transmission lines at the same voltage class. The typical wave impedance and propagation constant of EHV/UHV AC transmission lines at various voltage classes are shown in Table 2.3. It follows that the propagation constant varies slightly for transmission lines at different voltage classes.

Table 2.3: Characteristic Impedance and Propagation Constant of EHV/UHV AC Transmission Lines.

Voltage Class (kV)	500	750	1000
Z_C (Ω)	$270.1\angle -2.64°$	$250.1\angle -1.34°$	$242.9\angle -0.86°$
γ (rad/km)	$1.056\ 0\times10^{-3}\angle -87.36°$	$1.077\ 0\times10^{-3}\angle -88.66°$	$1.066\ 8\times10^{-3}\angle -89.14°$
α (nepers/km)	0.0486×10^{-3}	0.0252×10^{-3}	0.0161×10^{-3}
β (rad/km)	1.0549×10^{-3}	1.0767×10^{-3}	1.0667×10^{-3}

Table 2.4: Calculated Equivalent Parameters for a 300 km Transmission Line at 1000 kV.

Resistance (Ω)		Reactance (Ω)		Susceptance (S)	
R	*R'*	*X*	*X'*	*B*	*B'*
2.266	2.345	76.38	77.70	1.329×10^{-3}	1.318×10^{-3}

4. **Methods for calculating parameters of Π-type equivalent circuits**

   There are two methods for calculating the impedance and admittance of the Π-type equivalent circuit of a transmission line: (i) using distributed parameters and (ii) multiplying parameters of unit length by line distance.

   The calculation procedure of the first method is as follows:

1. Calculate the resistance, reactance, conductance, and susceptance per kilometer based on the configuration and phase-to-phase spacing of conductor bundles.
2. Calculate the wave impedance Z_C and propagation constant γ.
3. Calculate the impedance Z and admittance Y by substituting the line length into Eqs. (2.38) and (2.39), respectively.

   The calculation procedure of the second method is as follows:

1. Calculate the impedance Z_0 and admittance Y_0 per kilometer.
2. Multiply Z_0 and Y_0 by the line length, respectively, to obtain the lumped parameters of the Π-type equivalent circuit.

   The second method is used to approximately calculate the line parameters. For a 300-km transmission line at 1000 kV, the equivalent resistance R, reactance X, and susceptance B calculated using the first method, and the equivalent resistance R', reactance X', and susceptance B' calculated using the second method are shown in Table 2.4. It follows that the resistance and reactance calculated using the second method are somewhat larger, whereas the calculated susceptance is somewhat smaller. The calculation errors increase with the line length.

2.2 Transmission Characteristics of UHV AC Transmission Lines

2.2.1 Surge Impedance Load

Figure 2.4 shows the Π-type equivalent circuit for a UHV AC transmission line, where $Z = R + jX$, $Y = G + jB$. In calculation, the power loss attributable to corona and leakage current of insulators are neglected. Let the parallel conductance $G = 0$. $\dot{U}_1$ and $\dot{U}_2$ are the sending and receiving-end line voltages, respectively. $\dot{I}_1$ and $\dot{I}_2$ are the sending and receiving-end line currents, respectively. $\tilde{S}$ is the three-phase complex power, $\tilde{S} = P + jQ$.

Given $\dot{U}_2$, $\dot{I}_2$, and $P_2 + jQ_2$, we can draw the phasor diagram for $\dot{U}_1$, $\dot{I}_1$, $\dot{U}_2$, and $\dot{I}_2$, as shown in Figure 2.5.

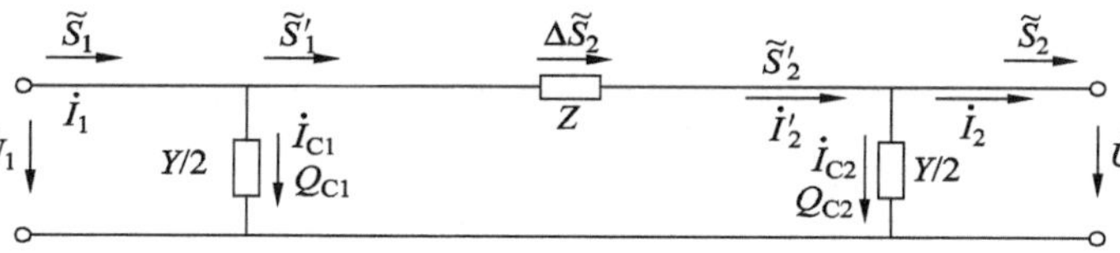

Figure 2.4
Π-type equivalent circuit for calculating power and voltage of a transmission line.

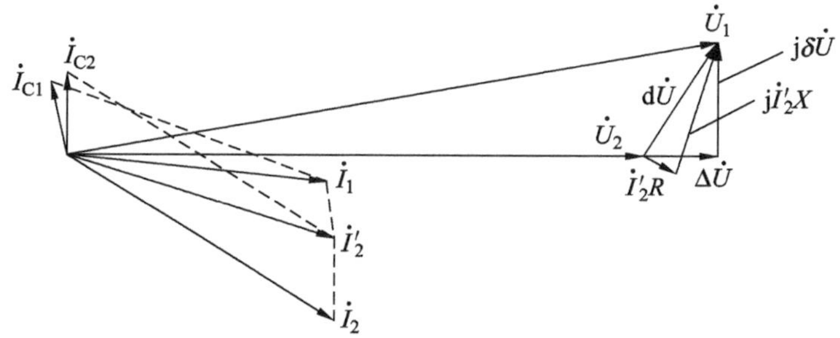

Figure 2.5
Phasor diagram for voltage and current of a transmission line.

The power loss $\Delta\tilde{S}_2$ of a transmission line is:

$$\Delta\tilde{S}_2 = I_2'^2 Z = \left(\frac{S_2'}{U_2}\right)^2 Z = \frac{P_2'^2 + Q_2'^2}{U_2^2}(R + jX) = \frac{P_2^2 + (Q_2 - Q_{C2})^2}{U_2^2}(R + jX) \tag{2.40}$$

The active power loss ΔP_2 and reactive power loss ΔQ_2 attributable to the line impedance Z are, respectively:

$$\left.\begin{aligned} \Delta P_2 &= \frac{P_2^2 + (Q_2 - Q_{C2})^2}{U_2^2} R \\ \Delta Q_2 &= \frac{P_2^2 + (Q_2 - Q_{C2})^2}{U_2^2} X \end{aligned}\right\} \tag{2.41}$$

where

Q_{C2}—charging power or reactive power due to line susceptance

$$\begin{aligned} \tilde{Q}_{C2} &= \dot{U}_2\left(\frac{Y}{2}\dot{U}_2\right)^* = \frac{1}{2}(0 - jB)U_2^2 = -\frac{1}{2}jBU_2^2 \\ Q_{C2} &= \frac{1}{2}BU_2^2 \end{aligned} \tag{2.42}$$

The natural power, also known as surge impedance load (SIL), indicates that a surge impedance of a line $Z_C = \sqrt{Z_0/Y_0}$ is respectively connected to each phase at the receiving end of a transmission line. If a surge impedance of a line Z_C is connected at the receiving end of a transmission line, then we have:

$$\frac{\dot{U}_2}{\dot{I}_2} = Z_C \text{ or } \dot{U}_2 = \dot{I}_2 Z_C \tag{2.43}$$

The wave impedance power $\tilde{S}_{ZC}$ is given by:

$$\tilde{S}_{ZC} = P_2 + jQ_2 = \frac{U_2^2}{Z_C^*} \tag{2.44}$$

Given the surge impedance $Z_C = 242.9\angle -0.86°$ as shown in Table 2.3, the wave impedance power of a 1000-kV transmission line $\tilde{S}_{ZC}$ is 4116.5-j61.8 (MVA). Considering the effects of the type and configuration of conductors, the impedance parameters of different types of UHV AC transmission lines may vary, causing the wave impedance power of the lines to change correspondingly. For example, if the delta arrangement of conductors in the UHV AC transmission line as shown in Table 2.3 is changed to horizontal arrangement, then the surge impedance Z_C will be changed to $253.9\angle -0.84°$, and the wave impedance power $\tilde{S}_{ZC}$ will become 3938.1-j57.7 (MVA).

Substituting Eq. (2.44) into Eq. (2.37) gives:

$$\begin{aligned}\dot{U}_1 &= \dot{U}_2\cosh\gamma L + \dot{U}_2\sinh\gamma L = (\cosh\gamma L + \sinh\gamma L)\dot{U}_2\\ \dot{I}_1 &= \dot{I}_2\sinh\gamma L + \dot{I}_2\cosh\gamma L = (\cosh\gamma L + \sinh\gamma L)\dot{I}_2\end{aligned}$$

Dividing the upper equation by the lower one, we have:

$$\frac{\dot{U}_1}{\dot{I}_1} = \frac{\dot{U}_2}{\dot{I}_2} = Z_C \text{ or } \dot{U}_1 = \dot{I}_1 Z_C \tag{2.45}$$

From the viewpoint of wave impedance power transmitted by a UHV AC line, the UHV AC transmission technology has the following characteristics:

1. The voltage and current at both the sending and receiving ends are in phase; the power factor and the magnitude of voltage and current along the line all remain unchanged.
2. The reactive power loss attributable to the line reactance is substantially equal to the reactive power generated by the line susceptance (line capacitance). Thus, reactive power is neither generated nor absorbed at both the sending and receiving ends of the transmission line.

The reactive power generated by the line capacitance is only related to U_1 and U_2, and is barely related to the transmitted active and reactive power, whereas the reactive power loss attributable to the line reactance is proportional to the square of transmitted power. If the transmitted power is higher than the wave impedance power, then the power source at the sending end has to inject reactive power into the line to maintain reactive power balance and voltage stability. As the transmitted power is increased, the reactive power to be injected will be increased correspondingly. If the transmitted power is lower than the wave impedance power, then the reactive power generated by the line capacitance will become higher than the reactive power consumed by the line reactance, resulting in an increase of voltage at both the sending and receiving ends. As a result, the power source at the sending end has to absorb reactive power.

2.2.2 Transmission of Active and Reactive Power

SIL is a function of voltage and impedance and admittance per unit length of a transmission line, regardless of the line length. The reactive power generated by the line capacitance and the reactive power loss attributable to the line reactance are both a function of the line length, that is, both the reactive power loss attributable to the line reactance and the reactive power generated by the line capacitance are increased with the line length, and vice versa. For an EHV AC transmission line and a UHV AC transmission line, the relationship between transmitted power and reactive power is identical.

Figure 2.6 shows the relationship between the transmitted power and the net reactive power (i.e., reactive power difference between the reactive power consumed by the line reactance

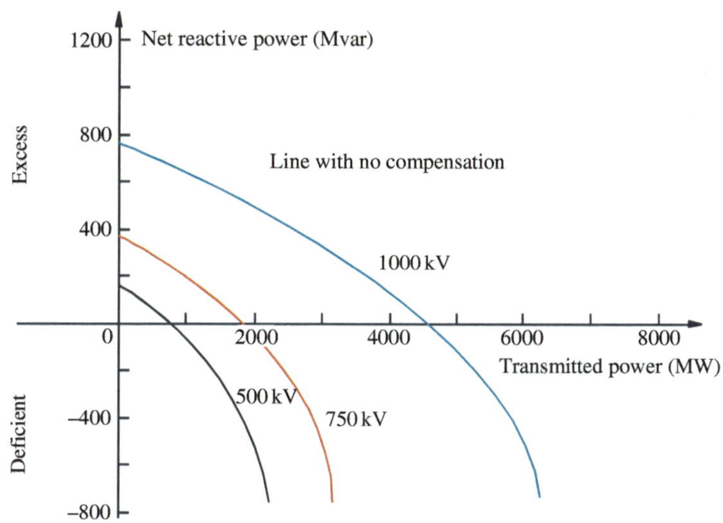

Figure 2.6
Net reactive power versus transmitted power.

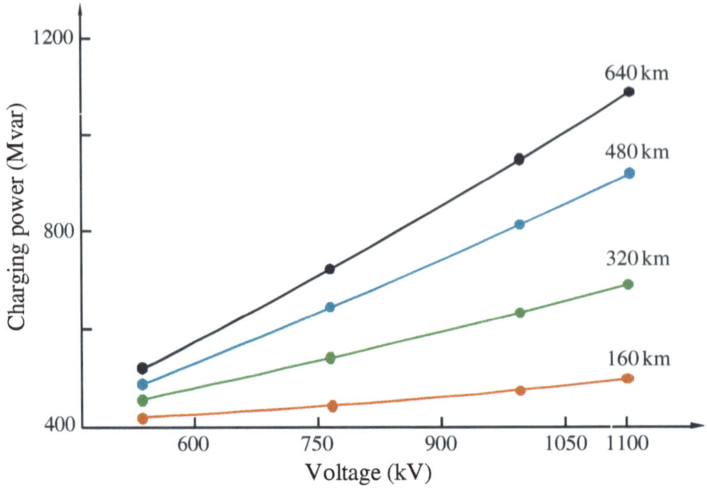

Figure 2.7
Reactive power generated by line capacitance versus line voltage and transmission distance.

and the reactive power generated by the line capacitance) for a 161-km transmission line at different voltage classes.

Figure 2.6 shows that the capacitance of a 1000-kV AC transmission line generates much more reactive power than that of 500- and 750-kV transmission lines. Therefore, it is more difficult to control reactive voltage for a UHV AC transmission line than an EHV AC transmission line.

Figure 2.7 shows the relationship between the reactive power generated by capacitance (i.e., charging power) and the transmission distance of transmission lines at different voltage classes. It follows that a 1000-kV UHV AC transmission line generates reactive power almost five-times that of a 500-kV line.

To limit the temporary overvoltage, shunt reactors are generally installed at the sending and receiving ends of EHV/UHV AC transmission line for reactive power compensation. For a 500-kV transmission line, the shunt compensation equipment includes HV shunt reactors and LV shunt reactors, which generally compensate 90% or more charging power for the lines. For the UHV AC transmission line, the shunt compensation capacity is determined by considering both the temporary overvoltage limit and reactive power regulation for different transmitted powers. Generally, the total capacity of HV and LV shunt reactors should allow the charging power to be substantially compensated. The capacity and withstand voltage class of HV and LV shunt reactors are determined according to the system conditions and their respective characteristics. Because shunt reactors are connected to a transmission line, the reactive power difference–power transmission characteristics curve tends to shift downward. When the transmitted power approaches or exceeds the wave impedance power, the self-balance of reactive power of the line would be upset and, as a result, the line needs to absorb a large amount of reactive power, which is increased with the increase in transmitted active power. This would make it more difficult to regulate the voltage and reactive power, and even compromise the power transmission capability of the UHV AC transmission line. If controllable reactors instead of fixed shunt reactors are used for compensation, then it is possible to satisfy the requirements for both temporary overvoltage limit and reactive power regulation, which is desirable for the operation of UHV AC power grid. The controllable reactors are regulated in such a manner that when a transmission line transmits a small power or is operated under no load, the compensation capacity reaches the maximum; when the power transmitted through the line is increased, the required compensation capacity is gradually reduced and the reactive power consumed by the line reactance is mainly offset by the reactive power generated by the line capacitance; and when three-phase tripping takes place for load shedding, the required compensation capacity is rapidly increased to limit the temporary overvoltage [1].

2.2.3 Power Loss and Voltage Decline

1. **Power loss of a transmission line**
   Equation (2.41) shows that the active power loss of a transmission line is proportional to the sum of squares of transmitted active and reactive power and is inversely proportional to the square of voltage. Therefore, for the same power transmitted, the increase in voltage of a transmission line can considerably reduce the active power loss. Reducing the transmitted reactive power can greatly reduce the active and reactive power loss, improve the cost effectiveness of line operation, and reduce the investment for shunt reactive power compensation at the receiving end.
   Equation (2.42) shows that the reactive power generated by the equivalent line capacitance is proportional to the square of the voltage. The susceptance per unit length of a 1000-kV UHV AC transmission line is approximately 1.1-times that of a 500-kV line. Therefore, the reactive power generated by capacitance of a 1000-kV UHV AC transmission line is approximately 4.4-times that of a 500-kV line.
2. **Comparison of power loss between EHV/UHV AC transmission lines**
   The power loss attributable to resistance of a transmission line is proportional to the square of current flowing through the line and is proportional to the resistance. The power loss attributable to resistance is a function of transmission distance, conductor resistivity, and voltage. The current flowing through the line ($I = S/U$) is inversely proportional to the voltage. Given a certain power transmitted, increasing the line voltage allows the current to be reduced, thereby considerably reducing the power loss due to line resistance. Additionally, increasing the conductor cross-section and reducing the resistivity of conductor materials can lower the line resistance, thereby reducing the resulting power loss.

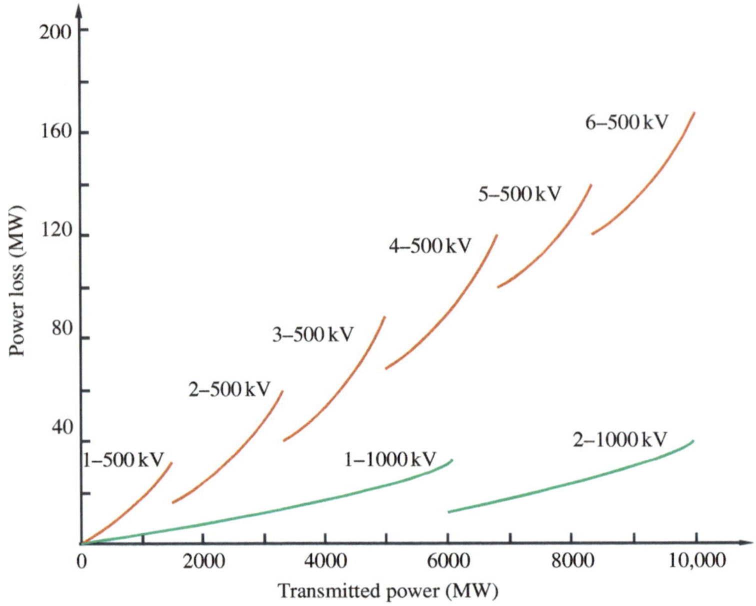

Figure 2.8
Power loss versus transmitted power for 1000- and 500-kV transmission lines.

Equation (2.42) shows that for a given power transmitted, the power loss attributable to line resistance is inversely proportional to the square of voltage while proportional to the resistance. Normally, the resistance per kilometer of a 1000-kV UHV AC transmission line is approximately 30% of that of a 500-kV line. If the current flowing through two transmission lines, respectively, at 1000 and 500 kV, then the power loss due to resistance of the former is only 30% of that of the latter. Thus, the application of UHV AC transmission line can significantly reduce the power loss due to line resistance. Figure 2.8 shows the relationship between power loss due to resistance and transmitted power of a 161-km transmission line, respectively, at 500 and 1000 kV.

In Figure 2.4, the sending-end power (i.e., the power fed into the line from the sending-end system or the power source) is:

$$\begin{aligned}\tilde{S}_1 &= \tilde{S}_1' - \mathrm{j}Q_{\mathrm{C1}} = \tilde{S}_2' + \Delta\tilde{S}_2 - \mathrm{j}Q_{\mathrm{C1}} = P_2 + \Delta P_2 + \mathrm{j}[Q_2 + \Delta Q_2 - (Q_{\mathrm{C1}} + Q_{\mathrm{C2}})] \\ &= P_1 + \mathrm{j}Q_1\end{aligned} \tag{2.46}$$

$$\left.\begin{aligned} P_1 &= P_2 + \Delta P_2 \\ Q_1 &= Q_2 + \Delta Q_2 - (Q_{\mathrm{C1}} + Q_{\mathrm{C2}}) \end{aligned}\right\} \tag{2.47}$$

where

$$Q_{C1} = \frac{1}{2} B U_1^2.$$

The power transmission efficiency $\eta = P_2/P_1 \times 100\%$ is the ratio of output power to input power of a line. Because of low power loss, the power transmission efficiency of a 1000-kV UHV AC transmission line is much higher than that of a 500-kV line.

Reducing the power loss due to line resistance gains considerable economic benefits in the following aspects:

a. The installed capacity of generating units in the power grid can be reduced. Equation (2.47) shows that, given a constant power demand, reducing power loss of the line can not only reduce the installed capacity but also the operating capacity. In other words, with the operating capacity of generating units kept unchanged, the saved power may be used for supply, thereby reducing the installed capacity of generating units in the power grid and, consequently, the construction costs of power plants.
b. The pressure of peak load regulation of the power grid can be mitigated. The power loss of a line is proportional to the square of transmitted power. During the peak load period, the transmitted power is increased, resulting in increased power loss of the line. Apparently, reducing power loss of the line can mitigate the pressure of peak load regulation of the power grid during the peak load period.
c. The power transmission efficiency can be increased and the electrical energy loss can be reduced, thereby reducing the operating costs of the power grid. Apart from the power loss due to resistance, corona discharge and leakage current of insulators also contribute to power loss of a UHV AC transmission line. The UHV AC transmission lines are designed to meet the environmental requirements, such as the audible noise limit. Therefore, the corona power loss of a UHV AC transmission line is comparable in magnitude with that of an EHV AC transmission line but accounts for a smaller percentage of transmitted power. If asymmetrical configuration of conductor bundles is used, then the corona loss can be reduced further. The power loss due to the leakage current of insulators is insignificant. Thus, from the perspective of economic benefits, under normal weather conditions, these two kinds of power losses of the UHV AC transmission line are negligible when compared with the power loss due to line resistance.

3. **Voltage decline across a transmission line**

As shown in Figure 2.5, let $\dot{U}_2 = U_2 e^{j0°}$. Then, the sending-end voltage of the transmission line is:

$$\begin{aligned}\dot{U}_1 &= \dot{U}_2 + \dot{I}'_2 Z = \dot{U}_2 + \left(\frac{\dot{S}'_2}{\dot{U}_2}\right)^* Z = \dot{U}_2 + \frac{P_2 - j(Q_2 - Q_{C2})}{U_2}(R + jX) \\ &= \left[\dot{U}_2 + \frac{P_2 R + (Q_2 - Q_{C2})X}{U_2}\right] + j\frac{P_2 X - (Q_2 - Q_{C2})R}{U_2} \\ &= (\dot{U}_2 + \Delta U) + j\delta U = \dot{U}_2 + d\dot{U}\end{aligned} \tag{2.48}$$

$$\left.\begin{aligned}\Delta U &= \frac{P_2 R + (Q_2 - Q_{C2})X}{U_2} \\ \delta U &= \frac{P_2 X - (Q_2 - Q_{C2})R}{U_2}\end{aligned}\right\} \tag{2.49}$$

$$U_1 \approx U_2 + \Delta U \tag{2.50}$$

where

$d\dot{U}$—voltage drop

ΔU—voltage loss, usually expressed as a percentage, that is,

$\Delta U\% = (U_1 - U_2)/U_N \times 100\%$

U_N—rated voltage of line.

It follows from Eq. (2.49) that the voltage loss is proportional to the transmitted reactive power and inversely proportional to voltage. Therefore, reducing the transmitted reactive power is helpful to regulate the system voltage, increase the receiving-end voltage class, and improve the voltage stability.

2.2.4 Power–Voltage Characteristics

1. **Voltage and current profiles for a transmission line open-circuited at receiving end**

With a transmission line open-circuited at the receiving end, $\dot{I}_2 = 0$, Eqs. (2.35) and (2.36) may be simplified to:

$$\dot{U} = \dot{U}_2 \cos \beta x \tag{2.51}$$

$$\dot{I} = j\frac{\dot{U}_2}{Z_C} \sin \beta x \tag{2.52}$$

Assuming that the line length is l, the voltage $\dot{U}_1$ and current $\dot{I}_1$ at the sending end are:

$$\dot{U}_1 = \dot{U}_2 \cos \beta l = \dot{U}_2 \cos \theta \tag{2.53}$$

$$\dot{I}_1 = \mathrm{j}\frac{\dot{U}_2}{Z_C} \sin \beta l = \mathrm{j}\frac{\dot{U}_1}{Z_C} \tan \theta \tag{2.54}$$

where

θ—line angle, $\theta = \beta l$, rad

The voltage $\dot{U}$ and current $\dot{I}$ of a transmission line may be expressed using the sending-end voltage $\dot{U}_1$:

$$\dot{U} = \dot{U}_1 \frac{\cos \beta x}{\cos \theta} \quad \text{or} \quad \frac{U}{U_1} = \frac{\cos \beta x}{\cos \theta} \tag{2.55}$$

$$\dot{I} = \mathrm{j}\frac{\dot{U}_1}{Z_C}\frac{\sin \beta x}{\cos \theta} \quad \text{or} \quad I/\frac{U_1}{Z_C} = \frac{\sin \beta x}{\cos \theta} \tag{2.56}$$

By using the transmission line parameters given in Table 2.3 and setting the sending-end voltage to the rated value and maintaining it constant, the voltage and current profiles along the transmission line can be calculated. Figure 2.9 shows the voltage and current profiles for a 300-km 1000-kV UHV AC transmission line open-circuited at the receiving end.

It follows from Figure 2.9 that for a 300-km 1000-kV AC transmission line open-circuited at the receiving end, the receiving-end voltage is 5.2% higher than the

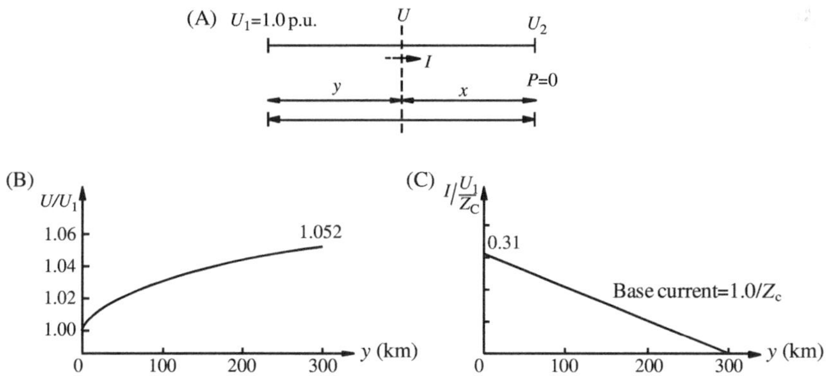

Figure 2.9
Voltage and current profiles for a 1000-kV AC transmission line open-circuited at receiving end: (A) schematic diagram, (B) voltage profile, and (C) current profile.

sending-end voltage; whereas for a 600-km transmission line open-circuited at receiving end, the steady-state voltage is 1.24 p.u., that is, the receiving-end voltage is 24% higher than the sending-end voltage. Thus, for long-distance UHV AC transmission lines, it is appropriate to set voltage supporting points at an interval of approximately 300 km.

2. **Voltage and current profiles for a transmission line connected with power sources at both ends and under no-load condition**
   Assuming that the voltages at the sending and receiving ends, respectively, are equal, for a lossless transmission line the voltage and current profiles may be given by:

$$\dot{U} = \dot{U}_1 \frac{\cos \beta(l/2 - 1)}{\cos(\theta/2)} \tag{2.57}$$

$$\dot{I} = -\mathrm{j}\frac{\dot{U}_1}{Z_C}\frac{\sin \beta(l/2 - x)}{\cos(\theta/2)} \tag{2.58}$$

   Figure 2.10 shows the voltage and current profiles along a 400-km UHV AC transmission line under no-load conditions where the voltages at the sending-end bus and at the receiving-end bus are constant.
   It follows from Figure 2.10 that, to allow stable operation of the UHV AC system, the generators at both the sending and receiving ends will be able to absorb the reactive power resulting from line charging. If the amount of reactive power exceeds the generator's reactive power capability limited by the underexcitation

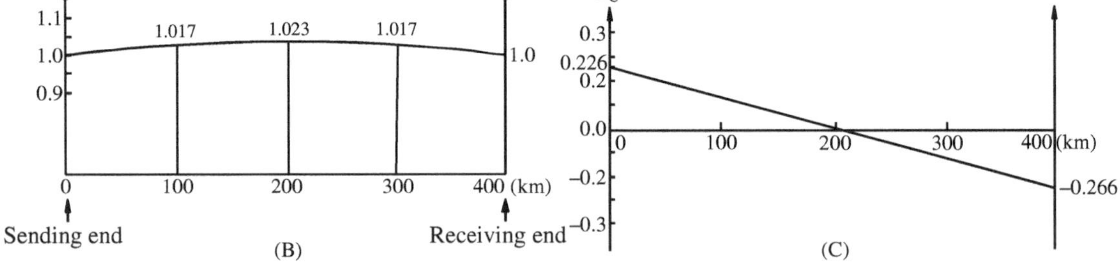

Figure 2.10
Voltage and current profiles for a 400-km line under no-load condition: (A) schematic diagram, (B) voltage profile, and (C) current profile.

condition or stability requirements, then reactive power compensation devices must be installed.

3. **Bus voltage variations during load increasing**

Corresponding to a load of $P_2 + jQ_2$ at the receiving end, we have:

$$\dot{I}_2 = \frac{P_2 - jQ_2}{U_2^*}$$

Substituting $\dot{I}_2$ into Eq. (2.42) and letting $x = l$ gives:

$$\dot{U}_1 = \dot{U}_2 \cos\theta + jZ_C \sin\theta \left(\frac{P_2 - jQ_2}{U_2^*} \right) \tag{2.59}$$

For any given receiving-end load and sending-end voltages, the receiving-end voltage can be derived from Eq. (2.59). Figure 2.11 shows the receiving-end voltage U_2 compared with load P_2 curve in the case of sending-end voltage being fixed.

From Figure 2.11, the following fundamental properties of AC transmission are evident:

a. There is an inherent maximum limit of power that can be transmitted at any load power factor.
b. Any value of power below the maximum can be transmitted at two different voltage classes. The normal operation corresponds to the upper value, within narrow limits at approximately 1.0 p.u. The operation at the lower voltage may lead to voltage instability.
c. The load power factor has a significant influence on U_2 and the maximum power that can be transmitted. When the load connected at the receiving end is inductive,

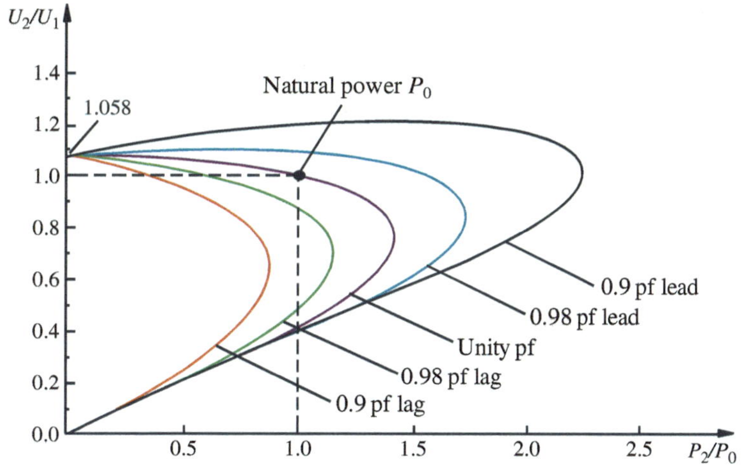

Figure 2.11
Receiving-end voltage versus power characteristic.

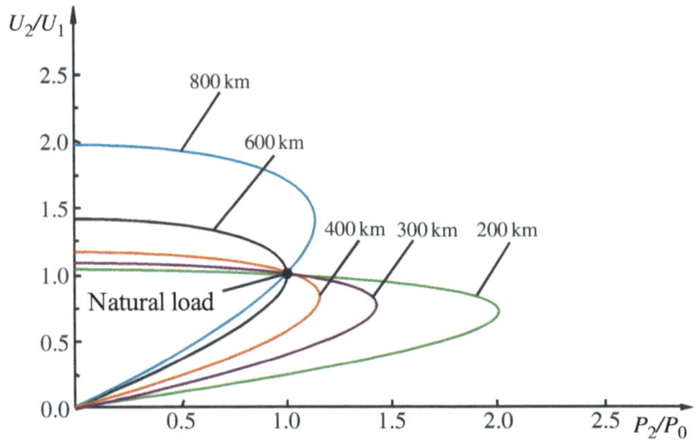

Figure 2.12
Receiving-end voltage versus line length and load.

the power limit and U_2 are both less. When the load connected at the receiving end is capacitive, the voltage U_2 is relatively stable, and the power limit is higher. This means that the receiving-end voltage can be regulated by the addition of shunt capacitive compensation or static var compensator (SVC) to increase the maximum limit of power that can be transmitted.

Figure 2.12 shows the relationship between receiving-end voltage, line length, and load. For long lines, the receiving-end voltage U_2 is very sensitive to the variations in load P_2.

2.3 Calculation Methods for Stability and Transmission Capability of UHV AC System

2.3.1 Basic Concept of Power System Stability

As an inherent attribute of the power system, stability is the ability of synchronous generators in the power system to maintain or restore synchronous operation following a disturbance. Maintaining stability is essential for normal operation of the power system.

To date, no consensus standard is available to classify the power system stability around the world. According to DL 755-2001 *Guide on Security and Stability for Power System*, power system stability is generally classified into power angle stability, frequency stability, and voltage stability, which can be further subdivided into a number of subcategories in terms of the magnitude of disturbance, characteristics of dynamic process, as well as the operating characteristics of the elements and control systems involved in operation. The classification of power system stability according to DL 755-2001 is shown in Figure 2.13.

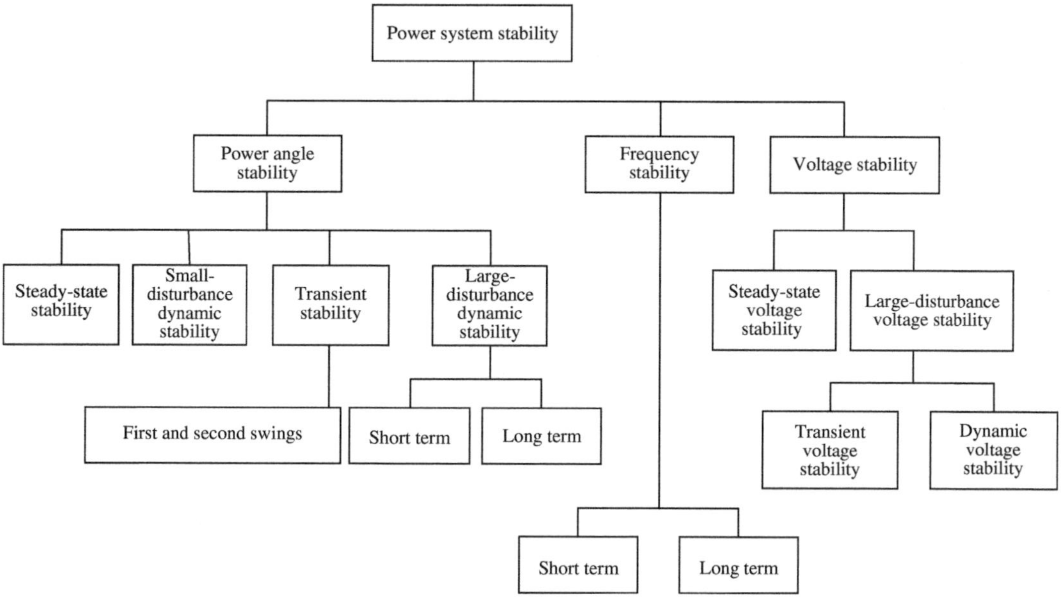

Figure 2.13
Classification of power system stability according to DL 755-2001.

2.3.1.1 Power angle stability

Power angle instability means that following a disturbance, the interconnected synchronous generators can no longer maintain synchronism. The ability of generators to maintain synchronism following a disturbance in the power system depends on their electromagnetic torques (including synchronizing torque and damping torque). The relationship between the change in electromagnetic torque T_e and the synchronizing and damping torques is given by:

$$\Delta T_e = T_s \Delta\delta + T_D \Delta\omega \tag{2.60}$$

where

T_s—synchronizing torque coefficient

T_D—damping torque coefficient

$T_s\Delta\delta$—synchronizing torque, in phase with power angle perturbation

$T_D\Delta\omega$—damping torque, in phase with speed deviation.

The power angle stability of the power system hinges on the magnitude and sign (positive, negative) of synchronizing and damping torques of each generator. Insufficient synchronizing torque would result in aperiodic drift in rotor angle, and insufficient damping torque would result in asynchronous oscillation.

The power angle stability of a power system can be classified into steady-state stability, transient stability, small-disturbance dynamic stability, and large-disturbance dynamic stability.

1. **Steady-state stability**

   Steady-state stability is concerned with the ability of the power system to avoid aperiodic instability following a small disturbance. Its physical characteristic is anti-small-disturbance dynamic stability in connection with the synchronizing torque. It is mainly used to define the steady-state stability reserve of the power system during normal operation and following a contingency.

   The connection diagram of a generator connected to the power system through the UHV transmission line is shown in Figure 2.14. The internal electromotive force of the generator is E_q and the internal impedance is X_d. The parameters of the system excluding the generator are equivalent to jX_s, $X_s = X_{T1} + 1/2X_L + X_{T2} + X_D$, where X_{T1} and X_{T2} are the short-circuit reactance of step-up and step-down transformers, respectively; X_L is the line reactance; and X_D is the equivalent reactance of remaining portions of the power system.

   Figure 2.15 shows the power-angle characteristic curve of the power system shown in Figure 2.14. Under normal operating conditions, there is an equilibrium between

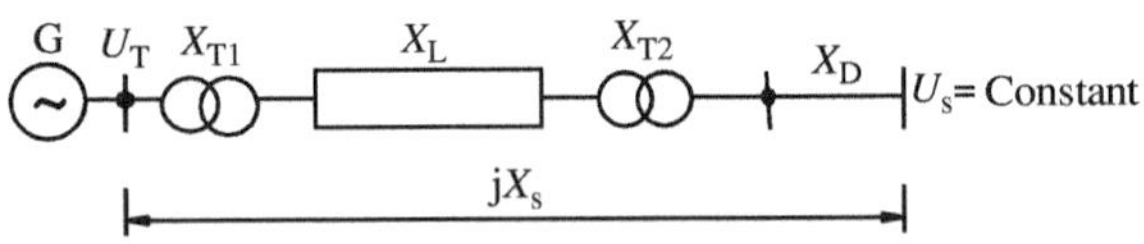

Figure 2.14
Connection diagram of a generator connected to the power system through a UHV transmission line.

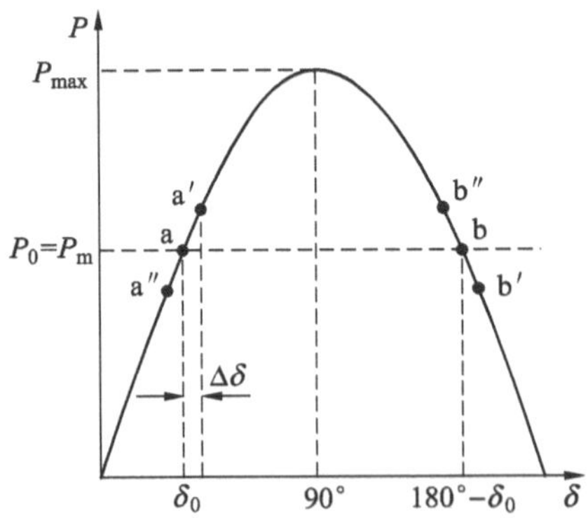

Figure 2.15
Power-angle characteristic curve of the power system.

the mechanical input power P_m from the prime mover (such as a steam turbine) to the generator and the electrical output power P_0 of the generator at the corresponding rotor angle δ_0, that is, $P_m = P_0$ at point a in Figure 2.15. If a small disturbance occurs in the power system, causing the power angle δ to increase by $\Delta\delta$, then the electrical output power of the generator will reach point a′. Because the power of the prime mover remains unchanged, at point a′ the electrical power is larger than the prime mover power, resulting in deceleration of the generator rotor. As a result, the power angle δ will decrease and start to shift toward the original point a. After several small swings of the power angle, the system will return back to operating point a. Similarly, if a small disturbance causes the power angle δ to decrease by $\Delta\delta$, then the generator will be at operating point a″, where the electrical power is smaller than the prime mover power. As a result, the rotor will accelerate and the power angle δ will increase and shift toward point a. After several small swings of the power angle, the system will return back to the operating point a. Apparently, at point a, the power system is able to restore to the original equilibrium point after being subjected to any small disturbance. Therefore, the system is stable in the steady-state. It follows from Figure 2.15 that if $0° < \delta < 90°$, then the system operates stably.

At point b (corresponding to a power angle $180° - \delta_0$), as shown in Figure 2.15, under steady-state conditions there is also an equilibrium between the prime mover power and the electrical power of the generator. However, when a small disturbance occurs, the power angle δ is increased by $\Delta\delta$ and shifts to point b′. The decrease in electrical output power of the generator will cause the generator to accelerate and the rotor angle δ to increase further, unable to return back to point b. As a result, the generator will lose synchronism and the system will lose stability. Similarly, if a small disturbance results in decrease in the power angle δ, then the generator rotor will decelerate due to the electrical power of the generator becoming larger than the prime mover power, the generator rotor will decelerate, and the power angle δ will decrease further to δ_0 where a new equilibrium will be established. Therefore, at operating point b, the system is unstable in the steady-state. Figure 2.15 shows that if $90° < \delta < 180°$, then the system is unstable.

From the aforementioned analysis, it follows that the steady-state stable condition of the power system is $dP/d\delta > 0$. At $\delta = 90°$, $dP/d\delta = 0$. Thus, the operating point where $\delta = 90°$ is the critical point for the steady-state stability, at which the electrical power reaches its maximum. The steady-state stability limit is $P_{max} = (E_q U_s)/(X_d + X_s)$.

2. **Transient stability**

   Transient stability mainly indicates the ability of the power system to maintain stability in the first and second oscillatory periods after being subjected to a large disturbance. It may be used to define the transient stability limit and stability

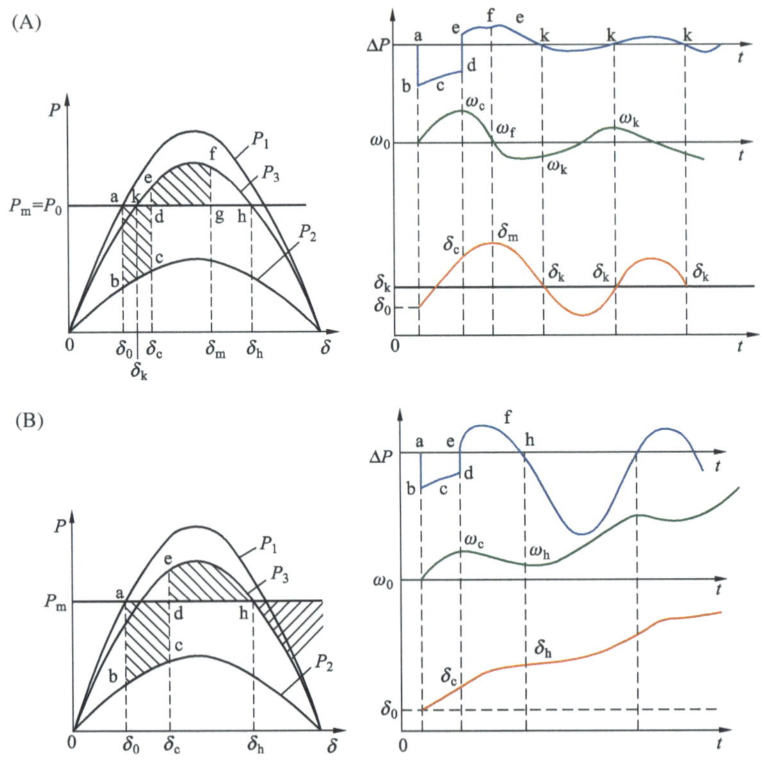

Figure 2.16
System electromechanical transient process during a fault of transmission line: (A) transient stability and (B) transient instability.

measures of the power system. Its physical characteristics are related to the synchronizing torque.

For the system shown in Figure 2.16, if a short-circuit fault occurs at the sending end of a transmission line, after a short period of time, the relay protection will operate to trip the circuit breakers on both sides of the transmission line to clear the fault. Figure 2.16 shows the electromechanical transient process during and following a fault.

Under normal operating conditions, the system operates at point a on the power-angle characteristic curve P_1. If a line fault occurs, then the system shifts from point a to point b on power-angle characteristic curve P_2 under the fault state. At this time, the electrical output power of the generator decreases, but the mechanical input power from the prime mover to the generator does not have enough time to follow the change. As a result, the generator rotor starts to accelerate and the power angle δ increases. At point c, where the power angle is δ_c, the fault is cleared and the operating point shifts from point c on power-angle characteristic curve P_2 to point e on postfault power-angle characteristic curve P_3. At this time, the electrical output power is larger than the prime mover power. As a consequence, the generator decelerates under the effects of braking torque. Due to the inertia of the rotor, the power angle δ increases further until it reaches the maximum at point f, where the rotor restores the synchronous speed. Then, the rotor angle starts to decrease under the effects of braking torque. After passing a new equilibrium point k, the rotor starts to accelerate once again. When the rotor finally reaches the synchronous speed, the power angle δ will not decrease any more. Afterward, the power angle δ oscillates along the power-angle characteristic curve P_3. Under the effects of damping torque, the generator finally reaches a new equilibrium state at operating point k and maintains synchronous operation, that is, the power system can maintain transient stability following a large disturbance.

If P_0 and δ_0 at the initial operating point are larger than the aforementioned values, then it is possible that following a fault the power angle becomes larger and larger, causing the generator to go out of step, as shown in Figure 2.16B, that is, the power system loses transient stability following a large disturbance.

This physical process of transient stability shows that from the instant when a fault occurs at the initial power angle δ_0 to the instant when the fault is cleared at power angle δ_c, the generator rotor is accelerated due to excess input power. The work performed by the excess power over the relative angular displacement is equal to the acquired additional kinetic energy during the relative displacement of the rotor. The work performed is the area enclosed by points a, b, c, and d, as shown in Figure 2.16, which is known as the accelerating area and is equal to $\int_{\delta_0}^{\delta_c}(P_m - P_2)d\delta$. Following clearance of the fault, when the power angle shifts from δ_c to δ_m, the electrical power is greater than the mechanical power, causing a braking torque. The work performed by the braking torque over the relative angular displacement is equal to the kinetic energy dissipated by the rotor. This work is the area enclosed by points d, e, f, and g, and is equal to $\int_{\delta_c}^{\delta_m}(P_3 - P_m)d\delta$. When the dissipated kinetic energy during rotor deceleration is equal to the kinetic energy acquired during rotor acceleration, the system is under transient stability; therefore, the condition for transient stability is:

$$\int_{\delta_0}^{\delta_c}(P_m - P_2)d\delta = \int_{\delta_c}^{\delta_m}(P_3 - P_m)d\delta \tag{2.61}$$

Equation (2.61) is known as the equal area criterion, that is, when the decelerating area equals the accelerating area, the rotor angular speed restores to the synchronous speed and the power angle δ starts to decrease. To maintain system stability, the rotor must restore to the synchronous speed before reaching point h. The stability limit occurs exactly when the rotor restores to the synchronous speed at point h. If the fault clearing time of relay protection and circuit breaker is definite, then it is theoretically possible to derive the transient stability limit for the power transmitted through the line under a certain system operating scheme based on the equal area criterion. Theoretically, there is a critical stability between transient stability and transient instability for operation of the transmission line. The initial operating power corresponding to such critical stability is the transient stability limit under a certain system operation scheme.

3. **Small-disturbance dynamic stability**

Small-disturbance dynamic stability indicates the ability of the power system to avoid oscillatory instability under small disturbance. Its physical characteristics are related to the damping torque. The small-disturbance dynamic stability is mainly used to analyze the damping characteristics of the power system under normal and postfault operating conditions.

Currently, the method for studying the dynamic stability of a complex power system is mainly based on Lyaponov's first approximation theorem. It is based on the principle that the dynamic characteristics of a power system may be expressed by a set of nonlinear differential equations and a set of nonlinear algebraic equations:

$$\left.\begin{aligned} \frac{dx_i}{dt} &= f_i(x_1, x_2, \ldots, x_n) \quad i = 1, 2, \ldots, m \\ 0 &= g_i(x_1, x_2, \ldots, x_n) \quad i = m+1, m+2, \ldots, n \end{aligned}\right\} \tag{2.62}$$

By linearizing the equations around equilibrium point x_0, expressing the variables by the sum of their initial values and microincrements ($x_i = x_{i0} + \Delta x_i$), expanding the resulting equation sets into Taylor series around equilibrium point x_0, and neglecting quadratic and higher-order terms of microincrements, we have:

$$\left.\begin{aligned} \frac{d\Delta x_i}{dt} &= \sum_{j=1}^{n} \frac{\partial f_i}{\partial x_j} \Delta x_j \quad i = 1, 2, \ldots, m \\ 0 &= \sum_{j=1}^{n} \frac{\partial g_i}{\partial x_j} \Delta x_j \quad i = m+1, m+2, \ldots, n \end{aligned}\right\} \tag{2.63}$$

The state and nonstate vectors are defined, respectively, as $\Delta X = [x_1, x_2, \ldots, x_m]^T$and $\Delta Y = [x_{m+1}, x_{m+2}, \ldots, x_n]^T$; Eq. (2.63) is expressed in matrix form:

$$\begin{bmatrix} \Delta \dot{X} \\ 0 \end{bmatrix} = \begin{bmatrix} J_A & J_B \\ J_C & J_D \end{bmatrix} \begin{bmatrix} \Delta X \\ \Delta Y \end{bmatrix} = J \begin{bmatrix} \Delta X \\ \Delta Y \end{bmatrix} \tag{2.64}$$

Eliminating nonstate variables from Eq. (2.64), we have:

$$\Delta \dot{X} = (J_A - J_B J_D{}^{-1} J_C)\Delta X = A \Delta X \tag{2.65}$$

Equation (2.65) is the state equation describing the linear system, where $\boldsymbol{A}$ is an $n \times n$ matrix, known as the state matrix of the system. For the linear system described by the state equation, its dynamic stability depends on all eigenvalues of the state matrix. Thus, the general process to analyze the dynamic stability of a power system can be summarized as follows: (1) calculate the steady-state values of individual variables under given steady-state operating conditions, that is, the values at equilibrium point; (2) linearize the differential and algebraic equations describing the transient process around the steady-state value; and (3) derive the state matrix $\boldsymbol{A}$ and determine the system stability based on the eigenvalues of the state matrix.

Assuming that the ith pair of eigenvalues of matrix $\boldsymbol{A}$ is λ_i, we have:

$$\lambda_i = \sigma_i \pm j\omega_i, \quad \zeta_i = -\sigma_i / \sqrt{\sigma_i^2 + \omega_i^2} \tag{2.66}$$

where

σ_i—attenuation coefficient, s^{-1}

ω_i—angular frequency of oscillation, rad/s

ζ_i—damping ratio.

If $\sigma_i > 0$, $\zeta_i < 0$, then the damping of the ith electromechanical oscillation mode is negative, and the corresponding time domain response curve is amplified; the larger σ_i is, the faster the curve grows; therefore, the system is unstable. If $\sigma_i = 0$, $\zeta_i = 0$, then the damping is zero, and the corresponding time domain response curve is of equal amplitude, representing a critical state in which the power system is also deemed unstable. If $\sigma_i < 0$, $\zeta_i > 0$, then the corresponding time domain response curve is attenuating; the larger the absolute value of σ_i is, the faster the curve attenuates, presenting a better damping performance and a better system stability.

4. **Large-disturbance dynamic stability**

Large-disturbance dynamic stability indicates the ability of the power system to maintain stability with the effect of system dynamic elements and control devices when subjected to a large disturbance. Its physical characteristics are related to the damping torque.

The method for calculating and analyzing the large-disturbance dynamic stability is not distinctly different from that of transient stability, mainly using digital simulation. The analysis of the large-disturbance dynamic stability mainly aims to determine the dynamic stability after the system becomes transiently stable. As compared with the calculation of transient stability, the calculation and analysis of large-disturbance dynamic stability must take into account detailed models for dynamic elements and control devices, such as excitation system and associated additional controls, prime mover governor, and power electronics controllers.

2.3.1.2 Voltage stability

Voltage stability indicates the ability of a power system to maintain or restore acceptable stable voltage and prevent voltage collapse after being subjected to a small or large disturbance. The main factor contributing to voltage instability is the inability of the power system to meet the demand for reactive power after being subjected to a disturbance, increase in load demand, or change in operating condition.

Voltage stability can be classified into steady-state voltage stability and large-disturbance voltage stability in terms of the magnitude of disturbance.

1. **Steady-state voltage stability**

   Steady-state voltage stability indicates a system's ability to control voltages following incremental changes in system load. It is mainly used to define the steady-state voltage stability reserve of the power system during normal operation and following a contingency.

   In cases in which a generator is equivalent to transient reactance X'_{d} and transient electromotive force E', and the line susceptance is fully compensated using line shunt reactors, the simplified equivalent circuit of the generator connected to the power system through an EHV/UHV transmission line is shown in Figure 2.17, where $X = X'_{\mathrm{d}} + X_{\mathrm{T1}} + 1/2X_{\mathrm{L}} + X_{\mathrm{T2}}$.

   The steady-state voltage stability is generally analyzed as a steady-state problem. For the system as shown in Figure 2.17, the characteristics of the power transmission system (i.e., $p-v$ curve and $v-q$ curve) are generally used to determine the voltage stability margin and analyze the factors influencing voltage stability.

   The active power and reactive power at the receiving end are, respectively, given by:

$$P = \frac{EU_{\mathrm{R}}}{X}\sin\delta = P_{\max}\sin\delta \tag{2.67}$$

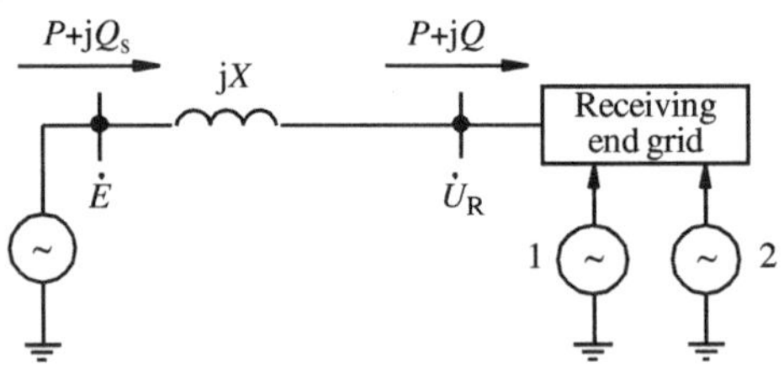

Figure 2.17
System diagram for voltage stability study.

$$Q = \frac{EU_R}{X}\cos\delta - \frac{U_R^2}{X} = P_{max}\cos\delta - \frac{U_R^2}{X} \tag{2.68}$$

$$P_{max} = \frac{EU_R}{X}$$

where

δ—power angle between $\dot{E}$ and $\dot{U}_R$.

The reactive power at the sending end is given by:

$$Q_s = \frac{E^2}{X} - \frac{EU_R}{X}\cos\delta = \frac{E^2}{X} - P_{max}\cos\delta \tag{2.69}$$

To obtain the general relationship between power and voltage and to determine the maximum transmissible power through a transmission line and corresponding critical voltage, the variables can be normalized using the short-circuit power E^2/X, that is:

$$p = P\Big/\left(\frac{E^2}{X}\right) = \frac{PX}{E^2},\quad q = Q\Big/\left(\frac{E^2}{X}\right) = \frac{QX}{E^2}$$

$$v = \frac{U_R}{E}$$

Then, we have:

$$p = \left(\frac{EU_R}{X}\sin\delta\right)\Big/\left(\frac{E^2}{X}\right) = v\sin\delta \tag{2.70}$$

$$q = \left(\frac{EU_R}{X}\cos\delta - \frac{U_R^2}{X}\right)\Big/\left(\frac{E^2}{X}\right) = v\cos\delta - v^2 \tag{2.71}$$

Trigonometric equation $\sin^2\delta + \cos^2\delta = 1$ is used to transform Eqs. (2.70) and (2.71), and then we have:

$$p = \sqrt{v^2 - (q + v^2)^2} \tag{2.72}$$

According to Eq. (2.67), given the transmission parameter X and allowable voltage decline (i.e., U_R/E is given), we can draw a family of q–p curves, as shown in Figure 2.18. The line connecting tangent lines of the family of curves is the power limit line of the transmission system, which shows the critical relationship among active power, reactive power, and voltage constraints to maintain the voltage

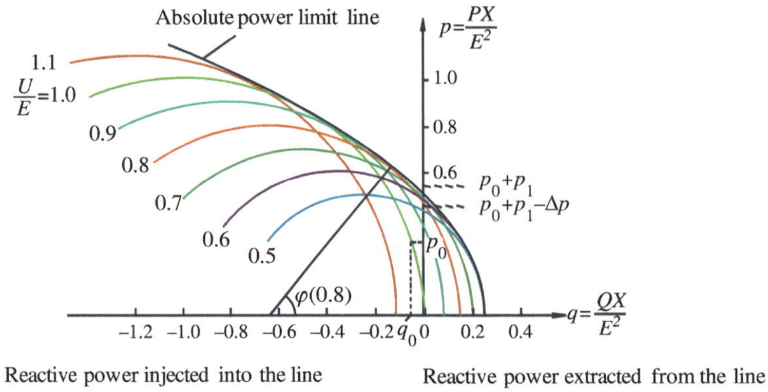

Figure 2.18
Receiving-end voltage and family of $q-p$ curves of a transmission line.

stability of the receiving-end system. When the system is heavily loaded, if the sending-end generator is required to feed more active power to the receiving end, then larger reactive power support must be made available at the receiving end to maintain the receiving-end voltage at a given level. As shown in Figure 2.18, if the receiving-end power is p_0 and the receiving-end voltage is required to maintain at 90% of the sending-end electromotive force, then the receiving-end system must offer reactive power support, namely, feed reactive power to the transmission line. To increase the active power transmitted through the transmission line, sufficient reactive power reserves or reactive power voltage support must be available at the receiving end to maintain stable operation of the system.

The transmission system reaches the steady-state stability limit at the power angle $\delta = 90°$ and the maximum possible power transmitted is EU_R/X. In this case, the sending end is required to generate reactive power E^2/X and the receiving end will feed reactive power U_R^2/X to the transmission line. At this time, the reactive power injected into the transmission line at each end is of the same order of magnitude as the active power transmitted through the line.

a. The $p-v$ curve of transmission system shows the change of the receiving-end voltage U_R as a function of the transmitted power at a given receiving-end load power factor, that is, the change of the receiving-end voltage U_R is a function of load at a given sending-end electromotive force.
   Assuming that the power factor angle is φ, then $Q = P \tan \varphi$, $q = p \tan \varphi$, Eq. (2.72) may be transformed to:

$$p^2 + (p\tan \varphi - v^2)^2 = v^2$$

$$v^4 - (1 + 2p\tan \varphi)v^2 + p^2(1 + \tan^2\varphi) = 0 \tag{2.73}$$

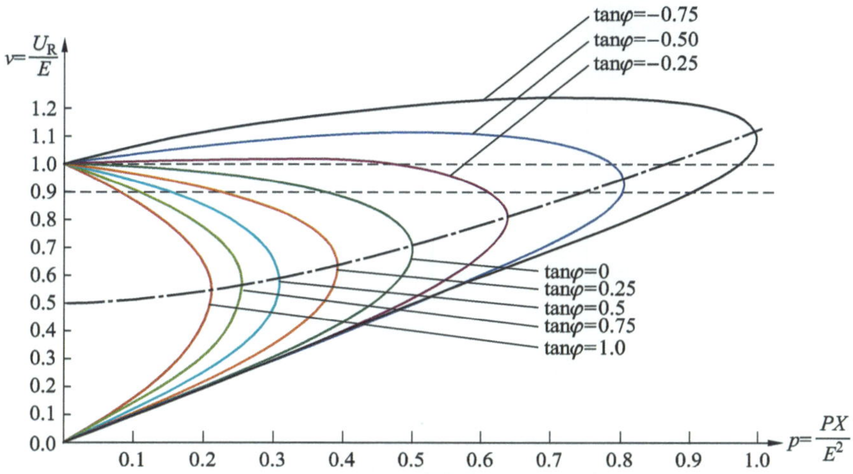

Figure 2.19

Family of $p-v$ curves for the receiving end of a transmission line.

Note: tan $\varphi = 1.0$, 0.75, 0.5, 0.25, and 0 correspond to a load power factor of 0.0707, 0.8, 0.0894, 0.97, and 1.0, respectively.

According to Eq. (2.73), for a given tan φ, we can draw a $p-v$ curve. By varying values of tan φ, we can draw a family of $p-v$ curves, as shown in Figure 2.19. From Figure 2.19 the following fundamental properties of a transmission system are evident:

i. There is an inherent maximum power that can be transmitted through a transmission line at any load power factor, that is, the power limit. The dotted line in Figure 2.19 shows the maximum power limit that can be transmitted at different load power factors.

ii. Any value of power below the maximum can be transmitted at two different values of voltage U_R. The normal operation corresponds to the upper value, within narrow limits of approximately 1.0 p.u., namely, within the boundary of dashed lines as shown in Figure 2.19. The operation at the lower voltage may lead to voltage instability.

iii. The load power factor has a significant influence on the maximum power that can be transmitted. When the load connected at the receiving end is inductive, that is, Q is positive, the power limit and U_2 are both less. When the load connected at the receiving end is capacitive, that is, Q is negative, the voltage U_2 is relatively high, and the power limit is higher.

b. The $v-q$ curve. Equation (2.73) may be transformed to:

$$q^2 + 2v^2 q = v^2(1 - v^2) - p^2 \tag{2.74}$$

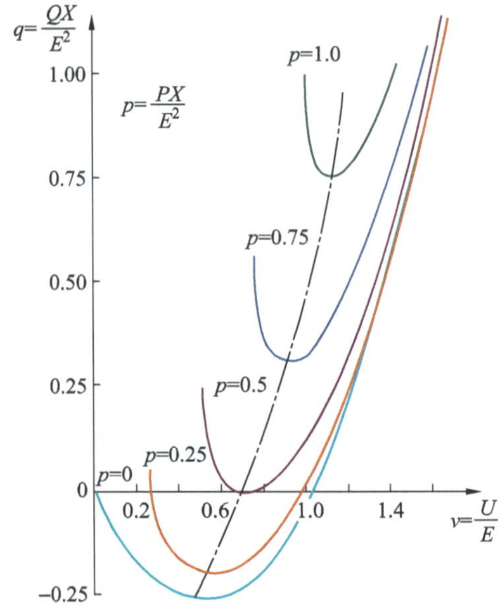

Figure 2.20
Family of $v-q$ curves for the receiving end of a transmission line.

According to Eq. (2.74) or Figure 2.18, for a given p, we can draw a $v-q$ curve. By varying the values of p, we can draw a family of $v-q$ curves, as shown in Figure 2.20.

Figure 2.20 shows that the reactive power increases with the increase of the receiving-end voltage. Therefore, the zone where the derivative $\mathrm{d}Q/\mathrm{d}U$ is larger than zero is the voltage stability zone. The point at which the derivative $\mathrm{d}Q/\mathrm{d}U$ is equal to zero is the voltage stability limit point, which corresponds to the critical voltage. Thus, the right side of the lowest point of the $v-q$ curve is the voltage stability zone, whereas the left side is the voltage instability zone.

1. **Large-disturbance voltage stability**

   Large-disturbance voltage stability (including transient voltage stability and dynamic voltage stability) indicates the ability of a power system to prevent voltage collapse after being subjected to a large disturbance. Transient voltage stability is mainly used to analyze rapid voltage collapse problems, whereas dynamic voltage stability is mainly used to analyze power system voltage stability under the effects of slower-acting dynamic elements and control devices, such as on-load voltage regulating transformer, generator stator and rotor overcurrent and underexcitation limiters, operable shunt capacitors, secondary voltage and frequency controls, and thermostatical loads.

For the transmission system, as shown in Figure 2.17, large disturbance refers to the situation when one circuit of a double-circuit transmission line is tripped due to faults and the power transmitted by it is suddenly transferred to the other circuit, or a generator in a weak system at the receiving end is tripped, or a large load is suddenly connected to the receiving-end system, or the reactive power compensation at the receiving end is suddenly switched out of service. The problems caused by reallocation of system power (especially reactive power) and deficit in reactive power due to such large disturbance is referred to as the large-disturbance voltage stability problems. Due to the interaction among nonlinear dynamic characteristics of system elements, the large-disturbance voltage stability is a long-term dynamic process, with the time frame ranging from a few seconds to more than 10 min. Therefore, the large-disturbance voltage stability is also known as the long-term voltage stability [2].

The mechanism and process of voltage instability are described, and a generator tripping in the receiving-end system is used as an example. As shown in Figure 2.17 and Figure 2.18, under normal operating conditions, the active power transmitted to the receiving-end system through a transmission line is p_0, and the bus voltage U_R is 1.0 p.u. To support power transmission, the receiving-end system must supply reactive power q_0 to the transmission line. Because the receiving-end system is weak, after generator 1 trips, the lost power p_1 must be offset by a remote generator. Thus, the transmitted power through the transmission line increases from p_0 to $p_0 + p_1$. As the power transmission is increased and the receiving-end system is under power control, the transmission line will consume more reactive power. To maintain almost unchanged U_R, the receiving-end system has to increase the reactive power supply to the line. However, the tripping of the generator 1 results in loss of some reactive power at the receiving end. Assuming that the receiving-end system is unable to supply newly increased reactive power demand, U_R will inevitably decrease to a low level. As a consequence, the system load at the receiving end has to be reduced by Δp. Thus, the transmitted power will be reduced from $p_0 + p_1$ to $p_0 + p_1 - \Delta p$. During such power swings, the generator excitation current is automatically regulated to the possible maximum level and the tap changer of the load-end transformer operates, causing the load bus voltage to increase. This will bring the load power back to the original value. The power transmitted through the transmission line may be restored to $p_0 + p_1$. Because the receiving-end system is unable to further increase reactive power support, the voltage and reactive power support from the remote generator are also limited. As a result, the receiving-end voltage U_R continues to decrease, tending to make it impossible for U_R to reach a new stable equilibrium point. Consequently, the system voltage will collapse and the remote generator will lose synchronism with the receiving-end unit.

2.3.1.3 Frequency stability

Frequency stability indicates the ability of a power system to maintain or restore steady acceptable frequency and to prevent frequency collapse after being subjected to a sudden active power disturbance. It is mainly used to study the effectiveness and rationality of the configuration of spinning reserves and underfrequency load-shedding devices in the power system, as well as the coordination between the power grid and power sources.

The frequency reflects the basic state of active power supply–demand balance in the power system. If the total active power output from all power plants in the power system meets the total power load demand of the whole power grid and can be duly adjusted in response to the load variation, then the average operating frequency of the power grid is maintained at the rated value. If the supply of active power is higher than the demand, then the average operating frequency will become higher than the rated value; if the supply of active power is lower than the demand, then the average operating frequency will become lower than the rated value.

The operating conditions of many electric facilities of consumers in the power system are closely related to frequency. The speed of asynchronous motors widely used in industrial and agricultural production is also related to the system frequency. Any variation in frequency will result in change in the speed of the motors, thereby affecting the quality of products. The variations in system frequency will also affect the operation of the equipment in the power system itself and the auxiliary systems in power plants. Therefore, modern large power generating units place strict requirements on the system frequency. In addition, if the system frequency decreases, then the excitation current passing through asynchronous motors and transformers will increase, thereby resulting in the increase of system reactive power. This will lead to voltage decline if the standby reactive power source is insufficient.

In a large power system, the frequency dynamic variations, frequency stability, power angle stability, and voltage stability are not independent of each other. Instead, they are different sides of a uniform physical phenomenon that are mutually induced and correlated, and they are affected by the operating conditions of the power grid.

Any imbalance in active power will lead to a series of dynamic responses and readjustment of power generation to reach a new balance between the power generation and load. This process is generally divided into the following four stages:

Stage 1: Immediately after a disturbance occurs (after 0 s), the power angle of the generator will not suddenly change, and the change in output power is proportional to the synchronizing power P_s ($P_s = dp/d\delta$). The power variation is allocated among all the generators in the system according to their respective synchronizing coefficients and depends on factors such as the magnitude of power imbalance, the initial operating point of generator, as well as the electrical distance from the location of power variation to individual generators. Generally, the closer a unit is to the disturbance, the more shares of power deficiency that will be allocated to it, which is given by:

$$\Delta P_i = \frac{\mathrm{d}P_i/\mathrm{d}\delta_i}{\sum_{j\in i}^{N}(\mathrm{d}p_i/\mathrm{d}\delta_j)} \tag{2.75}$$

In some cases, due to the voltage decrease, some loads sensitive to voltage will reduce their power consumption, thereby partially offsetting the generation imbalance. However, the power imbalance caused by the decline in power generation capacity is mostly compensated by the remaining generators that are close to the location of power generation capacity decline.

Stage 2: Within 0.5–2 s following the disturbance, the power will be reallocated among the generating units according to their rotating inertia. Due to the deficit in power generation, the frequency starts to decrease at a rate that is limited by the energy dissipated from the inertia of generator rotors. Within 1–2 s, the frequency decreases in an oscillatory manner at an average rate. At this time, the power deficiency will be reallocated among the generators according to their relative inertia, regardless of their locations. The generators with larger inertia will be allocated with more power deficiency, which is expressed by the following equation:

$$\Delta P_i/\Delta P \approx M_i \Big/ \sum_{j\in i}^{N} M_j \tag{2.76}$$

Stage 3: After 2–3 s, the generator governors gradually start to respond and change the output power of the prime mover, thereby increasing the output power of the generators. The increase rate and maximum increase amount of output power of generators are closely related to the available spinning reserve of individual generators as well as characteristics of the governors and units (fossil-fired or hydropower units). The shares of power deficiency allocated to individual units are proportional to the unit capacity, spinning reserve, and characteristics of the speed governing system.

Stage 4: Within dozens of seconds to a few minutes after the disturbance, the automatic generation control (AGC) will correct the deviations in both tie-line power and system frequency and will restore them within the spinning reserve controllable by the AGC. Thereafter, the operators' manual regulation will be initiated.

In case of any deficit in active power, the magnitude and decline rate of the power system frequency mainly depend on four factors: (i) the amount of generation power deficiency; (ii) total inertia of remaining generators; (iii) load frequency characteristics; and (iv) spinning reserve.

2.3.2 Power System Security and Stability Standard and Stability Criterion

1. **Classification of power system capabilities to withstand large disturbance**

The capability of a power system to withstand large disturbance can be classified into the following three levels:

Level 1: Following a large disturbance, the power system maintains stable operation and the grid continues to provide normal service.

Level 2: Following a large disturbance, the power system maintains stable operation with partial load loss.

Level 3: Following a large disturbance, the power system cannot maintain stable operation and measures must be taken to avoid system collapse and to minimize load loss.

a. Level 1: If, during normal operation, the power system experiences a disturbance caused by the fault of a single component, the protections, switches, and reclosing operate correctly, and additional measures are not taken, then the power system maintains stable operation, the grid continues to provide normal service, no other components work beyond their specified emergency overload capability, and no interlock trip occurs. Such single-component faults mainly include successful reclosing following any single-phase transient ground fault of the line, any transformer out of service at the receiving end due to fault, and failure of any circuit of AC tie-line that was tripped due to fault or without fault.
However, for three-phase fault on the AC send-out line of a power plant, monopolar fault on DC send-out line of a power plant, and the single circuit at higher voltage in an electromagnetic loop network being disconnected due to fault or without fault, it may be necessary to trip units or rapidly reduce the output of generating units.

b. Level 2: If the power system is subjected to any of the severe disturbances described during normal operation, then protections, switches, and reclosers operate correctly, the power system is able to maintain stable operation, and, if necessary, stability control measures such as tripping generating unit and shedding load are allowed. Such disturbances mainly include failure to reclose following a single-phase permanent fault on a single circuit, failure to reclose following three-phase disconnection without fault, fault of any bus section, failure to reclose following a simultaneous single-phase ground fault on two differently designated phases of double circuits on the same tower, and three-phase simultaneous tripping of two circuits.

c. Level 3: If the power system loses stability due to any of the faults described, then necessary measures must be taken to prevent system collapse, widespread and extended outage, and disastrous outage to key customers (including auxiliary power), and to minimize the load loss. The power system will restore normal operation as soon as possible. Such faults mainly include failure of the switch to trip upon fault, unwanted operation or refusal to operate upon fault by relay protection and automatic devices, and multiple faults.

2. **Main criteria for various types of power system stability**
   a. Criteria for steady-state power angle stability. For different power systems under normal operation, the steady-state power angle stability reserve coefficient calculated by using the power angle stability criteria will be 15–20%. Under postfault or special operating conditions, the coefficient will not be smaller than 10%.

      The steady-state power angle stability reserve coefficient K_p is given by:

$$K_p = (P_j - P_z)/P_z \times 100\% \tag{2.77}$$

      where

      P_j, P_z—maximum and normal transmitted power of a line or section, respectively.

   b. Criteria for transient stability. The generating units do not go out of step during the first and second oscillatory periods after the power system experiences a large disturbance, which results in the relative increase in the power angle between individual units.
   c. Criteria for small-disturbance dynamic stability. During normal operation, the damping ratio of the inter-area oscillation modes and oscillation modes that are strongly correlated to large power plants and large units generally will be more than 0.03. Under postfault operating conditions, the damping ratio will be at least 0.01–0.02.
   d. Criteria for large-disturbance dynamic stability. In the dynamic swing process after the power system is subjected to a disturbance, the relative power angle of generator, the active power supplied by the generator, and the active power transmitted through the transmission line decrease in an oscillatory manner, the power system is able to restore the voltage and frequency to the allowable range, with a damping ratio no smaller than 0.01.
   e. Criteria for steady-state voltage stability. Given the maximum regional load or maximum section power flow, the active power margin of a regional load during normal operation or maintenance will be more than 8%, and both the active power margin of a regional load and the reactive power margin of a bus load following an $N-1$ fault will be more than 5%.

      The active power margin K_p of a regional load is given by:

$$K_p = \frac{P_{max} - P}{P} \times 100\% \tag{2.78}$$

      where

      P, P_{max}—active power at the initial and critical operating points, respectively.

The reactive power margin K_q of a bus load is given by:

$$K_q = \frac{Q_{max} - Q}{Q} \times 100\% \tag{2.79}$$

where

Q, Q_{max}—reactive power at the initial and critical operating points, respectively.

f. Criteria for transient and dynamic voltage stability. Currently, the practical criterion for transient stability of a power system following a disturbance is: the power system is able to restore the load bus voltage to a value more than 0.80 p.u. in 10 s; for mid-term and long-term voltage stability, the power system is able to maintain or restore the load bus voltage to a value more than 0.90 p.u.
g. Criteria for frequency stability. The power system is able to rapidly restore to approximately the rated frequency for continuous operation without frequency collapsing, and to avoid the system operating at an excessively high or low value for a long time after the event.
h. Criteria for mid-term and long-term stability. The criteria for power angle stability, voltage stability, and frequency stability are applicable.

2.3.3 Calculating Methods for Transmission Capability of the UHV AC System

To calculate the transmission capability of the UHV AC system, it is necessary to calculate and analyze the steady-state stability, transient stability, small-perturbations/large-disturbance dynamic stability, steady-state voltage stability, transient/dynamic voltage stability, and frequency stability to derive the maximum power limit of the UHV transmission lines or sections that can meet the various stability constraints.

1. **Steady-state power angle stability**
   a. *Eigenvalue method*: The eigenvalue method for analyzing the steady-state power angle stability is as follows: (i) calculate the distribution of power flow and the steady-state values of individual state variables under given operating conditions; (ii) linearize the equation describing the transient process around the steady-state value; and (iii) derive the characteristic matrix and define the system steady-state stability based on the eigenvalues of the characteristic matrix.
   The criterion for steady-state power angle stability is that no eigenvalue is a positive real number.
   b. *Practical algorithm*: The practical algorithm for steady-state power angle stability involves the use of stability calculation program. By gradually increasing the power of the sending-end units and correspondingly decreasing the power of the receiving-end units or increasing the receiving-end load, calculate the maximum power that

can be transmitted through a transmission line or section, which is known as the steady-state power angle stability limit.

In the calculation process, the system frequency and voltage will be maintained within the normal range as practicable as possible. Therefore, it is essential to take into account the speed-governing system and excitation system, to ensure the substantial balance between power increase and decrease, and to consider the necessity of switching of reactive power compensation according to the actual conditions. It should also be noted that the power-changing scheme is to accommodate the actual power flows

2. **Transient stability**

   The calculation of transient stability requires the use of following mathematical models:

   – For dynamic components
     - Subtransient and transient electromotive force equations of the generator
     - Equations of motion of the generator rotor
     - Equations of the generator excitation control system (including the power system stabilizer (PSS))
     - Equations of the prime mover and speed-governing system
     - Dynamic characteristic equations of the motor load
     - Equations of DC transmission system control
     - Dynamic characteristic equations of other dynamic devices (such as flexible AC transmission system (FACTS) devices)
   – For static components
     - Power network equation
     - Generator voltage equation
     - Load static characteristic equation
     - DC line voltage equation

In the calculation and analysis of transient stability, the numerical integration-based time domain simulation program is preferred. Specially, the numerical integration method to derive the time domain solution to the equation describing the disturbed motion and then the system stability is judged based on the variation in the relative angle between individual generator rotors and changes of system voltage and frequency.

3. **Dynamic stability**

   The power system dynamic stability can be classified into small-disturbance dynamic stability and large-disturbance dynamic stability. The former is usually calculated with the eigenvalue method based on linear models of the power system, and the latter is usually calculated with the time domain simulation method based on numerical integration.

a. *Small-disturbance dynamic stability calculation*: This calculation mainly involves the following aspects: (i) calculate the eigenvalues; (ii) analyze the damping ratio of the dominant oscillation mode; (iii) perform modal analysis (participation factor analysis, eigenvector analysis, eigenvalue sensitivity analysis, etc.) for the oscillation modes with a negative or weak damping ratio; (iv) select the configuration scheme for the PSS based on the modal analysis; (v) determine parameters for the PSS; and (vi) check the damping effect of the PSS.

b. *Large-disturbance dynamic stability calculation*: The computing time of large-disturbance dynamic stability shall be up to 10–15 oscillatory periods. The large-disturbance dynamic stability of a power system may be determined based on the power angle swing curve, active power oscillatory curve, and the central/key point voltage variation curve. In the analysis, the effects of transient components will be eliminated. Commonly used tools are based on progression analysis, for example, the Prony analysis may be used to analyze both the power angle curve of the generating unit and the power curve of the transmission line derived from the time domain simulation to obtain the frequency and damping ratio of the oscillation mode of interest. If the Prony analysis tool is unavailable, then the approximate calculation equation for the damping ratio of the sinusoidal oscillation curve may be used, which is given as:

$$\zeta = \ln(A_I/A_{I+N})/2N\pi \tag{2.80}$$

where

ζ—damping ratio

A_I—magnitude of the Ith oscillation

A_{I+N}—magnitude of the $(I+N)$th oscillation.

If the time domain simulation curve is a nonstandard damped sinusoidal curve, then this equation may be used to obtain the average damping ratio of N oscillations.

4. **Voltage stability**

a. *Steady-state voltage stability calculation*: The loads are gradually increased (the loads may be increased proportionally with the initial power factor held constant, if appropriate) to obtain the critical point at which voltage loses stability and then the voltage stability margin at the current operating point is estimated.

The load increase shall be simulated using the constant power load model with the initial power factor held constant. The output of generators will be allocated by taking into account the direction of power flow from or to the area undergoing study, which will be consistent with the actual direction of power flow on the grid as much as possible.

b. *Large-disturbance transient and dynamic voltage stability calculation*: The mathematical models for this calculation are substantially similar to that of transient stability. The conventional time domain simulation program may be used.

The calculation of transient and dynamic processes involves the use of mathematical models for load dynamic characteristics, generators and associated excitation systems and speed-governing systems, generator overexcitation limit characteristics and forced excitation operating characteristics, reactive power compensation, and DC transmission system, as well as undervoltage load shedding and other components and controls.

c. *Mid-term and long-term voltage stability calculation*: In mid-term and long-term dynamic processes, apart from the components required for calculating the transient voltage stability, the mathematical models for transformer on-load voltage regulating (OLTC), generator stator and rotor overcurrent and overexcitation/underexcitation limiters, automatically switched shunt capacitors and reactors, secondary voltage and frequency controls (underexcitation limiter, automatic voltage control (AVC, AGC, etc.), thermostatically controlled loads, and other relevant components will also be taken into account.

Special mid-term and long-term dynamic simulation programs or extended transient stability programs (which can simulate the dynamic processes of aforesaid components) can be used for this calculation.

5. **Frequency stability**

When calculating the frequency stability, the possible maximum power imbalance will be taken into account. If the system is split into several islanding subsystems, then the possible maximum deficiency or surplus of power in individual subsystems caused by switching out of the largest or several generating units, open circuit of the system tie-line or long-distance transmission line, or DC blocking will also be considered.

Frequency stability calculation may use the time domain simulation program. The system modeling will use detailed models of generators, prime mover governors, excitation systems, and load frequency characteristics, and it will be necessary to simulate the automatic underfrequency load shedding, underfrequency disconnection, overfrequency generator tripping, hydro generator underfrequency self-startup, fossil unit over speed (including overacceleration) protection, and other frequency-related automatic devices. For long-term frequency stability problems, the dynamic characteristics of the prime mover and its associated power system and the long-term dynamic characteristics of secondary voltage and frequency controls (underexcitation limiter, AVC, AGC, etc.) and other components will also be simulated.

6. **Mid-term and long-term dynamic processes**

Apart from the component models required for calculating the transient stability, the calculation of mid-term and long-term dynamic processes will also take into account the mathematical models of the dynamic components, including: (i) protection and control of the generator and the associated excitation system; (ii) protection and control of the grid; (iii) protection and control of the prime mover/energy supply system of the power plant; and (iv) control operations of the operators.

For the mid-term and long-term dynamic processes calculations and analyses, the variable step-size numerical integration algorithm-based time domain simulation programs suitable for the rigid dynamic system will be used. Specially, the numerical integration method is used to derive the time domain solution to the equation describing the disturbed motion, and then the system stability is determined based on the variation of the relative angle between individual generator rotors and changes in the system voltage and frequency.

2.4 Influence of System Parameters on Transmission Capability of the UHV AC System

2.4.1 Transformer Reactance/Line Reactance Ratio of UHV System

There are two ways to connect a UHV transmission line to a power system:

1. The transmission line directly connects power plants to the receiving-end grid, allowing power plants to be connected to the receiving-end grid separately on a point-to-point basis. Such a connection method incorporates the design philosophy in which the power transmission line is regarded as the associated supporting project of power plants. It is widely used in UHV transmission engineering in China.
2. The transmission line connects the sending and receiving-end systems, allowing interconnection and bulk power transmission between them. To use this connection method, it is necessary to plan for both power grid and power sources first, according to the requirements of future load development. The grid planning should be before appropriately planning the power source and should serve to guide the planning and construction of power sources. Proper and reliable grid configuration at the sending end should be developed to connect the power plants to the sending-end grid reasonably. Such a connection method enables the sending-end system (including power source bases or station clusters) to flexibly transfer bulk power to the receiving-end system.

No matter what connection scheme is used for UHV power transmission, both ends of the transmission line should be equipped with step-up and step-down transformers. The parameters such as reactance of these transformers will affect the UHV transmission capability. The reactance of transformers is a function of their structures and insulating material and generally increases with the increase of the rated voltage.

To obtain the relationship between transformer reactance and line reactance, it is necessary to calculate the reactance parameters of transformers based on their technical data first.

Table 2.5: Typical Reactance-Related Parameters of 500-, 750-, and 1000-kV Transformers Used in China.

Voltage of transformer (kV)	500	750	1000
Capacity of step-up transformer (MVA)	750	780	1120
Turns ratio of step-up transformer	22/525	18/800	27/1050
Reactance of step-up transformer (p.u.)	0.120 (0.016)	0.145 (0.019)	0.158 (0.014)
Capacity of step-down transformer (MVA)	750	2100	3000
Turns ratio of step-down transformer	525/230	800/363	1050/525
Reactance of step-down transformer (p.u.)	0.123 (0.016)	0.179 (0.009)	0.180 (0.006)

Note: The base voltage corresponding to transformer reactance is the mean of the individual rated voltage classes, the base capacity is the capacity of the transformer itself, and the base capacity corresponding to the values in brackets is 100 MVA.

Table 2.6: Ratio of Transformer Reactance to Total Line Reactance.

Transmission line length (km)		100	200	300	400	500	600
Voltage class (kV)	500	0.688	0.524	0.423	0.355	0.306	0.268
	750	0.765	0.619	0.520	0.448	0.394	0.351
	1000	0.778	0.636	0.538	0.467	0.412	0.368

Note: When calculating the ratio of transformer reactance to total line reactance, the step-up transformer reactance, step-down transformer reactance, and line reactance shall be converted to the values for the same capacity.

Table 2.5 shows the typical reactance-related parameters of 500-, 750-, and 1000-kV transformers used in China.

Based on the transformer data shown in Table 2.5 and the transmission line parameters in Table 2.2, we can derive the proportion relationship between the step-up transformer reactance X_{T1} and step-down transformer reactance X_{T2} and the total line reactance, including the transmission line reactance X_L, that is:

$$b = \frac{X_{T1} + X_{T2}}{X_{T1} + X_{T2} + X_L}$$

Table 2.6 shows the ratio of transformer reactance to the total line reactance when the generator is directly connected to 500-, 750-, and 1000-kV transmission systems through the step-up transformer.

It is shown in Table 2.6 that the higher the transmission voltage, the greater influence the transformer reactance has on the transmission capability.

2.4.2 Ratio of Generator Reactance to UHV Transmission Line Reactance

Generally, the 600-MW or higher generator and its associated step-up transformer form a unit connection that is directly connected to the 500-kV or higher bus, then the power is

Table 2.7: Typical Parameters of 300-, 600-, and 1000-MW Generators Used in China.

Generator power (MW)	300	600	1000
Generator capacity (MVA)	333	667	1111
Terminal voltage (kV)	20	22	27
d-axis synchronous reactance (p.u)	1.682 (0.505)	1.894 (0.284)	2.611 (0.235)
d-axis transient reactance (p.u.)	0.206 (0.062)	0.273 (0.041)	0.267 (0.024)
d-axis subtransient reactance (p.u.)	0.160 (0.048)	0.200 (0.030)	0.222 (0.020)
q-axis synchronous reactance (p.u.)	1.752 (0.526)	1.894 (0.284)	2.489 (0.224)
q-axis transient reactance (p.u.)	0.243 (0.073)	0.273 (0.041)	0.356 (0.032)
q-axis subtransient reactance (p.u.)	0.160 (0.048)	0.200 (0.030)	0.244 (0.022)
Negative sequence reactance (p.u.)	0.163 (0.049)	0.220 (0.033)	0.211 (0.019)
Rotor inertia time constant (s)	2.319	5.547	10.637
Field winding time constant with open-circuited stator (s)	4.350	8.450	8.860
d-axis damper winding time constant with open-circuited stator (s)	0.051	0.050	0.036
g-axis damper winding time constant with open-circuited stator (s)	1.000	0.900	2.500
q-axis damper winding time constant with open-circuited stator (s)	0.051	0.050	0.200

Note: The base voltage corresponding to generator reactance is the mean of the individual rated voltage classes, the base capacity is the capacity of the generator itself, and the base capacity corresponding to the values in brackets is 100 MVA.

sent to the load center (i.e., the sending-end system) through the step-up substation buses, EHV/UHV step-up transformers, and UHV transmission lines. Therefore, when analyzing the transmission capability of an EHV/UHV transmission line, the influence of internal reactance of the generator should be taken into account. Table 2.7 shows the typical parameters of 300-, 600-, and 1000-MW generators used in China.

When planning for the power system, the generator is generally equivalent to transient electromotive force E' and *d*-axis transient reactance X'_d. Therefore, when studying the ratio of generator reactance to line reactance, X'_d is generally used to compare with the line reactance X_L. It should be noted that as the power send-out capability of the power plant is increased, the capacity of parallel-connected units and step-up transformers has to be increased correspondingly.

When a generator (power plant) is connected to the power system through an EHV or UHV transmission line, the *d*-axis transient reactance of generator X'_d/transmission line reactance X_L ratio is $\theta = X'_d/X_L$. Table 2.8 shows the proportion of the transient reactance of a 600-MW generator to the reactance of transmission lines at different voltage classes.

It is shown in Table 2.8 that the higher the transmission voltage, the greater influence the internal reactance of the transformer has on the transmission capability. For long-distance and short-distance EHV and UHV transmission lines, X'_d and $X_{T1} + X_{T2}$ are the main factors limiting the transmission capability of the entire transmission project.

Table 2.8: Ratio of 600-MW Generator Transient Reactance to Transmission Line Reactance.

Transmission line length (km)		100	200	300	400	500	600
Voltage class (kV)	500	2.600	1.300	0.867	0.650	0.520	0.433
	750	2.750	1.375	0.917	0.688	0.550	0.458
	1000	2.800	1.400	0.933	0.700	0.560	0.467

Note: When calculating the ratio of generator reactance to line reactance, the generator reactance and line reactance shall be converted to the values for the same capacity.

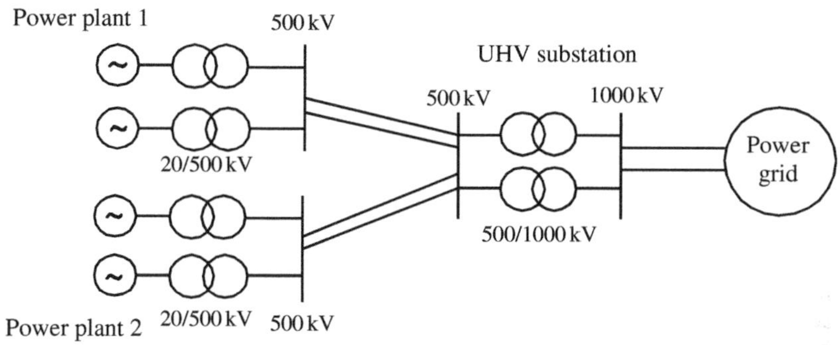

Figure 2.21
Schematic diagram of generators (power plants/stations) connected to UHV power grid through two levels of step-up transformers.

2.4.3 Influence of Connection Scheme of Generators (Power Plants/Stations) on UHV Transmission Capability

Generally, generators (power plants/stations) can be connected to the UHV power grid in two schemes. Let us take a 500-kV system as an example: in the first scheme, as shown in Figure 2.21, the generators (power plants/stations) are connected to 500-kV step-up substation buses through 500-kV step-up transformers to bring the power together. Then, the power is connected to the UHV power grid through 500/1000-kV step-up transformers. In the second scheme, the generators are directly connected to the UHV power grid through UHV step-up transformers, as shown in Figure 2.22.

The first connection scheme has the advantage that the manufacturing of 500/1000-kV UHV step-up transformers is less technically challenging and is thus widely used. However, the use of two-stage step-up transformers would lead to additional tie-reactance between the power plant and the grid, which would compromise the power send-out capability of the power plant. By contrast, the second scheme can reduce the tie-reactance between the power plant and the power grid, thus contributing to the increased power send-out capability of the power plant.

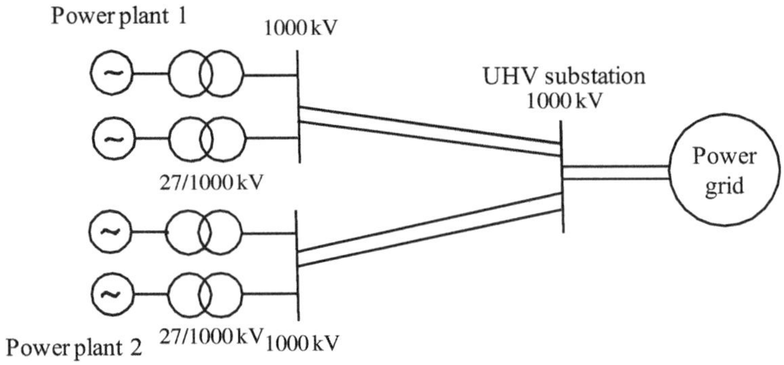

Figure 2.22
Schematic diagram of generators (power plants/stations) connected to UHV power grid through UHV step-up transformers.

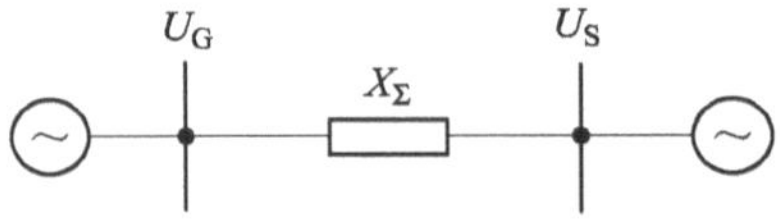

Figure 2.23
Two-generator equivalent system.

The first connection scheme has the advantage that the manufacturing of 500/1000-kV UHV step-up transformers is less technically challenging and is thus widely used. However, there will be an additional tie-reactance between the two levels of transformers, which would compromise the available output of the power plant. In December 2010, China successfully developed the first UHV AC step-up transformer that can directly boost the terminal voltage of 1000 MW generating units from 27 to 1000 kV. This makes the second scheme technically feasible. The third phase of Pingwei Power Plant at the sending end of the Huainan–Shanghai UHV AC Transmission Link plans to adopt the second scheme.

The power send-out system of a power plant may be equivalent to a two-generator system, as shown in Figure 2.23, where U_G is the internal electromotive force of the generator, U_S is the internal electromotive force of the system, and X_Σ is the total system impedance, consisting of generator reactance, step-up transformer reactance, line reactance, system internal impedance, and so on.

The steady-state stability limit of the two-generator system is given by:

$$P = \frac{U_G U_S}{X_\Sigma}$$

The influence of these two different connection schemes on the power transmitting capability (steady-state stability limit) of a power plant by using a specific calculation example is described. A power plant has two 1000-MW generating units that are connected to a nearby UHV substation and further to the power system through a 300-km double-circuit UHV line. Assuming that the system's short-circuit current is 50 kA and that the thermal stability problems caused by $N-1$ fault of generator transformers is neglected. If the first connection scheme is used, then the power plant is connected to the UHV bus through three 750-MVA, 27/500-kV step-up transformers and one 3000-MVA, 500/1000-kV step-up transformer. If the second scheme is used, then the power plant is connected to the UHV bus through two 1120-MVA, 27/1000-kV step-up transformers.

Using 100 MVA as the base value, the *d*-axis transient reactance of each unit $X'_{\mathrm{d}} = 0.024$, the reactance of each 27/500-kV step-up transformer $X_{\mathrm{T1}} = 0.016$, the reactance of each 500/1000-kV step-up transformer $X_{\mathrm{T2}} = 0.006$, the reactance of each 27/1000-kV step-up transformer $X_{\mathrm{T3}} = 0.014$, the reactance of a 300-km UHV single circuit $X_{\mathrm{L}} = 0.007\ 77$, and the system reactance $X_{\mathrm{S}} = 0.001\ 15$.

In the first connection scheme, the total system reactance is:

$$X_{\Sigma 1} = \frac{X'_{\mathrm{d}}}{2} + \frac{X_{\mathrm{T1}}}{3} + X_{\mathrm{T2}} + \frac{X_{\mathrm{L}}}{2} + X_{\mathrm{S}} = 0.0284$$

The steady-state stability limit is 3521 MW.

In the second connection scheme, the total system reactance is:

$$X_{\Sigma 1} = \frac{X'_{\mathrm{d}}}{2} + \frac{X_{\mathrm{T3}}}{2} + \frac{X_{\mathrm{L}}}{2} + X_{\mathrm{S}} = 0.0240$$

The steady-state stability limit is 4167 MW.

It follows that the application of two-stage step-up transformers would increase the equivalent impedance of step-up transformers and reduce the power send-out capability of the power plant. In contrast, the application of one-stage UHV step-up transformers will increase the power send-out capability of the power plant.

2.4.4 Influence of System Parameters on Transmission Capability of UHV AC System

This section mainly deals with the UHV transmission capability between two grids. When analyzing the factors that affect the UHV transmission capability of an interconnection system with two large grids during planning and designing of the UHV transmission system, the sending-end system and receiving-end system may be represented by two equivalent reactance and electromotive forces. The resultant equivalent two-machine system is used to study the influence and effects of the system parameters on the UHV transmission capability.

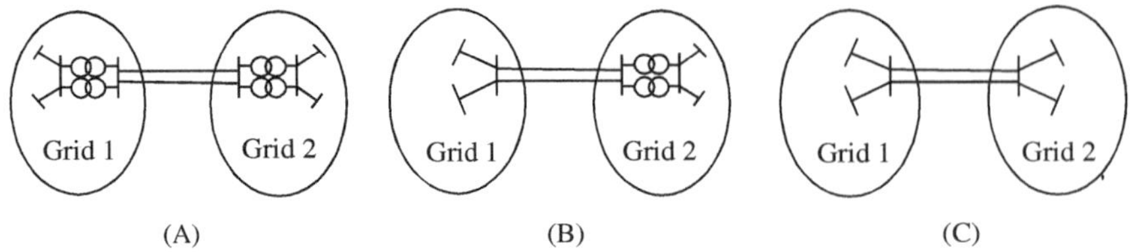

Figure 2.24
Three schemes of connecting two power grids through a UHV tie-line: (A) two ends of the UHV tie-line are connected to the system through transformer, (B) one end of the UHV tie-line is connected to the system through a UHV grid and the other through transformer, and (C) both ends of the UHV tie-line are connected to the system through UHV grid.

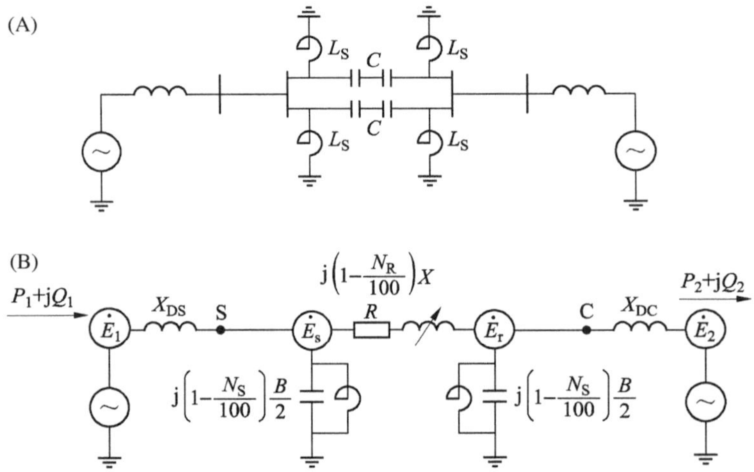

Figure 2.25
Schematic diagram of UHV two-generator system: (A) schematic diagram and (B) equivalent circuit diagram.

There are mainly three schemes to connect two power grids through a UHV tie-line, as shown in Figure 2.24:

Scheme 1: Two ends of the UHV tie-line are connected to the system through the transformer.
Scheme 2: One end of the UHV tie-line is connected to the system through a UHV grid and the other is connected through the transformer.
Scheme 3: Both ends of the UHV tie-line are connected to the system through the UHV grid. The three schemes can be analyzed in a similar manner. Scheme 3 is used as an example to analyze the aforementioned influence and effects without loss of generality.

1. **UHV equivalent two-generator system**
   The UHV equivalent two-generator system is shown in Figure 2.25, where R is the line resistance; C is the line series compensation capacitance; L_S is the line shunt

compensation inductance; X is the line reactance; B is the line susceptance; X_{DS} and X_{DC} are the equivalent reactance, respectively, at the sending end and receiving end, and their magnitudes represent the robustness of the sending- and receiving-end systems, respectively; $\dot{E}_1$ and $\dot{E}_2$ are the equivalent electromotive forces of the sending and receiving-end systems, respectively; and $\dot{E}_s$ and $\dot{E}_r$ are the voltages at two ends of the line, respectively.

When the capacitor C is connected to the line in series, the equivalent reactance is given by:

$$X_{L\text{-}C} = \mathrm{j}X - \mathrm{j}\frac{1}{\omega C} = \mathrm{j}X\left(1 - \frac{X_C}{X}\right) = \mathrm{j}\left(1 - \frac{N_R}{100}\right)X \tag{2.81}$$

where $X_C = 1/\omega C$ is the reactance of series compensation capacitor and $N_R = [(1/\omega C)/X] \times 100 = X_C/X \times 100$ is the rate of series compensation of capacitors.

When the shunt reactance X_S is connected to both ends of the line, the equivalent susceptance is given by:

$$B_{L\text{-}S} = \mathrm{j}\frac{B}{2} - \mathrm{j}\frac{1}{X_S} = \mathrm{j}\frac{B}{2}\left[1 - \left(\frac{1}{X_S}\right)\Big/ B/2\right] = \mathrm{j}\left(1 - \frac{N_S}{100}\right)\frac{B}{2} \tag{2.82}$$

where X_S is the shunt reactance and $N_S = \left[\left(\frac{1}{X_S}\right)/\frac{B}{2}\right] \times 100$ is the rate of shunt reactance compensation.

Given the network configuration and operating conditions of both the sending and receiving-end systems, the short-circuit current calculation program may be used to calculate the three-phase short-circuit current I_K at the connection points S and C, respectively, at both ends. Then, the equivalent reactance X_D of the sending-end and receiving-end systems is given by:

$$X_D = \frac{U}{\sqrt{3}I_K}$$

where

U—normal operating voltage at short-circuit point; for simplicity, the average system voltage may be used.

2. **Criterion and calculation method for UHV transmission capability**

   This section mainly considers the steady-state power angle stability constraints, using 15% as active power reserve margin for calculation and 85% steady-state stability limit as the transmission capability of the system. The practical algorithm for steady-state power angle stability is used for calculation.

3. **Influence of robustness of sending- and receiving-end systems on UHV transmission capability**

   The robustness of the sending and receiving-end systems is represented by the short-circuit current at the connection point of UHV transmission lines. Table 2.9

Table 2.9: Robustness of Sending- and Receiving-End Systems at Different Voltage Classes.

Rated Voltage (kV)	Robustness of Sending- and Receiving-End Systems	
	S_K (MVA)	$X_K = 100/S_K$
500	54,558	0.00183
750	81,837	0.00122
1000	109,116	0.00092

Note: The equivalent reactance at the sending and receiving ends is based on 100-MVA capacity.

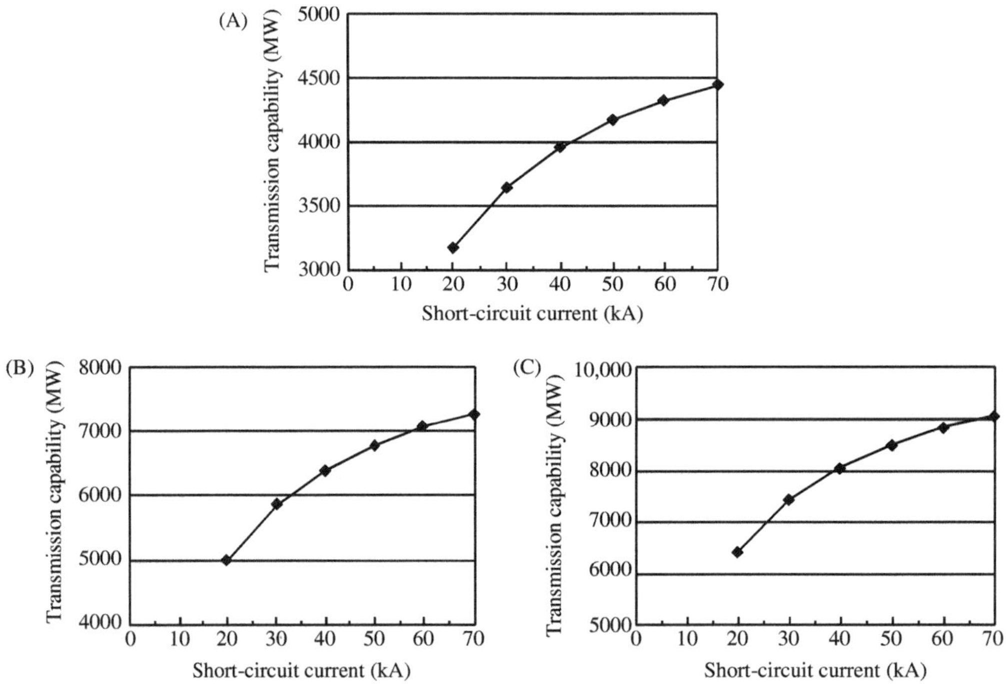

Figure 2.26
Influence of robustness of sending and receiving-end systems on EHV/UHV transmission capability: (A) 300-km double-circuit line at 500 kV, (B) 400-km double-circuit line at 750 kV (HV reactor compensation degree 80%), and (C) 600-km double-circuit line at 1000 kV (HV reactor compensation degree 80%).

shows the system robustness S_K and system equivalent reactance X_K when a short-circuit current of 63 kA flows through the connection point of EHV/UHV transmission lines.

Figure 2.26 shows the influence of the robustness of sending and receiving-end systems on the EHV/UHV transmission capability. It can be seen that the transmission capacity increases significantly with the increase in the robustness of the sending and receiving-end systems. And the higher the voltage class, the more significantly the transmission capability is increased. This is because for the same transmission distance, the higher the voltage class, the smaller the proportion of line impedance to the total system impedance, the larger the proportion the grid impedance will account for and, consequently, the more robust the sending and receiving-end systems become and, ultimately, the more significantly the UHV transmission capability is able to vary. However, with the increase in the transmission distance, the amplitude of variation tends to decrease. Such variation rules indicate that during the initial stage of UHV backbone construction when the 500-kV power grids at the sending and receiving-end are not robust enough, the inherent transmission capability of a UHV system would be considerably constrained.

4. **Influence of shunt reactor on UHV transmission capability**

   Due to the presence of large capacitor charging current, shunt reactors are generally installed at both ends of EHV/UHV transmission lines to compensate capacitive current to limit the temporary overvoltage. Generally, the shunt reactor can affect the wave impedance power of a transmission line and, hence, the maximum transmission capability. How seriously it affects the transmission capability depends on the voltage class and the system reactance at the sending and receiving ends. Figure 2.27 shows the influence of line shunt reactor on the EHV/UHV transmission capability. The line shunt reactor affects the transmission capability to some degree. Physically, it can be seen from the star-to-delta impedance transformation equations that the connection of the shunt reactor is equivalent to increasing the reactance of the transmission line, which causes the SIL to decrease, thereby indirectly compromising the transmission capability.

   The installation of shunt reactors with high compensation degree can address the power frequency overvoltage, secondary arc current, and other issues. However, when the system is heavily loaded, this would increase the reactive power loss and reduce the system voltage level, thus affecting the system transmission capability. The controllable reactor can effectively solve the contradictions between reactive power compensation and overvoltage limit. Specially, when the power flow in the system varies, the compensation capacity is regulated according to the requirements of voltage regulation. Once a transient process occurs, its capacity is rapidly increased to enhance the compensation effects greatly, thereby limiting the switching overvoltage and power frequency overvoltage. In addition, the controllable reactor is helpful for enhancing the system stability. It can not only provide reactive power support upon fault to raise the system dynamic voltage but also increase the system damping to suppress oscillation.

5. **Influence of series capacitor compensation on UHV transmission capability**

   The series capacitor compensation may further improve the UHV transmission capability significantly. Figure 2.28 shows the effects of different rates of series capacitor compensation on improving the transmission capability (for a 600-km

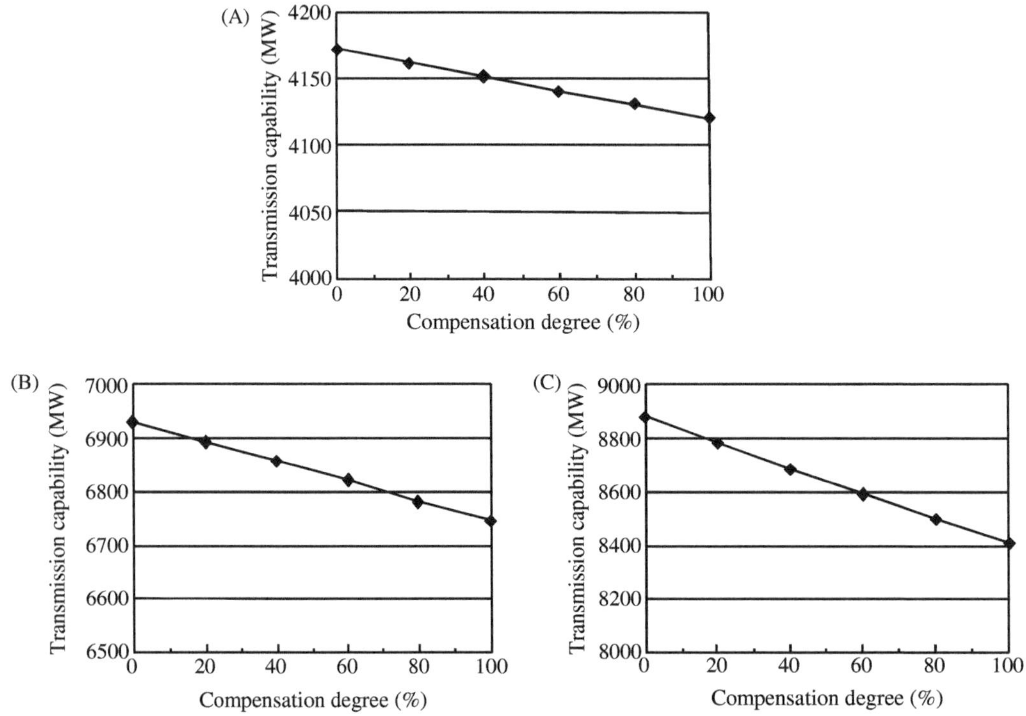

Figure 2.27
Influence of shunt reactors on EHV/UHV transmission capability: (A) 300-km double-circuit line at 500 kV, (B) 400-km double-circuit line at 750 kV, and (C) 600-km double-circuit line at 1000 kV.

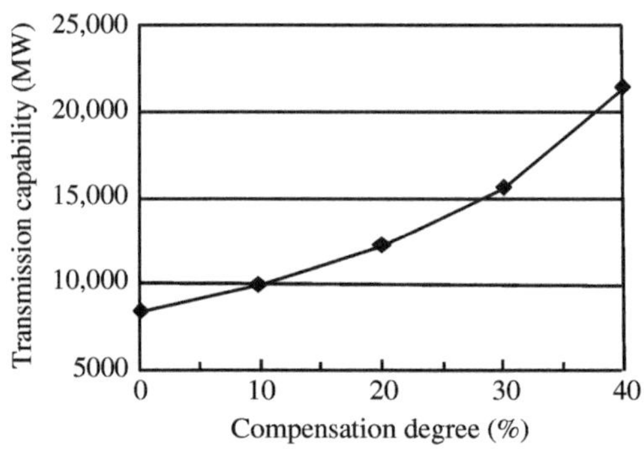

Figure 2.28
Influence of different compensation degrees of series capacitor on transmission capability (600-km double-circuit line at 1000 kV).

double-circuit line at 1000 kV). The application of series capacitor compensation is equivalent to reducing the line length. Even if the sending and receiving-end systems are weak, the series capacitor compensation is also able to improve the transmission capability satisfactorily.

The controllable series compensation allows smooth, continuous, and rapid adjustment of the equivalent capacitive reactance, thereby changing the line reactance. Therefore, the controllable series compensation not only plays the same role as the conventional fixed series compensation, namely, compensating the line reactance to improve the transmission capability, but also is helpful to control the line power flow, improve the system transient stability, and dampen the low-frequency oscillation. Moreover, it is also capable of suppressing subsynchronous oscillations and eliminating the restrictions of subsynchronous oscillation upon the rate of series compensation. Thus, the controllable series compensation is technically more superior to the fixed series compensation. China is performing a feasibility study of the application of the UHV controllable series compensation system. It shows that to accommodate the need for long-distance bulk power transmission using UHV power grids in the future, to enhance the operating stability level of grids, and to reduce the transmission corridor, it is feasible to equip the transmission channels with controllable series compensation.

References

[1] Z. Qinyong, G. Qiang, B. Guangquan, et al., Application of controllable reactors in China's power grid at extra and ultra voltage class, Proc. CSEE 27(7) (2007) 1–6.

[2] M. Shiying, Y. Yonghua, T. Yong, et al., Simulation and evaluation for short term and mid/long term voltage stability, Power Syst. Technol. 30(19) (2006) 14–20.

[3] S. Huadong, T. Yong, M. Shiying, A commentary on definition and classification of power system stability, Power Syst. Technol. 30(17) (2006) 31–35.

[4] W. Xiaogang, Y. Yonghua, B. Liangeng, et al., Summary of commissioning test on the 1000kV UHV AC demonstration project, Proc. CSEE 29(22) (2009) 12–18.

[5] Z. Jian, Z. Wenchao, X. Yang, et al., Simulation and verification of system commissioning test for 1000kV AC pilot project, Power Syst. Technol. 33(16) (2009) 29–32.

[6] Z. Liang, G. Qiang, T. Qin, et al., Analysis on stability characteristics of UHV synchronous power grid, Power Syst. Technol. 28(34) (2008) 47–51.

[7] S. Yinbiao, Z. Wenliang, Z. Xiaoxin, et al., Security evaluation of UHV synchronized power grid, Proc. CSEE 27(34) (2007) 1–6.

[8] Z. Wenliang, Z. Xiaoxin, Y. Yonghua, et al., Composition and analysis of "North China-Central China-East China" UHV synchronous power grid, Proc. CSEE 30(16) (2010) 1–5.

CHAPTER 3

Characteristics of UHV DC Transmission System

Chapter Outline

3.1 Basic Principles of HVDC Transmission System

3.1.1 Basics of HVDC Conversion Technology

The application of DC power transmission entails conversion of alternating current to direct current and vice versa. In particular, to allow electric energy to be transmitted from the power source to the load center, alternating current must first be converted into direct current at the sending end, referred to as rectification; direct current must then be converted

Ultra-high Voltage AC/DC Grids.
DOI: http://dx.doi.org/10.1016/B978-0-12-802161-3.00003-2

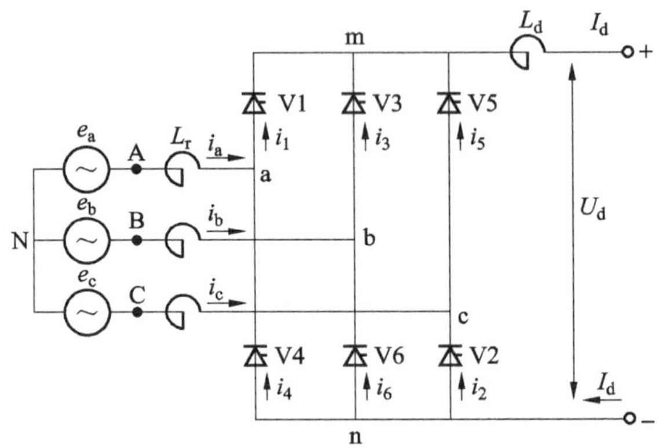

Figure 3.1
Schematic diagram of a six-pulse rectifier.

back into alternating current at the receiving end, referred to as inversion. The facility where the rectification occurs at the sending end is called a rectifier station, and the facility where the inversion occurs at the receiving end is called an inverter station. Rectifier stations or inverter stations are collectively known as converter stations. The devices that enable rectification and inversion operation are, respectively, called rectifier and inverter, and are collectively known as converter (converter unit).

Currently, converter used for HVDC and UHV DC power transmission generally comprise thyristor-based converter bridges and converter transformers that supply the commutation voltage. Thyristor is a technically proven power electronic device that can withstand high voltage and has high through-current capability. The on and off characteristics of thyristor are the basis for analyzing the thyristor conversion technology. A thyristor is characterized by the following: (i) current can only flow from anode to cathode, although a very small reverse current (reverse leakage current) may occur (unless the thyristor is damaged); (ii) it can be turned on if the anode is at a positive voltage with respect to the cathode, namely, a commutating voltage must be present, and a gate pulse is applied; and (iii) once turned on, it cannot be turned off until the current flowing through it decreases to zero. After the thyristor is turned off, sufficient deionization time when the anode is at a negative voltage with respect to the cathode is necessary; otherwise, even if no gate pulse is applied, the thyristor will be turned on when a positive voltage is imposed at the anode.

3.1.2 Six-Pulse Converter

1. **Rectifiers**
   Figure 3.1 shows the schematic diagram of a six-pulse rectifier. Figure 3.2 depicts the relevant voltage and current waveforms of a rectifier during normal operation. The six-pulse rectifier works in such a manner that the six bridge arms (valves) V1 through

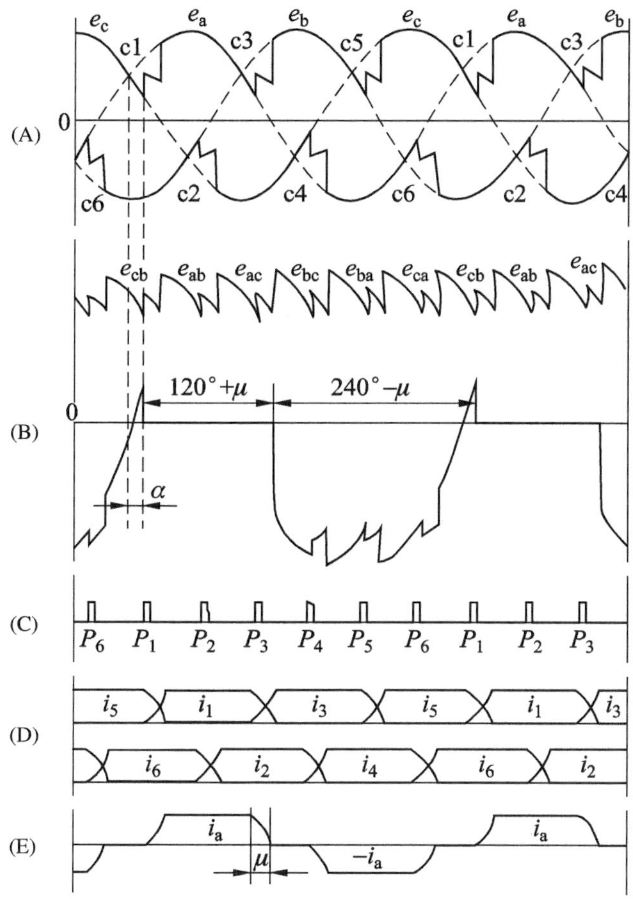

Figure 3.2

Voltage and current waveforms of a six-pulse rectifier: (A) waveforms of AC electromotive force and voltages at *m* and *n* points on DC side with respect to the neutral point; (B) waveforms of DC voltage and voltage across valve 1; (C) sequence and phase of firing pulses; (D) current waveform of valves; and (E) waveform of current through phase A on AC side.

V6 connected in a three-phase bridge configuration are turned on and off sequentially, thereby converting the alternating current into direct current. The numerals 1 through 6 indicate the sequence in which the valves are turned on. Generally, each valve consists of a number of thyristors connected in series; therefore it has the characteristics of thyristors and can meet the DC voltage design requirement. In Figure 3.1, e_a, e_b, and e_c represent, respectively, equivalent electromotive force of individual phases of an AC system and are generally supplied by the converter transformer; L_r represents the equivalent commutation reactance of each phase and mainly consists of the leakage reactance of the converter transformer; L_d represents the inductance of the smoothing reactor. The equivalent line voltages e_{ac}, e_{bc}, e_{ba}, e_{ca}, e_{cb}, and e_{ab} are the commutating voltages of valves. It is specified that the zero crossing point where the commutating voltage changes from negative to positive is the timing start point of the valve's firing angle α. Under idealized conditions, the three-phase AC system has a symmetrical configuration; therefore, the firing angles of the six valves are equal and evenly spaced and the firing pulses are spaced by 60°.

The area of the positive half of the commutating voltage waveform ($\alpha = 0-180°$) is the region where the valves can conduct in forward direction. The instant when a firing pulse is applied (α angle) is the conduction instant of valves. For rectifiers, when α falls within 0–90° (90° not included), the converter outputs a positive DC voltage. Within one power frequency cycle, a valve from the valve group sharing the common anode (namely, V2, V4, and V6) and another different phase from the valve group sharing the common cathode (namely, V1, V3, and V5) will conduct, allowing the AC current to flow through the rectifier into the DC circuit. In this way, the AC current is converted into a DC current.

A description of the commutation process by using an example of commutating the current from V1 to V3 follows. Once V3 is turned on, phases A and B of the converter transformer are short-circuited through V1 and V3; at this moment, the current flowing through V3 is equal to the phase-to-phase short-circuit current that increases from zero, whereas the phase-to-phase short-circuit current through V1 is in the reverse direction to its original current and, consequently, the resulting current through V1 is equal to the difference between the two. When the phase-to-phase short-circuit current through V1 becomes equal to the original current, the resultant current through V1 decreases to zero, causing V1 to turn off. At this time, all DC current will flow through V3 and the commutation process completes. Because of the presence of inductance, the current through the valves that are, respectively, turned on and off cannot be varied instantly. The time taken to commutate the current from one valve to another is called "overlap angle" μ.

For a six-pulse rectifier, the mean DC voltage U_{d1} may be expressed as:

$$U_{d1} = U_{di01}\cos\alpha - \frac{3}{\pi}X_{r1}I_d = U_{di01}\cos\alpha - d_{r1}I_d$$
$$U_{di01} = 1.35U_1 \tag{3.1}$$
$$d_{r1} = \frac{3}{\pi}X_{r1}$$

where

U_{di01}—ideal no-load DC voltage, kV

U_1—RMS line voltage on the rectifier side of the converter transformer, kV

α—firing angle of rectifier, (°)

I_d—DC current, kA

X_{r1}—equivalent commutation reactance, Ω

d_{r1}—specific commutation voltage drop, that is, the DC voltage drop caused by per-unit DC current during commutation, Ω.

Equation (3.1) gives the rectifier's DC voltage as a function of its DC current, also known as the rectifier's voltage–current characteristic.

Overlap angle μ is an important parameter of converters and directly affects their reactive power and harmonic performance.

The overlap angle of a rectifier is:

$$\mu_1 = \text{arc cos}\left(\cos\alpha - \frac{2X_{r1}I_d}{\sqrt{2}U_1}\right) - \alpha \tag{3.2}$$

It is shown in Eq. (3.2) that, during rectifier operation, μ_1 decreases with the increase of α ($\alpha < 90°$) but increases with the increase of X_{r1}. When α and X_{r1} are constant, μ_1 increases with the increase of I_d or the decrease of U_1.

2. **Inverters**

Inverters discussed here are active inverters; they need an AC system to supply commutating voltage and current. Figures 3.3 and 3.4, respectively, show the schematic diagram and the voltage and current waveforms of a six-pulse inverter.

It is shown in Eq. (3.1) that when $\alpha < 90°$ and the internal voltage drop of a converter is ignored, the DC output voltage is positive and the converter operates in rectifier mode; when $\alpha = 90°$, the DC output voltage is zero; when $\alpha > 90°$ (but $\alpha < 180°$), the DC output voltage is negative and the converter operates in inverter mode. When a converter operates as an inverter on its own, it is impossible for it to supply a DC current to an external load. However, when combined with a rectifier, the inverter serves as load for the rectifier and can operate as intended, driven by the positive DC voltage output from the rectifier.

Like the rectifier, a six-pulse inverter also consists of six valves connected in a three-phase bridge configuration and a converter transformer that serves to supply commutation voltage. Each valve contains a number of thyristors connected in series

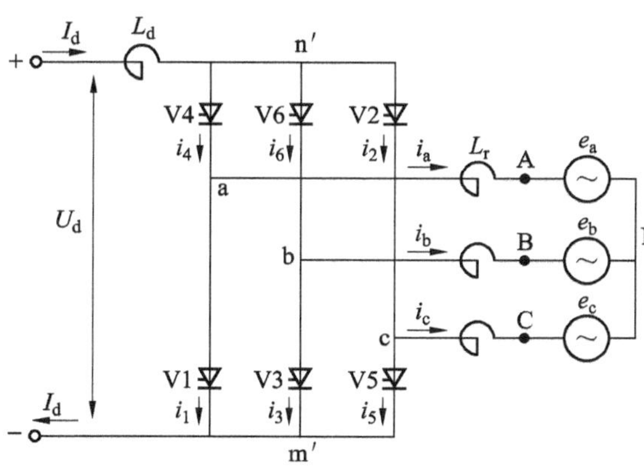

Figure 3.3
Schematic diagram of a six-pulse inverter.

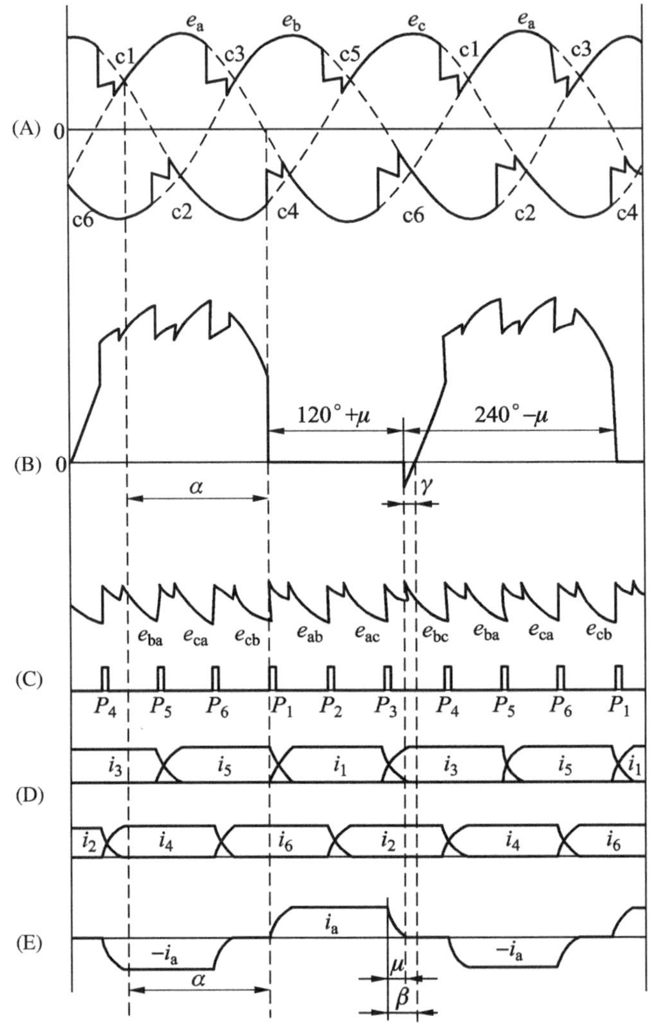

Figure 3.4

Voltage and current waveforms of a six-pulse inverter: (A) waveforms of AC electromotive force and voltages at m' and n' points on DC side with respect to the neutral point; (B) waveforms of DC voltage and voltage across valve 1; (C) sequence and phase of firing pulses; (D) waveform of valve current; and (E) waveform of current through phase A on AC side.

and has the characteristics of thyristors. Six valves, V1 through V6, are in the same sequence as that of a rectifier. Also, the current is commutated from one valve to another through the phase-to-phase short-circuit current with the help of the converter transformer. Within one power frequency cycle, one valve connected to the anode and another connected to the cathode that are of different phase will conduct, allowing the DC current to flow into the three-phase windings of the converter transformer. In this way, the DC current is converted to AC current.

The mean DC voltage U_{d2} across the inverter may be expressed by:

$$\begin{aligned} U_{d2} &= -(U_{di02}\cos\gamma - d_{r2}I_d) \\ &= -(U_{di02}\cos\beta + d_{r2}I_d) \\ U_{di02} &= 1.35U_2 \\ d_{r2} &= \frac{3}{\pi}X_{r2} \\ \beta &= 180° - \alpha = \gamma + \mu \end{aligned} \quad (3.3)$$

where

U_{di02}—ideal no-load DC voltage through inverter, kV

U_2—RMS line voltage on the inverter side of the converter transformer, kV

d_{r2}—commutating voltage drop of inverter, Ω

X_{r2}—equivalent commutation reactance of inverter, Ω

β—advance angle of inverter, (°)

γ—extinction angle, expressed in (°), a measure of time interval; in this interval, after the valve is turned off and DC current drops to zero, the valve suffers from negative voltage.

The overlap angle of an inverter is:

$$\mu_2 = \arccos\left(\cos\gamma - \frac{2X_{r2}I_d}{\sqrt{2}U_2}\right) - \gamma \quad (3.4)$$

During inverter operation, the overlap angle μ_2 is a function of I_d, U_2, γ, and X_{r2}. Given $\gamma = \beta - \mu_2$, if β is kept constant or nearly constant, then an increase of μ_2 will lead to the decrease of γ. If the extinction angle γ is decreased to the point when the time for the negative voltage is applied across the valve is less than that required for it to recover its blocking capability, the valve would conduct again once a positive voltage is applied. This phenomenon is called commutation failure. Commutation failure will result in periodic short-time short-circuiting across the inverter, causing the DC voltage to decrease instantaneously and the DC current to increase instantaneously. To avoid commutation failure, it is specified that γ should be larger than or equal to γ_0. Generally, γ_0 is taken as 17–18° to allow sufficient time for the valve to recover its blocking capability, with due regard to the asymmetry of the three-phase voltage and parameters of the AC system and other factors such as margin reserved for the response time of discrete control. In the case of a rectifier valve, once it is turned off, it will be in a reverse blocking state for a long time; therefore, no commutation failure would result. By comparing Figures 3.2 and 3.4, it is apparent that during the commutation process, the current waveforms of inverter valve and rectifier valve are different because of different firing instants. The rate of increase in valve current shows an increasing tendency for rectifier valves but a declining tendency for inverter valves.

3. **Reactive power**

There is a phase displacement between AC voltage and current in each phase of the converter. It may be approximated to the power-factor angle φ of the converter. Both inverter and rectifier need to absorb reactive power from the AC system during operation; therefore, the converter station should be equipped with reactive compensation devices.

The power factor of a rectifier station is:

$$\cos\varphi_1 \approx \frac{1}{2}\left[\cos\alpha + \cos(\alpha + \mu_1)\right] \tag{3.5}$$

The power factor of an inverter station is:

$$\cos\varphi_2 \approx \frac{1}{2}\left[\cos\gamma + \cos(\gamma + \mu_2)\right] \tag{3.6}$$

The reactive power absorbed by a rectifier is:

$$Q_{\mathrm{C1}} = P_{\mathrm{d1}}\tan\varphi_1 = P_{\mathrm{d1}}\sqrt{\left(\frac{U_{\mathrm{di01}}}{U_{\mathrm{d1}}}\right)^2 - 1} \tag{3.7}$$

The reactive power absorbed by an inverter is:

$$Q_{\mathrm{C2}} = P_{\mathrm{d2}}\tan\varphi_2 = P_{\mathrm{d2}}\sqrt{\left(\frac{U_{\mathrm{di02}}}{U_{\mathrm{d2}}}\right)^2 - 1} \tag{3.8}$$

where

P_{d1}—active power on the rectifier side, MW, $P_{\mathrm{d1}} = U_{\mathrm{d1}}I_{\mathrm{d}}$

P_{d2}—active power on the inverter side, MW, $P_{\mathrm{d2}} = U_{\mathrm{d2}}I_{\mathrm{d}}$.

It is shown in Eqs. (3.7) and (3.8) that thyristor valves absorb a large amount of reactive power during operation, and the amount of reactive power absorbed by converters is directly proportional to the DC power transmitted.

The current flowing from the converter to the AC side is not perfectly sinusoidal; therefore, the waveform of resulting DC voltage is not smooth. Through Fourier analysis of the waveforms, a six-pulse converter will produce characteristic harmonics of orders $(6k \pm 1)$ and $6k$, respectively, on the AC side and DC side (*k is a positive integer*). Therefore, it is necessary to provide AC filters of $(6k \pm 1)$ orders on the AC side, and DC filters of $6k$ orders on the DC side. The harmonic and reactive power characteristics of thyristor converters have a direct bearing on the requirements of performance and capacity of the converter transformer.

For six-pulse converters, the AC-side current and the parameters of the converter transformer are given here.

a. The RMS line current on AC side may be approximated to:

$$I_a \approx \sqrt{\frac{2}{3}} I_d = 0.816 I_d \tag{3.9}$$

The RMS fundamental component of line current on the AC side of a rectifier or inverter may be approximated to:

$$I_{a1} \approx \frac{\sqrt{6}}{\pi} I_d = 0.78 I_d \tag{3.10}$$

b. The rated capacity of a three-phase converter transformer may be expressed by:

$$S = \frac{\pi}{3} U_{\mathrm{di0N}} I_{\mathrm{dN}} \tag{3.11}$$

where
U_{di0N}—ideal rated no-load DC voltage, kV
I_{dN}—rated DC current, kA.

3.1.3 Twelve-Pulse Converter

A 12-pulse converter consists of two six-pulse converters. The DC sides of the two six-phase converters are connected in series and the AC sides are connected in parallel. On the valve side, one converter transformer is star-connected and the other is delta-connected, giving rise to a 30° phase displacement between the commutating voltages of two six-pulse converters. A 12-pulse converter may be operated with two double-winding converter transformers or one three-winding converter transformer. Figure 3.5 shows the schematic diagram of a 12-pulse converter operated with two double-winding converter transformers.

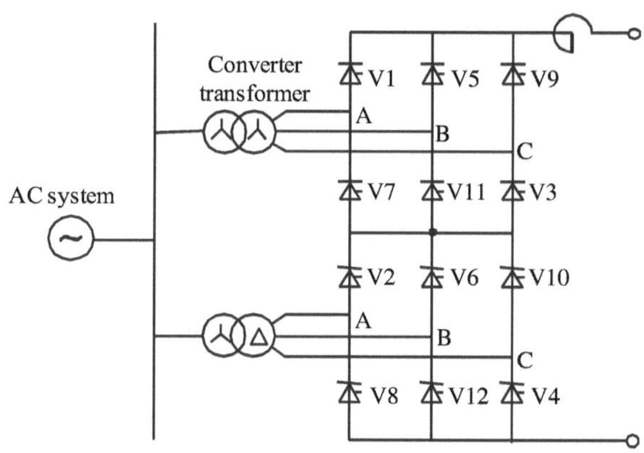

Figure 3.5
Schematic diagram of a 12-pulse converter.

The 12-pulse converter consists of 12 valves numbered from V1 to V12. Numerals of valves shown in Figure 3.5 correspond to the situation in which the voltage in the star configuration leads that in the delta configuration. Within one power frequency cycle, 12 sequential firing pulses in-step with AC system are required and the pulse interval should be 30°. A 12-pulse converter works in the same way as that of a six-pulse converter. Also, it commutates current from one valve to the next through the phase-to-phase short-circuit current of the AC system. The DC voltage across a 12-pulse converter is twice that across a six-pulse converter.

A 12-pulse converter will produce characteristic harmonics of orders $12k \pm 1$ and $12k$, respectively, on the AC side and the DC side. Therefore, apart from the ability to increase the DC voltage, a 12-pulse converter is advantageous in the following ways:

a. The DC voltage contains less harmonic components. The DC voltage across the 12-pulse converter is the sum of DC voltage across two six-pulse converters. Within one power frequency cycle, 12 pulses occur and each six-pulse converter produces harmonics of orders 6 and $6(2k+1)$ (e.g., 18, 30, 42,...) in the DC voltage, which are part of the harmonics produced by the two six-pulse converters, are 180° out of phase with each other, and thus cancel out. As a result, the DC voltage contains only harmonics of orders $12k$; therefore, only filters of orders $12k$ are required on the DC side.
b. The AC voltage contains less harmonic components. For a 12-pulse converter operated with two double-winding converter transformers, the harmonics of orders $6(2k-1) \pm 1$ (e.g., 5, 7, 17, 19,...) in AC current simply circulate on the grid side windings between the two transformers, rather than flowing into the AC system. For a 12-pulse converter operated with one three-winding converter transformer, no harmonic of orders $6(2k-1) \pm 1$ would be contained in the transformer's grid side windings because the harmonics of these orders on the two valve side windings are 180° out of phase and thus cancel each other. Therefore, only filters of orders $(12k \pm 1)$ are required.

The use of 12-pulse converters can effectively improve the harmonic performance on both the AC and DC sides, simplify the filtering devices, reduce the land occupation, and lower the construction costs of converter stations, which are the main reasons why most HVDC transmission systems adopt 12-pulse converters as the basic converter units.

3.2 Characteristics of UHV DC Transmission System

3.2.1 System Composition

The DC transmission systems can be classified into two categories in terms of system composition, namely, the two-terminal and multiterminal system, which is also the case with UHV DC transmission systems.

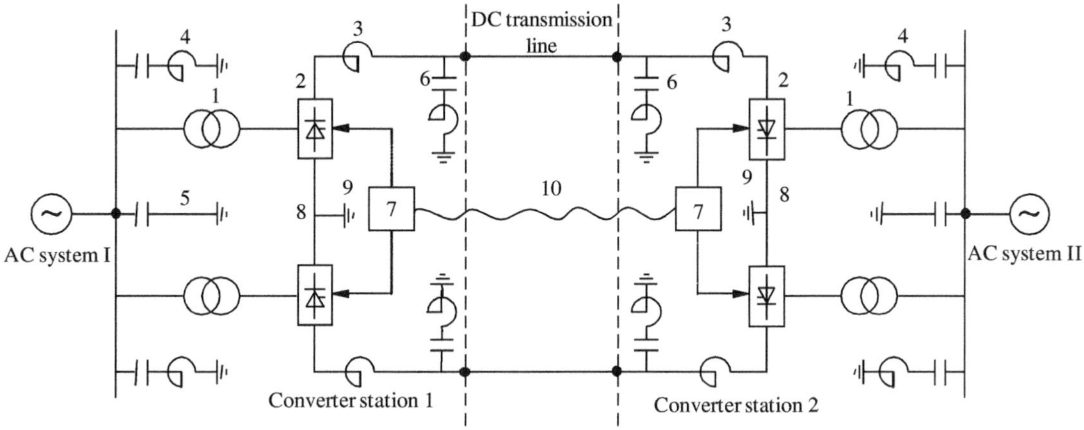

Figure 3.6
Schematic of a bipolar two-terminal system: 1, converter transformer; 2, converter valve; 3, smoothing reactor; 4, AC filter; 5, capacitor bank; 6, DC filter; 7, control and protection system; 8, neutral bus; 9, earth electrode line/earth electrode; 10, telecommunication system.

1. **Two-terminal system**

A two-terminal long-distance DC system generally comprises one rectifier station and one converter station built at different geographical location, a transmission line connecting the two terminals, earth electrodes, and earth electrode lines. Figure 3.6 shows the schematic diagram of a typical bipolar configuration.

The HVDC system is a DC system connected between different AC buses and comprises the minimum converter units that can be operated independently. The DC system within converter stations may consist of a number of converters and DC and AC switchyard equipment. A converter mainly comprises converter bridges, control and protection systems, and converter transformers. DC switchyard equipment mainly includes smoothing reactors, DC filters, and various types of DC switchgears. AC switchyard equipment mainly includes the filtering and reactive power compensation devices such as AC filters and capacitor banks. The DC secondary system mainly includes the DC control and protection system and associated communication systems.

a. The converter valve is the core component of a DC transmission system to allow AC/DC conversion. Mercury-arc valves were once used in early DC transmission systems. However, they were found to have many disadvantages, such as complicated manufacturing, high costs, high failure rate attributable to arc-back, low reliability, and difficult maintenance, that hinder the application and development of DC transmission. With the rapid development of power electronic and microelectronic technologies, the high-voltage and large-power thyristor becomes a mature technology. Characterized by no arc-back failure, easy manufacturing, easy testing, easy operation, and easy maintenance, thyristors have significantly improved the operating performance and reliability of DC transmission. Since the 1970s, thyristor valve-based converters have become the mainstream products used for DC transmission and greatly promoted the development of DC technology. The thyristor is a low-frequency semiconductor device with no self-turn-off capability. Such thyristors comprise the line-commutated converter in which the valves are commutated through use of the phase-to-phase short-circuit current on the valve side of the converter transformer. There are two types of thyristors used in DC systems, respectively, the electrically triggered thyristor (ETT) and light-triggered thyristor (LTT). The increasing capacity of the DC system requires the development of thyristor technology with higher voltage and current ratings. The high-capacity 6-inch thyristor first developed by China for UHV DC applications is a leading technology around the world.

b. The converter transformer serves to connect the AC system and the converter bridges by realizing the voltage transformation and electrical isolation between the AC and the DC sides and limiting the short-circuit current. There are four available types of converter transformers, the three-phase three-winding, the three-phase two-winding, the single-phase three-winding, and the single-phase two-winding. In contrast to conventional AC power transformers, the valve side windings are subject to not only the combined load stress of DC and AC voltage but also the stress caused by polarity reversal. In addition, a series of harmonic currents is present in windings on both sides.
c. The smoothing reactors and DC filters function together to smooth the current on the DC side, protect the converter station from the line's steep-front overvoltage, avoid interruption of DC current, and reduce the probability of commutation failures.
d. The DC and AC filters mainly serve to filter the characteristic harmonics generated on both AC and DC sides as the converter acts as a harmonic current source for the former and as a harmonic voltage source for the latter.
e. The reactive power compensation devices are used to provide the reactive power needed by the converter stations. The thyristor-based converter absorbs a large amount of reactive power during its operation, up to 40–60% of the active power delivered by the DC system. To this end, apart from the use of AC filters to supply reactive power, additional compensation devices are required, such as capacitors, condensers, or static VAR (Volt Ampere Reactive) compensators.
f. The DC control and protection system enables the normal start and stop of the system, change of operating modes, automatic regulation of steady state and dynamic state, troubleshooting, and protection of system and equipment, thus playing an important role in guaranteeing the performance and reliability of the DC system. Because of the development of computer technology, all the control and

protection equipment today incorporate the advanced high-performance multiple-microcomputer parallel processing technology, standard bus technology, network communication technology with standardized protocols, and the hierarchical redundant structure, which have dramatically improved the control and protection system's real-time performance and reliability.

g. The earth electrode and earth electrode line are connected to the ground or sea to form a DC return circuit. In contrast to the safety earthing equipment in converter stations, it is necessary to take into account several factors for earth electrode and earth electrode line, such as the electro-corrosion caused by ground current to the earth electrodes and the nearby underground metallic conduits, and the saturation of nearby neutral grounding transformers because of the increase in DC bias. When designing the earth electrodes, it is necessary to determine their allowable operating current and operating time.

Two-terminal DC transmission systems generally fall into three configurations: monopole (either positive or negative); bipolar (positive and negative poles); and back-to-back systems (without DC transmission lines).

A monopole system usually uses monopolar ground return configuration or monopolar metallic return configuration as shown in Figure 3.7. Both of the configurations use a high-voltage transmission line. The difference lies in that the former uses the ground or seawater as the return path, whereas the latter uses a low-voltage line as the return path. In the monopolar ground return configuration, a DC current flows through the earth electrodes, whereas in the monopolar metallic return configuration there is no DC current through the ground and the earthing point on the DC side is of protective earthing.

Most bipolar systems adopt neutral bus grounding at both terminals, as shown in Figure 3.6. A bipolar system consists of two independently operated monopolar ground return systems in the same converter station. The two poles share the same earth

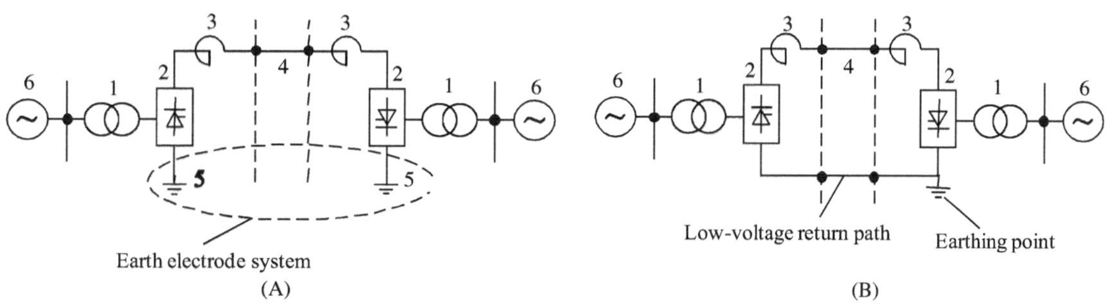

Figure 3.7
Schematic of monopolar DC system wiring configurations: (A) monopolar ground return system and (B) monopolar metallic return system. 1, converter transformer; 2, converter bridge; 3, smoothing reactor; 4, DC transmission line; 5, earth electrode system; 6, AC system.

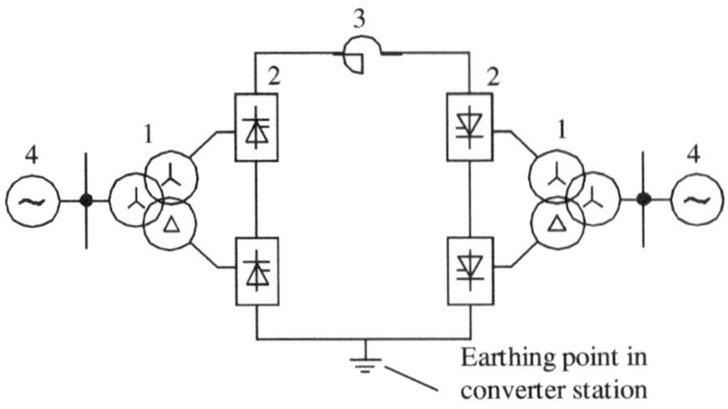

Figure 3.8
Back-to-back wiring configuration: 1, converter transformer; 2, converter bridge; 3, smoothing reactor; 4, AC system.

electrode line and earth electrode, and the current through the ground is equal to the difference between the two pole currents. During normal balanced bipolar operation, only a small unbalanced current (generally smaller than 1% of the rated current) caused by measuring and control errors flows through the ground. If a pole is out of service due to faults, then the bipolar system can be automatically switched to monopolar operation mode with the loss of, at most, 50% of the total capacity, thus improving the system reliability. Moreover, this configuration allows phased construction of a transmission project. A bipolar system may also adopt the configuration in which only the converter station at one terminal is grounded or the bipolar configuration has a metallic neutral line. However, these two configurations are rarely used in practice.

As shown in Figure 3.8, the back-to-back system is a two-terminal system without a DC transmission line and an earth electrode line/earth electrode. In a back-to-back DC system, the rectifier and inverter are located in the same converter station (usually known as a back-to-back converter station) and the two terminals are, respectively, connected to different AC systems. The back-to-back system is mainly characterized by: (i) the DC-side voltage is low but the current is high; (ii) it allows full use of the through-current capability of large-section thyristors; (iii) DC filters and earth electrodes are eliminated, thus lowering the construction cost; and (iv) a number of back-to-back units independent of each other can be constructed within the same converter station and the coordinated control of them may contribute to higher interconnection benefits and stability of AC systems.

2. **Multiterminal system**

A multiterminal system generally consists of three or more converter stations and can interconnect three or more AC grids. Of all the DC transmission projects that have been commissioned around the world, only a few are multiterminal systems. A multiterminal system allows DC power to be collected from multiple sources or transmitted to multiple receiving terminals. Also, it enables a number of AC systems located in different regions

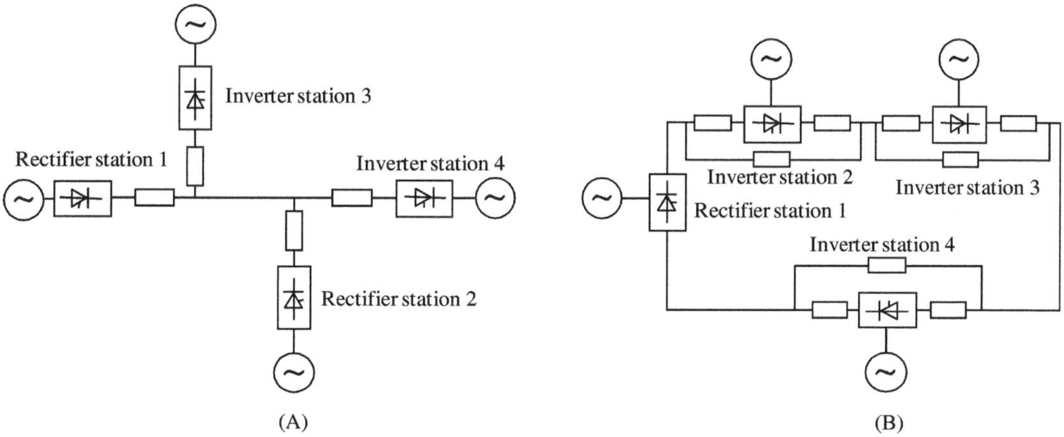

Figure 3.9
Wiring schematic diagram of multiterminal HVDC transmission systems: (A) parallel multiterminal system and (B) series multiterminal system.

or isolated grids to be interconnected. Each converter station can act as either a rectifier station or an inverter station provided that the total power, when it operates under the rectifier mode, is equal to that when it operates under the inverter mode. As shown in Figure 3.9, the converter stations may be connected either in parallel (DC lines may be arranged in the form of branches or a loop) or in series. For multiterminal systems, in-depth studies should be performed for the main circuit and control and protection system because of the complexity in power allocation among converter stations during dynamic and transient processes, system fault management, and measures to improve the system reliability. The HVDC Italy–Corsica–Sardinia (three terminals, small size) and the Quebec–New England system (a five-terminal link with three terminals in practical use) are the examples of multiterminal systems that have been put into operation. In addition, the Nelson River Bipole 1 and Bipole 2 in Canada and the Pacific DC Intertie in the United States are also featured with the operating performance of a multiterminal system.

3. **Configuration of main circuit and characteristics of main equipment of a UHV DC transmission system**
   a. *Configuration of the main circuit of DC systems*: Currently, most UHV DC systems are two-terminal systems. To accommodate the requirement of higher transmission capability, higher DC system voltage and current may be considered. The main circuit of the DC system may adopt a number of converters connected in parallel or series. Through in-depth study of and comparison between the options where one, two, or three 12-pulse converters are connected in series (uniformly or nonuniformly distributed voltage) or in parallel on each pole at each terminal, it follows that operating the power transmission at higher voltage is an economic option. Considering the greatly increased transmission capacity and difficulty in

manufacturing and transportation of converter transformers for ±800 kV UHV DC systems, it was determined that the option in which two 12-pulse converters carrying the same DC voltage are connected in series on each pole at each terminal is given priority for China's UHV DC systems. In such an option, each 12-pulse converter has its separate valve hall and double-valve configuration is used. With this option, the voltage can be boosted to the required level, thereby increasing the transmission capacity and reducing power loss. In addition, in the event of outage of one converter, the remaining converters can still operate normally. This ensures the system reliability while increasing the transmission capacity. This typical configuration has been successfully used in practice.

The configuration of the main circuit of a UHV DC link also differs from that of a conventional HVDC link in that each converter is connected in parallel with a bypass circuit breaker and associated disconnector so that a 12-pulse converter can be switched in and out readily as well as isolated from the other converters that are in service.

In UHV DC systems, dry type smoothing reactors are used and evenly arranged on the pole bus and neutral bus. This can reduce the investment and allow easy switching in and out of 12-pulse converters without causing adverse effect on the steady-state and transient characteristics of the DC system.

b. *Characteristics of main equipment*: In a UHV DC system with two 12-pulse converters connected in series, although the voltage across DC terminals of each converter does not change, an increase in the DC voltage and power of each pole necessitates increased insulation for the valves and converter transformers connected at the high voltage side of the converter, smoothing reactors, and DC filters connected at the pole bus. This places higher requirements on the voltage-dependent equipment in the converter station, such as the high-voltage end (±800 kV) converter transformers and associated bushings and DC wall through bushings and arresters, and necessitates a novel design for the external insulation of station equipment, such as porcelain post insulators, intended to withstand a voltage of ±800 kV.

If it is necessary to increase the DC current while boosting the DC voltage, the through-current capability of thyristor-based converters is required to be increased. The DC current in Xiangjiaba–Shanghai and Jinping–Sunan UHV DC transmission systems is as large as 4000 and 4500 A, respectively; therefore, 6-inch thyristors and associated converter valve cooling systems are used.

3.2.2 Operation of DC Transmission System

3.2.2.1 Wiring configurations

One of the major operating characteristics of DC transmission systems, including the UHV DC systems, is that multiple operating modes are available, and manual or automatic switching between different modes is possible. How a DC system operates depends on factors such as the configuration of DC circuit, power flow direction, DC voltage, transmitted power, and reactive power control mode, as well as any combination thereof.

By considering the system equipment conditions and requirements of the AC systems at both terminals, it is possible to reasonably select the optimal operating mode of the DC system to improve the security, stability, reliability, and equipment availability of both the AC and DC systems.

1. **Monopolar configurations**
   Three available monopolar configurations that have varying operating performance and different requirements on equipment are described.
   a. *Monopolar ground return*: It requires that the relevant equipment in converter stations, the DC line, and the earth electrode systems at both terminals are all in sound condition. During operation, all the DC current will flow through the ground return path. With this configuration, it is necessary to take into account the constraints imposed by the design of the earth electrode on the current flowing through it and its operating duration. Otherwise, the service life of the earth electrode will be reduced. The line loss in this operating mode is somewhat larger than that of one pole operated under balanced bipolar mode, which is attributed to the resistance of the earth electrodes and earth electrode lines at both terminals added to the DC circuit resistance.
   b. *Monopolar metallic return*: In this configuration, it is required that not only the equipment in converter stations at both terminals and the DC line of the operating pole are in sound condition but also the DC line of the pole out of service reaches the same insulation level as the metallic return path. The operating current is simply limited by the overload capability of a monopole, regardless of the earth electrode system. During operation, the resistance of the DC circuit is approximately equal to the sum of the resistance of lines with positive and negative polarity. Therefore, the line loss is approximately twice that of one pole operated under the bipolar mode. This configuration may be used in cases in which the converters or earth electrode system of one pole requires repair after faults or scheduled maintenance or the current to the ground is limited.

      For a DC system, it is possible to transfer from the monopolar ground return to the monopolar metallic return and vice versa via the ground return transfer switch (GRTS) and metallic return transfer breaker (MRTB) without shutting down the system. This can reduce the power loss due to system outage and contribute to the increased system reliability and availability.
   c. *Monopolar ground return system with two DC lines in parallel*: In this configuration, the equipment in converter stations and the earth electrode systems at both terminals of the operating pole and the DC lines of both poles are required to be in sound

condition. The line loss is approximately one-half that of one pole operated under the bipolar operation mode. However, the magnitude of operating current and the operating duration are constrained by the overload capability of a single pole and the design of earth electrodes. Moreover, the metallic return conductors in the DC switchyards at both terminals are required to be fully insulated.

2. **Bipolar configurations**
   Both the bipolar metallic return configuration (also called bipole with metallic neutral wire or bipolar three-wire system) and the bipolar configuration with only one converter station grounded (also called bipolar two-wire system) are rarely used in practice.

As shown in Figure 3.6, the bipolar configuration with neutral-point earthing at both terminals is commonly used in DC projects. This configuration requires that all the equipment is in sound condition and allows flexible operation and high reliability. During the balanced bipolar operation, only a small unbalanced current flows through the grounding systems at both terminals and thus has little effects on the earth electrodes. It is believed to be the optimum operating mode of a bipolar system. When one pole is out of service because of faults, the DC system can automatically switch to monopolar ground return mode. The healthy pole may compensate the loss of the other pole by increasing its output utilizing its short-term, long-term, or inherent overload capacity to minimize impact on AC systems.

For a bipolar system with two 12-pulse converters connected in series at each terminal, each converter consists of one 12-pulse converter valve and one group of associated converter transformers and can operate independently. Thus, each pole has a total of four 12-pulse converters at both terminals. By varying the number of converters put into operation, selecting a monopolar or bipolar system and ground return or metallic return, a total of 45 different configurations are available for a UHV DC system built with two poles, namely, one bipolar configuration with all converters in operation, eight bipolar configurations with an unbalanced number of converters in operation for both poles (i.e., one pole has only one converter in operation while the other pole has two), 16 bipolar configurations with only one converter in service per pole, four monopolar configurations with two converters in service, and 16 monopolar configurations with one converter in service. Added to these is the deicing configuration, in which the two high-voltage end 12-pulse converters are connected in parallel to increase the line current for deicing purposes.

3.2.2.2 Direction of power flow

With the aid of the control system, a DC system can flexibly reverse its power flow direction, which is known as power reversal. There are two kinds of power reversals, namely, normal power reversal and emergency power reversal. Generally, the latter is automatically completed by the control system and the former can be either manually or automatically achieved.

Because of the inherent unidirectional conduction of thyristor valves, the direction of the current through a DC circuit cannot be changed. Therefore, the reversal of power flow direction has to resort to reversing the polarity of pole line voltage. For instance, when a bipolar system transmits power in a forward direction, the voltage polarity of pole 1 is positive and that of pole 2 is negative. After reversal, the voltage polarity of pole 1 becomes negative and that of pole 2 becomes positive.

A power reversal necessitates the change of operating conditions of converter stations at both terminals. Particularly, the rectifier station is changed to operate under inverter mode while the inverter station operates under rectifier mode. Therefore, it is required that the control and protection systems of converter stations at both terminals are able to satisfy the control requirements for both the rectifier and inverter operations. Usually, the inverter station as seen along the prevailing transmission direction has a capacity slightly smaller but a reactive power compensation capacity slightly larger than the rectifier station. If the transmission capacity in both directions is required to be the same, then appropriate adjustment and changes have to be made to the converters at both terminals, regulating the range of converter transformer tap changers and reactive power compensation configuration.

The normal power reversal may be achieved manually by operators or automatically under predefined conditions. To reduce the impact of power reversal on the AC systems connecting both terminals, the process should be controlled at a relatively slow pace, usually taking a few seconds or longer.

If a fault occurs in the AC system that requires urgent active power support from the DC system, then an emergency power reversal may be used. In this case, the quicker the power reversal is, the more effective the support to the AC system is. Because the power reversal involves a change in voltage polarity, the reversal speed mainly depends on the charging and discharging times of equivalent capacitance of DC lines. For overhead lines, this can be completed within a dozen cycles. For DC cable lines, to avoid damages to cable insulation caused by fast voltage polarity reversal, the reversal speed must be limited.

UHV DC systems may be provided with bidirectional power flow transmission capability as required.

3.2.2.3 Operation at rated or reduced voltage

The DC system voltage is defined as the mean voltage at the line side of the pole bus with respect to the neutral bus in the rectifier station:

$$U_{\mathrm{dr}} = nU_{\mathrm{d1}} \tag{3.12}$$

where

n—number of six-pulse converters connected in series per pole.

The DC system can be selected manually or automatically to operate at rated or lower voltage. Operating at reduced voltage is an operation that could be used during harsh weather conditions or severe pollution to reduce the probability of fault occurring, thereby increasing the system reliability and availability. The DC voltage can be regulated by quickly varying the firing angle through the control system and changing tap changer positions of converter transformers. Too small a reduction in voltage may not have the desired effect of improving system availability. However, excessive reduction would result in the system operating at a large firing angle, thereby compromising the operating conditions of the system. Generally, the preferred voltage for reduced voltage operation should be 70–80% of the rated voltage with the firing angles reaching up to 40–50°.

As long as the system can be operated at the rated voltage, the operation at reduced voltage should be avoided. This is because, for a given transmitted power, reduction in the DC voltage will cause the DC current to increase proportionally, contributing to higher power loss and operating cost. When the system operates at reduced voltage, the operating current is generally limited by the rated current. Therefore, when the system is operated at 70% of the rated voltage, the maximum transmitted power will be reduced to 70% of the rated power. However, to avoid unstable operation of the DC system after a disturbance and to allow desirable recovery characteristics, the low-voltage current limiter may be used to protect the system from operating under low-voltage large-current conditions when the system voltage drops abnormally.

In addition, another operating mode is to operate the system with a single converter, reducing the DC voltage by a factor of two.

3.2.2.4 Active power control

The active power transmitted by a DC system can be controlled rapidly. There are two control modes, power control and current control.

1. **Current control**
   The DC system current is defined as the operating current through each pole:

$$I_{\mathrm{d}} = \frac{U_{\mathrm{dr}} - U_{\mathrm{di}}}{R} \tag{3.13}$$

   where
   U_{di}—voltage of the inverter-side pole bus with respect to the neutral bus, $U_{\mathrm{di}} = nU_2$
   R—resistance of the DC circuit, Ω, mainly consisting of resistance of DC lines, smoothing reactors, earth electrode lines, and earth electrodes. It varies from one system configuration to another.

Under the current control mode, the DC current is maintained constant via the current regulator in the rectifier station. If the inverter station is operated at the extinction-angle γ control mode and the rectifier station is operated at the current control mode, then as the DC voltage experiences small changes, the transmitted power will vary as well. In contrast, if the inverter station is operated at the voltage control mode and the rectifier station is operated at the current control mode, then the transmitted power will remain constant.

2. **Power control**

The DC system power is defined as the output power of the rectifier station. The power of a single pole is given by:

$$P_{\mathrm{dm}} = U_{\mathrm{dr}} I_{\mathrm{d}} \tag{3.14}$$

The power of a balanced bipole is given by:

$$P_{\mathrm{db}} = 2P_{\mathrm{dm}} \tag{3.15}$$

Generally, the DC voltage is controlled by the regulator installed in the inverter station, whereas the current or power is controlled by the regulator installed in the rectifier station. Under the power control mode, the transmitted power is maintained constant via the power regulator installed in the rectifier station. During operation, when the DC voltage increases, the power regulator will operate to reduce the DC current correspondingly; conversely, when the DC voltage decreases, the power regulator will operate to increase the DC current (up to the real-time permissible maximum limit). In this way, the transmitted power can be maintained at or close to the set value. Under the power control mode, the transmitted power is set manually by the operator or automatically according to the predefined load curve. During operation, the DC current changes as a function of the DC voltage.

In a coordinated control, the inverter station usually adopts extinction angle or voltage control to stabilize the DC voltage. During operation, the DC voltage changes with the AC system voltage or the DC current. If extinction angle control is adopted, then the extinction angle should be set to the minimum safe value. If voltage control is adopted instead, then the voltage regulator on the inverter side usually maintains constant DC voltage (for operation at either rated or reduced voltage).

Normally, the power control mode should be used. If there is a fault occurring to the communication system, then fluctuations in current may cause mismatches in control parameters used by the rectifier and inverter. To avoid this, control mode at the rectifier station may be switched to current control while maintaining the prefault operating current value.

3.2.2.5 Balanced and unbalanced bipolar operation

Balanced bipolar operation is an operating mode in which the two poles have the same voltage, current, and, hence, the same transmitted power. In contrast, unbalanced bipolar operation is an operating mode in which the two poles have different voltage or current. Therefore, the balanced and unbalanced bipolar operations correspond to different combinations of the aforementioned DC voltage and power control modes.

The balanced bipolar operation is characterized by the following: (i) both poles are operated at the same rated or reduced voltage; and (ii) power transmitted by both poles is equal because the same current flows through them and the current through the earth electrodes is generally less than 1% of the rated DC current. Normally, this mode is selected for a bipolar system. It has many advantages, including the following: (i) the system's design capacity can be fully exploited; (ii) the system equipment can operate under desirable conditions; (iii) the system loss is small; and (iv) the operating cost is lower and the reliability is higher.

The unbalanced bipolar operation has three categories: unbalanced voltage, unbalanced current, and unbalanced voltage and current operations. The unbalanced voltage operation means that one pole is operated at rated voltage while the other is operated at reduced voltage. In this case, it is desirable to ensure that both poles operate at the same current, thereby reducing the current through the earth electrodes to the minimum. With different operating voltages, the two poles would transmit different power.

If the pole that is operated at reduced voltage is also required to reduce its DC current, then both poles will be operated at different voltages and different currents. In this case, the differential current between the poles will flow through the earth electrodes.

If both poles use the same voltage control mode but have different current settings, then the system operates at unbalanced current. In this case, the differential current between the poles flows through the earth electrodes.

To minimize the effects of unbalanced operation on both the AC and DC systems, the pole operating under the design conditions should usually be placed under power control mode. Thus, irrespective of whether the other pole is operated at reduced voltage or increased/ reduced current, it will try to maintain constant transmitted power for both poles.

Moreover, each pole of a bipolar system can operate with a single 12-pulse converter, giving rise to more bipolar operation modes with balanced or unbalanced voltage and power combination.

3.2.2.6 Reactive power control

Regarding the reactive power interchanged between the converter station and AC system, the reactive power supplied from the AC system to the converter station is positive and that from the converter station to the AC system is negative.

1. Rectifier station (Q_{S1})

$$Q_{S1} = Q_{C1} - Q_{F1} - Q_{RC1} \tag{3.16}$$

2. Inverter station (Q_{S2})

$$Q_{S2} = Q_{C2} - Q_{F2} - Q_{RC2} \tag{3.17}$$

where

Q_{F1}, Q_{F2}—reactive power supplied by the AC filters at rectifier station and inverter station, MVAr, respectively

Q_{RC1}, Q_{RC2}—reactive power, respectively, supplied by other reactive power compensation devices at rectifier station and inverter station, MVAr.

It is vital to control reactive power at converter stations. This can reduce its effects on the reactive power or voltage of the AC systems, contributing to increased stability of the DC system. Such controllability is achieved through various reactive power control modes provided by the control system. These control modes mainly include the control over the reactive power exchange between converter stations and AC system, and the AC voltage control. In the former works, the reactive power exchange between the converter stations and the AC system should be limited within a certain range, whereas the latter maintains the converter bus voltage within a certain range. Generally, the DC transmission system uses the reactive power control mode. However, when the converter station is connected to a weak AC system or when the AC voltage is somewhat high, the AC voltage control mode may be used instead. The control range of reactive power at converter stations is limited and is restricted by the configuration of reactive power compensations at converter stations, design conditions of equipment at converter stations, system transmission capacity, and control modes.

The reactive power control at converter stations is achieved mainly by means of: (i) switching in/out of the AC filter banks or capacitor banks to change the reactive power supplied by the converter stations; and (ii) regulating the firing angles to change the reactive power absorbed by the converters. Switching in/out of filter banks will cause the reactive power to change in a stepped manner. Also, when the DC system is operated under light load conditions, it is necessary to switch-in the absolutely minimal number of filter banks to satisfy the requirement of filter capacity limit. As a result, excess reactive power could potentially be injected into the AC system, causing the AC bus voltage to increase. Therefore, installation of reactive power absorption equipment may be required. In contrast, regulating the firing angles can change the reactive power absorbed by the converters quickly and smoothly and, therefore, is desirable. Quickly increasing the firing angle α or the extinction angle γ can also reduce the dynamic overvoltage on the AC side of converter stations. However, the regulation range of firing angle α is restricted by the DC system performance and equipment stress. In practice, usually a combination of these two is used to achieve satisfactory performance. Also, synchronous condensers or static VAR compensators may be an alternative to control the reactive power at converter stations.

As a UHV DC link transmits more power, the converter stations at both terminals need more reactive power, leading to an increased number of reactive power compensation equipment and more complicated switching and control logics.

3.2.3 Characteristics and Applications of UHV DC Transmission

3.2.3.1 Advantages and applications

The UHV DC transmission is an efficient way to achieve bulk power transmission over long distances.

In addition to having all the characteristics of the conventional HVDC transmission, UHV DC transmission can take full advantage of bulk power transmission over long distances, which remains an urgent problem in China's power industry development.

A DC overhead line requires only two conductors, one positive and the other negative, and features a simpler tower structure, narrower line corridor, lower construction cost, and line loss. In addition, a DC system may use the ground (or sea) instead of a dedicated conductor as the return path and may take advantage of the low resistivity and lower loss of the ground return. On a DC line, no capacitive current is present and the voltage is uniformly distributed along the line, eliminating the need for installation of shunt reactors. Moreover, it has a large transmission capacity and, hence, is particularly suited for bulk power transmission over long distances. A DC cable line can withstand high voltage, has a large transmission capacity, and has high power density, low line loss, and long service life. In addition, its transmission distance is not restricted by the capacitive current. This is why most long-distance cross-sea and underground transmission lines (e.g., supplying power to cities that are constrained by insufficient line corridor) use DC cables.

Increasing the transmission voltage can reduce the power loss while increasing the transmission distance. Because the line loss of a DC overhead line is mainly caused by resistance, UHV DC is more desirable for power transmission over a long distance. With a certain current, increasing the UHV DC transmission voltage is able to increase the transmitted power, which can be up to the maximum power as limited by the thermal limit of DC lines and equipment in converter stations at both terminals, thus making full use of the transmission capacity of UHV DC systems. When it is necessary to increase the DC current to allow a larger transmission capacity, the cross-sectional area of the subconductors of transmission lines may be increased to 1250 mm^2 to reduce the line loss. To this end, it is required to improve or enhance some aspects, such as the conductor type, bundle configuration, insulation coordination, insulator string, distance from conductors to the ground, distance from DC lines to other transmission lines that they cross, towers and loads applied thereon, as well as electromagnetic effects.

The UHV DC transmission system has the following advantages:

1. Isolated by the DC system, interconnection can be made between AC grids operating at different frequencies (e.g., one at 50 Hz and the other at 60 Hz), and each grid can maintain independent operation at its own frequencies and voltages.
2. Because the AC systems interconnected through the DC system are not required to operate synchronously, the transmission capacity and distance of the DC system are not limited by the requirements of the synchronous operation of the AC systems at both terminals.
3. The short-circuit capacity of the connected AC systems is not increased, thus limiting the increase in short-circuit current of AC systems.
4. By virtue of its quick and accurate regulation function, the UHV DC transmission system can increase the stability and transmission capacity of the AC systems, and can serve as the tie line between regional grids to increase the operational stability, reliability, and flexibility of the entire interconnected system. Because of these important characteristics, the UHV DC systems permit bulk power transmission over long distances.

China started the study of UHV DC technology in 2005, and results of the study indicate that the conventional HVDC system is a preferred option for power transmission over a distance less than 1300 km, and UHV DC system is more cost-effective for power transmission over a distance more than 1300 km. To date, China has built the Xiangjiaba–Shanghai ±800-kV UHV DC transmission link with a rated transmission capacity of 6400 MW and a transmission distance of 1891 km, and the Jinping–Sunan ±800-kV UHV DC transmission link with a rated transmission capacity of 7200 MW and a transmission distance of 2059 km; both have already been put into service. The operating practices with these links show that the construction cost per unit capacity of a ±800-kV UHV DC link is approximately 28% lower than that of a ±500-kV HVDC link. Apparently, from the techno-economic perspective, UHV DC system is more competitive when used for bulk power transmission over long distances.

Reliability is of great importance for UHV DC systems delivering bulk power over long distances. Today, China's UHV DC systems use a configuration with two 12-pulse converters that are series-connected per pole and have as many as 45 operating modes in normal conditions. In addition, the unique control strategy of switching a single 12-pulse converter online makes it possible to isolate and repair a single 12-pulse converter after a fault occurs. Moreover, the control and protection philosophy aiming at avoiding forced outage of a pole/bipole and the interpole power compensation function intended to minimize the power loss all contribute to greatly increased reliability of UHV DC systems.

Because two series-connected 12-pulse converters are provided for each pole, once a single converter is out of service because of a fault, the remaining ones can still operate. Despite the loss of one-half power of the pole, the complete outage of the pole can be avoided.

By using additional connections in the converter stations at both terminals and surge arresters and the primary switches in DC switchyard, it is possible to change the connections of primary equipment and achieve the operating mode whereby the high-voltage end of the two poles is connected in parallel. In this case, the DC line current will be much higher than the rated current, considerably enhancing the deicing capability. Moreover, the operating mode whereby the two poles transmit power in opposite directions and the associated control protection strategy can also protect the line from icing. All these measures can effectively handle the snow and cold weather, consequently greatly reducing the occurrence of outages of the DC transmission system and improving the system reliability.

The control and protection system of UHV DC systems has a reasonable hierarchical and distributed structure, as well as a complete redundancy configuration and a self-diagnosing function. The DC protections all use triple configuration and 2-out-of-3 voting logic. The bipolar protection is, respectively, incorporated into the pole protection, with the output strategy of removing the pole that is in trouble. Moreover, the application of the state-of-the-art technology of the converter maintenance key allows one converter of a monopole to be isolated and repaired while maintaining the other in normal operation. All these features can further improve the reliability of UHV DC systems. Therefore, the annual design fault outage rate for a single pole is reduced from five times to two times per year, and that for a bipole is reduced from 0.1 times to 0.05 times per year.

In addition, other important UHV DC features include its controllability of the transmitted active and reactive power, rapid dynamic response, and ability to conduct staged commissioning. To accommodate the bulk power transmission over long distances, the UHV DC technology has greatly evolved and advanced in terms of the system research, main circuit configuration, equipment parameters, control and protection strategy, as well as the operation technology, which have already been applied successfully in actual projects.

3.2.3.2 Limitations and development trends of HVDC transmission technology

The thyristor-based HVDC transmission technology still has with a number of problems:

a. Compared with an AC system, an HVDC converter station requires more equipment, the converters are expensive to manufacture, and the system configuration and operation are more complicated.
b. As the thyristor converters generate harmonics on both the AC and DC sides during operation and consume a large amount of reactive power, it is essential to install AC filters, DC filters, and reactive power compensation devices in converter stations at both terminals, leading to larger land occupation, construction cost, and operation cost.
c. Because of the inherent features of thyristor converters, commutation failure is inevitable when the AC and DC systems encounter a fault or disturbance. Although the

state-of-the-art technology can reduce the probability of commutation failure to a minimum, it cannot be eradicated completely.

d. The use of the ground or sea return path may cause electro-corrosion of buried metallic parts and conduits adjacent to the earth electrodes, and other problems such as the bias magnet and vibration due to DC current flowing through the neutral earthing transformers. Moreover, it is difficult to site the earth electrodes when the ground soil resistivity is high.
e. The inherent electrostatic adhesion of DC current causes more severe contamination to DC lines and equipment in converter stations than in AC systems, thus requiring higher external insulation level.
f. Because the DC current wave has no zero crossing point, it is difficult for HVDC circuit breakers to perform arc extinction, which creates difficulty for intermediate tapping and hinders the development of multiterminal DC system. The HVDC circuit breakers developed so far need to be further improved in terms of technical maturity and economy to accommodate the requirements of UHV DC applications.

The UHV DC systems are characterized by large transmission capacity, more equipment, and multiple operating modes, thus necessitating DC control and protection strategies with higher performance and reliability. To accommodate bulk power transmission of UHV DC links, great efforts should be made to develop the main equipment with larger capacity, such as converter valves and transformers, and to improve and update the control and protection system to increase the stability and reliability of the DC system itself. However, it is essential to study how to guarantee the safe and stable operation of UHV DC systems, focusing on the analysis and research of the interaction between the AC and DC systems. Moreover, developing novel technology to address problems currently facing the thyristor-based conversion technology is required.

With the advent of insulated gate bipolar transistor (IGBT), a novel metal-oxide semiconductor turn-off device, the voltage source converter (VSC) based on the IGBT and pulse width modulation (PWM) was applied in DC transmission systems. The world's first test VSC-HVDC system (±10 kV, 3 MW, and 10 km) was constructed in 1997 in Sweden. Then, in 1999 in Sweden, the first commercial VSC-HVDC project (±80 kV, 50 MW, and 140 km) was put into operation. In 2011, China's first flexible DC transmission project—Shanghai–Nanhui flexible DC demonstration project (±30 kV, 18 MW)—was completed. The Zhoushan flexible DC transmission project (±200 kV, 1000 MW) under construction is a five-terminal link. The application of VSC technology will be further expanded with the increase in the capacity of IGBTs.

The VSC-HVDC technology has the following main advantages: (i) it allows simultaneous control of the active and reactive power and requires no reactive power compensation; (ii) VSC converters only generate a small amount of harmonics of high orders; (iii) the VSC converter is self-commutated and operates in passive inverter mode, making it suitable for power feeding to passive networks; and (iv) the controllable turn-off capability avoids problems such as commutation failures. The VSC-HVDC transmission system needs less equipment, has a simpler structure, and occupies less land. However, because of the constraint imposed by the manufacturing technology of IGBTs, the power loss caused by a VSC converter is approximately double that of a thyristor-based converter. The IGBT's voltage and current ratings are both lower than that of thyristors. At present, the capacity of the largest IGBT-based project under construction is only 1000 MW. In addition, it is impossible for a VSC-HVDC link to clear faults on the DC side by virtue of its own control capability, which results in relatively low reliability for long-distance overhead lines. For this reason, it is more suitable to be used in transmission links using cable lines with a lower fault rate. Nevertheless, the VSC-HVDC is still a highly promising technology.

To deal with the problems with conventional HVDC technology, such as reactive power compensation and commutation failure, the capacitor commutated converter (CCC) technology was developed. It works in such a manner that capacitors are series-connected between the converter and the converter transformer to mitigate the effects of AC system voltage disturbance on the converter and reduce the reactive power compensation needed by the converter station. For CCCs to be applied in actual projects, it is necessary to investigate the overall performance of DC systems and technical requirements on capacitors.

3.3 Safety, Stability, and Operation of UHV DC Transmission System

3.3.1 Role of AC Systems in Supporting UHV DC Systems

Considering the uneven distribution of energy sources and load centers and the geographic features of China, DC transmission technology plays an important role in transmitting bulk power from large energy bases over long distances. However, the conventional thyristor converter-based DC systems require the voltage support from AC grids when commutation occurs. Hence, the system capability of AC grids is crucial to the normal operation of the DC system. China has commissioned the most DC transmission links with the largest power transmission capacity around the world. The receiving ends of those links are concentrated at the load centers in the eastern and southern parts of the country. During China's 12th 5-year plan, Jinping–Suzhou, Hami–Zhengzhou, and Xiluodu–Zhejiang UHV DC transmission links were completed and commissioned successively, making the feature of a multi-DC infeed network in the Eastern China more distinct. Such a dense multi-DC infeed network is unique to China and requires robust AC grids to ensure its normal operation. Therefore, building a robust hybrid grid integrating AC and DC systems that complement and support each other is essential to guarantee the safety and economical efficiency of the grids.

The UHV DC system has a large transmission capacity and its security and stability are closely related to the robustness of AC systems at the sending and receiving ends. This mainly lies in the following aspects:

1. The AC systems at the sending and receiving ends, respectively, supply commutation voltage to the rectifiers and inverters of the DC system, creating a prerequisite for conversion.
2. The AC systems at both ends are essential for DC transmission. The sending-end AC system serves as the power source to supply the power to be transmitted, whereas the receiving-end AC system serves as the load to consume and accommodate the transmitted power.
3. The sending-end AC system must be sufficiently robust for because it should be able to supply sufficient reactive power and voltage support to the rectifier station. However, it is more capable of withstanding both the active and reactive power surge resulting from faults in the DC system, which is essential in reducing the number of generating units that have to be intertripped under such faults.
4. The robustness of the receiving-end AC system is essential for the safe and stable operation of the DC system, particularly when multiple DC transmission lines are densely terminated at the receiving system. If the receiving-end AC system is robust enough, although there is potential for the commutation to fail at multiple inverter stations when a severe fault occurs in the AC system, its voltage can be recovered rapidly once the fault is cleared and the DC system can also restore to normal operation rapidly. Therefore, when a commutation failure occurs at an inverter station, immediate blocking protection is not required. Instead, if appropriate steps are taken, then the aforementioned recovery process can be accelerated, avoiding consequent commutation failures. In contrast, if the AC system is weak, once a severe fault occurs, its voltage cannot be restored to the normal value and commutation failure will occur at multiple inverter stations, which will further exacerbate the operation of the AC system and cause it to lose stability. In addition, if a fault in the DC system results in blocking of a single pole or bipole, a robust receiving-end AC system is able to withstand the surge caused by abrupt power change and the operating voltage and frequency of the system can be maintained within the normal range, occurrence can prevent load rejection and reducing the load loss to the minimum.

3.3.2 Connection of UHV DC Transmission Systems

As the number of DC transmission systems grows, interconnections between AC and DC systems are becoming more and more complicated, from the simple interconnection of a single-circuit DC system with two asynchronous AC systems to the parallel AC/DC power transmission, from the connection of the DC system to a large AC grid at the sending end to the islanded operation at the sending end, and from single infeed to multiple infeeds. A large hybrid AC/DC system combining several connection modes also appears. These diversified interconnections will inevitably make the system secure and stable, and make the relevant controls more complicated.

A DC system can be connected to an AC system mainly in the following modes:

1. *Through a single DC circuit*: The sending-end grid and the receiving-end grid are interconnected through just a DC link. If this mode is used, then it is necessary to consider the effects of severe fault in the DC system on the AC grids.
2. *Single DC circuit, islanded operation*: The sending-end power source is directly connected to a DC system through simple electrical connection to transmit power. Although the islanded system at the sending end has a simple configuration, there are some prominent concerns about system stability control, subsynchronous oscillations, overvoltage suppressing measures, and start-up operation.
3. *Parallel AC/DC power transmission*: This has the following advantages: (i) offers a new transmission mode that enables the system to better accommodate different operating requirements; (ii) increases the AC system robustness and rotational inertia under certain conditions to improve the system damping behavior; and (iii) allows the AC and DC systems to support each other by virtue of fast controllability of the DC system, thus enhancing the transmission capability of the parallel AC/DC system. However, when this mode is used, the AC power network is required to be robust enough to deal with the great amount of power transferred from the DC system and maintain the system voltage at a proper level after the fault.
4. *Multi-send-out mode*: In this mode, the rectifier stations of multiple DC systems are situated at the same place and are electrically in close proximity to each other. It is mostly used in ultralarge energy bases from where the power is sent out through multiple DC systems to receiving ends at different locations. In China, this mode is widely used in the extra-large energy bases in the northwestern and southwestern regions. To use this mode, it is necessary to perform an in-depth study of the interaction between DC systems and their coordinated control under fault conditions.
5. *Multi-infeed mode*: In this mode, the inverter stations of multiple DC systems are situated at the same place and are electrically in close proximity to each other. It is commonly used in different regions, such as eastern and southern China, and it will be used more widely as more and more DC systems are built in China. Therefore, it is necessary to perform an in-depth study of the stability and security of the multi-DC infeed network.

Apart from these modes, a combination of these modes has also been used in practice. For instance, the hydropower produced in southwest China is sent to east China through a hybrid AC/DC system using a combination of multi-send and multi-infeed modes. In this case, the system stability behavior is more complicated and the operation control becomes more difficult.

Because a UHV DC system has a large transmission capacity, the voltage class of the AC grid to which the system is connected is a key factor affecting system security and stability. Based on the experiences gained in connecting the UHV DC systems to the 500-kV AC grid, it is expected that in the future it is possible to connect UHV DC systems directly to 1000-kV UHV AC grids and connect the high-voltage and low-voltage converters of a UHV DC system, respectively, to two grids at different voltage classes.

3.3.3 Stability Evaluation Methods for Interconnected UHV DC–AC System

A conventional DC transmission system comprises thyristor-based converters that require voltage support from the grid for commutation. Therefore, a conventional DC system can only deliver power to AC grids that are supported by stable power sources, and its transmission capacity is closely related to the robustness and stability of the connected AC system. The transmission capacity of a DC system is restricted by the robustness of the connected AC system, which lies in two aspects: (i) the equivalent impedance of the AC system as seen from the converter bus is related to the fault recovery capability of the AC power network; and (ii) the mechanical inertia (rotational inertia) of the AC system is related to the supporting capability offered by the power sources in it. The robustness of the interconnected AC/DC system is mainly evaluated in terms of the following two indexes.

1. **SCR and ESCR**

The relative robustness of the AC system compared with the connected DC system can be evaluated using the short-circuit ratio (SCR). SCR is defined as the ratio of the short-circuit capacity of the AC system to the rated power of the DC system, namely:

$$\mathrm{SCR} = \frac{S_{\mathrm{SC}}}{P_{\mathrm{d}}} \tag{3.18}$$

where

S_{SC}—the short-circuit capacity of the connected AC system, MVA

P_{d}—rated power of the DC converters, MW.

After considering the effect of the reactive compensation devices (such as filters and shunt capacitors) on the system robustness, the effective SCR (ESCR) concept is used to give a more practical evaluation of the total system robustness and it is defined by:

$$\mathrm{ESCR} = \frac{S_{\mathrm{SC}} - Q_{\mathrm{C}}}{P_{\mathrm{d}}} \tag{3.19}$$

where

Q_{C}—capacitive reactive compensation, MVAr.

In the case of high SCR/ESCR systems, the variation of active/reactive power in the converter station leads to small AC voltage changes of the converter bus and nearby AC systems. Therefore, the additional transient voltage control at the bus is normally not required. The reactive power balance between the AC system and the converter station can be achieved by switching the reactive power equipment in/out.

In the case of low SCR/ESCR systems, the variations in the AC system or in the HVDC power could lead to voltage oscillations and require special control measures, such as additional static VAR compensators or synchronous condensers. The concept of SCR/ESCR offers a simple and visualized way to assess the relative robustness of the AC and DC systems. In practical applications, the SCR/ESCR can be closely correlated with the technical issues such as the voltage stability of the interconnected AC/DC system, AC system overvoltage caused by blocking of the DC system, possible resonance frequency of the interconnected system, and the maximum transmitted power allowed by the AC system. Based on the operating experiences, the interconnected AC/DC systems of the single-infeed configuration can be classified into three categories in terms of robustness: strong system, $\text{ESCR} > 3$; moderate and weak system, $2 < \text{ESCR} < 3$; and very weak system, $\text{ESCR} < 2$. Other classification criteria described herein are based on SCR are similar to these.

2. **OSCR and OESCR**

Due to the constraints imposed by the configuration or load variations of the AC system, the DC system usually cannot be operated at full load, that is, actual power transmitted is lower than the rated power. Consequently, the use of SCR/ESCR to evaluate the relative robustness of AC system cannot reflect its effects on the stability of the interconnected system. In view of this, the concepts of operation SCR (OSCR) and operation effective SCR (OESCR) are introduced to indicate the actual SCR and ESCR corresponding to the actual power transmitted.

OSCR is defined as:

$$\text{OSCR} = \frac{S_{\text{SC}}}{P_{\text{d\_operation}}} \tag{3.20}$$

where

$P_{\text{d\_operation}}$—actual power of DC system, MW.

After taking into account the reactive power compensation capacity of the filters and shunt capacitors, the OESCR is defined as:

$$\text{OESCR} = \frac{S_{\text{SC}} - Q_{\text{C\_operation}}}{P_{\text{d\_operation}}} \tag{3.21}$$

where

$Q_{\text{C\_operation}}$—actual net capacitive and reactive compensation, MVAr.

3. **AC system constrained maximum DC power [1]**
For an interconnected AC/DC system, the DC system may be deemed as a special load of the AC system, as seen from the converter bus, that is, the rectifier side is a load and the inverter side is a power source. In this case, there occurs an instability problem of the AC system under critical condition that is related to the maximum power transmitted through the DC system as constrained by the AC system stability. This maximum limit may be calculated using the following set of equations:

$$\left.\begin{array}{l} P_{\mathrm{d}} = CU^2[\cos 2\gamma - \cos(2\gamma + 2\mu)] \\ Q_{\mathrm{d}} = CU^2[2\mu + \sin 2\gamma - \sin(2\gamma + 2\mu)] \\ I_{\mathrm{d}} = KU[\cos\gamma - \cos(\gamma + \mu)] \\ U_{\mathrm{d}} = P_{\mathrm{d}}/I_{\mathrm{d}} \\ P_{\mathrm{ac}} = [U^2\cos\theta - EU\cos(\delta + \theta)]/|Z| \\ Q_{\mathrm{ac}} = [U^2\sin\theta - EU\sin(\delta + \theta)]/|Z| \\ Q_{\mathrm{C}} = B_{\mathrm{C}}U^2 \\ P_{\mathrm{d}} - P_{\mathrm{ac}} = 0 \\ Q_{\mathrm{d}} + Q_{\mathrm{ac}} - Q_{\mathrm{C}} = 0 \end{array}\right\} \tag{3.22}$$

where
γ—extinction angle, (°)
μ—overlap angle, (°)
C, K—parameters of converter transformer and constant related to the reference ratings of DC system
Z—equivalent impedance of AC system, Ω
θ—equivalent impedance angle of AC system, (°)
δ—phase angle of converter bus voltage, (°)
B_{C}—equivalent susceptance of AC filters, S
Q_{C}—reactive power generated by AC filters, MVar
P_{ac} and Q_{ac}—active and reactive power of the equivalent circuit of AC system, MW and MVar.

Equation (3.22) contains nine equations and 12 variables. Assuming that the equivalent electromotive force of the AC system remains constant, there will be 11 independent variables. Depending on the objects to be analyzed, two of the 11 variables can be defined as independent variables, and the others are then determined accordingly. Generally, given the SCR in conjunction with the operating conditions and parameters of the DC system, the maximum transmitted power of the DC system under certain interconnection conditions can be obtained using Eq. (3.22).

To ensure that the critical stability conditions of the interconnected system can be met when the DC system operates at rated power, it is required that the AC system to which the DC system is connected must be strong enough to provide supports. The constraint relationship described in Eq. (3.22) shows that blindly increasing the transmission capacity of the DC system without reinforcing the AC grid structure may lead to instability of the power network once the critical operating condition of AC grid at either end is reached. Therefore, coordinated development of both the AC and DC transmission to build hybrid grids interconnecting is essential to maintain the security and stability of the power system.

4. **Relative mechanical inertia constant**

The AC system stability is closely related to its mechanical inertia constant. For the AC system at the receiving end of the DC transmission, its robustness is of vital importance when the DC infeed represents a major part of the supply. An insufficient mechanical inertia indicates the AC system's poor capability to withstand disturbances. As the UHV DC system transmits an enormous amount of power, a single pole fault or bipole fault will cause a large power deficiency in the AC system. To measure the ratio of the DC power to the equivalent rotational inertia of the AC system, an index of equivalent mechanical inertia of the DC system is introduced.

The relationship between the variations of system frequency, mechanical power input, and electrical power output can be represented by:

$$df = \frac{(P_m - P_e)f_0}{2H}dt \tag{3.23}$$

where

P_m—generator mechanical power, MW

P_e—generator electrical power, MW

f_0—system nominal frequency, Hz

H—inertia constant of the generator, s.

After the DC system is connected to the AC system, the mechanical inertia constant of the AC system relative to the DC system is

$$H_{dc} = H\frac{S_{GE}}{P_d} \tag{3.24}$$

where

S_{GE}—system equivalent capacity, MVA.

In a simple single-DC infeed system, the variation of AC system frequency as a function of the DC power can be represented by

$$df = \frac{f_0 S_{GE}(P_m - P_e)}{2H_{dc}P_d}dt \tag{3.25}$$

Equation (3.24) shows that the larger the proportion of power fed from the DC system, the weaker the receiving-end AC system relative to the DC system. Consequently, the smaller the mechanical inertia of the system, the poorer the capability of the interconnected system is to withstand disturbances. Therefore, the DC infeed proportion should be maintained at a proper level.

For asynchronous interconnection through a DC system, by using the concept of relative mechanical inertia in conjunction with Eq. (3.25), it is possible to approximately estimate the frequency variation range of the grids in cases when the DC system experiences a single pole or bipole fault at the sending and receiving ends. This method essentially establishes a mathematical relationship between the generator inertia constant, DC transmitted power, and AC system frequency. Thus, the AC system fault clearing time can be calculated based on the estimated system frequency fluctuation, providing a basis for designing the control parameters of generator frequency adjustment.

5. **MSCR**

Although the SCR can reflect the strength of the AC system relative to the DC system satisfactorily in the case of the single-DC infeed configuration, it cannot work effectively for multiple DC infeed system. This is because a multiple DC infeed system has a number of sending ends and the various DC systems interact and affect one another significantly. Simply using SCR to assess a multi-DC infeed system would cause a large deviation and yield an excessively optimistic outcome.

By reference to the definition of traditional SCR, International Council on Large Electric systems (CIGRE) has introduced the concept of multi-infeed SCR (MSCR) to evaluate the robustness of the AC system relative to the DC system in a multi-DC infeed configuration. Specifically, MSCR is the ratio of the three-phase short-circuit capacity of the AC system as seen from the converter bus of the *i*th DC system to the equivalent DC power fed into the interconnection point [2,3], expressed as

$$\mathrm{MISCR}_i = S_{\mathrm{ac}i}/P_{\mathrm{deq}i} \tag{3.26}$$

where

MISCR_i—MSCR corresponding to the *i*th DC system

i—*i*th DC system

$S_{\mathrm{ac}i}$—three-phase short-circuit capacity as seen from the converter bus of the *i*th DC system

$P_{\mathrm{deq}i}$—equivalent DC power of the *i*th DC system after allowing for the influence of other DC systems, MW.

It is obtained from the characteristics of multi-infeed DC/AC system, taking into account factors such as electrical distance, and is transmitted from other DC systems

When simply considering the electrical coupling of the AC system and using the multi-port Thevenin equivalent at the converter bus side of the multiple DC systems, the MSCR can be defined as:

$$\mathrm{MISCR}_i = \frac{S_{\mathrm{ac}i}}{P_{\mathrm{deq}i}} = \frac{1}{|Z_{\mathrm{eq}ii}|P_{\mathrm{d}i} + \sum_{j=1,j\neq 1}^{n}|Z_{\mathrm{eq}ij}|P_{\mathrm{d}j}} \tag{3.27}$$

where

Z_{eqii}—self-impedance of the ith converter bus in the equivalent impedance matrix, Ω

Z_{eqij}—mutual impedance between the ith and jth converter buses in the equivalent impedance matrix, Ω

P_{di}—rated power of the ith DC system, MW

P_{dj}—rated power of the jth DC system, MW

With the MSCR concept, it is possible to evaluate the robustness of the AC system relative to the DC system in multi-infeed configuration. Meanwhile, through analyzing the factors [4] that affect the magnitude of MSCR of individual DC systems, the network structure of the receiving-end AC system can be optimized. This acts as the basis for developing a reasonable strategy that serves to optimize the selected infeed ends for multiple DC systems [5], thus enabling the AC network to provide stronger capability of supporting the UHV DC system.

3.3.4 Interaction Between UHV DC System and AC System

1. **Voltage stability**
   Numerous studies show that when considering the stability of a DC infeed system, irrespective of a long-distance transmission system or a back-to-back system, the voltage stability is always the greatest concern. Essentially, the voltage instability results from the fact that the reactive power supplied by the power system cannot meet demand or that the power transmission over a long distance reduces the system voltage to an unacceptable level. The DC system absorbs reactive power during commutation. When other conditions remain unchanged, the larger the DC power, the more reactive power the DC system will absorb and, consequently, the larger the voltage drop that the bus will experience. During normal operation, the reactive power consumed by the DC system is mainly supplied by reactive power compensation devices in converter stations, such as the AC filters and capacitors. In the event of a fault, transient voltage fluctuation will occur. Such change in operating conditions will cause variations in the output of the compensation devices. Whether these devices can supply sufficient reactive power needed by the DC system has a direct bearing on the magnitude of the reactive power exchanged between the AC and DC systems. If the AC system fails to support the dynamic reactive power variations of the DC system, then the AC system will suffer voltage instability. Because UHV DC systems generally transmit an enormous amount of power to load centers, they place higher requirements on the reactive power

balancing capability and voltage stability of the receiving-end AC system. In addition, the blocking of a single pole or bipole may cause a large amount of power shift to AC interconnection circuits, leading to a significant increase in the amount of reactive power consumed by the AC system, and further compromising the AC system voltage stability. Therefore, during the power network planning and operation, the UHV DC system is required when possible to connect to the robust AC systems at the receiving end and to avoid a situation with too many DC systems terminating in the same region. When necessary, the dynamic reactive power capability of the AC system should be enhanced.

2. **Frequency stability**
Because the DC system transmits a large amount of power that could account for a relatively large proportion of the demand at the receiving-end AC system, any faults occurring in the DC system, such as commutation failures, could lead to significant fluctuation in the transmitted DC power, causing a large shock to both sending and receiving AC systems and, consequently, compromising the system frequency stability. The power that can be transmitted by a DC system has an upper limit that depends on the mechanical inertia constant of the connected AC system. For a parallel AC/DC transmission system, the AC system is required to have a robust structure, large transmission capacity, and high stability margin. When a fault occurs in the DC system, a certain amount of power would be transferred to the AC interconnection lines, thus reducing the amount of generators to be tripped at the sending end and the demand to be disconnected at the receiving end. However, the AC interconnection lines and transformers are likely to be overloaded. Moreover, the AC system voltage may drop because of the increased power flow; therefore, measures should be taken to maintain it at a reasonable level to endure normal operation of the AC system and maintain commutation of the DC system. After a fault on the AC interconnection lines, it is possible to take advantage of the ramp-up, mid-term, and long-term overload capability of the DC system to reduce the load loss and improve system frequency stability.

3. **Subsynchronous oscillation**
The experience gained in the system operation and theoretical analysis show that subsynchronous oscillation is more likely to occur in cases when steam turbine generator units are located near to the rectifier station than in cases when hydro turbine generator units are close to the inverter station. This is particularly the case when the DC system and steam turbine generator units have similarly rated capacities and are close to each other. For an islanded DC transmission system whose sending-end power sources mainly consist of steam turbine generator units, the subsynchronous oscillation problem will be more prominent and is a critical problem affecting the islanded operation. It is recommended in IEC 60919-3:1999 *Performance of High-Voltage*

Direct Current (HVDC) Systems—Part 3: Dynamic Conditions that the unit interaction factor (UIF) should be used in screening analysis of the subsynchronous oscillations. The UIF is calculated using the following equation:

$$\mathrm{UIF} = \frac{P_d}{S_G(i)}\left[1 - \frac{S_{SC}(i)}{S_{SC}}\right]^2 \tag{3.28}$$

where

$S_G(i)$—rated capacity of the ith generator, MVA

$S_{SC}(i)$—three-phase short-circuit capacity at the converter bus at the rectifier side excluding the ith generator, MVA

S_{SC}—three-phase short-circuit capacity at the converter bus at the rectifier side including the ith generator, MVA.

If the UIF is less than 0.1, then the probability of subsynchronous oscillation is considered to be low.

The UIF method applies to circumstances in which the generators connected to the same bus have varying shaft mechanical characteristics, that is, they have different inherent torsional frequencies. In cases when multiple generators have the same shaft mechanical characteristics and inherent torsional frequency, the generators of the same type must be treated as an equivalent generator before the UIF method is applied.

References

[1] X. Zheng, Dynamic Behavior Analysis of AC/DC Power Systems, China Machine Press, Beijing, 2004.

[2] CIGRE Working Group B4.41, Systems with Multiple DC Infeed. CIGRE, 2008.

[3] P.F. de Toledo, B. Bergdahl, G. Asplund, Multiple Infeed Short Circuit Ratio-Aspects Related to Multiple HVDC into One AC Network. Transmission and Distribution Conference and Exhibition, IEEE Power Engineering Society, Dalian, China, 2005 1–7.

[4] G. Xiaojiang, T. Yong, G. Qiang, et al., Influence factors and theory for CIGRE MSCR index, Power Syst. Prot. Control 40(9) (2012) 69–74.

[5] G. Xiaojiang, G. Jianbo, M. Shiying, et al., A method for multi DC terminal location selection based on multi-infeed short circuit ratio, Proc. CSEE 33(10) (2013) 36–42.

CHAPTER 4

Internal Overvoltages in UHV Grid and Their Suppression

Chapter Outline

Ultra-high Voltage AC/DC Grids.
DOI: http://dx.doi.org/10.1016/B978-0-12-802161-3.00004-4

4.1 Classification of Internal Overvoltages and Overvoltage Level in UHV System

The internal overvoltage of a power system refers to the increase of voltage due to electromagnetic energy conversion or transfer inside the system caused by the change in system parameters as a result of switching of circuit breaker, fault, or other reasons. It may be classified as temporary overvoltage, switching overvoltage (or slow-front overvoltage), or very fast-front overvoltage.

The temporary overvoltage usually has low amplitude but lasts for a long time at a frequency equal or close to the power frequency. The causes of temporary overvoltage include the Ferranti effect, voltage increase on the healthy phase as a result of an asymmetrical grounding fault, sudden changes in load, resonance, and ferroresonance. In the insulation coordination design of an EHV or UHV system, the first step should be to calculate the temporary overvoltage because:

1. It is often followed by switching overvoltage and its amplitude has a direct impact on that of the switching overvoltage.
2. It is an important basis for determining the rated voltage of surge arresters, and thus the overvoltage protection level of the system.
3. It lasts for a long time and threatens the operational security of the equipment. Therefore, it may dictate the design of the internal and external insulation of the equipment.

Switching overvoltage is characterized by large amplitude, high-frequency oscillation, strong damping, and short duration. It is caused by switching of circuit breakers or system fault, including the overvoltages after normal switching operations, such as closing of lines, transformers, reactors, and post-fault reclosing of a line, and the overvoltages as a result of tripping, a fault, or fault clearing. Switching overvoltages may have different degrees of effects on the insulation of electrical equipment and protection equipment, primarily depending on their amplitude, wave shape, and duration.

Very fast transient overvoltage (VFTO) usually occurs in gas-insulated metal-enclosed switchgear (GIS). For instance, an overvoltage with a high frequency (300 kHz–100 MHz) and a short wave front (3–100 ns) may occur as a result of switching of a disconnector, a grounding switch, or a short-circuit fault, which is known as VFTO. Its frequency is much higher than that of a lightning overvoltage. Therefore, metal oxide arresters (MOA), although commonly used in the power system, are seldom used to suppress the VFTO because they cannot reduce the residual voltage to an acceptable level under high frequency. VFTOs may jeopardize the security of the GIS and its adjacent equipment, particularly the interturn insulation of transformers. They may also cause high-frequency oscillations in transformers.

Similarly, internal overvoltages of a DC transmission system may also be classified into switching overvoltage and temporary overvoltage. These overvoltages are primarily caused by various faults and switching operations in the AC and DC systems on both sides of the converter station. Therefore, they may also be classified into the following types based on their causes: (i) temporary overvoltage and switching overvoltage on the AC side, which are caused by switching operations and faults in the AC grid and (ii) temporary overvoltage and switching overvoltage on the DC side, which are caused by switching operations and faults on both the AC and DC sides. The amplitude, wave shape, and duration of an internal overvoltage of the DC system depend on a variety of factors, such as the type of switching operation (or fault), DC system structure, configuration and protection level of arresters, DC control and protection, and DC power transmitted. The action sequence of DC control and protection system has a particularly significant impact on the internal overvoltage.

Secondary arc current and recovery voltage are also discussed in this chapter. The secondary arc current, although not an overvoltage, is an important electromagnetic transient phenomenon. In EHV AC and UHV AC systems, single-phase auto-reclosing is commonly used for clearing a single-phase to ground fault. After an instantaneous single-phase to ground fault, the circuit breakers at both ends of the faulty phase will open to isolate the fault. However, because of the inductive and capacitive coupling between the faulty phase and healthy phases, an inductive current will flow through the fault arc path, known as the secondary arc current. This current is a key factor affecting the success of single-phase auto-reclosing. Because AC systems rated at 330 kV and higher are characterized by high voltage level, large capacity, and long transmission lines, the secondary arc will last for a long time or may not be able to self-extinguish, which will cause failure of auto-reclosing. Therefore, measures need to be taken to suppress the secondary arc current and recovery voltage.

A UHV AC system has electromagnetic transient and overvoltage problems similar to those of an EHV AC system. However, because of its larger capacity, longer transmission lines, and large reactive power, the UHV AC system faces more prominent overvoltage problems, as detailed here:

1. Load rejection may cause serious temporary overvoltage.
2. Load variation during normal operation may cause some problems to reactive power regulation and voltage control.
3. There is no room to improve the switching impulse to withstand voltage by increasing the air gap, and the internal insulation level of the UHV equipment is limited. Therefore, measures need to be taken to suppress the switching overvoltages. Otherwise, the equipment would require an oversized and complex insulation structure, which will increase the costs and impair the operational security.

4. Compared with an EHV system, the secondary arc current and recovery voltage on faulty lines are higher and the duration of secondary arc is longer. Failure of auto-reclosing might occur if no proper measures are taken, and this will affect the system reliability.

For the EHV AC systems in China, the allowable switching overvoltages are not higher than 2.2 p.u. for 330-kV systems, 2.0 p.u. for 500-kV systems, and 1.8 p.u. for 750-kV systems. For the UHV AC systems, the values are not higher than 1.6 p.u. for substations and not higher than 1.7 p.u. for lines. This shows that a higher rated system voltage poses higher requirements for limiting the switching overvoltage.

For a UHV DC system, the switching overvoltage is one of the key considerations in determining the insulation level of the equipment. Considering the limited insulation performance offered by air gap, high-altitude installation, and difficulties in equipment manufacturing, more stringent requirements are imposed on the overvoltage suppression and protection of the UHV DC system.

4.2 Temporary Overvoltage and Its Suppression

4.2.1 Temporary Overvoltage Caused by Load Rejection and Its Suppression

Temporary overvoltage caused by load rejection is one of the major concerns for EHV and UHV power systems. This is because it has a direct effect on the amplitude of the switching overvoltage and might endanger the security of the equipment and of the system. It is important for determining the rated voltage of arresters and affects the overvoltage protection level of the system. In a UHV system, temporary overvoltage usually occurs as a result of load rejection during normal operations or a single-phase to ground fault. As specified in GB/Z 24842-2009 *Overvoltage and Insulation Coordination of 1000 kV UHV AC Transmission Project*, the temporary overvoltage caused by load rejection should be less than 1.3 p.u. (1 p.u. = $Um/\sqrt{3}$), except in the case of three-phase load rejection during a single-phase to ground fault, when the overvoltage on the line side can exceed 1.3 p.u. for a short period (no more than 0.5 s) but should be lower than 1.4 p.u.

4.2.1.1 Main causes of temporary overvoltage

1. **The Ferranti effect**
   For a long transmission line, when its receiving end has no load, the impedance at the sending end of the line is capacitive. Taking into account the internal inductive impedance of the power source, due to the Ferranti effect [1], the voltage at the receiving end is higher than that at the sending end, and the voltages at both ends are higher than the electromotive force of the power source.

Given a no-load, loss-free transmission line with the receiving-end current $\dot{I}_2 = 0$ and a length of l, as shown in Figure 4.1, the sending-end voltage $\dot{U}_1$, receiving-end voltage $\dot{U}_2$, and voltage $\dot{U}_x$ at a given point are given by:

$$\left.\begin{aligned} \dot{U}_1 &= \dot{U}_2 \cos(\alpha l) \\ \dot{I}_1 &= \mathrm{j}\frac{\dot{U}_2}{Z_C}\sin(\alpha l) \end{aligned}\right\} \tag{4.1}$$

$$\dot{U}_x = \dot{U}_2\cos(\alpha x) = \frac{\dot{U}_1}{\cos(\alpha l)}\cos(\alpha x) \tag{4.2}$$

where

$\alpha = \omega\sqrt{L_0 C_0}, Z_C = \sqrt{L_0/C_0}$

L_0 is the inductance per unit length of the line

C_0 is the capacitance per unit length of the line

x is the distance from the given point to the receiving end of the line.

Propagation coefficient K_{12} for the receiving-end voltage to the sending-end voltage of an unloaded line is given by:

$$K_{12} = \frac{\dot{U}_2}{\dot{U}_1} = \frac{1}{\cos(\alpha l)} \tag{4.3}$$

It is shown in Eq. (4.3) that the voltage begins to increase from the sending end and peaks at the receiving end in the shape of a cosine curve, as shown in Figure 4.2. When $\alpha l = 90°$, it is equivalent to a series resonance occurrence as viewed from the sending end, with $K_{12} \to \infty$ and $\dot{U}_2 \to \infty$, when the line length is equal to one-quarter the power frequency wavelength, approximately 1500 km.

At the same time, the capacitive current in the unloaded line also causes increase of voltage on the reactance of the power source. As a result, the sending-end voltage is higher than the electromotive force of the power source and the temporary overvoltage is exacerbated. After taking into account the reactance of the power source, Eq. (4.4) can be derived from Eq. (4.1):

$$\dot{E} = \dot{U}_1 + jX_S\dot{I}_1 = \left[\cos(\alpha l) - \frac{X_S}{X_C}\sin(\alpha l)\right]\dot{U}_2 \tag{4.4}$$

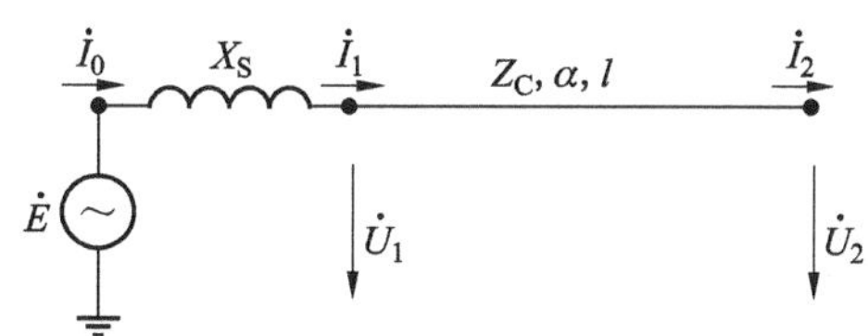

Figure 4.1
Impact of impedance of power source on a long transmission line without load.

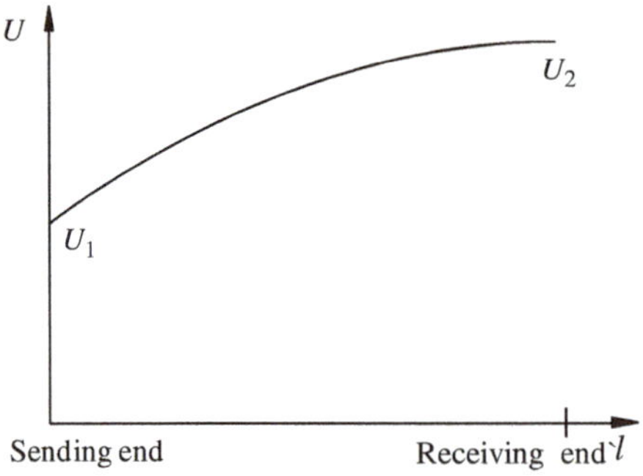

Figure 4.2
Voltage profile for a long transmission line without load.

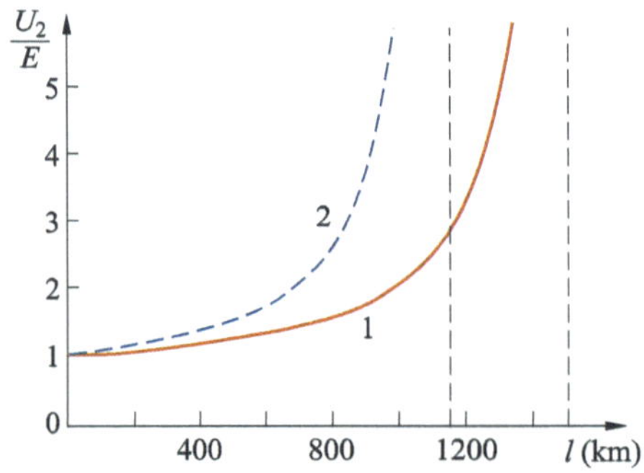

Figure 4.3
Increase of receiving-end voltage as a function of length of unloaded long transmission line.

Let $\varphi = \tan^{-1}(X_S/Z_C)$, then the propagation coefficient K_{02} of the receiving-end voltage to the electromotive force of the power source is given by:

$$K_{02} = \frac{\dot{U}_2}{\dot{E}} = \frac{1}{\cos(\alpha l) - (X_S/Z_C)\sin(\alpha l)} = \frac{\cos\varphi}{\cos(\alpha l + \varphi)} \tag{4.5}$$

From Eq. (4.5), the impact of the reactance of the power source can be expressed by angle φ. When $\alpha l + \varphi = 90°$, $K_{02} \to \infty$ and $\dot{U}_2 \to \infty$. At this point, $\alpha l = 90° - \varphi$ and the line length is less than 1500 km. The introduction of the reactance of the power source is equivalent to the increase of the line length, which causes the resonance to occur earlier, as shown in curve 2 (Figure 4.3). Curve 1 would occur when the impedance of the power source is zero.

2. **Load rejection**
When the transmission line is operating under a heavy load, temporary overvoltage may also occur when the load is rejected as a result of sudden opening of the circuit breaker at the receiving end, usually known as load rejection effect [1]. In this case, factors

affecting the overvoltage amplitude include the power flow on the line before load rejection, capacity of in-feed power source, and line length. Added to these are generator speed increase and operations of automatic voltage regulator and governor.

Assuming that during normal operation, the bus voltage is $\dot{U}_1$, sending-end current is $\dot{I}_1$, power factor is cos φ, transmitted active power is $P = 3U_1I_1 \cos\varphi$, transmitted reactive power is $Q = 3U_1I_1 \sin\varphi$, and reactance of the power source is X_S, the generator's electromotive force $\dot{E}$ is given by:

$$\dot{E} = \mathrm{j}X_S\dot{I}_1 + \dot{U}_1 \tag{4.6}$$

$$E = \sqrt{(U_1 + I_1X_S \sin\varphi)^2 + (I_1X_S \cos\varphi)^2} \tag{4.7}$$

Before load rejection, the line carries certain active power or inductive reactive power; therefore, the electromotive force of the power source is necessarily higher than the bus voltage, that is, $E > U_1$. After load rejection, according to the constant flux linkage theorem, the transient electromotive force of the power source could be deemed to remain approximately the same, that is, $E'_d \approx E$. The circuit breakers at the receiving end will open to reject load, forming a scenario in which the power source feeds a long, non-loaded line. Because the temporary voltage on a long transmission line will increase due to the Ferranti effect after the load at the receiving end is rejected, the sending-end voltage will be higher than the electromotive force of the power source, but the overvoltage is even worse at the receiving end.

3. **Single-phase to ground fault**

Asymmetrical short-circuit is the most common fault on transmission lines. The zero sequence component of the short-circuit current will cause the power frequency voltage on the healthy phases to increase, known as the asymmetrical effect [1]. Single-phase to ground fault is one of the most common asymmetrical short-circuit faults.

Assuming that a single-phase to ground fault occurs on phase A and with a boundary condition of $\dot{U}_A = 0$, $\dot{I}_B = \dot{I}_C = 0$, we have:

$$\left.\begin{aligned} \dot{U}_1 + \dot{U}_2 + \dot{U}_0 = 0 \\ \dot{I}_1 = \dot{I}_2 = \dot{I}_0 \end{aligned}\right\} \tag{4.8}$$

where

$\dot{U}_1, \dot{U}_2, \dot{U}_0$—positive, negative, and zero sequence components of the voltage at the faulty point

$\dot{I}_1, \dot{I}_2, \dot{I}_0$—positive, negative, and zero sequence components of the current at the faulty point.

Based on these conditions, a complex sequence network in the case of single phase grounding can be formulated, as shown in Figure 4.4. From this network, the sequence currents and voltages of the healthy phases are given by:

$$\left.\begin{aligned} \dot{I}_1 = \dot{I}_2 = \dot{I}_0 &= \frac{\dot{E}_A}{Z_1 + Z_2 + Z_0} \\ \dot{U}_B &= \frac{(a^2-1)Z_0 + (a^2-a)Z_2}{Z_0 + Z_1 + Z_2}\dot{E}_A \\ \dot{U}_C &= \frac{(a-1)Z_0 + (a-a^2)Z_2}{Z_0 + Z_1 + Z_2}\dot{E}_A \end{aligned}\right\} \tag{4.9}$$

where

a—operator, $a = e^{j120°}$

Z_1, Z_2, Z_0—positive, negative, and zero sequence input impedance of the network as viewed from the faulty point.

By using $K^{(1)}$ to represent the increase of voltage on the healthy phases after a single-phase to ground fault, Eq. (4.9) may be simplified to $\dot{U} = K^{(1)}\dot{E}_A$, where:

$$K^{(1)} = -\frac{1.5Z_0}{Z_0 + Z_1 + Z_2} \pm j\frac{\sqrt{3}(2Z_2 + Z_0)}{2(Z_0 + Z_1 + Z_2)}$$

For a large capacity power source, $Z_1 \approx Z_2$. By neglecting the resistive components of the positive, negative, and zero sequence impedance, it may be further simplified to:

$$K^{(1)} = -\frac{1.5(X_0/X_1)}{2 + (X_0/X_1)} \pm j\frac{\sqrt{3}}{2} \tag{4.10}$$

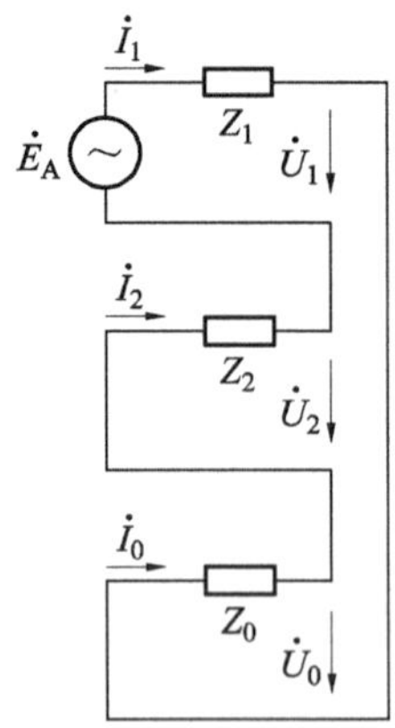

Figure 4.4
Complex sequence network in the case of a single-phase to ground fault.

From Eq. (4.10), this sort of temporary overvoltage is closely related to X_0/X_1 (zero sequence reactance/positive sequence reactance, as viewed from the faulty point). If X_0/X_1 increases, then the temporary overvoltage due to load rejection during a single-phase to ground fault tends to increase as well.

4. **Generator speed increase**
   After load rejection, the governor and brake cannot function immediately because of their inertia. As a result, the generator speed will increase, and so will its electromotive force and frequency, which aggravates the temporary overvoltage on the non-loaded line. Furthermore, the automatic voltage regulator will also influence the duration and amplitude of the temporary overvoltage.
   Therefore, in the case of load rejection due to single-phase to ground fault, all these factors will jointly result in a higher temporary overvoltage. However, this overvoltage will only last for a short period because of the presence of ground fault.

In general, because of its high capacitive reactive power, large capacity, and long length (particularly in China), the UHV line will face serious temporary overvoltage problems. If no or improper measures are taken, then the temporary overvoltage will affect the operational security of the UHV system.

4.2.1.2 Suppression of temporary overvoltage caused by load rejection

1. **Installing fixed high-voltage shunt reactors**
   To transmit electricity from west China to east China and to enable power exchange between north China and south China, many of the UHV transmission lines have to be very long. As such, the charging power on a line section is huge, and compensation using high-voltage shunt reactors is required to suppress the temporary overvoltage, as shown in Figure 4.5.
   The inductive reactive power produced by these reactors partially compensates the capacitive reactive power of the line, as if the line is shortened. This way, the temporary overvoltage on the line is suppressed. Under the condition of commonly used undercompensation, the input impedance at the sending end is still capacitive, as viewed from the sending-end of line, but with higher amplitude, resulting in a smaller capacitive current in the non-loaded line. Thus, the voltage increase at the sending end

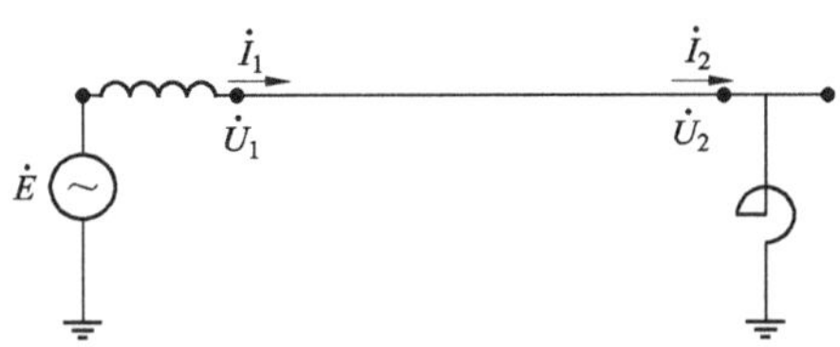

Figure 4.5
Line with high-voltage shunt reactors.

is suppressed, with the reactance of the power source remaining unchanged. Therefore, the temporary overvoltage at both ends of the line is reduced. However, it should be noted that the compensation should not be excessive so that difficulties are not caused for reactive compensation and voltage control during heavy load operation. For early UHV grids, the compensation degree provided by high-voltage shunt reactors could be up to 80–90%, which may be reduced for strong grids or short UHV lines. In the 1000-kV Jindongnan–Nanyang–Jingmen UHV pilot system, with the existing configuration of shunt reactors, the temporary overvoltage caused by load rejection on the lines is not more than 1.4 p.u.

2. **Installing controllable high voltage shunt reactors**

For a long, heavy-loaded transmission line, a compensation degree of 80–90% provided by high-voltage shunt reactors may cause great difficulties for reactive power compensation and voltage control, or even impair the transmission capacity. A preferred solution to this problem is to use controllable high-voltage shunt reactors to provide low compensation (for instance, 60%) during heavy-loaded operation. This way, the reactive power supplied by the power source to the line can be reduced and the electromotive force of the power source can be reduced to an acceptable level to maintain the reactive power balance and increase the transmission capacity. Also, the compensation degree provided by the high-voltage shunt reactors can be promptly increased to a higher value (for instance, 90%) when temporary overvoltage occurs.

For a scenario in which five generating units send 3500-MW power through 2×350-km lines (with an intermediate switchyard), the temporary overvoltage due to load rejection during a single-phase to ground fault on the line side may reach 1.47 p.u. if 90% of compensation is provided by fixed shunt reactors. The value may be reduced to 1.39 p.u. if controllable shunt reactors are used. This demonstrates the strong suppressing effect of controllable shunt reactors on temporary overvoltage on a long-distance, large-capacity UHV transmission system.

Other alternative solutions include using good conductor ground wire or optical fiber composite overhead ground wire (OPGW) and intertrip at both ends of the line or overvoltage relay protection.

4.2.1.3 Duration of temporary overvoltage due to three-phase load rejection during a single-phase to ground fault

According to studies, short-duration temporary overvoltages higher than 1.3 p.u. usually occur due to three-phase load rejection during a single-phase to ground fault. Previously, this duration was specified as 0.3–0.5 s based on the understanding that the standby circuit breaker will operate within 0.3 s after the circuit breakers of three phases fail. For a UHV system, this duration will be a major consideration for selecting metal oxide surge arresters (MOA). Long-duration overvoltages occur only when one end of the line trips and circuit breakers at the other end fail. In the past, circuit breakers failed for two main reasons.

First, relay protection fails to trip. However, all EHV and UHV lines are provided with different types of main protections and statistics show that they are unlikely to fail concurrently. Second, after the circuit breakers trip at one end of the line, the circuit breakers at the other end fail to trip. However, all EHV and UHV lines are operated on a phase-segregated basis and statistics show that circuit breakers of all three phases are unlikely to fail concurrently. In the design of 500-kV series compensation devices, it has been found that only the failure of the circuit breakers on one phase needs to be considered. If the circuit breakers on the faulty phase fail and those on the healthy phases trip, then the circuit breakers at the other end will trip within 0.1 s, which is the true duration of the overvoltage due to three-phase load rejection during a single-phase to ground fault. If the circuit breakers on the faulty phase do not fail, then the short-circuit fault would have been cleared after the circuit breakers on the faulty phase properly operate. In this case, only the overvoltage due to single-phase trip without fault occurs.

Calculations show that when the overvoltage due to load rejection during a single-phase to ground fault is not high, and when the overvoltage level is close to or lower than that due to three-phase load rejection without fault, there will be no major difference in overvoltage whether the circuit breakers on one phase or three phases fail; if the former is high, or much higher than the latter, then there will be a major difference in their overvoltage. Specifically, for the temporary overvoltage due to three-phase load rejection during a single-phase to ground fault, the value in cases in which the circuit breakers on a single phase fail is not higher than 1.3 p.u., which is much lower than when the circuit breakers on three phases fail. Therefore, this overvoltage can be considered as an overvoltage due to load rejection without fault.

The actual duration of the temporary overvoltage of 1.3–1.4 p.u. due to three-phase load rejection during a single-phase to ground fault is not more than 0.1 s.

4.2.2 Resonance Overvoltage and Its Suppression

In systems of different structures and voltage classes, different types of resonance overvoltages may occur. Resistive and capacitive components in power systems are generally considered to have linear parameters, whereas inductive components have three types of characteristic parameters. Under the joint effects of these three types of characteristic parameters and certain capacitance values and other conditions, the following three types of resonance phenomena may occur.

1. *Linear resonance*. In a circuit with linear resonance, as shown in Figure 4.6A, the inductance is a constant that does not vary with the voltage and current across the element. Such inductive elements include the core-free inductive elements (such as the inductance of a transmission line and leakage inductance of a transformer) and

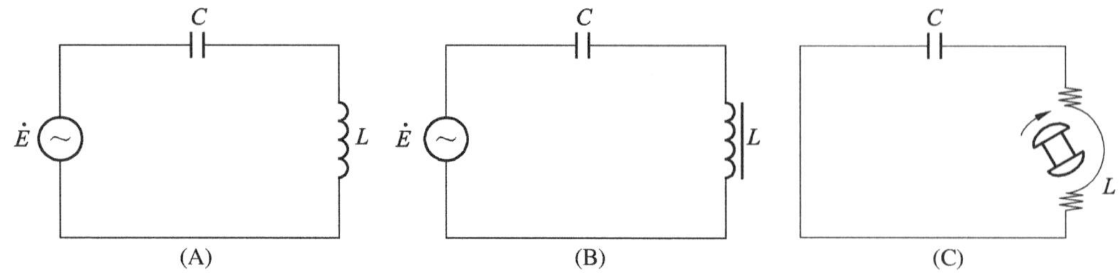

Figure 4.6: Three types of resonance circuits.
(A) circuit with linear resonance; (B) circuit with ferroresonance; and (C) circuit with parametric resonance.

core-equipped inductive elements with almost linear excitation characteristics (such as an arc suppression coil, which usually has an air gap in the core). These inductive elements, together with given parameters, will give rise to intense linear resonance.

2. *Ferroresonance.* In a circuit with ferroresonance, as shown in Figure 4.6B, the inductance is not a constant; it varies with the current or flux because the inductive element in this circuit has a core subject to saturation. In this circuit, which contains a non-linear inductive element, ferroresonance may occur when certain conditions are met. In practice, ferroresonance usually occurs in a circuit consisting of a non-loaded transformer, a voltage transformer (normally operating under near-zero load), and a capacitor.
3. *Parametric resonance.* In a circuit with parametric resonance, as shown in Figure 4.6C, the inductance parameters of some elements change periodically because of external effects. For instance, the inductance changes periodically when the generator is rotating, and parametric resonance may occur when the generator is connected to a capacitive load (such as a non-loaded line) and the parameters do not fit. The overvoltage due to parametric resonance caused by the generator is also known as self-excitation overvoltage.

In EHV and UHV systems, power frequency resonance overvoltage, which is a linear resonance overvoltage, is most commonly seen; overvoltage due to switching on non-loaded transformer is a ferroresonance overvoltage; and generator self-excitation overvoltage is a parametric resonance overvoltage.

1. **Power frequency resonance overvoltage and its suppression**
   The open-phase operation of an EHV or UHV system means that the circuit breakers on one or two phases are closed while the circuit breakers on the other two or one phase are open at both ends. These operations may occur during normal switching or single-phase auto-reclosing. Figure 4.7 shows a typical circuit with shunt reactors and open phases. In this figure, shunt reactors X_R are connected to the receiving end of the line, the circuit breakers on phases B and C at the sending end are closed, and the

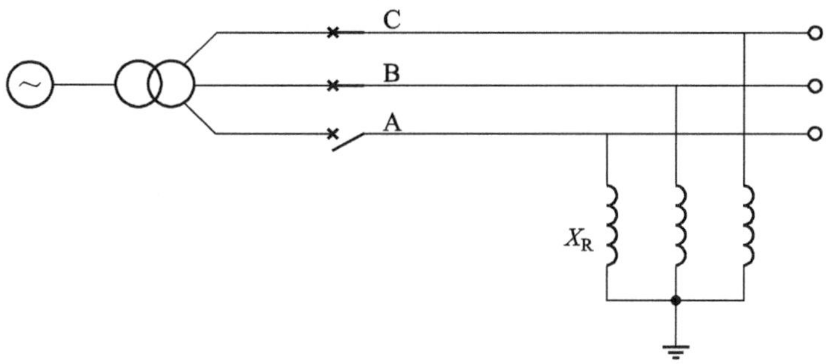

Figure 4.7
Circuit with shunt reactors and open phases (without neutral grounding reactor on the neutral of high-voltage shunt reactors).

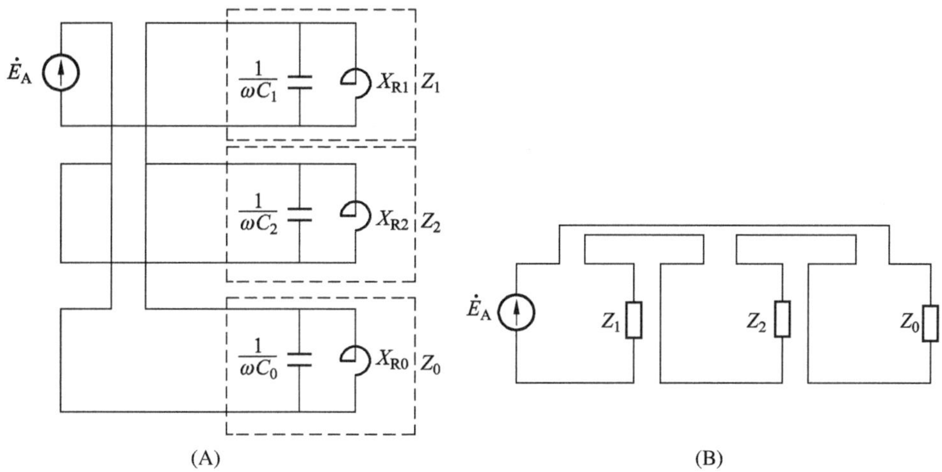

Figure 4.8: Complex sequence network diagram of the circuit with shunt reactors and open phase(s). (A) With one open phase and (B) with two open phases.

circuit breaker on phase A is open. A power-frequency capacitive transfer voltage is generated on the open phase (phase A) because of the coupling of voltages on phases B and C by inter-phase capacitance C_{12}, forming a resonant circuit. In this circuit, high-power frequency resonance overvoltages will occur on the open phase, which may cause insulation failure of the shunt reactors. Figure 4.8 shows the complex sequence network diagram of the circuit with shunt reactors and one or two open phases.

To simplify analysis, internal impedance of the power source and resistance and inductance of the line are neglected and an equivalent circuit is developed, which consists of Z_1, Z_2, and Z_0, as shown in Figure 4.8, representing the positive, negative, and zero sequence impedance at the sending end of the line, respectively, with $Z_1 = Z_2$.

In the case of one open phase, as shown in Figure 4.8A, the open phase A—ground voltage—is:

$$\dot{U}_A = \dot{E}_A \frac{Z_1 - Z_0}{Z_1 + 2Z_0} \tag{4.11}$$

In the case of two open phases, as shown in Figure 4.8B, the open phases B and C—ground voltage—are:

$$\dot{U}_B = \dot{U}_C = \dot{E}_A \frac{Z_0 - Z_1}{Z_0 + 2Z_1} \tag{4.12}$$

The line is usually undercompensated; therefore, the positive sequence impedance is capacitive. If

$$\frac{1}{\sqrt{L_1 C_1}} < \omega < \frac{1}{\sqrt{L_1 C_0}} \tag{4.13}$$

then the zero sequence impedance Z_0 is overcompensated, and Z_1 and Z_0 have opposite signs. In this case, power frequency resonance may occur if certain parameters are met. If no shunt reactors are connected to the line, then Z_1 and Z_0 will have the same sign and apparently no resonance will arise.

If the zero sequence impedance is also undercompensated, then both Z_1 and Z_0 will be capacitive and no resonance will occur in the circuit. This can be achieved by installing a suitably sized reactor on the neutral of the shunt reactors. Studies show that this is an effective measure to prevent power frequency resonance and reduce the power frequency transfer overvoltage on the open phase. It can also reduce the secondary arc current resulting from a fault or the recovery voltage on the faulty point, and thus help improve the success rate of single-phase auto-reclosing. For instance, in the UHV AC transmission pilot project with this configuration of reactors, the voltage on the open phase(s) is not higher than 30 kV.

2. **Overvoltage as a result of switching on a non-loaded transformer and its suppression**
When switching on a non-loaded transformer, large-excitation inrush currents may be produced because of non-linear excitation of the transformer. In the waveform of excitation inrush current, in addition to the fundamental wave, odd harmonics such as third, fifth, and seventh harmonics exist; the excitation inductance will change at a frequency of even multiples of the power source frequency, and the transformer thus becomes a harmonic source containing odd and even harmonics. If the system's parallel natural frequency or the excitation inductance's variation frequency is close to the frequency of a harmonic of a certain order, high-amplitude resonance overvoltage may occur. Because the resistance of the circuit is low, when switching-on a non-loaded transformer, such harmonic overvoltage may last for a long time, thus leading to

damage of the MOA because of absorption of excessive energy. If a non-loaded transformer connected to a bus is switched on, then resonance overvoltage may occur on other buses in the grid and cause the protection to operate when its amplitude and duration exceed the overvoltage protection settings of the given bus.

The main factors influencing the resonance overvoltage as a result of switching-on an unloaded transformer include:

a. *No-load excitation characteristic and remanence of the transformer.* These are key parameters influencing the amplitude of the overvoltage and excitation inrush current.
b. *Natural frequency of the system.* The amplitude of overvoltage is related to the natural frequency of the system as viewed from the transformer. The natural frequency depends on the structure of the system. The closer the natural frequency approaches the integral multiples of the power frequency, the higher the resonance overvoltage might become. The amplitude of resonance overvoltage is also related to the system damping. However, it is difficult to accurately calculate the damping at frequencies higher than the power frequency.
c. *Closing phase angle.* Considering that the closing phase angles are random, the overvoltage should be calculated on a statistical basis to identify the worst conditions.

If the overvoltage is high and lasts for a long time, then the following suppression measures must be taken: (i) install a closing resistor in the transformer circuit breaker; (ii) reduce the bus voltage before closing; (iii) switch-on low-voltage reactors before closing; and (iv) if the resonance overvoltage will last for a long time, then use a relay protection to disconnect the non-loaded transformer from bus when the overvoltage exceeds the protection settings (1.25–1.30 p.u. (rms) for 0.5 s or 1.40 p.u. (rms) for 0.1 s) [2].

3. **Generator self-excitation overvoltage and its suppression**

If a capacitive load (for instance, a non-loaded line) is connected to a synchronous generator, even if the excitation current is small or zero (rise of voltage from zero), self-excitation of the generator will still occur because the capacitive current tends to assist the magnetic excitation of the generator. As a result, the current and terminal voltage of the generator will increase sharply, leading to an extremely high self-excitation overvoltage. In this case, the generator will not be able to operate in parallel with other generators. From the perspective of physics, the self-excitation of the generator is caused by parametric resonance in the circuit due to periodical changes in the inductance while the generator rotates.

The parametric resonance occurring during the synchronous operation of the salient pole generator is called synchronous self-excitation and it should meet:

$$X_d > X_C > X_q \quad (4.14)$$

where

X_d—direct-axis synchronous reactance

X_q—quadrature-axis synchronous reactance

X_C—capacitive reactance of the transmission line.

Apparently, synchronous self-excitation will not occur on a non-salient pole generator. The parametric resonance occurring during asynchronous operation of the generator is called asynchronous self-excitation and it should meet:

$$X_q > X_C > X'_d \tag{4.15}$$

where

X'_d—direct-axis transient reactance.

If expression (4.16) is met, then no self-excitation will occur when a long, non-loaded line is connected to the generator:

$$X_C > X_d \tag{4.16}$$

Usually, the generator is connected to the system via a step-up transformer. Therefore, as we take the leakage reactance into account, this condition is corrected to:

$$X_C > X_d + X_T \tag{4.17}$$

As specified in DL/T 5429-2009 *Technical Code of Design for the Electric Power System*, if expression (4.18) is met, then no self-excitation will occur when a long, non-loaded line is connected to the generator.

$$W_h > (Q_L - Q_R)X_d^* \tag{4.18}$$

where

W_h—rated capacity of the generator, MVA

Q_L—charging power of the line, Mvar

Q_R—capacity of high-voltage or low-voltage reactor, Mvar

X_d^*—equivalent synchronous reactance of generator (including the per-unit value of the leakage reactance of step-up transformer, on the basis of generator capacity).

Expressions (4.17) and (4.18) are equivalent. Therefore, the self-excitation overvoltage may be suppressed by installing shunt reactors on a long, non-loaded EHV and UHV line, which can compensate the capacitance of the line and thus increase the equivalent capacitive reactance in the circuit to beyond the range where self-excitation of generator may occur.

4.3 Secondary Arc Current and Its Suppression

4.3.1 Secondary Arc Current and Recovery Voltage

Single phase auto-reclosing is usually used for the EHV and UHV transmission lines in China to improve system stability. To increase the success rate of single-phase auto-reclosing, the secondary arc current and recovery voltage must be taken into consideration. As shown in Figure 4.9, after a single-phase to ground fault in the line, the circuit breakers at both ends of the faulty phase (phase A) will open while the healthy phases (phases B and C) maintain the operating voltage. Under the effects of the inter-phase capacitance C_{12} and inter-phase inductance M, there is a current flowing through the faulty point, known as the secondary arc current. This current consists of a capacitive component (latitudinal component) and an inductive component (longitudinal component). The capacitive component is the current supplied by the voltage on the healthy phases through the inter-phase capacitance C_{12}. At the same time, the load current on the healthy phases produces an electromotive force on the faulty phase through inter-phase inductance, which together with the phase to ground capacitance and high-voltage shunt reactors forms a circuit supplying a current to the faulty point, which is the inductive component. Usually, the capacitive component is dominant for most uncompensated lines. After the secondary arc extinguishes, C_{12} and M will produce a recovery voltage across the arc path, which may make the arc reignite. This makes the auto-extinction more difficult and therefore may cause failure of auto-reclosing.

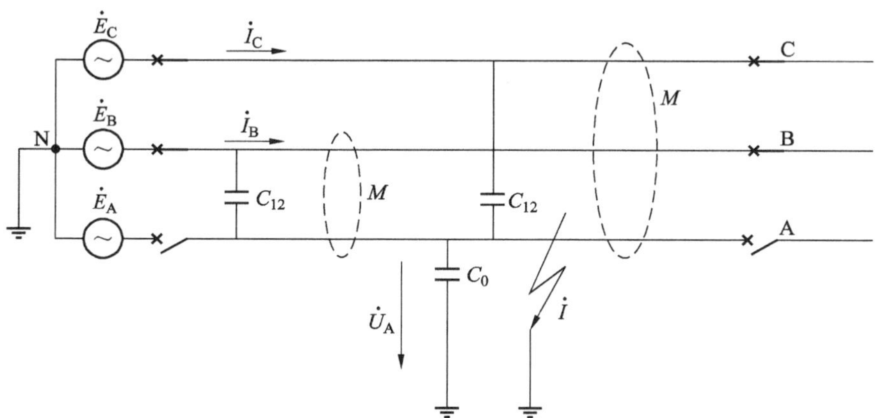

Figure 4.9
Schematic of secondary arc current.

4.3.2 Suppression of Secondary Arc Current

For an EHV system, its high-voltage level, large capacity, and long transmission distance will cause the secondary arc to last for a long time or even to be unable to self-extinguish. Therefore, suppression measures are required. For a UHV system, it is more difficult to suppress the secondary arc current as a result of its even higher voltage class and transmission power. The following are the main suppression measures used for EHV and UHV systems:

1. **Installing a neutral grounding reactor**
   For lines compensated with shunt reactors, the secondary arc current and recovery voltage can be reduced by installing a neutral grounding reactor, as shown in Figure 4.10A. By compensating the inter-phase capacitance and phase to ground capacitance, particularly by fully compensating the inter-phase capacitance, the capacitive component of the secondary arc current can be reduced. Furthermore, the phase-to-ground impedance can be increased, and thus the inductive component can be reduced. These measures have been widely used in 500-kV systems in China.
   To simplify calculation, the four-reactor circuit, as shown in Figure 4.10A, is converted to an equivalent six-reactor circuit, as shown in Figure 4.10B. This equivalent circuit consists of three-phase star-connected reactors with the neutral grounded and delta-connected rectors. The relationship between the equivalent ground reactance X_{L0}, equivalent reactance between phases X_{L12}, high-voltage shunt reactance X_R, and neutral grounding reactance X_N is given by:

$$X_{L0} = X_R + 3X_N \tag{4.19}$$

$$X_R = \frac{(X_{L12}/3)X_{L0}}{X_{L12}/3 + X_{L0}} \tag{4.20}$$

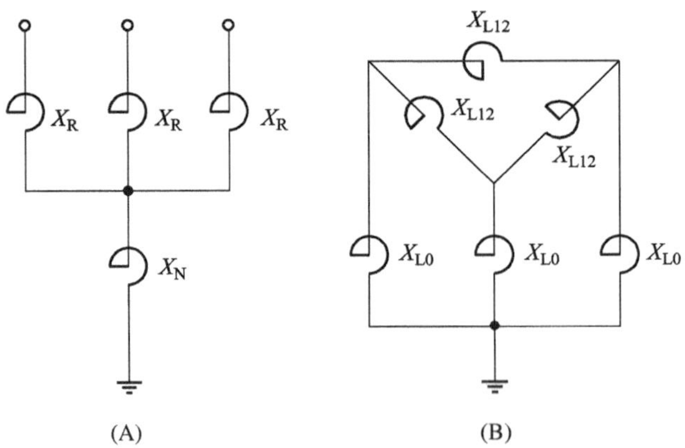

Figure 4.10: Installing a neutral grounding reactor.
(A) Four-reactor circuit and (B) six-reactor circuit.

When a single phase fault occurs in the system, the circuit breakers on both sides of the faulty phase will open. To simplify analysis, only the capacitive component, the major contributor to secondary arc current, is considered, whereas the inductive component is left out. Then, a Thevenin equivalent circuit can be developed, as shown in Figure 4.11. In this equivalent circuit, E is the RMS of normal phase voltage. As viewed from the faulty phase, the Thevenin equivalent electromotive force of the healthy phases is $\dot{E}/2$. The internal impedance (inter-phase impedance) of the equivalent source is the impedance of the parallel connected $X_{L12}/2$ and $X_{C12}/2$.

By choosing a properly sized reactor to achieve full inter-phase compensation, that is, $|X_{L12}| \approx |X_{C12}|$, a very high inter-phase impedance is produced and, thus, both the secondary arc current and recovery voltage can be reduced.

2. **Installing a high-speed grounding switch (HSGS)**
As shown in Figure 4.12, HSGSs are used to provide a path to divert the secondary arc current at the faulty point. In the event of a single-phase to ground fault, the HSGSs will operate in the sequence shown in Figure 4.13: (i) immediately after the circuit breakers at both ends of the faulty phase opened, the HSGSs, which have a low resistance, are quickly closed so that the secondary arc current is diverted to them and (ii) the HSGSs open and the arc is forcibly extinguished by their arc extinguishing capability. After the secondary arc is extinguished, the faulty phase will be reclosed.

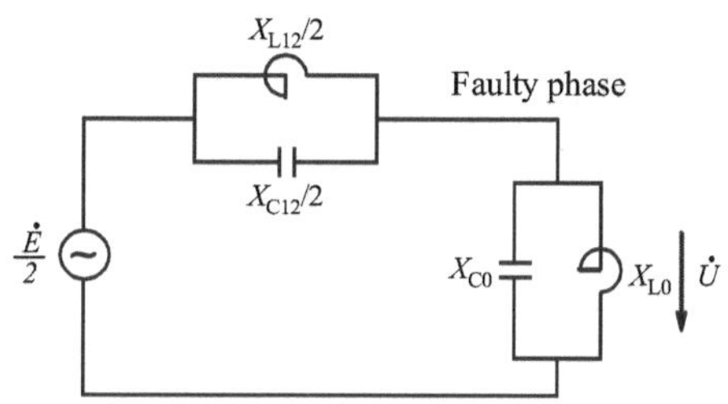

Figure 4.11
Equivalent circuit of the line with an open phase and shunt reactors grounded via neutral grounding reactor.

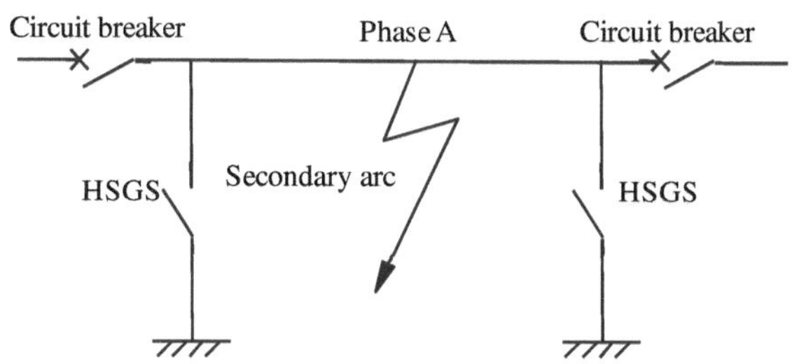

Figure 4.12
Schematic of HSGS.

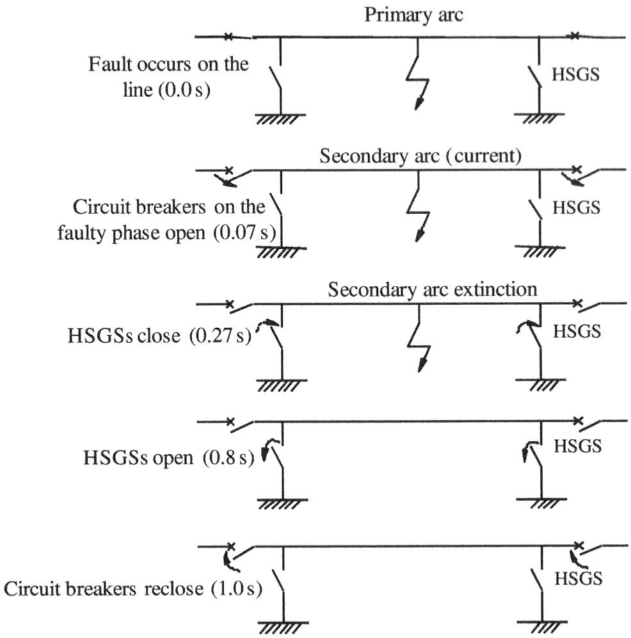

Figure 4.13
Operation sequence of HSGS.

4.3.3 Self-Extinguishing Characteristics of Secondary Arc

A secondary arc, also known as free arc, is an open arc that may freely move in the atmosphere. Its duration depends on many factors, including secondary arc current, recovery voltage, and environmental conditions (mainly wind speed). Only a few countries, such as Japan, Germany, and the Soviet Union, have performed tests and studies of the self-extinguishing characteristics of secondary arc. In China, some organizations have also studied secondary arc from aspects of calculation methods, suppression measures, and self-extinguishing characteristics. China Electric Power Research Institute (CEPRI) used an equivalent single-phase circuit in the laboratory to simulate the secondary arc when the circuit is uncompensated, undercompensated, and overcompensated under different wind speeds, recovery voltage gradients, and secondary arc currents. In the test circuit, as shown in Figure 4.14, $\dot{E}$ is the voltage of power source, $\dot{I}_L$ is the inductive current, $\dot{I}_C$ is the capacitive current, $\dot{I}$ is the secondary arc current, and $\dot{U}_a$ is the voltage of arc. CEPRI also studied the self-extinguishing characteristics of secondary arc for EHV/UHV transmission lines to obtain the recommended self-extinguishing durations (probability confidence level: 90%) for secondary arc in a circuit, which is either compensated with high-voltage shunt reactors or not compensated, under different secondary arc currents and recovery voltages [3,4], as shown in Figures 4.1 and 4.2. The arc path is exposed to natural wind.

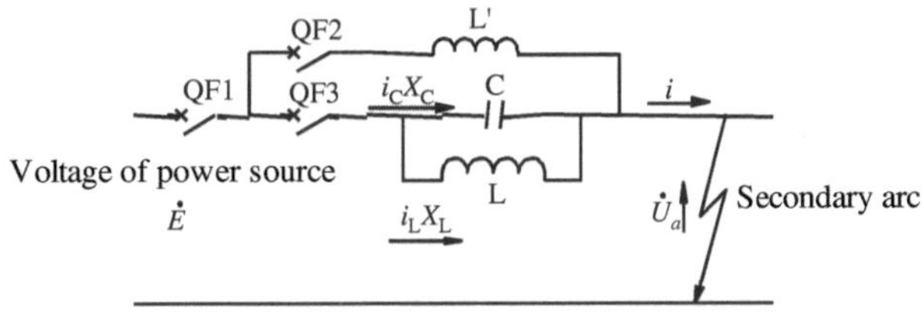

Figure 4.14: Test circuit for self-extinguishing characteristics of secondary arc.
QF1, main circuit breaker; QF2, arc ignition circuit breaker; QF3, compensation circuit breaker; L', arc inductance; C, capacitance; L, compensation inductance.

The durations of the secondary arc are highly scattered at a wind speed of 1.12 m/s and less, and are inversely proportional to the wind speed. When the wind speed exceeds 3 m/s, the durations are less scattered, but this rarely occurs in actual transmission projects. Therefore, a wind speed of 1.5 − 2.5 m/s was used in studies. The single-phase reclosing time means the time period from the occurrence of a single fault in the system and line protection operations to successful reclosing, which generally includes the time for the protections and switches to operate, time for arc path ionization and dielectric strength recovery, and duration of the secondary arc (also known as self-extinguishing time). Studies suggest that after the extinction of the secondary arc, the time for recovery of dielectric strength of arc path should be at least 0.04 s and generally 0.1 s, and the zero current interval should be generally taken as 0.1 s (Tables 4.1 and 4.2).

Studies suggest that under the same or approximate test currents, the duration of the secondary arc in a compensated circuit is much shorter than that in an uncompensated circuit. It should be noted that the test results are obtained based on short air gaps in the laboratory. For UHV lines, the air gap of which is long, the test results are subject to adjustment based on the operation experience. Usually, if the secondary arc currents are the same, the secondary arc in a longer air gap is more easily to extinguish.

4.3.4 Selection of Neutral Grounding Reactor for Fixed Shunt Reactors

1. **Single-circuit transmission line**
   Studies show that the secondary arc current and recovery voltage can usually be reduced to a low level by installing suitable line shunt reactors and associated neutral grounding reactor. Table 4.3 shows the calculated secondary arc currents and recovery voltages in the Changzhi−Nangyang and Nanyang−Jingmen lines of Changzhi−Nangyang−Jingmen UHV AC Transmission Pilot Project, with the reactance of the neutral grounding reactor calculated via Figure 4.11. From this table, we can see that the secondary arc currents of the line can be reduced to less than 10 A, and the recovery voltage gradients can be reduced to less than 10 kV/m, allowing successful single-phase reclosing after approximately 1 s.

Table 4.1: Recommended Duration Limits for Secondary Arc in Uncompensated Circuits.

Recovery Voltage Gradient (kV/m)	Secondary Arc Current (A)	Secondary Arc Duration (s)
10.0	12.0	0.10–0.15
	15.0	0.15–0.25
	18.4	0.25–0.40
	21.0	0.31–0.49
	24.0	0.37–0.54
	30.0	0.42–0.58
	42.0	0.52–0.65
16.8	12.0	0.30–0.50
	15.0	0.40–0.60
	18.4	0.50–0.70
	21.0	0.60–0.78
	24.0	0.66–0.85
	30.0	0.74–0.94
	42.0	0.90–1.11
20.0	18.4	0.60–0.80
	21.0	0.81–0.96
	24.0	0.82–1.02
	30.0	0.85–1.16
	42.0	1.17–1.42
23.0	18.4	0.70–1.04
	21.0	0.74–1.16
	24.0	1.09–1.30
	30.0	1.39–1.51
	42.0	1.25

2. **Double-circuit transmission line**

For double-circuit transmission lines, there are electrostatic and electromagnetic coupling between the three phases of one circuit and between the two circuits. When a single-phase to ground fault occurs in any circuit, the coupling between the faulty phase and the healthy phases, as well as between other conductors of parallel circuits, will induce secondary arc current and recovery voltage across the faulty point. Therefore, the coupling among conductors of a double-circuit line is stronger than that of a single-circuit line.

In addition to the normal double-circuit operation, single-circuit operation is also possible for a double-circuit line. To ensure reliability, the conditions for single-phase reclosing should be satisfied on both circuits. Therefore, for a double-circuit line compensated by high-voltage reactors, the selected neutral grounding reactor should be suitable for the double-circuit operation and the single-circuit operation. However, because the coupling effects between phases of one circuit and between conductors of

Table 4.2: Recommended Duration Limits for Secondary Arc in Compensated Circuits.

Compensation Mode	Recovery Voltage Gradient (kV/m)	Secondary Arc Current (A)	τ or τ'[a] (ms)	Wind Speed (m/s)	Secondary Arc Duration (s)	Remarks
Undercompensated	12.0	28.0	0.88	1.5–2.0	≤0.2	
		35.0	1.20	1.5–2.0	≤0.2	
		40.0	1.48	1.5–2.0	0.64–0.83	
	20.0	28.0	0.88	1.5–2.0	≈0.2	
		35.0	1.20	1.5–2.0	≈0.2	
		40.0	1.48	1.5–2.0	0.70–0.83	
		48.0	2.03	2.0	1.26	
	25.0	48.0	2.03	1.5–2.0	1.30	Duration: 2 s at a wind speed lower than 1.0 m/s
	30.0	65.0	3.45	1.1	>2	
		80.0	2.32	2.0–2.5	0.99–1.04	Duration: 2 s at a wind speed lower than 1.5 m/s
	45.0	80.0	2.32	1.5–2.5	Cannot be extinguished by itself	
Overcompensated	12.0	18.4	0.84	0.9	≤0.2	
	15.0	40.0	0.65	0.9	0.40	
			1.46	1.5–2.0	0.64–0.72	
			1.49	0.9	0.81	
	20.0	18.4	0.84	0.9	0.25	
	25.0	40.0	1.46	1.5–2.0	0.85–0.94	

[a]τ is defined as the difference between the 1/4 cycle duration of compensation circuit frequency and that of power frequency (5 ms). τ and τ' are respectively used for undercompensated and overcompensated circuits.

Table 4.3: Secondary Arc Current and Recovery Voltage for Changzhi–Nangyang and Nanyang–Jingmen Lines.

Line Name	Secondary Arc Current (A)		Recovery Voltage (kV)		Highest Voltage on Neutral Grounding Reactor (kV)
	Sending End	Receiving End	Sending End	Receiving End	
Changzhi–Nanyang line I	9.8	6.9	49.4	35.7	87.6
Nanyang–Jingmen line I	3.6	3.6	20.4	20.8	93.0

parallel circuits are different, providing different neutral grounding reactors for these two operation modes is recommended. Therefore, the reactors must be carefully chosen.

Take a 300-km, 1000-kV UHV double-circuit line for example, where 600-Mvar shunt reactors are installed at both ends of each circuit. Figure 4.15 shows the variation

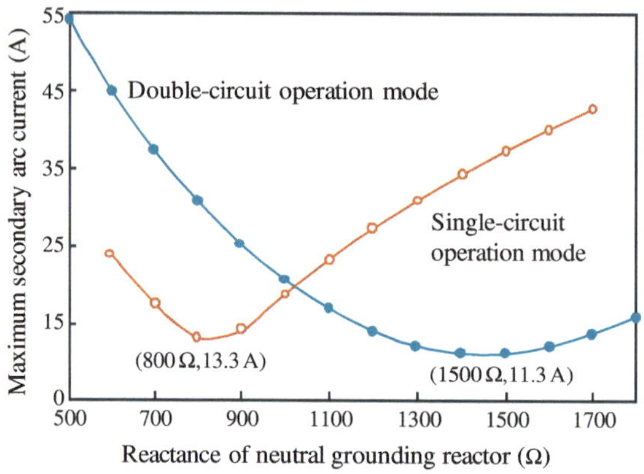

Figure 4.15
Variation of secondary arc current with neutral grounding reactor reactance.

of the secondary arc current with the reactance of the neutral grounding reactor, respectively, for double-circuit and single-circuit operations. The figure shows that the secondary arc current is the smallest when the neutral grounding reactor has a reactance of 1500 Ω for double-circuit operation and when the neutral grounding reactor has a reactance of 800 Ω for single-circuit operation. As a compromise for the two operation modes, the reactance is determined as 1000 Ω. With this reactance, the maximum secondary arc current and recovery voltage are 21 A and 53 kV and 19 A and 58 kV, respectively, for single-circuit and double-circuit operations, allowing successful single-phase reclosing within approximately 1 s.

4.3.5 Selection of Neutral Grounding Reactor for Controllable Shunt Reactors

For a line compensated with controllable shunt reactors, when it is operating under a heavy load, the compensation degree can be adjusted to 60%. After the circuit breakers on both sides trip after a single-phase to ground fault, two scenarios are considered: the capacity of the line shunt reactors is adjusted or not adjusted. In the former scenario, a maximum compensation degree of 90% is considered and neutral grounding reactor should be selected on this basis. Table 4.4 shows the calculated secondary arc currents and recovery voltages when the line carries 3500 MW.

This table shows that when the compensation cannot be or is not allowed to be adjusted to provide a higher compensation, the secondary arc currents will reach a high level, up to 57 A for a 500-km line and 83 A for a 600-km line, so that the secondary arc will last a long time or cannot self-extinguish, which may cause failure of single-phase reclosing within approximately 1 s.

Table 4.4: Secondary Arc Currents and Voltages on a Line with Controllable Shunt Reactors.

No.	Line Length (km)	Capacity of Shunt Reactor Adjusted or Not	Compensation Degree		Secondary Arc Current (A)	Recovery Voltage (kV)
			Normal Operation	Faulty Phase Trip-Out		
1	500	Not adjusted	60%	60%	57.0	58.4
2		Adjusted	60%	90%	17.6	62.8
3	600	Not adjusted	60%	60%	82.8	70.7
4		Adjusted	60%	90%	26.9	72.4
5	2 × 350	Not adjusted	60%	60%	26.6	41.1
6	(with switchyard)	Adjusted	60%	90%	7.7	45.4

4.3.6 Selection of HSGS

Studies show that on a long UHV transmission line, the secondary arc current may be reduced to a low level by installing a grounding reactor on the neutral part of the line shunt reactors to ensure successful single-phase reclosing. This is demonstrated by an operating single-circuit UHV AC transmission pilot project in China. In this project, two sections of the transmission line (respectively, 281 and 359 km) are installed with shunt reactors and a neutral grounding reactor, whereby the secondary arc current is effectively reduced. Therefore, HSGSs are not necessary for China's UHV lines.

However, with the development of grids, short UHV transmission lines without shunt reactors may appear. For lines of the same length, the secondary arc current in an uncompensated UHV line is much higher than that in a 500-kV line. Compared with a compensated line, it is more difficult to extinguish the arc in an uncompensated line under the same secondary arc current, which may cause failure of single-phase reclosing. HSGSs may be required on these lines. For a 100-km, 1000-kV untransposed double-circuit line that carries 10,000 MW, if no shunt reactors are installed, then the secondary arc current and recovery voltage are 57 A and 147 kV (voltage gradient: approximately 15 kV/m), which are far beyond the recommended values in Table 4.1. If HSGSs are installed at both ends of the line, when the ground resistance is $<1.0\ \Omega$ and the resistance of the secondary arc is 10–50 Ω, then the secondary arc current and recovery voltage will be 21.7 A and 1.3 kV, respectively, allowing successful single-phase reclosing within approximately 1 s.

4.3.7 Impact of Series Compensation Capacitor on Transient Secondary Arc Current

For a line with series compensation capacitor, the capacitor may not be bypassed during a single-phase to ground fault if the faulty point is far from the capacitor or the short-circuit capacity on both sides of the line is low, because the metal oxide varistor (MOV) has a low current and energy absorption capacity. After the circuit breakers of the faulty phase at both ends of the line trip, the capacitor, high-voltage reactors, and resistance of the faulty point arc path will form a low-frequency oscillation circuit, as shown in Figure 4.16. The residual charge in the series compensation capacitor will be discharged through the circuit, giving rise to a low-frequency transient component superimposed on the power frequency component of the secondary arc current at the faulty point, which is detrimental to single-phase reclosing.

Figure 4.17 shows the simulated wave shapes for the secondary arc current in Nanyang–Jingmen UHV AC transmission line, which is provided with series compensation. It shows that when the series compensation capacitor is not bypassed, the secondary arc current attenuates slowly, with a peak value of approximately 62.3 A in 0.5 s after the line circuit breakers tripped. Moreover, the secondary arc current wave shape contains a low-frequency (approximately 6 Hz) oscillation component, which reduces the number of zero crossings of the secondary arc current and therefore is detrimental to self-extinguishing of the secondary arc. To suppress the secondary arc current, the circuit

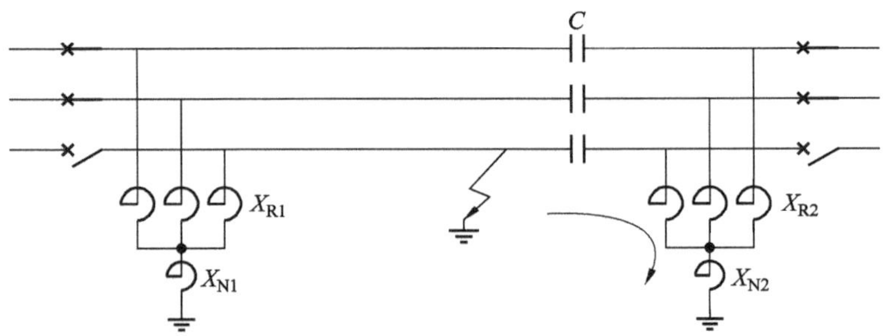

Figure 4.16
Low-frequency oscillation circuit formed in single-phase to ground fault on a line with series compensation.

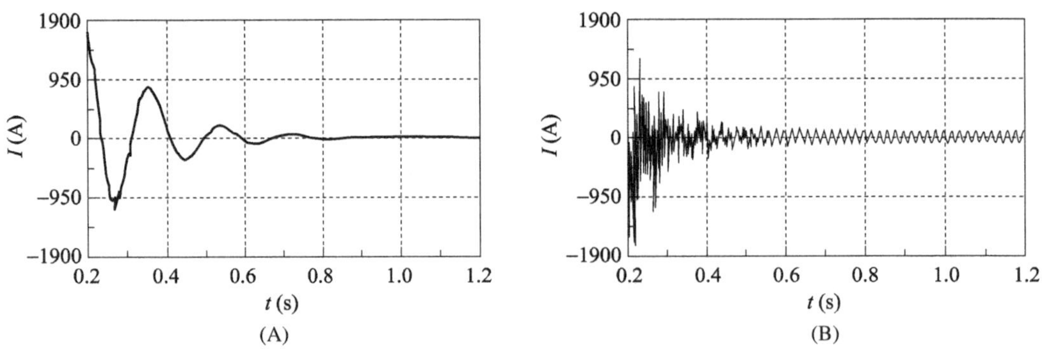

Figure 4.17
Simulated wave shapes of the secondary arc current on a 1000-kV line with series compensation: (A) capacitor not bypassed and (B) capacitor bypassed.

breaker and capacitor bypass may be linked, that is, although the line protection gives trip signals to the circuit breaker, the bypass breaker of the series compensation capacitor is ordered to close by phase. Figure 4.17B shows the wave shape of the secondary arc current during such a process. It indicates that when the capacitor is bypassed, the secondary arc current attenuates rapidly, with a peak of approximately 11.3 A in 0.5 s after the line circuit breaker trips, which is significantly lower than when the capacitor is not bypassed.

4.3.8 Impacts of Phase Sequence on Secondary Arc Current in Double-Circuit Lines

For EHV or UHV double-circuit lines, inverse-phase sequence (i.e., ABC/CBA) arrangement is often used. This arrangement has the advantage of a low ground electric field. However, the same phase sequence (ABC/ABC) and different phase sequence (ABC/BAC) offer lower audible noise and corona loss. Figure 4.18 shows the transposition modes in different phase sequence arrangements in a complete transposition cycle. When two or more transposition cycles are involved, the phases are transposed in the following modes: (i) mode A, phases in each transposition cycle are arranged in the same direction and (ii) mode B, phases in two adjacent transposition cycles are arranged in opposite directions. Figure 4.19 shows modes A and B in inverse-phase sequence arrangement in two transposition cycles.

Take a 340-km, 1000-kV double-circuit line with 720-MVar shunt reactors at both ends of each circuit as an example. The calculated secondary arc currents and recovery voltages

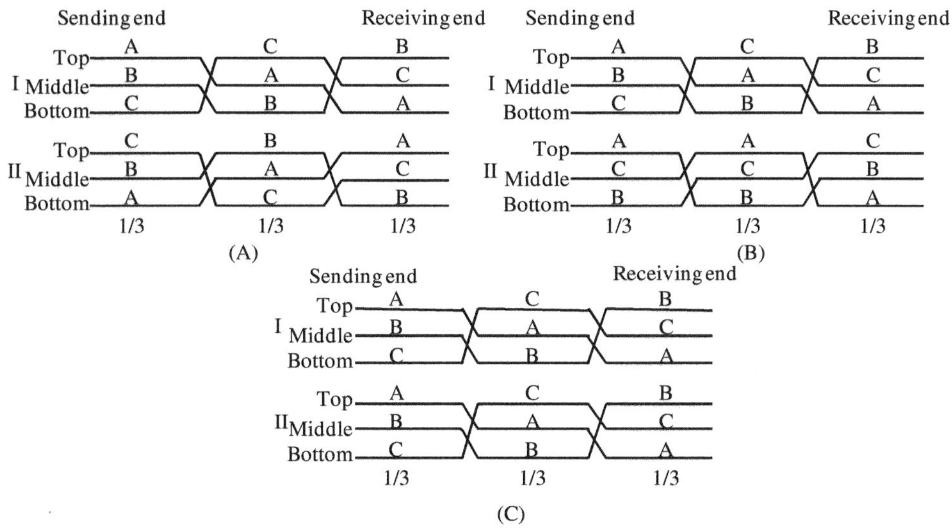

Figure 4.18: Transposition modes in different phase sequence arrangements in a transposition cycle. (A) Inverse phase sequence; (B) other phase sequence; and (C) same phase sequence.

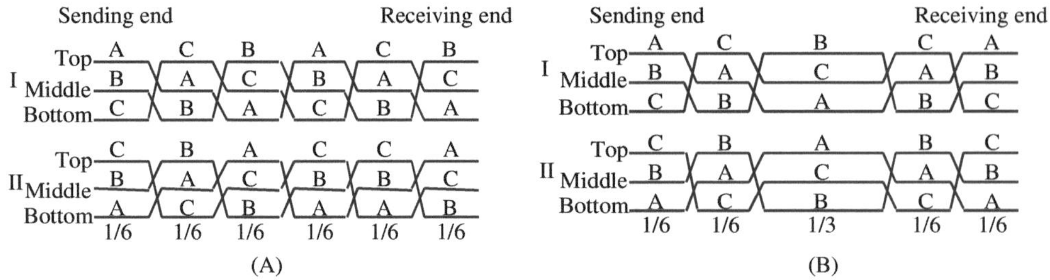

Figure 4.19: Transposition modes in inverse phase sequence in two transposition cycle. (A) Transposition mode A and (B) transposition mode B.

Table 4.5: Secondary Arc Currents and Recovery Voltages in Different Phase Sequence Arrangements.

Phase Sequence Arrangement and Transposition			Neutral Grounding Reactor (Ω)	Secondary Current (A)	Recovery Voltage (kV)
Inverse phase sequence	In a transposition cycle	—	700	26.7	70.5
	In two transposition cycles	Mode A	800	22.1	56.7
		Mode B	800	17.0	43.6
Other phase sequence	In a transposition cycle	—	500	33.5	97.3
	In two transposition cycles	Mode A	500	30.6	88.9
		Mode B	500	28.7	83.6
Same phase sequence	In a transposition cycle	—	500	37.7	109.4
	In two transposition cycles	Mode A	500	30.9	89.6
		Mode B	500	24.4	71.0

across the line in different phase sequence arrangements are shown in Table 4.5. It can be seen that the secondary arc currents and recovery voltages in the inverse-phase sequence are the lowest. Mode B outperforms mode A in suppressing the secondary arc current and recovery voltage.

4.4 Switching Overvoltages and Its Suppression

4.4.1 Closing Overvoltage and Its Suppression

The most common switching overvoltages in an EHV or UHV system are closing (including the single-phase reclosing) overvoltage and opening overvoltage (including the overvoltage caused by fault clearing). Generally, the maximum statistic phase-to-ground switching overvoltage should be no more than 1.7 p.u. along a 1000-kV line and no more than 1.6 p.u. in a substation, and the maximum statistic phase-to-phase switching overvoltage should be no more than 2.9 p.u. The statistic switching overvoltage means the

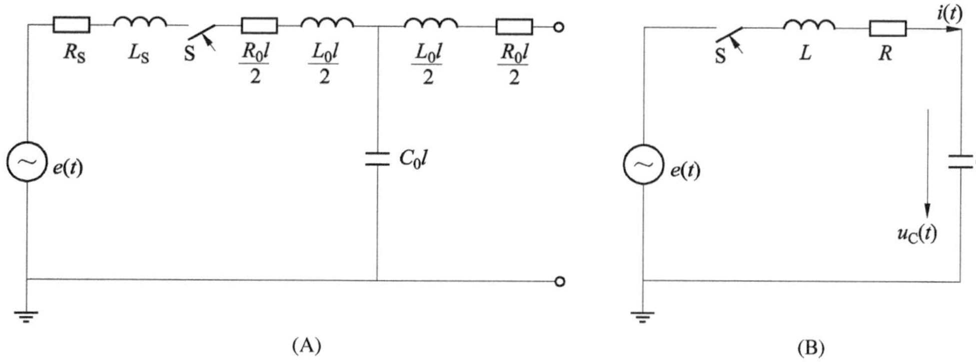

Figure 4.20: Equivalent circuit for switching on an unloaded line.
(A) Equivalent circuit and (B) simplified equivalent circuit.

switching overvoltage with an occurrence probability of no more than 2% and a base value of $1100\sqrt{2}/\sqrt{3}$ kV. If, in rare cases, the switching overvoltage exceeds these limits, then it should be analyzed according to its probability and consequences.

Switching-on an unloaded line is a type of operation in the power system, and it can be classified into two situations: normal closing and auto-reclosing.

4.4.1.1 Physical process of closing overvoltage

A single-circuit transmission line is used as an example for analyzing the process of closing overvoltage. Before normal closing, the initial line voltage is zero. Assuming that the three-phase lines are symmetrically arranged and their circuit breakers are closed simultaneously, the lines may be considered as a single phase line.

To simplify analysis, an equivalent circuit is used to represent the physical process of a closing overvoltage. As shown in Figure 4.20, R_S, L_S, and $e(t) = E_{sn}\sin(\omega t + \varphi_0)$ are, respectively, the equivalent resistance, inductance, and phase electromotive force of the source; R_0, L_0, and C_0 are the resistance, inductance, and capacitance per unit length of the line; l is the line length; and R, L, and C are the resistance, inductance, and capacitance, respectively, of the simplified equivalent circuit, $R = R_S + (R_0 l/2)$, $L = L_S + (L_0 l/2)$, and $C = C_0 l$.

Before normal switching, the line has no residual charge and its initial voltage is zero, that is, $u_C(0) = 0$. Closing switch S is equivalent to applying a sinusoidal electromotive $e(t)$ on the series oscillating circuit consisting of L and C, as described by:

$$u_C = RC\frac{du_C}{dt} + LC\frac{d^2u_C}{dt^2} = e(t) \tag{4.21}$$

By solving the differential equation, we have:

$$u_C(t) = U_{Cm}\left\{-\cos(\omega t+\varphi)+\sqrt{\left(\frac{\omega}{\omega_0}\right)^2\sin^2\varphi+\cos^2\varphi}\times\left[e^{-\delta t}\cos(\omega_0 t-\theta)\right]\right\} \quad (4.22)$$

where

$$\varphi=\varphi_0-\arctan\frac{\omega L-(1/\omega C)}{R},\ \tan\theta=\frac{\omega}{\omega_0}\tan\varphi,\ \theta=\frac{R}{2L},\ \text{and}\ \omega_0=\sqrt{\frac{1}{LC}-\delta^2}$$

Let the closing phase angle $\varphi_0 = 90°$, considering that $\omega L < (1/\omega C)$, and by neglecting the resistance R, we have:

$$u_C(t) = U_{Cm}(\cos\omega t - \cos\omega_0 t) \quad (4.23)$$

where

$$U_{Cm}=\frac{(1/\omega C)}{(1/\omega C)-\omega L}E_m=\frac{1}{1-(\omega^2/\omega_0^2)}E_m\ \text{and}\ \omega_0=\frac{1}{\sqrt{LC}}$$

Analysis results show that closing is an oscillatory process in which the electromotive force $e(t)$ charges C through L. Both Eqs. (4.22) and (4.23) contain two components, a steady-state component with an angular frequency the same as the power source's angular frequency of ω and a free component with an angular frequency of ω_0. The inequality $U_{Cm} > E_m$ in nature reflects the Ferranti effect of an unloaded line. The maximum possible overvoltage during the oscillation may be estimated based on the steady-state voltage during the oscillation and the initial voltage.

$$|u_{C.max}| = |2U_s - U_i| \quad (4.24)$$

It is obvious that under the conditions described by Eq. (4.23), we have $|u_{C.max}| = 2U_{Cm}$. The transient component ω_0 varies with system parameters. For EHV lines, the value of ω_0/ω is generally in the range of 1.5–4.0. For low-voltage lines, the value of ω_0 is usually high.

However, in practice, switching-on an unloaded line is much more complex than the equivalent circuit shown. First, the transient component may have several oscillation frequencies. Second, the simultaneity bias of the three-phase circuit breakers should be considered. Third, the transient overvoltages on three phases may interact with each other. Because of these reasons, the overvoltage as a result of switching on an unloaded line will be higher than $2U_{Cm}$. During the single-phase reclosing, there is no transient inter-phase interaction because the other two healthy phases work normally. As a result, three-phase closing usually produces higher overvoltages than single-phase reclosing. Therefore, three-phase reclosing is not adopted in EHV and UHV transmission systems in China.

4.4.1.2 Measures for suppressing closing overvoltages

To suppress the switching overvoltage of the UHV system as required in the Chinese national standard GB/Z 24842-2009 *Overvoltage and Insulation Coordination of 1000 kV UHV AC Transmission Project*, the following measures may be considered in addition to those given in Section 4.2.1.

1. **MOA**
   In recent years, because of its improved performance, the MOA has become one of the major means for suppressing switching overvoltages across the world. In recent studies of UHV transmission, MOAs with a rated voltage of 828 kV are used for both substations and lines.
2. **Closing resistor for circuit breaker**
   This is an effective measure to suppress switching overvoltages. As shown in Figure 4.21, the circuit breaker will operate in the following two stages: the auxiliary contact closes and the main contact closes after a period of time. In the first stage, higher resistance will result in lower overvoltage. In the second stage, after the resistor is short-circuited, the larger the resistor is, the higher the overvoltage will be. Generally, the closing resistor is determined as 350–700 Ω.
   In the EHV/UHV grids of China, switching overvoltages have been effectively suppressed by wide use of MOA. Therefore, circuit breakers on 330-kV lines are generally not equipped with closing resistors; on 500-kV lines, closing resistors are needed only when the switching overvoltages exceed permitted values; whereas in 750- and 1000-kV systems, closing resistors are generally required because switching overvoltages have to be suppressed to a relatively lower level.
3. **Closing phase control**
   To minimize the closing overvoltage, the closing phase of the circuit breaker should approach the voltage's zero crossing point as far as possible. Broadly speaking, closing

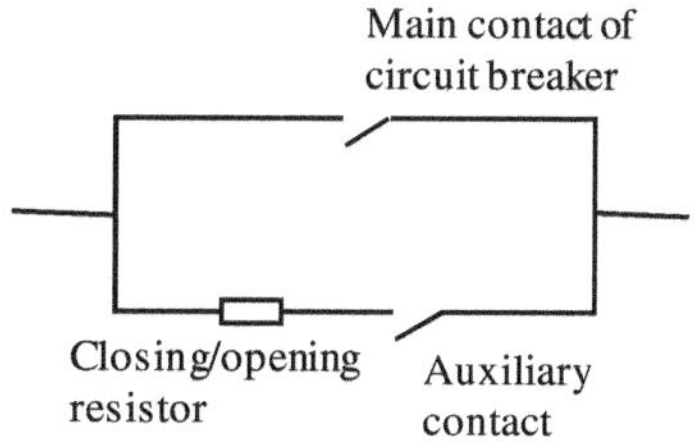

Figure 4.21
Schematic diagram of a circuit breaker with closing/opening resistor.

phase control is to control the closing phase according to bus/line voltage, thereby suppressing the inrush currents and overvoltages in grids [5]. The phase control circuit breaker developed based on this theory has a simple structure and a low cost. However, its control accuracy in actual application is doubtful. In 1998, the International Council on Large Electric Systems (CIGRE) discussed the strengths and weakness of the phase control circuit breaker and concluded that its effectiveness can be proved by analyses, calculations, and field tests. Such circuit breakers are usually used for the switching of lumped elements, such as capacitor banks and transformers, but are rarely used on transmission lines. American Electric Power (AEP) conducted tests to suppress the closing overvoltage in 1500-kV UHV systems by combining phase-control closing and other measures.

The amplitude of switching overvoltages is closely related to the system structure and operating conditions. Proper operating conditions may help reduce the overvoltage. Taking the power transmission from an isolated power plant to the grid via an EHV or UHV line, for example, the switching overvoltage caused by switching-on an unloaded line will be high on the plant side and low if the switching is on the grid side (at substation or switchyard). This is mainly because there are other lines connected to the bus in operation, which will absorb the oscillation energy generated from switching-on the line from the grid side, thus reducing the switching overvoltage.

In a UHV system, the closing overvoltage usually cannot be reduced to an acceptable level only using MOA. Therefore, other measures may be required, such as installing a closing resistor in circuit breakers and phase-control circuit breakers. Among them, the former is more often used. Figure 4.22 shows the phase-to-ground closing overvoltage as a function of the closing resistance under a certain system operation mode. In this case, the shunt closing resistance is 200–900 Ω. This figure shows that the overvoltage is lowest when the closing resistance is 300–400 Ω. However, lower resistance will result in higher energy

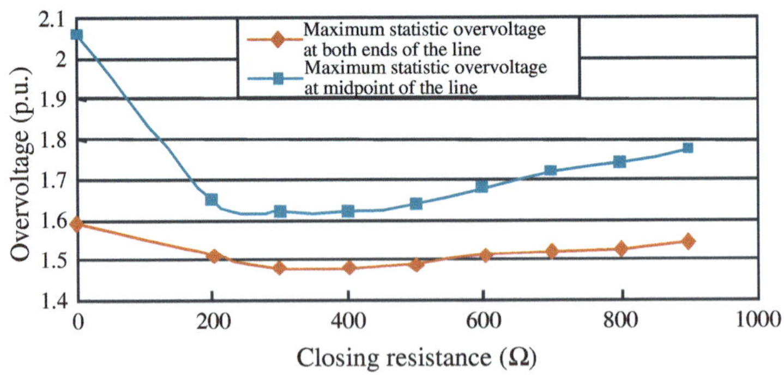

Figure 4.22
Statistic closing overvoltage as a function of closing resistance.

consumption. When the working time of the closing resistor is more than 10 ms, the statistic closing overvoltage decreases as the working time of the closing resistor increases, but the energy consumption of the closing resistor increases significantly. Comprehensively considered, the closing resistance of the line circuit breaker is chosen as 400–600 Ω, and the working time is chosen as (9.5 ±1.5) ms.

For the UHV AC pilot project in China, when the closing resistance is 400, 500, and 600 Ω, the statistic switching overvoltages are, respectively, 1.59, 1.62, and 1.64 p.u. at the midpoint of the line and 1.54, 1.55, and 1.56 p.u. at the line ends, which are all within the allowable range. In this case, the energy to be consumed by the closing resistor is no more than 2.5 MJ and the energy consumption of the MOA is also significantly reduced. If a closing resistance of 700 Ω is used, then the closing overvoltage will be higher. For instance, switching on the Jindongnan–Nanyang lines from the Nanyang side will produce a statistic phase-to-ground switching overvoltage of up to 1.72 p.u., which exceeds the aforesaid limit (1.70 p.u.), and the statistic phase-to-phase switching overvoltage is up to 2.91 p.u., exceeding the limit (2.9 p.u.) as well. These results demonstrate the poor applicability of a 700-Ω closing resistor. Therefore, the closing resistance is determined to be 400–600 Ω.

According to available data, the American Bonneville Power Administration (BPA) uses a closing resistance of 300 Ω, the Soviet Union uses a closing resistance of 378 Ω, Italy uses a closing resistance of 500 Ω (for both opening and closing resistance), Chinese UHV AC pilot project uses a closing resistance of 600 Ω, and Japan uses a closing resistance of 700 Ω (for both opening and closing resistors). It can be seen that most counties use a small closing resistance (300–600 Ω), which is suitable for long transmission lines. Only Japan uses a large closing resistance because transmission lines are relatively short there.

4.4.2 Opening Overvoltage and Its Suppression

For UHV systems, two main opening overvoltage scenarios should be considered:

1. Load rejection overvoltages including the overvoltage attributable to three-phase load rejection by fault tripping while a single-phase fault or three-phase fault occurs, and overvoltage attributable to three-phase load rejection by unexpected tripping without faults.
2. Overvoltage after fault clearing. This is the overvoltage that occurs on other healthy lines in parallel (including the other circuit on the same tower) after clearing of a fault on one line.

4.4.2.1 Load rejection overvoltages

As mentioned, the load rejection overvoltages include two major types: one is a three-phase load rejection after single-phase to ground fault and the other is three-phase load rejection without faults. The following discussions focus on the effects of different suppression measures.

1. **MOA only**

   For transmission lines with conventional fixed shunt reactors, the line with a longer length and larger transmitted power will lead to higher overvoltage and higher energy consumption of MOAs during load rejection. In the UHV AC transmission pilot project of China, when only MOA is used, the overvoltage after a load rejection without fault or attributable to phase-ground fault is 1.60 p.u. at the sending end and 1.70 p.u. at the midpoint of the line, both of which are within permitted limits.

2. **Circuit breaker equipped with opening resistor and MOA**

   In addition to MOA, the circuit breaker equipped with an opening resistor is also an option to suppress the overvoltage. The schematic diagram of the circuit breaker with an opening resistor is shown in Figure 4.21. The circuit breaker opens in the following two stages: the main contact opens first and, after a period of time (known as switching-on time), the auxiliary contact opens. In the first stage, higher resistance will result in higher overvoltage. In the second stage, the overvoltage will decrease as the resistance increases. For the circuit breakers in 330- to 750-kV lines, opening resistors are generally not necessary for suppressing the load rejection overvoltage. On the 1000-kV UHV line in Japan, circuit breakers are equipped with opening resistors of 700 Ω, which also serve as closing resistors.

   Figure 4.23 shows the overvoltage as a function of opening resistance during the switching at the sending end after a single-phase to ground fault. This figure shows that the overvoltages at the midpoint and opposite side are higher than that at the opening side. In general, when the opening resistance exceeds 200 Ω, the higher the opening

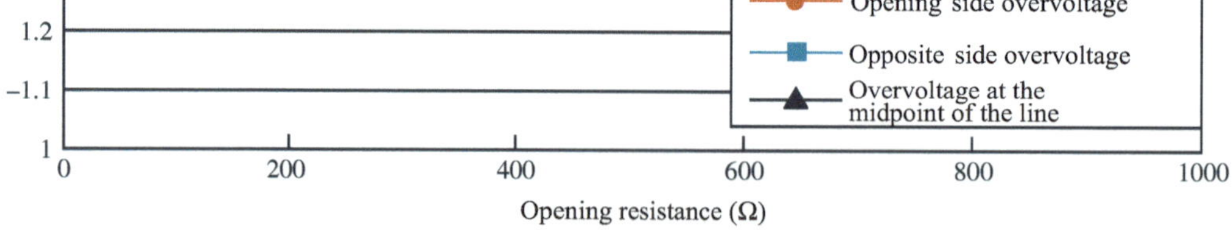

Figure 4.23
Opening overvoltage as a function of opening resistance.

resistance is, the better the effect in suppressing the overvoltage at the midpoint of the line will be. It takes a longer time for the opening resistor to take effects than it does for the closing resistor. In studies, the working time of the opening resistor is considered to be 30 ms.

3. **Controllable high-voltage shunt reactors**
For a line compensated with controllable shunt reactors, the compensation degree should be adjusted to 60% under heavy load and to approximately 90% within 50–100 ms after load rejection. For comparison, Table 4.6 shows the switching overvoltages attributable to load rejection during normal operation on a 600-km line transmitting 3500-MW power compensated with fixed and controllable shunt reactors. This table shows that with controllable shunt reactors, the overvoltage at the midpoint increases from 1.68 to 1.73 p.u. This is because it takes time for the controllable shunt reactors to increase the compensation. In contrast, the overvoltage at both ends only slightly increases because of the existence of MOAs. Therefore, to reduce the overvoltage at the midpoint, a MOA may be installed there for slow-responding controllable shunt reactors.

4.4.2.2 Overvoltage after fault clearing

The amplitude of overvoltage after clearing a phase-to-ground fault depends on the system structure and location and type of the fault. The following paragraphs discuss the suppression of overvoltage using only MOA.

Table 4.7 shows the maximum overvoltages on other healthy lines in parallel after clearing a single-phase to ground fault on a typical UHV line when only MOAs and shunt reactors are used. This table shows that the maximum overvoltage is 1.56 p.u. at the substation or switching substation and 1.70 p.u. on the line.

Table 4.8 shows the maximum statistic overvoltages after the clearing of a three-phase to ground fault. It can be seen that the overvoltage remains no higher than 1.6 p.u. at the

Table 4.6: Controllable Shunt Reactor Versus Fixed Shunt Reactor in Effect on Opening Overvoltage.

Item	Opening Side Overvoltage (p.u.)	Overvoltage at the Midpoint of the Line (p.u.)	Opposite Side Overvoltage (p.u.)	Maximum Energy Consumption of Arrester (MJ)
Fixed shunt reactor (90% compensation)	1.43	1.68	1.51	2.5
Controllable shunt reactors (50 ms)	1.44	1.73	1.54	6.1
Controllable shunt reactors (100 ms)	1.44	1.73	1.54	7.6

Table 4.7: Overvoltage After Clearing a Single-Phase to Ground Fault on a Typical UHV Line.

Project	Faulty Line	Statistic Phase to Ground Overvoltage on Other Healthy Conductors of Parallel Circuits (p.u.)	
		Maximum Overvoltage on the Line	Maximum Overvoltage at Both Ends of the Line
UHV AC transmission pilot project of China	Changzhi–Nanyang line I	1.70	1.56
	Nangyang–Jingmen line I	1.61	1.55
Expansion project of UHV AC transmission pilot project of China	Changzhi–Nanyang line I	1.68	1.51
	Nangyang–Jingmen line I	1.61	1.47
Huainan–Shanghai UHV AC transmission project	Huainan–Wannan line	1.50	1.46
	Wannan–Zhebei line	1.64	1.55
	Zhebei–Huxi line	1.60	1.50

Table 4.8: Overvoltage After Clearing a Three-Phase to Ground Fault on a Typical UHV Line.

Project	Faulty Line	Statistic Phase to Ground Overvoltage on Other Healthy Conductors of Parallel Circuits (p.u.)	
		Maximum Overvoltage on the Line	Maximum Overvoltage at Both Ends of the Line
UHV AC transmission pilot project of China	Changzhi–Nanyang line I	1.76	1.59
	Nangyang–Jingmen line I	1.95	1.60
Expansion project of UHV AC transmission pilot project of China	Changzhi–Nanyang line I	1.88	1.59
	Nangyang–Jingmen line I	2.05	1.57
Huainan–Shanghai UHV AC transmission project	Huainan–Wannan line	1.83	1.55
	Wannan–Zhebei line	1.85	1.57
	Zhebei–Huxi line	1.86	1.58

substation or switching substation. However, it may reach up to 2.05 p.u. on some towers in the middle of healthy lines.

In summary, the overvoltage after clearing a fault has the following three major characteristics if only MOAs and shunt reactors are used.

1. The overvoltage after clearing a single-phase to ground fault is no higher than 1.6 p.u. at both ends of the line and 1.7 p.u. along the line, which are all within the limits. Single-phase to ground fault is more likely to occur than three-phase to ground fault. However, because of its low amplitude, the overvoltage after clearing a single-phase to ground fault could rarely cause failure of the system.

2. The maximum statistic overvoltage at the substation or switching substation after clearing a three-phase to ground fault rarely exceeds 1.6 p.u., which is far less than the impulse withstand level of the electrical equipment. Therefore, such an overvoltage is unlikely to cause system failure.
3. After clearing a three-phase to ground fault, excessive overvoltage may occur near the midpoint of the line, with an amplitude of up to 2.0 p.u. or even higher, which exceeds the limit (1.7 p.u.) and may cause flashover or even line failure. However, it should be pointed out that in most cases such high overvoltages only occur on one phase and the overvoltage on other phases is low.

It can be seen that only the overvoltage along the line might exceed the limit and may only cause flashover. Moreover, a certain flashover rate is permitted because the insulation of towers can restore itself. Studies show that the overvoltage of this type has a very low probability of causing derivative fault. Taking the Huainan–Wannan lines as an example, its failure rate because of this reason is approximately 9.0×10^{-5} times per year.

The overvoltage may be reduced by installing an opening resistor in circuit breakers. However, the opening resistor, which requires a large amount of energy, is costly and unreliable. If the overvoltage after clearing a three-phase to ground fault, which rarely occurs, must be suppressed, then an effective measure is to install a 828-kV MOA at the midpoint of the line, which outperforms opening resistors in reliability and economy.

In conclusion, it is neither desirable nor feasible to install an opening resistor for the circuit breaker to suppress the overvoltage because such overvoltage only influences the insulation of towers and can rarely cause system failure [6].

4.4.2.3 Necessity of installing an opening resistor in circuit breaker

During the switching process, the closing/opening resistor will absorb a large amount of energy. In determining the energy required, the following factors should be considered:

1. Because of manufacturing difficulties, only one resistor should be used for both opening and closing purposes. As previous analysis suggests, the closing resistance is preferably 400 Ω and the opening resistance is preferably 800 Ω. As a compromise, the resistance is determined to be 600 Ω.
2. The major portion of energy required by the closing/opening resistor is not the energy consumed in the process of suppressing the switching overvoltage, but the energy required by the circuit breaker for other switching operations. According to the Chinese and IEC test standards, there are mainly two energy requirements for the circuit breaker:
   a. Able to complete such an operation cycle at the rated short-circuit capacity: O-CO-CO.
   b. Able to complete such an operation cycle after an out-of-phase fault: CO.

Based on these requirements, the estimated energy consumption of the closing/opening resistor is shown in Table 4.9. The energy consumed by the resistor when suppressing the

Table 4.9: Desired Capacity of the Closing/Opening Resistor in Circuit Breakers.

Type	O-CO-CO (MJ)	CO after an Out-of-Phase Fault (MJ)
400 Ω closing resistor	21	41
400 Ω closing/opening resistor	90	228
600 Ω closing resistor	14	30
600 Ω closing/opening resistor	63	164

switching overvoltage is always lower than the values given in this table. If a resistor is used both for closing and opening, then the higher value of the two should apply. A 600-Ω closing/opening resistor requires 164 MJ. For reference, the energy consumption of a 700-Ω closing/opening resistor for a 1000-kV circuit breakers in Japan is estimated to be 133 MJ.

The opening overvoltage may be reduced by installing an opening resistor in line circuit breakers. However, the opening resistor consumes much energy and is very expensive. Taking a 600-Ω shunt resistor, for example, if it is only used for closing, then the energy consumption will be approximately 30 MJ; if it is used both for closing and opening, then the value may be up to 164 MJ. As a result, an additional cost of RMB 10 million will be required for each circuit breaker. In addition, it is difficult to reach an agreement regarding a reasonable closing and opening resistance; higher resistance will affect the performance in suppressing the opening overvoltage, whereas lower resistance means consuming a large amount of energy.

Reliability is another problem with an opening resistor. Because of the lack of actual operating data of using an opening resistor in UHV circuit breakers, its reliability can only be analyzed here based on the data of the closing resistor in 500-kV circuit breakers. Closing resistors used to be widely applied for 500-kV circuit breakers in China. However, they have been phased-out since the mid 1990s because of their high failure rate. Even so, from 2001 to 2005, there was, on average, one fault in closing resistors in circuit breakers each year. In 2003, three phases/times of opening resistor explosions occurred in two imported circuit breakers in a substation. For a combined opening/closing resistor, the failure rate is expected to be even higher.

Studies show that most load rejection overvoltages may be reduced to an acceptable level only with the MOA at line ends. Although the overvoltage after clearing a fault, while possibly high, is very unlikely to occur and will not damage the equipment at line ends. Whether it is feasible to use an expensive, vulnerable component to suppress this unlikely overvoltage calls for much deliberation. Therefore, the opening resistor is generally not a recommended option.

4.5 VFTO and Its Suppression

4.5.1 VFTO and Its Impact

VFTO occurs during the switching of circuit breakers, disconnectors, and grounding switches, as well as short-circuit faults in GIS. VFTO causes a threat to GIS and the equipment connected to it, particularly the interturn insulation of a transformer, and may lead to high-frequency oscillation in the transformer. In China, there used to be some VFTO events on 500-kV systems that damaged large transformers.

GIS has high dielectric strength and an even electric field. In GIS, the striking distance is short, discharge develops quickly, and the step wave produced during spark is steep. The back-and-forth refraction and reflection of the voltage wave produced during switching take only a very short time because of short duct and low capacitance of the GIS, and the VFTO oscillation frequency is very high (from 300 kHz to 100 MHz). The front time of overvoltage wave is in the range of 3–100 ns.

In GIS, switching of the disconnector is a major cause of VFTOs. In the switching process, the moving contact moves at a low speed, leading to multiple pre-strikes or re-strikes between the contacts. After a spark, residual voltage on the load side of the disconnector will persist until next spark occurs. This process will repeat until the end of switching. The amplitude of VFTO depends on the difference between the voltages on both sides of the disconnector before spark and the power frequency voltage phase during the spark. Theoretically, the VFTO will peak during the spark process when the voltage on the source side and residual voltage on the load side reach the peak power frequency voltage, but in opposite phases [7–9].

Existing standards have not specified the standard test wave shape for VFTO. It is generally believed that for a GIS, the VFTO withstand level should be slightly higher than lightning impulse withstand voltage (LIWV). Therefore, the insulation used to withstand VFTO is usually designed based on the LIWV of the equipment. As the nominal voltage of the system increases, the ratio of the LIWV to the nominal voltage gradually decreases but the ratio of the VFTO withstand level to the nominal voltage is basically unchanged. This reduces the safety margin between the VFTO withstand level and LIWV. Therefore, a UHV GIS will have more prominent VFTO problems [10].

4.5.2 VFTO Characteristics

1. **Characteristics test**
   Laboratory and field tests could provide characteristics of VFTO similar to those in actual operation. The IEC standards recommend a typical test circuit for the simulation of switching of an operating disconnector where severe VFTO may occur. Japan

develops a laboratory test circuit for the simulation of switching of a 1000-kV disconnector with damping resistor. ABB performs relevant tests as well. In the UHV AC test base in China, two UHV GIS test circuits are developed for the simulation of VFTO characteristics, one using a high-speed GIS and the other using a low-speed GIS, as illustrated in Figure 4.24, in accordance with *IEC 62271-102: High-Voltage Switchgear and Controlgear—Part 102: Alternating Current Disconnectors and Earthing Switches*. In addition, based on engineering needs, each circuit is provided with a branch bus with adjustable length, which makes the circuits valuable for typical engineering applications. They have been used for extensive switching tests of the disconnector to obtain the basic characteristics of VFTO.

Through tests and analysis, the full wave shape characteristics of the VFTO (as illustrated in Figure 4.25), number of sparks, frequency spectrum, residual voltage distribution (as illustrated in Figure 4.26) oscillation factor distribution, and VFTO amplitude characteristics with or without advance DC voltage application are obtained. The suppression effect of the damping resistor on VFTO is shown in Figure 4.27. The test results are the basis for future simulation and insulation coordination studies for UHV GIS.

2. **Simulation of VFTO characteristics**
Compared with the field test, it is easier to study VFTO under various operating conditions by using digital simulation, which is also a primary means for predicting VFTO in substations. The simulation is based on a circuit model, mainly consisting of a

Figure 4.24
UHV GIS VFTO test circuit layout.

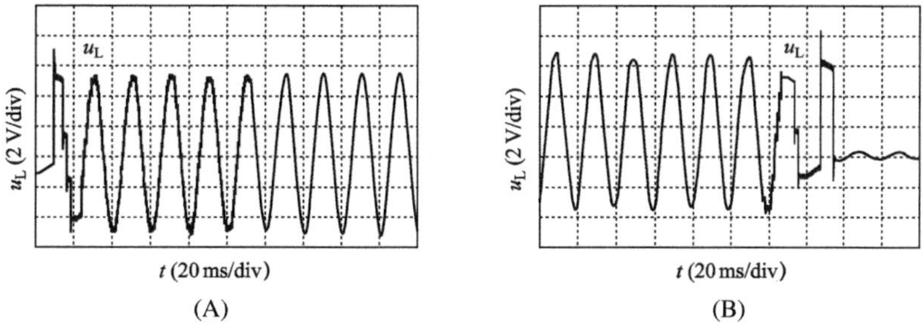

Figure 4.25
VFTO on load side during switching operations of disconnector: (A) closing and (B) opening.

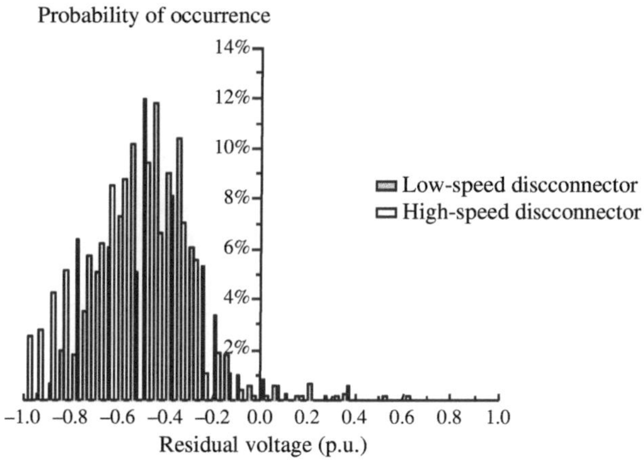

Figure 4.26
Distribution of residual voltage in last spark during opening.

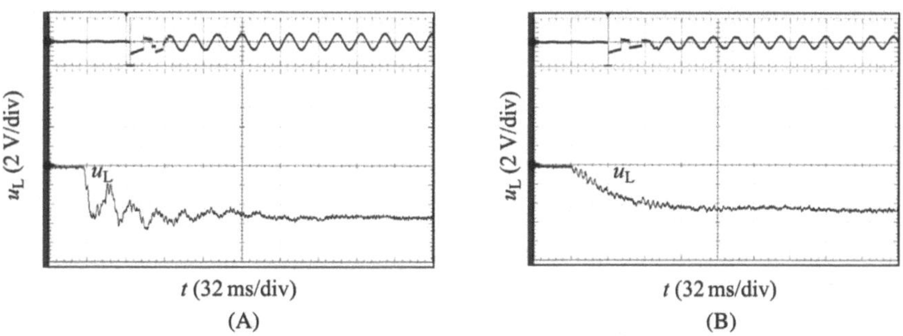

Figure 4.27: VFTO wave shapes for closing of high-speed disconnector.
(A) Without damping resistor and (B) with damping resistor.

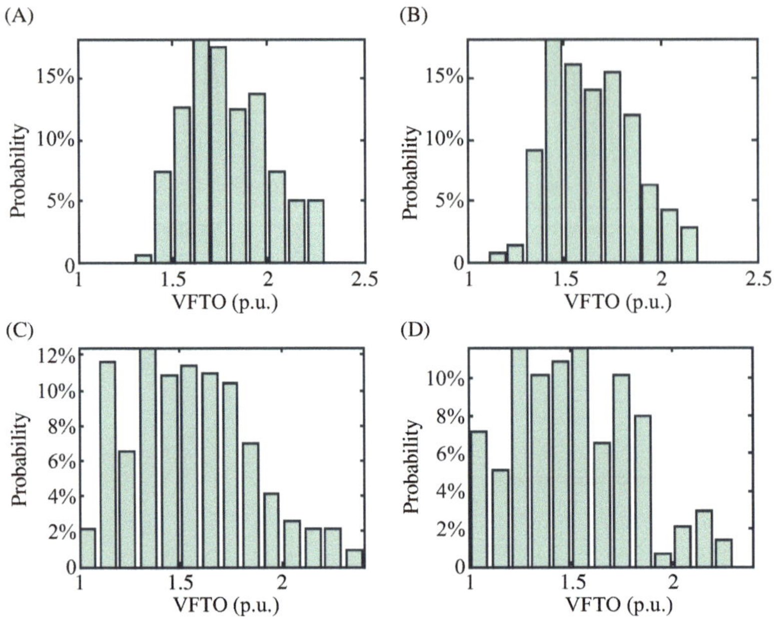

Figure 4.28: Simulation results versus test results for peak VFTO distribution during opening operation of disconnector.

(A) Simulation with low-speed disconnector; (B) test with low-speed disconnector; (C) simulation with high-speed disconnector; and (D) test with high-speed disconnector.

distributed transmission line and lumped elements, such as capacitive element and inductive element. For instance, the elements of GIS can be modeled by the Electromagnetic Transient Program (EMTP) for the computation of VFTO. By comparing the results from test and simulation, the latter is found to be accurate.

In engineering studies, simulation is usually based on a model in which the voltage on the source side and residual voltage on the load side reach the peak power frequency voltage, but in opposite phase during a spark, to obtain the peak VFTO. China has computed the VFTO in substations of the UHV AC pilot project and the Huainan-Shanghai UHV AC transmission project to determine the VFTO withstand level and suppression measures.

The VFTO distribution characteristics for multiple sparks during switching of the disconnector are usually simulated using the gap spark voltage distribution characteristics obtained by measurements and theoretical analysis, simulated dynamic characteristics during switching operations of the disconnector, and conditions required for the ignition and extinguishing of sparks.

By simulating the same circuits as mentioned for the VFTO characteristics test for UHV GIS, the peak VFTO distribution during switching is obtained. Figures 4.28 and 4.29 illustrate the simulation results and test results for comparison.

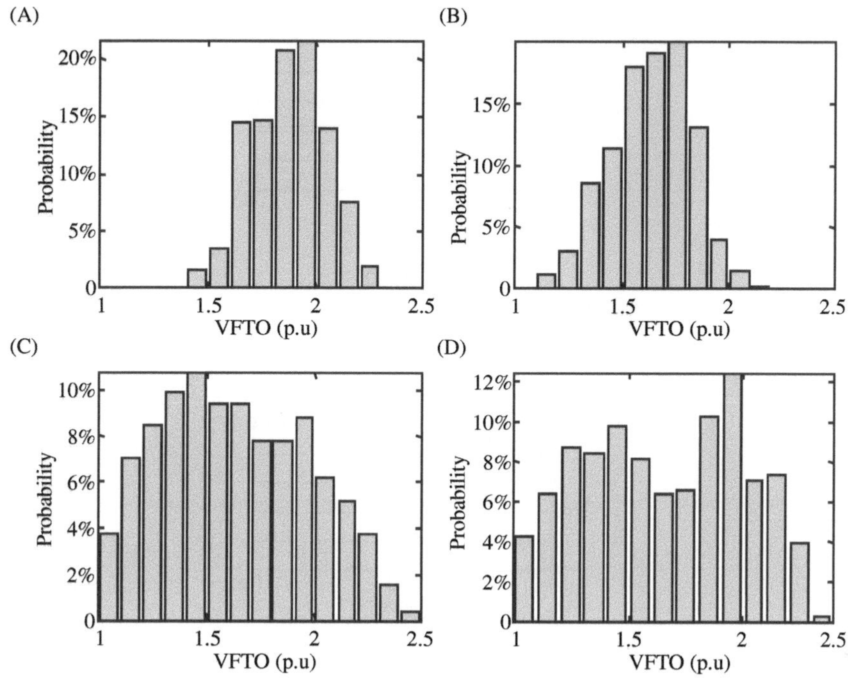

Figure 4.29: Simulation results versus test results for peak VFTO distribution during closing operation of disconnector.
(A) Simulation with low-speed disconnector; (B) test with low-speed disconnector; (C) simulation with high-speed disconnector; and (D) test with high-speed disconnector.

As illustrated, the simulation and test results show approximately the same trend that the VFTO distribution during the switching operations of the high-speed disconnector is scattered, although it is less scattered for the low-speed disconnector.

4.5.3 Suppression of VFTO

VFTOs are primarily suppressed by installing a damping resistor in the disconnector. This damping resistor is series-connected to the discharge circuit during the opening/closing of the disconnector to suppress the electromagnetic oscillation due to repeated gap sparks, thus reducing the amplitude and frequency of VFTO [10,11].

Figure 4.30 depicts the curve of VFTO amplitudes versus damping resistance under typical disconnector operating conditions. These amplitudes are obtained from damping resistors of different resistances. This figure shows that the damping resistor can significantly reduce the VFTO, that is, a higher resistance will lead to a lower VFTO. With a resistance of 100–500 Ω, all VFTOs are less than 1.5 p.u. When the resistance exceeds 500 Ω, VFTOs

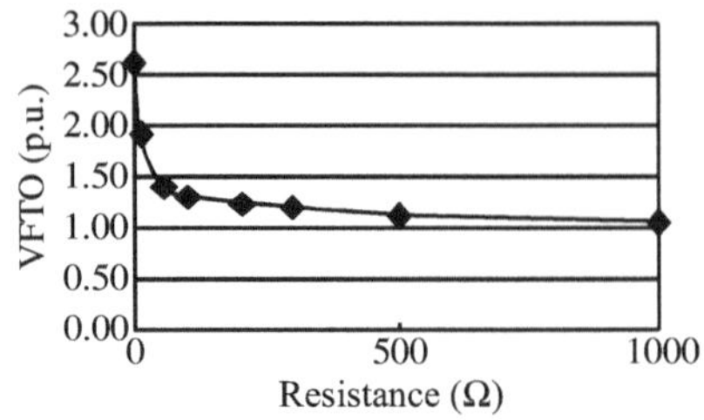

Figure 4.30
Curve of VFTO amplitude versus damping resistance.

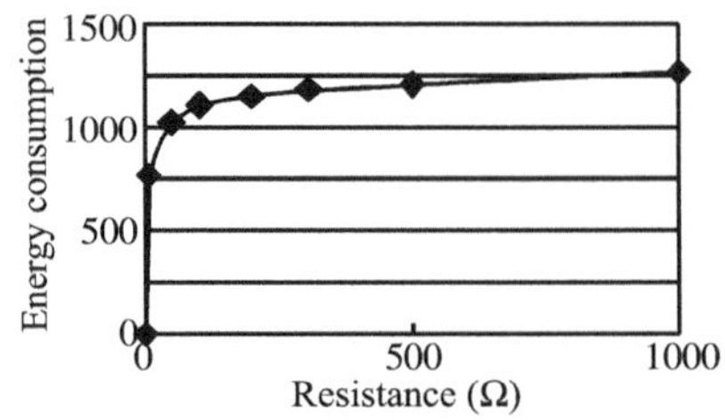

Figure 4.31
Curve of the energy consumption of damping resistance versus damping resistance.

decrease slightly. Figure 4.31 depicts the curve of the energy consumption of the damping resistor versus resistance during a single spark. It can be seen that higher resistance results in higher energy consumption. When the resistance is in the range of 0–100 Ω, the energy consumption sharply increases as the resistance increases. As the resistance increases from 100 to 500 Ω, the energy consumption increases from 1100 to 1200 J. After the resistance exceeds 500 Ω, the energy consumption only slightly increases.

As analyzed, when the damping resistance is in the range of 100–500 Ω, all VFTOs are less than 1.5 p.u., so the insulation coordination requirements are met. After the damping resistance exceeds 500 Ω, VFTOs slightly decrease. Considering this perspective, the resistance should be in the range of 100–500 Ω. Also, higher resistance will cause higher energy consumption, more manufacturing difficulties, and higher cost. Therefore, the damping resistance should be determined considering various factors such as VFTO limits, manufacturing, and cost.

In addition, theoretical and test studies show that switching speed of the disconnector has a major influence on VFTO, that is, slower switching may help reduce VFTO. Therefore, a proper switching speed and disconnector structure may contribute to reduced VFTO [11].

In recent years, other measures, such as MOA and magnetic rings on the GIS bus duct, have attracted wide attention. However, these measures are subject to further tests and simulation studies before being used.

4.6 Internal Overvoltage of DC Transmission System and Its Suppression

4.6.1 Causes

The internal overvoltage of an HVDC transmission system is caused by various switching operations or faults in the AC and DC systems, respectively, on both sides of the converter station, including temporary overvoltage and switching overvoltage. The temporary overvoltage may last from a few to hundreds of power frequency cycles. The switching overvoltage generally maintains its peak for only half of a cycle and then attenuates rapidly. The internal overvoltage of an HVDC system differs from that of an AC system in the forming mechanism, frequency of occurrence, amplitude, and wave shape. In addition to the types of switching operation and fault, the DC system configuration, arrangement and protection level of MOA, time sequence of control and protection actions, and transmitted power will all affect the amplitude, wave shape, and duration of the internal overvoltage. Moreover, overvoltage on the AC side of the converter station will dictate the overvoltage protection level and insulation level of the AC side. Also, it may be propagated to the converter valve and the MOA of the valve by the converter transformer based on the turns ratio, and it produces overvoltage on the DC side through the conducting valve, thus affecting the insulation coordination on the DC side.

For a UHV DC transmission system in which each pole has multiple series-connected 12-pulse converters, the internal overvoltage may be a primary concern influencing the equipment's insulation level. Similar to AC systems, internal overvoltages of a DC system are also classified into temporary overvoltage, switching overvoltage, and steep-front overvoltage, based on their duration.

1. **Temporary overvoltage**
   Temporary overvoltage refers to overvoltages lasting from a few to hundreds of cycles. It directly acts on equipment, particularly the MOAs, placing a higher requirement on their energy absorption capacity. In addition, the temporary overvoltage may initiate further faults, therefore causing the switching overvoltage to increase.
   a. Temporary overvoltage on the AC side of converter station. The most typical temporary overvoltage occurs on the AC bus of converter stations, which directly influences the MOA on the AC bus and will be propagated to the valve side by the converter transformer, therefore affecting the valve's MOA. Temporary overvoltages on the AC bus are mainly caused by load rejection and switching on transformers.
   b. Temporary overvoltage on the DC side of converter station. This type of overvoltage mainly includes temporary overvoltage propagated from the AC side and overvoltage as a result of faults on converters. The temporary overvoltages occurring on the AC bus of converter stations due to various reasons will be

propagated to the DC side by the converter transformer, causing accumulation of a large amount of energy on the valve's MOA. Faults such as loss of control pulse and commutation failures in converters may result in the AC fundamental voltage intruding the DC side.

2. **Switching overvoltage**

   Switching overvoltage refers to overvoltages with a front time of 30–1000 μs or more. The switching overvoltages in DC systems are different from those in an AC system in many aspects, such as the causes, mechanism of development, amplitude, and wave shape.

   a. Switching overvoltage on the AC side of converter station. This overvoltage occurs due to switching operations and faults on the AC side, such as closing and reclosing after a line-to-ground fault, switching on and re-switching on AC filters/shunt capacitors, and disconnection of all or most of the AC lines from an inverter station. It may affect the insulation level of the equipment connected to the AC bus and the energy absorbed by the AC-side MOAs. Also, it can be propagated to the converter valves by the converter transformer, causing faults within the valves.
   b. Switching overvoltage on the DC side of converter station. This overvoltage mainly includes the switching overvoltages transmitted from the AC side and overvoltages as a result of faults on converters. Such overvoltages on the AC side may propagate to the converter by the converter transformer. However, they will not impose overlarge stresses on the DC equipment because of the protection provided by MOAs on the AC bus. When a short circuit occurs in the converter, the consequent discharge of DC filter capacitors and inrush of alternating current will usually produce switching overvoltages on the converter itself and the neutral bus of the DC system.
   c. Switching overvoltage on DC transmission lines. Given the static and electromagnetic interactions between the bipolar transmission line, the short circuit between one transmission line pole and the ground under bipolar operation will produce switching overvoltages on the healthy pole, which will propagate along the transmission lines to the DC switchyard at the converter station. This switching overvoltage will influence the design of the tower for DC transmission lines as well as the overvoltage protection and insulation coordination of the DC switchyard at both converter stations. The overvoltage amplitude depends on the transmission line parameters and the impedance at both ends.

3. **Steep-front overvoltage**

   The steep-front overvoltage refers to overvoltage waves with a voltage ramp rate of no smaller than 1200 kV/μs. When any point between the valve-side outlet of the converter transformer at high potential and the converter valve is short-circuited to the ground, the pole-to-ground voltage generated by the converter's stray capacitance will be directly applied on the converter valves, producing steep-front overvoltage there.

Also, when all valves of one converter in the two or more series-connected converters are turned-on or a bypass pair is inadvertently activated, the remaining non-conducting converters will withstand all the pole-to-ground voltage, resulting in steep-front overvoltage.

4.6.2 Suppression Measures

Like all other power systems, an HVDC transmission system also requires overvoltage protection devices such as MOAs to suppress overvoltages and protect the equipment, thus increasing the system reliability and lowering the equipment cost.

1. **Installing MOAs in converter stations**
   MOAs installed in the converter station can suppress overvoltages acting on the electrical equipment. Because the rated operating voltage of a UHV DC system is much higher than that of an EH VDC system, the insulation level of electrical equipment in the UHV DC system should be higher than that in EH VDC system. Like an EH VDC system, the UHV DC converter station also uses MOA for overvoltage protection; therefore, MOA's configuration and parameters play an important role in the insulation coordination and construction cost of a UHV DC project.

   The MOAs should be arranged following the principle that overvoltages on the AC and DC sides, respectively, are suppressed by MOAs installed on each side while the key equipment is protected from overvoltages by MOAs arranged in close proximity to it. Generally, the converter transformer's valve side windings are protected by connecting several types of MOAs used for other equipment in series. The valve side windings of the converter transformer at the highest potential can be directly protected by MOAs installed immediately closest to them.

   MOAs installed in a UHV DC converter station have diversified features and a complicated configuration. Their arrangement and location are determined based on the layout of the main electrical equipment and the overvoltage protection requirements. They may be classified as AC arrester, valve arrester, converter unit arrester, DC bus arrester, DC line arrester, neutral bus arrester, and DC filter arrester according to their mounting positions and their respective protection targets. The MOA configuration should be as simple as possible while ensuring equipment security.

   Considering the difficulties in manufacturing and transportation of UHV DC converter transformers, China's operating ±800-kV Xiangjiaba–Shanghai, Jinping–Sunan, and Yunnan–Guangdong links all have a configuration in which each monopole has two series-connected 12-pulse converter units at the same voltage rating at each end. Figure 4.32 shows the schematic of a typical arrangement of MOAs of a single pole. Table 4.10 shows the designation and symbol of MOAs. Of course, the configuration of multiple series-connected converter units of a monopole may vary with the requirements and conditions of different DC links. Table 4.11 gives several

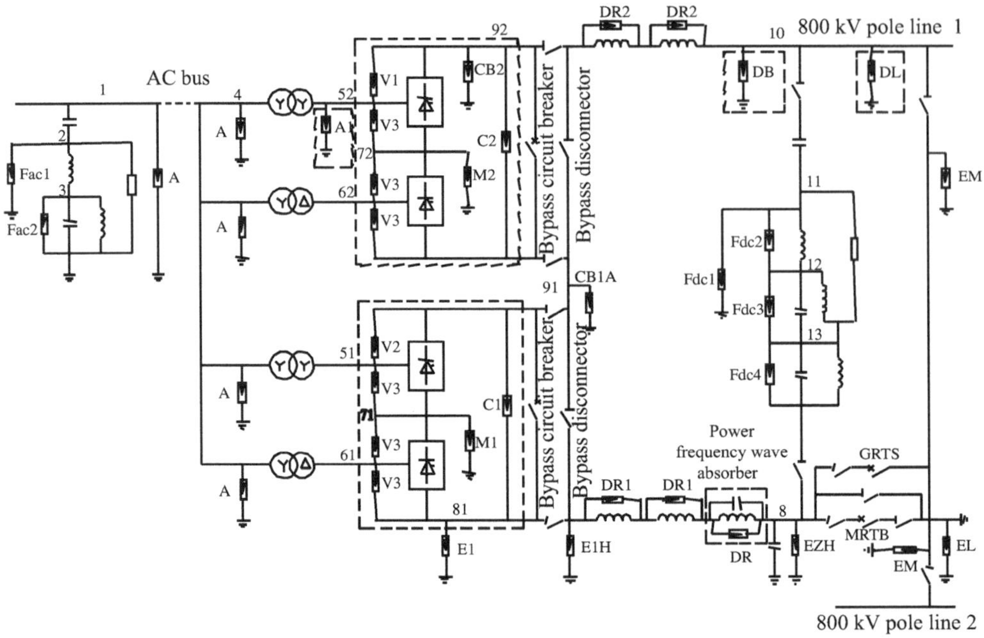

Figure 4.32
Configuration of MOAs of a monopole having two series-connected 12-pulse converter units in a UHV DC converter station.

Table 4.10: Designations and Symbols of MOAs in a UHV DC Converter Station.

Designation	Symbol	Designation	Symbol
MOA for AC bus	A	MOA for DC bus	DB
MOA on valve side of converter transformer	A1	MOA for DC line	DL
MOA for AC filter	Fac1, Fac2	MOA for smoothing reactor (power frequency absorber)	DR1, DR2 (DR)
MOA for valves	V1, V2, V3	MOA for neutral bus	E1, E2
MOA for six-pulse bridge	M1, M2	High energy-absorption MOA for neutral bus	E1H, E2H
MOA for 12-pulse bridge	CB1A	MOA for earth electrode line	EL
MOA between 12-pulse bridges	C1, C2	MOA for metallic return	EM
MOA for converter unit DC bus	CB2	MOA for DC filters	Fdc1, Fdc2, Fdc3, Fdc4

configurations of series-connected converter units that were once considered during the feasibility study of UHV DC projects at different voltage classes. Accordingly, the installation mode and parameters of smoothing reactors and associated MOAs also vary due to different series connection types of the converter units. In addition, different design philosophies of different projects and contractors result in different arrangements

Table 4.11: Different Configurations of Series-Connected Converter Units for UHV DC Projects.

Voltage Class (kV)	Quantity of Series-Connected Converter Units of a Monopole	Configuration of Multiple Series-Connected 12-Pulse Converter Units		
±800	Two 12-pulse converter units	400 kV + 400 kV	500 kV + 300 kV	600 kV + 200 kV
±1000	Two 12-pulse converter units	500 kV + 500 kV	–	–
	Three 12-pulse converter units	400 kV + 400 kV + 200 kV	333 kV + 333 kV + 333 kV	–
±1100	Two 12-pulse converter units	550 kV + 550 kV	–	–

Note: 400, 500, 300, 600, 333, 200, and 550 kV are the voltage ratings of 12-pulse converter units.

of MOAs in UHV DC converter stations, mainly reflected in the selection of MOAs C1 and C2 between 12-pulse bridges and the MOA A1 installed on the valve side of the high-voltage end Y/Y converter transformer. For instance, in the ±800-kV Xiangjiaba–Shanghai transmission line project and the ±800-kV Jinping–Sunan project, both were not equipped with MOAs C1, C2, and A1, whereas in the ±800-kV Yunnan–Guangdong project MOAs C1, C2, and A1 are installed, but without MOA CB2 for the DC bus.

2. **Thyristors with forward direction-protective firing function**
   The thyristor valve overvoltage protection is composed of the forward direction-protective firing system of thyristors and MOAs. When the valves are subject to overvoltages with a fast or steep front beyond the protection level of the MOAs, some of the thyristors series-connected to the valve will suffer breakdown due to the nonlinear voltage distribution. Adding a protective firing system can protect the thyristors from being damaged as a result of forward direction overvoltage.
   The protection level of the protective firing system should be selected following the principle that clearing a fault in the AC system will not trigger the protective firing system for inverter valves; therefore, they should be determined based on the protection level of the valve's MOA during faults.
3. **Installing closing/opening resistor or phase selector on the circuit breaker**
   The switching overvoltages in a UHV DC system may also be suppressed by installing a closing/opening resistor or phase selector on the circuit breaker. The circuit breaker of the converter transformer is normally equipped with a closing resistor to limit the inrush current on its closing and to prevent harmonic resonance overvoltages in the AC system, thereby avoiding overload on internal components on the low-voltage side due to the harmonic current injecting into the AC filters. The circuit breakers of the AC filter banks and capacitor banks are equipped with a phase selector for limiting the inrush current on closing to reduce system disturbances caused by switching operations.
4. **Installing damping devices**
   A damping circuit may be used to reduce the amplitude of oscillating overvoltages in DC systems. For instance, the oscillation frequency and amplitude of oscillating

overvoltages occurring on the turn-off of valves can be reduced by connecting a damping circuit in parallel. Because of its long DC transmission lines, a UHV DC system has a resonant frequency approximating power frequency or two-times the power frequency, and the resonance overvoltage can be suppressed by installing absorbers.

5. **Suppression measures for overvoltages in control system**

   Some measures for suppressing overvoltages in the control system are as follows:

   a. Using the fast reactive power control incorporated in converter station can suppress the temporary overvoltages on the AC bus. The control strategy is to restrict the duration of temporary overvoltages by switching-on the least filters, thus facilitating the recovery of both the AC and DC systems after faults and ensuring the quick recovery of the DC-transmitted power after faults on the equipment in converter stations.
   b. Providing positive damping to suppress the resonance overvoltage on the DC transmission system via the control system.
   c. After putting the upper or lower 12-pulse converter unit into operation, as shown in Figure 4.32, in the rectifier station or inverter station, the bypass circuit breaker connected in parallel to its two terminals needs to be opened. Because the bypass circuit breaker cannot interrupt the DC current, its opening should be coordinated with the control program.
   d. Restricting a firing angle can be accomplished by using converter blocking or bypass pair to prevent the overvoltages produced on the outage of the DC system.

4.6.3 Internal Overvoltage Suppression Effects in DC Transmission System

1. **Overvoltage caused by AC side**

   The switching overvoltage and temporary overvoltage on the AC side of the converter station is transferred to the valve side through the converter transformer based on the turns ratio and produces overvoltage on the DC side through the conducting converter valve. Such overvoltage can be effectively suppressed by MOA on the DC side.

   Underrated bipolar operating conditions, a simulation was conducted for the overvoltage and MOA's performance during a three-phase fault on rectifier station's 500-kV AC line in the Xiangjiaba–Shanghai UHV DC project. The simulation result shows: (i) the maximum overvoltages on the DC transmission line pole and the valve top at the rectifier station are 1274 and 1138 kV, respectively, with smoothing reactors split on the line pole and the neutral bus; (ii) MOA DL on the DC line side, MOA E1H at the bottom of valve on the rectifier side, and MOA E1 and E1H at the bottom of valve on the inverter side all operate and their performances are shown in Figure 4.33; and (iii) the maximum current and energy consumption of MOA for the valve on the

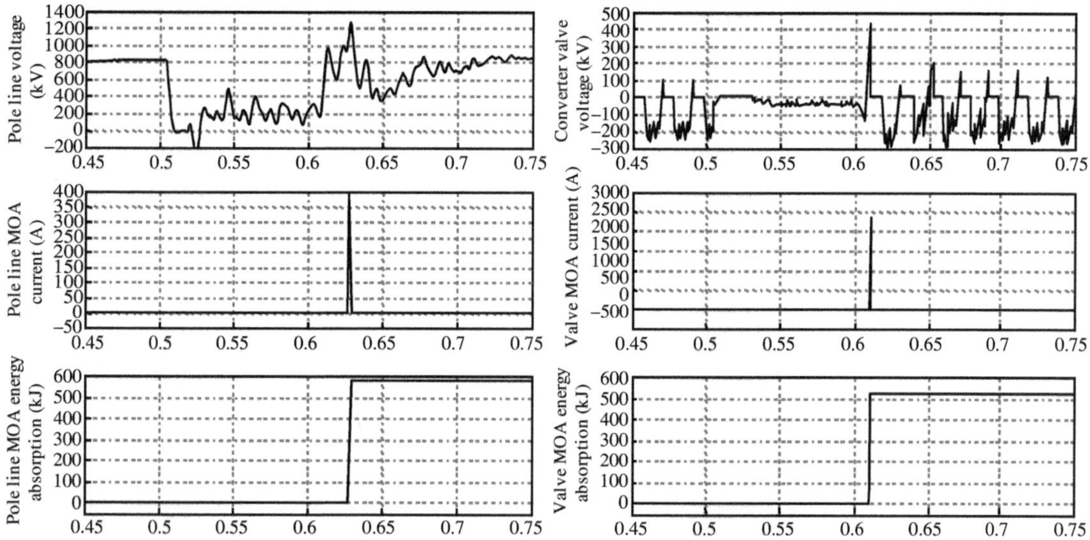

Figure 4.33

Calculated wave shapes for voltage and current and energy consumption of MOA on line pole and valve on the Xiangjiaba side in the case of AC three-phase ground fault of rectifier station.

Table 4.12: Overvoltage Level During Switching of Filter HP24/36 on the Rectifier Side.

System Conditions	Bus Voltage Before Switching (kV)	Measured Location			Overvoltage (Peak Value, kV) Random Closing/Phase Selection Closing
Switching of the first filter bank	544	AC bus (1)			692 (1.54 p.u.)/ 458 (1.02 p.u.)
		Component	Across the component	C1 (1–2)	720/461
				L1 (2–3)	330/9
				C2 (3)	35/2
			To ground	L1H (2)	330/10

rectifier side are 2374 A (peak value) and 520 kJ, respectively, effectively reducing the overvoltage level on the DC side.

The switching of AC filters or capacitors will produce overvoltage on the 500-kV AC bus, filters, or capacitors at the converter station.

Table 4.12 compares the overvoltage level and MOA conditions between switching in filter or capacitor banks by random closing and phase selection closing [12]. In the case of random closing, the maximum switching overvoltage on the 500-kV bus is 1.54 p.u. In the case of phase selection closing, the resultant impacts and overvoltage on individual components of filters or capacitor banks and the disturbance to the system will be reduced. In this case, the maximum switching overvoltage on the 500-kV bus is only 1.02 p.u.

2. **Emergency shutdown of DC system**
   When a fault occurs in an operating DC system, its protection devices will act and make the system stop operation. This is called an emergency shutdown and it includes the following two scenarios:
   a. *Emergency shutdown activated by the rectifier station*: The rectifier station blocks the converter after restricting the firing angle by more than 160°, whereas the inverter station blocks the converter by relevant protection. During such an emergency shutdown, overvoltage will not occur on the line pole but rather on the DC filter circuit.
   b. *Emergency shutdown activated by the inverter station*: The inverter station first switches-on the bypass pair and then blocks the converter; after a delay, the rectifier station blocks the converter by relevant protection after restricting the firing angle. Because the switching of bypass pair is equivalent to a short circuit at the DC transmission line end, the DC filter will experience a high-amplitude overvoltage. If the converter blocking at the inverter station precedes the rectifier station, the AC/DC side equipment will also experience an overvoltage. The overvoltage at individual places of the converter station is suppressed by relevant MOAs.
3. **Commutation failure (loss of control pulse)**
   The failure of converter control or valve conduction, such as conduction failure, commutation failure, or complete loss of control pulse for a valve, will produce additional components (mainly of power frequency) on the DC side and overvoltage on the DC bus, DC filters, and neutral bus. The asymmetrical faults in the AC grid near the converter station will produce additional components (mainly of second harmonic) on the DC side. When the natural frequency of oscillation on the DC side approaches the power frequency, second harmonic, or third harmonic, the aforementioned faults may result in a high-amplitude overvoltage or overcurrent under the amplification effects. Therefore, in the design stage, DC main circuit parameters should be properly selected to avoid the presence of resonance conditions; otherwise, other measures must be taken. For example, the natural frequency of oscillation on the DC side of Xiangjiaba–Shanghai project is close to the power frequency and second harmonic frequency. To avoid resonance, triple-tuned DC filters with a tuning frequency of 100/600/2000 Hz are used, and a power frequency absorber is installed on the neutral bus (Figure 4.32). According to the site test, in cases when the converter suffers the loss of control pulse, or when the AC line of converter station suffers a single-phase instantaneous ground fault, the system experiences no resonance, indicating that these measures are effective in resonance suppression.

   If a fault on the converter control system or other reasons result in the complete loss of valve control pulses in the inverter, neither the switching of bypass pair nor the commutation can be conducted, and the conducting valves before the fault will maintain the conduction. In this case, a high-amplitude overvoltage may occur on the DC bus before the bypass breaker is closed (Figure 4.32) or the incoming line circuit breaker of converter transformer is tripped. The overvoltage amplitude and duration may be reduced by blocking the rectifier as soon as possible through inter-station communication. In addition, the complete loss of control pulse in an inverter may cause current interruption only in relevant valves, resulting in serious overvoltage across the valve. Such overvoltage is suppressed by relevant MOAs and valve-protective firing.

4. **Switchover between DC connection modes**
The switchover between DC earth return mode and metallic return mode is common during monopolar operation. This is performed by the metallic return transfer breaker (MRTB) or ground return transfer switch (GRTS) in the DC circuit breaker, which is only installed on the neutral bus of the rectifier station (Figure 4.32). During such switchover, the DC system maintains normal on-load operation. To switch between the two connection modes, the ground and metallic return circuits are first put into parallel operation. Because of its small impedance, the ground return circuit carries approximately 90% of the total current. In case of a switchover from ground return operation to metallic return operation, the MRTB connected to the ground return circuit in series is opened to interrupt the DC current in the ground return circuit, or the GRTS connected to the metallic return circuit in series is opened to interrupt the DC current in the metallic return circuit. Because partial current is forced to change the path during the switching operation, overvoltage occurs across the transfer switch, the amplitude of which is proportional to the switching current. Thus, for the switchover from ground return operation to metallic return operation, the overvoltage across the MRTB is more serious. In addition, when the grounding of neutral bus at the rectifier station is changed from through the earth electrode at the rectifier station to through the earth electrode at the opposite station, overvoltage will occur on the neutral bus, DC line pole, and DC equipment, the amplitude of which is proportional to the switching current (varies slightly with the voltage class of the DC system). During such switchover, the MOVs parallel-connected to both ends of the circuit breakers are used to suppress the overvoltage across the MRTB or GRTS and to absorb the corresponding energy (for a large amount of energy, more parallel MOVs are required), whereas the overvoltage at other places at the converter station is suppressed by relevant MOAs.

Taking the Xiangjiaba–Shanghai UHV DC system as an example, because the neutral bus at the rectifier station is changed from grounded to ungrounded during the switchover from ground return to metallic return, a high-amplitude overvoltage will appear on the DC line pole, neutral bus, and DC filters at Fulong Converter Station (the rectifier station). Figure 4.34 shows the results of the site test. From the top to the bottom, respectively, are the voltage wave shapes for the DC filter reactors L1, L2, and

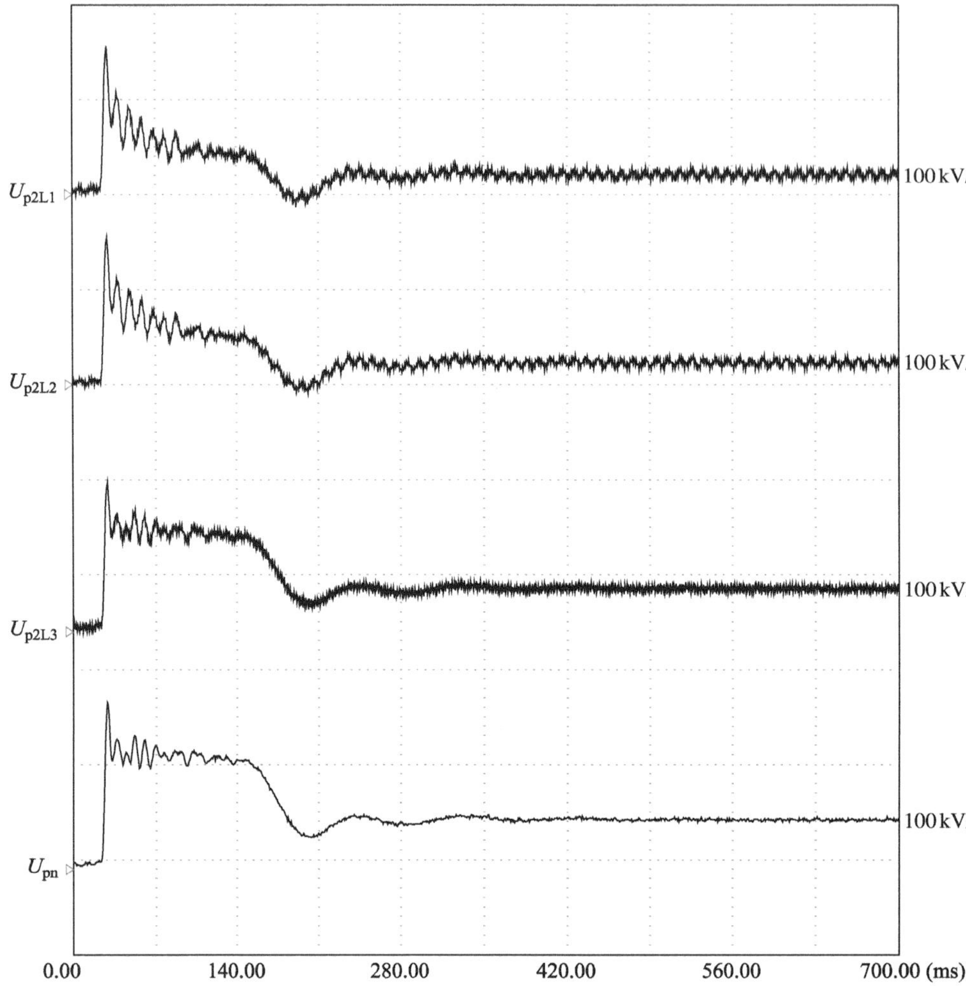

Figure 4.34
Wave shapes of terminal-to-ground overvoltage on DC equipment at Fulong Converter Station caused by switchover from ground return to metallic return (forward transmission, pole 2 HV converter unit in service).

L3 and the neutral bus of pole 2. The maximum overvoltage on the neutral bus at Fulong Converter Station caused by such switchover is 178.0 kV. The measured overvoltages on DC equipment at both stations are lower than the impulse insulation level of the equipment.

5. **Live switching of DC filter**
During the operation of the DC system, live switching of DC filters is sometimes required. Such switching will cause overvoltage on the DC line pole and the DC filter being switched, as well as the other DC filters in service. In addition, under bipolar

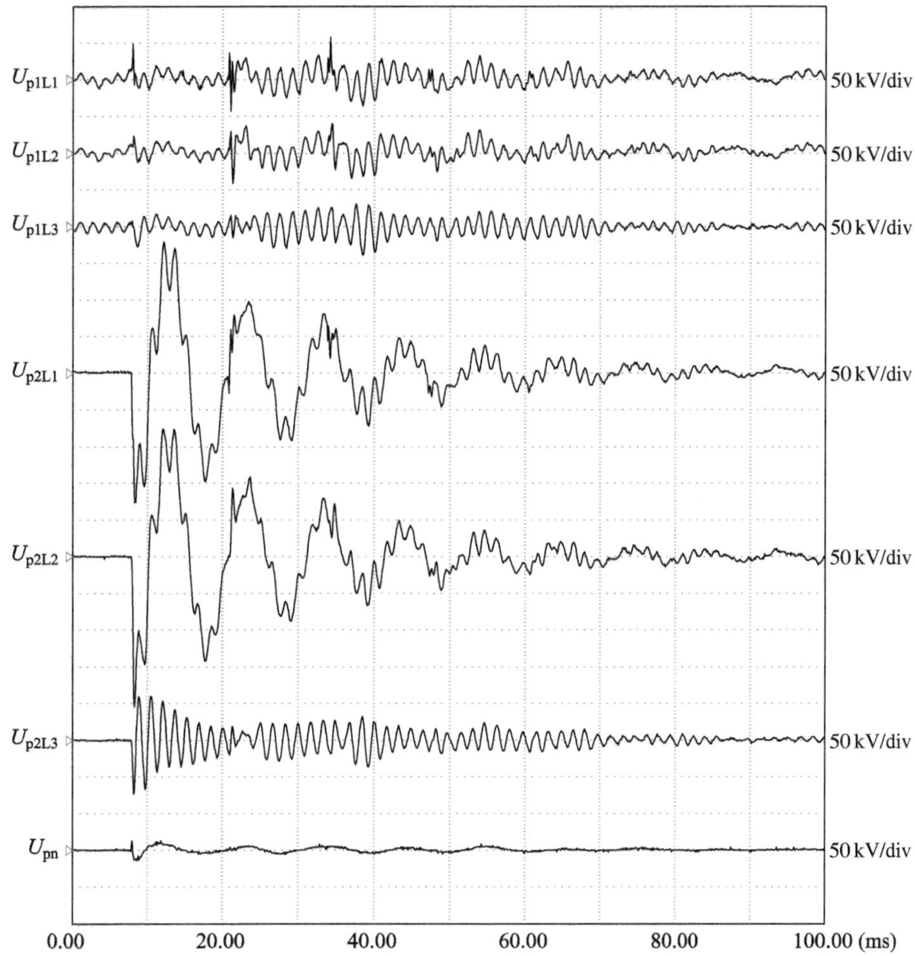

Figure 4.35
Wave shapes of terminal-to-ground overvoltage on DC equipment caused by live switching of DC filter on pole 2 at Fulong Converter Station (±800 kV, forward power transmission).

operation, the live switching of DC filter on one pole will cause overvoltage on the DC equipment of the other pole. These overvoltages are suppressed by relevant MOAs.

Figure 4.35 shows the typical overvoltage wave shapes caused by the live switching of DC filter at Fulong Converter Station Xiangjiaba–Shanghai project under bipolar operation. From the top to the bottom of the figure, respectively, are the voltage wave shapes for the DC filter reactors L1, L2, and L3 of pole 1 and the DC filter reactors L1, L2, and L3 as well as the neutral bus of pole 2. The measured maximum terminal-to-ground overvoltages of the DC line pole, DC filter reactors L1, L2, and L3 as well as the neutral bus at Fulong Converter Station are 911, 179.2, 205.8, 73.7, and 15.0 kV, respectively. Such values of DC filter reactors L1, L2, and L3 and the neutral bus at

Fengxian Converter Station (the inverter station) are 139.6, 91.4, 44.8, and 12.1 kV, respectively, lower than the impulse insulation level of the equipment.

6. **Short-circuit fault on converter**

If a short-circuit fault occurs within the converter, then the DC filter discharge and AC current inrush will generally result in overvoltage on the converter itself and the DC equipment at the neutral bus. Such grounding and short-circuit faults include: valve top-to-ground fault, valve top-to-lead (on the valve side of converter transformer) short-circuit fault; valve top-to-neutral bus short-circuit fault; and single phase-to-ground short-circuit fault on the valve side of converter transformer. The most typical short-circuit fault is the short-circuit to ground between the valve side outlet of converter transformer and the converter valve. The aforementioned ground faults and short-circuits would produce steep-front overvoltages similar to lightning impulse overvoltage. Generally, short-circuit is not expected to occur between terminals of converter. However, switching-on bypass pair or closing bypass circuit breaker is similar to a short-circuit between terminals. For multiple 12-pulse converter units connected in series, such an operation would produce overvoltage on other converter units.

Taking the valve top-to-ground fault on the rectifier side as an example, when the fault occurs, the DC current on the neutral bus is larger than that on the DC transmission line. Generally, the valve differential protection operates to block the rectifier station and issues blocking signals to the inverter side. The calculation results of the maximum transient overvoltage at various places on the AC and DC sides in the converter station indicate that the overvoltage across the smoothing reactor of the DC line pole is 785 kV, whereas the overvoltage at other places is not so high, as shown in Table 4.13 [12]. Under such a fault, only the MOAs on the neutral buses at the rectifier station and inverter station operate to reduce the overvoltage level.

Figure 4.36 shows the calculation results of the overvoltage across the smoothing reactor on the DC line pole (left) and neutral bus (right) at the rectifier station during the valve top-to-ground fault on the rectifier side. The two columns (from top to bottom) give the voltage wave shapes across the corresponding smoothing reactor of pole 1 and pole 2.

Table 4.13: Energy Consumption and Current of MOA During Valve Top-to-Ground Fault at Rectifier Station.

MOA Name	Symbol	Rectifier Side/Inverter Side Energy Consumption (kJ)	Rectifier Side/Inverter Side Current (A)
Neutral bus MOA	E1	38/24	46/56
	E1H	372/43	652/167
	EL	−/37	−/238

Note: "−" means that MOA does not operate.

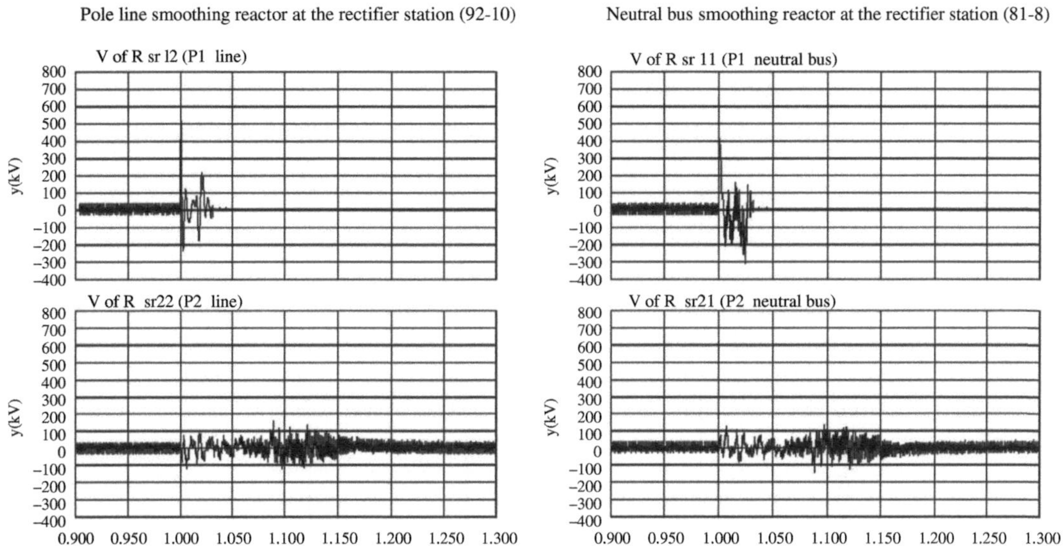

Figure 4.36
Voltage wave shape across the smoothing reactor on the DC line pole and neutral bus at the rectifier station during valve top-to-ground fault on the rectifier side.

7. **DC transmission line ground fault**
   If a pole-to-ground fault occurs outside the smoothing reactor at the converter station or on the DC transmission line, then the faulty pole (mainly the equipment on the neutral bus and the DC filters and reactors) would be subject to overvoltage, and under bipolar operation the healthy pole will also experience an induced overvoltage.
   a. *Overvoltage distribution along the DC transmission line.* A number of engineering studies show that when a ground fault occurs along the DC line, the amplitude of the overvoltage on the line is related to many factors such as the location of the fault, the length and parameters, and terminal impedance of the line. Generally, the overvoltage level in the middle of the line is the highest, whereas the overvoltage near the converter station depends on the characteristics of MOAs on the DC line and thus is relatively low. For the ±800-V DC transmission line 2071 km in length divided into 20 sections, if, underrated bipolar operation, a ground fault occurs at different places along the line, then the overvoltage profile along the line, as shown in Figure 4.37 [12].
   b. *Overvoltage at the midpoint of the line.* Studies show that if a ground fault occurs at the midpoint of DC line, the maximum overvoltage at the midpoint of the line on the healthy pole is generally as high as 1.6–1.7 p.u., and the amplitude of overvoltage is reduced with the increase of ground resistance of the transmission tower. Figure 4.38 shows the calculated overvoltage wave shape at the midpoint of the line on the healthy pole when a flashover to ground occurs on the pole at the

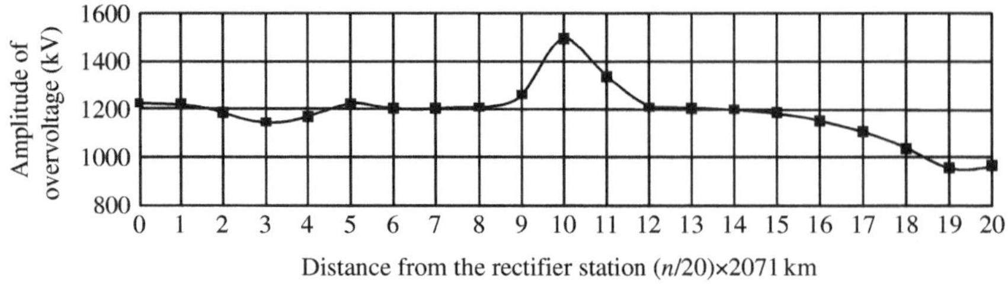

Figure 4.37
Overvoltage profile when a ground fault occurs at different places along the DC line.

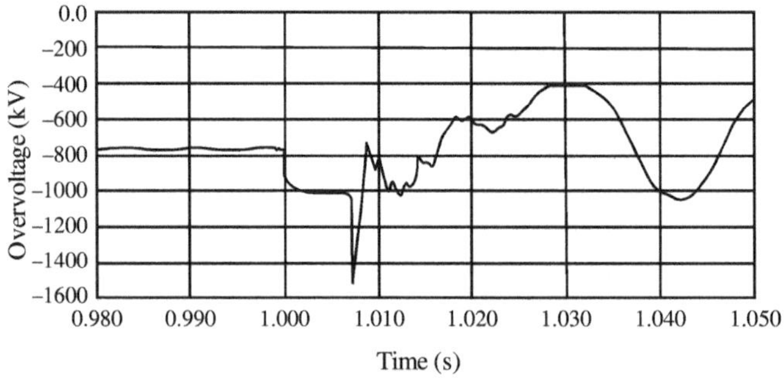

Figure 4.38
Overvoltage wave shape at the midpoint of the line on a healthy pole when a ground fault occurs at the midpoint of the DC line.

midpoint of the ±800-kV DC line. It follows that the normal operating voltage is 786 kV before fault, and that there are two voltage steps after fault, with the first being caused by the instant electromagnetic coupling between poles on occurrence of the ground fault, and with the second being caused by the superposition of the transient electromagnetic waves propagated to the faulty point after refraction and reflection at the converter stations; the transient component was generated by the discharge of the DC filter at the end of converter stations. The front time of the second step is approximately 250 μs. Based on this analysis, the transient disturbance caused by ground fault can be detected by the DC control and protection device only after it is propagated to the rectifier station and subjected to refraction and reflection. Therefore, the characteristics of DC system control have nothing to do with the overvoltage level and therefore cannot suppress its amplitude.

Studies show that reducing the capacity of the DC filter's HV capacitor can reduce the amplitude of overvoltage at the midpoint of line, and that the capacity is closely related to the overvoltage distribution along the DC line. The DC filter's HV capacitor affects the overvoltage in the following respects: the change of parameters and structure of DC filter causes variation in the end impedance of converter station and, further, the amplitude of refraction and reflection components of electromagnetic wave propagated from the faulty point to the converter station; and, for the same voltage, a smaller capacity of the HV capacitor results in less energy storage and smaller and shorter peak discharge current, thereby reducing the electromagnetic transient components on the healthy pole caused by coupling. Therefore, reducing the capacity of the HV capacitor through optimizing the parameters of DC filters while meeting the requirements of harmonic filtering can reduce the overvoltage at the line midpoint and further the required insulation level. Considering the non-uniform distribution of actual line parameters and other factors, the location of maximum overvoltage may deviate from the midpoint.

8. **Online switching of 12-pulse converter unit**

   To improve the reliability and operational flexibility of the power system, UHV DC transmission projects are provided with the online switching of a single 12-pulse converter unit and the following optimized control strategies to reduce the overvoltage risk.

   1. *Switching-on 12-pulse converter unit.* The converter unit is switched-on, first on the rectifier side and then on the inverter side. After the associated bypass circuit breakers are opened, the control system will adjust the firing angle to increase the DC voltage to the set value. Field measurements show that, in practice, switching-on a 12-pulse converter unit online has a small influence on the operating converter units on both the pole of interest and the other pole. No overcurrent occurs and the voltage on the healthy pole varies slightly.
   2. *Switching-off a 12-pulse converter unit.* Under the monopolar with two converter units connected in series, switching-off one converter unit online is conducted in the following two ways:
      a. *Switching-off by manual operation.* The converter unit is manually switched-off, first on the inverter side and then on the rectifier side. When the voltage across the converter unit to be switched-off is close to zero, the bypass pair is put into operation and at the same time a closing command is issued to the bypass circuit breaker. Once the bypass circuit breaker is closed, the converter unit of interest is blocked immediately. At this time, the DC voltage on the associated pole is reduced to half of the voltage under normal operation. If the DC system operates in the constant current control mode, then the DC current on the associated pole remains unchanged and that on the healthy pole is not affected. In the joint control mode, the control system will automatically coordinate the switching-off process of the converter unit between two converter stations. Such a process causes a small disturbance to the entire DC system.
      b. *Switching-off by protection.* After a fault on one converter unit, the faulty converter unit is switched-off by the corresponding protection. If the inter-station

communication is functional, then after receiving the protection actuation signals the opposite station will switch-off the corresponding converter unit through sequence control. If the communication fails, then the opposite station first switches off the LV converter unit through unbalance control of the converter. If the converter unit at the rectifier station is first switched-off because of fault, then the current through the DC line is interrupted temporarily, until the corresponding converter unit at the inverter station is switched-off. If the converter unit at the inverter station is first switched-off because of fault, then the current through the DC line will increase sharply, which will be reduced by the shifting phase of the firing angle of the converter on the rectifier side, and the current will recover to the normal value until the corresponding converter unit at the rectifier station is switched-off. Field test shows that the amplitude of overvoltage and overcurrent caused by switching-off the converter unit by relevant protection is lower than that caused by the ground fault on the DC line.

References

[1] Z. Weobo, G. Yuming, Power Systems Overvoltage and Insulation Coordination, Tsinghua University Press, Beijing, 1988.

[2] L. Jiming, W. Xiaogang, B. Liangeng, X. Zutao, Resonance overvoltage caused by energizing no-load transformers in 1000 kV AC transmission system, Power Syst. Technol. 31(2) (2007) 4–10.

[3] C. Rongjiang, G. Nihong, S. Yong, Test study on self-extinction behavior of secondary arc on power system, Proc. CSEE 16(2) (1996) 73–78.

[4] H. Yanmiao, S. Gao, C. Rongjiang, Test research of secondary arc in 1000 kV UHV double-circuit transmission lines, Proc. CSEE 31(16) (2011) 138–143.

[5] Q. Jiali, et al., Discussion on switching overvoltage in 1000 kV power networks and phase controlled circuit breaker, Power Syst. Technol. 29(10) (2005) 1–4.

[6] L. Jiming, C. Weijiang, H. Bin, et al., Failure rate calculation and necessity discussion on opening-resistor of ultra high voltage circuit breakers, Proc. CSEE 32(7) (2012) 161–166.

[7] Z. Zhaohua, L. Jiming, B. Liangeng, et al., Very fast transient overvoltages (VFTO) and a study of a practical case, Power Syst. Technol. 20(3) (1996) 10–14.

[8] H. Shanqing, Fast transient phenomena in GIS, High Voltage Apparatus 1(2) (1995) 41–46.

[9] Y. Yamagata, K. Tanaka, S. Nishiwaki, I. Miwa, N. Takahashi, T. Kokumai, T. Komukai, et al., Suppression of VFT in 1100 kV GIS by adopting resistor-fitted disconnector, IEEE Trans. Power Deliv. 11(2) (1996) 872–880.

[10] C. Weijiang, Y. Xianglian, W. Shaowu, et al., Recent progress in investigations on very fast transient overvoltage in gas insulated switchgear, Proc. CSEE 31(31) (2011) 1–11.

[11] H. Bin, L. Jiming, C. Weijiang, et al., Analysis on the influence of the switching speed of disconnector on very fast transient overvoltage, Proc. CSEE 31(31) (2011) 12–17.

[12] J. Weiping, Z. Yiying, et al., Study on Overvoltage and Insulation Coordination for ±800 kV DC Project, China Electric Power Research Institute, Beijing, 2007.

CHAPTER 5

Lightning Overvoltage and Protection of UHV Grid

Chapter Outline

5.1 Lightning and Its Main Parameters

5.1.1 Lightning Mechanism

Lightning is a massive electrostatic discharge phenomenon of cloud to ground, cloud to cloud, or within a cloud cluster. It is mainly caused by the thermal effect of damp air convection. It is generally believed that the thundercloud charges are the charges carried on

Ultra-high Voltage AC/DC Grids.
DOI: http://dx.doi.org/10.1016/B978-0-12-802161-3.00005-6

numerous distributed waterborne particles (i.e., water droplets, ice crystals, and snowflakes), rather than free ions or electrons. As these waterborne particles change their states (e.g., from water to ice), collide, split (or blend), and absorb electrons (or ions), intense charging occurs. Under the effects of intense air flow and the Earth's gravitational field, the waterborne particles carrying charges of different polarity will separate because of different aerodynamic characteristics and accumulate in different areas within the thundercloud. Oppositely charged areas create an electric field that, in turn, polarizes the waterborne particles in the air between them. The polarizing effect may catalyze the charging and heterocharge separation process in the thundercloud, thus creating two major charged central areas in a vertical direction, that is, the lower negatively charged area and the upper positively charged area in the thundercloud. The former is typically 1.5–5 km (2–3 km at the center) above the ground and the latter is 4–10 km above the ground, with both of them having a horizontal width of 1–10 km.

Electric charges in the thundercloud are not distributed evenly; several concentrated charge areas usually form in the thundercloud. Discharges of a thundercloud usually occur between one cloud to another or within a cloud cluster, and only a small fraction occurs between the cloud and the ground. The electric potential from a thundercloud to the ground can be as high as 10^5–10^6 kV.

In most cases, the lightning discharge channels are linear, and sometimes sheet lightning and, very rarely, sphere lightning can be seen. The linear lightning that may initiate from a cloud and that travels into the ground through atmosphere is called downward lightning (which can be subdivided into positive and negative ones), whereas linear lightning that initiates from a sharp projection on the ground and travels upward is called upward lightning. Both downward and upward lightning can be subdivided as positive or negative. Negative downward lightning is the most common type of linear lightning.

The linear discharge process of the negative downward lightning usually consists of three stages: the leader discharge; main discharge; and residual discharge. When the electric field strength reaches 2500–3000 kV/m in a densely charged region of a thundercloud, a downward-developed discharge occurs first, which is referred to as leader discharge. A downward negative leader discharge may split and form branches in a tree-like pattern because of the randomly distributed plasma present in the air. As the downward negative leader discharge approaches the ground, one or more upward discharges will be initiated from some projections on the ground, which is referred to as upward head-on leader discharges(s). When a head-on leader discharge meets a downward leader branch, an intense neutralization occurs, accompanied by enormous current (10–1000 kA), flickers, and thunder; this is the main discharge stage. The main discharge process develops upward through the downward negative leader discharge channel and ends when it reaches the thundercloud. Then, the residual charge in the cloud travels downward through the main discharge channel, which is referred to as the residual discharge stage. Because of the high resistance of the cloud, the current amplitude of the residual discharge is low (100–1000 A), but it lasts for a longer period of time (0.03–0.15 s).

Because there usually are several densely charged central areas in a single thundercloud cluster, the discharge from the thundercloud at one time is usually composed of multiple discharge impulses; therefore, it is called multi-impulse discharge. According to the statistics of 6000 measurements, the time interval between two discharge impulses is generally 30–50 ms, with the minimum of 15 ms and the maximum of 700 ms, and the time interval increases with the number of discharge impulses. Each discharge impulse is called one lightning stroke. The main discharge mechanism for subsequent strokes is the same with the first stroke except for generally smaller current (approximately one-third to one-half the current of the first stroke). However, the current wave-front time of the subsequent strokes is much less than the first stroke. Therefore, the maximum current increasing the steepness of subsequent strokes is three- to five-times greater than that of the first stroke, inducing a high overvoltage at an inductive impedance object struck by the lightning. Figure 5.1 shows the whole process of negative downward lightning. Figure 5.1A is a high-speed photograph showing a negative downward lightning and Figure 5.1B shows the changing lightning current during discharge [1].

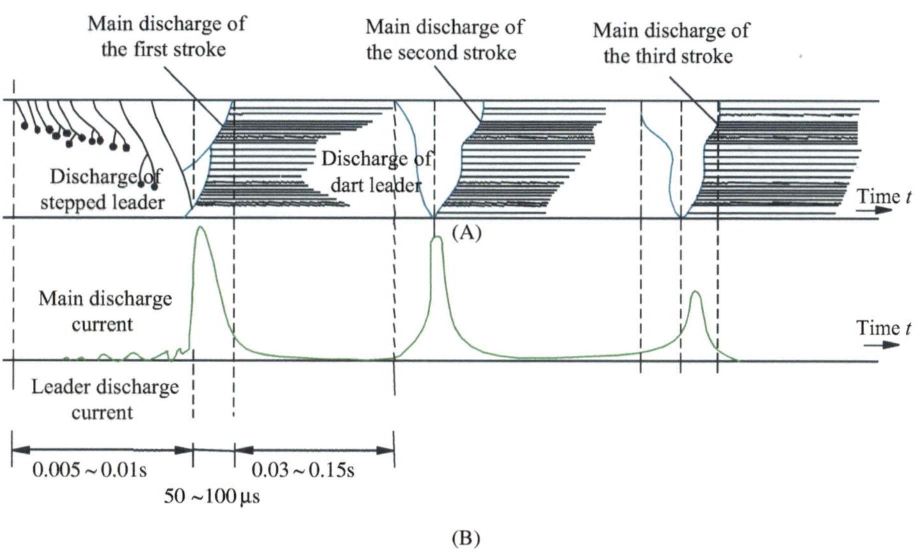

Figure 5.1: Negative downward lightning.
(A) High-speed photographs of negative downward lightning and (B) changing lightning current during discharge.

5.1.2 Lightning Parameters

The cloud-to-ground (CG) lightning is a quick discharge of electric charge stored in the cloud to the ground. The breakdown of atmosphere as a prerequisite for the formation of lightning leader and main discharge can only occur when the initial potential of the thundercloud is very high. However, the potential of a ground object struck by the lightning is not determined by this initial potential, but rather the product of the lightning current and the impedance of this object (which is the impedance between the striking point and the zero-potential reference point on ground). Therefore, this process can be treated as the action of a current source from the point of view of the power source property.

Generally, the following two types of lightning parameters have significant impact on grid operation: (i) lightning overvoltage is characterized by lightning current waveform, polarity, and wave impedance and (ii) lightning overvoltage level is characterized by lightning frequency, such as lightning current amplitude probability distribution, CG lightning flash density, and upward lightning probability

1. *Lightning current waveform*: The CG lightning current waveforms measured around the world show basically the same characteristics. Most lightning impulses are repetitive impulses of the same polarity and a few are small negative overshoot impulses. The discharge process at one time usually involves multiple leaders, main discharge processes, and follow-up currents.

   A typical lightning current waveform is usually expressed using a double-exponential equation as shown:

$$i = I(e^{\alpha t} - e^{\beta t}) \tag{5.1}$$

   where
   I—lightning current amplitude, kA
   α, β—constants
   t—time period, s.

   According to the observations around the world, the front wave of approximately 85% lightning currents is in the range of 1–5 μs and 2.5 μs, on average. The front value of 2.6 μs is recommended for the power system projects in China. The wavelength of lightning currents is in the range of 20–100 μs and 50 μs, on average.
2. *Lightning current polarity*: When the thundercloud is positively charged, both the lightning discharge and lightning current are positive. When the thundercloud is negatively charged, the lightning current is negative. According to the statistics of actual measurements, 75–90% of lightning currents are negative.
3. *Wave impedance of lightning current channel*: Figure 5.2 shows the equivalent circuit of the transmission line struck by lightning. If the line conductor's operating voltage is

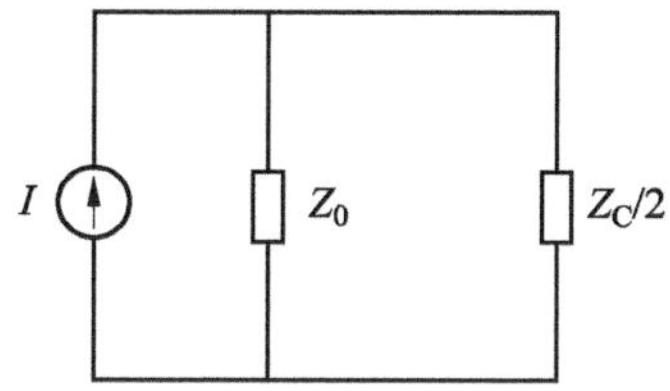

Figure 5.2
Equivalent circuit of transmission line (conductor) struck by lightning.

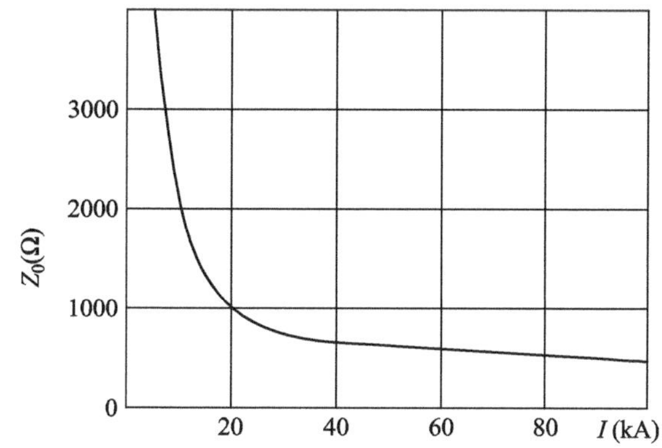

Figure 5.3
Wave impedance of lightning curve channel Z_0 versus lightning current amplitude I curve.

ignored, then the voltage U_C on the line conductor when struck by lightning may be calculated using:

$$U_C = \frac{IZ_C}{2 + Z_C/Z_0} \tag{5.2}$$

where

Z_0—wave impedance of lightning current channel, Ω

Z_C—wave impedance of line conductor.

It is shown in Eq. (5.2) that the wave impedance of lightning current channel has impact on the calculation of the lightning withstand level of transmission line. Therefore, modeling of the wave impedance of the lightning current channel is required for the calculation [2].

The study shows that the wave impedance of lightning current channel is in the range of 300–3000 Ω and is closely related to the lightning current amplitude. Through observation and calculation, the scientists of the former Soviet Union have worked out the wave impedance of the lightning current channel Z_0 versus the lightning current amplitude I curve, as shown in Figure 5.3 [3,4].

4. *Probability distribution of lightning current amplitude*: The lightning current amplitude of a lightning stroke is random. For a given region, through statistical analysis of extensive data of the lightning current amplitudes measured, a lightning current amplitude probability distribution curve can be established. The lightning current amplitude probability distribution varies from one region to another and is mainly dependent on the latitude, landform, weather, and thunderstorm intensity of the region. Because of varying geographical and climate conditions along the UHV transmission line corridor, the distribution of lightning activities changes considerably. China has obtained data regarding the characteristics of lightning activities along transmission lines through observations. However, the accuracy of the available data needs to be improved. Currently, for purpose of the lightning protection assessment, the probability distribution of lightning current amplitude is calculated using the following equations recommended in DL/T 620-1997 *Overvoltage Protection and Insulation Coordination for AC Electrical Installations*.

   For the regions with more than 20 thunderstorm days during a year, the probability distribution of lightning current amplitude is calculated using:

$$\lg P = -\frac{I}{88} \tag{5.3}$$

   where

   P—probability of the lightning current amplitude exceeding I.

   For the regions with no more than 20 thunderstorm days per year, its probability distribution of lightning current amplitude will decrease and the following formula is recommended:

$$\lg P = -\frac{I}{44} \tag{5.4}$$

5. *CG lightning flash density*: Both the CG lightning flash density and thunderstorm days can reflect the frequency and intensity of CG lightning. CG lightning flash density means the number per year of CG lightning flashes per square kilometer in a region. Thunderstorm days mean the days in a year when lightning (between a cloud and the ground, from one cloud cluster to another cloud cluster, or within one cloud cluster) occurs and at least one thunder can be heard. The CG lightning flash density and thunderstorm days vary significantly between years. Therefore, when lightning performance is calculated and assessed, average CG lightning flash density and thunderstorm days over years (usually referred to as CG lightning flash density and as thunderstorm days for short) are used, which are represented by N_g and T_d. In China, it is recommended to use the following empirical conversion equation proposed by the International Council on Large Electric Systems (CIGRE) in

Table 5.1: Typical Values of Thunderstorm Days and CG Lightning Flash Density.

Thunderstorm days T_d (d/a)	15	31	40	63	90	115
CG lightning flash density N_g [times/(km^2 a)]	0.78	2.0	2.78	5.0	7.98/8.0	11.0

Table 5.2: Classification of Regions with Lightning Activities.

Classification		CG Lightning Flash Density N_g [Times/(km^2 a)]	Thunderstorm Day T_d (d/a)
Description	**Code**		
Less thunderstorm region	A	$N_g < 0.78$	$T_d < 15$
Middle thunderstorm region	B_1	$0.78 \leq N_g < 2.00$	$15 \leq T_d < 31$
	B_2	$2.00 \leq N_g < 2.78$	$31 \leq T_d < 40$
More thunderstorm region	C_1	$2.78 \leq N_g < 5.00$	$40 \leq T_d < 63$
	C_2	$5.00 \leq N_g < 7.98$	$63 \leq T_d < 90$
Strong thunderstorm region	D_1	$7.98 \leq N_g < 11.00$	$90 \leq T_d < 115$
	D2	$11.00 \leq N_g$	$115 \leq T_d$

1980 to express the relationship between the CG lightning flash density and thunderstorm day [5]:

$$N_g = 0.023 T_d^{1.3} \tag{5.5}$$

Table 5.1 shows the typical thunderstorm days and CG lightning flash densities calculated using Eq. (5.5).

This empirical equation allows thunderstorm days to be converted into CG lightning flash density for ease of calculation. This equation indicates that the CG lightning flash density and thunderstorm days used for analyzing the frequency and intensity of the CG lightning exhibit some regular correlation function. If the available data of CG lightning flash density are scientific and reliable, then no data regarding thunderstorm days are required. However, when data are incomplete and unreliable, the data of thunderstorm days may be used as an effective alternate tool to calculate the CG lightning flash density.

For early lightning warning, regions with lightning activities may be classified into different levels according to their CG lightning flash density or thunderstorm days, as shown in Table 5.2.

6. *Upward flash probability*: Upward flash is a special type of lightning discharge initiated from an elevated object on the ground. The lightning current amplitude of upward flash is low whether any return stroke occurs. The relationship between the upward flash probability and the height of the ground object is illustrated in Figure 5.4. The available probability distribution data of lightning current amplitude essentially for downward flashes (excluding upward flashes) were obtained based on the measurements on towers of average height.

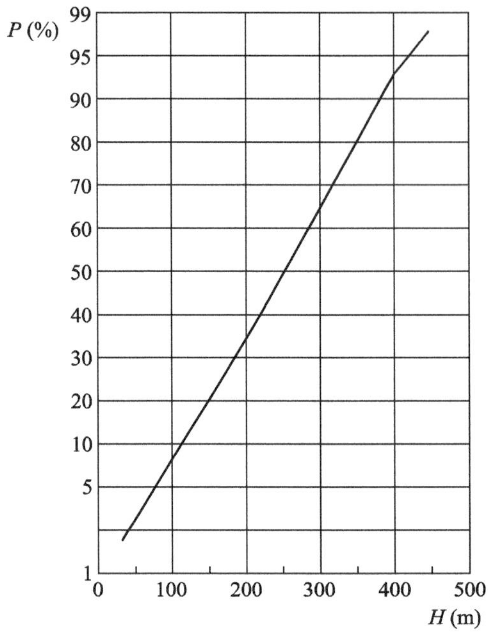

Figure 5.4
Upward flash probability *P* versus object height *H* curve.

For tall UHV transmission line towers, the probability of the towers struck by lightning increases with the tower height and, hence, leads to a larger probability of upward flashes. Therefore, in the simulation calculation for lightning performance of UHV transmission line, the probability of upward flashes initiated from the tower should be taken into account and the calculations should be performed separately for upward flashes and downward flashes. Otherwise, the simulation calculation error would be large.

5.1.3 Lightning Overvoltage

The lightning discharge is a complex physical process. However, from the viewpoint of lightning protection, it can be considered as a process in which an electromagnetic wave travels to the object on the ground through the lightning current channel with wave impedance. The lightning overvoltage is formed on the transmission line conductors or electrical equipment because of discharge from a thundercloud.

1. **Direct lightning strike overvoltage**
   Direct lightning strike overvoltage is the overvoltage occurring on any conductor or electrical equipment when directly struck by lightning. It may do harm to the line and equipment at various voltage classes.

Depending on the position of line struck by lightning, the direct lightning overvoltage may occur in the following two scenarios. First, when the lightning strikes line tower or ground wire, the point of strike's potential to ground is considerably increased because of the lightning current flowing through its impedance. When the potential difference between the point of strike at tower or ground wire and the conductor exceeds the impulse discharge voltage of the line insulation, flashover will occur on the conductor, giving rise to an overvoltage on the conductor. This phenomenon is referred to as back strike because the potential of the tower or ground wire (absolute value) is higher than that of the conductor. Second, when the lightning directly strikes the conductor due to absence of ground wire or due to shielding failure, overvoltage will occur on the conductor, and the lightning strike is referred to as shielding failure.

2. **Lightning-induced overvoltage**
   Lightning-induced overvoltage is the overvoltage occurring on the conductor or electrical equipment due to lightning strike on the ground or other objects nearby. Essentially, it is generated because the lightning leader discharge development takes a long time, during which the charged leader channel causes electrostatic induction in a nearby conductor, and the main discharge takes a very short time, during which charges in the lightning leader channel are neutralized rapidly. This means that the electric field generated from the lightning leader disappears in a very short time, causing slow release of the induced charge on the conductor or electrical equipment near the object struck by lightning. In this way, an overvoltage occurs.

5.2 Lightning Protection for UHV Overhead Transmission Line

5.2.1 Characteristics of Lightning Protection

The lightning protection performance of UHV overhead transmission line has the following two characteristics: good back strikes withstand capability against back strikes and poor back strikes withstand capability against shielding failure. The insulation level of UHV overhead transmission line is very high. Therefore, compared with EHV overhead transmission line, breakdown of insulator strings or air gap is less likely to occur in the event of a lightning strike on ground wire or tower top, which ensures a good back strike withstand capability. Generally, in China, the back strike withstand level is no less than 158 kA for a 500-kV overhead transmission line and no less than 200 kA for UHV overhead transmission line. For the overhead transmission lines rated 220 kV or less, the shielding failure trip-out rate and back strike trip-out rate are almost the same. For 500- and 750-kV overhead transmission lines, most lightning trips are due to shield failure strikes. UHV overhead transmission lines are more susceptible to shielding failures because their towers are much higher, and almost all lightning trips are caused by shielding failures.

Table 5.3: Operation Experience Data from Former Soviet Union and Japan with Regard to Lightning Performance of UHV Overhead Transmission Line.

Line Location	Line Type	Operation Year	Total Trips	Lightning Trips		
				Total Trips	Trip-Out Rate [Times/ (100 km a)]	Unsuccessful Reclosing Attempts
Soviet Union	1150 kV single-circuit	1985–1994	19	16 (84%[a])	0.6	6 (32%)
Japan	1000 kV double-circuit (on the same tower)	1993–2007	68	67 (98%[a])	0.9	2 (3%[b])

[a]Proportions (%) of lightning trips to total trips.
[b]In Japan, arcing horns are used for UHV transmission lines and significantly reduce the probability of unsuccessful reclosing attempts.

Therefore, shielding failure is major concern for lightning protection. However, it should be noted that despite of a high probability of shielding failures, the trip-out rate that is caused is not high because of the high insulation level of the UHV overhead transmission line. Therefore, the total lightning trip-out rate of the UHV overhead transmission lines is lower than that of EHV overhead transmission lines.

According to the experiences obtained by the former Soviet Union and Japan in operating UHV overhead transmission lines, the number of lightning trips accounts for more than 80% of the total trips of UHV overhead transmission line, which means that lightning strike is the main contributor to the trips of UHV overhead transmission lines. Table 5.3 shows the operation experience data with regard to lightning performance. It follows that lightning tripping poses the key risk to the safe operation and system stability of the UHV overhead transmission line.

China has placed more stringent requirements than Japan has on the lightning trip-out rate of the UHV transmission line. For 1000-kV UHV AC transmission lines, the design limit is 0.1 times/(100 km a), which is lower than that for 500-kV lines.

Considering that the lightning trip-out rates in actual operation of the transmission lines in former Soviet and Japan are much higher than the design limit, the lightning protection design of the UHV overhead transmission line in China has the following challenges: (i) develop better simulation and calculation models and methodology for assessment and calculation of lightning performance and (ii) develop economically viable and effective lightning protection measures and determine the relevant design parameters using applicable lightning performance simulation assessment methodology to ensure that the lightning trip-out rate is within the design limit.

5.2.2 Methods of Calculating Lightning Trip-Out Rate

In 2009, the Chinese national standard GB/Z 24842-2009 *Overvoltage and Insulation Coordination of 1000 kV UHV AC Transmission Project* was published, which is formulated based on the engineering and operating experiences and research findings from Jindongnan–Nanyang–Jingmen 1000-kV UHV AC Pilot Project and other UHV transmission projects across the world. This standard recommends the methods for the calculation and analysis of the lightning trip-out rate, that is, electromagnetic transient program (EMTP) or traveling wave method for back strike trip-out rate calculation and electric geometrical model (EGM) for shielding failure trip-out rate calculation.

5.2.2.1 Methods for calculating back strike trip-out rate

To calculate the back strike trip-out rate with EMTP or traveling wave method, proper parameters and models are required for the modeling of the line condition and the lightning strike process.

In China, the lightning current wave shape used for calculating the back strike trip-out rate of UHV lines is considered to be 2.6/50 μs, and towers are modeled with wave impedance and wave velocity. As compared with the simulation of lightning performance of EHV transmission lines, greater attention should be given to the impacts of the following factors on UHV transmission lines:

1. *Induced voltage*: A lightning strike on the top of a tower will induce a voltage component on the conductor. This concept is also accepted in Europe, the United States, and Japan, and they have also put forward some equations for calculating the induced voltage component; however, their calculated results are much lower than those accepted by China [6]. In actual calculation of lightning protection, they consider the induced voltage to be negligible, which differs dramatically from the practices of the former Soviet Union and China.

   In the United States, the induced voltage component U_i is given by the following approximate calculation equation [7]:

$$U_i = 6h_s I/(10 + I^{0.8}) \tag{5.6}$$

   where

   h_s—average height of ground wire above ground, m.

   By contrast, the Wuhan High Voltage Research Institute proposes the following equation that is obtained through the fitting calculation from a set of calculated curves:

$$U_i = [(1.771 + 1.754h_c - 0.010\,88h_c^2 + 1.935 \times 10^{-5}\,h_c^3) \times I^{0.170\,6h_c^{0.222}}](1 - k_0) \times 1.33 \tag{5.7}$$

where

h_c—average height of conductor above ground, m

k_0—geometric coupling factor between conductor and ground wire.

The following equation is proposed jointly by the Wuhan University and Wuhan High Voltage Research Institute through cooperative study [8]:

$$U_i = 2.2I^{0.4}h_c\left(1 - \frac{h_s}{h_c}k_0\right) \tag{5.8}$$

They indicate that the induced voltage component of the lightning overvoltage and the injected lightning current peak at different times. When the latter reaches its peak value, the former only reaches 70% of its peak value.

As proposed by the China Electric Power Research Institute, the induced voltage component U_i is given by:

$$U_i = \frac{60ah_{c.t}}{k_\beta c}\left[\ln\frac{h_t + d_R + k_\beta ct}{(1 + k_\beta)(h_t + d_R)}\right]\left(\frac{1 - h_s}{h_c}k_0\right) \tag{5.9}$$

where

a—steepness of lightning current, kA/μs

$h_{c.t}$—attachment height of the conductor, m

h_t—tower height, m

k_β—ratio of the main discharge speed to the speed of light c, $k_\beta = \sqrt{I/(500 + I)}$

d_R—length of the head-on leader discharge when a lightning strikes on tower, $d_R = 5I^{0.65}$, m.

Figure 5.5 shows the curves of the induced voltage components on the conductor as a function of lightning current when a lightning strikes the top of the tower calculated using different equations under the same conditions.

2. *Power frequency voltage*: In China, when the lightning protection of transmission lines at lower voltage levels is calculated, the power frequency voltage impact is not considered. However, for 1000-kV lines, the power frequency voltage can reach up to approximately 20% of the lightning impulse discharge voltage of the insulation, which may cause great impact; therefore, it should be taken into account in the calculation. The power frequency voltage peak value is generally taken as 898 kV for the calculation.
3. Criteria for flashover of air gap

a. *Intersection method*: At the point where the overvoltage waveform intersects the insulation impulse discharge volt–second characteristic curve, a flashover is deemed to occur, as shown in Figure 5.6.
b. *Criteria for long air gap flashover based on the leader discharge propagation model*: It is also recommended in IEC 60071-4:2004: *Insulation Coordination—Part 4:*

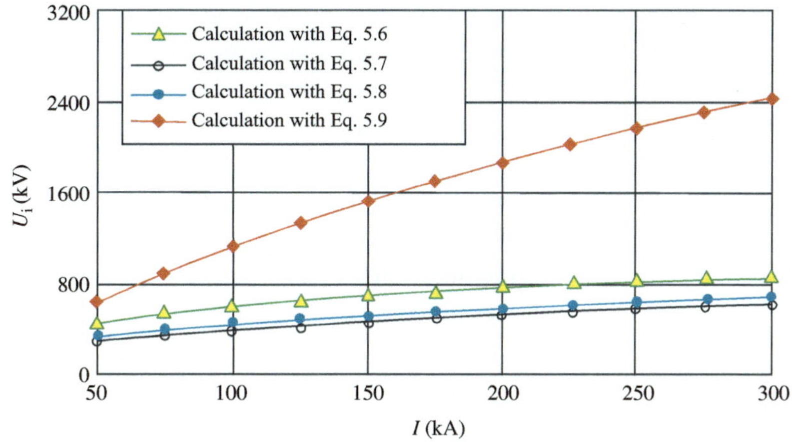

Figure 5.5
Induced voltage components on the conductor as a function of lightning current when a lightning strikes on top of the tower.

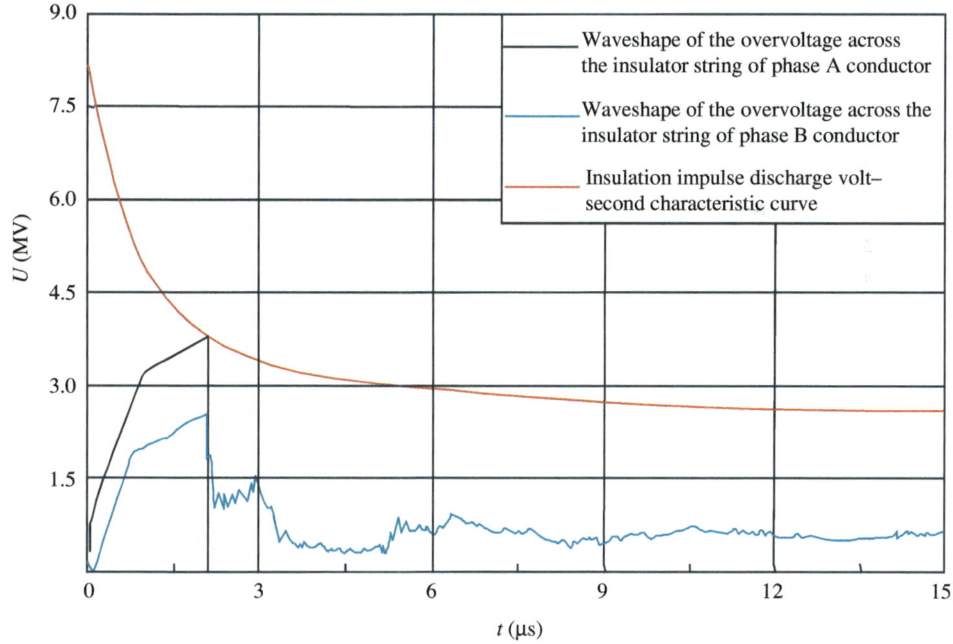

Figure 5.6
Wave shape of the overvoltage across the insulator string in case of a lightning strike and relevant volt—second characteristic curve.

Computational Guide to Insulation Coordination and Modeling of Electrical Network. The volt–second characteristic curve of air gap discharge voltage is obtained from the test with standard lightning impulse. However, during the lightning back strike processes, the lightning overvoltage across the air gap (such as insulator strings) of the line actually involves a short tail wave rather than a standard lightning wave due to the effects of the reflection wave from adjacent towers. With the leader discharge propagation model, the effects of a short tail wave on the air gap discharge can be reflected by taking into account the amplitude change and duration of overvoltage during the entire lightning flashover process.

When the voltage $u(t)$ applied across the air gap exceeds a certain value, the leader discharge will be initiated and propagate. If the leader discharge penetrates through the air gap or the leader discharge initiated from the electrodes at both ends meet together, then air gap breakdown will occur. The propagation velocity of the leader discharge $v(t)$ is given by:

$$v(t) = 170d_g[u(t)/(d_g - l_1) - E_0]e^{[0.0015u(t)/d_g]} \tag{5.10}$$

where

d_g—length of air gap, m

l_l—leader discharge length, m

E_0—a constant, dependent on air gap structure and voltage polarity, generally 500 kV/m (without regard to the impacts of altitude).

Because the leader discharge propagation process is closely related to the electric field strength and electrode shape and because it is complicated, few test data are available for a suitable model leader discharge propagation process.

In the UHV transmission study, some parameters in equation is calculated using the method of leader propagation model (LPM) to infer backward from the volt-second characteristic curve under the standard lightning impulse. Then, these parameters are used to calculate the discharge process of line insulator strings under the effects of a short tail wave for comparison. By comparing the back strike trip-out rate calculated using these criteria, it can be found that the results are similar.

5.2.2.2 Methods for calculating shielding failure trip-out rate

The shielding failure trip-out rate of transmission lines is calculated mainly using two methods: (i) the method recommended in DL/T 620-1997 *Overvoltage Protection and Insulation Coordination for AC Electrical Installations* and (ii) EGM method. Some scholars are exploring methods for calculating the shielding failure trip-out rate based on the LPM. The calculation equation recommended in DL/T 620-1997 is an empirical equation developed based on the experiences with the lightning protection of transmission lines rated 220 kV or less. However, because of the limitation of the technology at that time, the mechanism of the shielding failure cannot be clearly expressed; therefore, it is not applicable to the lines with high-voltage classes and on high towers.

1. **Electric geometrical model**

   The EGM is used to relate the electrical variables, such as lightning current, with striking distance (the critical breakdown distance between the head of the discharge channel of the leader lightning from cloud and the struck object) to derive the striking distance calculation equation that is used for calculation of the shielding failure trip-out rate. However, the equation is based on theoretical analysis model and correction based on field observation. Such statistical nature of observations obtained from operation experiences of transmission lines dictates that the EGM can only reflect the average shielding failure performance of transmission lines. For the transmission lines of higher voltage class or new tower types, some equations and parameters of the EGM have to be corrected based on the operating experiences with new transmission lines along with related simulation calculations.

   The EGM of a line directly struck by the lightning due to shielding failure is shown in Figure 5.7. If the lightning leader head falls on arc AB, then the lightning strikes on the ground wire and the conductor is protected from lightning strike. Therefore, the curve AB is referred to as the protection arc. If the lightning leader head falls on arc BD, then the lightning will strike on the conductor due to shielding failure of the ground wire. Therefore, the arc BD is referred to as the exposure arc. If the lightning leader head falls on the plane DE, then the lightning will strike on the ground. Therefore, the plane DC is referred to as the lightning capture surface. As the lightning current amplitude increases, the exposure arc BD diminishes. When the lightning

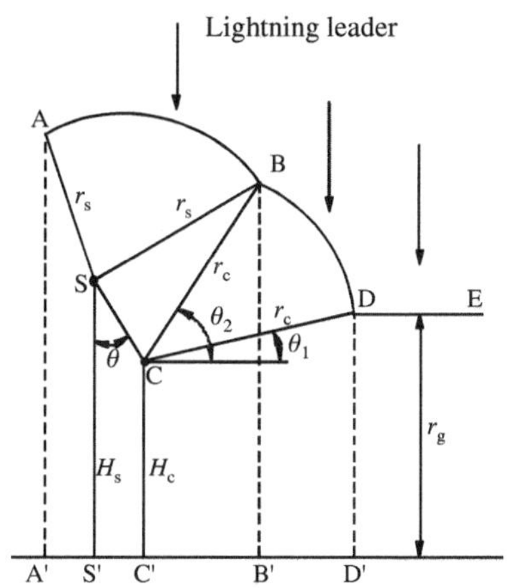

Figure 5.7
EGM of the line struck by lightning due to shielding failure.

current amplitude increase up to I_{max} (the maximum current for shielding failure on the conductor), the arc BD vanishes, which means that no shielding failure will occur.

Not all shielding failures cause insulation flashover. This occurs only when the lightning current exceeds the lightning withstand level of the line.

In the lightning protection calculation for the UHV lines in China, the shielding failure trip-out rate is calculated using an improved EGM that has taken into account the following factors:

a. Distribution probability of lightning leader angle of incidence θ, which is calculated by:

$$P(\theta) = K_m \cos^3 \theta \tag{5.11}$$

where

K_m—coefficient, generally 0.75.

b. Considering that the striking distance to ground differs from that to conductor, a striking distance correction factor β is introduced, which is the ratio of the striking distance to ground versus the striking distance to conductor, and its value varies with the height of the conductor above ground. In different countries, the correction factor β might be calculated using different methods [7,9]. For instance, the calculation equation recommended in the IEEE standard is as follows:

$$\beta = \begin{cases} 0.36 + 0.17\ln(43 - H_c) & H_c < 40\text{ m} \\ 0.55 & H_c \geq 40\text{ m} \end{cases} \tag{5.12}$$

where

H_c—average height of conductor above ground, m.

In previous studies performed in Japan, β was taken as 0.7, which means that the striking distance to ground is shorter than that to conductor, so that the lightning capture capability of ground is weaker than that of conductor and, consequently, the shielding capability of the ground for the conductor is reduced. However, this coefficient value cannot explain the field observations of shielding failure on transmission lines in recent years. Through analysis, it is believed that this value may be applicable only to a short air gap (4–5 m) rather than a long air gap. Japan has corrected the striking distance coefficient based on test results. According to the latest study results, the data obtained from field observation can be explained when the coefficient is taken as 1.1 for high towers. For Chinese UHV projects, the coefficient is taken in the range of 0.7–0.8. The study of the correction factor for the striking distance in case of a long air gap is underway.

c. *Impact of power frequency voltage on insulation flashover*: Because the power frequency voltage on conductors of a UHV transmission line accounts for a proportion of the insulation flashover voltage and cannot be negligible, it is necessary to simulate its impact on insulation flashover.

   d. *Topographical impact*: In the calculation of shielding failure trip-out using the EGM method, the topographical impacts are usually taken into account in two ways. First, in the EGM, an inclination angle of ground equivalent to the topographical impacts is used. When a line is routed over a mountain ridge, the conductor on the outside of the slope has a greater height above the ground than the conductor over flat ground and therefore is more susceptible to shielding failure. In this case, the maximum lightning current may be calculated by rotating the coordinate axis of the EGM by a certain degree. Second, for the lines over flat areas, the height of conductors will not vary greatly. Therefore, the average height of conductors may be used for calculation. However, for lines over mountainous areas and for large-span lines, the height above ground of conductors and ground wires varies greatly, and sometimes the variation is even greater than the tower height. In this case, the modeling of topographical impact should be based on the actual height of the conductor, and then the shielding failure current at different geometrical positions is calculated using this EGM.
2. **Leader propagation model**

   Based on the similarities between long air gap discharge and lightning discharge, combined with the new achievement made in long air gap discharge tests and lightning observation, the proposed LPM gives more physical connotation of the lightning shielding analysis model because it has taken into account the development of and competition between the upward leaders initiated from the grounded objects. The LPM simplifies the process of lightning attachment to overhead transmission lines into four physical processes, as shown in Figure 5.8.
   a. *Development of downward leader toward to ground*: Lightning downward leaders find their way to ground randomly, and are almost totally immune to the effects of ground objects. However, statistically, it can be approximately considered that the lightning discharge trajectory is a vertical channel and when its tip reaches 200–300 m above the ground (others believe that after the upward leader is initiated), the lightning leader will propagate along the maximum electric field flux direction.
   b. *Initiation and development of upward leader*: As the downward leader travels toward the ground, the electric field on the surface of ground objects (including overhead lightning wire and conductor) increases continuously until the electric field of any ground object is strong enough to launch an upward leader from the grounded object. Afterward, the upward leaders freely advance toward the maximum electric field direction at a certain velocity, that is, it is assumed that the upward leaders initiated from the grounded wire and phase conductor, respectively, will not interact with each other.
   c. *Connection of downward leader and upward leader*: Assuming that both the downward leader and upward leader propagate along their own maximum electric field direction, respectively, their relative movement speed is a key factor for determining which object will be attached by the downward leader.

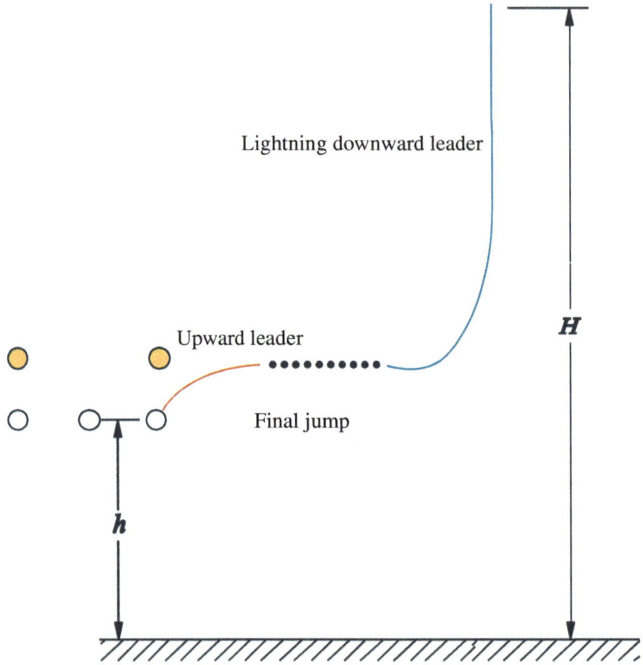

Figure 5.8
Physical process of lightning downward and upward leaders in LPM.

d. *Final jump*: The downward leader will connect with the first upward leader, where the final jump condition is satisfied. Sometimes, no upward leader is initiated from the target object; however, once the average electric field between the downward leader and the target object exceeds the critical breakdown electric field, the final jump will occur and downward leader will attach to the object.

As compared with the EGM, the LPM analyzes the physical process of lightning development in more detail while taking into account a variety of complicated factors that may have an impact on the attachment process. Therefore, using LPM allows a more accurate calculation of the shielding failure strike trip-out rate. However, it is difficult to determine all the parameters and criteria required for the calculation of the physical process of lightning development. Moreover, when different calculation conditions and parameters (such as different upward leader inception and velocity value of the upward leader) are used, the calculation results differ greatly. Limited by the test conditions, these calculation parameters and conditions are difficult to determine directly from the tests conducted in laboratories. Therefore, the shielding failure strike trip-out rate for the large air gap is generally extrapolated from the test results for the short air gap, and hence its confidence level is a question that necessitates extensive research efforts before it can be applied to practical engineering.

Currently, no unified international guidelines have been established for the LPM calculation method; therefore, few countries use the LPM for engineering design.

As compared with the LPM, the EGM considers fewer factors and uses fewer parameters. Moreover, different countries across the world use nearly the same parameter values in the EGM, and the calculation results are similar to each other and consistent with the actual operation experience. Therefore, various countries and international organizations, such as IEC and IEEE, recommend the use of the EGM in the engineering design because it is more mature than the LPM.

In China, the lightning protection design is mainly based on the calculation results with EGM, whereas the calculation results with LPM are taken as the reference. However, checking the calculation results with LPM against the operating experiences of 500-kV lines is required.

5.2.3 Application of Lightning Protection for UHV Overhead Transmission Line

5.2.3.1 Lightning protection for 1000-kV AC single-circuit line

As shown in Figure 5.9, cat-head-type tower and cup-type tower are the typical tower types used for the single-circuit line of the UHV AC transmission demonstration project in China. Cat-head-type towers are used in plains areas, and cup-type towers are used in mountainous areas. I-string insulators are used for the outside phase conductor and V-string insulators are used for the center phase conductor.

For 1000-kV single-circuit lines, the back strike trip-out rate is very low because of the high insulation level of the line. Therefore, the lightning protection of the line should focus on how to prevent trip-out caused by shielding failures.

The lightning protection for UHV transmission lines in the former Soviet Union is used as an example. Because single-circuit configuration is used, shielding failure is the main cause for lightning trip-out and the shielding failure trip-out rate is excessively high because of the excessive shielding angle (more than 20°) of the ground wire.

Studies show that with a reduced lightning shielding angle, the shielding failure trip-out rate can be significantly reduced; the landform has great influence on the shielding failure trip-out rate. Therefore, different lightning shielding angles may be selected for the ground wires of the UHV transmission lines according to landforms along the line. For the UHV single-circuit transmission line in China, the lightning shielding angle is required to be less than 5° for cat-head-type towers in plains areas; for cup-type towers in mountainous areas, it is required to be less than $-5°$. Table 5.4 shows the corresponding calculated shielding failure trip-out rates.

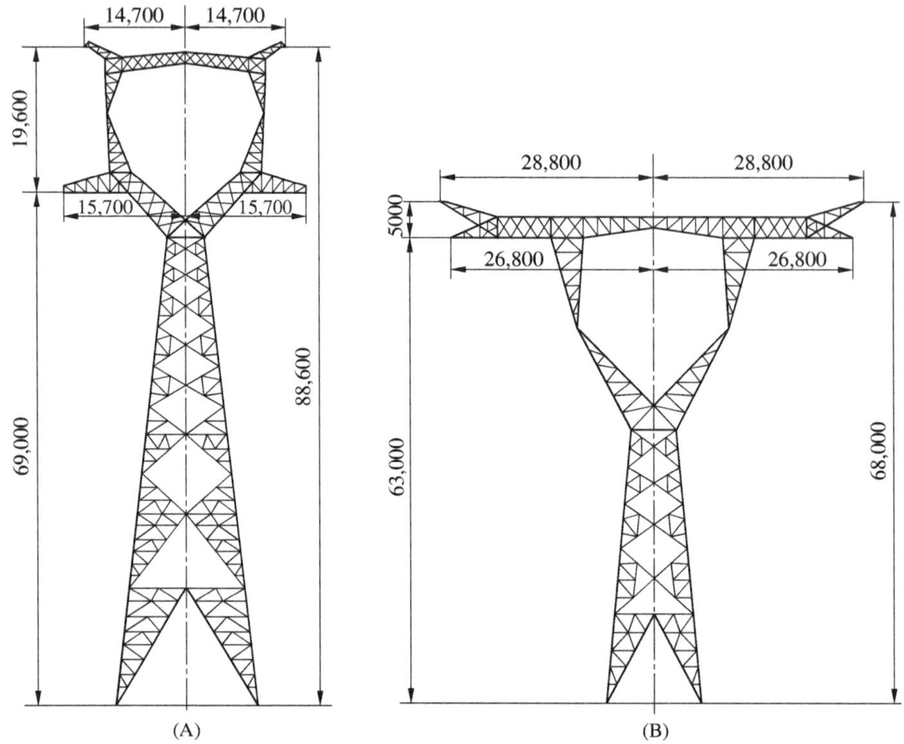

Figure 5.9
Typical tower for 1000-kV AC single-circuit line: (A) cat-head-type tower (ZMP2) and (B) cup-type tower (ZBS2).

Table 5.4: Shielding Failure Trip-Out Rates for Single-Circuit Lines on Typical Towers [Times/(100km · a)].

	Ground Inclination Angle θ (°)			
Tower Type	**0**	**10**	**20**	**30**
ZBS2	0	0	4.8×10^{-9}	0.019
ZMP2	0	0.006	0.108	0.618

Theoretically, as the lightning shielding angle is reduced, the spacing between two ground wires has to be increased, which reduces the shielding effect of both ground wires on the center phase conductor. This will increase the risk of insulation flashover and trip-out of the line due to shielding failure on the center phase conductor. To verify this, the trip-out rate of the line due to shielding failure on the center phase conductor is analyzed with EGM. Using cup-type tower (ZBS2) as an example, assume that the spacing between two ground wires is 57.6 m and the vertical distance (Δh) between the ground wire and the conductor is

Table 5.5: Maximum Lightning Current Possibly Striking the Center Phase Conductor (EGM).

Tower	Spacing Between Ground Wires (m)	Vertical Distance Between Conductor and Ground Wire Δh (m)	Maximum Lightning Current Possibly Striking Center Phase Conductor (kA)
ZBS2	57.6	15	7.9
		13	8.8
ZMP2	29.4	13	2.7
		11	2.8

13 m. The calculation results, as shown in Table 5.5, indicate that there is potential for shielding failure on the center phase conductor only when the lightning current is low ($I < 9$ kA), such that lightning can penetrate between ground wires. The lightning current of such magnitude is not large enough to cause insulation flashover and, hence, line trip-out. In contrast, lightning with a larger current will strike the ground wire because it cannot penetrate between ground wires and strike the center phase conductor. Therefore, even if the spacing between two ground wires is increased after the lightning shielding angle is properly reduced, no line trip-out will occur because of shielding failures on the center phase conductor.

Calculation results based on either EGM or LPM show that increasing the vertical distance between the ground wire and the conductor could reduce the line trip-out rate due to lightning strike on the center phase conductor. To ensure safety, the height of the ground wire for ZBS2 type towers is increased by 2 m compared with the original design.

The UHV AC transmission pilot project has a total line length of 641 km and was put into operation in January 2009. By July 2013, no lightning-induced line trip-out occurred in operation for more than 4 years. This demonstrates that the lightning protection design of this UHV AC transmission pilot project is effective and successful, fully meeting the design requirement of lightning trip-out rate being less than 0.1 times/(100 km a).

5.2.3.2 Lightning protection of 1000 kV AC double-circuit line sharing a tower

As land resources are becoming increasingly scarce, it is essential to properly adopt double-circuit or multicircuit transmission lines sharing a tower to increase the transmission capacity per unit width of the corridor and thereby improve the utilization efficiency of the transmission line corridor. However, according to experience with the design and operation of UHV projects across the world, erecting double-circuit transmission lines poses a risk of high lightning trip-out rates. For instance, in contrast to the practices of erecting single-circuit transmission lines adopted by the former Soviet Union, Japan erected double-circuit transmission lines. Although Japan uses a negative lightning shielding angle to reduce the lightning trip-out rate due to shielding failure after learning from the experience of the former Soviet Union in the lightning protection of UHV transmission lines, the lightning trip-out rate due to shielding failure is still much higher than expected. Therefore, more exact lightning protection measures are required for the Huainan–Shanghai 1000-kV UHV AC double-circuit transmission line to reduce the lightning trip-out rate to an acceptable level.

Through investigation and analysis, it can be found that the UHV double-circuit lines in Japan have the following characteristics: (i) most lines run through mountains exposed to frequent lightning and towers for these lines are much higher than those for single-circuit lines; (ii) lines are provided with arcing horns that considerably reduce the lightning impulse insulation level; and (iii) a large number of tension towers are used in mountainous areas (accounting for 74% of the total) and the tension towers where line flashover occurs account for 94% of the total number of towers with line flashover. By incorporating Japan's experience in the operation of the UHV double-circuit line, China has taken the following measures for lightning protection in the design of a 1000-kV UHV AC double-circuit line [10]:

1. Optimize the tower type for UHV double-circuit line. For selection of a suitable tower type, China has made comparisons between umbrella-type towers and drum-type towers by taking into account a wide range of factors, including the lightning protection performance, insulation, ice-shedding-induced line jumping, and electromagnetic environment. Finally, the umbrella-type tower was selected. Figure 5.10 shows the umbrella-type tower and drum-type tower.
2. Adopt different lightning shielding angles and lightning impulse insulation levels for the lines in the regions with different landforms. The shielding angle of ground wire for a double-circuit line is generally no more than $-3°$ in plains areas and rolling country and $-5°$ in mountainous areas. The shielding angle for an overhead incoming feeder should be less than $-4°$ in plains areas and should be reduced in mountainous area when conditions permit. At 500 m above sea level, the required minimum air gap of the double-circuit line to withstand lightning impulse should be 6.7 m in plains areas and 7.0 m in mountainous areas.
3. Take special lightning protection measures for the jumper of the tension angle tower. The shielding angle of the ground wire with respect to the jumper on the tower of the UHV transmission line is no more than 0°. The shielding angle of the ground wire with respect to the jumper of the tension angle tower for an overhead incoming feeder to a substation should be less than 4°. Other alternative measures may be taken as well. For instance, in case of a tension angle tower of an incoming feeder where the shielding angle of the outside ground wire with respect to the hard jumper is larger, a horizontal lightning rod along the jumper direction or a ground jumper wire may be installed at the top of the ground wire support.
4. When conditions permit, use its footing's natural grounding electrode instead of artificial grounding conductor for UHV double-circuit towers. For example, in the Huainan–Shanghai UHV AC double-circuit transmission line, the artificial horizontal

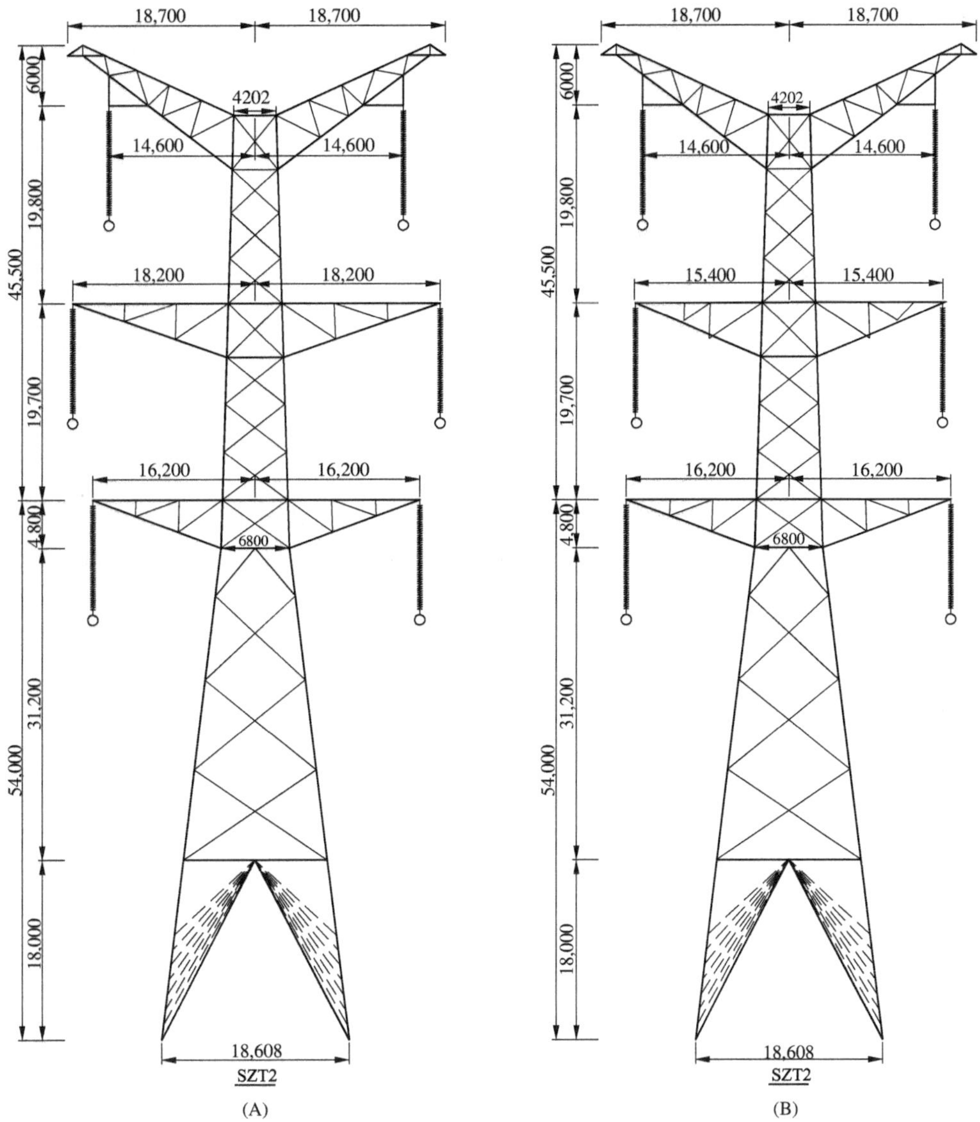

Figure 5.10: Tower types for 1000-kV AC double-circuit line.
(A) Drum type and (B) umbrella type.

grounding conductor may be eliminated if the use of its footing's natural grounding electrode of the tower can lower the power frequency grounding resistance to less than 25 Ω.

5. Use differentiated design and evaluation methods of lightning protection for UHV double-circuit line towers. Different lightning protection methods may be used for different UHV double-circuit line sections according to the tower types, landform, and

lightning activities. The shielding failure trip-out rate is assessed and calculated for each tower so that measures can be reinforced and lightning protection measures can be taken for the points vulnerable to lightning strikes.

6. Because the distance of large-crossing UHV line is relatively short, the lightning protection performance of single-circuit and double-circuit lines in terms of safe operating years is estimated rather than the number of lightning trip-outs per 100 km during a year. It is required that the lightning safe operating years is no less than 50 years. For large-crossing lines, as the height of tower increases, the insulation level of the transmission line should be increased accordingly, focusing on increasing the air gap between conductors and tower based on lightning overvoltage calculations. The lightning shielding angle for the large-crossing lines should be less than that for other lines.

5.2.3.3 Lightning protection of UHV DC transmission line

The lightning protection of UHV DC transmission lines should be designed based on transmission capability of DC power in conjunction with past operational experience, the intensity of lightning activities, the topographic and geomorphologic characteristics, and the soil resistivity in the local area. After the lightning withstand level and flashover rate of the entire line are calculated, a techno-economic analysis should be performed to choose rational differentiated lightning protection modes.

The UHV DC electrode lines usually operate at a low-voltage class and the conductors are arranged on both sides of the tower. Regarding the importance of the electrode line, ground wire is required to route along the electrode line. To increase the coupling coefficient between the ground wire and the conductor and to improve the lightning withstand level, the shielding angle of the ground wire should not exceed 30°. Because the insulation level of the electrode line is low, a lightning strike on either the ground or the line may cause flashover on the electrode line. Because DC current has no zero-crossing point, stable arcing may be established. A portion of DC current will be directed to the ground through the tower. During monopolar ground operation mode or during bipolar ground mode but with one failed ground pole, the DC follow-up current at the point of strike will be tremendous and cause difficulty in extinguishing the arc by itself. To prevent dropping of insulator strings due to DC arc burnout of insulators, installing arcing horns connected in parallel with the insulators of electrode line is required.

In China, the lightning withstand performance of a UHV DC line is not inferior to that of a ±500-kV DC line. In the Xiangjiaba–Shanghai ±800-kV UHV DC Pilot Transmission Project, the following lightning protection scheme is used:

1. Provide dual ground wires. The shielding angle of ground wires is controlled within −10° in mountainous areas and that for the outside conductor is 0° for plains areas. The horizontal distance between the two ground wires is no more than five-times

the vertical distance from the ground wire to the conductor. The ground wires on different towers are continually grounded at the top of each tower, and the tower footing grounding resistance should meet the design requirement.

2. Provide dual ground wires for the large-crossing line sections. The allowable horizontal displacement of the suspension points between the ground wire and the conductor is no less than 3 m at the tower. The shielding angle of the ground wire for the conductor on the tower is $<0°$. The distance between the conductor and the ground wire at the midpoint of spans should be increased to more than 20 m to ensure that the lightning withstand level at the midpoint of spans can reach 200 kA. The footing grounding resistance of the tower is designed to be $<5\ \Omega$.

UHV DC lines have almost the same lightning performance as UHV AC lines. According to line operation experience, the lightning flashover rate is approximately 0.1 times/(100 km a). Preliminary analysis results show that almost all lightning flashovers are caused by shielding failures on a positive polar conductor. Operation experiences show that the positive polar conductor of a UHV DC line is more susceptible to shielding failures than the negative polar conductor.

5.2.3.4 Lightning protection of hybrid EHV/UHV AC multicircuit line sharing a tower

In the planned Huainan–Nanjing–Shanghai transmission project, the hybrid EHV/UHV AC multicircuit transmission line sharing a tower will be used in areas where the line corridors are very difficult to obtain. As compared with the conventional UHV double-circuit line, the towers of hybrid EHV/UHV multicircuit lines sharing a tower are much higher and unfavorable to prevent lightning back strikes and shielding failures, thus requiring special consideration for lightning protection design.

1. The multicircuit line towers are more likely to suffer lightning strikes because of their larger height and lightning-attractive area. In the event of a lightning strike on the tower or ground wire, the negative reflection wave from the footing ground resistance will take a long time to reach the top of tower and the amplitude of overvoltage on the tower is high. Therefore, the EHV circuits are more likely to suffer a back flashover, and there is a higher probability that both EHV circuits will suffer a flashover simultaneously than will the conventional EHV double-circuit line. In contrast, the probability of flashover on UHV circuits is very low because most lightning energy has been dissipated after flashover on EHV circuits. Therefore, one of the major concerns for erecting the hybrid EHV/UHV AC multicircuit line is the increased back strike trip-out rate of the circuit at a lower voltage class. The most effective solution to this is to increase the insulation strength of the line at a lower voltage class.
2. The calculation based on the EGM shows that for a hybrid EHV/UHV multicircuit line, the height difference between the above-ground wire and the EHV line is up to 60–70 m, so it is difficult for the ground wire to provide effective shielding for the EHV line. However, the lower phase conductor of the UHV circuit has some shielding effect on the EHV line.

3. Using I-string insulators or V-string insulators has different effects on lightning performance of UHV and EHV lines. Specifically, if V-string insulators are used, then the shielding failure flashover rate of the UHV line is lower and that of the EHV line is higher than that of the conventional EHV line. If I-string insulators are used, then the shielding failure flashover rate of UHV circuits is higher than that when V-string insulators are used. This rate of hybrid-type EHV line is lower than that of the conventional EHV line.

Figure 5.11 shows the typical hybrid EHV/UHV AC multicircuit line tower used in the Huainan–Shanghai EHV transmission line. Table 5.6 presents the simulation calculation results of the lightning trip-out rate of the multicircuit line section of Huxi–Zhebei.

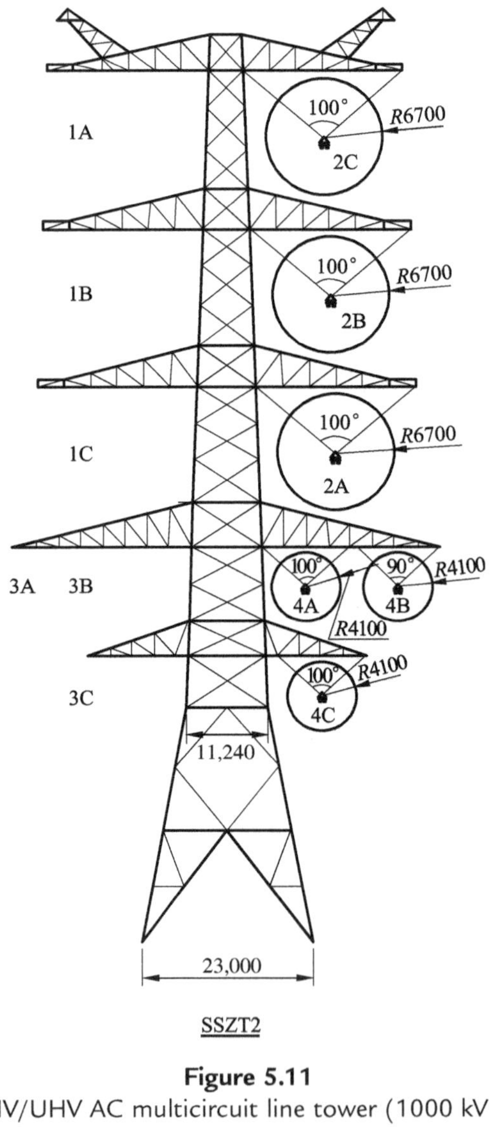

Figure 5.11
Hybrid EHV/UHV AC multicircuit line tower (1000 kV + 500 kV).

Table 5.6: Lightning Trip-Out Rate of Multicircuit Line [Times/(100km · a)].

Tower Type	Line (kV)	Back Flashover Rate	Shielding Failure Flashover Rate	Total
Hybrid 1000 and 500 kV line, with I-string insulators used for 1000 kV circuit	1000	0.012	0.081	0.093
	500	0.119	0.017	0.136
Hybrid 1000 and 500 kV line, with V-string insulators used for 1000 kV circuit	1000	0.012	0.012	0.024
	500	0.119	0.295	0.414

5.3 Lightning Protection of UHV Substation and Converter Station

5.3.1 Simulation on Lightning Protection of UHV Substation and Converter Station

1. **Lightning protection contents**

   The damage caused by lightning strike to UHV substations and converter stations originates from two aspects: (i) direct lightning strike, which is when lightning strikes the station equipment directly to cause equipment damage, and (ii) lightning intruding wave, which is the lightning wave resulting from flashover on the transmission line (also known as back strike) when lightning strikes the tower or ground wire, or from shielding failure when lightning strikes the conductors directly and is propagated along the transmission line to the equipment within the station to cause overvoltage. Therefore, the lightning protection of the substation and converter station is designed from the two aspects correspondingly.

   The direct lightning strike protection is usually achieved using lightning rods or wires. The calculation of the protection range of lightning rods or wires for UHV substations and converter stations may be performed using the same method as that for 500-kV substations. This design method can provide substations with high reliability against direct lightning strike. For more details regarding this method, refer to DL/T 620-1997 *Overvoltage Protection and Insulation Coordination for AC Electrical Installations.*

2. **Simulation method for lightning intruding overvoltage**

   The lightning wave that intrudes into the UHV substation or converter station through the UHV transmission line will cause complex refraction and reflection waves on the connections and electrical equipment and, consequently, produce overvoltage on the insulation of equipment.

   The lightning-intruding overvoltage may be calculated with the deterministic method or Monte Carlo method, each of which has its own advantages. Generally, for substations, both methods are applicable; for the converter station, only the deterministic method is applicable.

The deterministic method is based on the worst-case operation conditions along with the most stringent boundary conditions to obtain the maximum overvoltage level in the substation, and then the lightning impulse withstand level of equipment and arrester configuration are determined to meet the safety requirements. With this method, the safety margin of the internal insulation of the equipment should be no less than 15% during usual operation mode and no less than 10% during special operation mode; for the external insulation, it should be no less than 5%.

When the Monte Carlo method is used, the amplitude of lightning current, point of lightning strike on the incoming feeder, operating voltage of the conductor, lightning impulse withstand level, and operation mode are taken as the interval random variables that are used for statistical calculation, and analysis is based on interval combination to obtain the safe operation years of a substation (mean time between failures, MTBF). The MTBF of a 1000-kV substation affected by lightning intruding overvoltage should not be less than 1500 years.

In engineering practices, both methods can be used to assess the lightning protection performance of substations through simulation modeling. In China, EMTP or traveling wave method is used for simulation modeling of the lightning-intruding overvoltage in a UHV substation. Specifically, the transmission line (including the tower, ground wire, and conductors) and the bus, connecting lines, and electrical equipment in the substation are assumed to be equivalent to a network that allows high accuracy and compliance with the actual situation. In this network, the electrical equipment is simulated using the equivalent input capacitance of the equipment, and the incoming feeder and connecting lines are simulated using distributed parameters.

The amplitude of the lightning-intruding overvoltage is related to factors such as lightning strike point, substation operation mode, incoming feeder parameter, amplitude, and amplitude probability distribution of lightning current.

Usually, the closer the lightning strike point is to the substation inlet, the higher the lightning-intruding overvoltage will be. However, when the point of strike is too close to the substation inlet, in case of a back lightning strike, the lightning-intruding overvoltage will be reduced because the overvoltage on the ground wire will be mitigated by the wave of opposite polarity reflected from the terminal tower close to the substation. Therefore, in case of shielding failure, the lightning-intruding overvoltage will be high when the lightning strikes the conductor on the tower closest to the substation. By contrast, in case of a back strike, the maximum lightning-intruding overvoltage might occur when the lightning strikes the second closest tower or its ground wire. These should be given more attention in the simulation. Accordingly, the calculation of lightning-intruding overvoltage due to a back strike should be based on the scenario that the lightning strikes the top of the tower (in which case the worst-case overvoltage occurs) rather than the ground wire in the middle of the span. If the lightning strikes the ground wire in the middle of the span, then no flashover will occur between ground wire and conductor as long as the relevant standards are observed. Even if flashover occurs on the insulator string, the overvoltage will be lower than that when the lightning strikes the top of the tower. Likewise, in the calculation of the lightning intrusion wave overvoltage resulting from shielding failure, only the scenario of the lightning strikes occurring on the conductor at the tower position should be considered, because here the conductor and the ground wire have their highest positions and the largest shielding angle; therefore, the potential lightning current is the largest.

In the calculation of lightning-intruding wave overvoltage, the operation mode to be considered for UHV substation is different from that for a 500-kV substation in China. Both the normal operation mode and special operation mode are considered for a UHV substation, whereas only normal operation mode is considered for a 500-kV substation. The special operation mode means the mode in which the circuit breaker on the line side is open. In such case, the overvoltage is the most severe but occurs at a very low probability.

The special operation mode may occur in the following three scenarios: the line experiences multiple consecutive lightning strikes; the line is ready for connection to the grid; or the line is in a hot standby state. The UHV line is usually not in the hot standby state, and this state can be avoided and precluded by manual control. Therefore, in the simulation of this mode, only the former two scenarios need to be considered.

The normal operation modes mean all the possible combinations of the lines, buses, and transformers that can be put into operation. Among these modes, the overvoltage is usually higher when a fewer number of lines and transformers are in operation (e.g., a single-circuit outgoing feeder and a single transformer connected).

The amplitude of the lightning current through the incoming feeder is another key factor influencing the amplitude of the lightning-intruding wave overvoltage. In case of back lightning strikes, the larger the lightning current amplitude used in the simulation analysis, the higher the amplitude of the resulting overvoltage but the less likely that lightning current will occur. For the method of calculating the probability distribution of lightning current amplitude, see the previous section. If the lightning intruding overvoltage generated as a result of back lightning strike in a UHV substation is calculated using the deterministic method, then the largest lightning current should be assumed to be 230 kA for the special operation mode, and 250 kA is the largest that should be assumed for the normal operation mode. In case of shielding failures, the largest lightning current for use in simulation is determined using the EGM.

5.3.2 Lightning Protection of UHV Substations

5.3.2.1 Direct lightning strike protection

The main measures for protecting substations from direct lightning strike include the use of lightning rod or ground wire or a combination of these. Japan and some Western European countries often use ground wire, whereas China uses a combination of lightning rods and lightning shielding wire for UHV substations.

Specifically, in a UHV substation in China, lightning rods and/or ground wires are used for outdoor equipment. The lightning rod on the framework should be connected with grounding grid and a centralized grounding device should be provided in the vicinity thereof. The calculation and analysis show that the air gap between the grounded part and the energized part of the framework installed with lightning rod should be no less than 7 m. The location where the lightning rod at the framework (except for the gantry structure of transformer) is connected to the main grounding grid should be no less than 15 m apart (along the grounding conductor) from the location where the ground wire at the transformer is connected to the main grounding grid. Outdoor Gas-insulated metal-enclosed switchgear (GIS) requires no direct lightning strike protection but should have its enclosure reliably grounded. Moreover, the lightning rod or wire and the protected equipment should be spaced apart far enough to ensure that the protected equipment is protected from back lightning strike when lightning strikes the lightning rod or wire.

5.3.2.2 Electric equipment protection against lightning-intruding overvoltage

The probability of lightning directly striking a line is far higher than that of lightning striking a substation. When a lightning strikes ground wire or a tower, causing back flashover and lightning electric current to enter the conductor, or when lightning directly strikes the conductor by puncturing its shield, lightning overvoltage traveling waves will be formed and intrude into the substation through the line. The lightning-intruding overvoltage may exceed the insulation level of the equipment and endanger the safe operation of the substation. Therefore, controlling lightning-intrusion wave overvoltage is a focal point in UHV substation lightning protection.

In contrast to EHV substations, UHV substations are protected from lightning strikes not only by taking measures inside the substations to suppress the intruding overvoltage but also by suppressing its origin by improving the design of the incoming feeder to provide integrated protection on both the substation and its incoming feeder.

1. **Improvement of incoming feeder of the substation for lightning protection**
   The main way of suppressing the lightning-intruding overvoltage is improving the lightning protection performance of the incoming feeder, including reducing the

possible largest lightning current through the line in case of shielding failures, and improving the back strike withstand ability. The details are as follows:

a. Reduce the lightning shielding angle for the incoming feeder as far as possible, thereby reducing the largest possible lightning-intruding current along the line and the amplitude of the lightning-intruding wave due to shielding failure.

   The lightning shielding angle of the ground wire for overhead incoming feeders should be $<4^\circ$ in plains areas and should be further reduced in mountainous areas if conditions permit. The shielding angle of the ground wire with respect to the jumper on tension angle towers should be $<0^\circ$ and other alternative measures may be taken instead, such as installing lightning rods along the jumper direction or installing ground wire jumpers.

b. Improve the insulation strength of the incoming feeder and reduce the grounding resistance of the tower, thereby reducing the back strike trip-out rate and the likelihood of a lightning wave intruding into the substation due to back strikes on the line.

   For UHV substations in China, the lightning withstand level of the incoming feeder section (2 km) to the substation is required to be no less than 250 kA to prevent back flashover on the line.

c. Install an additional ground wire. In the UHV pilot project, a single-circuit transmission line is erected and the conductors are arranged horizontally. Therefore, reducing the lightning shielding angle would lead to increased spacing of two ground wires, compromising the shielding effect of ground wires on the center phase conductor. Although the shielding failures on the center phase conductor have low current amplitude, they can still pose risks to the substation if the lightning strike occurs in the vicinity. Therefore, if the spacing of both ground wires on the incoming feeder tower is more than four-times the vertical distance between the conductor and the ground wire, then an additional ground wire should be installed to prevent shielding failure on the center phase conductor.

2. **Improvement in the configuration of metal oxide arrester in substation**

Installing metal oxide arresters (MOAs) in the substation is another major measure to limit the lightning intruding overvoltage. MOAs installed on either the bus side or the line side in a 1000-kV substation all have a rated voltage of 828 kV, nominal discharge current of 20 kA, and lightning protection level of 1620 kV. Usually, MOAs are installed at a distance from the equipment. The larger the distance between MOAs and equipment, the less capable the MOAs are of protecting the equipment from lightning strikes. Therefore, in a 1000-kV substation, the configuration (locations and number) of MOAs has a significant impact on their capability of suppressing the lightning-intruding overvoltage.

Installing MOAs in the vicinity of the key equipment in a UHV substation can effectively limit the lightning-intruding overvoltage on the UHV equipment and thereby prevent the equipment from damage. Generally, each piece of key equipment (such as the main transformer and high-voltage reactor), bus, and outlet of the line all require one set of MOAs.

However, an increased number of MOAs will result in increased construction investment and O&M costs of the project. Therefore, the optimal configuration of MOAs should be developed according to the results of the calculations and analysis of the lightning-intruding wave in the substation. The study results show that in some UHV substations, no MOA is required for the GIS bus, and the high-voltage reactors and the line outlet may share one set of MOAs, provided that the configuration of such MOAs meets the requirements of relevant engineering design. The specific configuration of MOAs should be finalized based on the simulation and calculation results of the lightning-intruding wave in the substation.

3. **Improvement in the lightning protection design for the series compensation platform and disconnectors**
Generally, after a series compensation device is additionally installed in a substation during the extension phase, the highest lightning-intruding overvoltage will occur at the series compensation devices, although the amplitude of the lightning-intruding overvoltage in the existing substation is not high. Therefore, provisions must be made to suppress the lightning-intruding overvoltage on the series compensation devices.

For this purpose, MOAs are still used. Because the series compensation device itself is large and the total length of the series compensation platform and associated equipment is longer than 77 m, the MOAs may exhibit low protection capability for the equipment at the farthest end. Great attention should be given to this. According to the study results, the distance between the MOAs and the nearest end of the series compensation platform should not be more than 80 m.

5.3.3 Lightning Protection of UHV Converter Station

5.3.3.1 Direct lightning strike protection

The direct lightning strike protection of a converter station is substantially the same as that of a substation. Both use lightning rods and wire to reduce the likelihood of direct lightning strikes with certain intensity to an acceptable level. A converter station includes a variety of outdoor yards that are used for different purposes and at different voltage levels, such as AC switchyard, high-voltage DC switchyard, AC filter yard, and DC neutral bus yard. The protection configuration against a direct lightning strike for those yards should be tailored according to the actual conditions.

In China, as with ±500-kV converter stations, UHV converter stations are protected from direct lightning strikes primarily by use of lightning rods and wires whose lightning protection mechanisms and design methodology are similar to those for substations.

The direct lightning strike protection of DC yard, AC filter area, and converter transformer area is achieved primarily by use of ground wires. In the DC filter area and DC neutral bus area, because the facilities have a low insulation level and occupy a large area of land, higher shielding is required. If lightning rods are still used, then their height and density have to be increased. This will greatly increase the construction costs for lightning rods and their foundations, jeopardize the overall visual appeal of converter stations or cause difficulties in equipment arrangement, and increase land occupation. Therefore, it is advisable to route ground wires between the lightning rods on the frameworks and standalone lightning rods on the boundary of the station area. In a UHV DC yard, ground wires should be routed at a height of approximately 40 m.

5.3.3.2 Lightning-intruding overvoltage protection for electric equipment

A UHV DC converter station comprises three major parts: AC switchyard, DC switchyard, and converter area. The lightning-intruding overvoltage in a converter station is mainly caused by the back strike and shielding failure on the incoming feeders from the AC and DC switchyards.

1. **Lightning-intruding overvoltage**
   a. *Lightning overvoltage on AC switchyard equipment*: It is propagated from the AC transmission line and is generated in the same mechanism as that in a substation. Because an AC switchyard has more outgoing feeders and is connected with multiple AC filters and capacitor banks, all these can damp the lightning overvoltage to some degree. Moreover, the lines, auxiliary transformer, converter transformer, and filter bus are all provided with surge arresters that will limit the lightning overvoltage. Therefore, the lightning overvoltage on the AC equipment in a converter station is lower than that in a substation.
   b. *Lightning overvoltage in converter area*: The lightning-intruding wave is shielded by the converter transformer on the AC side and by the smoothing reactor on the DC side of the converter. Therefore, the lightning overvoltage propagated to the converter through the interwinding capacitance of the converter transformer and the interturn capacitance of the smoothing reactor has low amplitude and its wave shape resembles that of a slow-front overvoltage.
   c. *Lightning overvoltage in DC switchyard*: In contrast to a substation, a converter station can operate in both bipolar mode and monopolar mode. The monopolar operation mode may be further classified into metallic return operation mode and ground return operation mode (not commonly used). In different operation modes, the lightning-intruding overvoltage has different characteristics.
      i. *Lightning overvoltage transmitted from DC line*: The polarity of the DC incoming feeder has a significant impact on the lightning-intruding overvoltage in the DC yard. Because a majority of lightning strikes are negative, most back

strikes and shielding failures occur on the conductor of positive polarity under the effect of the DC operating voltage of the DC switchyard. Shielding failures can also occur on the conductor of negative polarity. In this case, the negative lightning voltage superimposes the operating voltage to induce a much higher overvoltage.

The overvoltage generated as a result of lightning wave intruding into the DC yard through the pole conductor is usually lower because the equipment installed at the DC yard, such as DC filter and smoothing reactor, can damp the lightning wave, and the DC pole bus, metallic return conductor, electrode line, and neutral bus are all provided with surge arresters.

ii. *Lightning overvoltage on neutral bus*: The overvoltage on the neutral bus is a result of the lightning-intruding overvoltage generated on the metallic return conductor, electrode line, or DC line being coupled to the neutral bus through DC filters.

The electrode line operates at a lower voltage level and arcing horns are installed at both ends of insulators. In the monopolar ground return operation mode, due to the surge arresters installed on the neutral bus and the impulse energy storage capacitors installed at the inlet of the neutral bus, the lightning intruding overvoltage is low. The impulse energy storage capacitors perform well in suppressing the lightning-intruding overvoltage.

In the monopolar metallic return operation mode, the amplitude of lightning-intruding wave in case of either back strikes or shielding failures is much higher than that in the ground return operation mode because the tower carrying the metallic return conductor is much taller than the tower carrying the electrode line and has a high insulation level. Therefore, special attention must be given to the lightning-intruding overvoltage in the monopolar metallic line return operation mode when designing the lightning protection of the neutral bus.

2. **Lightning-intruding overvoltage protection**

UHV converter stations are protected from lightning-intruding overvoltage primarily by use of MOA that are provided according to the following principles:

a. The lightning overvoltage on the AC side should be suppressed using the MOAs installed on the AC side as practically as possible.
b. The lightning overvoltage on the DC side should be suppressed via the MOAs for the DC line, DC bus, and neutral bus.
c. Key components, such as valves, and the low-voltage components of AC and DC filters should be protected via the MOAs in close proximity to them.
d. As with the AC side, the MOAs installed in the DC yard should be arranged in close proximity to the protected equipment. If the arrester and the protected equipment are far apart, then effective measures for lightning protection should be determined through simulation calculations according to the specific engineering design.

References

[1] R.H. Golde, Lightning, Water Resources and Electric Power Press, Beijing, China, 1983.

[2] Mirinicov, et al., Electrical Design of 330–500 kV Overhead Transmission Line, Water Resources and Electric Power Press, Beijing, China, 1977, translation version.

[3] Terestchenko, Design and Construction of 500 kV Transmission Line, CIGRE, 1978.

[4] Anderson, et al., EHV Transmission Line, Water Resources and Electric Power Press, Beijing, China, 1979.

[5] R.B. Anderson, A.J. Erisksson, Lightning parameters for engineering application, Electra 39(69) (1980) 65–102.

[6] Г.Н. Александров, Design of EHV Transmission Lines, Water Resources and Electric Power Press, Beijing, China, 1987.

[7] CRIEPI, Guide to lightning protection design of power stations, substations and underground transmission lines, CRIEPI Report No. 40, 1995 (12).

[8] L. Qiang, G. Dingxie, Z. Yuanfang, Sensitivity analysis of lightning performance calculations for transmission line, High Voltage Eng. 31(3) (2005) 31–33.

[9] Estimating lightning performance transmission lines II—updates to analytical models, IEEE Working Group Report 8 (3) (1993).

[10] G. Dingxie, Z. Peihong, D. Min, et al., Comparison and analyses on over-voltage and insulation coordination of UHV AC transmission system between China and Japan, High Voltage Eng. 35(6) (2009) 1248–1253.

[11] EPRI, Transmission Line Reference Book 345 kV and Above, second ed., Califonia, 1982.

[12] Guide to procedures for estimating the lightning performance of transmission lines. CIGRE report, Reference 63, Working Group 33-01, 1991–10.

CHAPTER 6

External Insulation Characteristics and Insulation Coordination of UHV Transmission System

Chapter Outline

Ultra-high Voltage AC/DC Grids.
DOI: http://dx.doi.org/10.1016/B978-0-12-802161-3.00006-8

6.1 Discharge Characteristics of External Insulation

6.1.1 Classification of External Insulation

1. **Insulation of overhead transmission lines**
   The major insulating medium of UHV overhead transmission lines is air. In addition, insulators are used for insulation between the conductor and tower. Thus, the insulation of UHV overhead transmission lines can be classified into air gaps and insulators.
   Air gaps can be further classified into four types:
   a. Air gap between the conductor and tower
   b. Air gap between conductors
   c. Air gap between the conductor and ground in the middle of the span
   d. Air gap between the conductor and vehicles or driving machinery on the ground in the middle of the span
2. **Insulation of substations and converter stations**
   As with UHV overhead transmission lines, the major insulating medium of UHV substations (with open HV switchgear) and converter stations is air, and insulators are also used for insulation between the conductor and structure.

6.1.2 Discharge Characteristics of Air Gaps of UHV Overhead Transmission Lines

During operation, the air gap under AC/DC operating voltage, switching overvoltage, and lightning overvoltage will show different discharge characteristics. The amplitude of discharge voltage depends on factors such as the voltage type (operating voltage, switching overvoltage, or lightning overvoltage), polarity, waveform (front time of switching overvoltage), shape of electrodes forming the air gap, length of the air gap, and local atmospheric conditions.

Over the years, the former Soviet Union, Italy, the United States, Canada, the United Kingdom, and other countries have conducted several studies of discharge characteristics of air gaps of EHV/UHV overhead transmission lines, including various typical electrodes, tower air gaps, and air gaps between neighboring conductors in the middle of the span.

In the early studies and engineering design of UHV AC/DC transmission projects, tests of discharge characteristics of rod–rod and rod–plane gaps and air gaps for full-scale towers, conductors, and fittings used in projects were performed in China and fruitful achievements have been made. Also, the world's leading air gap study facilities such as the test device for simulating long wave-front overvoltage that may occur in UHV AC systems have been built in China.

6.1.2.1 Discharge characteristics of air gaps for AC lines

1. **Discharge characteristics of air gap between conductor and tower**
   a. Power frequency voltage discharge characteristics

      For UHV AC single-circuit or double-circuit towers with I-string insulators, the minimum safety distance from the conductor to tower body depends on the operating voltage. To properly determine the distance, the power frequency voltage discharge tests have been performed on the air gap between the outside-phase conductor and tower body of the full-scale UHV cat-head tower, cup tower, and double-circuit tower, respectively, in China. Through these tests, the discharge characteristics curves for air gap of these tower types were obtained, providing a basis for engineering design. Figure 6.1 shows the power frequency voltage discharge test on a UHV cat-head tower with a gap varying from 2.0 to 4.5 m, and Figure 6.2 shows the corresponding discharge characteristics curve [1]. Figure 6.3 shows the power frequency voltage discharge test on a UHV cup tower with a gap varying from 2.9 to 3.3 m, and Figure 6.4 shows the corresponding discharge characteristics curve.

      Tests on the full-scale cat-head tower and cup tower of the 1000-kV Jindongnan–Nanyang–Jingmen UHV AC pilot project show that the relationship

Figure 6.1
Power frequency voltage discharge test for an outside-phase conductor of a UHV cat-head tower.

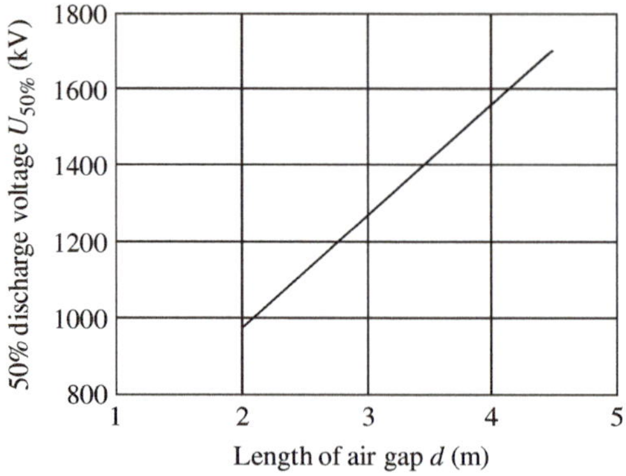

Figure 6.2
Discharge characteristics curve.

Figure 6.3
Power frequency voltage discharge test for an outside-phase conductor of a UHV cup tower.

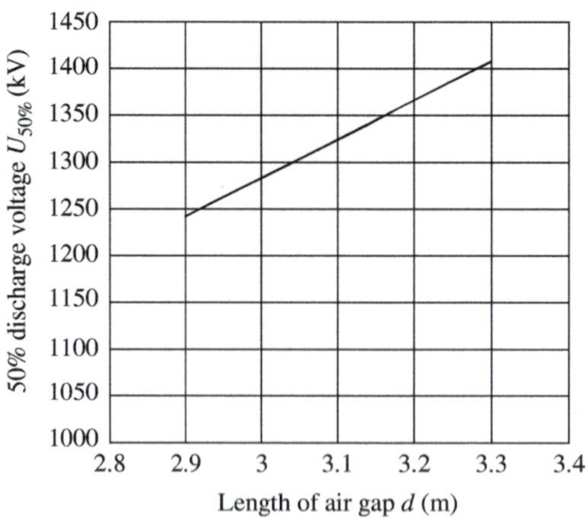

Figure 6.4
Discharge characteristics curve.

between 50% power frequency discharge voltage $U_{50\%}$ and the air gap length can be expressed as $U_{50\%} = kS$, where S represents the length of air gap and k represents the gradient factor. When $2\text{ m} \leq S \leq 4.5\text{ m}$, k ranges from 392 to 485 kV/m and decreases with the increase of the air gap length.

In the power frequency voltage discharge tests of the conductor-to-tower air gap of a full-scale UHV double-circuit tower, because the top phase and the middle phase have the same frame structure, their discharge characteristics are also the same. However, the bottom phase, because of a very different frame structure, has different discharge characteristics. Therefore, the center phase and bottom phase can represent the insulation characteristics of air gaps across the tower.

During the tests, the distance from the middle-phase conductor (with I-string insulators) to the tower body varies from 2.7 to 3.5 m, and that from the bottom-phase conductor (with I-string insulators) to the tower body varies from 2.5 to 3.5 m, and the tower widths at the points nearest to the conductors are 5.8 and 7.5 m, respectively. The test results are shown in Figure 6.5.

Figure 6.5 shows that within the given range of gaps, the power frequency discharge voltage of both the middle-phase and bottom-phase conductors increases almost linearly with the increase of air gap. Within the gap range from 2.5 to 3.5 m, the average discharge voltages of the center phase and bottom phase are 417 and 405 kV/m, respectively. The former is 3.0% greater than the latter, which is mainly caused by the difference in the tower frame structure and width of tower body.

b. Switching impulse voltage discharge characteristics
   i. *Development of UHV long wave-front switching impulse voltage generator.* For long-distance, high-capacity EHV transmission systems, the switching

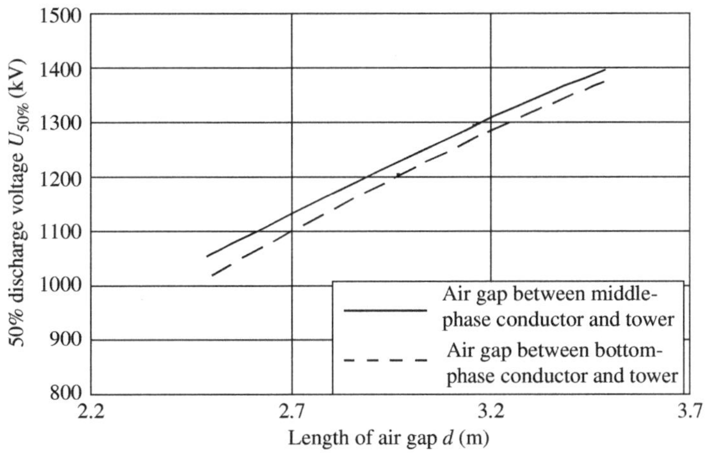

Figure 6.5
Power frequency voltage discharge characteristics curve.

overvoltage waveforms measured and calculated by simulation with a transient network analyzer all have a front time longer than 1000 μs. For UHV transmission systems, the front time of switching overvoltage is even longer, usually up to 2000–3000 μs. UHV systems have a higher voltage level and, therefore, require higher insulation strength. If the waveform of the switching impulse voltage in the test can be made as close to the overvoltage waveform within the system as possible, the operational reliability and construction cost efficiency of UHV lines can both be considered. So, an impulse voltage generator for simulating long wave-front overvoltage in a UHV system is necessary to external insulation tests and studies. No countries have ever succeeded in developing a generator capable of generating impulse voltage with a front time of 2000–3000 μs used for testing UHV voltage grade. To determine the proper air gaps between conductors and towers for UHV AC systems, a program to develop such a generator was launched in China.

In the development process, a number of key technical challenges and problems (such as synchronized firing for multiple stages in series, low-voltage output efficiency, and severe long wave-front loop oscillation) were overcome, and finally a 7500-kV-long wave-front switching impulse voltage generator was successfully developed. This generator has high discharge efficiency and can generate switching impulse voltage above 2500 kV, with a front time longer than 2500 μs, meeting the test requirements for EHV/UHV transmission lines.

- *Structural parameters of the generator.* This generator, used in the UHV AC test base, consists of 25 stages and has a nominal voltage of 7500 kV. It is equipped with a ±150 kV DC charger. The charging voltage is 300 kV (±150 kV) at each stage, and each main capacitor is a 1.35 μF pulse capacitor. This generator can generate both standard switching impulse voltage and 1000 to 2500 μs long wave-front switching impulse voltage whose waveforms and amplitudes can meet test requirements of EHV/UHV systems. In particular, the peak value of 2500 μs long wave-front switching impulse voltage could reach above 2500 kV. The generator casing has a diameter of approximately 6 m and a height of 30 m. An air conditioning system is arranged inside it. The casing is assembled using epoxy resin.
- *Charging switch.* The charging switch closes to charge the reservoir capacitor at each stage with the same voltage, which reduces the non-uniformity occurring in common circuits; it opens when the capacitors discharge, thus reducing the energy loss caused by charging resistance and increasing the discharge efficiency. The opening and closing of this switch are controlled by a high-precision rack-and-pinion driving module, which is simply and properly structured with good manufacturability.
- *Multi-stage firing device.* The generator is equipped with a multi-stage firing device as the firing switch. The firing device fires each stage by first breaking down a

small gap and then making the main gap conductive. The firing impulse is passed from one stage to the next via a transmission R-C installation. For voltage grading purposes, a circular copper tube is arranged in each small gap of a multi-stage main gap, and a copper strip and epoxy capacitance bar are used to form an equalizing capacitor. This achieves voltage grading for multi-stage gaps during firing, makes the gaps more easily conductive, and thus enhances synchronization among multiple stages. The success in developing this device makes it possible for China to perform studies of the characteristics of external insulation under long wave-front switching impulse voltage in EHV AC/UHV AC systems.

At present, the design of air gaps controlled by switching overvoltage for UHV AC transmission lines and substations in China are based on the test data of switching impulse with a 1000-μs front time. So, the studies are focused on the discharge characteristics of air gaps under switching impulse with a 1000-μs front time.

ii. Switching impulse voltage discharge characteristics of air gaps of full-scale UHV single-circuit towers. During the switching impulse voltage test, the distance between the outside-phase conductor and tower body varied from 4.5 to 8.2 m (as shown in Figure 6.6). Figure 6.7 shows the discharge characteristics curve of the air gap under standard switching impulse voltage. The tower width at the point nearest to the conductor is 6.6 m.

To obtain the long wave-front switching impulse discharge characteristics of the air gap between the outside-phase conductor (with I-string insulators) and UHV cat-head tower, discharge tests were performed on an air gap of 5.6 m between the conductor and tower leg under switching impulses with front times of 500, 1000, and 5000 μs, respectively, to obtain the 50% discharge voltage. The test results indicate that,

Figure 6.6
Switching impulse voltage discharge test for the outside-phase I-string of a UHV cat-head tower.

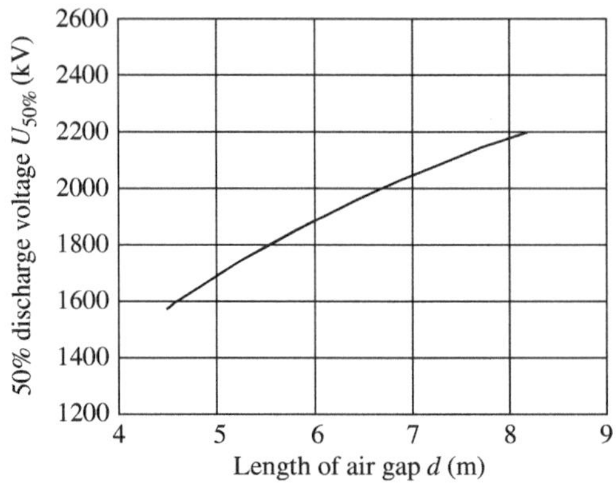

Figure 6.7
Discharge characteristics curve.

Figure 6.8
Switching impulse voltage discharge test for the outside-phase I-string of a UHV cup tower.

compared with the discharge voltage under standard switching impulse, the discharge voltage under switching impulses with front times of 500, 1000, and 5000 μs is increased by 5.1%, 7.0%, and 18.8%, respectively.

For a UHV cup tower, discharge tests were performed on the air gap between the outside-phase conductor and tower body with a typical distance of 6.4 m under switching impulse voltages with front times of 120, 250, 500, 1000, and 5000 μs, respectively. Figure 6.8 shows the test configuration. Figure 6.9 shows the 50% discharge voltage ($U_{50\%}$) curve. The test results indicate that compared with 50% discharge voltage under standard switching impulse, 50% discharge voltage under switching impulses with front times of 500, 1000, and 5000 μs is increased by 4.4%, 8.0%, and 18.2%, respectively. The test data provide a basis for waveform correction [2].

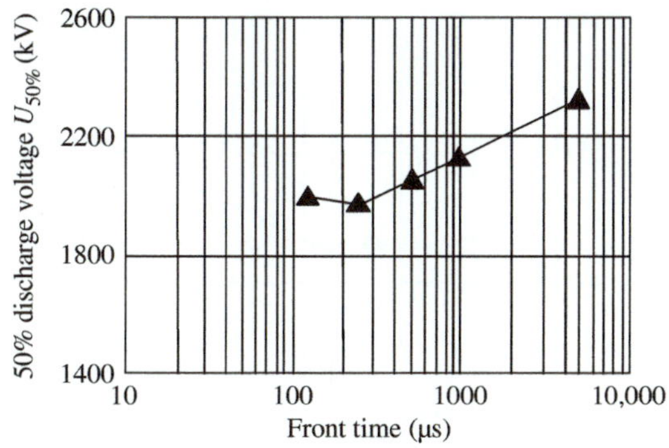

Figure 6.9
Discharge characteristics curve under switching impulses with various front times.

Figure 6.10
Switching impulse voltage discharge test on the gap between the middle-phase conductor and tower window of a UHV cup tower.

Figure 6.10 shows the picture of switching impulse voltage discharge test on the gap between the middle-phase conductor and tower window of a UHV cup tower, where the distance from the middle-phase conductor to the cross-arm and to the diagonal brace is 8.1 and 7.7 m, respectively. The discharge characteristics curve under switching impulses with different front times are shown in Figure 6.11, which can be used as a basis for determining the gap length and correcting waveforms.

iii. Switching impulse discharge characteristics of air gaps of a full-scale UHV double-circuit tower. In the studies of switching impulse discharge characteristics of air gaps of a full-scale UHV double-circuit tower, tests were performed on air gaps between a conductor and various parts of the center phase and bottom phase under switching impulse with a 1000-μs front time. The test results had been applied in the design of double-circuit towers in the Power Transmission Project

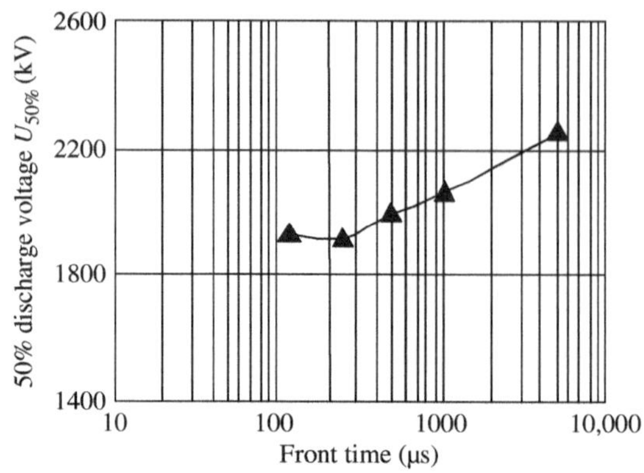

Figure 6.11
Discharge characteristics curve under switching impulses with different front times.

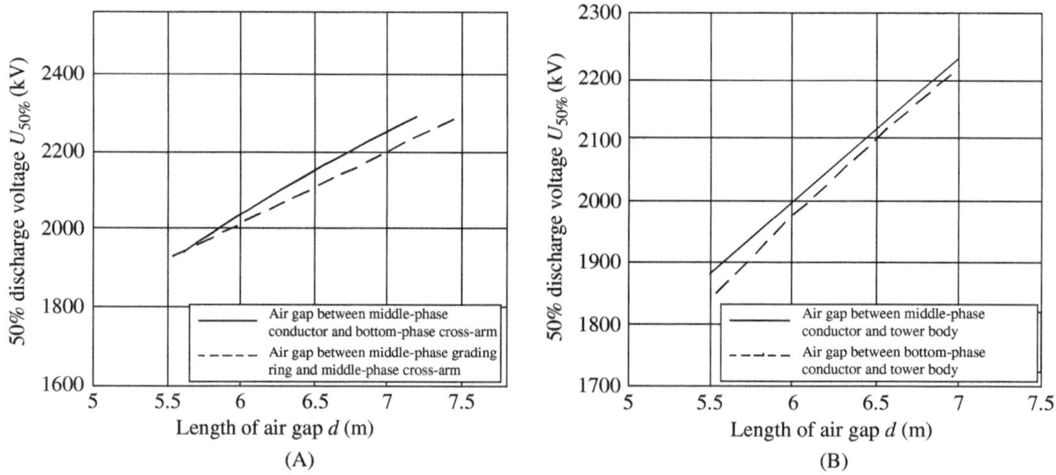

Figure 6.12: Switching impulse discharge characteristics curves of the air gap between the I-string and tower in a UHV double-circuit line.
(A) Air gap between middle-phase I-string and middle-phase/bottom-phase cross-arm; (B) air gap between middle-phase/bottom-phase I-string and tower body.

from Anhui to East China. In the tests, an insulating rope was used to pull the conductor and make the I-string insulators form a wind deflection angle of 20°, and then the relative position of the conductor to tower's different parts was adjusted to form main gaps for test. Figure 6.12A shows the discharge characteristics curves of the air gap between the grading ring on the HV side of the middle-phase insulator and middle-phase cross-arm, and that between the middle-phase conductor and bottom-phase cross-arm. Figure 6.12B shows the

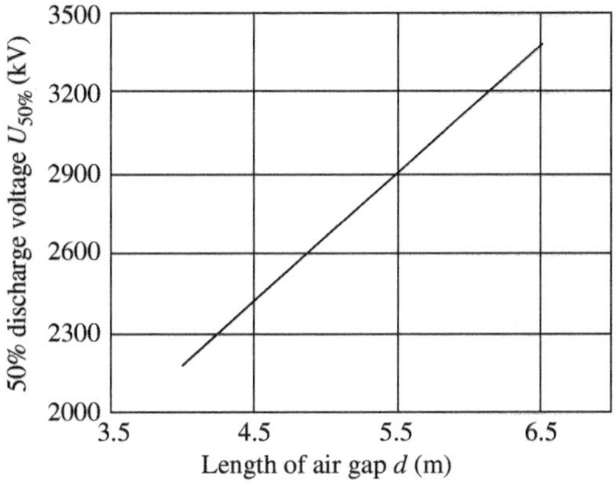

Figure 6.13
Lightning impulse discharge characteristics curve of the air gap between outside-phase conductor and tower window of a UHV cat-head tower.

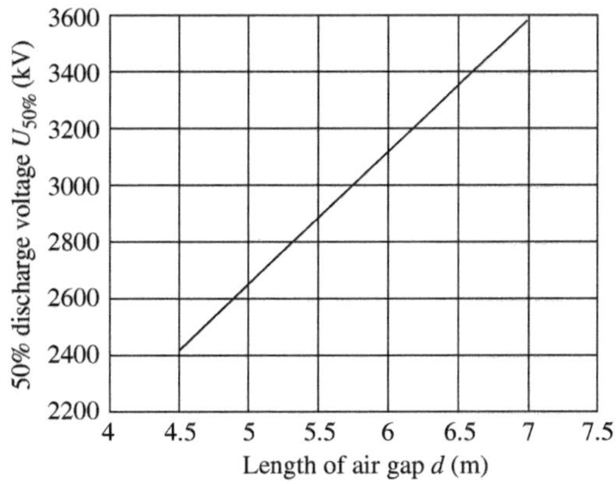

Figure 6.14
Lightning impulse discharge characteristics curve of the air gap between outside-phase conductor and tower window of a UHV cup tower.

discharge characteristics curves of the air gap between the middle-phase conductor and tower body, and that between the bottom-phase conductor and tower body.

c. Lightning impulse voltage discharge characteristics
   i. Lightning impulse discharge characteristics of air gaps of a full-scale UHV single-circuit tower. Standard lightning impulse discharge tests were performed on outside-phase air gaps of the UHV cat-head tower and cup tower, respectively. Figures 6.13 and 6.14 show the lightning impulse discharge characteristics curves of the air gap between the outside-phase conductor and tower window for the UHV cat-head tower and cup tower, respectively.

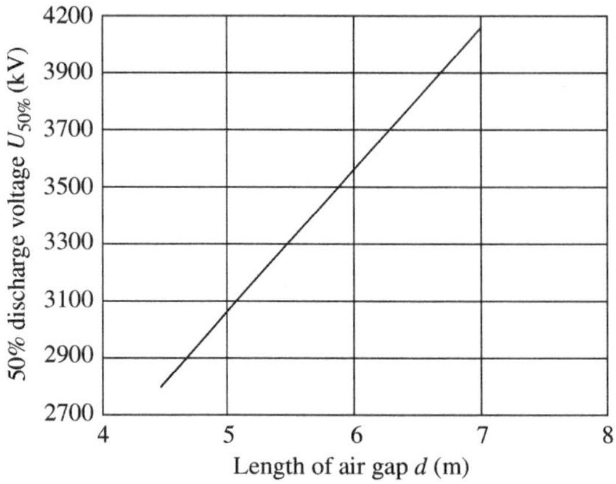

Figure 6.15
Lightning impulse discharge characteristics curve of the middle-phase (with V-string insulators) of a UHV cat-head tower.

For UHV cat-head towers, standard lightning impulse discharge tests were performed on the air gap between the middle-phase conductor (with V-string insulators) and cross-arm. In the test, the gap length was adjusted with a template parallel to the cross-arm. Figure 6.15 shows the relation curve between the lightning impulse discharge voltage and length of the air gap.

For UHV cup towers, standard lightning impulse discharge tests were performed on the air gap between the middle-phase conductor (with V-string insulators) and cross-arm. In the test, the gap length was changed within the range from 5.0 to 7.0 m by a template, whereas the distances from the conductor to the side and to the bottom of the tower window were both maintained at 7.7 m. Standard lightning impulse was applied on the conductor for test. Figure 6.16 shows the test results.

The test results indicate that the lightning impulse discharge voltage gradient of the air gap between the outside-phase conductor (with I-string insulators) and tower body is 587 kV/m, and that between the middle-phase conductor (with V-string insulators) and tower window is 575 kV/m. The frame structure of the tower has a minor effect on the lightning impulse discharge voltage.

ii. Lightning impulse discharge characteristics of air gaps of a full-scale UHV double-circuit tower. Figure 6.17 shows the lightning impulse discharge characteristics curves of the air gap between the middle-phase conductor and top surface of the bottom-phase cross-arm and the air gap between the bottom-phase conductor and tower body.

As shown in Figure 6.17, the 50% lightning impulse discharge voltage increases linearly with the increase of gap length. For a gap length from the middle-phase conductor to top surface of the bottom-phase cross-arm varying from 5.5 to 7.0 m,

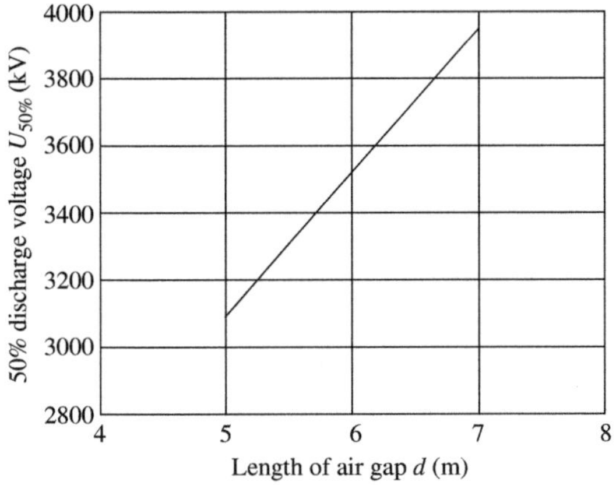

Figure 6.16
Lightning impulse discharge characteristics curve of the middle-phase (with V-string insulators) of a UHV cup tower.

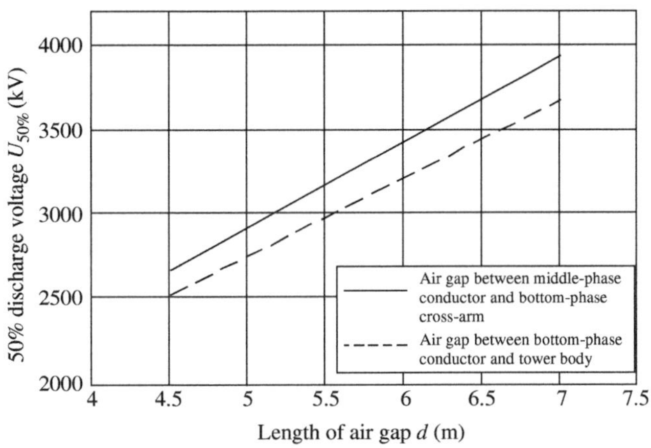

Figure 6.17
Lightning impulse discharge characteristics curves of gaps with I-string on a UHV double-circuit tower.

the discharge voltage gradient ranges from 562.8 to 580.9 kV/m. And for a gap length from the bottom-phase conductor to the tower body varying from 4.5 to 6.5 m, the gradient ranges from 545 to 580 kV/m.

2. **Discharge characteristics of air gaps between phase conductors**
During line switching operation or when a ground short-circuit fault occurs on a UHV AC transmission line, if the overvoltage amplitude between two phase conductors is beyond the insulation level, then flashover may occur between them. As such, the test and simulation results of phase-to-phase overvoltage are usually referenced in design of

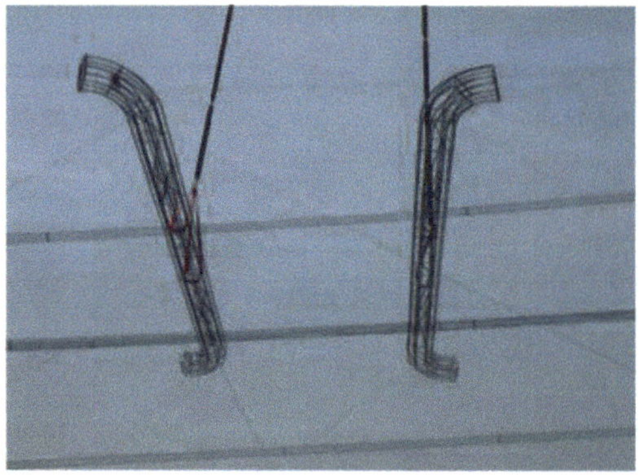

Figure 6.18
Switching impulse voltage discharge test on the air gap between eight-bundle conductors.

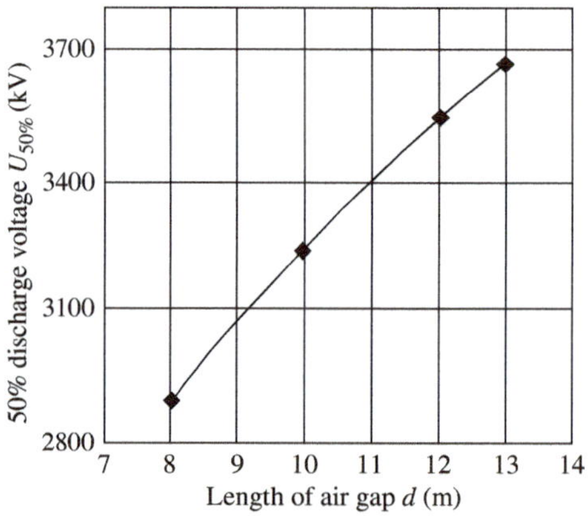

Figure 6.19
Discharge characteristics curve under switching impulse with a 1000-μs front time.

transmission lines to determine the minimum safe and economic air gap between phases.

The required insulation strength between phases of a UHV line depends on the switching overvoltage. To clarify the discharge characteristics of phase-to-phase air gaps, two sets of impulse voltage generators were used to produce positive and negative (respectively) switching impulse voltages to apply to the air gap between conductor bundles or between fittings. Figures 6.18 and 6.19, respectively, show configuration of the discharge test on the air gap between eight-bundle conductors and the corresponding discharge characteristics curve.

The test results show that when the gap varies from 8 to 13 m, the 50% discharge voltage gradient ranges from 282 to 362 kV/m, and the gradient decreases with the increase of the gap. The relation between 50% discharge voltage $U_{50\%}$ of the air gap between eight-bundle conductors under switching impulse with a 1000-μs front time, and the air gap length S is given by:

$$U_{50\%} = 1047.2S^{0.49}(8\text{m} \leq S \leq 13\text{m})$$

6.1.2.2 Air gaps of DC lines

1. **DC voltage discharge characteristics**

   Because the DC discharge voltage of a unit length of air gap is very high, its required air gap length is much less than that required by switching or lightning impulse voltage. DC discharge tests were performed on rod–plane and rod–rod gaps, which can be used as references for evaluation of DC discharge characteristics of power transmission projects. The tests were performed at an altitude of 50 m (Beijing) and 4300 m (Tibet), respectively. Figures 6.20 and 6.21 show the discharge characteristics curves obtained from the tests. For comparison purpose, test data from other countries are also presented in the figures.

   The test results show that, except the rod–plane gap under negative voltage, the rod–plane gap and rod–rod gap under positive voltage and rod–rod gap under negative voltage all have low discharge dispersity and a variation coefficient smaller than 0.9%. In addition, in Figures 6.20 and 6.21 it can be seen that with the increase of

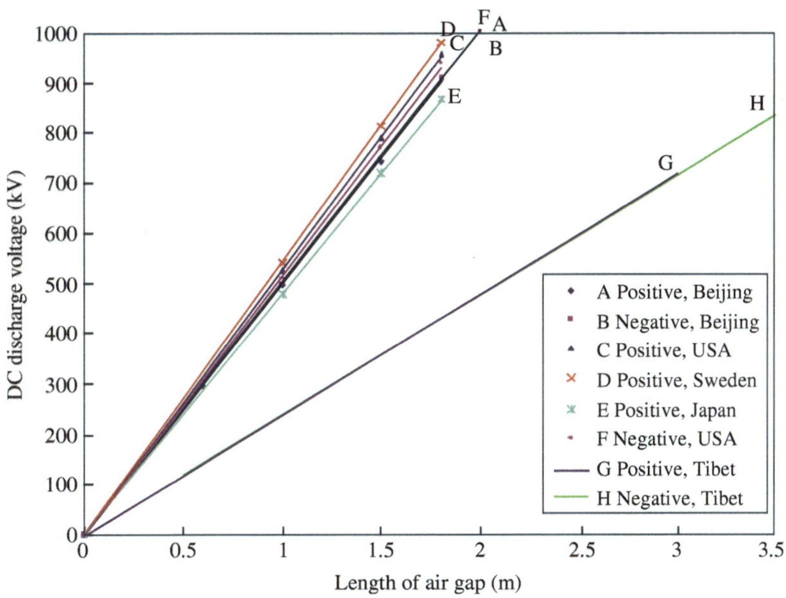

Figure 6.20
DC voltage discharge test results of rod–rod gap in different places.

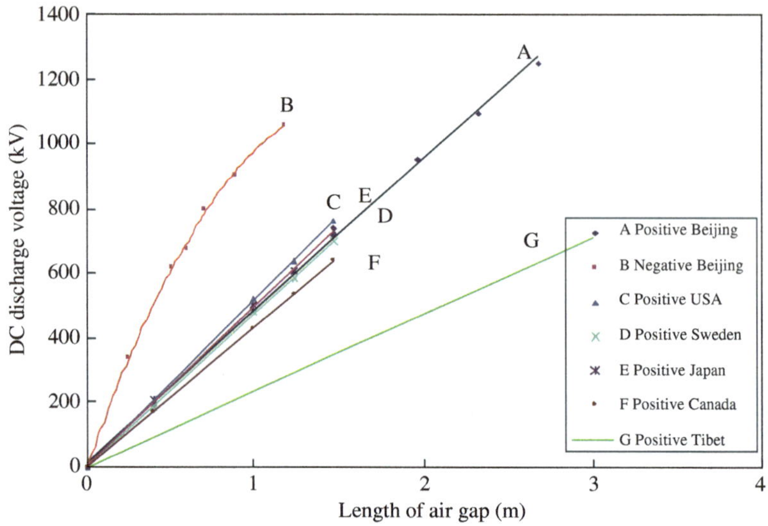

Figure 6.21
DC voltage discharge test results of rod–plane gap in different places.

altitude, the DC discharge voltage of rod–plane/rod–rod gap decreases much faster than switching impulse discharge voltage.

2. **Switching impulse voltage discharge characteristics**
In ±800- and ±1100-kV HVDC transmission lines, the conductors are suspended by V-string insulators, and the distance between the conductor and tower body is determined by the switching overvoltage. And the 50% switching impulse discharge voltage is sensitive to the shape of electrodes (i.e., the shape and size of the tower head and width of the tower body). To determine a proper air gap, switching impulse voltage discharge tests need to be performed on a full-scale tower.

For UHV DC single-circuit lines, switching impulse discharge tests were performed on transmission lines rated ±1100 kV or less in the outdoor test field of the UHV DC test base and the high-altitude outdoor test field of the Tibet test base. Tower heads of the same structure and size were used for tests at different altitudes to make the test results comparable [3].

Figures 6.22 and 6.23, respectively, show the configuration of the tower head for the tests in Beijing and Tibet. The voltage applied is positive standard switching impulse voltage with a front time of 250/2500 μs.

Through such tests, the relation curve between 50% switching impulse discharge voltage at the tower head (with V-string insulators) and the air gap length was obtained for the altitudes of 50 and 4300 m, as shown in Figure 6.24. According to these test results, the relation curve between the positive switching impulse discharge voltage and

Figure 6.22
Configuration of the top hamper of the tower in the outdoor test field of Beijing test base.

Figure 6.23
Configuration of the top hamper of the tower in the outdoor test field of Tibet test base.

air gap length for other altitudes was derived with the interpolation method, as shown in Figure 6.25.

The test results indicate that the switching impulse discharge dispersion is relatively small in Beijing and Tibet. The variation coefficient of test data is smaller than 3.7% in Beijing and smaller than 4.2% in Tibet (both are below the common variation coefficient of 5–6%). Also, the variation coefficient does not increase sharply with the increase of altitude.

In the Xiangjiaba–Shanghai ±800-kV UHV DC demonstration project, because the line corridor is narrow, F-shape towers and V-string insulators are used and the upper and lower conductors are arranged vertically in the eastern section of the line. This differs greatly from the I-string configuration typically used in ±500-kV DC lines and

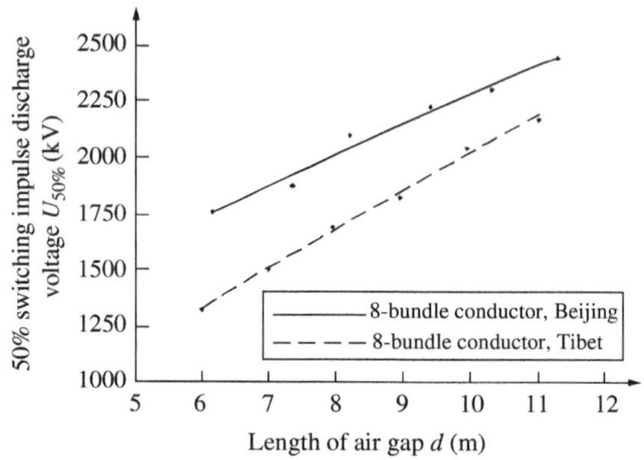

Figure 6.24
Switching impulse discharge characteristics curves.

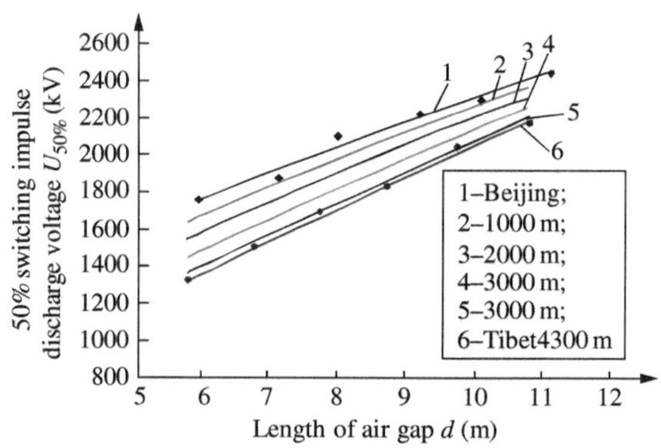

Figure 6.25
Switching impulse discharge characteristics curves for different altitudes.

the horizontal V-string configuration typically used in ±800-kV single-circuit lines in China. Figure 6.26 is a single-line diagram of an F-shape single-circuit tower for a ± 800-kV UHV DC transmission line. From the figure it can be seen that, for the tower head of a single-circuit F-shape tower with vertical V-string configuration, the width of the tower body corresponding to the lower conductor is greater than that when the single-circuit V-string conductor is arranged horizontally, and thus the switching impulse discharge voltage for the air gap is relatively lower; because of the lower cross-arm, the upper air gap of the tower head is greatly different from the air gap when the conductors are horizontally arranged. Thus, tests are required to determine the discharge characteristics. The impulse discharge tests on air gaps of an F-shape tower were performed in the outdoor test field of SGCC's UHV DC test base. Figure 6.27 shows a picture of a simulated tower head manufactured for such tests. The distance from the grading ring to the cross-arm is almost the same as that from the grading ring

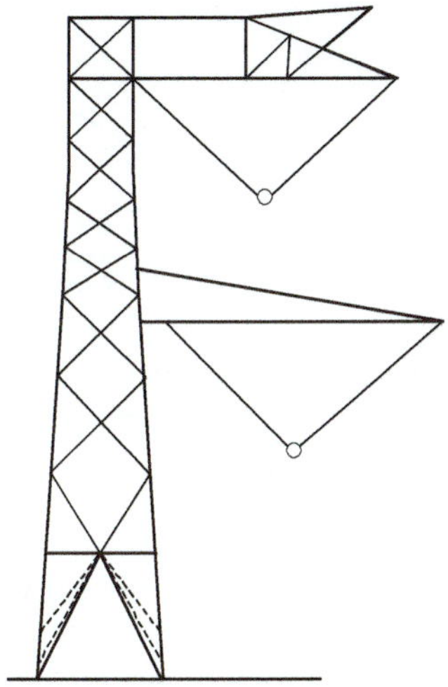

Figure 6.26
Single line diagram of an F-shape tower.

Figure 6.27
Arrangement for test on the upper air gap of an F-shape tower.

to the tower body, approximately 1.22-times the distance from the conductor to the lower cross-arm.

Figure 6.28 shows the relation curve between 50% positive switching impulse discharge voltage at upper section of the F-shape tower and the length of air gap. For comparison purposes, the 50% positive switching impulse discharge characteristics curve for the tower head with V-string conductors horizontally arranged in ±800-kV UHV DC transmission lines is also given in the figure. It can be seen that for the same distance from the grading ring to the upper cross-arm, the presence of a lower cross-arm leads to a decreased 50% switching impulse discharge voltage at the upper section of the F-shape

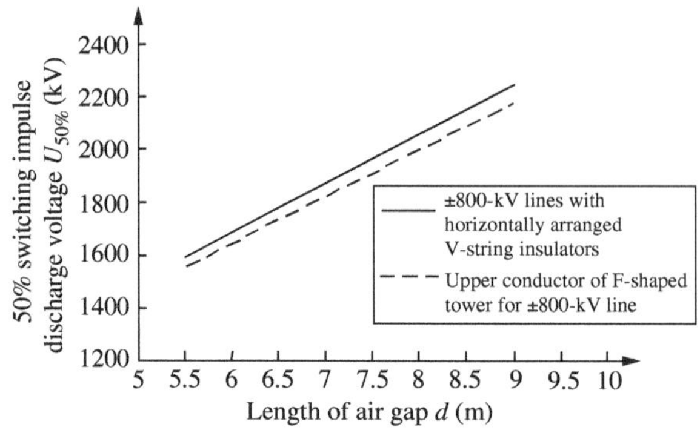

Figure 6.28
Switching impulse discharge characteristics curve of upper section of an F-shape tower.

tower. In this test, the 50% switching impulse discharge voltage decreases by approximately 3.7% when the effects of the height above the ground are excluded.

The arrangement of the lower conductor on an F-shape tower is similar to that of single-circuit conductors in horizontal arrangement, except that the corresponding tower body is wider. According to the test results of ±800-kV UHV DC transmission line towers with V-string insulators, for the same air gap length, the switching impulse discharge voltage reduces by 3.5% approximately every 1 m that the width of the tower body is increased by. In the actual design of an F-shape tower, the width of the lower cross-arm is 4.5 m, whereas for ±800-kV towers with V-string insulators horizontally arranged, the width of the tower body is 3.4 m. It can be deemed that with the same air gap length, the 50% switching impulse discharge voltage $U_{50\%}$ at the lower section of F-shape tower is 3.85% lower than the discharge voltage at the tower head in a ±800-kV UHV DC transmission line with conductors horizontally arranged and suspended by V-string insulators. Figure 6.29 shows the switching impulse discharge characteristics curve of tower head for ±800-kV UHV DC transmission lines with horizontally arranged V-string insulators, and the switching impulse discharge characteristics curve at the lower section of an F-shape tower derived from the relation mentioned [4].

The vertical arrangement of the double-circuit conductors on the same tower differs greatly from the traditional arrangement of the horizontally arranged single-circuit conductors. The additional lower cross-arm has some impacts on the discharge characteristics of the upper section. To determine the reasonable design values, switching impulse discharge tests are performed on the air gap of a full-scale simulated tower head. Figures 6.30 and 6.31 show the 50% switching impulse discharge voltage curve.

3. **Lightning impulse discharge characteristics**

Lightning impulse discharge tests were performed on air gaps of a DC line with V-string insulators in Beijing and Tibet, respectively, and the tested air gaps meet the

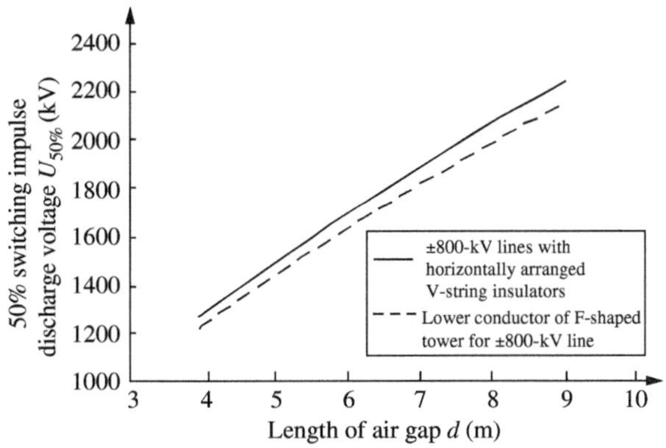

Figure 6.29
Switching impulse discharge characteristics curve of lower section of an F-shape tower.

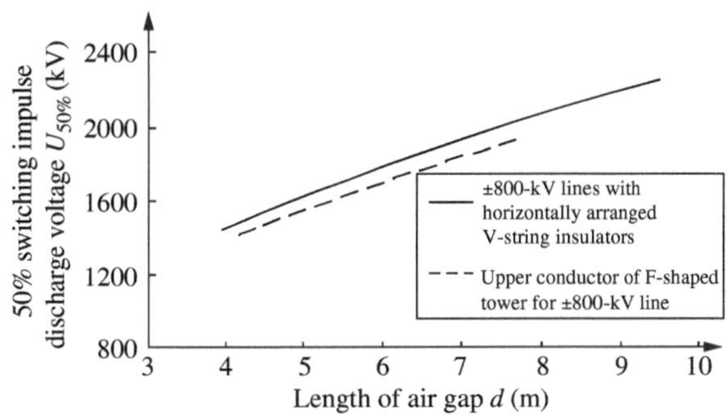

Figure 6.30
Switching impulse discharge characteristics curve of upper section of the tower head of a ±800-kV tower.

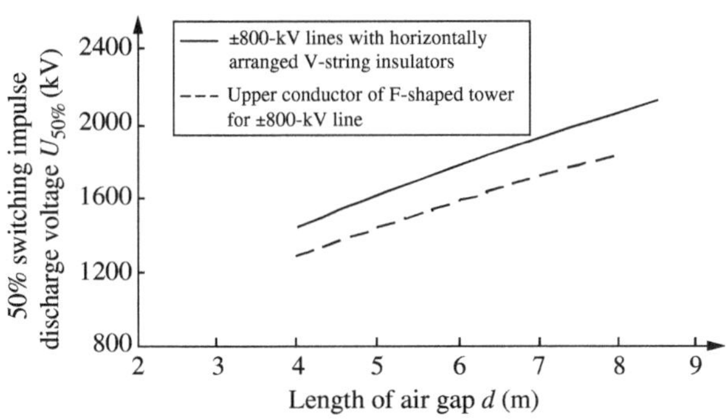

Figure 6.31
Switching impulse discharge characteristics curve of lower section of the tower head for a ±800-kV tower.

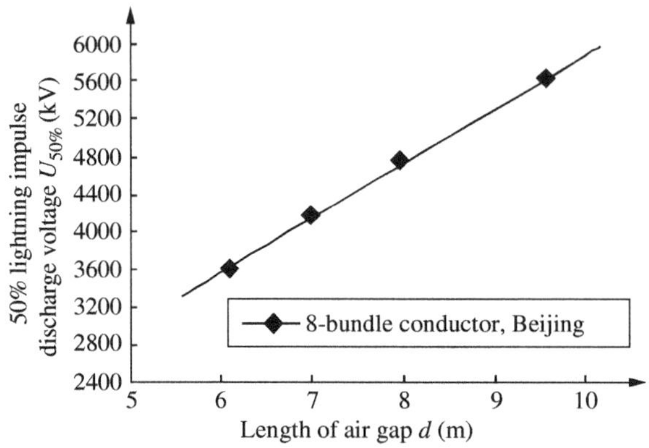

Figure 6.32
Lightning impulse discharge characteristics curve for tower head air gaps in Beijing.

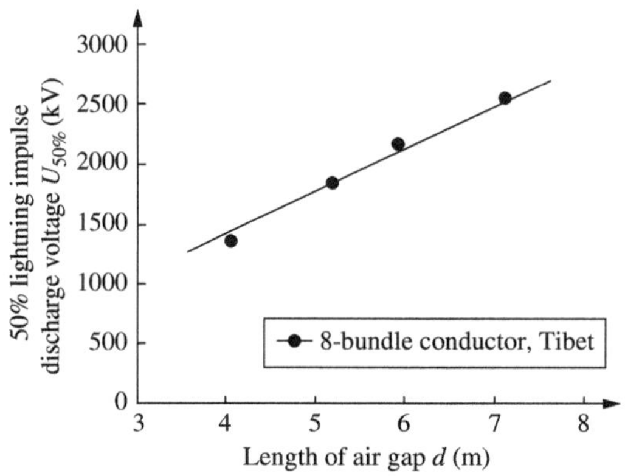

Figure 6.33
Lightning impulse discharge characteristics curve for tower head air gaps in Tibet.

design requirements of ±800- to ±1100-kV transmission projects. Figures 6.32 and 6.33, respectively, show the relation curve between the 50% positive lightning impulse discharge voltage and air gap length obtained in Beijing and Tibet.

The test results indicate that the lightning impulse discharge voltage and the air gap length have a good linear relationship, and the discharge dispersion or the variation coefficient is small: the variation coefficient of the test data from Beijing is smaller than 2% and that from Tibet is smaller than 3%. In addition, during the test, it was observed that the lightning impulse discharge always takes the shortest path from the HV electrode to the ground electrode, and the flashover arc appears as a bright straight line. It is shown in Figures 6.32 and 6.33 that the lightning impulse discharge voltage gradients in Beijing and Tibet is 591 and 354 kV/m, respectively, and that the ratio of the former to the latter is approximately 1.67.

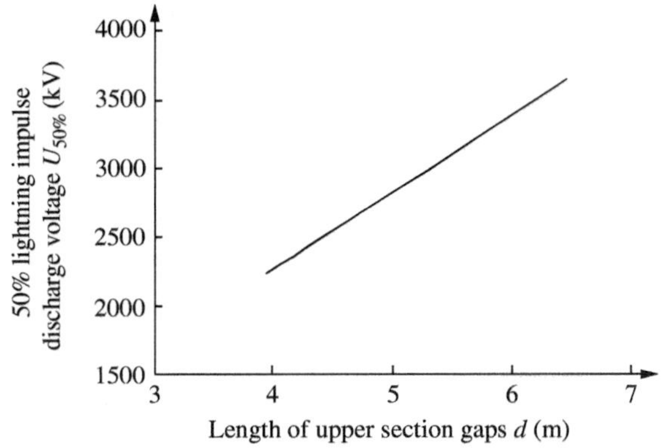

Figure 6.34
Lightning impulse discharge characteristics curve at upper section of a tower for ±800-kV double-circuit system.

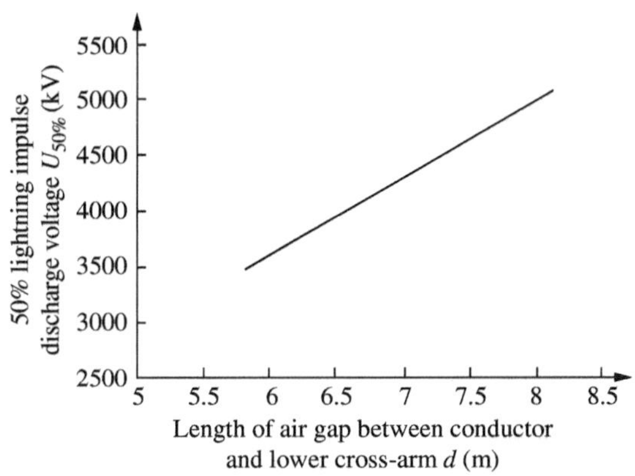

Figure 6.35
Lightning impulse discharge characteristics curve at lower section of a tower for ±800-kV double-circuit system.

For ±800-kV double-circuit lines, the 50% lightning impulse discharge voltage curves at the upper section and lower section of a tower are, respectively, shown in Figures 6.34 and 6.35.

6.1.3 Discharge Characteristics of Air Gaps in UHV Substations and Converter Stations

Systematic tests and studies of the discharge characteristics of air gaps in UHV substations and converter stations have been performed in China.

6.1.3.1 Discharge characteristics of typical air gaps in UHV substations

In studies of typical electrodes in the UHV AC substation, switching and lightning impulse voltage discharge tests were performed on the air gaps between the bundle conductor, tube bus, grading ring, and the gantry structure. Also, standard and long wave-front switching impulse voltage discharge tests were performed on typical phase-to-phase air gaps such as those between bundle conductors, between tube buses, between equipment grading rings, and between shielding rings at the ends of conductors. A large number of discharge characteristics curves were obtained through these studies and tests.

1. **Switching impulse voltage discharge characteristics of the air gap between four-bundle conductor and gantry structure**
   The simulation structure in substation used for tests has a girder (35 m high and 4 m wide) and two columns (5 m wide). The four-bundle conductor is simulated with rigid galvanized iron pipes at a 1:1 scale, spaced at 400 mm. Figure 6.36 shows the test arrangement, and Figure 6.37 shows the test results with varying distances between the outer side of the conductor and the column.
2. **Switching impulse discharge characteristics of the air gap between tube bus and gantry structure**
   Figure 6.38 shows the arrangement for the switching impulse voltage discharge test on the gap between the tube bus and structure leg, where the simulated structure in substation used for test has a girder (35 m high and 4 m wide) and two columns (5 m wide) and the tube bus is 17 m above the ground. The test results are shown in Figure 6.39.

Figure 6.36
Arrangement for the discharge test on the air gap between four-bundle conductor and gantry structure.

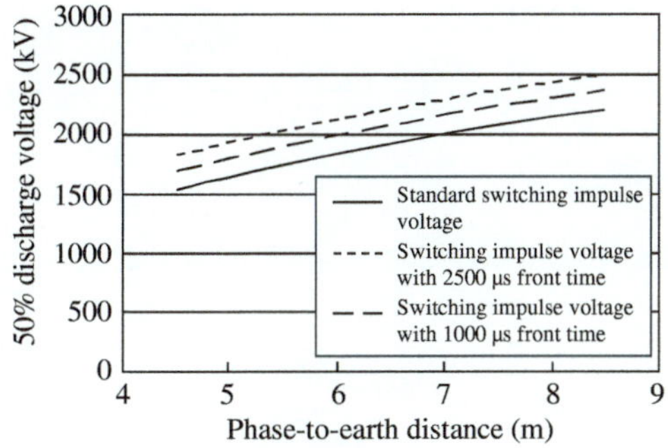

Figure 6.37
Discharge characteristics curves.

Figure 6.38
Arrangement for the discharge test on the air gap between tube bus and gantry structure.

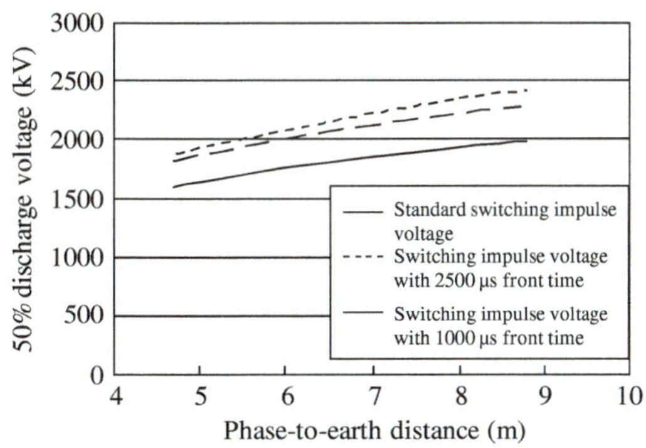

Figure 6.39
Discharge characteristics curves.

Figure 6.40
Arrangement for the discharge test on the air gap between grading ring and gantry structure.

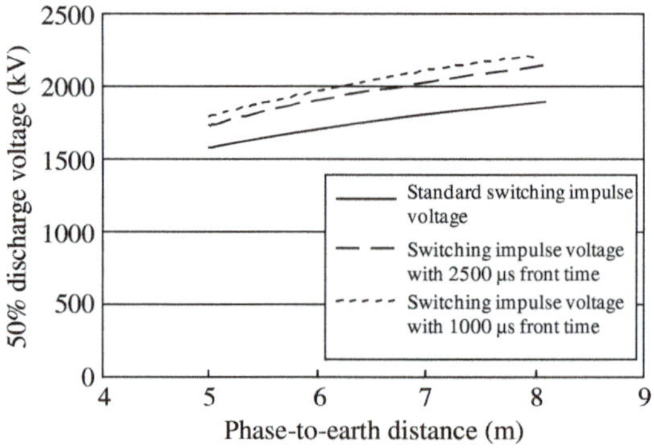

Figure 6.41
Discharge characteristics curves.

3. **Switching impulse voltage discharge characteristics of the air gap between grading ring and gantry structure**

Figure 6.40 shows the arrangement for the discharge test, where the arrester grading ring is 16.5 m above the ground, the structure column is 5 m wide, and the girder is 35 m high. The test results are shown in Figure 6.41.

The test results show that the discharge voltage under switching impulse voltage with a 2500-μs front time is 14.8%–16.0% higher than that under the standard switching impulse voltage.

In addition, discharge tests under switching impulse with a 1000-μs front time were performed on typical phase-to-phase air gaps as well, and the discharge parameters under various air gap lengths, voltage distribution factors, and heights above ground were obtained. The study of the effects of voltage distribution factor α on discharge voltage shows that when α is in the range of 0.3–0.5, it basically has a linear relationship with the discharge voltage.

6.1.3.2 Discharge characteristics of typical air gaps in UHV converter stations

For ±800-kV converter stations, DC voltage discharge tests and switching impulse discharge tests were performed on air gaps between typical electrodes (such as the terminal ball and grading ring) and grounding bodies. Corresponding discharge characteristics curves were obtained, which were used to support insulation design and selection of equipment in converter stations.

The equipment in the valve hall of a UHV converter station mainly consists of the valve side bushing of converter transformer, converter valve bank, disconnector, surge arrester, wall bushing for outgoing DC line, as well as the pole bus connected to the main circuit. The equipment in outdoor DC switchyard mainly includes smoothing reactor, filter, disconnector, surge arrester, optical current transformer, DC voltage divider, and pole bus for HVDC leading out, so a variety of air gaps can be found on the equipment. In the research of discharge characteristics of air gaps in a UHV converter station, simulation tests were conducted on the typical air gaps whose minimum length must be controlled, including the air gaps between grading ball and plane, between tube bus and plane, and between tube buses.

1. **Switching impulse voltage discharge characteristics of the air gaps between tube bus and wall/ground**
   For the test, an aluminum tube bus with a diameter of 0.3 m was used, and a metallic ball with a diameter of 1.5 m was installed at both ends of the bus. Metal wire nets were connected to simulate the wall of the valve hall, and the ground was covered with galvanized iron plates to simulate the ground or a wall of the valve hall. Figures 6.42 and 6.43, respectively, show the arrangement for the tests in Beijing and Tibet, and Figure 6.44 shows the switching impulse discharge test results.

Figure 6.42
Arrangement for the discharge test in Beijing.

Figure 6.43
Arrangement for the discharge test in Tibet.

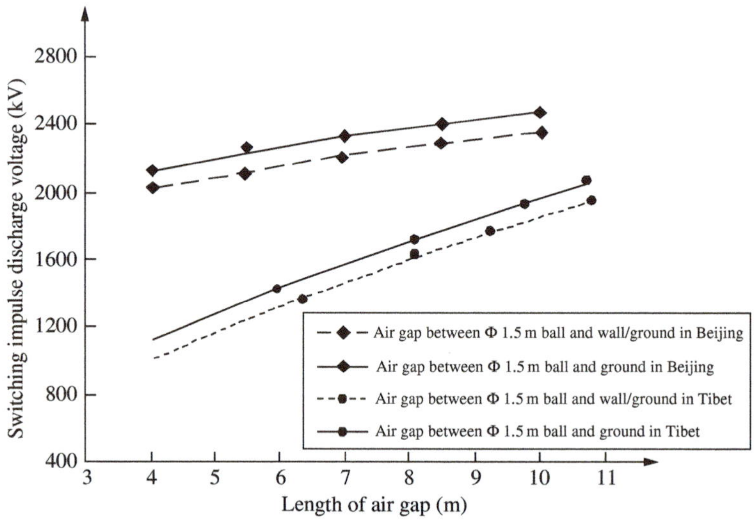

Figure 6.44
Discharge characteristics curves of the air gaps between tube bus and wall/ground.

2. **Switching impulse voltage discharge characteristics of the air gaps between tube bus and barrier/ground**
   For the test, an aluminum tube bus with a diameter of 0.3 m was used and a metallic ball with a diameter of 0.5 m was installed at both ends of the bus; beneath the bus was a 1.9 m high simulated metallic barrier, and the ground was covered with galvanized iron plates.

Figure 6.45
Arrangement for the discharge test in Beijing.

Figure 6.46
Arrangement for the discharge test in Tibet.

Figures 6.45 and 6.46, respectively, show the arrangement for the discharge tests in Beijing and Tibet.

The switching impulse discharge test results are shown in Figure 6.47. The test results indicate that as the electrodes have different structural shapes, in the positive switching impulse test no discharge happened between the tube bus and barrier. Hence, to determine the air gap distance between the pole bus and ground, the effects of the metallic barrier on the discharge voltage value may be disregarded.

6.1.4 Altitude Correction

According to the method for atmospheric correction of the withstand voltage of external insulation at altitudes up to 2000 m given in IEC 60071-2—1996 *Insulation*

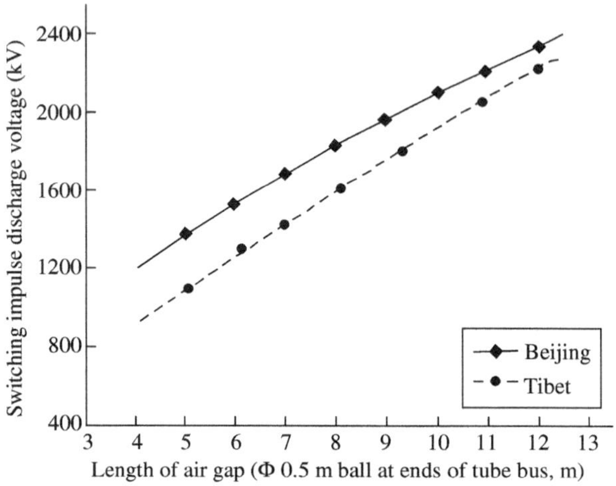

Figure 6.47
Discharge characteristics curves of the air gaps between tube bus and barrier/ground.

Coordination—Part 2: Application Guide, for external insulation at an altitude of H (m), the withstand voltage U_{PH} under standard atmospheric conditions can be corrected using Eq. (6.1):

$$U_{P0} = k_a U_{PH} \tag{6.1}$$

where

$$k_a = e^{m(H/8150)}$$

where
U_{P0}—withstand voltage at the altitude of 0 m, kV
k_a—altitude correction factor
m—a correction factor varying with the voltage type and gap structure, which is taken as 1.0 for lightning impulse withstand voltage and short-duration power frequency withstand voltage. For switching impulse withstand voltage, see Figure 6.48.

Because UHV AC/DC transmission lines are likely to pass through areas at an altitude above 2000 m, an equation of altitude correction factor k_a applicable to air gaps of towers at high altitudes according to the results of discharge tests performed at various altitudes was developed in China.

$$k_a = \frac{1}{1.0 - mH \times 10^{-4}} \tag{6.2}$$

where
H—altitude, m
m—a correction factor varying with the voltage type and electrode structure.

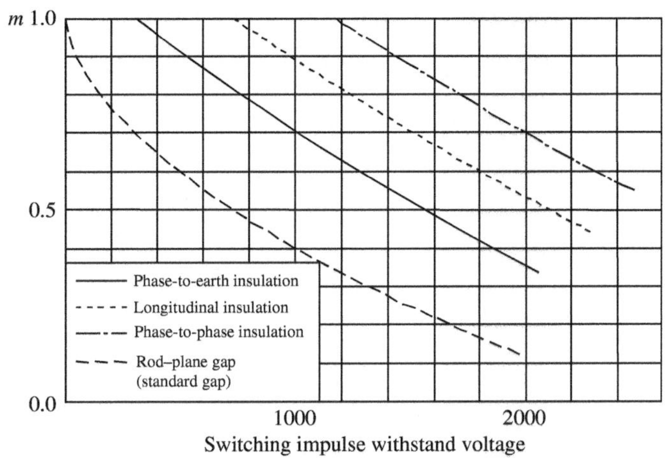

Figure 6.48

Correction factor m versus switching impulse withstand voltage.

According to the power frequency voltage discharge tests on the bundle conductor-to-tower air gap performed in Wuhan (at an altitude of 23 m) and Xining (at an altitude of 2260 m), the correction factor m is 1 for power frequency discharge voltage and $2.73-0.0015U$ (U refers to 50% switching impulse discharge voltage, kV) for standard switching impulse discharge voltage. Also, positive switching impulse discharge tests were performed on rod–plane gaps, respectively, in Beijing, Guiyang, Xining, and Yangbajing, which are at different altitudes. Different parameters were used as the variable of the altitude correction equation. After comparing and discussing the results obtained from these equations, it was concluded that the calculation result is the closest to the actual data when m is taken as $0.137L+1.161$ (L refers to the length of the air gap). The altitude correction factor k_a decreases as L increases [5].

According to the lightning impulse discharge tests on air gaps of AC/DC transmission lines performed in Wuhan, Kunming, and Xining, the correction factor m is 1 for lightning impulse discharge voltage.

According to the positive DC discharge voltage tests on rod–plane gaps and rod–rod gaps performed in Beijing and Tibet, the correction factor m is 1.2 for DC discharge voltage.

Table 6.1 gives the values of altitude correction factor for various altitudes and voltage types.

6.1.5 Surface Flashover Characteristics of Insulators in UHV Power Grids

6.1.5.1 Surface flashover characteristics of AC insulators

For the pollution problem in external insulation of transmission lines, the 50% power frequency withstand voltage tests on artificially polluted long string insulators of various

Table 6.1: Altitude Correction Factor.

Altitude (m)	0	1000	1500	2000	2500	3000	3500	4000	4300
Power frequency voltage[a]	1	1.11	1.18	1.25	1.33	1.43	1.54	1.67	1.75
DC voltage	1	1.14	1.22	1.32	1.43	1.56	1.73	1.92	2.06
Switching impulse	1	Depending on the amplitude of switching impulse discharge voltage or the length of the air gap							
Lightning impulse[b]	1	1.11	1.18	1.25	1.33	1.43	1.54	1.67	1.75

[a]The values for altitudes of 2500 m and above are extrapolated.
[b]The values for altitudes above 2260 m are extrapolated.

types used in UHV AC transmission lines are performed in China, including common insulators, double-shed insulators, tri-shed insulators, bell insulators, and composite insulators arranged in "I," "II," and "V" strings. In these tests, the effects of non-uniform pollution distribution on the upper and lower surface of insulators on voltage withstand capability are also considered. The test results show that composite insulators have the best performance against pollution flashover, followed by tri-shed insulators, double-shed insulators, bell insulators, and common insulators. The pollution flashover voltage varies linearly with the length of insulator string, but non-linearity correction is necessary for light pollution. The pollution flashover characteristics of V-strings are similar to those of I-strings, but the difference in pollution accumulation over the long-term must be considered. The pollution flashover characteristics of multiple parallel strings spaced at 600 mm are similar to those of I-strings, but the difference in the occurrence probability of flashover must be considered.

For the pollution problem on external insulation of substations, 50% power frequency withstand voltage tests were performed on artificially polluted post insulators and bushings used in UHV substations. The test results show that compared with insulators (with an average diameter of 665 mm) used in EHV substations, those used in UHV substations have a larger diameter (average 900–1000 mm), which leads to a lower performance against pollution flashover. The diameter correction factor k_D can be taken as 1.04. The pollution withstand voltage is somewhat non-linear with the dry arcing distance, and the 50% pollution flashover voltage $U_{50\%}$ calculated through curve fitting is 2.5–8.6% greater than test results.

For the problem at high altitudes in China, high-altitude pollution tests were performed on insulators used in UHV transmission line in a large-scale artificial climate chamber, and the high-altitude AC pollution withstand voltage and altitude correction factor of insulators used in transmission lines and substations were obtained. The test results show that the AC discharge voltage U of a polluted insulator decreases as the atmospheric pressure P decreases, and their relationship can be expressed as:

$$U = U_0(P/P_0)^n \tag{6.3}$$

where
U_0—pollution flashover voltage under standard atmospheric pressure, kV
P_0—standard atmospheric pressure, MPa
n—character index of altitude, the magnitude of which reflects the extent to which the atmospheric pressure affects the pollution flashover voltage.

For ease of use in engineering, the relation between the altitude and flashover voltage can also be expressed as:

$$\frac{U}{U_0} = 1 - k_1 H \tag{6.4}$$

where
k_1—rate of decrease, the magnitude of which reflects the extent to which the altitude affects the pollution flashover voltage
H—altitude, m.

The pollution flashover voltage decreases with the increase of altitude. The higher the altitude becomes, the more quickly the pollution flashover voltage decreases. Laboratory test results show that the effects of altitude on pollution flashover voltage are related to the type of insulator and pollution degree other than the length of insulator string. The pollution flashover voltage of common porcelain insulators decreases most quickly with the increase of altitude, followed by double-shed insulators, tri-shed insulators, and composite insulators. Consequently, it is highly recommended that composite insulators should be used in heavily polluted environments at high altitudes, and tri-shed insulators should be used in lightly and moderately polluted environments.

Because insulators are also subject to icing, icing flashover tests were performed on insulators used in UHV transmission in a large-scale artificial climate chamber, and the AC icing flashover voltage of insulators used in transmission lines and substations was obtained. The test results show that, under heavy icing conditions, the icing flashover voltages of insulator strings of different shed types but of same length are approximately the same. The icing flashover voltage of an insulator string is somewhat non-linear in length, that is, the gradient of discharge voltage decreases with the increase of length of insulator string. Therefore, a major means to increase the icing flashover voltage in heavy icing regions is to prevent the formation of an ice bridge on the string by using V-string insulators or composite insulators with ice-proof sheds.

6.1.5.2 Characteristics of flashover on surface of DC insulators

The pollution flashover on insulators under DC voltage differs from that under AC voltage in that it is more severe and that the DC pollution flashover voltage under negative polarity is lower than that under positive polarity. This is because, with the electrostatic precipitation effect, more contaminants deposit on the surface of a DC insulator; however, the current does not cross the zero point before DC arc occurs, making it easier for the DC arc to develop across the insulator.

China's UHV DC transmission lines are at voltages of ± 800 and ± 1100 kV and their external insulation faces similar issues to UHV AC lines. Because the UHV DC insulators are different from those used in ± 500-kV DC lines in terms of disc diameter, string length, and parallel configuration, tests are required on various types of insulators to clarify their dust accumulation characteristics, pollution flashover characteristics, linear relation between pollution flashover voltage and string length, and effects of string type on pollution flashover voltage. For UHV DC switchyards, to increase the mechanical and electrical strength, the diameter and height of the insulators must be increased. At the same time, a larger diameter will give rise to a lower pollution withstand voltage, and it has not been determined whether the pollution withstand voltage of insulator string varies linearly with the spacing between insulators. Similar to UHV AC lines, the insulation coordination of UHV DC lines are also facing high altitudes and severe icing problems.

For pollution problem on external insulation of transmission lines, the 50% power frequency withstand voltage tests on artificially polluted insulators of various types used in UHV DC transmission lines were performed in China, including double-shed insulators, tri-shed insulators, bell insulators, and composite insulators of various shed shapes arranged in "I", "II" and "V" strings; in the tests, their difference in dust accumulation characteristics were also considered. The test results show that composite insulators have the best performance against pollution flashover, followed by bell insulators, double-shed insulators, and tri-shed insulators. The pollution flashover characteristics of V-strings are similar to those of I-strings, but the difference in pollution accumulation over the long-term must be considered. The pollution flashover characteristics of multiple parallel insulators spaced at 600 mm or above are similar to those of I-strings, but the difference in the occurrence probability of flashover must be considered.

Outdoor DC switchyard is adopted for the UHV DC transmission projects in China. Considering the restrictions of outdoor DC switchyard on the height of the insulators, composite external insulation and porcelain insulation coated with room temperature vulcanizing silicone (RTV) are recommended. The test results indicate that the 50% pollution withstand voltage of a post insulator used in a UHV DC converter station varies linearly with the height of insulators.

The external insulation of a UHV DC transmission line in China is also subject to high-altitude and polluted environments, so high-altitude artificial pollution tests were performed on the insulators used in UHV DC projects in a large-scale artificial climate chamber, and the high-altitude DC pollution withstand voltage and altitude correction factor of line and substation insulators were obtained. The test results show that the DC discharge voltage of a polluted insulator decreases with the decrease of atmospheric pressure, and their relation can be expressed by Eqs. (6.4) and (6.5). The artificial pollution tests also indicate that the effects of altitude on the pollution flashover voltage are related to the type of insulator and pollution degree other than the length of insulator string, and that the atmospheric pressure has a smaller impact on the pollution flashover voltage under DC voltage than that under AC voltage.

Because external insulation of DC lines is subject to pollution, icing, and other complicated environmental conditions at high altitudes, icing flashover characteristics tests were performed on insulators at various altitudes in a large-scale artificial climate chamber. The DC icing flashover voltages under low atmospheric pressures were obtained. According to the test results, under heavy icing conditions, the icing flashover voltages of insulator strings of different shed shapes but the same length are approximately the same; the icing flashover voltage of a single insulator is somewhat non-linear with its string length; the discharge voltage per unit length of string decreases with the increase of string length, and the icing flashover voltage decreases with the decrease of atmospheric pressure. Therefore, V-strings, tension strings, or composite strings with ice-proof sheds are used in areas subject to heavy icing to increase the icing flashover voltage of insulators.

6.2 Air Gaps of UHV Overhead Transmission Lines

6.2.1 Conductor-to-Tower Air Gap Under Operating Voltage

1. **AC lines**

   For AC transmission lines, the operating voltage has a bearing on the air gap between the conductor and tower. The 50% power frequency discharge voltage $U_{rw.p}$ of the air gap (taking into account the effect of wind deflection) at 0 m and under standard atmospheric conditions should meet the following requirement:

$$U_{rw.p} \geq k_s k_c (\sqrt{2} U_m / \sqrt{3}) \tag{6.5}$$

   where

   U_m—maximum operating voltage of the system, kV

   k_s—safety factor, taken as 1.05

   k_c—coordination factor, generally taken as 1.1 for UHV transmission lines.

   The wind pressure non-uniformity factor used for calculating the wind deflection angle of a suspension insulator string should be determined in accordance with Appendix B to DL/T 620—1997 *Overvoltage Protection and Insulation Coordination of AC Electric Installations*, while the wind speed used for calculating the wind deflection should be taken with the design maximum wind speed of the transmission line.

   Based on Eq. (6.5) and considering atmospheric condition correction, Table 6.2 gives the minimum conductor-to-tower air gap of 1000-kV overhead transmission lines under the operating voltage at different altitudes.

2. **DC lines**

In UHV DC transmission lines, V-strings are commonly used, and the operating voltage has no bearing on the conductor-to-tower air gap. The minimum air gap under DC operating voltage can be calculated using the formula given in DL/T 436—2005 *Technical Guide for HVDC Overhead Transmission Lines*, where the wind deflection angle should be calculated with the design maximum wind speed of the transmission line.

The relationship between the required 50% DC discharge voltage $U_{\text{rw.p}}$ and the rated DC operating voltage U_{N} is given by:

$$U_{\text{rw.p}} = U_{\text{N}}k_2k_3/[(1-3\sigma)k_1] \tag{6.6}$$

where

U_{N}—rated operating voltage

k_2—humidity correction factor for the discharge voltage of the air gap under DC voltage

k_3—safety factor, ranging from 1.1 to 1.15

σ—coefficient of variation of the discharge voltage of the air gap under DC voltage, taken as 0.009

k_1—air density correction factor for the discharge voltage of the air gap under DC voltage.

Table 6.3 gives the recommended air gaps of ±800-kV UHV DC overhead transmission lines under operating voltage.

Assuming a rated operating voltage of 800 kV, a safety factor of 1.15, and a coefficient of variation of 0.01, then a clearance of 2.1 m from the conductor to tower is justified for such a voltage level [6].

Table 6.2: Required Minimum Conductor-to-Tower Air Gap of 1000-kV Overhead Transmission Line Under Operating Voltage (m).

Altitude	1000	1500
Air gap	2.9	3.1

Table 6.3: Recommended Conductor-to-Tower Air Gap for ±800-kV UHV DC Overhead Transmission Line Under Operating Voltage.

Altitude	1000	2000
Air gap	2.1	2.4

6.2.2 Conductor-to-Tower Air Gap Under Switching Overvoltage

1. **AC lines**

   The air gap of a UHV AC overhead transmission line under switching overvoltage is determined using the statistical method. Assuming that a 1000-kV line is 400 km long and has a span of 500 m, the highest statistical switching overvoltage along the line is 1.6 p.u. and occurs only over half of the entire line length. The overvoltage is in Gaussian distribution with a standard deviation σ_0 of 0.12 $\overline{u}_0$ ($\overline{u}_0$ refers to the average overvoltage). The middle-phase conductor is suspended by V-string insulators while the two outside-phase conductors are suspended by I-string insulators, and the flashover probability of the air gap between the middle-phase conductor and tower is calculated. The total number of insulators Z is 400, and the standard deviation of discharge voltage σ_f is 0.05 $\overline{u}_f$ ($\overline{u}_f$ refers to the average discharge voltage). Table 6.4 shows the calculation process and results. The results indicate that the flashover probability P'_z of multiple insulators is closely related to the statistical coordination factor k_c. When k_c is 1.1, P'_z is 0.52, which is too high to be acceptable. And the value of P'_z for $k_c = 1.27$ is 10% of that when $k_c = 1.2$. Under switching overvoltage, the two outside phases of a 1000-kV line are very unlikely to experience flashover and, thus, the flashover probability of the line is dictated by the center phase. For a 1000-kV UHV overhead transmission line, if one to three switching operations are scheduled for a year, then the occurrence times of flashover under the switching overvoltage will be only 0.15–0.45 times in 30 years.

Table 6.4: Flashover Probability of the Air Gap Between the Middle-Phase Conductor and Tower Under Switching Overvoltage for a 1000-kV Transmission Line.

Switching overvoltage	Multiple (reference value: $\sqrt{2}U_m/\sqrt{3}$)	1.6		
	Statistical overvoltage U_0 (kV)	1437		
	Coefficient of variation $\frac{\sigma_0}{\overline{u}_0}$	0.12		
	50% discharge voltage $\overline{u}_f$ (kV)	1825	1724	1581
	Coefficient of variation of discharge voltage $\frac{\sigma_f}{\overline{u}_f}$	0.05		
Statistical coordination factor	$k_c = \frac{\overline{u}_f}{U_0}$	1.27	1.20	1.10
Standard variation λ'		−4.05	−3.50	−2.68
Flashover probability of a single insulator P_s		0.013×10^{-3}	0.12×10^{-3}	1.84×10^{-3}
Flashover probability of multiple insulators P_z'		0.005	0.047	0.52

As stipulated in GB/Z 24842—2009 *Overvoltage and Insulation Coordination of 1000-kV UHV AC Transmission Project*, the 50% switching impulse discharge voltage of the conductor-to-tower air gap of a UHV transmission line $U_{rw.s}$ is given by:

$$U_{rw.s} = k_s k_c U_0 \tag{6.7}$$

where

k_s—safety margin, taken as 1

k_c—coordination factor, ranging from 1.1 to 1.26 for I-strings and taken as 1.27 for V-strings

U_0—statistical phase-to-ground switching overvoltage of the line, kV.

Table 6.5 gives the required air gaps for both single-circuit and double-circuit UHV overhead transmission lines under switching overvoltage, as specified in GB/Z 24842—2009. For air gaps subject to wind deflection, the wind speed used for calculating the wind deflection angle should be 50% the design maximum wind speed.

Table 6.6 gives the air gaps under both long and short wave-front switching impulse voltages. The transmission line is at an altitude of 1000 m, the middle-phase insulator string is of V configuration, and the insulation coordination principles and methods are the same for both long and short wave-front switching impulses. For ease of comparison, all results are given based on the width of the tower window $W = 1.2$ m. According to Table 6.6, using the test data under short wave-front switching impulse

Table 6.5: Required Conductor-to-Tower Air Gap Under Switching Overvoltage.

Type of Line		Air Gap (m)		
		At Altitude of 500 m	At Altitude of 1000 m	At Altitude of 1500 m
Single circuit	I-strings on outside phases	5.9	6.2	6.4
	V-string on center phase	6.7 (7.9)	7.2 (8.0)	7.7 (8.1)
Double circuit		6.0	6.2	6.4

Note: The figures outside brackets refer to the distance between the middle-phase conductor and diagonal brace, while those inside brackets refer to the distance between the middle-phase conductor and upper cross-arm.

Table 6.6: Air Gaps Under Switching Impulses with Different Front Times.

Switching Overvoltage (p.u.)	Air Gap (m)	
	1000 μs Long Wave-Front Switching Impulse	250 μs Short Wave-Front Switching Impulse
1.6	5.71	8.19
1.7	6.33	9.47

will lead to an excessive air gap, whereas for UHV transmission lines the front time of a switching impulse is generally longer than 1000 μs. Taking into account the impacts of the waveform and tower width, it is reasonable and feasible to adopt the required minimum gap under long wave-front switching impulse in practical projects.

2. **DC lines**

As compared with ± 500-kV lines, ± 800-kV UHV DC lines have a much higher voltage level; therefore, the air gap has to be increased accordingly. Regarding minimizing the width of the line corridor, V-strings are universally used in ± 800-kV UHV DC lines and the air gap is dependent on the switching overvoltage. In east China, where land for line corridor is limited, the V-string conductors are arranged not only arranged horizontally but also, in some cases, vertically. The insulation coordination can be designed differently for various sections along the line based on the distribution of the overvoltage having a 2% probability of not being exceeded. For example, this overvoltage in the middle section of a DC line is much higher than that in other sections, and thus the air gap may be determined separately to improve the construction economy.

With the simplified statistics method, the required value of 50% positive switching impulse discharge voltage $U_{rw.s}$ of the conductor-to-tower air gap can be calculated with the equation given in DL/T 436—2005:

$$U_{rw.s} = \frac{k_2 k_3}{(1 - 2\sigma)k_1} U_m \tag{6.8}$$

where

k_1—air density correction factor for the switching impulse discharge voltage
k_2—humidity correction factor for the switching impulse discharge voltage
k_3—multiple of the overvoltage having a 2% probability of not being exceeded, determined according to simulation results
σ—coefficient of variation of switching impulse discharge voltage, taken as 0.05
U_m—highest operating voltage.

For a maximum switching overvoltage of 1.7 p.u. along a ± 800-kV UHV DC line, Table 6.7 [6] and Table 6.8, respectively, give the recommended air gaps based on the

Table 6.7: Recommended Air Gap for a Tangent Tower with Conductors in Horizontal Arrangement in a ± 800-kV UHV DC Line with a Maximum Switching Impulse Overvoltage of 1.7 p.u.

Type of Tower	Air Gap (m)	
	At Altitude of 1000 m	At Altitude of 2000 m
Tangent tower with conductors in horizontal arrangement	6.0	6.7

Table 6.8: Recommended Air Gaps for an F-Shape Tower with Conductors in Vertical Arrangement in a ±800-kV UHV DC Line with a Maximum Switching Impulse Overvoltage of 1.7 p.u.

Type of Tower		Air Gap (m)	
		At Altitude of 0 m	At Altitude of 1000 m
F-shape tower with conductors in vertical arrangement	Upper conductor	5.7	5.8
	Lower conductor	6.2	6.3

Table 6.9: Required Minimum Air Gap Under Lightning Overvoltage for UHV AC Transmission Lines.

Type of Tower	Air Gap (m)		
	At Altitude of 500 m	At Altitude of 1000 m	At Altitude of 1500 m
Single circuit	Not specified		
Double circuit	6.7	7.1	7.6

discharge characteristics curves of air gaps of tower head with V-string insulators arranged both horizontally and vertically.

In determining the air gap between the conductor and tower with the statistics method, the flashover probability of the insulation across the line on the healthy pole under switching impulse can be determined as 0.001 times per switching.

6.2.3 Conductor-to-Tower Air Gap Under Lightning Overvoltage

1. **AC lines**
   For a UHV AC transmission line, the air gap under lightning overvoltage depends on the lightning flashover probability. For a single-circuit line, because the air gap between the conductor and tower is large under windless condition, it is unnecessary to be specified for lightning overvoltage. For a double-circuit line, to achieve the expected lightning trip rate, a sufficient distance must be maintained between the conductor and the cross-arm below it. Table 6.9 specifies the required minimum air gap under lightning overvoltage for UHV AC transmission lines.
2. **DC lines**
   For a DC transmission line, the size of tower head is usually not determined by the air gap for preventing lightning flashover. This is because when a flashover occurs on a

DC line, the rectifier station will change to inverter operation mode immediately, the voltage and current quickly decrease to zero, and then after deionization the station will restart and resume power transmission. This process lasts only for 100–300 ms and, hence, will not lead to operation interruption in most cases. When lightning flashover occurs, the fault current flowing through insulator strings is much smaller in amplitude and duration than the short-circuit current in an AC line. Hence, damage to the insulators is much less severe as well, and the problem with the AC circuit breaker requiring repair due to repeatedly breaking the short-circuit current will not occur.
A DC line has inherent unbalanced insulation due to its operating voltage, that is, when lightning flashover occurs on a pole, the other pole can maintain normal operation and is capable of withstanding an overstress of 1.4- to 1.5-times the rating value at 80 ms after the faulty pole is blocked. Therefore, for a ± 800-kV UHV DC transmission line with V-string insulators, the air gap length may not be specified to prevent lightning overvoltage [7].

6.2.4 Recommended Conductor-to-Tower Air Gap for UHV Overhead Transmission Lines

1. **AC lines**
   Based on this analysis, Table 6.10 specifies the recommended conductor-to-tower air gap of a UHV AC overhead transmission line at an altitude of 1000 m in China. For ease of comparison, the corresponding data from the Soviet Union and Japan are presented as well.

Table 6.10: Recommended Conductor-to-Tower Air Gap for UHV AC Overhead Transmission Lines (m).

Country	Operating Voltage	Switching Overvoltage		Lightning Overvoltage	
		V-string	I-string	V-string	I-string
China	2.9/3.1[a]	7.2/8.0[b]	6.2/6.4[a]	7.1	7.1
The Soviet Union	2.5	8.0–9.0[c]	6.0–7.0[c]	—	—
Japan	3.09/3.21[d]	—	6.0/6.55[e] 5.69/6.75	—	6.62[f] 6.2

[a]The numerator and denominator, respectively, correspond to the altitudes below 1000 m and below 1500 m.
[b]The numerator and denominator, respectively, refer to the distance from the middle-phase conductor to the diagonal brace and to the upper cross-arm.
[c]The maximum switching overvoltage along a line in the Soviet Union is 1200 kV (1.8 p.u.). The smaller values and greater values in the "V-string" column and "I-string" column under the category of "Switching Overvoltage," respectively, refer to the distance between the conductor and tower body and that between the conductor and tower window.
[d]The numerator and denominator, respectively, correspond to the altitudes below 1800 m and below 2100 m.
[e]The figure in the first line corresponds to I-string insulators while that in the second line corresponds to tension insulator strings. And the numerator and denominator, respectively, refer to the distance from the conductor to the lower cross-arm and that from the jumper to the tower body.
[f]The figure in the first line corresponds to I-string insulators while that in the second line corresponds to tension insulator strings.

Table 6.10 shows that:

1. The highest system voltage is 1100 kV in China and Japan. The air gaps under operating voltage are 2.9 and 3.1 m at altitudes of 1000 and 1500 m, respectively, in China, and are 3.09 and 3.21 m at altitudes below 1800 and below 2100 m, respectively, in Japan. Taking into account factors such as the altitude and maximum design wind speed, the values of the two countries are close. In the Soviet Union, the air gap required under operating voltage is only 2.5 m, which is smaller than that in China and Japan.
2. In China, the air gaps between the outside-phase conductor and tower are 6.2 and 6.4 m at altitudes of 1000 and 1500 m (with a switching overvoltage of 1.7 p.u.). In Japan, the gaps are 6.54 and 6.66 m at altitudes below 1800 and below 2100 m, respectively (with a switching overvoltage of 1.6 p.u.). In China, the air gap under switching overvoltage is determined based on the distance that has to be maintained under the wind deflection angle at 50% the maximum design wind speed. The maximum design wind speed is 40 m/s in Japan, whereas the air gap under switching overvoltage is determined based on the distance that has to be maintained under the wind deflection angle at a wind speed of 15 m/s. After considering this factor, the values of the two countries are still close.

2. **DC lines**

The conductor-to-tower air gap required for ± 800-kV UHV DC overhead transmission lines is specified in Table 6.11. Because the lines often pass through high-altitude areas, the values need altitude correction.

6.3 Air Gaps in UHV Substations and Converter Stations

6.3.1 Required Air Gaps Under Operating Voltage

1. **Substations**

The electrical equipment in a substation should be able to withstand continuous operating voltage and temporary overvoltage of a given amplitude for a given duration.

Table 6.11: Required Conductor-to-Tower Air Gap for ±800-kV UHV DC Overhead Transmission Lines (m).

Altitude	Operating Voltage	Switching Overvoltage (1.7 p.u.)
1000	2.1	6.0 (6.2/6.3)
2000	2.4	6.7

Note: The figures inside the bracket respectively refer to the minimum air gap between the upper conductor and tower and that between the lower conductor and tower (for F-shape towers).

The required 50% power frequency discharge voltage of a phase-to-earth gap $U_{\mathrm{rw.p}}$ is given by Eq. (6.9):

$$U_{\mathrm{rw.p}} = k_s k_c U_{\mathrm{pg}} \tag{6.9}$$

where

k_s—safety margin, taken as 1.05

k_c—coordination factor, taken as 1.06

U_{pg}—maximum phase-to-earth temporary overvoltage, taken as 1.4 p.u.

The required 50% power frequency discharge voltage of a phase-to-phase gap $U_{\mathrm{rw.p}}$ is given by Eq. (6.10)

$$U_{\mathrm{rw.p}} = k_s k_c U_{\mathrm{pp}} \tag{6.10}$$

where

U_{pp}—maximum phase-to-phase temporary overvoltage on the pole bus, taken as $1.3\sqrt{3}$ p.u.

The pollution flashover withstand capability of external insulation of electrical equipment should be suitable for continuous operating voltage under the pollution degree of the area where the substation is situated.

Table 6.12 specifies the required minimum air gaps under power frequency voltage for UHV substations at altitudes up to 1000 m.

2. **Converter stations**

For the DC side of a converter station, the main air gaps are those for the equipment in the valve hall and DC switchyard, mainly considering the air gaps required by the equipment under rated DC operating voltage. An open circuit fault occurring in the inverter station will cause a temporary DC overvoltage, and when the valve at the converter station is deblocked with a very small delay angle, theoretically, a switching overvoltage up to twice the normal voltage may appear at the open-circuit end due to the total reflection of traveling wave at this end. And once the switching overvoltage attenuates, a temporary DC overvoltage with high amplitude will be generated on the equipment connected to the DC pole in the converter station. Because the control and protection system for the DC system acts merely milliseconds after a fault or switching occurs, this temporary overvoltage can basically be avoided, so it may be disregarded in the design of insulation coordination.

Table 6.12: Required Minimum Air Gaps Under Power Frequency Voltage for UHV Substations (m).

Minimum air gap between conductor and gantry structure	4.2
Minimum air gap between equipment and gantry structure	4.2
Minimum air gap between phases	6.8

6.3.2 Required Air Gaps Under Switching Overvoltage

6.3.2.1 Substations

As stipulated in GB/Z 24842—2009, the required switching impulse discharge voltage of air gaps in a UHV substation can be calculated with either the simplified statistical method or the semi-statistical method. The simplified statistical method, as stated in IEC 60071-2—1996, takes into account the Gaussian distribution characteristics of switching overvoltage and flashover voltage of air gaps, and it calculates the required 50% switching impulse discharge voltage based on the acceptable insulation failure rate. This method facilitates differentiated insulation design and thus ensures economy and safety of a project. The semi-statistical method, as introduced in DL/T 620—1997, is commonly used to determine the air gaps of 500 kV substations in China, but the value of the coordination factor is different for UHV substations.

1. **Simplified statistical method**
   a. *Required phase-to-earth discharge voltage*: The required 50% switching impulse discharge voltage of the phase-to-earth gap $U_{rw.s}$ is given by

$$U_{rw.s} = \frac{U_{rw}}{1-1.28\sigma} = \frac{k_s k_{cs} U_{rp}}{1-1.28\sigma} \tag{6.11}$$

where

σ—coefficient of variation of switching impulse discharge voltage, taken as 0.06

U_{rw}—required 90% switching impulse voltage

U_{rp}—representative overvoltage, taken as the statistical phase-to-earth switching overvoltage

k_{cs}—statistical coordination factor, taken as 1.15 for an insulation failure rate of 0.01%

k_s—safety factor, taken as 1.05.

The discharge voltage is obtained through discharge tests under standard wave-front voltage (250 μs). In determining the minimum air gaps, such data need to be corrected to a 1000-μs wave-front voltage. This can be accomplished by dividing the required 50% switching impulse discharge voltage by a factor of 1.13.

   b. *Required phase-to-phase discharge voltage*: The required 50% switching impulse discharge voltage of the phase-to-phase gap $U_{rw.s}$ is given by:

$$U_{rw.s} = \frac{U_{rw}}{1-1.28\sigma} = \frac{k_s k_{cs} U_{rp}}{1-1.28\sigma} \tag{6.12}$$

where

σ—coefficient of variation of switching impulse discharge voltage, taken as 0.03

U_{rw}—required 90% switching impulse voltage

U_{rp}—representative overvoltage, taken as the statistical phase-to-phase switching overvoltage
k_{cs}—coordination factor, taken as 1.15 for an insulation failure rate of 0.01%
k_s—safety factor, taken as 1.05.

The waveform correction factor is taken as 1.13 for correcting the test data to a 1000-μs wave-front voltage.

2. **Semi-statistical method**
   a. *Required discharge voltage for phase-to-earth air gaps*: The required 50% discharge voltage for the insulator string and the phase-to-earth air gap under positive switching impulse should conform to Eq. (6.13), and it should not be smaller than the corresponding discharge voltage of the disconnector and post insulator used for electrical equipment:

$$U_{rw.s} = \frac{U_{ps}}{1 - 3\sigma} = k_1 U_{ps} \tag{6.13}$$

where
U_{ps}—switching impulse protective level (SIPL) of surge arrester, kV
σ—coefficient of variation of discharge voltage of the air gap under switching overvoltage, taken as 0.06
k_1—switching overvoltage coordination factor for the insulator string, taken as 1.22.

   b. *Required discharge voltage for phase-to-phase air gaps*: The required 50% switching impulse discharge voltage of the phase-to-phase gap is given by Eq. (6.14):

$$U_{rw.s} = \frac{k_2 U_{ps}}{1 - 3\sigma} = k_3 U_{ps} \tag{6.14}$$

where
σ—coefficient of variation of switching impulse discharge voltage, taken as 0.035
k_2—ratio of phase-to-phase overvoltage to phase-to-earth overvoltage, taken as 1.81
k_3—switching overvoltage coordination factor of a phase-to-phase gap, taken as 2.0.

Table 6.13 specifies the required minimum air gaps under switching impulse for UHV substations at altitudes up to 1000 m.

Table 6.13: Required Minimum Air Gaps Under Switching Impulse for UHV Substations (m).

Minimum air gap between the conductor and gantry structure	6.8
Minimum air gap between the equipment and gantry structure	7.5
Minimum air gap between phases	10.1 (between grading rings) 9.2 (between 4-bundle conductors) 11.3 (between tube buses)

6.3.2.2 Converter stations

In determining the air gaps of equipment on the DC side of a UHV converter station, mainly the combination of DC voltage, commutation pulse, lightning impulse, and switching impulse are considered. Because the energized conductors on the DC side mostly use fixed electrodes, the air gaps are therefore mainly determined by lightning and switching impulses. Generally, the positive DC withstand voltage and positive impulse withstand voltage of an air gap are lower than negative withstand voltages, and the positive lightning impulse breakdown voltage is higher than the positive switching impulse breakdown voltage. Therefore, switching impulse, rather than lightning impulse, is decisive in determining the air gap under a certain DC voltage. The combination of standard switching impulse and DC voltage can be replaced by the result obtained from a simplex switching wave test, where the switching wave amplitude is the sum of the DC voltage and slow-front overvoltage peak. The air gaps of the DC switchyard equipment are dependent on the protective level of the surge arrester. The required 50% switching impulse discharge voltage of an air gap $U_{\mathrm{rw.s}}$ is given by:

$$U_{\mathrm{rw.s}} = \frac{k_1 U_{\mathrm{ps}}}{(1-2\sigma)k_2} \tag{6.15}$$

where

k_1—margin coefficient, which is 1.2 for switching impulses
U_{ps}—SIPL of the surge arrester
σ—coefficient of variation of switching impulse discharge voltage, taken as 0.06
k_2—altitude correction factor of switching impulse discharge voltage, taken as 1 for altitudes less than 1000 m.

A large amount of test data shows that, for an air gap within 10 m, the lightning impulse flashover voltage varies linearly with the length of the air gap, whereas the switching impulse flashover voltage tends to be non-linearly saturated with increase of the air gap. For example, on the curve of 50% flashover voltage of a DC voltage plus switching impulse versus rod-plate gap, a turning point appears at 1600 kV. The minimum air gap between equipment and earth can be determined based on either the switching impulse discharge characteristics curves of air gaps (inclusive of air gaps between the tube bus/grading ring of live equipment and the gantry structure) formed by full-scale electrodes of various shapes or the minimum phase-to-earth air gaps, as specified in Table A.2 in IEC 60071-2—1996, according to the required 50% switching impulse discharge voltage of the air gap between the equipment and earth.

6.3.3 Air Gaps Under Lightning Overvoltage

6.3.3.1 Substations

As stated in GB/Z 24842—2009, the required lightning impulse discharge voltage of an air gap in a UHV substation can be calculated with two methods. In method A, insulation coordination is based on the representative overvoltage derived through simulation, whereas in method B, insulation coordination is based on the rated residual voltage under the nominal lightning current of the surge arrester. The latter is commonly used to determine the air gaps under lightning impulse for 500-kV substations in China.

1. **Method A**
   a. *Required discharge voltage for phase-to-earth air gaps*: The required 50% lightning impulse discharge voltage of the phase-to-earth gap is given by Eq. (6.16):

$$U_{rw.l} = k_s k_{cs} U_{rp} \tag{6.16}$$

   where
   k_s—safety factor, taken as 1.05
   k_{cs}—statistical coordination factor; $k_{cs} = 1/(1-3\sigma)$. σ is taken as 0.03 and k_{cs} 1.1
   U_{rp}—representative phase-to-earth overvoltage.

   b. *Required discharge voltage for phase-to-phase air gaps*: The minimum phase-to-phase gap under lightning impulse can be taken as 1.1-times the minimum phase-to-earth gap.
2. **Method B**
   a. *Insulator strings*: The required 50% positive lightning impulse discharge voltage of an insulator string $U_{rw.l}$ is given by Eq. (6.17), which should not be lower than the corresponding discharge voltage of the disconnector and post insulator used for electrical equipment:

$$U_{rw.l} = k_1 U_{pl} \tag{6.17}$$

   where
   k_1—lightning overvoltage coordination factor for the insulator string, taken as 1.45
   U_{pl}—lightning impulse protective level (LIPL) of the surge arrester (kV).
   b. Required discharge voltage for phase-to-earth air gaps. The required 50% discharge voltage of the phase-to-earth gap under positive lightning impulse is given by Eq. (6.17).

Table 6.14: Required Minimum Air Gaps Under Lightning Impulse for UHV Substations (m).

Minimum air gap between conductor and gantry structure	5.0
Minimum air gap between equipment and gantry structure	5.0
Minimum air gap between phases	5.5

c. Required discharge voltage for phase-to-phase air gaps. The required 50% discharge voltage of the phase-to-phase gap under lightning impulse is given by Eq. (6.18):

$$U_{\mathrm{rw.l}} = k_1 U_{\mathrm{pl}} + \frac{\sqrt{2}U_{\mathrm{m}}}{\sqrt{3}} \tag{6.18}$$

Table 6.14 specifies the required minimum air gaps under lightning impulse for UHV substations at altitudes up to 1000 m.

6.3.3.2 Converter stations

As with switching overvoltage, the discharge voltage under lightning overvoltage for an air gap of equipment on the DC side of a UHV converter station should be commensurate with LIPL of the surge arrester. The required 50% lightning impulse discharge voltage $U_{\mathrm{rw.l}}$ is given by Eq. (6.19):

$$U_{\mathrm{rw.l}} = \frac{k_1 U_{\mathrm{pl}}}{(1 - 2\sigma)k_2} \tag{6.19}$$

where
k_1—margin factor, which is 1.25 for lightning impulses
U_{pl}—LIPL of the surge arrester
σ—coefficient of variation of lightning impulse discharge voltage, which is taken as 0.03
k_2—altitude correction factor of lightning impulse discharge voltage.

Similarly, the minimum air gap between equipment and earth can be found from the lightning impulse discharge characteristics curves of air gaps between various shapes of real electrodes (e.g., tube bus/grading ring of live equipment and the gantry structure) or via the correlation between the standard lightning impulse withstand voltage and minimum air gaps, as illustrated in Table A.1 in IEC 60071-2—1996, according to the required 50% lightning impulse discharge voltage of the air gap between the equipment and earth.

6.3.4 Recommended Air Gaps for a UHV Substation

Recommended air gaps for a 1000-kV UHV substation under operating voltage, switching overvoltage, and lightning overvoltage, respectively, are made available

Table 6.15: Air Gaps in UHV Substations (m).

Type of Applied Voltage	Phase-to-Earth Gap A_1		Phase-to-Phase Gap A_2
	A_1'	A_1''	
Temporary overvoltage	4.2		6.8
Switching overvoltage	6.8	7.5	10.1 (between grading rings) 9.2 (between 4-bundle conductors) 11.3 (between tube buses)
Lightning overvoltage	5.0		5.5

Table 6.16: Minimum Air Gap Between the Pole Bus and Barrier in the Outdoor DC Switchyard of a ±800-kV Converter Station (m).

Altitude	Rigid Bus-Barrier	Flexible Bus-Barrier	Post Insulator
0	5.90	6.12	5.28
2000	7.58	7.72	6.89

through the aforementioned analysis. Table 6.15 shows the selected air gaps between the conductor and gantry structure (A_1'), between the equipment and gantry structure (A_1''), and between phases (A_2) for substations at an altitude of 1000 m or less. In this table it can be seen that the air gaps in a UHV substation are also dictated by switching overvoltage [8].

6.3.5 Recommended Air Gaps for DC Switchyard of a UHV Converter Station

To determine air gaps in a converter station, the lightning impulse and switching impulse discharge characteristics curves of the full-scale air gaps (inclusive of air gaps between the tube bus/grading ring of live equipment and the gantry structure or wall, or combinations thereof) in the DC switchyard and valve hall are required. For this reason, standard switching impulse discharge tests were conducted on the air gaps between the tube bus and wall, the pole bus insulator, the pole bus and barrier, and corresponding discharge characteristics curves were obtained.

Table 6.16 specifies the recommended minimum air gap for the pole bus and minimum height of the post insulator in the outdoor DC switchyard for ±800-kV converter stations at altitudes of 0 and 2000 m, respectively [9].

Table 6.17 specifies the minimum air gap between the pole bus and wall in the valve hall and in the indoor DC switchyard for ±400- and ±800-kV converter stations at altitudes of 0 and 2000 m, respectively.

If the switching impulse discharge characteristics curve of a full-scale air gap is unavailable, then the electrode shape-varying gap factor specified by CIGRE, EPRI, and in technical specifications for equipment used in converter stations in China may be used to determine the relationship between the air gap and 50% discharge voltages under lightning impulse and switching impulse, and thus determine the air gap.

Table 6.18 specifies the minimum air gaps at various locations of a converter station, which are calculated with different gap factors *k* as outlined in IEC 60071-2—1996 according to the insulation level of the converter station.

Table 6.17: Minimum Air Gap Between the Pole Bus and Wall in the Valve Hall and in the Indoor DC Switchyard for ±400- and ±800-kV Converter Stations, Respectively (m).

	Pole Bus-Single Wall		Pole Bus-Corner of Wall	
Voltage Level (kV)	±400	±800	±400	±800
At altitude of 0 m	3.59	7.41	3.87	8.58
At altitude of 2000 m	4.36	9.00	4.79	10.62

Table 6.18: Minimum Air Gaps in Converter Stations.

Location	Equipment	LIWV (kV)	SIWV (kV)	Air Gap (mm)		
				$k=1.15$	$k=1.3$	$k=1.4$
Valve hall in a ±800-kV converter station	Between terminals of a valve	426	450	878	736	656
	Between terminals of the upper 12-pulse valve group	916	927	2856	2395	2117
	Between terminals of the lower 12-pulse valve group	903	825	2364	1962	1734
	Between the valve side of HV Y/Y converter transformer and earth	1745	1594	6922	5690	5067
	Between the valve side of HV Y/Δ converter transformer and earth	1367	1295	4897	4042	3628
Valve hall in a ±400-kV converter station	Between the valve side of LV Y/Y converter transformer and earth	1059	1029	3367	2803	2510
	Between the valve side of LV Y/Δ converter transformer and earth	1019	973	3080	2583	2290
DC switchyard	Between the line side of smoothing reactor installed on pole bus and earth	1800	1600	6706	5415	4749
	Between the line side of smoothing reactor installed on neutral bus and earth	574	452	771	730	694

Table 6.19: Main Parameters of Porcelain/Glass Cap and Pin Type Suspension Insulators Used in UHV AC Transmission Projects.

Type	Model	Shape of Sheds	Rated Mechanical Loads (kN)	Height (mm)	Disc Diameter (mm)	Creepage Distance (mm)
Porcelain insulator	U300BP/195D	Double-shed	300	195	330	485
	U300BP/195T	Tri-shed	300	195	360	635
	U420BP/205D	Double-shed	420	205	380	550
	U420BP/205T	Tri-shed	420	205	400	635
	U550B/240	Common	550	240	380	700
Glass insulator	FC550p/240	Common	550	240	320	700

6.4 Selection of UHV Insulators

6.4.1 Selection of Type and Number of Insulators for Overhead Transmission Lines

1. **Insulators for 1000-kV UHV AC transmission lines**
   a. *Length of a single I-string*: Open-type insulators (double-shed or tri-shed) have a high capability of self-cleaning. Compared with common types, open-type insulators suffer less dust accumulation when exposed to the same pollution environment, and thus have better performance against pollution flashover. Therefore, double-shed or tri-shed insulators are used in suspension strings. Tension strings suffer less dust accumulation on their surface than suspension strings; therefore, depending on actual requirements, common insulators or insulators with two sheds or three sheds may be used. In addition, considering the required mechanical strength for bearing line load, 300, 420, and 550 kN porcelain and glass insulators of cap-and-pin type are used in 1000-kV lines. Table 6.19 gives the major parameters of such insulators.

   The length of an insulator string used in 1000-kV UHV AC pilot project is determined based on the pollution withstand voltage to minimize the need for cleaning while ensuring operational safety of the line. The insulation configuration under pollution conditions is based on the artificial pollution tests on long insulator strings and the insulator's pollution withstand voltage. The configurations of suspension I-strings at altitudes of 1000 m and less are as follows: 54 double-shed insulators for a pollution class of II[1] or below; 65 insulators for a class III; and

[1] At the early stage of construction of UHV projects, environmental pollution was classified into classes 0, I, II, III, and IV by pollution severity based on GB/T 16434-1996 *Environmental Pollution Classification and External Insulation Selection for High-Voltage Transmission Line, Power Plant and Substation*. Later, this standard was replaced by GB/T 26218.1-2010 *Selection and Dimension of High-Voltage Insulators Intended for Use in Polluted Conditions—Part 1: Definitions, Information and General Principles*, and environmental pollution was then classified into classes A, B, C, D, and E by pollution severity. For more information, see GB/T 26218.1-2010.

70 insulators with a length up to 14.5 m for class IV. On this basis, composite insulator strings with a length of 9.75 m are used in the 1000-kV UHV AC pilot project in the polluted areas of class III or IV to improve the insulation level of the line and minimize the lengths of insulator strings. In later UHV AC transmission projects in China, taking into account the effects of pollution and icing on flashover characteristics of insulators, the length of composite insulator strings is reduced to 9.0 m for polluted areas of class III (now class D) or below, and to 9.7 m for class IV (now class E) areas at altitudes less than 1000 m.

In view of the geographical conditions across China, UHV transmission lines often pass through areas at altitudes above 1000 m. A large number of studies indicate that the flashover voltage of an insulator at a high altitude is lower than that in a plains area. Hence, if porcelain or glass insulators are used at a high-altitude and in severely polluted areas, then the tower has to be large due to long strings. This will significantly increase the costs of the transmission line. To prevent occurrence of pollution flashover and to limit the tower size, a large number of composite insulators are also used in UHV AC transmission lines where they pass through severely polluted areas at high altitudes.

b. *Effects of insulator string type and selection of insulation configuration*: In a 1000-kV transmission line, V-string insulators are used on the center phase and I-string insulators on outside phases; in the case of a double-circuit line, I-string insulators are used. Using V-string insulators on the center phase can not only effectively reduce the air gaps but also eliminate the effects of wind deflection, thereby reducing the size of the tower head remarkably, saving land for line corridors, and decreasing the costs of the line. However, if V-string is also used on the outside phases, then the width of the tower will be increased and the investments will increase accordingly.

When cap-and-pin type suspension insulators are used in V-strings, less dust accumulation will occur on the surface of porcelain/glass isolators than in I-strings, resulting in a smaller insulator number and string length and, hence, a smaller size of tower head. And for the ease of manufacture and installation, all composite insulators are designed to have the same structural height, whether intended for use in a V-string or an I-string.

2. **Insulators for ±800-kV UHV DC transmission lines**

Because of the differences in environment, altitude, and geographical and meteorological conditions, international criteria on selection of insulators for DC transmission projects, especially the test data and methods used in design of external insulation, are not applicable for the projects in China. As a result, the insulators intended for use in ±800-kV UHV DC transmission projects in China have to be selected based on the specific atmospheric conditions of China.

a. *Determination of the length of a single I-string*: Double-shed insulators, tri-shed insulators, bell insulators, as well as composite insulators are mainly used in DC transmission projects. The artificial pollution tests of heavy insulator strings of various lengths conducted in the large-scale artificial climate chamber show that the DC pollution flashover voltage of insulators varies in proportion to the length of string. The test data from other countries also support this conclusion.

According to the result of DC pollution tests on long insulator strings conducted in the large-scale artificial climate chamber, bell insulators of at least 65 pieces in each string may be used in a lightly polluted area (assuming an equivalent salt deposit density (ESDD) of 0.05 mg/cm^2) at an altitude up to 1000 m.

The UHV DC transmission lines in China also pass through high-altitude areas. The length of an insulator string depends on its performance against pollution flashover. In a high-altitude and severely polluted area, if porcelain or glass insulators are used, then a single string may comprise more than 100 insulators. To prevent occurrence of pollution flashover and limit the tower size in such an area, heavy composite insulators are used in China.

Different from AC lines, the composite insulators used in DC lines are exposed to risks such as ablation of the silicone rubber due to DC arcing, corrosion of metallic fittings at ends due to leakage current, and impacts on the core due to ion transfer under DC voltage. All these problems should be solved.

According to artificial pollution test results, it is determined that in moderately and heavily polluted areas (with an ESDD >0.10 mg/cm^2), rod suspension composite insulator strings with a length of 11–12 m should be used for UHV DC transmission projects.

b. *Effects of insulator string types and selection of insulation configuration*: In ± 800-V transmission lines, suspension insulators are arranged in double I-string or V-string types, whereas tension insulators are arranged in multiple parallel string types. Studies show that under the same pollution class, the flashover characteristics of a single I-string are almost the same as those of a single V-string. However, considering that the dust deposits on a V-string are approximately 56.5–96.7% of those on an I-string, which is taken as 80% in insulation design, the number of bell insulators in a V-string is no less than 56 pieces in a lightly polluted area and the length of a composite insulator string is approximately 10 m in areas of a high ESDD. Because strings used in ± 800-kV UHV DC transmission comprise 60 or even more insulators, consideration should be given to the influence of the spacing between strings on the DC pollution flashover voltage when two or more strings are arranged in parallel. Artificial pollution tests on long insulator strings indicate that when the spacing (center-to-center) between strings in parallel is >600 mm, the pollution flashover voltage of these strings is approximately the same as that of a single string. Therefore, the spacing between UHV insulator strings is determined to be 600 mm. However, consideration should be given to the difference in the occurrence probability of flashover between multiple parallel strings and a single string.

3. **Coordination between length of insulator string and air gap under switching overvoltage**
   Insulators used in UHV AC/DC overhead transmission lines should be able to withstand not only the operating voltage but also the switching overvoltage. In particular, when V-string insulators are used on the center phase in the tower window, both the string and air gap are subject to the effects of switching overvoltage; thus, their discharge strength needs to be coordinated. Because the air gap of the middle phase has been determined with the statistical method as stated, the 50% switching impulse discharge voltage of insulators (including the I-string insulators on outside phases) should not be lower than that of the air gap of the middle phase.

6.4.2 Selection of Insulators Used in Substations and Converter Stations

1. **Insulators used in substations**
   Pollution on the surface of post insulators and bushings used in 1000-kV UHV AC transmission systems in China is a major problem that needs to be addressed. If the pollution withstand voltage varies linearly with the dry arcing distance, then the pollution withstand voltage of a 1000-kV post insulator or bushing should be twice that of a 500-kV post insulator or bushing, and thus the dry arcing distance or height of the former should also be twice that of the latter. If the relation is non-linear, then the pollution withstand voltage and dry arcing distance or structural height of the 1000-kV post insulator or bushing need to be more than twice those of the 500-kV post insulator or bushing. To achieve the desired mechanical strength, the diameter of insulators should be increased, which, however, will lead to a lower pollution withstand voltage. The performance of an insulator against pollution is affected by the creepage distance, dry arcing distance, shape of sheds, and structural height and diameter. This means that it is impractical to improve the performance of an insulator with a given height simply by increasing the creepage distance to an unlimited measure.

   Studies were performed on the effects of shape of sheds, average diameter, and structural height to ground of a post insulator on its pollution withstand voltage, and artificial pollution tests were performed on 1000-kV porcelain post insulators with a structural height of 8.8 m and 1000-kV hollow porcelain bushings with a structural height of 8 m, respectively, in the large-scale megavolt artificial climate chamber.

   Because bushings used in UHV transmission have a large diameter, dust deposits on them may result in flashover during heavy rain. Studies show that an effective means for preventing flashover is to block rain with partitions or to increase the length of air gap. To this end, a water shed may be installed every 700–1000 mm (insulation distance) along the axis of a large bushing to prevent water bridging between sheds and occurrence of flashover.

In China, the structural height and creepage distance of 1000-kV porcelain post insulators used in the areas with a pollution class of C and below are 10,000 and 30,250 mm, respectively. In 1000-kV UHV AC transmission, hollow porcelain and composite insulators (bushings) are used. They are mainly installed (vertically or with an inclination) on electrical devices such as outdoor transformers, reactors, GISs, circuit breakers, current/voltage transformers, and surge arresters. The bushings used in UHV substations have a specific creepage distance no less than 25 mm/kV and spacing between sheds no less than 70 mm, meeting the requirements of external insulation in areas with a pollution class of C and below. The hollow porcelain insulators used on transformers, reactors, current/voltage transformers, and zinc oxide arresters have a height of 10 m, whereas the hollow composite insulators used on GISs have a structural height of 10.8 m.

2. **Insulators used in converter stations**
The design of external insulation for converter stations mainly depends on the anti-pollution performance of insulators, including their natural dust accumulation characteristics and pollution flashover performance. Compared with an AC insulator, a DC insulator suffers more dust accumulation on its surface, and because the current does not cross the zero point before DC arc occurs, it is easier for the DC arc to develop across the insulator. To obtain the dust accumulation characteristics of insulators of various types used in converter stations under AC voltage and DC voltage, two test stations were created near Fulong Converter Station and Fengxian Converter Station, respectively. The test stations were equipped with automatic meteorological instruments to keep records of meteorological conditions and analyze their effects on dust accumulation. In addition, remote monitoring devices were also installed on the test stations to monitor their operation in real time. To obtain the pollution flashover characteristics of insulators, artificial pollution tests were performed on all-porcelain post insulators and porcelain post insulators coated with RTV.

According to artificial pollution test results, it was determined that, for moderately (with an ESDD smaller than 0.05 mg/cm^2) and heavily polluted areas, the specific creepage distance of post insulators and bushings with equal diameter and deep under-rib sheds used in the DC switchyard should be 48.4–59.3 and 55.9–62.3 mm/kV, respectively, and that of post insulators and bushings with alternating sheds should be 54.9–60.4 and 71.5–78.7 mm/kV, respectively.

Demonstration of the outdoor DC switchyard design scheme showed that the costs of DC insulators are the primary constraint in DC switchyard construction. If the external insulation design scheme of conventional ±500-kV projects is adopted, then the height of a ±800-kV post insulator will exceed 11 m in a heavily polluted area. This makes it difficult to meet the required mechanical strength and aseismic requirements, and it makes the design and manufacture of ±800-kV bus equipment much more difficult and costly. By analyzing the prospects, operational experience, and economy and technology of various external insulation schemes, including outdoor all-porcelain insulators, composite insulators with porcelain core, hollow composite insulators, and RTV-coated insulators, it has been determined that porcelain post insulators coated with RTV, composite bushings, and hollow composite insulators will be used in the DC switchyard in a moderately polluted area. The use of composite materials in UHV DC converter stations greatly reduces the height of post insulators and facilitates the design of external insulation.

The wall bushing is a key device in connecting the equipment in the valve hall with that in the DC switchyard. Its sheds can effectively prevent the flashover caused by pollution or uneven distribution of rain in the event of heavy rain, and can allow for convenient cleaning. The wall bushings used in UHV DC converter stations are made of composite materials and have either alternating sheds or equal-diameter sheds without under-ribs.

6.5 Insulation Level of UHV Electrical Equipment

6.5.1 Parameters of Surge Arrester

1. **Parameters of surge arresters used in substations**
   The protective level of a surge arrester is related to its rated voltage, which is then mainly determined based on the amplitude and duration of the temporary overvoltage. The rated voltage of surge arrester on the bus side is taken as 828 kV considering a maximum temporary overvoltage of 1.3 p.u. (826 kV). Although the maximum temporary overvoltage could reach 1.4 p.u. (889 kV) on the line side, it persists only for <0.5 s; hence, the rated voltage can also be taken as 828 kV. This can significantly reduce the system's overvoltage level. Table 6.20 gives the major parameters of surge arresters used in a 1000-kV substation.
2. **Parameters of DC surge arresters used in converter stations**
   The crest value and peak value of the continuous operating voltage (CCOV and PCOV, respectively) of a surge arrester should be higher than the maximum operating voltage of the system at the installation location. In addition, the combination of operating voltage, harmonic voltage, and high-frequency transient overvoltage should be considered to avoid accelerated aging and reduced reliability of the arrester due to absorption of energy.

   The rated voltage (AC side) or reference voltage (DC side) of a surge arrester should be determined by comprehensively considering factors such as the applied voltage ratio, CCOV, PCOV, temporary overvoltage, LIPI, SIPI, and energy absorption capability. The reference voltage U_{ref} is generally defined as the DC voltage when the surge arrester conducts 1 mA (i.e., the voltage at which the arrester begins to act). It is a major parameter that dictates the material property, dimension, and number of the

Table 6.20: Major Parameters of Surge Arresters Used in a 1000-kV Substation (kV).

Nominal System Voltage	Installation Location	Rated Voltage (RMS Value)	Continuous Operating Voltage (RMS Value)	Residual Voltage at 30/60 μs and 2 kA Switching Impulse Current	Residual Voltage at 8/20 μs and 20 kA Lightning Impulse Current
1000	Line side, bus side, and transformer side	828	638	≤1460	≤1620
500	Transformer side	420	318	≤858	≤1046
110[a]	Bus, and reactor for reactive compensation	158	126	≤301	≤363[b]
	Capacitor for reactive compensation	174	138	≤331	≤400[b]

[a]110 kV isolated neutral system.
[b]Residual voltage at 8/20 μs and 10 kA lightning impulse current.

resistors connected in series or parallel. Most small-diameter, single-column arresters will act under a reference voltage at 1 mA, whereas for large-diameter arresters, this voltage may be determined as the reference voltage at 5 mA. The reference voltage is related to the current density of the resistor. The applied voltage ratio of a DC surge arrester characterizes the voltage load applied on a resistor, expressed as the ratio of the CCOV and PCOV to the reference voltage U_{ref}, after accounting for the non-uniform distribution of voltage across the arrester. It must be properly determined through stability tests (including ageing test) and pollution tests and must take into account factors such as stability, magnitude of leakage current, CCOV, DC voltage component, installation location (indoor or outdoor), effects of temperature on voltage–current characteristics, and effects of pollution on potential distribution over the porcelain or silicon rubber housing. It has a great influence on aging of arresters. With a smaller applied voltage ratio, the resistive leakage current under continuous operating voltage will be reduced, the resulting loss can be easily balanced with the heat dissipation capacity, and no thermal runaway will occur. However, a higher applied voltage ratio will reduce the protective level of the arrester, thereby significantly reducing the insulation level of equipment. For a DC arrester used in the ± 800-kV UHV DC valve hall, the voltage distribution non-uniformity factor is smaller than that of an AC arrester; therefore, such a DC arrester does not require housing. This facilitates heat dissipation and prevents explosion in the event of a thermal runaway, and increases the applied voltage ratio.

In Figure 4.32, the surge arrester V is subject to valve voltage only when the valve is off in an AC cycle, and the heat generated by the leakage current under the valve voltage in a cycle is very low, on average. Hence, a higher applied voltage ratio ranging from 0.95 to 1 may be selected. The surge arrester A1 is subject to a high voltage only when the valve is on in an AC cycle, where the voltage steps from 600 to 800 kV, but persists only for 10 ms. Therefore, an applied voltage ratio of approximately 0.95 may also be selected. The surge arresters C2, M1, M2, and CB1A are subject to DC voltage plus harmonic voltage. Part of the current resulting from the harmonic voltage is discharged because of the stray capacitance of the arresters and, especially, the commutation overshoot generates less heat on the resistors than that of the DC voltage component. In addition, as these surge arresters are installed indoors, the effects of pollution and ambient temperature may be disregarded. Hence, an applied voltage ratio of approximately 0.9 may be selected. The surge arresters DB and DL are subject to a very high, pure DC voltage. If they are installed outdoors, then pollution may lead to non-uniform distribution of potential over the porcelain or silicon rubber housing, causing local overheating on the resistors. In addition, the ambient temperature has a great impact on heat dissipation and voltage–current characteristics of the arresters. Consequently, a lower applied voltage ratio of 0.8–0.9 seems more reasonable. The PCOV of surge arresters E1H, E2, E2H, EL, and EM is rather low, and their applied voltage ratio is generally not a concern. The surge arresters A1, M1, and M2, which have a higher applied voltage ratio, may be provided with temperature and total current detection devices.

In the case of a short DC line, the operating voltage in the inverter station is only slightly lower than that in the rectifier station. Therefore, the DC side's reference voltage could be the same in both the inverter station and rectifier station, thereby reducing the variety of arresters and spare parts and making manufacture and test of arresters simpler. In contrast, for a long DC line, the reference voltage in the inverter station can be determined based on the operating voltage to reduce the protective level of arresters and insulation level of equipment in the station.

The SIPL, LIPL, and steep-front impulse protective level (STIPL) of a surge arrester are determined against the residual voltage under 30/60 μs switching impulse, 8/20 μs lightning impulse, and steep-front current impulse with a wave-front of 1 μs, respectively.

The energy of the arrester on the DC side is closely related to the type and duration of faults occurring in the converter station, and the response speed and delay time of the control and protection system. The amplitude and duration of the discharge current during a single surge or continual surges should be specified in determining the arrester energy. Repeated actions of a surge arrester due to DC control or operating sequence, such as the restart of DC system after an earth fault, could be deemed as a single discharge. The energy discharged is the sum of energy discharged during the repeated actions. In finalizing the energy of an equivalent single discharge, the energy of short-time surge discharge reducing the energy absorption capacity of the arrester should be considered.

Arresters with matching characteristics may be connected in parallel to accommodate the energy of single discharge and reduce the residual voltage of the arresters. For the parallel connection, multiple resistors could be arranged in a porcelain housing or multiple arresters could be connected in parallel. In general, the arrester manufacturer should consider the distribution non-uniformity of discharge current among the several resistors of an arrester or among multiple arresters.

A higher reference voltage (U_{ref}) could reduce the required specific energy (kJ/kV) of the arrester and make the manufacture easier.

6.5.2 Insulation Level of UHV AC Electrical Equipment

6.5.2.1 AC test voltage

AC withstand voltage tests on UHV electrical equipment are intended to examine the long-time operational reliability of equipment under operating voltage stress by monitoring partial discharge in the voltage rise process, and to examine the equipment's capability of withstanding temporary overvoltage that may occur during operation. Therefore, temporary overvoltage should be fully considered in insulation design.

Tables 6.21–6.23 provide the requirements of the capability of 1000-kV UHV transformers, shunt reactors, and MOAs in China to withstand a temporary overvoltage with a given duration.

Table 6.21: Requirements on the Capability of 1000-kV Transformers to Withstand Temporary Overvoltage.

Multiple of rated voltage under the in-service tap changer (p.u.)	1.05	1.10	1.25
Duration	Continuous under rated loads	Continuous under 80% rated loads	20 s under rated loads

Table 6.22: Requirements on the Capability of 1000 kV Shunt Reactors to Withstand Temporary Overvoltage.

Multiple of rated voltage (p.u.)	1.15	1.20	1.30	1.40	1.50
Duration	60 min	20 min	3 min	20 s	8 s

Table 6.23: Requirements on the Capability of 1000-kV MOAs to Withstand Temporary Overvoltage.

Multiple of rated voltage (p.u.)	1.00	1.10	1.15	1.20
Duration	20 min	10 s	1 s	0.2 s

Table 6.24: Withstand Voltage of Internal Insulation of Transformers in Japan and the Former Soviet Union (kV).

Country	Highest Voltage of Equipment	Lightning Impulse	Switching Impulse
Japan	1100	1950	1425
The Former Soviet Union	1200	2400	1950

6.5.2.2 Switching/lightning impulse withstand voltage (SIWV/LIWV)

In view of differences in insulation coordination principles and in performance of surge arresters, the insulation level of electrical equipment in Japan and the former Soviet Union differs greatly. Table 6.24 provides the respective withstand voltage of internal insulation of transformers of the two countries. It indicates that proper insulation coordination principles (i.e., to obtain desirable reliability level at a reasonable cost) are an essential prerequisite for determining the insulation level of electrical equipment [10].

The withstand voltage of internal insulation of electrical equipment is usually determined based on the SIPL and LIPL of the surge arrester multiplied by a given coordination factor (safety margin).

1. **SIWV**
   1. *Method A*

i. *Internal insulation of electrical equipment*: The required SIWV (RSIWV) $U_{\text{rw.s}}$ is given by Eq. (6.20)

$$U_{\text{rw.s}} = k_s k_{\text{cd}} U_{\text{rp}} \tag{6.20}$$

where

k_s—safety factor, taken as 1.15

k_{cd}—deterministic coordination factor, taken as 1.05

U_{rp}—representative overvoltage, taken as the maximum switching overvoltage derived from statistical calculation (kV).

ii. *External insulation of electrical equipment*: The RSIWV between phase and earth or between phases is given by Eq. (6.21)

$$U_{\text{rw.s}} = k_s k_{\text{cs}} U_{\text{rp}} \tag{6.21}$$

where

k_s—safety factor, taken as 1.05

k_{cs}—statistical coordination factor, taken as 1.15 for an insulation failure rate of 0.01%

U_{rp}—representative overvoltage, taken as the statistical switching overvoltage between the phase and earth or between phases (kV).

2. *Method B*: For both the internal and external insulation of electrical equipment, the RSIWV between phase and earth is given by Eq. (6.22):

$$U_{rw.s} = 1.15U_{ps} \tag{6.22}$$

where

U_{ps}—SIPL of surge arrester (kV).

2. **LIWV**
   1. *Method A*: The required LIWV (RLIWV) is given by Eq. (6.23)

$$U_{rw.1} = k_s U_{rp} \tag{6.23}$$

where

k_s—safety factor. For internal insulation, it is taken as 1.15 under common operation and 1.1 under special operation; for external insulation, it is taken as 1.05

U_{rp}—representative overvoltage, determined based on lightning intruding overvoltage on each piece of equipment in the substation.

$U_{w.1}$, the specified full-wave LIWV of internal insulation of equipment in the substation, should be equal to or greater than $U_{rw.1}$, the RLIWV.

2. *Method B*: The required full-wave LIWV of internal and external insulation of a transformer and shunt reactor is given by Eq. (6.24):

$$U_{rw.1} = 1.33U_{pl} \tag{6.24}$$

where

U_{pl}—LIPL of surge arrester (kV).

The coordination factor of 1.33 is determined by taking into account a safety margin of 1.15 and an equipment aging factor of 1.15.

$U_{rw.1}$, the required full-wave LIWV of insulation of HV electrical equipment, current transformer, separately tested bushing, bus post insulator, and cable and its fittings are given by Eq. (6.25):

$$U_{rw.1} = 1.45U_{pl} \tag{6.25}$$

The coordination factor of 1.45 is determined by taking into account a safety margin of 1.15, an equipment aging factor of 1.15, and a distance factor of 1.1.

Table 6.25: Specified Withstand Voltage of Insulation of 1000-kV Equipment (kV).

Equipment	Lightning Impulse	Switching Impulse	Power Frequency Voltage
Transformer and reactor	2250 (2400 for chopped lightning impulse)	1800	1100 (5 min)
GIS (circuit breaker and disconnector)	2400	1800	1100 (1 min)
Post insulator and disconnector (open type)	2550	1800	1100 (1 min)
CVT	2400	1800	1200 (5 min)
Bushing (transformer and reactor)	2400 (2760 for chopped lightning impulse)	1950	1200 (5 min)
Bushing (GIS)	2400	1800	1100 (1 min)
Longitudinal insulation of switchgear	2400 + 900	1675 + 900	1100 + 635 (1 min)

The required chopped-wave LIWV of insulation of a transformer, shunt reactor, and current transformer is taken as 1.1-times the required full-wave LIWV.

Overall, the insulation level of UHV equipment in China is lower than that in the former Soviet Union and higher than that in Japan. This is determined by the overvoltage level of UHV systems and technical capabilities of China [8].

Table 6.25 gives the specified withstand voltage of insulation of various 1000-kV equipment in China, which may be optimized based on actual project conditions.

6.5.3 Insulation Level of UHV DC Electrical Equipment

1. **Principles and methods for determining insulation level**

   As with AC equipment, the insulation strength and characteristics of DC equipment, including the characteristics of surge arresters and control and protection devices in converter station, are also determined based on temporary overvoltage, switching overvoltage, lightning overvoltage, and steep-front overvoltage that may act on the equipment, considering the cost of equipment, maintenance, and accidents, to achieving safe, economic, and reliable operation of the DC equipment.

   In design of insulation coordination, the connection of the two 12-pulse converters needs to be specified first, the Y/Y converter transformer is installed at the HV end while the Y/Δ converter transformer is installed at the LV end, and shunt reactors are installed both on the pole bus and neutral bus. Then, a model is established based on the AC/DC system data (including data of equipment on the AC side and DC side, such as the transformer and line), characteristics of surge arrester, valve control and protection strategy for the converter station (including response and delay of the valve protection firing circuit), and operating conditions to determine the representative overvoltage, current, and energy of the surge arrester through calculation and analysis. This is an iterative process of optimizing the coordination between requirements for surge arrester and insulation level of equipment.

Insulation coordination involves the following:

a. Determine the representative voltage and overvoltage U_{rp} through systematical calculation and analysis
b. Determine the insulation that meets the performance criteria
c. Determine the coordination current and then the coordination withstand voltage U_{cw}
d. Determine the factor of difference between the type test condition and actual operating condition
e. Determine the required withstand voltage U_{rw}
f. Determine the specified withstand voltage U_w

2. **SIWV/LIWV**

The representative overvoltage is the calculated switching overvoltage, lightning overvoltage, or steep-front overvoltage. For the equipment directly protected by a surge arrester, the representative overvoltage is the protective level of the arrester.

The protective level of a surge arrester can be determined as follows:

a. The protective level of a single surge arrester is determined based on the residual voltage across the arrester at the coordination current.
b. In cases when the equipment (valve side of the converter transformer, for example) in the converter station is protected with two serially connected surge arresters, and when the SIPL is determined based on the sum of the respective SIPL of the two arresters, the protective level may be determined with any of the following three methods (the first is preferred):
   i. The coordination current corresponding to the SIPL of the two surge arresters is determined based on the maximum coordination current flowing through one of the arresters after a fault occurs, whereas the discharge currents, respectively, flowing through the two arresters are different, thus producing a certain margin.
   ii. The coordination current corresponding to the SIPL of the two arresters is determined based on the maximum coordination currents flowing through the arresters under different faults, whereas the maximum coordination current will never occur on the two arresters simultaneously after the same fault, thereby producing a greater margin than that of method 1.
   iii. The protective level is determined based on the maximum switching overvoltage derived from simulation without considering the coordination current. This method does not produce any margin, which poses a risk that the protective level of the surge arresters is higher than the withstand capability of the protected equipment. This method is optional when the fault giving rise to the maximum overvoltage has a very low probability.

c. The LIPL of the two arresters can be determined with method 1, that is, by adding the residual voltage across each arrester at the same coordination current. This method is practical in engineering because the lightning intruding overvoltage can be disregarded for equipment in the valve hall. However, for the DC bus connecting the valve side of smoothing reactor and wall bushing, the LIPL of the two serially connected surge arresters should be calculated by establishing a comprehensive lightning model, which includes the inductance and distributed capacitance to ground of the line between the two arresters.

For equipment on the DC side, the coordination withstand voltage U_{cw} is derived with the deterministic method by multiplying the representative overvoltage U_{rp} by the deterministic coordination factor k_{cd}, that is, $U_{cw} = k_{cd}U_{rp}$ (see GB/T 311.2—2002 *Insulation Coordination—Part 2: Application Guide for Insulation Coordination for High-Voltage Transmission and Transformation Equipment*). The following conditions are considered for the coordination factor k_{cd}:

1. Limitations of data and model used for overvoltage calculation; effects of strong non-linearity of the arrester on the coordination current.
2. Difference in the waveform and duration between the overvoltage and standard test voltage.
3. The coordination factor of lightning overvoltage is taken as 1, considering that the representative lightning overvoltage is calculated without considering the effects of occurrence probability but under stringent conditions.

   GB 311.1—2012 *Insulation Coordination—Part 1: Definitions, Principles and Rules* gives the equation for calculating the required withstand voltage of insulation with the deterministic method:

$$U_{rw} = U_{cw}k_s k_a \tag{6.26}$$

where
k_a—atmospheric correction factor
k_s—safety factor, depending on the type of internal and external insulation.

k_s is determined considering the life of insulation, changes in characteristics of surge arrester, and the dispersion of product quality.
k_s is taken as 1.15 for internal insulation and as 1.05 for external insulation. For calculation of the atmospheric correction factor, see IEC 60071-2.

Insulation coordination of ±800-kV UHV DC converter stations is also designed with the deterministic method. According to experience with ±500-kV projects and GB/T 311.3—2007 *Insulation Coordination—Part 3: Procedures for High-Voltage Direct Current (HVDC) Converter Stations*, for HVDC converter stations at altitudes lower than 1000 m, the withstand voltage of equipment protected directly by the surge arrester in its immediate vicinity can be derived by multiplying the protective level of the surge arrester by a ratio. This ratio accounts for the insulation coordination factor, safety margin, and atmospheric correction factor of external insulation for an altitude of 1000 m. Table 6.26 gives the ratio of required impulse withstand voltage to protective level of surge arrester.

For AC and DC switchyard equipment not directly protected by a surge arrester, its insulation coordination should be determined considering effects of distance under transient lightning overvoltage and steep-front overvoltage.

Because the thyristors of a converter valve are provided with a monitoring device, it is easy to identify and replace the failed thyristors; therefore, the thyristors have no aging problem. It is generally considered that the withstand voltage of a converter valve restores to its original value after each maintenance. In addition, the converter valve is protected directly by arresters, and the cost and loss of the converter valve are approximately proportional to its insulation level. Reducing the insulation level also contributes to a lower height of the valve and valve hall. Therefore, the ratio of the required withstand voltage of insulation of converter valve to the protective level of surge arrester, as specified in Table 6.26, including RSIWV/SIPL and RLIWV/LIPL, and required steep-front impulse withstand voltage (RSFIWV)/STIPL can be selected from ranges of 1.10–1.15, 1.10–1.15, and 1.15–1.20 through techno-economic comparison.

Table 6.26: Ratio of the Required Withstand Voltage of Insulation to the Protective Level of Surge Arrester.

Equipment	RSIWV/ SIPL	RLIWV/ LIPL	RSFIWV/ STIPL
AC switchyard (including bus, outdoor insulation, and other common equipment)	1.2	1.25	1.25
AC filter components	1.15	1.25	1.25
Converter transformer (oil-insulated equipment)			
Line side	1.20	1.25	1.25
Valve side	1.15	1.20	1.25
Converter valve	1.10–1.15	1.10–1.15	1.15–1.20
DC equipment in valve hall	1.15	1.15	1.25
Equipment in the DC switchyard (outdoor, including DC filter and smoothing reactor)	1.15	1.20	1.25

Notes:

1. The ratio may be increased or decreased depending on the insulation criteria of the specific equipment. For example, the ratio of the converter transformer bushing may be increased.
2. The ratio applies only to the equipment directly protected by the surge arrester in its immediate vicinity.
3. STIPL applies to valve arresters.

The specified withstand voltage should be equal to or greater than the required withstand voltage. For AC equipment, the specified withstand voltage is determined as the value following the voltage equal to or higher than the same type test voltage in the series standard values of specified power frequency and impulse withstand voltage given in GB 311.1-2012 according to the required withstand voltage. For HVDC equipment, no series standard values of withstand voltage are given; instead, the specified withstand voltage is determined to be an applicable value.

Because the lightning intruding overvoltage on the valve side of converter transformer is not considered, the ratio of specified switching impulse withstand voltage (SSIWV) to the specified lightning impulse withstand voltage (SLIWV) may be determined based on the manufacturing and operation experience, generally 0.83–0.95. The CIGRE working group recommends a ratio of 0.95, but does not explain why. Operation experience with converter transformers and AC transformers indicates that a smaller ratio will increase the failure rate of the transformer. To prevent voltage that is too high on the AC winding on the line side, applying switching impulse on one terminal of the valve-side winding while connecting the other terminal to ground is not allowed. Instead, the switching impulse should be applied on the main insulation of the valve-side winding with its terminals connected and the line-side winding short-circuited and earthed. Lightning impulse tests should be performed to examine the longitudinal insulation of winding. The test voltage should not be too low; SSIWV/SLIWV ratio of 0.83 is recommended. Considering that increasing the SSIWV/SLIWV ratio of the HV converter transformer will cause manufacturing difficulties, a ratio of approximately 0.90 is recommended.

Table 6.27 provides the recommended values of SIPL and SIWV for major nodes of the Fulong Converter Station in Shanghai–Xiangjiaba transmission project.

Table 6.27: SIPL and SIWV in Fulong Converter Station.

Directly Protected Equipment	SIPL	SIWV
Line side of DC bus smoothing reactor	1389	1600
Valve side of DC bus smoothing reactor	1384	1600
Between terminals of HV converter	761	950
Between terminals of LV converter	725	850
Valve side of HV Y/Y converter transformer	1384	1600
Valve side of HV Y/Δ converter transformer	1128	1300
Valve side of LV Y/Y converter transformer	900	1050
Valve side of LV Y/Δ converter transformer	827	970
Between terminals of pole bus smoothing reactor	739	850
Between terminals of neutral bus smoothing reactor	424	550
Between phases on valve side of converter transformer	461	550

References

[1] G. Dingxie, Z. Peihong, C. Yong, et al., Selection of air clearance of 1000 kV transmission line tower, High Voltage Eng. 33(11) (2007) 15–19.

[2] L. Zhenya, Overvoltage and Insulation Coordination of UHV AC Transmission Systems, China Electric Power Press, Beijing, 2008.

[3] UHV DC Transmission Technology Study Collection (2009), China Electric Power Press, Beijing, 2010.

[4] D. Yujian, L. Weiming, S. Zhaoying, S. Zhiyi, Impulse discharge characteristics tests and altitude correction research on air gaps of ±1000 kV DC transmission line tower, Proc. CSEE 34 (2011) 156–162.

[5] G. Chen, Z. Wenliang, F. Jianbin, et al., Altitude correction research on switching impulse voltage of rod–plane gap, High Voltage Eng. 38(1) (2012) 140–146.

[6] China Electric Power Research Institute, UHV Transmission Technology—DC Transmission, China Electric Power Press, Beijing, 2012.

[7] Z. Peihong, X. Muhong, Gu Dingxie, et al., Study on overvoltage and insulation coordination of ± 800 kV DC systems, High Voltage Eng. 32(12) (2006) 125–132.

[8] G. Dingxie, Z. Peihong, X. Muhong, et al., Study on overvoltage and insulation coordination of AC 1000 kV transmission systems, High Voltage Eng. 32(12) (2006) 1–6.

[9] L. Zhenya, Overvoltage and Insulation Coordination of UHV DC Transmission Systems, China Electric Power Press, Beijing, 2009.

[10] G. Dingxie, Overvoltage and insulation coordination of UHV transmission systems in China, China Electr. Power 32(4) (1999) 65–68.

CHAPTER 7

Electromagnetic Environment in UHV Transmission Projects

Chapter Outline

Ultra-high Voltage AC/DC Grids.
DOI: http://dx.doi.org/10.1016/B978-0-12-802161-3.00007-X

7.1 Overview

The environmental impacts of power transmission projects generally include the impacts on ecological environment, soil erosion and water loss, land occupation by line corridors, conformity and compatibility between routing/siting and planning, electromagnetic environmental effects, and even the landscape impact. After the projects are put into operation, the electromagnetic phenomena will become the major environmental concern. Today, with the increasingly growing environmental quality awareness, the public becomes more and more concerned about the electromagnetic environmental effects in the vicinity of power transmission facilities.

For AC transmission projects, the electromagnetic environmental problems mainly include the power-frequency electric field and electrostatic induction, power-frequency magnetic field and electromagnetic induction, radio interference (RI) (including television interference (TVI)), and audible noise [1]. For DC power transmission, there are also problems resulting from ion current and the total electric field [2,3]. All the problems occur in a frequency range of 50 Hz up to 1 GHz, and also include 0 Hz. Because the electric and magnetic fields and RI may affect the living environment and quality of life for human beings, proper technical measures should be taken to mitigate the impacts of these electromagnetic environmental problems. How to control the electromagnetic environment (EME) in UHV transmission projects is one of the major concerns for UHV grid construction and operation.

During the development of UHV transmission technology, China has given a great deal of attention to the electromagnetic environmental impacts and performed numerous specialized studies regarding this subject. Several tests at UHV AC and UHV DC test bases have been conducted, aiming at ensuring that the EME parameters of the completed UHV AC and UHV DC demonstration projects fully meet the applicable national standards on environmental protection.

7.2 Electric and Magnetic Fields of UHV Transmission Projects

7.2.1 Electric and Magnetic Fields of UHV AC Transmission Projects

In China, the AC grids operate at a frequency of 50 Hz, thus producing 50-Hz power-frequency electric and magnetic fields. The magnitude of the power-frequency electric and magnetic fields is defined by electric field strength and magnetic flux density, respectively.

The power-frequency electric and magnetic fields are quasi-static fields [4]. Such quasi-static characteristics allow the electric and magnetic fields to be discussed separately. As shown in Figure 7.1, by applying a voltage U across two conductors that are d apart, an

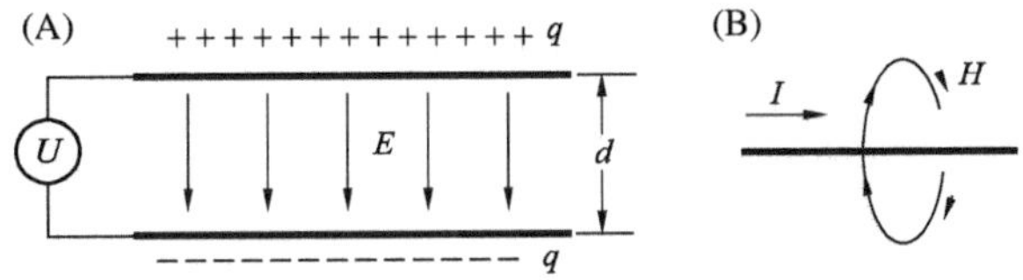

Figure 7.1: Generation mechanism of power-frequency electric and magnetic fields. (A) Electric field and (B) magnetic field.

electric field E is generated between them; when current I flows through a conductor, a magnetic field H is generated around the conductor.

The power-frequency electric and magnetic fields generated by transmission lines are at extremely low frequency (ELF) and have a wavelength as high as 6000 km. According to the antenna theory, it is unlikely for a transmission link to become an effective radiation source; therefore, the power-frequency electric and magnetic fields must be distinguished from the high-frequency radiation. The International Commission on Non-Ionizing Radiation Protection (ICNIRP) and other involved international organizations scrupulously differentiate between the terms electric, magnetic, and electromagnetic fields instead of using a general term—electromagnetic radiation (EMR). They only use the term electromagnetic field (EMF) when referring to fields at frequencies higher than 100 kHz.

7.2.1.1 Power-frequency electric fields produced by UHV AC lines

The power-frequency electric fields produced by UHV AC lines have the following characteristics. Firstly, the electric field at any point in space is a rotating elliptical field. However, when approaching the ground level, the electric field becomes perpendicular to the ground. Within the range up to 2 m above the ground, the vertical component of the electric field strength is distributed substantially uniformly, whereas the horizontal component can be negligible. Secondly, the power-frequency electric fields are liable to be shielded by trees and buildings. As a result, their strength is greatly reduced. Thirdly, regardless of the variation of transmission capacity, the voltage of the line is kept relatively stable; even the line operates under no load. Therefore, the strength of the power-frequency electric field is also relatively stable.

1. **Calculation of the electric field strength**
   To control the power-frequency electric field of UHV transmission lines, it is essential to understand its characteristics and laws first. There are two means available for analyzing the electric field of transmission lines, measurement and calculation.
   The electric field of an AC line mainly results from the charges on it. For a given conductor configuration and voltage, it is necessary to calculate the conductor charges first and then use the results to calculate the strength of the spatial electric field.

Figure 7.2
Field test for cup towers of 1000-kV Changzhi–Nanyang line I.

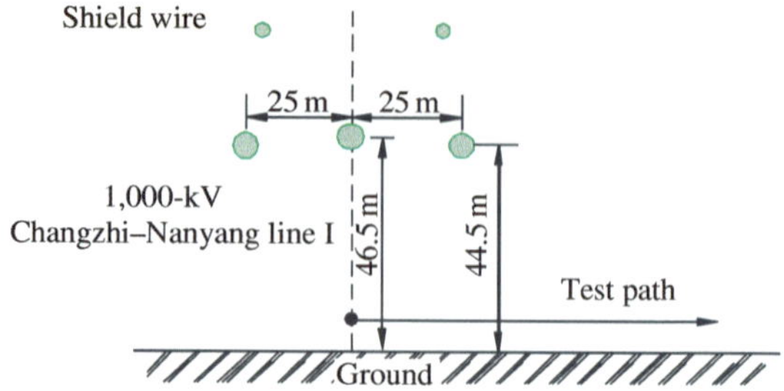

Figure 7.3
Test points for lateral distribution of electric fields produced by conductors carried by cup towers.

The CIGRE working group 36.01 proposed an equivalent charge method for calculating the strength of power-frequency electric fields produced by AC transmission lines. This method uses an equivalent single conductor instead of bundle conductors, thereby simplifying the calculation process.

Because the equivalent radius of the bundle circle of an overhead line is much smaller than the line height, the concern normally lies with the strength of the electric fields near the ground. Therefore, it may be assumed in calculation that the equivalent charges are located at the line's geometric center, that is, the equivalent charges are linear.

2. **Calculation and measurement of the power-frequency electric field strength distribution of a sample UHV AC line**
China's 1000-kV Jindongnan–Nanyang–Jingmen UHV AC pilot and demonstration project mainly uses two types of tower, the cup tower (with conductors horizontally arranged) and the cat-head tower (with conductors triangularly arranged), respectively.

Figure 7.2 shows the field test of the lateral distribution of electric fields produced by conductors carried by cup towers. Figure 7.3 shows the locations of the test points. The test path starts from the projective point on the ground of the center phase conductor in the middle of span, and the electric field strength is tested in the direction perpendicular to the conductor. Figure 7.4 shows the comparison between the calculated and measured values of the power-frequency electric field strength for conductors carried by cup towers.

Figure 7.5 shows the field test of the lateral distribution of the electric field produced by conductors carried by cat-head towers. Figure 7.6 shows the test points. The test path starts from the projective point on the ground of the center phase conductor in the middle of span, and the electric field strength is tested in the direction perpendicular to the conductor. Figure 7.7 shows the comparison between the calculated and measured values of the power-frequency electric field strength for conductors carried by cat-head towers.

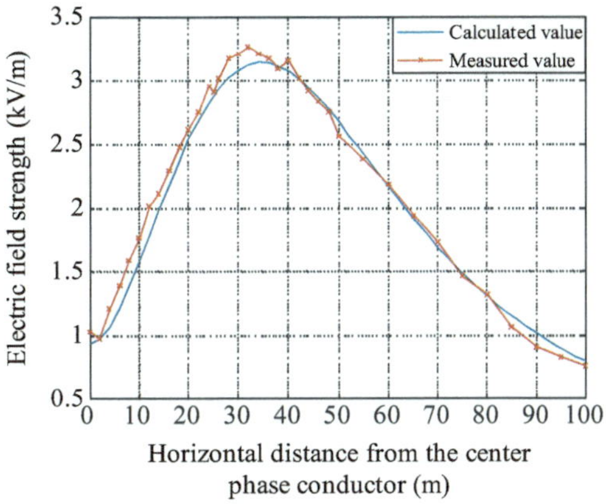

Figure 7.4
Calculated values versus measured values of the power-frequency electric field strength for conductors carried by cup towers.

Figure 7.5
Field test for cat-head towers of 1000-kV Changzhi–Nanyang line I.

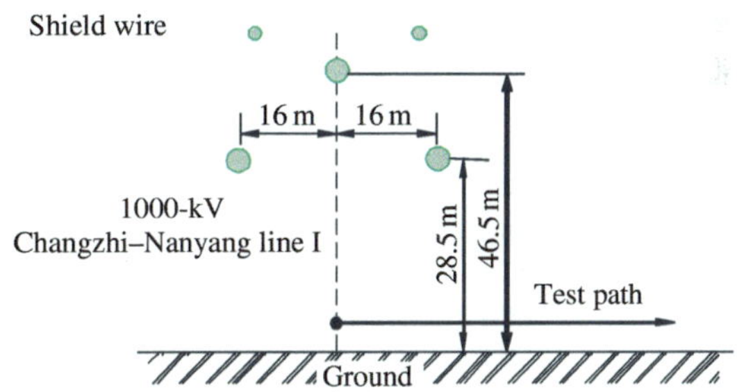

Figure 7.6
Test points for lateral distribution of electric field produced by conductors carried by cat-head towers.

Figures 7.2–7.7 show that the calculated and measured values of the power-frequency electric field strength match well, and that the maximum strength occurs outside of the outer phase conductor.

7.2.1.2 Power-frequency electric fields of UHV substations

UHV substations mainly comprise power transformers, circuit breakers, disconnectors, voltage transformers, current transformers, surge arresters, overhead bus, incoming line and outgoing line, as well as the interconnection overhead lines within the substation. Similar to the overhead transmission lines, charges on and currents through these conductors exposed to air will produce power-frequency electric and magnetic fields in the substation.

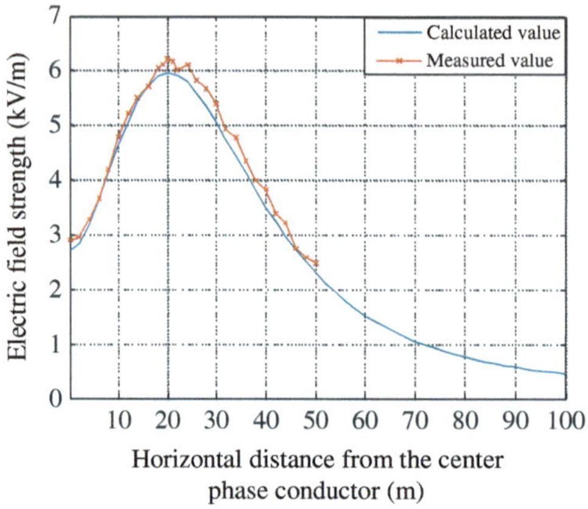

Figure 7.7
Calculated values versus measured values of the power-frequency electric field strength for conductors carried by cat-head towers.

Table 7.1: Maximum Electric Field Strength in Typical Areas of Changzhi Substation (2011).

Area	Maximum Electric Field Strength (kV/m)
1000 kV GIS	7.10
1000 kV high-voltage reactor	9.10
1000 kV main transformer	9.57
500 kV high-voltage reactor	7.82

The measured data of the electric field in UHV substations at various voltage levels indicate that the electric field strength at the ground level is usually approximately 10 kV/m, and the electric field strength within a substation is not uniformly distributed.

After the 1000-kV Jindongnan–Nanyang–Jingmen UHV AC pilot and demonstration project was put into operation, comprehensive measurements of power-frequency electric fields were performed in typical areas of the substations.

1. In the 1000-kV Changzhi substation, the test points are mainly located in the areas of the 1000-kV reactor, 1000-kV GIS, 1000-kV main transformer, and 500-kV GIS. Table 7.1 shows the maximum electric field strength measured in individual typical areas.
   The test results show that the maximum electric field occurs in the vicinity of the main transformer and is 9.57 kV/m (<10 kV/m).
2. In the 1000-kV Nanyang switching station, the test points are mainly located at the areas of 1000-kV high-voltage reactor and 1000-kV GIS. Table 7.2 shows the maximum electric field strength measured in individual typical areas.

Table 7.2: Maximum Electric Field Strength in Typical Areas of Nanyang Switching Station (2011).

Area	Maximum Electric Field Strength (kV/m)
1000-kV high-voltage reactor of Changzhi–Nanyang line	10.60
1000 kV GIS	8.60
1000-kV high-voltage reactor of Nanyang–Jingmen line	10.74

Table 7.3: Maximum Electric Field Strength in Typical Areas of Jingmen Substation (2011).

Area	Maximum Electric Field Strength (kV/m)
1000-kV high-voltage reactor	10.16
1000 kV GIS	7.88
1000-kV main transformer	10.44
500 kV GIS	7.14

The test results show that the maximum electric field occurs in the high-voltage reactor area in Nanyang switching station, at 10.74 kV/m (slightly larger than 10 kV/m). In the high-voltage reactor areas of both the Changzhi–Nanyang and Nanyang–Jingmen lines, there are areas where the electric field strength is slightly larger than 10 kV/m. The electric field strength in the GIS area is <10 kV/m.

3. In the 1000-kV Jingmen substation, the test points are mainly located at the 1000-kV reactor, 1000-kV GIS, 1000-kV main transformer, and the 500-kV GIS. Table 7.3 shows the maximum electric field strength measured in individual typical areas.

   The test results show that the maximum electric field of 10.44 kV/m occurs in the 1000-kV main transformer area, slightly larger than 10 kV/m but smaller than 11 kV/m. In the 1000-kV high-voltage reactor area, there is only a small area where the electric field strength is slightly larger than 10 kV/m.

7.2.1.3 Power-frequency magnetic fields produced by UHV AC lines

The power-frequency magnetic fields produced by UHV AC lines have the following characteristics: (i) as the line current varies with the power load, the power-frequency magnetic field strength varies correspondingly; (ii) the magnetic field strength decreases rapidly with the distance away from the line, at a larger rate than the electric field strength; (iii) the magnetic field distribution can only be influenced by ferromagnetic material and, hence, the power-frequency magnetic fields around transmission lines is not as susceptible to be perturbed as electric fields (the trees and buildings can barely shield the magnetic fields); and (iv) the power-frequency magnetic field of transmission lines is a rotating elliptical field.

1. **Calculation of magnetic field**
   The Ampere's Circuital Law is used to calculate the magnetic field produced by current flowing through each conductor, and then the results are superimposed to obtain the strength of magnetic field around the bundle conductor.
   For a three-phase AC transmission line, the spatial magnetic field is the composition of the magnetic fields produced by each phase. Similar to the power-frequency electric field, the space trajectory of the total magnetic field vector of a three-phase AC line is in the shape of an ellipse.
2. **Calculation and measurement of the power-frequency magnetic flux density distribution of a sample UHV AC line**
   Figures 7.8 and 7.9 show the comparison between the calculated and measured values of the magnetic flux density of conductors carried by the cup tower and cat-head tower. The test path starts from the projective point on the ground of the center phase conductor in the middle of the span, and the magnetic flux density is measured in the direction perpendicular to the conductor.

Figures 7.8 and 7.9 show that the calculated and measured values of the power-frequency magnetic flux density match well, and that the maximum density occurs at the center of the line.

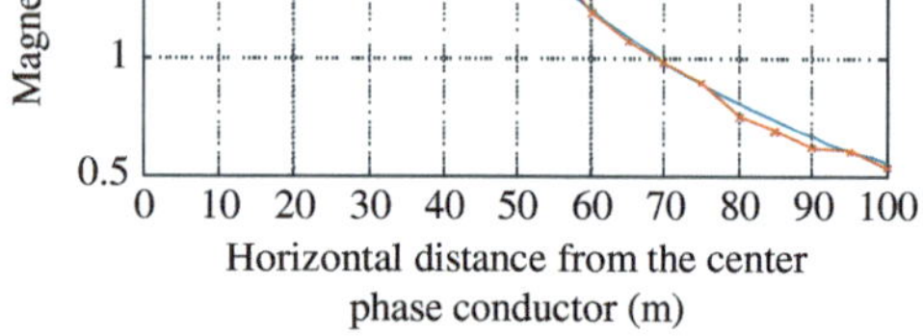

Figure 7.8
Calculated values versus measured values of the magnetic flux density produced by conductors carried by cup towers.

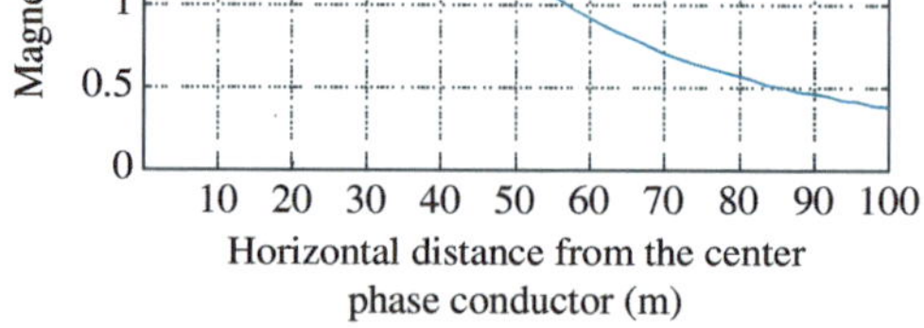

Figure 7.9
Calculated values versus measured values of the magnetic flux density produced by conductors carried by cat-head towers.

7.2.1.4 Power-frequency magnetic fields in UHV substations

In a substation, numerous current-carrying conductors are arranged in a criss-cross manner and there is a wide variety of energized equipments. As a result, the power-frequency magnetic field at any point in space is a three-dimensional field rather than a two-dimensional one. Generally, the magnitude of magnetic field at the point is represented by the power-frequency magnetic field strength at 1 m above the ground. The distribution and magnitude of the power-frequency magnetic fields in substation are a function of the arrangement of current-carrying conductors and the magnitude of current flowing through them. The current-carrying conductors in a UHV substation have higher voltage than those in an EHV substation, which has a great impact on the distribution and magnitude of the electric fields but has no substantial effect on the distribution of magnetic fields. The measurements of the power-frequency magnetic fields at different locations of the 500- and 330-kV substations show that in an EHV substation, the maximum magnetic field occurs in the vicinity of the 35-kV shunt reactors. However, there are few references available regarding power-frequency magnetic fields in UHV substations.

After the 1000-kV Jindongnan–Nanyang–Jingmen UHV AC pilot and demonstration project was put into operation, comprehensive measurements of the power-frequency magnetic fields were performed in typical areas of substations.

1. In the 1000-kV Changzhi substation, the test points are mainly located at the 1000-kV high-voltage reactor, 1000-kV GIS, and 1000-kV main transformer, and the 500-kV GIS. Table 7.4 presents the maximum magnetic flux density measured in various typical areas.

   The test results show that:

   a. The maximum magnetic flux density of 31.25 μT in the 1000-kV areas occurs in the vicinity of the GIS (<100 μT).
   b. The maximum magnetic flux density in the 500-kV areas is 21.57 μT (<100 μT).
2. In the 1000-kV Nanyang switching station, the test points are mainly located in the 1000-kV high-voltage reactor and GIS areas. Table 7.5 presents the maximum magnetic flux density measured in various typical areas.

Table 7.4: Maximum Magnetic Flux Density in Typical Areas of Changzhi Substation (2011).

Area	Maximum Magnetic Flux Density (μT)
1000 kV GIS	31.25
1000 kV high-voltage reactor	22.80
1000 kV main transformer	26.41
500 kV high-voltage reactor	21.57

Table 7.5: Maximum Magnetic Flux Density in Typical Areas of Nanyang Switching Station (2011).

Area	Maximum Magnetic Flux Density (μT)
1000 kV high-voltage reactor of Changzhi–Nanyang line	8.21
1000 kV GIS	15.63
1000 kV high-voltage reactor of Nanyang–Jingmen line	8.50

Table 7.6: Maximum Magnetic Flux Density in Typical Areas of Jingmen Substation (2011).

Area	Maximum Magnetic Flux Density (μT)
1000 kV high-voltage reactor	9.98
1000 kV GIS	18.41
1000 kV main transformer	13.35
500 kV GIS	22.60

The test results show that the maximum flux density in the Nanyang switching station is 15.63 μT (<100 μT).

3. In the 1000-kV Jingmen substation, the test points are mainly located in the areas of the 1000-kV high-voltage reactor, 1000-kV GIS, 1000-kV main transformer, and the 500-kV GIS. Table 7.6 presents the maximum magnetic flux density measured in various typical areas.

   The test results show that:

   a. The maximum magnetic flux density of 18.41 μT in the 1000-kV areas occurs in the vicinity of the GIS area (<100 μT).
   b. The maximum magnetic flux density in the 500-kV areas is 22.60 μT (<100 μT).

7.2.2 Limits of Power-Frequency Electric and Magnetic Fields of UHV AC Lines

1. **Limit of power-frequency electric field**

   The limits on the power-frequency electric fields are set to prevent the adverse health effects caused by exposure to power-frequency electric fields. Generally, such limits are determined by considering three aspects, preventing transient electric shock causing discomfort, steady-state electric shock current larger than the let-go threshold, and harmful ecological effects. In addition, some international safety regulations are also available for limiting the steady-state electric shock current caused by large objects or vehicles beneath transmission lines.

a. *ICNIRP and IEEE limits*. ICNIRP (formerly IRPA/INIRC) is a reputable international body that is recognized by the World Health Organization (WHO). In April 1998, ICNIRP officially put forward the *Guidelines for Limiting Exposure to Time-Varying Electric, Magnetic and Electromagnetic Fields (300 GHz and below)* (referred to as the Guidelines). In 2010, ICNIRP published a new version in which the limits of power-frequency electric fields remain the same. The Guidelines specify different limits for the general public and the occupational exposure, i.e. 5 kV/m and 10 kV/m separately. The public exposure limit is the primary concern in environmental protection.

   It is specified in IEEE Std.C95.6—2002 that the exposure limit of the 50-Hz power-frequency electric field is 5 kV/m for the general public and 20 kV/m for a controlled environment. A controlled environment refers to an area with restricted access.

b. *International limits*. CIGRE has also summarized the limits of electric fields in the vicinity of transmission lines used in some countries. Although such limits vary for different countries or utilities, there are still some consistent requirements: the maximum electric field strength value of 10–15 kV/m is used for the area under lines (or in agricultural regions); 10 kV/m is used for the location where lines cross roads; and <5 kV/m is used for public places or near residential houses.

c. *Limits in China*. It is recommended in HJ/T 24—1998 *Technical Regulations on Environmental Impact Assessment of Electromagnetic Radiation Produced by 500 kV Ultrahigh Voltage Transmission and Transfer Power Engineering* that 4 kV/m be tentatively taken as the assessment criterion for power-frequency electric fields in residential areas. It is specified in clause 16.0.5 of DL/T 5092—1999 *Technical Code for Designing 110–500 kV Overhead Transmission Line* that where a 500-kV transmission line crosses buildings not permanently occupied by people or runs close to residential buildings, the maximum unperturbed electric field at 1 m above the ground around them should not exceed 4 kV/m.

   In China, the limits of power-frequency electric field strength for UHV AC lines are the same as those for the 500-kV AC lines, that is, the limit of electric field at 1.5 m above the ground under an AC line is 7 kV/m for general regions (e.g., the regions readily accessible to the public and the roads where a line crosses them) and is 10 kV/m for the farmland where a line crosses it. When the line runs close to the residential buildings, the maximum unperturbed electric field at 1 m above the ground around the buildings should not exceed 4 kV/m [5].

2. **Limit of power-frequency magnetic field**

   a. *ICNIRP and IEEE limits*. The European Union recommended that its member countries should use the ICNIRP Guidelines as a mandatory standard to guarantee the safety of occupational personnel and general public.

The 1998 edition of the Guidelines specifies that the magnetic field exposure limits for the general public and the occupational personnel are 100 and 500 μT. In the 2010 edition of the Guidelines, the public exposure limit is increased to 200 μT.

IEEE Std.C95.6—2002 sets different exposure limits for various parts of a human body: 904 μT for head and 2710 μT for limbs.

b. *Limits in foreign countries*. Most foreign countries follow the ICNIRP guidelines. Germany legislated the use of ICNIRP guidelines. In 2004, the National Radiological Protection Board (NRPB) suggested that limits given in the ICNIRP guidelines should be used as the national standard in electromagnetic protection in the United Kingdom. Australia uses the same limits of power-frequency magnetic field as those specified in the ICNIRP guidelines.

c. *Limits in China*. HJ/T 24—1998 recommends that the public exposure limit of 100 μT proposed by ICNIRP should be used as the assessment criterion for magnetic flux density.

Although the limits used by different countries are different, most countries tend to use the limits given in the ICNIRP guidelines. In developing UHV transmission projects, China uses the stricter limit of 100 μT specified in the ICNIRP guidelines as the limit of power-frequency magnetic fields produced by transmission lines.

7.2.3 Total Electric Field and DC Magnetic Field in UHV DC Transmission Projects

During the normal operation of DC transmission lines, certain corona discharges occur on the conductor surface [6]. The spatial ions distribution resulting from corona discharge on a DC line is quite different from that on an AC line. As the polarity of AC voltage reverses periodically, when coronas occur on an AC line, all the charged ions generated by corona discharge within the voltage's positive half cycle will be drawn back to the conductor in the negative half cycle. In this way, the charged ions just move back and forth within a small range around the conductor and no charged ions are present between the phase conductors or between the phase conductor and the ground. In contrast, the polarity of DC voltage does not reverse, so when coronas occur on a DC line, among the charged ions generated by coronas on a pole conductor, the ions with opposite polarity will be drawn back to the conductor while those with the same polarity will be repelled from the conductor and continue to move along the electric field lines. Thus, charged ions will occupy the entire space between the two pole conductors and between the pole conductor and the ground. Figure 7.10 shows the electric field lines of a bipolar transmission link and the distribution of charged ions. The entire space shown in Figure 7.10 can be approximately divided into three regions: the region between the positive pole conductor and the ground, which is full of positive ions; the region between the negative pole conductor and the ground, which is

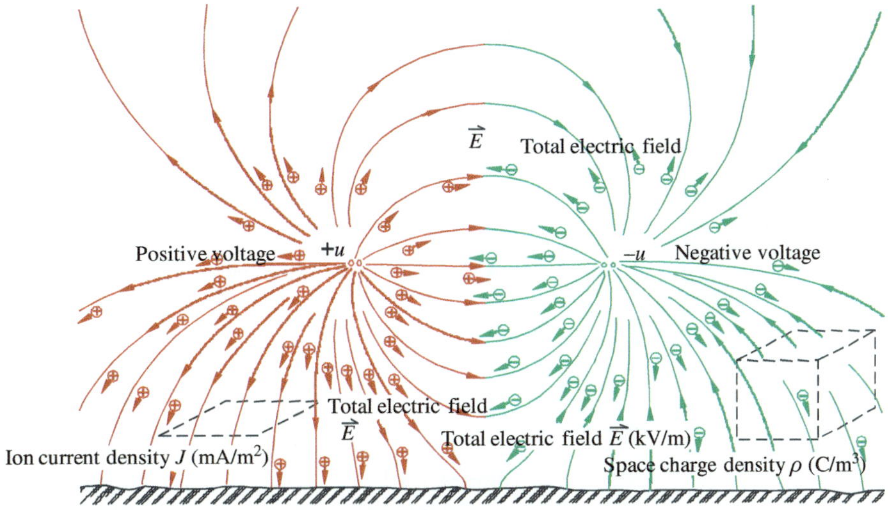

Figure 7.10
Lines of electric force of a bipolar transmission line and the distribution of charged ions.

full of negative ions; and the region between the positive and negative pole conductors, where positive and negative ions co-exist.

The current generated by movement of charged ions in space is called ion current. The ion current through a unit area is referred to as the ion current density J, expressed in mA/m^2. To measure the quantity of charged ions in a space, the concept of space charge density ρ, expressed in C/m^3, is introduced. For DC lines, the electric fields originate from two kinds of charges: charges on the line conductor, which generate a static or nominal electric field; and charged ions in space, which generate an ionic electric field. The superposition of both forms a total electric field E (kV/m) Also, in converter stations, the presence of charged ions in space results in a total electric field and ion current.

1. **Total electric field of UHV AC lines**
   Figure 7.11 shows the distribution of the total and nominal electric fields on the ground under a ±800-kV DC line, with their calculated strength values represented by the solid and dotted lines, respectively. The points of different shapes, square and round, in this figure represent the maximum and minimum measured values, respectively. The maximum total electric field occurs at 1−2 m outside the pole conductor and the minimum, zero value, generally occurs at the center between the two pole conductors. It should be particularly noted that Figure 7.11 shows the idealized scenario in which no wind blows. Because the migration speed of positive and negative ions under the effect of electric field is of the same order of magnitude as the wind speed, even a soft wind at a low speed of 1 m/s will change the distribution of the total electric fields. Soft winds blowing in a direction perpendicular to the line will cause the positions of

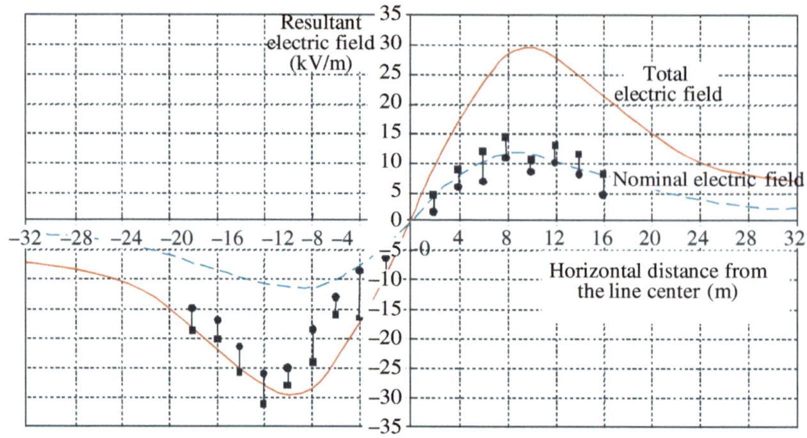

Figure 7.11
Distribution of total and nominal electric fields on the ground under a ±800-kV DC line.

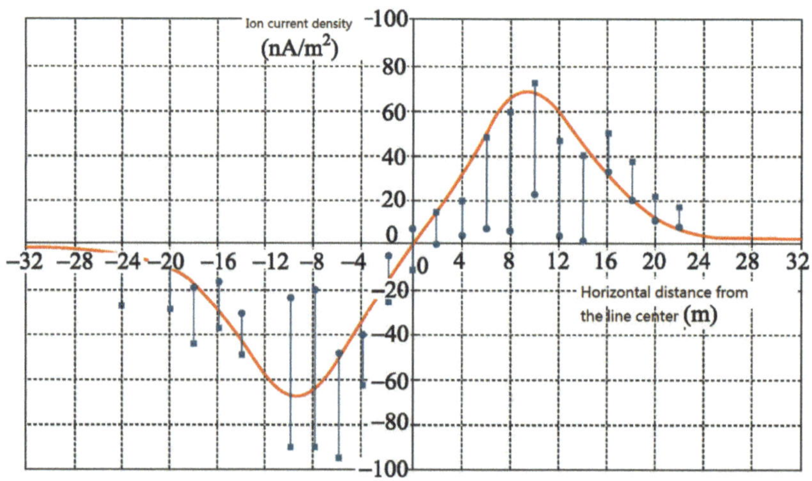

Figure 7.12
Distribution of ion current density on the ground under a ± 800-kV DC line.

the maximum and minimum field to move downwind, and relatively strong winds will largely change the distribution of total electric fields.

Figure 7.12 shows the distribution of ion current density on the ground under a ±800-kV DC line. The curve represents the calculated values and the solid lines represent the variation range of the measured values. Because the migration rate of negative ions (1.8(cm/s)/(V/cm)) is larger than that of positive ions (1.4(cm/s)/(V/cm)) under the effect of electric field, the measured value of the negative ion current density is larger than that of the positive ion current density.

The magnitudes of the total electric field and ion current density are a function of the field strength on the conductor surface and the corona onset field strength. The smaller the corona onset field strength and the larger the field strength on the conductor surface, the larger the total electric field and ion current density will be. Therefore, both increasing the corona's onset field strength and decreasing the field strength on the conductor surface can reduce the total electric field strength and the ion current density.

The field strength of the conductor surface depends on the conductor configuration (including the number and radius of subconductors) and the conductor-to-ground height. For conductors having the same total cross-sectional area, the field strength of the surface of a conductor containing more subconductors is generally smaller than that with less subconductors. For conductors having the same number of subconductors, the field strength on the surface of a conductor that contains subconductors with a larger cross-sectional area is smaller than that with smaller cross-sectional area. Therefore, increasing both the number and cross-sectional area of the subconductors leads to reduced total electric field and ion current density. Increasing the pole conductor height can reduce the field strength on the conductor surface and expand the ion diffusion range, thereby reducing the total electric field strength and ion current density on the ground. Based on these laws, it is possible to reduce the total electric field on the ground under a well-designed UHV DC line to the same level as a ±500-kV EHV DC line.

The corona onset field strength is related to the conductor surface condition and weather conditions such as humidity. The corona onset field strength of a conductor with a rough surface is lower than that with a smooth surface. For a given type of conductor, the corona onset field strength under humid conditions is lower than that under dry conditions and, consequently, the total electric field of a DC line under humid conditions is higher than that under dry conditions.

Wind blowing can significantly affect the total electric fields of a DC line because wind blowing expands the diffusion range of ions and, hence, reduces the ion current density. This will usually reduce the maximum total electric field on the ground.

The frequency and amplitude of the corona discharge current on DC lines vary with the pole polarities. The air humidity has different effects on the corona discharges of positive and negative pole conductors. In addition, insects' different preferences to perch on positive or negative polarity conductors result in different surface conditions, thereby causing different effects on the corona discharge. All these contribute to considerable differences between the total electric fields under the positive and negative pole conductors of a DC line.

2. **Total electric field in converter stations**

   The electric fields in a converter station originate from a combination of the charges on various energized conductors (such as the DC bus, various switchgears, and fittings) and the space charges generated by coronas on conductors. The energized DC bus is the major source of electric fields. In contrast to the bundle conductors used in DC lines, DC buses in the converter station generally consist of hollow rigid metallic tube with a large radius. Their height above the ground is usually smaller than the transmission line conductor. A well-designed tubular bus with a large radius almost experiences no corona on sunny days, in which case the electric fields under the bus are dominated by nominal electric field.

In a ±800-kV converter station (outside the valve hall), the absolute value of the measured maximum total electric field and ion current density on the ground is smaller than 30 kV/m and 25 nA/m^2, respectively.

According to the engineering experiences and analysis, the limit of total electric field strength is the primary factor dictating the model and height of the tubular bus above the ground. Figure 7.13 depicts the lateral distribution curves of the electric field strength on the ground under a −800-kV ϕ300-mm tubular bus under dry conditions when the heights are 14.8, 16.6, and 18.5 m above the ground. The ground electric field strength can be reduced by increasing the radius and height of the tubular bus.

The limits of the electric field strength and the ion current density in a ±800-kV converter station are generally determined with reference to those for a DC line.

3. **DC magnetic field of UHV DC lines**

The current through an operating DC line generates magnetic fields in space whose strength mainly depends on the line geometric configuration and the magnitude of current. The maximum magnetic field generated by a UHV DC line is comparable with the earth's magnetic field and is much smaller than the public exposure limit recommended by ICNIRP. Therefore, such limits will impose no constraint on the design of UHV DC lines.

The magnetic fields produced by a DC line are mainly related to the line configuration and current. In China, the rated current of ±800 kV DC links falls within 3000−5000 A. Figure 7.14 depicts the lateral distribution of magnetic fields on the ground under a ±800-kV DC line with an inter-pole spacing of 22 m and operating

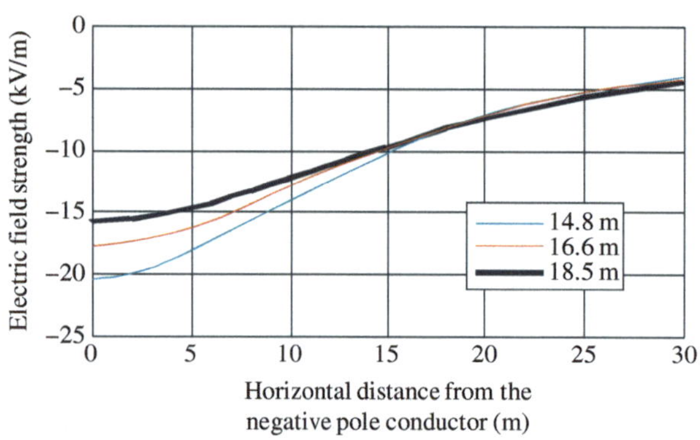

Figure 7.13
Lateral distribution of electric field strength on the ground under a ϕ300-mm tubular bus.

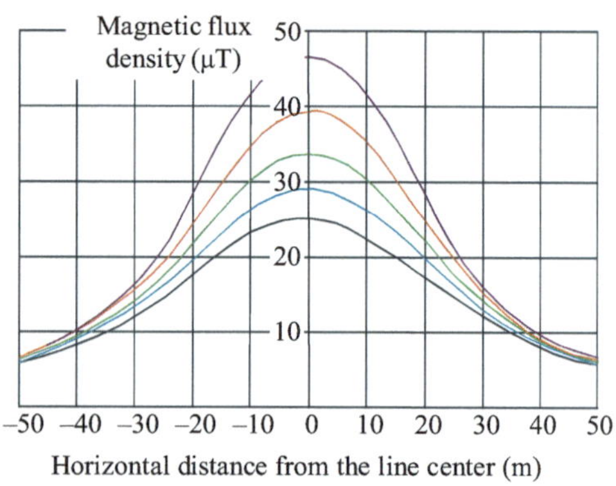

Figure 7.14
Lateral distribution of magnetic fields on the ground under a ± 800-kV DC line.

current of 4000 A under bipolar operation. The curves from top to bottom in this figure correspond to different pole conductor-to-ground heights within the range of 17–25 m. For ±800-kV DC lines in China, the pole conductor-to-ground height is no less than 18 m. Figure 7.14 shows that with such a height, the maximum magnetic field on the ground under the DC line is <45 μT. The earth's magnetic field on the earth surface also falls within 40–60 μT, which is comparable with the DC magnetic field produced by the DC lines.

In China, the DC links are mainly operated under bipolar mode, and in normal operation only a very small unbalance current (<1% of the rated operating current) occurs and flows into the ground through the earth electrodes. When analyzing the magnetic fields at locations with a certain distance from the DC line, the unbalanced current can be assumed to flow into the ground through the earth electrodes in the receiving-end (sending-end) converter station and then back to the sending-end (receiving-end) converter station through the ground return path. If a DC line is assumed to be straight and infinitely long, then the magnetic fields produced by it may be calculated using Ampere's Circuital Law. However, a DC line generally has a route of broken line whose magnetic fields can be calculated using integral method according to the Biot–Savart law.

4. **DC magnetic fields in converter stations**

Under different operation modes, the tubular bus in the DC switchyard of the converter station may sometimes carry a DC current up to several thousand amperes that will generate magnetic fields under the tubular bus.

Figure 7.15 shows the lateral distribution of magnetic flux density on the ground under a tubular bus that carries a current of 4000 A and is 13–18 m high above the ground. Figure 7.16 depicts the maximum magnetic flux density on the ground versus

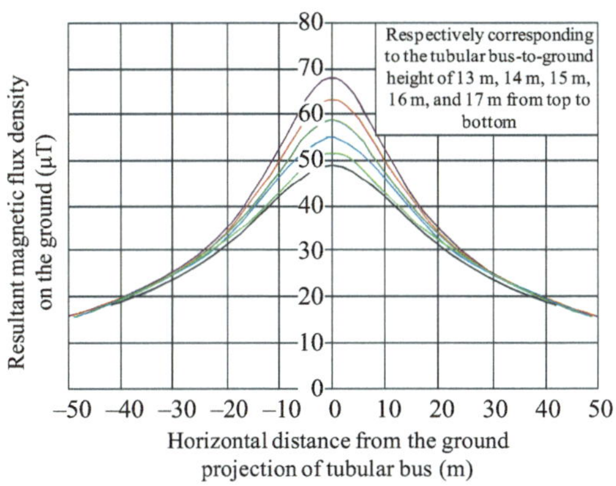

Figure 7.15
Lateral profile of magnetic flux density on the ground under the tubular bus.

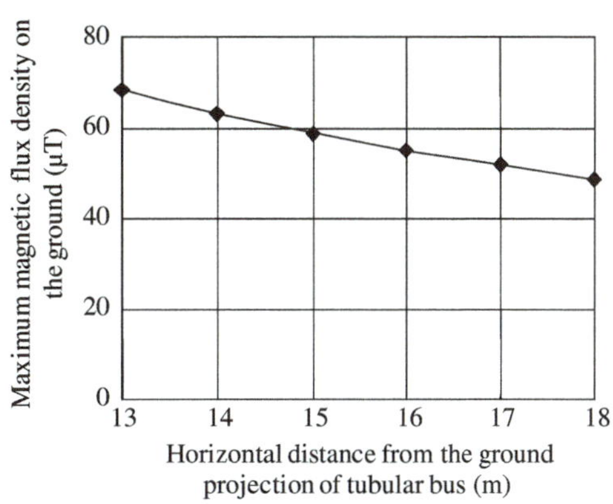

Figure 7.16
Maximum magnetic flux density versus bus height.

the height of tubular bus curve. Figures 7.15 and 7.16 show that the magnetic flux density on the ground under a tubular bus is up to 70 μT, which is comparable with the earth's magnetic field in northern China. Therefore, the magnetic field limit imposes no constraint on determining the height of tubular bus, and it is not necessary to check the magnetic flux density when designing the bus in the converter station.

7.2.4 Limits of Total Electric Field and DC Magnetic Field for UHV DC Line

1. **Limits of total electric field**
ICNIRP stated that for biological and health effects of electric and magnetic fields with frequency ranges up to 100 kHz, the major mechanism of interaction is induction of currents in tissues. Although the derived reference levels for general public exposure provided in these guidelines do not contain the reference level for exposure to electric fields with frequency ranges up to 1 Hz, relevant notes indicate, "No electric field value is provided for frequencies <1 Hz, which are effectively static electric fields. Perception of surface electric charges will not occur at field strengths <25 kV/m. Spark discharges causing stress or annoyance should be avoided."
Below are the limits of the electric field, ion current density, and static electric field for a DC line that are used by some countries.
   a. The United States Department of Energy specifies that the strength of the total electric field on the ground under a DC line should be less than 30 kV/m.
   b. In Canada, the strength of the maximum total electric field under a DC line should be less than 25 kV/m and the ion current density should be less than 100 nA/m^2.
   c. The allowable maximum strength of the total electric field on the ground under the ±600-kV DC line used in the Itaipu Dam Project, Brazil, is 40 kV/m.
   d. In the late 1980s, when building its first DC transmission project—the Gezhouba–Shanghai ± 500-kV HVDC transmission project, China proposed a limit of 30 kV/m as the allowable maximum strength of total electric field under a DC line. After the project was completed and commissioned, China formulated an industrial standard DL 436—1991 *Technical Guide for HVDC Overhead Transmission Lines*, which specifies that the limit of the total electric field under a DC line is 30 kV/m. The investigation results of the environmental effects caused by China's four operating ± 500-kV DC links show no adverse health consequence. It is specified in DL/T 1088—2008 *Electromagnetic Environment Parameter Limit for ± 800 kV UHV DC Transmission Lines* that the total electric field strength in the residential areas shall be lower than 25 kV/m, 80% of the measured values during the measurement should not exceed 15 kV/m, and in the area where a transmission line crosses farmland/road and other places accessible to people the maximum strength of total electric field should not exceed 30 kV/m.

2. **Limits of DC magnetic field**
ICNIRP officially issued the *Guidelines on Limits of Exposure to Static Magnetic Fields* in 1994 and in its revised edition in 2009. These guidelines stated that the highest exposure for members of the general public occurs in patients undergoing diagnostic examination by magnetic resonance imaging (MRI) or spectroscopy (MRS). In MRI procedures, magnetic flux densities range from 0.15 to 3 T, and the exposure is usually limited to <1 h. Exposures can also occur during other medical applications. In addition to staff involved in the use of MRI equipment, a few other occupations are associated with exposure to strong magnetic fields. Strong fields are produced in high-energy technologies such as thermonuclear reactors, magnetohydrodynamic systems, superconducting generators, particle accelerators and isotope separation units, and electrolytic processes such as aluminum production. In addition, magnetic flux densities on the order of 20 μT are produced under HVDC transmission lines.
Table 7.7 shows the limits of exposure to static (DC) magnetic field given in the ICNIRP guidelines.
The limit of public exposure to static magnetic field is 400 mT, which is much higher than the limit of exposure to power-frequency magnetic field (0.1 mT).

7.2.5 Effects of Power-Frequency Electric and Magnetic Fields

1. **Effects of power-frequency electric field on animals and plants**
Studies of the biological effects of the power-frequency electric field began in 1960s. After several decades of studies around the world, some deterministic conclusions have been drawn.
   a. Tests and studies conducted abroad on the ecological effects of UHV pilot transmission lines show that the electric fields have no adverse effects on trees, shrubs, grass, crops, or wild animals under the line.
   b. In China, Wuhan High Voltage Research Institute and Tongji Medical University conducted exposure tests using animals in 1980s. White rats were exposed to power-frequency electric fields with a strength of 40 kV/m for 1000 h, and no detrimental biological effects were found on those rats' hemograms, biochemical indexes, and organ coefficients.

Table 7.7: Limits of Exposure to Static (DC) Magnetic Field.

Exposure Characteristics	Magnetic Flux Density
Head and trunk	2 T
Limbs	8 T
Any part of the body	400 mT

Transient and steady-state electric shocks to human or animals in an electric field can occur due to contact with any objects that are insulated from the ground or poorly grounded. The transient electric shock is a phenomenon that occurs when a person touches a conductor in the electric field and the charges accumulated on the conductor or the person due to electrostatic induction are released through the human body to the ground in the form of spark discharges at that instant. Electric shocks that occur under a transmission line are mostly transient. The steady-state electric shock is a phenomenon that occurs when a person grasps a conductor in an electric field and, due to capacitive coupling, the energized conductor generates a continuous power-frequency current through the human body and results in an electric shock. If the power-frequency electric fields under a UHV transmission line are controlled within the limits given in IEEE C95.6—2002, the transient electric shocks, if they occur, will simply cause a feeling similar to the scenario when a person walks on carpet and the friction between the feet and the floor leads to an accumulation of charges on the body that is released when the person touches a grounded object, resulting in a shock. Such electric shocks may make people feel uncomfortable but they will not hurt them.

2. **Effects of power-frequency magnetic fields on animals and plants**
Since the 1990s, numerous tests and studies have been conducted on the ecological impacts of power-frequency magnetic field. These involve the studies of the interaction mechanism between cellular/molecular structure, animal tissues, and power-frequency magnetic fields, as well as numerous tests of the effects of magnetic field exposure on breast cancer, brain tumor, septicemia, melatonin, immunity, and neurobiological characteristics. The tests of cellular/molecular structure have detected the subtle changes in biological characteristics of animals when exposed to high-level magnetic fields.

The China Electric Power Research Institute and the State Grid Electric Power Research Institute conducted an ecological effect test in Wuhan UHV AC Test Base for 2 years (2009–2011). Figure 7.17 shows the test scene. During the test, Sprague-Dawley (SD) rats were exposed to 100 μT power-frequency magnetic fields for

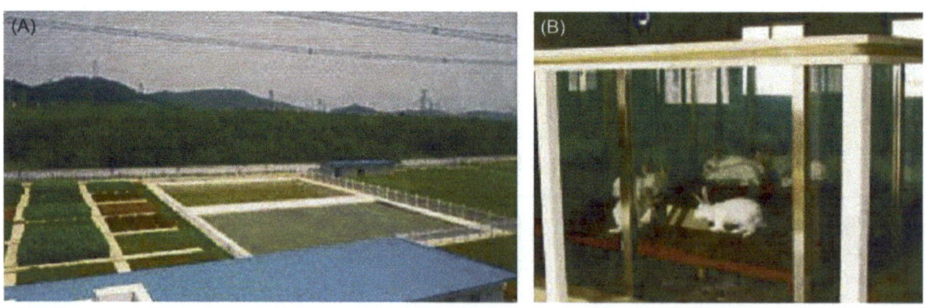

Figure 7.17: Ecological effect test of ELF electric/magnetic fields.
(A) UHV AC test base and (B) test of ecological effects on animals.

12 weeks. The behavioral, biochemical, and histological tests demonstrated that exposure to power-frequency magnetic fields has no significant effects on the SD rats. Also, such exposure has no effect on Alzheimer's disease in rats.

Test of the Japanese long-eared rabbits: Rabbits were exposed to 0, 30, 100, and 200 μT power-frequency magnetic fields for 12 months. The test results show that exposure to ELF magnetic fields has no significant effects on the weight of the Japanese long-eared rabbits. Although their routine examination results of blood tests and indices of hepatorenal function slightly fluctuated within the normal range, these fluctuations did not indicate any statistical discrepancy and clinical significance.

7.3 Noise from UHV Transmission Lines

7.3.1 Physical Measurement and A-Weighted Sound Level of Audible Noise

Audible noise control is an important technology for UHV transmission projects. During operation, the main equipments in substations and converter stations, such as transformers, reactors, filters, and capacitors, will produce low-frequency noise. Because the noise travels over a long distance, it may have certain impact on noise-sensitive locations in the surrounding area. The magnitude of noise produced by substations and converter stations generally relate to the operating load.

The audible noise from UHV transmission lines refers to the audible noise produced by corona and spark discharges around conductors.

The noise is generally measured using sound pressure level and sound power level [7]. It may be considered as a composition of the components with different frequencies. The loudness perceived by the human ear is different for sounds under different frequencies, even if they have the same sound pressure level. The human ear is most sensitive to sounds of 1000–5000 Hz. To allow the objective measurement to approach subjective perception, sound level meters are usually fitted with a filter, which is known as the A-weighing network. The sound pressure level obtained through the A-weighted network is called the A-weighted sound level. Practices have shown that the noise expressed in A-weighted sound level conforms to that perceived by the human ear.

Equivalent continuous A-weighted sound level is defined as the average energy of A-weighted sound level in a period of time T. It is referred to as equivalent sound level for short and denoted by L_{eq}. For steady-state noise, the equivalent sound level is its A-weighted sound level. For noise that is not steady-state, the equivalent sound level cannot be simply assessed with its A-weighted sound level, but it is expressed using statistical method instead, denoted by the probability or cumulative probability of sound level. L_N is used to assess the noise, which means that the sound pressure levels are higher than the L_N

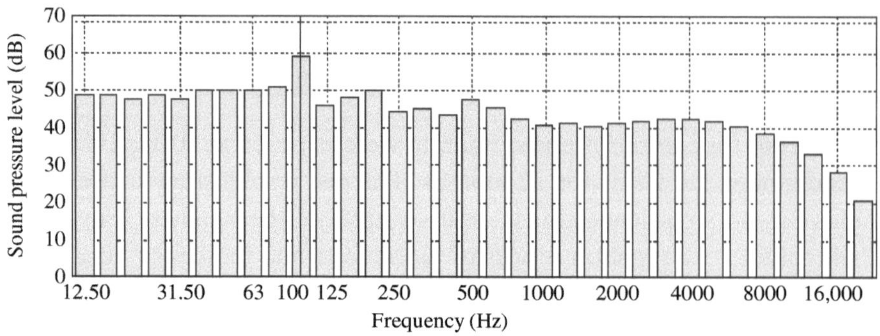

Figure 7.18
Typical spectrum characteristics of audible noise from UHV AC line.

level for N percent of the time span. For example, $L_{10} = 70$ dB(A) means that the sound or noise has a sound pressure level above 70 dB(A) for 10% of the time span. L_{50} is used generally to assess the corona noise. It represents the median of the noise levels.

7.3.2 Characteristics and Influencing Factors of Audible Noise from Overhead Transmission Lines

1. **Spectrum**
   The corona is the primary audible noise source of transmission lines. Audible noise can be divided into two parts in terms of the characteristics of frequency components for AC transmission lines: broadband noise component and pure tone component at multiple times the power-frequency.
   The broadband noise is formed by disorganized pulses randomly emitted from corona discharges on the conductor surface. Significantly different from the general environmental noise, the broadband noise is perceivable as a crackling and hissing noise, giving rise to the subjective annoyance.
   The so-called pure tone noise is perceivable as a buzzing noise caused by the back-and-forth motion of charged ions near the conductor due to the periodic change of the voltage. For AC transmission lines, as the voltage changes alternately between positive and negative, the conductor acts as a positive and negative corona electrode alternatively, and the positive ions and negative ions are repelled by and attracted to the conductor alternatively at twice the power-frequency. Therefore, the frequency of such noise is multiple times the power-frequency. If the power-frequency is 50 Hz, then the most obvious noise is at 100 Hz.
   Figure 7.18 shows the typical spectrum of audible noise from China's UHV AC pilot and demonstration project under rainy conditions. The noise is prominent at 100 Hz.
   Figure 7.19 shows the comparison of spectrum characteristics between the audible noise from UHV AC line and the environmental noise. The low-frequency pure tone

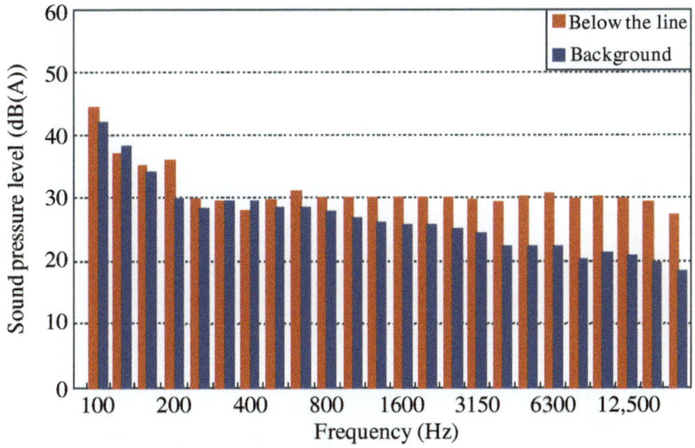

Figure 7.19
Spectrum of audible noise from UHV AC line versus that of environmental noise.

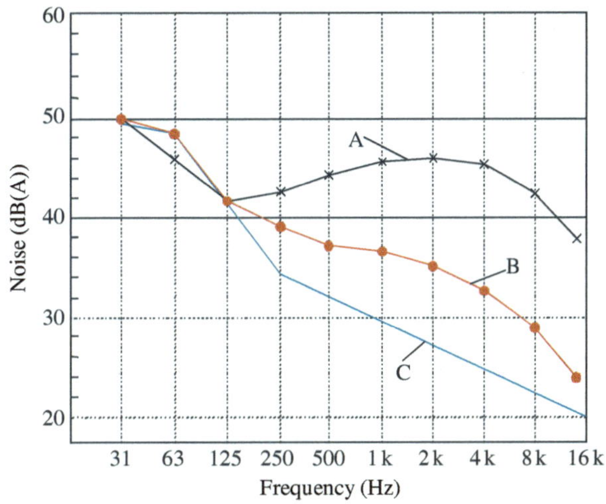

Figure 7.20
Spectrum of audible noise from a DC line versus that of environmental noise.

components at 100 and 200 Hz from the AC transmission line are apparent in the figure. In addition, with the increase of the frequency, the spectrum characteristics of the environmental noise and the audible noise from the line differ significantly. So, in situations with low environmental noise, the high-frequency noise from the AC transmission line can be easily identified. It is this characteristic that makes the corona noise from transmission lines acoustically abnormal.

Figure 7.20 shows the comparison of spectrum characteristics between the audible noise from a DC line and the environmental noise. Curves A and B denote the high corona noise and low corona noise under good weather conditions, respectively, and

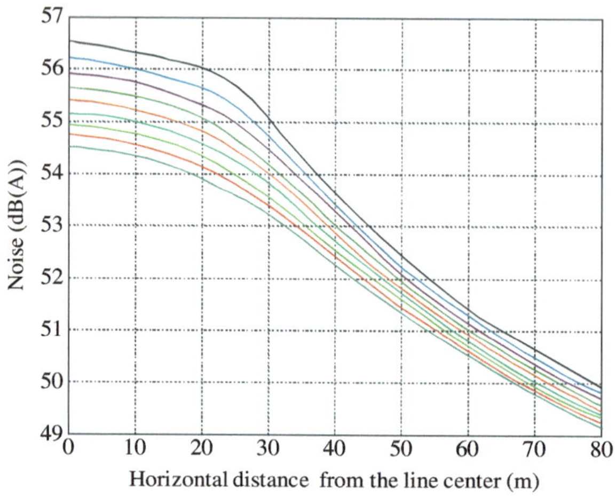

Figure 7.21
Lateral profile of A-weighted sound levels of audible noise from typical UHV AC transmission lines.

curve C denotes the background noise. The environmental noise attenuates significantly from 100 Hz, whereas the audible noise produced by corona on the DC line does not attenuate until it reaches a high frequency. Therefore, in situations with low environmental noise, the corona noise can be easily identified.

2. **Lateral profile of noise**
   Figure 7.21 shows the lateral profiles of A-weighted sound levels L_{50} of audible noise from typical UHV AC transmission lines under rainy conditions, with the individual profiles from top to bottom, respectively, corresponding to the minimum height of the conductor to ground in a range of 15–23 m. It follows that:
   a. The audible noise is gradually attenuated with distance from the ground projection of the line center in the direction perpendicular to the line. The audible noise attenuates at a slow rate beneath the line and at a higher rate away from the side conductor.
   b. The noise is reduced with the increase of the conductor height, but not greatly.

   The audible noise from a DC transmission line is mainly sourced from the positive polarity conductor, and its lateral attenuation characteristic is bilaterally symmetrical relative to the positive conductor, as shown in Figure 7.22. Therefore, when assessing the audible noise from DC transmission lines, the reference point is generally selected outside the positive conductor. The audible noise is attenuated with the increase of distance from the line at a slower rate than that of the electric field and RI.
3. **Influence of conductor type on audible noise**
   The audible noise may vary with the types of conductors. Using UHV AC transmission line as an example, the sound power levels of audible noise from eight-bundle

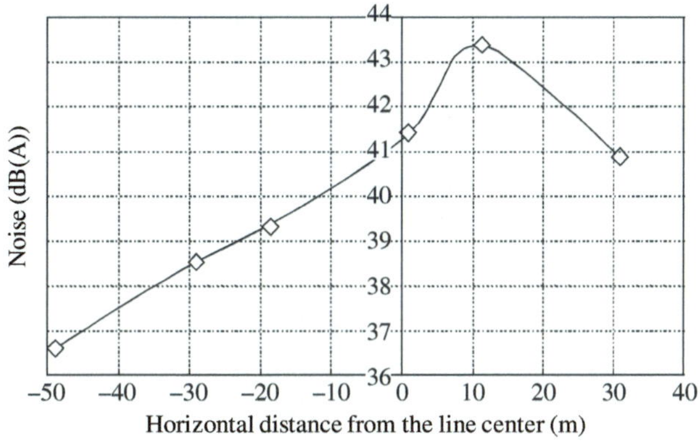

Figure 7.22
Lateral profile of A-weighted sound levels of audible noise from typical UHV DC transmission lines.

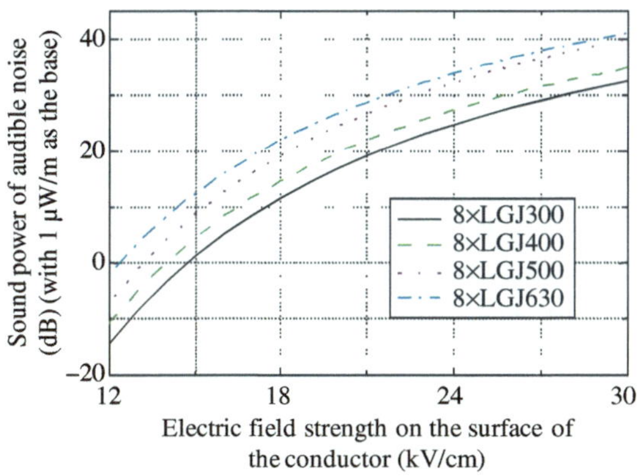

Figure 7.23
Sound power levels of audible noise from different types of eight-bundle conductors of a UHV AC line.

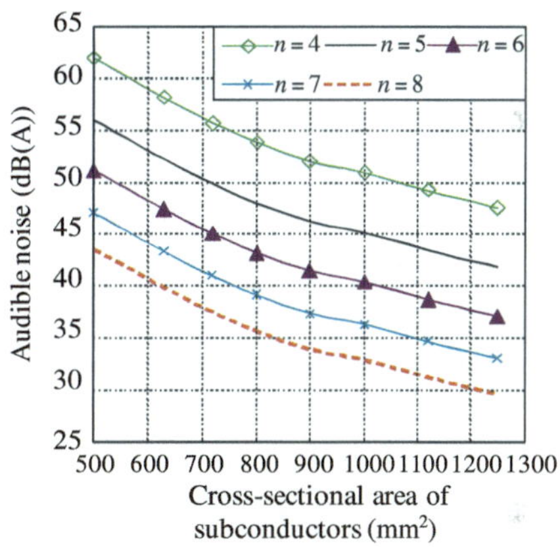

Figure 7.24
Audible noise from a ±800-kV DC line as a function of number and cross-sectional area of subconductors.

conductors are shown in Figure 7.23, where the spacing of subconductors is 400 mm and the cross-sectional areas of subconductors are 300–630 mm^2. Given the same electric field strength on the surface of the conductor, the larger the cross-sectional area of the subconductor, the higher the sound power of audible noise that is produced by the conductor.

Figure 7.24 shows the audible noise from the ±800-kV DC transmission line at 20 m away from the ground projection of the positive conductor as a function of both the number (4–7) and cross-sectional area (500–1250 mm^2) of subconductors under

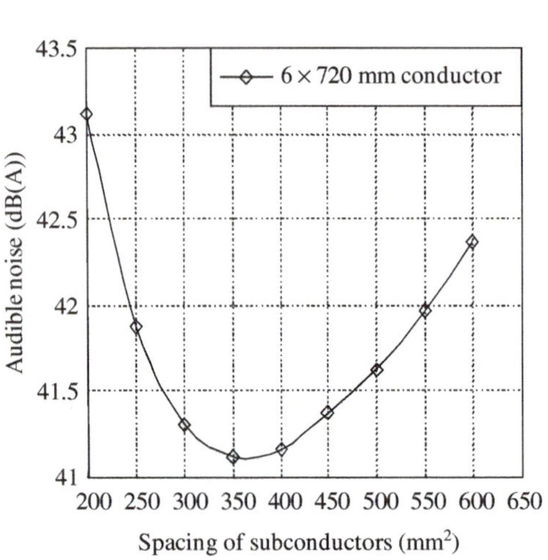

Figure 7.25
Audible noise from a ±800-kV DC line as a function of spacing of subconductors.

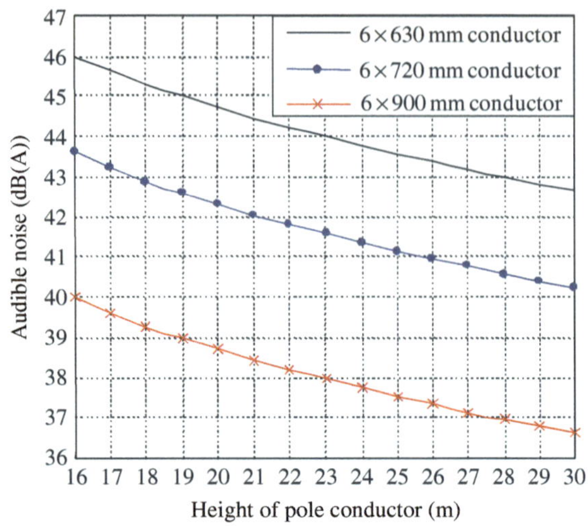

Figure 7.26
The ±800-kV DC line audible noise versus height of pole conductor above ground.

dry conditions. In this figure, the spacing of subconductors is 45 cm, the spacing of pole conductors is 22 m, and the height of the pole conductor to ground is 18 m.

The audible noise decreases with the increase of number and cross-sectional area of subconductors. Increasing the number of subconductors can significantly reduce the audible noise. The main reason is that for a given cross-sectional area of the subconductor, the increase of the number of subconductors leads to the increase of the total surface area of the conductor and the decrease of charge density at the surface of the conductor. Because the electric field strength at the surface of the conductor is proportional to the charge density, it will cause the electric field strength at the surface of the conductor to decrease, which in turn leads to weakened corona discharge, thereby reducing the corona effects (such as audible noise).

Figure 7.25 shows the audible noise from a ± 800-kV DC transmission line as a function of the spacing of subconductors. In this figure, the conductor type is 6 mm × 720 mm, the spacing of pole conductors is 22 m, and the height from the pole conductor to the ground is 18 m. It shows that the audible noise changes in a "U"-shape profile with the variation of the spacing of subconductors, and there is a minimum value in the profile. From the perspective of controlling the electromagnetic environmental impacts, the optimal spacing of subconductors should be determined according to the subconductor type in the design of UHV DC transmission lines to minimize the electric field strength at the surface of the conductor.

Figure 7.26 shows the audible noise from a ± 800-kV DC transmission line at 20 m away from the ground projection of the positive polarity conductor as a function of the

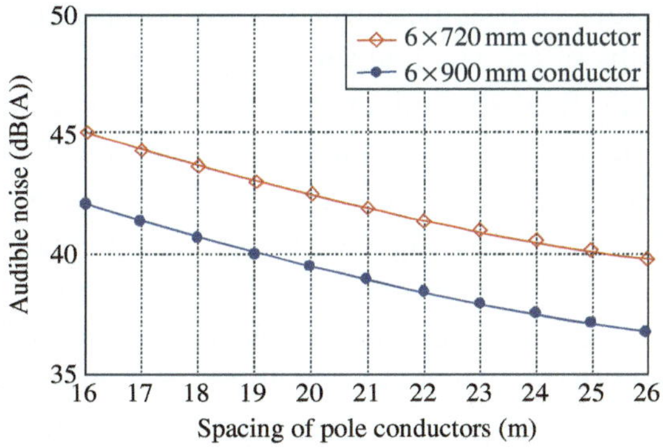

Figure 7.27
Audible noise at 0 m elevation versus spacing of pole conductors.

type of conductor (6 mm × 630 mm, 6 mm × 720 mm, 6 mm × 900 mm) and the height of the pole conductor from ground.

The rule of conductor type selection based on the transmission capacity results in low-level audible noise in the existing EHV DC transmission lines in China, causing no significant audible noise problem. But Figure 7.26 shows that the audible noise level will certainly increase because of voltage upgrade for UHV DC transmission lines, which requires special attention. Generally, the audible noise level is controlled by selecting an appropriate conductor, and the variation of the height of pole conductor from ground is used as a supplemental method under certain conditions.

Figure 7.27 shows the audible noise at 0 m elevation as a function of the spacing of pole conductors. It shows that the audible noise will be reduced with the increase in spacing of pole conductors, because the electric field strength at the surface of the conductor is reduced. The audible noise is decreased with the increase of both the spacing of pole conductors and the height of pole conductor above the ground, and the former measure is more effective in reducing the audible noise than the latter one.

4. **Influence of weather on audible noise from overhead transmission lines**

The degree of the corona discharge on overhead transmission lines and the resulting audible noise are considerably dependent on weather conditions.

For AC transmission lines, noticeable audile noise occurs under bad weather conditions like rain, fog, and snow, when substantial corona discharge occurs randomly along the conductor length due to collision and accumulation of water droplets on the conductor surface. Therefore, the corona noise under rainy conditions is generally used to assess and represent the characteristics of audible noise from AC transmission lines. However, because the formation of corona noise on rainy days is a complicated process, it is difficult to theoretically deduce an accurate corona noise prediction equation. As such, the audible noise prediction methods used in different countries are all studied with the results of tests performed in the corona cage or statistically analyzed with the results of the test line for a long time.

Given that the electric field strength at the surface of the conductor is in the range of 15–20 kV/cm, using the audible noise under heavy rain condition as the reference value, the audible noise under other weather conditions can be reduced by:

a. Approximately 5–10 dB under light rain, fog, or wet conditions
b. Approximately 15–20 dB under good weather conditions.

For DC conductors under rainy conditions, the corona onset field strength of the conductor is lower than that under sunny condition, and more ions are present around the conductor. At the very beginning of rain, the ion concentration on the surface of conductor is lower and the intensity of corona discharge is slightly greater than that under sunny conditions. After a period of rain, the corona onset field strength is further reduced and the ion concentration at the surface of the conductor is increased, causing the irregular surface of the conductor to be surrounded by charges. This will reduce the intensity of corona discharge and the audible noise will be less than that under sunny conditions. The attachment of falling objects to the conductor will increase the electric field strength on partial surface and cause increased audible noise. Such falling objects may vary with seasons, and they are in greater quantity in summer. When determining the limit of audible noise from DC transmission lines, primary consideration should be given to the audible noise under good weather conditions in summer.

7.3.3 Calculation of Audible Noise from UHV Transmission Lines

1. **Calculation of audible noise from UHV AC transmission lines**

By comparing the audible noise calculation formulas generalized by organizations in different countries, the formula recommended by Bonneville Power Administration (BPA) is generally believed to be representative and accurate. It is derived from the long-term data measured on actual test lines with different voltage levels and bundle configurations and is used to calculate the 50% exceeded level L_{50} under rainy conditions. The formula for predicting audible noise from UHV AC transmission lines as recommended by BPA is:

$$SL_A = 10\lg \sum_{i=1}^{Z} \lg^{-1}\left[\frac{PWL(i) - 11.4\lg(R_i) - 5.8}{10}\right] \tag{7.1}$$

where

SL_A—A-weighted sound level, dB(A) unified modified for dB

Z—number of phases

$PWL(i)$—sound power level on the i phase conductor, dB(A)

R_i—distance from the test point to the measured i phase conductor, m.

$PWL(i)$ is given by:

$$PWL(i) = -164.6 + 120\lg E + 55\lg d_{eq} \tag{7.2}$$

$$d_{eq} = 0.58n^{0.48}d$$

where

E—field strength at the surface of the conductor, kV/cm

d_{eq}—equivalent radius of the conductor, mm

n—number of subconductors

d—diameter of the subconductor, mm.

This prediction formula is applicable to conventional symmetrically spaced bundle conductors when the spacing of subconductors is 30–50 cm and the electric field strength at the surface of the conductor is 10–25 kV/cm. Because the air breakdown strength at high altitudes is lower than that at low altitudes, more severe corona discharge will occur and audible noise will be more noticeable. Generally, the audible noise produced by corona discharge increases by approximately 1 dB for every 300-m increase in altitude.

A 3-year observation was conducted on the test line at the UHV AC test base and the single circuit of the UHV AC pilot and demonstration project in China. In the statistical comparison of audible noise at 15 m from the outside phase from January to May performed at Henan Long-Term Observation Station, the L_{50} under rainy condition calculated using BPA's formula was 3.1 dB(A) greater than the measured value. In the analysis of audible noise data under rainy conditions from June to August performed by Hubei Long-term Observation Station, the calculated L_{50} audible noise was 1.84–2.48 dB(A) greater than the measured value, and the difference was 1.9–2.1 dB(A) at 20 m beyond the outside conductor. In 1982, the calculated L_{50} audible noise from the UHV test line (Project UHV 8) under rainy conditions obtained by the American Electric Power Research Institute (EPRI) was approximately 1.8 dB(A) greater than the measured value. These studies show that the audible noise from a 1000-kV UHV single circuit (eight-bundle conductor) under rainy conditions calculated using BPA's formula is approximately 2 dB greater than the measured value. Therefore, it is recommended that the calculation results of audible noise from China's UHV transmission lines (eight-bundle conductor, 1000 kV, single circuit) using BPA's formula can be corrected by reducing 2 dB.

2. **Calculation of audible noise from UHV DC transmission lines**

   The audible noise produced by corona discharge on DC transmission lines may be calculated using empirical formulas. Such formulas are derived by analyzing the data obtained from tests in corona cages and on test lines. The United States, Canada, Germany, Japan, and other countries have their respective formulas. The formulas derived by EPRI and BPA in the United States are more representative.

EPRI's formula for calculating audible noise produced by corona discharge on DC transmission lines is:

$$P_{dB} = 56.9 + 124\lg\frac{g_{max}}{25} + 25\lg\frac{d}{4.45} + 18\lg\frac{n}{2} - 10\lg R_P - 0.02R_P + k_n \quad (7.3)$$

where

g_{max}—maximum electric field strength at the surface of the conductor, kV/cm

d—diameter of the subconductor, cm

n—number of subconductors

R_P—distance from the positive conductor to the test point, m

k_n—correction term, if $n \geq 3$, $k_n = 0$ dB(A); if $n = 2$, $k_n = 2.6$ dB(A); and if $n = 1$, $k_n = 7.5$ dB(A).

Formula (7.3) applies to the circumstance when the electric field strength at the surface of the conductor is 15–30 kV/cm, the diameter of the subconductor is 2–5 cm, and the number of subconductors is <6. The result obtained from Formula (7.3) is the L_{50} audible noise under sunny conditions in summer; in spring and autumn, it should be corrected by reducing by 2 dB(A), and in winter it should be corrected by reducing by 4 dB(A).

BPA's formula for calculating audible noise produced by corona discharge on DC transmission lines is:

$$P_{dB} = -133.4 + 86\lg g_{max} + 40\lg d_{eq} - 11.4\lg R \quad (7.4)$$

where

$$d_{eq} = d \times 0.66n^{0.64}(n > 2) \quad (7.5)$$

$$d_{eq} = d(n = 1, 2) \quad (7.6)$$

where

g_{max}—maximum electric field strength at the surface of the conductor, kV/cm

R—distance from the positive conductor to the test point, m

d—diameter of the subconductor, mm

n—number of subconductors.

The results obtained from BPA's formula is the L_{50} audible noise under good weather conditions in spring and autumn; in summer and winter, it should be corrected by increasing or decreasing by 2 dB(A) as appropriate, and under bad weather it should be corrected by reducing by 6–11 dB(A).

7.3.4 Limits of Audible Noise for UHV Overhead Transmission Lines

Extensive international studies show that there are almost no public complaints when the audible noise from transmission lines is less than 52.5 dB(A), few public complaints when the audible noise is 52.5–59 dB(A), and significant public complaints when the audible noise is more than 59 dB(A). Worldwide, the audible noise limit for UHV AC transmission lines is 50–60 dB(A). Japan poses stricter requirements for environmental protection and sets the limit to 50 dB(A). In the United States, the Soviet Union, and Italy, the audible noise limits are no less than 55 dB(A) and are even as high as 60 dB(A).

Because continuous audible noise would be produced by DC transmission lines under sunny conditions, it should be generally controlled within the range when almost no complaints will be incurred. Worldwide, the audible noise limits for DC transmission lines are generally 40–45 dB(A). For example, it is 40–45 dB(A) in the United States, 40 dB(A) in Japan, and 40 dB(A) in Brazil at the edge of the DC transmission line corridors.

Currently, no audible noise limit standard for 1000-kV UHV AC transmission lines is available in China. Considering that the noise at noise-sensitive targets along the line should comply with GB 3096—2008 *Environmental Quality Standard for Noise*, the limit on L_{50} audible noise (under rainy conditions) from UHV AC transmission lines at 20 m away from the projection of the outside phase is set to 55 dB(A) during the engineering design. Generally, the audible noise from AC transmission lines under sunny conditions is 15–20 dB lower than that under rainy conditions.

For the audible noise from UHV AC overhead transmission lines, according to DL/T 1088—2008 *Electromagnetic Environment Parameters Limit for ± 800 kV UHV DC Transmission Lines*, the L_{50} audible noise produced by corona discharge under sunny conditions should be no higher than 45 dB(A) at 20 m away from the projection of the positive conductor on the ground; for non-residential areas with altitude higher than 1000 m, the audible noise limit should be corrected by increasing 3 dB for every 1000-m increase in altitude. Based on these results, the audible noise produced by corona discharge on UHV DC transmission lines in China should, in principle, comply with the noise limit for Category 1 area during nighttime. In the non-residential area at high altitudes, the corona noise should comply with the limit for Category 2 area during nighttime; for nearby houses, the limit for Category 1 area during nighttime should still apply.

7.3.5 Limits of Noise for UHV Substations and Converter Stations

1. **Environmental quality standard for noise**
   The noise level at noise-sensitive targets around substations and converter stations should comply with the environmental noise limit requirements of different functional zones specified in GB 3096—2008.
   For substations and converter stations located in cities or urban planning areas, the noise level should comply with the noise limits for five types of environment noise functional zones specified in GB 3096—2008. For substations located in rural areas not

regionalized in terms of environmental noise, the type of the environmental noise functional zone should be determined through consultation with local competent environmental protection authority above the county level.

2. **Noise emission standard for industrial enterprises at boundary**
The noise at the boundary of substations and converter stations should comply with GB 12348—2008 *Emission Standard for Industrial Enterprises Noise at Boundary*. For substations and converter stations located in cities or urban planning areas, the noise at the boundary of them should comply with the limits for five types of environmental noise function zones specified in GB 12348—2008. For substations and converter stations located in rural areas not regionalized in terms of environmental noise, if there are noise-sensitive buildings outside the plant boundary, then the environmental noise quality requirements outside the boundary should be determined by reference to GB 3096—2008 and GB/T 15190—1994 *Technical Specifications to Determinate the Suitable Areas for Environmental Noise of Urban Area* and the corresponding environmental noise limit should be followed.
It should be noted that the emission limits specified in GB 12348—2008 only refer to the noise emission (contribution) at the boundary of industrial enterprises, not including the background noise. Therefore, in the actual engineering survey, the results should be corrected according to the actual conditions.

7.3.6 Audible Noise Reduction Measures for UHV Transmission Lines

Transmission lines at 220 kV or less generally produce low audible noise that is barely noticeable. However, with the increase in the voltage level, especially for UHV transmission lines, the noise produced by corona discharge will increase, which may affect the normal life and work of nearby residents if improperly handled. Therefore, China gives great attention to the audible noise control during UHV construction, inputting lots of manpower and material resources to ensure compliance of UHV projects with relevant environmental protection requirements.

Studies show that the following measures may be taken to reduce the audible noise from UHV transmission lines:

1. For bundle conductors with symmetrically spaced subconductors, the number of subconductors is increased and the spacing of subconductors is controlled to reduce the electric field strength at the surface of the conductor. For bundle conductors with asymmetrically spaced subconductors, provisions are made to cause the charges to uniformly distribute on subconductors as much as possible, thereby improving the electric field distribution at the surface of the conductor. Additional subconductors are included in symmetrically spaced subconductors to improve the charge distribution at the surface of each subconductor and to reduce the electric field strength at the surface

of the conductor. Hydrophobic paint is applied on the conductor surface to minimize the accumulation of water droplets under rainy conditions, thereby reducing the strength of corona discharge and, hence, the audible noise.

2. For AC substations, the noise produced by corona discharge is eliminated by properly selecting HV electrical equipment using conductors that are free of corona discharge under sunny conditions and other measures.

   For the corona noise produced by fittings, the following noise reduction measures may be taken: (i) developing an optimal design scheme for fittings and grading rings through calculating the electric field strength at their surface to reduce the electric field strength and the corona noise; and (ii) improving the fabrication process of fittings for the 1000-kV system.

   The noise produced by the transformer proper is mainly reduced by reducing the noise contribution from the core through (i) using a high-permeability silicon sheet to manufacture the transformer core; (ii) reducing the flux density, because the transformer noise is approximately linear with respect to the flux density, and each increase in the flux density by 0.1 will lead to increase in transformer noise by 2.5 dB(A); (iii) applying multistage welding on the transformer core; (iv) installing thin-type vibration damping rubber membrane between core laminations; (v) installing vibration-isolation rubber pads between the bottom of transformer tank and the core; and (vi) installing vibration-isolation rubber pads between the transformer tank and the mounting foundation.

   The noise produced by the cooling equipment may be reduced by reducing the noise contribution from the equipment itself and effectively blocking the propagation path through two ways. Firstly, self-cooling radiators instead of air cooling radiators or forced oil circulation air coolers should be used as practically as possible. The selection of cooler should sufficiently allow for the noise index while meeting the design requirements. If fans are necessary, then low-noise cooling fans should be used as practically as possible, vibration isolators are required between the fan and the support, and silencers are installed at the inlet and outlet of the fans. Secondly, the construction between the transformer tank and the cooling fins should be reinforced. All cooling fins of the radiator are integral with the transformer tank by means of welding to reduce the vibration of fins and the noise.

   For HV shunt reactors, the following design measures may be taken to control vibration and noise: (i) using high-permeability silicon sheets, applying multistage welding on the core to maintain the flux density at a low level; (ii) strictly controlling the core dimension to ensure sufficient pressing and clamping force; (iii) adding a vibration damping layer between the core and the oil tank; (iv) increasing the thickness of the steel plates of the oil tank; (v) increasing the capacity of coolers, reducing the number of fan groups, and using low-noise fans; and (vi) effectively securing the magnetic shield and controlling its design flux density.

Sound barriers and box-type sound insulation devices (Box-in) are provided in front of the 1000-kV HV shunt reactor and main transformer. Analysis shows that when the sound barrier in front of the 1000-kV HV shunt reactor is 10 m high, the noise loss resulting from insertion of the sound barrier is approximately 10 dB(A), presenting prominent noise reduction effects. The use of Box-in noise reduction measures can also reduce the noise produced by the proper of the 1000-kV HV shunt reactor and the main transformer, that is, the HV shunt reactor is housed in a removable acoustic enclosure that is provided with ventilation, heat radiation, and silencers, and the bushing, cooler, fans, and pressure relief devices are arranged outside the acoustic enclosure. According to the calculation and the experience with DC converter stations regarding noise reduction using Box-in, the noise is generally reduced by 20 dB(A).

3. For DC converter stations, the main noise-generating equipments, such as the AC filter, converter transformer, smoothing reactor, and DC switchyard equipment, are optimally arranged to minimize the influence on surrounding environment; low-noise equipment types and configurations are selected; and sound isolation and muffling measures should be taken, such as providing sound barriers at the battery limit of the converter station or using Box-in converter transformers.

7.4 RI and TVI of UHV Overhead Lines

7.4.1 RI and TVI Characteristics and Effects of Overhead Lines

Corona discharges on conductors, insulators, or line fittings are the main source of RI because they inject current pulses into the line conductors. These propagate along the conductors in both directions from the injection point, as shown in Figure 7.28, thus generating magnetic fields around the conductors. As corona discharges and current injection points uniformly occur along an overhead line conductor, due to the aggregation effects, a steady-state current with a high pulse repetition frequency (PRF) is generated in the conductor, producing high-PRF steady-state RI fields around the overhead line.

Corona on the transmission line and discharges at some points may cause interference to the reception of radio and television broadcasting [8,9]. In addition, the reflection, refraction, and secondary radiation of lines and iron towers may produce unwanted signals that may cause interference to the radio reception. Corona discharges mainly cause

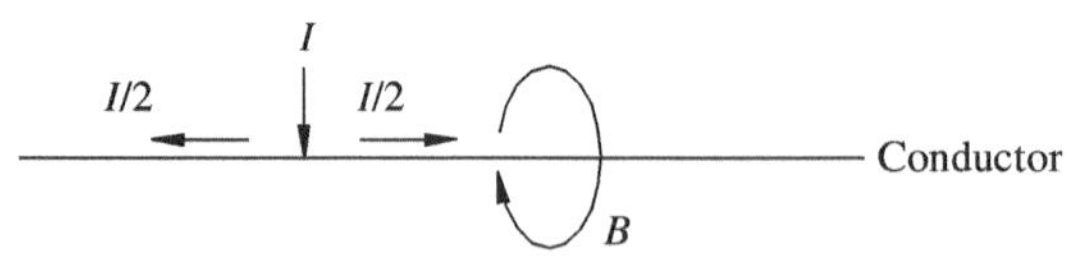

Figure 7.28
RI produced by corona.

interference to the short and medium wave band, whereas spark discharges mainly cause interference to the television frequency band. This section deals with the RI and TVI due to coronas on the transmission line.

The main characteristics of RI caused by corona are represented by the spectrum characteristics, lateral attenuation characteristics, and statistical distribution characteristics. The RI and TVI, respectively, refer to interferences to the radio broadcasting band and to the television broadcasting band of 48.5–958 MHz. The frequency of television signals is much higher than that of radio signals. The spectrum characteristics of interference signals caused by a transmission line show that it causes much less interference to television signals than to radio signals.

7.4.1.1 Radio interference

1. **Spectrum characteristics**
   The spectrum is the variation of the RI measured at a given point in the vicinity of a line, as a function of the measurement frequency. The reference frequency for measurements recommended by the International Special Committee on Radio Interference (CISPR) is 0.5 MHz. Figure 7.29 depicts the comparison between the typical RI spectrum characteristics of China's UHV AC pilot and demonstration project under heavy rain condition and the standard spectrum characteristics.
   When measuring the RI of a transmission line, the reference point is generally selected in the vicinity of the line. Three reference points are commonly used worldwide: 15 m from the ground projection of an outside phase conductor; 20 m from

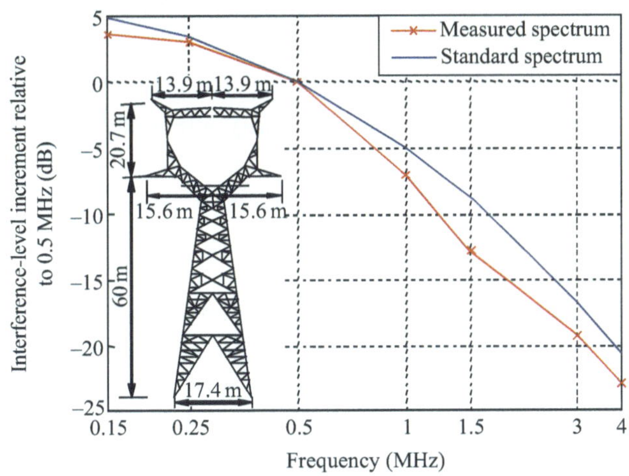

Figure 7.29
Typical spectrum characteristic of China's UHV AC pilot demonstration project under heavy rain condition as compared with 0.5 MHz.

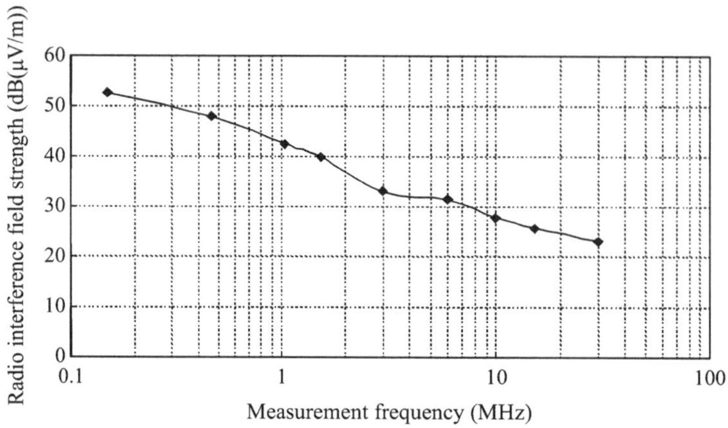

Figure 7.30
Spectrum characteristics of the RI of a DC line.

the nearest conductor; and edge of the line corridor. China uses 20 m from the ground projection of an outside phase conductor.

Figure 7.29 shows that the low-frequency band exhibits a high interference level. When the frequency rises above 4 MHz, the interference field strength is so low that it is negligible. Generally, of all the fields generated by corona discharges, considering those up to 30 MHz is sufficient.

Figure 7.30 shows the typical measured spectrum characteristics of a UHV DC line. It follows that the low frequency band exhibits a high interference level. With the increase of frequency, the interference field strength rapidly attenuates. Usually, when the frequency rises above 10 MHz, the interference field strength is so low that it is negligible.

2. **Lateral attenuation characteristics**

The RI fields generated by line coronas gradually attenuate with the distance from the line. Figure 7.31 depicts the lateral attenuation characteristic curve of the RI field strength under a 1000-kV AC line in fair weather. It shows that the RI field strength decreases rapidly with the distance from the line and approaches the background value when the distance is increased to 100 m.

Figure 7.32 shows the lateral profile of the RI under China's UHV AC pilot and demonstration transmission line. It follows that the measured RI field strengths on both sides are substantially symmetrical around the line center and the RI value gradually decreases with the distance from the line center.

The RI of a DC line is mainly generated by the positive pole conductor and the lateral attenuation characteristics of its field strength are similar to that of an AC line. Figure 7.33 depicts the lateral attenuation characteristic curve of the RI field strength under a bipolar DC transmission line. It follows that the lateral attenuation

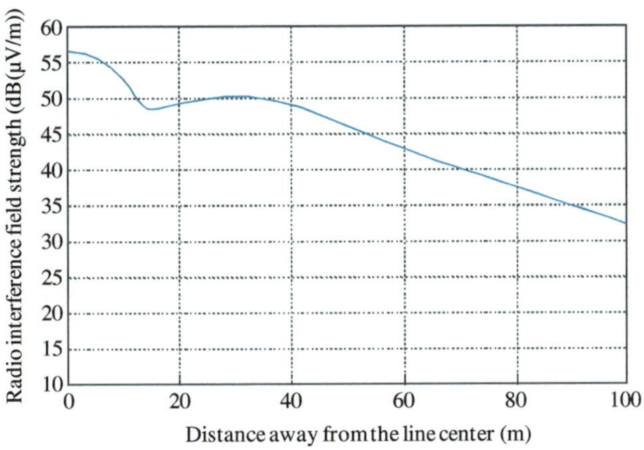

Figure 7.31
Lateral attenuation characteristics of RI field strength under a 1000-kV AC line.

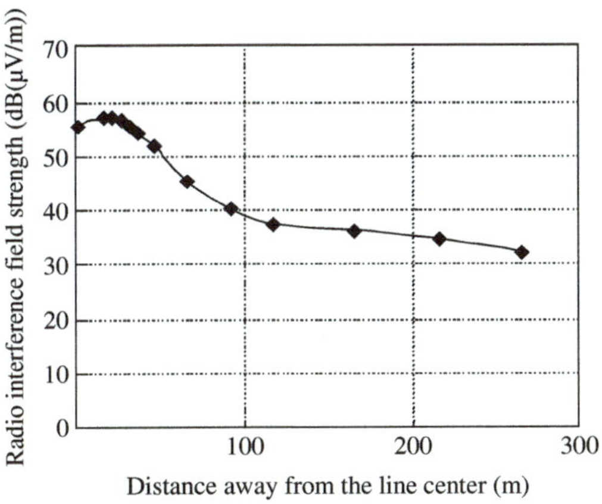

Figure 7.32
Lateral profile of RI field strength under a UHV AC pilot and demonstration transmission line.

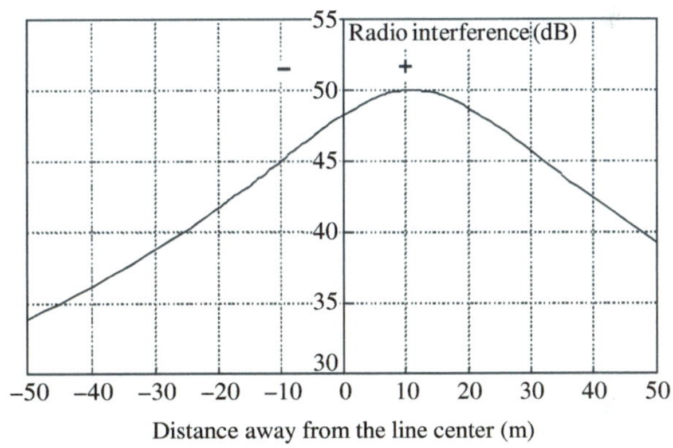

Figure 7.33
Lateral attenuation characteristics of RI field strength under a bipolar DC transmission line.

characteristics of the RI field strength are substantially symmetrical around the positive pole conductor. Therefore, when assessing the RI, the reference point is generally selected beyond the projection of the positive pole conductor on the ground.

Figure 7.33 also shows that the RI field strength decreases rapidly with the distance, which is different from the audible noise.

3. **Statistical distribution**

It is generally believed that the RI caused by corona discharges is an intrinsic feature of the HV overhead line and substantially falls within a frequency range up to 30 MHz.

Because the generation of corona discharge is affected by weather conditions, for example, AC conductor corona is more likely to occur in wet weather, and the RI level varies widely with weather conditions (e.g., it would be 20–30 dB higher on rainy days than on sunny days). In light of this, it is customary to define several characteristic levels, such as the average fair weather level, the 80% level, and the 95% level (under heavy rain). The 80% level is chosen as the characteristic value worldwide (i.e., the limit that is not exceeded for 80% of a year with 80% confidence). Moreover, the RI also depends on conductor parameters such as the conductor height, interphase distance, cross-sectional area of conductor (subconductor), and the number of subconductors.

The RI of a DC transmission line varies with the weather and season, but within a smaller range as compared with an AC line. Moreover, it is higher on sunny days than on rainy days; it is lower in late autumn and early winter, and it reaches the peak in summer.

When measuring the corona-induced RI, even at a fixed measuring point and frequency, the measuring results vary with a number of factors. The measuring results of RI caused by line coronas are of statistical nature due to the following reasons: (i) the random nature of the interference phenomenon; (ii) variations of the meteorological conditions along the line; (iii) changes in the surface state of the conductors, which is affected not only by weather conditions, such as rain and frost, but also by deposits of dust, insects, and other particles. These results are often presented according to statistical analysis methods in the form of cumulative distributions that usually include the fair weather, bad weather, and all-weather cumulative probabilities.

Figure 7.34 shows the typical RI cumulative probability curves measured for a transmission line of China's UHV AC pilot and demonstration project. In this figure,

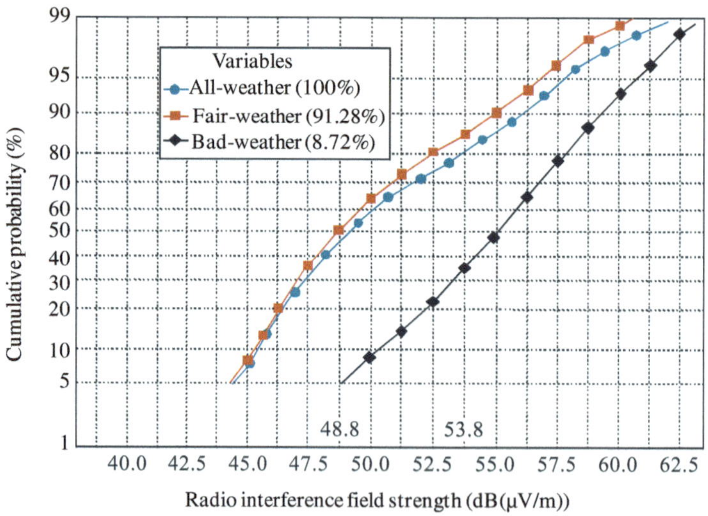

Figure 7.34
Typical RI cumulative probability curves for China's UHV AC pilot demonstration project (2010–2011, Wenxian of Henan).

"fair weather" refers to the statistical values during fair weather, "bad weather" refers to statistical values during drizzling, light rain, heavy rain, and snow, and "all-weather" refers to statistical values measured for all test times. The horizontal axis indicates the RI field strength expressed in decibels with 1 μV/m as the base. The vertical axis indicates the percent of number of measurements not higher than the corresponding field strength on the horizontal axis in the total number of measurements. As shown in Figure 7.34, on the "fair weather" curve, the RI field strength corresponding to a cumulative probability of 50% is 48.8 dB(μV/m); on the "all weather" curve, the RI field strength corresponding to a cumulative probability of 80% is 53.8 dB(μV/m), denoting that 80% of all values measured in various weather conditions do not exceed 53.8 dB(μV/m). Because the project undergoing testing is performed in three provinces where the weather conditions and the percent of fair and bad weather (including rain, snow, etc.) are different, the RI cumulative probability curve also varies.

7.4.1.2 Television Interference

The frequency of television signals is much higher than that of radio signals. It follows from the spectrum characteristics of interference signals caused by line coronas that the interference to television signals is much less severe than to the medium and short waves. The spark discharges across gaps are the main source of TVI. Considering the particularity of television signals, two aspects need attention. The electromagnetic noise as a result of spark discharges across gaps causes interference to the TV picture signals, jeopardizing the picture quality; the line conductors and towers may shield and reflect the TV signals, causing signal attenuation and ghosting. The TV audio works in frequency modulation mode and is intrinsically noise-proof, making it immune from the interference.

Presented here are the two main TVI sources:

1. *Intense discharge at the tip of fittings.* The frequency range of spark discharges extends up to the extremely high-frequency band, making them the major source of interference to television reception. However, such interference will only affect a limited area, and it is easy to locate and eliminate the source.
2. *Intermittent discharge resulted from rupture or damage of insulator's porcelain parts.* Rusting at the hinge joints of various metallic parts and between the steel pins and the iron caps of insulators produces insulation barriers where the electric field strength is very high, causing intermittent micro-arc discharges to occur under an alternating voltage.

7.4.2 Calculation of RI of Overhead Lines

When designing a transmission line, it is necessary to calculate the lateral profile of its RI to determine the line parameters according to the specified permissible interference level. The RI field strength of a transmission line may be calculated using two categories of formulas. The first is the empirical formulas based on numerous measured data and mainly apply to lines with single conductors and bundle conductors with up to four subconductors operating at voltages between 220 and 750 kV and at maximum voltage gradients between 12 and 20 kV/cm. The second is the semi-empirical formulas derived from the excitation function and mainly apply to lines using bundle conductors containing more than four subconductors. See CISPR 18-3 AMD 1−1996 *Amendment 1 Radio Interference Characteristics of Overhead Power Lines and High-voltage Equipment—Part 3: Code of Practice for Minimizing the Generation of Radio Noise.*

CISPR states that the 80% level of RI field strength may be obtained by subtracting 10−15 dB from the heavy rain level, and the fair weather level may be obtained by subtracting 6−10 dB from the 80% level. In other words, the difference between the heavy rain level and the fair weather level is recommended to be 16−25 dB.

The RI may also be predicated using empirical formulas. See CISPR 18-1:2010 *Radio Interference Characteristics of Overhead Power Lines and High-Voltage Equipment. Part 1: Description of Phenomena (second edition)* for the recommended formula applicable to the bipolar DC link. It provides the 50% level of RI for the reference frequency of 0.5 MHz and for the reference distance of 20 m from the positive pole conductor in fair weather. The 80% level can be obtained by adding 3−4 dB to the 50% level.

7.4.3 RI Limits for UHV Overhead Lines

According to CISPR, three technical requirements should be considered in determining the RI limit for transmission lines: (i) the lowest radio signal level that should be ensured; (ii) the minimum signal-to-noise ratio (SNG) that can ensure the satisfactory reception; and (iii) the protection distance (or belt), i.e., the minimum distance from the outside phase conductor to the location where the radio signal can be received satisfactorily.

Countries like Canada, Brazil, Poland, Czech Republic, Slovakia, Switzerland, and the former Soviet Union have all developed associated national standards. Of these standards, some specified only one limit regardless of the voltage levels of lines, whereas the others specified different limits corresponding to different voltage levels as well as different reference distances. For instance, Canada's national standard specifies that the RI limit should be based on a reference frequency of 0.5 MHz and at a reference distance of 15 m beyond the projection of the outside phase conductor on the ground. For lines rated 600 kV or more, the limit should be converted to the value at 20 m beyond the projection of the outside phase conductor on the ground. The 50% level is 53 dB(μV/m) and the 80% level is 61 dB(μV/m). However, in Brazil, the design criteria for 1000-kV lines are as follows: the RI during 50% of all time in fair weather is 42 dB(μV/m), which becomes 57 dB(μV/m) when converted to 0.5 MHz and 80% level, as in China.

It is specified in Q/GDW 305—2009 *Electromagnetic Environment Limits for 1000-kV Overhead Lines* that the RI limit for China's UHV lines is taken as 58 dB(μV/m) when measured at 20 m beyond the projection of the outside phase conductor on the ground, 2 m above the ground, and at the reference frequency of 0.5 MHz, thus ensuring that the measured RI level in fair weather does not exceed 55 dB(μV/m). The altitude correction is tentatively set to start from 500 m and the RI limit is increased by 1 dB for every 300-m increase in the altitude.

The average of the RI levels was measured at 20 m beyond the projection of the outside phase conductor on the ground for a UHV AC pilot and demonstration project and the results are as follows: from January to September in Jiaozuo, Henan Province, 48.99 dB(μV/m); and from June to August in Zhongxiang, Hubei Province, 47.8 dB(μV/m). Obviously, the RI levels in both regions meet the environmental protection requirements indicating that the measured RI level in fair weather should not exceed 55 dB(μV/m).

Table 7.8 lists the 80% RI level specified by the US Department of Energy for UHV DC transmission lines. For a ±800-kV DC line, the RI limit is measured at the edge of the line corridor and is 58 dB(μV/m) at the frequency of 0.5 MHz. When converted to the value for a reference distance of 20 m beyond the projection of pole conductor on the ground, it is approximately 63 dB(μV/m).

In Canada, the RI limits for transmission lines at different voltage levels vary while the limits for AC and DC lines at the same voltage level are identical. For transmission lines rated higher than 600 kV, the RI limit at 15 m beyond the projection of the outside phase conductor on the ground and 0.5 MHz is 63 dB(μV/m). When converted to the value at 20 m beyond the projection of the outside phase conductor on the ground, the limit is approximately 61 dB(μV/m).

With reference to the international limits and considering the characteristics of RI of DC lines, particularity China's UHV DC lines that run through high-altitude regions, it is specified in Q/GDW 145—2006 *The Control of the Electromagnetic Environment of ±800-kV DC Overhead Transmission Line* and DL/T 1088 *Electromagnetic Environment Parameters Limit for ±800 kV UHV DC Transmission Lines* that the 80% level of the

Table 7.8: 80% RI Level Specified by the US Department of Energy for UHV DC Transmission Lines.

Line Type	DC		Remarks
	±800 kV	±1200 kV	
80% RI level, dB(μV/m)	58	58	At the edge of line corridor and 0.5 MHz

Table 7.9: Effect of Number of Subconductors on the RI for a 1000-kV Transmission Line (with a Bundle Circle Radius of 1.05 m).

Number of Subconductors		6	8	10
500 mm^2 subconductor	Δg_{max}(kV/cm)	3.60	0	− 2.28
	$\Delta\Gamma$(dB)	8.57	0	− 7.69

Table 7.10: Effect of Cross-Sectional Area of Subconductor on RI for a 1000-kV Transmission Line (Eight-Bundle).

Cross-sectional Area of Subconductor (mm^2)	400	500	630
Δg_{max}(kV/cm)	1.45	0	− 1.31
$\Delta\Gamma$(dB)	1.62	0	− 1.88

corona-induced RI field strength at 20 m beyond the projection of the positive pole conductor on the ground and 0.5 MHz should not exceed 58 dB(μV/m) in general regions, which is a medium level around the world.

7.4.4 Measures to Reduce RI of UHV Overhead Lines

Both the number and diameter of subconductors have a direct effect on the maximum potential gradient on the conductor surface, which will then indirectly affect the calculation of RI level.

Table 7.9 lists the calculated results for a 1000-kV transmission line using bundle conductors that contain different number of subconductors. Δg_{max}(kV/cm) is the difference between the maximum surface field strength of an 8-mm × 500-mm conductor and that of others. $\Delta\Gamma$(dB) is the difference between the excitation function yield of an 8-mm × 500-mm conductor and that of others. These results reveal that increasing the number of subconductors can considerably decrease the RI field strength. Table 7.10 gives relevant calculation results showing the effects of cross-sectional area of the subconductor on the RI for a 1000-kV transmission line. Increasing the cross-sectional area of the subconductor can decrease the RI level.

Figure 7.35 depicts the curves of RI of a ±800-kV DC line as a function of the number and cross-sectional area of subconductors. The RI is measured at 20 m beyond the projection of the positive pole conductor on the ground and at 0.5 MHz; the number of subconductors ranges from four to eight, and the cross-sectional area of the subconductor ranges from 500 mm^2 to 1250 mm^2. It follows that increasing the number of subconductors can significantly reduce the RI. This is because for a given cross-sectional area of subconductors, an increased number of subconductors will contribute to a larger total

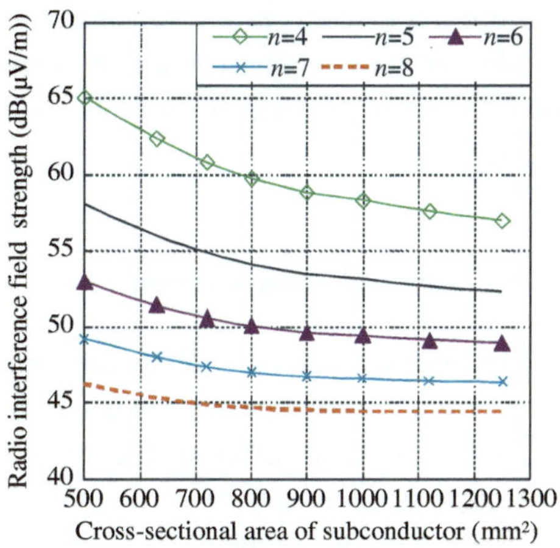

Figure 7.35
RI of a ±800-kV DC line with different number and cross-sectional area of subconductors.

surface area of the conductor, thus reducing the density of charges on conductor surface. Because the field strength on the conductor surface is directly proportional to the density of charges, the field strength is reduced. The intensity of corona discharges decreases with decreased field strength on the conductor surface, thus reducing the corona-induced RI.

7.5 Corona Losses of UHV Overhead Transmission Lines

7.5.1 Corona Performance of Overhead Transmission Lines

Corona generation on conductors is a major phenomenon of an operating HV overhead transmission line and is also the root cause of the RI and audible noise that concerns the public. Moreover, corona discharges cause losses in power delivered by lines and generate ozone.

The corona discharge features distinct polarity effects. Specifically, both the corona discharge process and corona current characteristics of a positive electrode are different from those of a negative electrode.

Corona discharges in gases are accompanied with the ionization and recombination processes, giving rise to audio, light, and heat effects that are reflected by sizzling sound, blue light, and temperature increase, respectively, in surrounding gases.

The corona generates a high-frequency pulse current that contains numerous harmonics of high orders, thereby causing RI. Also, corona causes audible noise and dissipates energy.

The main factors affecting the corona discharges on transmission lines include: (i) corona onset field strength on the conductor surface; (ii) conductor surface conditions; (iii) point particles in the vicinity of the conductor; (iv) water droplets on the conductor; and (v) air density, humidity, and wind conditions.

7.5.2 Corona Tests on UHV Overhead Transmission Lines

The use of conductors with different parameters, such as the number and diameter of subconductors, will result in different corona characteristics and corona losses. To visualize the corona characteristics of conductors with different parameters and provide experimental data for calculating corona losses, it is necessary to perform measurements, tests, and studies of visual corona and corona loss of overhead transmission line conductors.

When studying the corona loss of a transmission line, it is difficult to conduct accurate and intensive theoretical studies because too many factors are involved. Therefore, the corona phenomena are usually studied by test means, which mainly include the corona cages and test lines. In the early studies of the UHV transmission technology, it is economically impractical to erect long test lines and frequently replace conductors for corona tests conducted on conductors of varying dimensions. Usually, the UHV test bases and similar research institutes obtain data from corona tests on short test lines and then convert them to equivalent data for long three-phase lines.

1. **Measurement of corona loss of conductors in corona cage**
During the test, test conductors are usually placed at the center of a corona cage (Figure 7.36) to simulate the single or bundle conductors of transmission lines. A corona cage usually has two layers. The outer layer is the shielding cage and the inner layer is the corona measuring cage with a round or square cross-section. The shielding cage wall is grounded via a small resistor for measurement use. Because the conductors are placed near the cage wall, the corona cage can produce surface fields with strength comparable with those on an HV transmission line conductor when

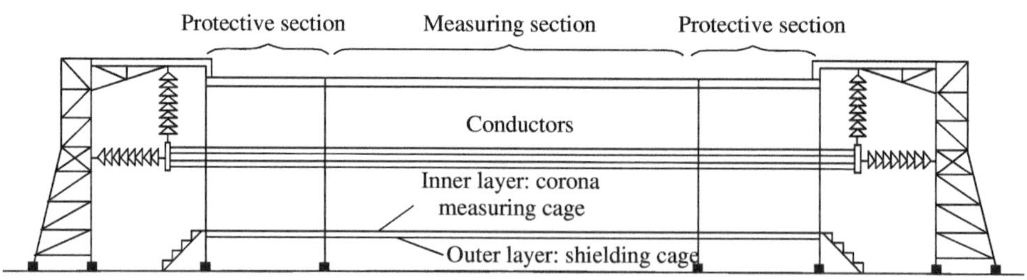

Figure 7.36
Schematic diagram of a corona cage.

applied with a lower test voltage to simulate the corona characteristics on HV conductors.

With the corona loss measuring system, it is possible to measure the conductor's corona loss in the corona cage. China developed a corona cage-based corona loss measuring system with the following features: (i) it incorporates a hybrid electronic current transformer that allows the safe and reliable measurement of the current through conductors in the corona cage; (ii) by using an outdoor high-accuracy capacitive voltage divider, it is possible to accurately measure the voltage; and (iii) the voltage signals are transmitted through coaxial cable and the current signals are transmitted through optical fiber. Based on the accurate voltage and current signals extracted, the system can measure the corona loss of conductors with the aid of the digital signal processing and virtual instrumentation technologies. Figure 7.37 shows the schematic diagram of the corona cage-based corona loss measuring system.

In the test with a corona cage, 8 × LGJ-400/35, 8 × LGJ-500/35, and 8 × LGJ-630/45 conductors are used and the spacing of subconductors is 400 mm. Figures 7.38 and 7.39, respectively, show the corona loss versus the applied voltage and the maximum (peak) field strength on conductor surfaces under both dry and wet (with a rainfall intensity of 20 mm/h) conditions.

2. **Corona loss test on UHV test lines**

China has performed some tests and studies of the UHV line corona loss with test lines at its UHV AC test base. The test base has a single-circuit test line that is approximately 1 km long and uses the 8 × LGJ-500/35 bundle conductor, with the spacing of subconductors being 400 mm. The test line is supported by two tension towers and two tangent towers arranged in a "tension-tangent-tangent-tension" configuration (three spans in total). All four towers have a structural type typical of that used by China's UHV transmission lines. To study the laws of corona losses of a UHV line, a long-term test was made on corona loss of the UHV test line under different meteorological conditions.

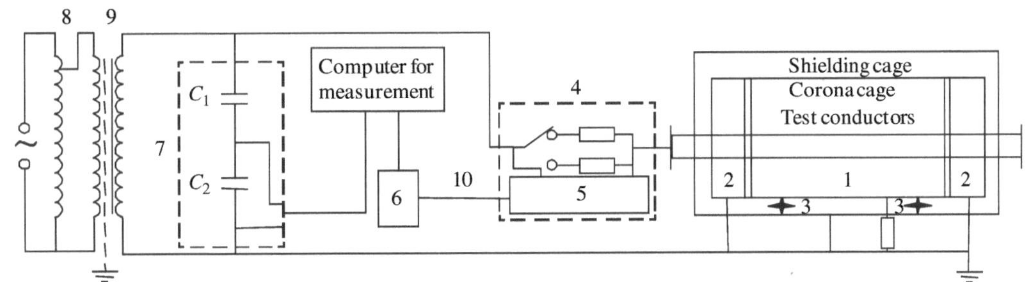

Figure 7.37: Schematic diagram of a corona cage-based corona loss measuring system.
1—Measuring section; 2—protective section; 3—insulating support; 4—measuring resistor switchover box; 5, 6—corona loss measuring module; 7—standard HV capacitive voltage divider; 8—voltage regulator; 9—transformer; 10—optical fiber.

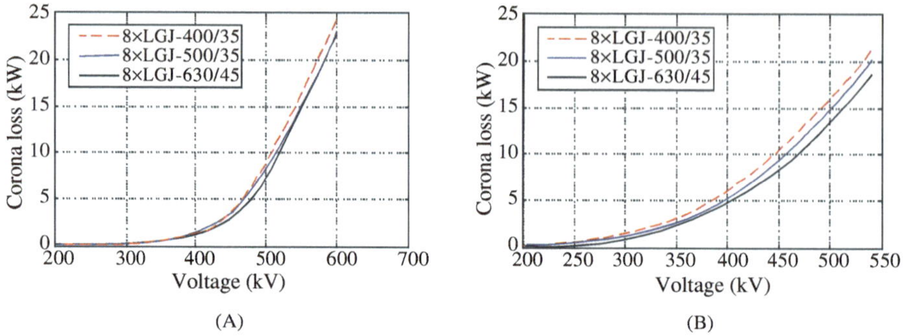

Figure 7.38: Corona loss measurements of UHV line conductors in the corona cage under different applied voltages.
(A) Dry and (B) wet (with a rainfall intensity of 20 mm/h).

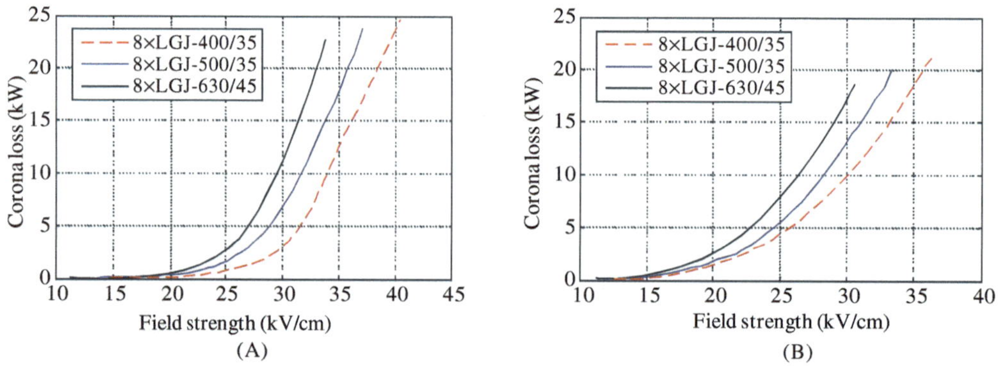

Figure 7.39: Corona loss measurements of UHV line conductors in the corona cage under different maximum field strengths on conductor surface.
(A) Dry and (B) wet (with a rainfall intensity of 20 mm/h).

Considering the symmetrical configuration of conductors, only phases A and B of the test line are tested. Figure 7.40 shows the corona loss measuring system.

Table 7.11 shows the corona loss power measured on sunny and cloudy days. P_a and P_{max} are, respectively, the average and maximum corona loss powers.

Test the corona loss of the UHV AC single-circuit test line under light, moderate, and heavy rain. Let the rainfall intensity be the independent variable and let the corona loss of phases A and B be the dependent variables. Perform fitting of the measured corona loss data using the least square method. Then, estimate the corona losses of the 1-km-long test line under light, moderate, and heavy rain according to the fitted curve. Table 7.12 shows the estimates.

Table 7.12 shows that the corona loss on rainy days ranges from 20 to 60 kW/km. Because the field strength on the center phase conductor surface is larger than that on the outside phase conductor surface, the corona loss of the former is also larger than that of the latter. Rainfall has a significant effect on the corona loss of the test line.

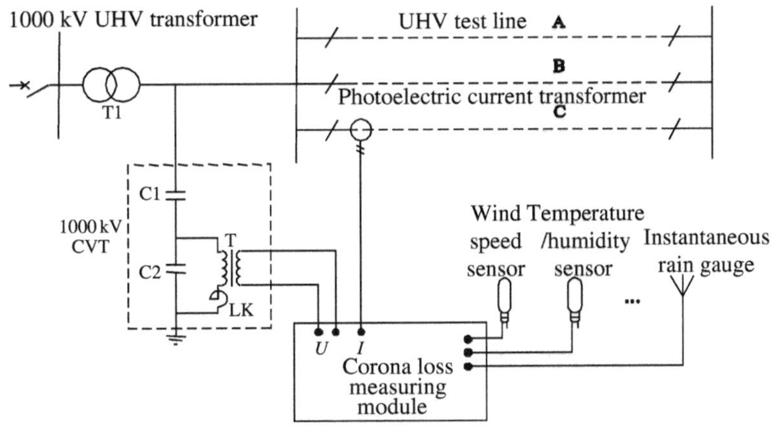

Figure 7.40
Schematic diagram of the corona loss measuring system for the test line at UHV AC test base.

Table 7.11: Corona Loss Power of the UHV AC Single-Circuit Test Line in Sunny and Cloudy Days.

Phase	Sunny Days (kW/km)		Cloudy Days (kW/km)	
	P_a	P_{max}	P_a	P_{max}
Outside phase	5.6396	6.7926	6.1657	7.1856
Center phase	6.5957	7.6406	7.2455	8.7543

Table 7.12: Corona Loss of the UHV AC Single-Circuit Test Line in Rainy Days.

Conductor Type	Meteorological Conditions	Corona Loss (kW/km)	
		Outside Phase	Center Phase
8 × LGJ-500/35	Heavy rain	36.50	50.61
	Moderate rain	26.40	36.71
	Light rain	19.10	26.62

Specifically, when it starts to rain, the corona loss increases rapidly; after it stops raining, the corona loss gradually decreases with the drying of conductor. Its attenuation time depends on the field strength on the conductor surface, wind speed, and air humidity. The corona loss increases with increasing rainfall intensity and is similar to a saturation curve. When the rainfall intensity is low, the corona loss increases rapidly with increasing rainfall intensity. As the rainfall intensity increases, the corona loss increases at a reduced rate. When the rainfall intensity increases to a certain level, the corona loss will gradually become saturated.

Figure 7.41 presents the corona loss measurements of a DC test line at China's UHV DC test base in fair weather. The corona loss under a voltage of 800 kV is approximately 10 W/m.

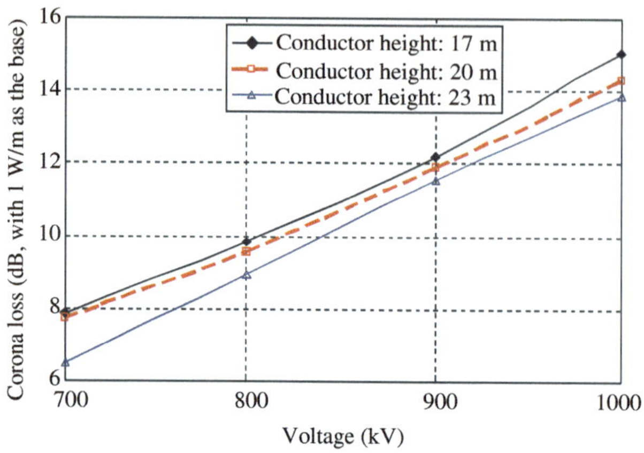

Figure 7.41
Corona loss measurements of a UHV DC line.

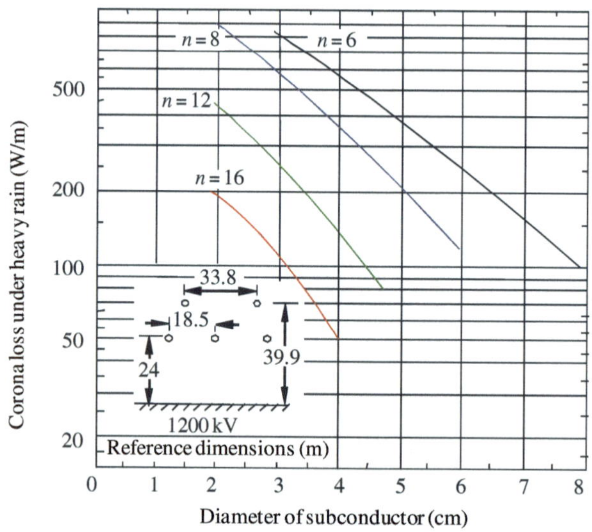

Figure 7.42
Corona loss of a 1200-kV line with reference geometric dimensions under heavy rain.

7.5.3 Corona Loss Calculation of AC Transmission Lines

When studying the corona loss of UHV transmission lines, the diagram method was proposed to simplify the calculation process. Figure 7.42 depicts the curve of the corona loss of a short test line with reference geometric dimensions under heavy rain (rainfall intensity: 12–25 mm/h) as a function of the diameter and number of subconductors. By consulting the curves, it is possible to obtain the corona loss per unit length of the short line under heavy rain.

Table 7.13: Climate Zones Divided by Corona Loss Pattern.

Climate Zone	Climate Type
1	Cold climate, no dry season, warm summer
2	Warm and rainy temperate climate, no dry season, hot summer
3	Steppe climate, dry winter
4	Cold climate, no dry season, warm summer, mountainous climate
5	Desert climate
6	Warm temperate climate, hot and dry summer
7	Warm and rainy temperate climate

Table 7.14: Corona Loss of Lines Expressed in Percentage of Corona Loss of the Short Test Line Under Heavy Rain.

Climate zone	1	2	3	4	5	6	7
Annual average corona loss	2.5%	1.7%	1.7%	3.2%	0.6%	1%	3.4%
Maximum corona loss	74%	36%	51%	74%	29%	33%	36%

To get the annual and maximum corona losses of a given commercial line, the United States divided the country into seven representative climate zones (Table 7.13) based on 10–15 years of meteorological data and proposed a criterion for classifying the statistical weather pattern. Specifically, the actual corona loss of a line is obtained by consulting Table 7.14 according to the specific climate zone to which the line belongs. Table 7.14 presents the annual average and maximum corona loss expressed as a percentage of the corona loss of the short test line.

It is worthy of note that the annual average corona loss for the seven climate zones is calculated according to the standard number of rainy days of each zone. Therefore, corrections should be made with the following formula according to the actual rainy days of these regions through which the line runs:

$$P_{\text{average}} = P_{\text{standard}} \times \frac{\Psi}{\Psi_{\text{s}}} \tag{7.7}$$

where

Ψ, Ψ_{s}—actual and standard number of rainy days during a year.

For the Huainan–Shanghai UHV AC transmission project, with the available line parameters and meteorological data, the project's corona loss is comprehensively assessed using the test data from China's UHV AC test base and the EHV/UHV transmission line design guides from the American EPRI (referred to as the design guides). Table 7.15 gives the assessment results, and the differences shown in the table are calculated using the estimates obtained from the design guides as the reference.

Table 7.15: Corona Loss Estimates of the Double-Circuit Line of Huainan–Shanghai Power Transmission Project.

Corona Loss	Year	Measurements-based Estimates	Design Guides-based Estimates	Difference (%)
Annual corona loss power in bad weather (kWh)	2007	161.235×10^6	153.113×10^6	–
	2008	138.218×10^6	131.859×10^6	–
	2009	131.12×10^6	124.924×10^6	–
Annual total corona loss power (kWh)	2007	161.535×10^6	173.313×10^6	− 6.80
	2008	138.522×10^6	152.378×10^6	− 9.09
	2009	131.428×10^6	145.804×10^6	− 9.86
	Annual average	143.828×10^6	157.165×10^6	− 8.49
Maximum corona loss power (kW/km)	–	444.60	459.07	− 3.15

This table shows that the annual and the maximum corona loss powers assessed with the test data are both lower than those obtained from the design guides.

7.5.4 Corona Loss Calculation of DC Transmission lines

Coronas on a DC transmission line also cause energy loss, leading to an increased annual operating cost. Actually, it is impossible to completely eliminate coronas because dust, insects, water droplets on conductor surface, and unevenness of the conductor surface itself all contribute to the generation of coronas. To ensure reasonable construction and annual operating cost of the transmission line, the designer should select appropriate conductor configurations to control the corona loss within an acceptable range.

The corona loss estimation methods for a bipolar DC link are summarized here.

Method 1: After numerous tests, the Hydro-Québec Research Institute (IREQ) derived a corona loss calculation formula for bipolar DC lines:

$$P = P_0 + k_1(g - g_0) + k_2 \lg\left(\frac{n}{n_0}\right) + 20\lg\left(\frac{d}{d_0}\right) \tag{7.8}$$

where

P—corona loss, dB (with 1 W/m as the base)
g—maximum field strength on conductor surface, kV/cm
d—diameter of subconductor, cm
n—number of subconductors
g_0, n_0, d_0—reference values; $g_0 = 25$ kV/cm, $n_0 = 6$, $d_0 = 4.064$ cm
k_1, k_2, P_0—empirical data, depending on seasons and weather conditions. Table 7.16 shows their values.

Table 7.16: Relevant Parameters of Formula (7.8).

Season	Weather Conditions	P_0 (dB)	k_1	k_2
Summer	Good	13.7	0.80	28.1
	Bad	19.3	0.63	9.7
Spring and autumn	Good	12.3	0.88	36.9
	Bad	17.9	0.72	12.8
Winter	Good	9.6	1.00	44.3
	Bad	14.9	0.85	10.2

Formula (7.8) applies to the voltage range of ± 600 to ± 1200 kV and 4–6 subconductors.

Method 2: Italy conducted intensive tests and derived two formulas for calculating the corona loss of the bipolar DC line in fair and bad weather:

Fair weather:

$$P = P_0 + 50\lg\left(\frac{g}{g_0}\right) + 30\lg\left(\frac{d}{d_0}\right) + 20\lg\left(\frac{n}{n_0}\right) - 10\lg\left(\frac{HS}{H_0 S_0}\right) \tag{7.9}$$

Bad weather:

$$P = P_0 + 40\lg\left(\frac{g}{g_0}\right) + 20\lg\left(\frac{d}{d_0}\right) + 15\lg\left(\frac{n}{n_0}\right) - 10\lg\left(\frac{HS}{H_0 S_0}\right) \tag{7.10}$$

where

P—corona loss, dB (with 1 W/m as the base)
g—maximum field strength on conductor surface, kV/cm
d—diameter of subconductor, cm
n—number of subconductors
H—height of conductor above the ground
S—spacing of poles, m
g_0, n_0, d_0, H_0, S_0—reference values; $g_0 = 25$ kV/cm, $n_0 = 3$, $d_0 = 3.05$ cm, $H_0 = 15$ m, $S_0 = 15$ m
P_0—empirical constant, 2.9 dB in fair weather and 11 dB in bad weather.

Formulas (7.9) and (7.10) do not account for seasons and apply to the voltage range of ± 150 to ± 1200 kV.

Method 3: This method was derived by Anneberg from Sweden through analyzing a great amount of test data on the DC test lines. The calculation formula for a bipolar DC line in fair weather is given by:

$$P = 2U(K+1)K_c nr \times 2^{0.25(g-g_0)} \times 10^{-3} \tag{7.11}$$

where

$$K = \frac{2}{\pi} \times \arctan\left(\frac{2H}{S}\right)$$

where

P—corona loss, kW/km
U—conductor-to-ground voltage, kV
K_c—conductor surface coefficient; 0.15 for smooth conductors and 0.35 for defective ones
n—number of subconductors
r—radius of subconductor, cm
g—maximum electric field strength on the conductor surface under operating voltage, kV/cm
g_0—reference values, $g_0 = 22$ kV/cm (where δ is the relative air density)
H—height of conductor above the ground
S—spacing of poles, m.

When comparing the corona loss results calculated by these three methods with the test data of the UHV DC test line, as shown in Figure 7.43, it is shown that for a conductor with an appropriate surface roughness coefficient, the annual average corona loss in fair weather calculated using the Anneberg formula fits best with the test data. The calculation results of the IREQ formula are approximately 2 dB higher than the average corona loss of the test data, and they fit with the maximum corona loss well. The calculation results of the Italy formula are 2–3 dB lower than the average corona loss of the test data.

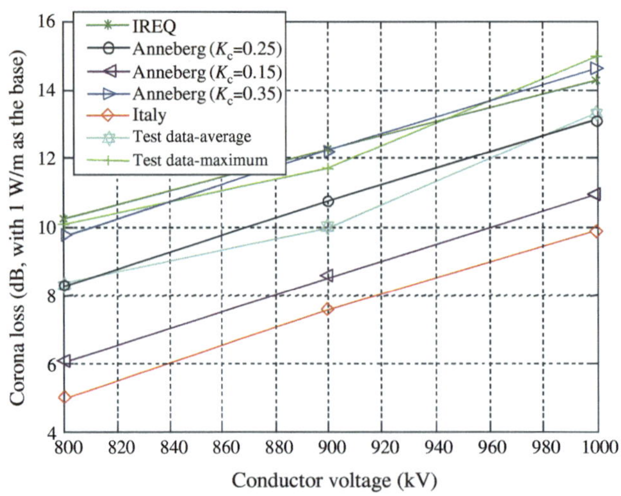

Figure 7.43
Comparison between the corona loss results of formulas used abroad under different voltages and the test data of the UHV DC test line.

References

[1] L. Zhenya, UHV DC Transmission Technology Study Collection (2009–2011), China Electric Power Press, Beijing, 2010–2012.
[2] L. Zhenya, UHV DC Transmission Technology Study Collection (2009–2010), China Electric Power Press, Beijing, 2011.
[3] S. Yinbiao, Z. Wenliang, Research of key technologies for UHV transmission, Proc. CSEE 27(31) (2007) 1–6.
[4] X. Guangrun, High Voltage Static Electric Fields, Shanghai Science and Technology Press, Shanghai, 1962.
[5] W. Xiong, W. Baoquan, Lu Yao, Study on electromagnetic environment for 1000 kV AC transmission line, High Voltage Eng. 32(12) (2006) 55–58.
[6] S. Fuheng, Environment Protection of High-Voltage Transmission, Water Resources and Electric Power Press, Beijing, 1989.
[7] M. Dongxing, H. Zonghui, Environmental Noise Control Engineering, second ed., Higher Education Press, Beijing, 2010.
[8] J.G. Anderson, Transmission Line Reference Book 345 kV and Above, second ed., Electric Power Research Institute, Palo Alto, CA, 1982.
[9] P.S. Maruvada, Corona Performance of High-Voltage Transmission Lines, Research Studies Press Ltd, 2000, ISBN 0 86380 254 0, 310 p.

CHAPTER 8

Equipment of UHV Overhead Transmission Lines

Chapter Outline

Ultra-high Voltage AC/DC Grids.
DOI: http://dx.doi.org/10.1016/B978-0-12-802161-3.00008-1

8.1 Towers

8.1.1 Types and Characteristics

8.1.1.1 Types

Ultrahigh-voltage (UHV) towers should meet electrical, mechanical, and economic requirements for transmission lines. Considering the service conditions, number of circuits, and local geographical conditions, and with reference to types of extra-high voltage (EHV) towers and UHV towers used worldwide, the following types of UHV towers are used in China:

1. *Guyed tower.* A guyed tower consumes less steel but occupies a large area. Because guyed towers need guy wires, they are usable only in plains areas or hilly areas. The cost of the tower itself is low, but the land cost is increasing; therefore, the total cost may be higher than that for self-supporting towers. For this reason, guyed towers are not recommended.
2. *Single-circuit self-supporting tower.* Single-circuit suspension self-supporting towers are mostly two types: cup-type or cat-head. Nearly all other types are evolved from these two types. In areas with narrow line corridors, cat-head towers or even cat-head towers with three V-strings are often used; however, in other areas, cup towers are preferred. In the 1000-kV Jindongnan–Nanyang–Jingmen UHV AC pilot and demonstration project, both types are used. Figures 8.1 and 8.2 show a cat-head tower and cup tower, respectively. In UHV DC transmission, claw-type suspension towers, as shown in Figure 8.3, are used.

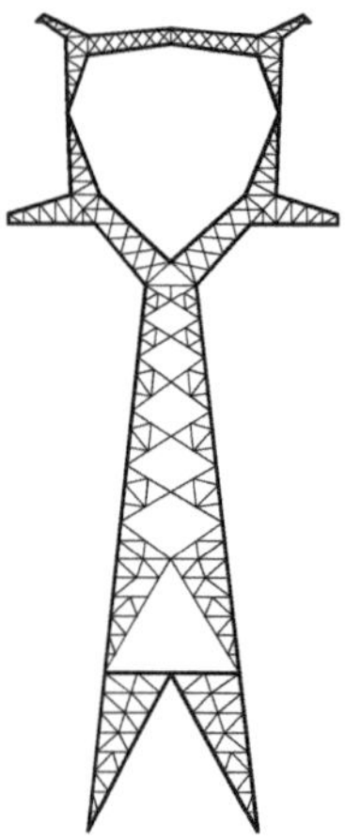

Figure 8.1
Schematic diagram of 1000-kV cat-head tower.

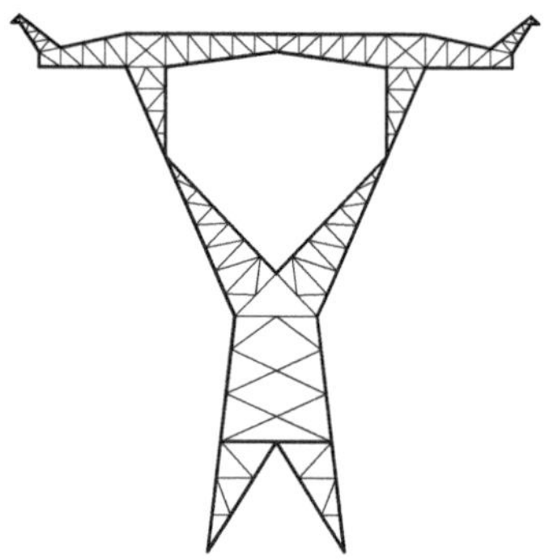

Figure 8.2
Schematic diagram of 1000-kV cup tower.

Most single-circuit angle towers are of a "干" type. This tower type has been widely used in transmission lines in various voltage classes for its simple structure, clear distribution of stresses, narrow line corridors, and ease of erection and maintenance. Abundant experience with this tower type has been accumulated. Single-circuit angle towers in the UHV AC pilot and demonstration project are arranged in a "干" form, as shown in Figure 8.4. The angle towers in the ±800-kV UHV DC demonstration project are shown in Figure 8.5.

3. *Double-circuit tower.* A double-circuit tower usually has three or four cross-arms arranged in an umbrella or drum form and can significantly narrow the line corridors by arranging the three-phase conductors vertically. This tower type is recommended for 1000-kV double-circuit transmission lines.

Figures 8.6 and 8.7, respectively, show the suspension tower and angle tower used in the Huainan–Shanghai UHV AC transmission project.

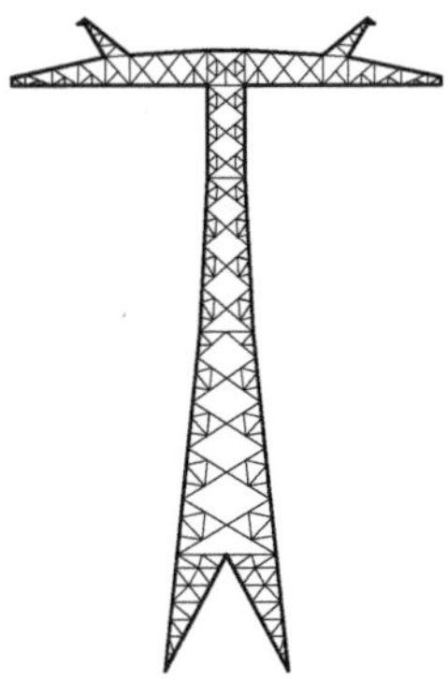

Figure 8.3
Schematic diagram of ±800-kV claw tower.

Figure 8.4
Schematic diagram of 1000-kV tower in a "干" form.

Figure 8.5
Schematic diagram of ±800-kV tower in a "干" form.

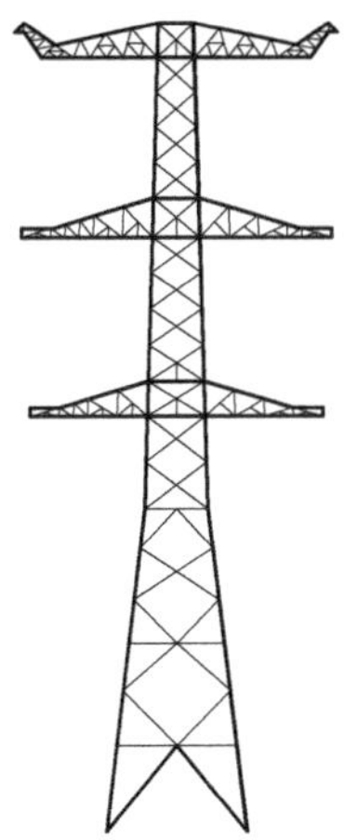

Figure 8.6
Schematic diagram of 1000-kV double-circuit suspension tower.

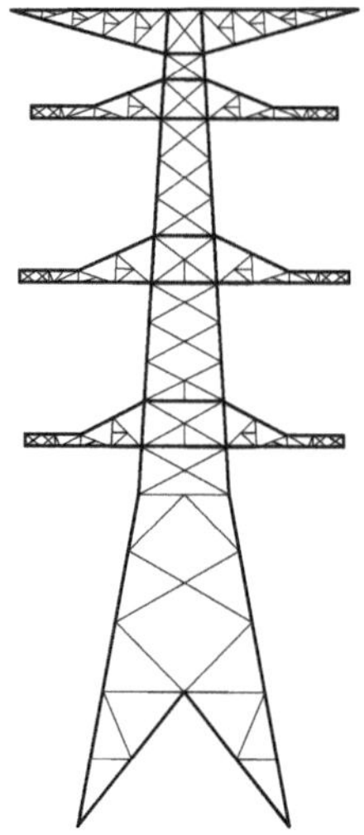

Figure 8.7
Schematic diagram of 1000-kV double-circuit angle tower.

8.1.1.2 Design principles and major technical characteristics

UHV towers should be designed following the principles given here:

1. Simple structural type, clear distribution, and transfer path of stress
2. Simple construction, proper arrangement of nodes to allow for easy fabrication and erection, and guaranteed operational safety
3. Compact arrangement of members to minimize the width of line corridors and save land resources
4. Reasonable arrangement of nodes and members to make full use of the bearing capacity of members
5. Use of proper materials to minimize consumption of steel and maximize the economy of the tower

In UHV AC single-circuit lines, cat-head and cup suspension towers are used, and the body and forks are made of Q420 high-strength hot-rolled angle steel (Q420 high-strength steel), accounting for approximately 20% of the total steel used for a whole tower; angle towers are in a "干" form, and the body and cross-arms are made of Q420 high-strength steel, accounting for approximately 25% of a whole tower. In UHV DC lines, suspension towers and angle towers are of claw-shape and "干"-shape respectively, and the body and cross-arms are made of Q420 high-strength steel, accounting for 25–35% of a whole tower.

In UHV AC double-circuit lines, both suspension towers and angle towers are mainly made of steel tube of Q235 and Q345, respectively. In the Huainan–Shanghai transmission project, the average tower height is more than 100 m, and average tower weight reaches

180 tons. All members are connected with standard forging hubbed flanges and inserting plates, making the design more efficient and reasonable. In mountainous areas where traffic is particularly difficult, angle steel may also be used for a UHV AC double-circuit tower for ease of construction.

8.1.2 Design and Optimization of UHV Towers

8.1.2.1 Determination of design loads

Experience with UHV transmission lines and pilot lines indicates that the load nature of a UHV tower is substantially the same as that of an EHV tower. However, a proper coefficient for importance of a structure and recurrence interval of meteorological loads should be determined according to the characteristics of UHV lines. Moreover, the increase of construction loads due to use of larger machinery should be considered for UHV towers.

1. **Determination of coefficient for importance of a structure**
   As stipulated in GB 50545—2010 *Code for Design of 110-kV–750-kV Overhead Transmission Line*, the coefficient for importance of a structure γ should be 1.1 for very important towers, 1.0 for common towers, and 0.9 for temporary towers. UHV towers should be defined as very important towers, for which γ should be 1.1. This is commensurate with requirements of the safety class of very important building structures set forth in GB 50068—2001 *Unified Standard for Reliability Design of Building Structures*.
2. **Recurrence interval of wind load**
   The longer the recurrence interval of wind load is, the greater the design wind speed is and the higher the resistance to wind of the tower has to be. However, this will increase the weight and cost of the tower. Therefore, determining a proper recurrence interval of wind load is vital for ensuring operational safety and reducing the cost of transmission line. For 500- and 750-kV transmission lines in China, the design recurrence interval of wind load has been prolonged from 30 to 50 years. Although for a UHV transmission line, the recurrence interval of wind load should be determined as 100 years because of its importance.

8.1.2.2 Optimization of structure design

1. **Material**
   Before appearance of UHV transmission, Q235 and Q345 steel specified in GB/T 700—2006 *Carbon Structural Steels* and in GB/T 1591—2008 *High-Strength Low-Alloy Structural Steels*, respectively, are commonly used for transmission towers, whereas in many other countries higher-strength steel is widely used. In Japan, the main legs of 1000-kV double-circuit towers are made of SS55 steel tube or STK55 section steel, and other members are made of SS41 angle steel or STK41 section steel.

The strength of low-strength steel used in China is close to the world average, whereas that of high-strength steel is 14–30% lower, and thus the steel consumption for transmission towers in China is much higher than that in other countries.

A type of high-strength steel Q420 is added in GB 50017—2003 *Code for Design of Steel Structures*, the strength of which is 21.7% higher than Q345. This new material could substantially reduce the amount of steel required for towers. Angle steel towers made of Q420 are used in UHV AC and UHV DC transmission projects in China, reducing the tower weight by 5–10%. Also, the use of this material simplifies the tower structure, reduces the weight of individual members, and reduces transport and installation costs. In conclusion, using Q420 can reduce the total cost by 2–6%.

Steel materials should be selected considering factors such as the importance and type of the structure, type of connection, thickness of steel, and ambient environment and temperature. Approximately 95% of members of a tower are mainly subject to compression, expressed by stability and strength under compression. Consequently, for a specific member, the strength class of steel can be selected based on its load characteristics and slenderness ratio λ:

a. If λ is smaller than 40, then strength is the major concern and Q420 high-strength steel should be used, which will greatly downgrade the specification of the member.
b. If λ ranges from 40 to 80, then stability is the major concern and Q420 high-strength steel can be used. However, this can downgrade the specification of the member by only one level. Due to limitations of specification series, in some cases the specification cannot be downgraded.
c. If λ is greater than 80, then stability is the major concern and Q420 high-strength steel should not be used.

2. **Cross-section of members**

The traditional towers in China are mostly connected with bolts, have a simple structure, and are easy to transport. They are widely used in EHV transmission lines. However, if an angle steel structure is used for UHV towers, which are subject to higher loads, then the main leg of a suspension tower has to be made of double-piece or four-piece combined angle steel. For an angle tower, even four-piece combined angle steel could barely provide the required bearing capacity and, thus, structural angle steel columns must be used. Such a complicated structure causes difficulty in design, fabrication, and erection and results in more safety hazards.

Compared with an angle steel tower, a steel tube tower has more technical and economical advantages, and it is more suited to UHV transmission lines [1].

a. *Load characteristics*. A steel tube tower has advantages such as low wind pressure, great bending stiffness of cross-sections, simple structure, clear stress transfer, and full use of bearing capacity of materials. This structure can reduce the tower weight and the force applied on the foundation, and can improve the tower's capability of withstanding natural disasters under extreme conditions.

In calculating of wind loads, the wind loads sustained by the tower body are generally assumed to be 40–50% of the total loads acting on the tower. A circular steel tube has good aerodynamic performance, and the shape factor of wind load is approximately half that of angle steel. Therefore, steel tube towers may be used to reduce the wind loads applied to the tower body, provided that the strength and stability requirements are satisfied.

b. *Characteristics of cross-section.* The cross-section of a steel tube tower is centrally symmetric and has the same characteristics in all directions. With the material uniformly distributed, it has great bending stiffness. When used for a tension member of a tower, the steel tube cannot exhibit its advantages over angle steel of the same sectional area; however, if used for a bending member, then the application of steel tube with a smaller sectional area and a larger radius of gyration can fully utilize the dynamic characteristics of the material and meet requirements on the bearing strength, rigidness, and stability for the member. In particular, when steel tube is used for a large tower with long members and is subject to heavy loads, the steel tube exhibits even higher stability.
c. *Comprehensive benefits.* Compared with an angle steel tower, a steel tube tower can reduce the weight by 15–20%, remarkably reduce the force applied to the foundation, save concrete by approximately 20%, reduce width of line corridors, minimize demolition of buildings and destruction to plants and trees, save resources, and create environmental benefits. High-strength steel tubes will produce even more notable benefits.

The shape of the cross-section of UHV transmission towers should be determined through techno-economic comparison. In China, angle steel towers are mainly used in UHV AC single-circuit lines and UHV DC lines, and steel tube towers are mainly used in UHV AC double-circuit lines.

3. **Hubbed flanges**

Most main legs of steel tube towers are connected with flanges, mainly rigid flanges and flexible flanges. In the Huainan–Shanghai transmission project, hubbed flanges were developed and used first [2]. Hubbed flanges are one-piece and forged, allowing for more compact metal texture and improvement in the mechanical properties of the material. They are superior to rigid flanges in that they have fewer weld seams, higher fabrication efficiency, and higher quality; they are superior to flexible flanges in their higher strength and rigidness. Regarding these merits, forging hubbed flanges were introduced to connect the main legs of steel tube towers for the Huainan–Shanghai transmission line. As shown in Figure 8.8, a hubbed flange with a butt-welded form, single-groove outside, straight wall inside, and a design similar to that of a flexible flange (i.e., thin flange plate permitting a slight deformation).

Through tests and studies, CEPRI proposes a standard specification sheet that includes 98 pairs of hubbed flanges. The flanges used for connecting pipes of different diameters are classified into four categories and should be designed based on the maximum pipe wall thickness without considering the difference in pipe wall thickness. The flange should have the same strength as the steel tube, and grade 8.8 common rough galvanized bolts with dual nuts and dual gaskets should be used.

4. **Inserting plate**

The diagonal bracings and main leg of a steel tube tower are jointed mainly by inserting U-shape, I-shape, C-shape, T-shape, or X-shape plates, as shown in Figure 8.9.

These types should be selected based on the member position, characteristics, and magnitude of loads, and on structural requirements [3]. Because I-shaped and T-shaped inserting plates are not widely applicable, series configurations are performed only for

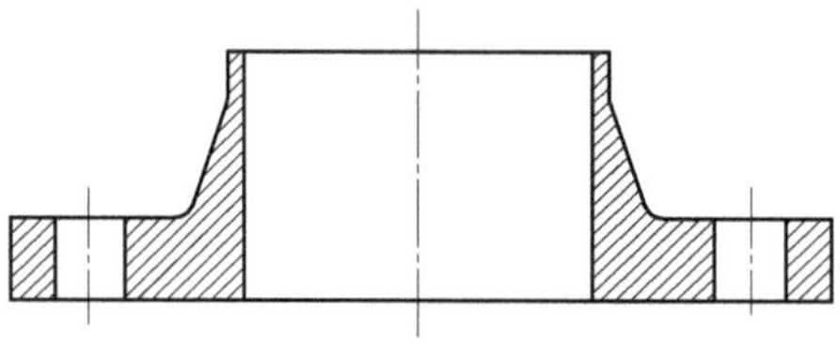

Figure 8.8
Schematic diagram of a forging hubbed flange.

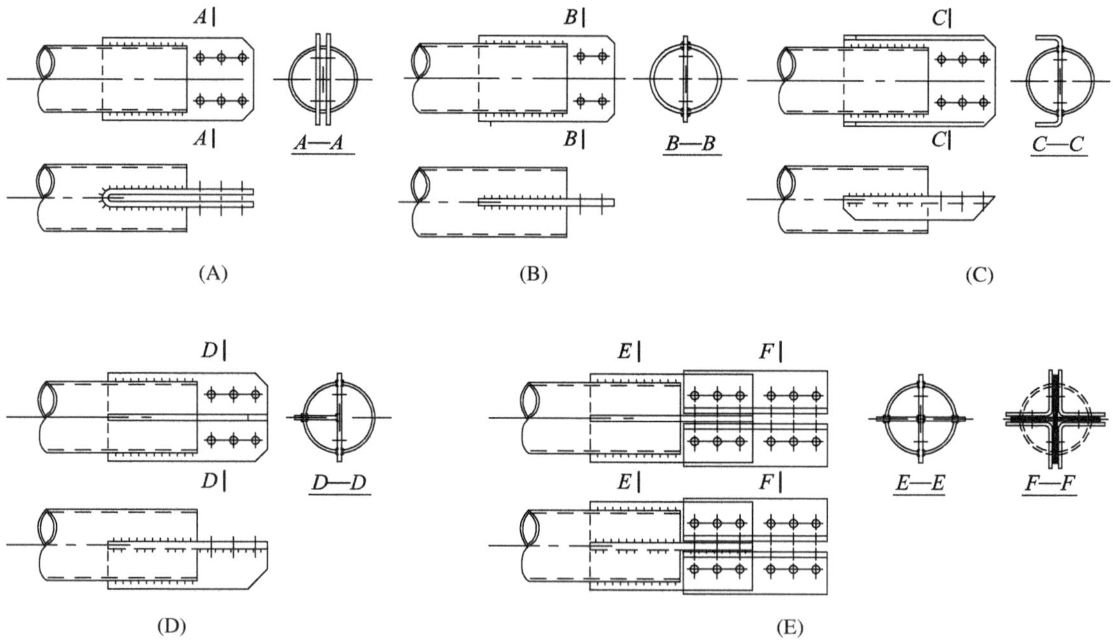

Figure 8.9: Schematic diagram of various types of inserting plates.
(A) U-shaped inserting plate; (B) I-shaped inserting plate; (C) C-shaped inserting plate; (D) T-shaped inserting plate; and (E) X-shaped inserting plate.

the other three types. The bearing capacity of the inserting plate should be designed at three levels, 100%, 70%, and 50%, of the strength of steel tube. In addition, the strength of the inserting plate, weld seams and bolts, layout, and stiffness of joints should all be considered. The model of an inserting plate should be determined based on the specifications and material of the steel tube because each model matches only one specification and material.

5. **Eiffel effect**

Wind speeds vary indefinitely with time in all the three dimensions and, because a tower structure occupies a huge space, the wind speed will never reach the peak value simultaneously over the tower height. If the internal force of diagonal bracings of the tower body is calculated based on the maximum wind load on the entire tower, then sometimes a very small force will be obtained; however, when the distribution of wind speed changes, the internal force will be much higher than the obtained value. This will jeopardize the safety of the tower structure. Such a feature of wind speed has a great impact on the stress acting on the diagonal bracings of a sloping tower. With multiple slopes, a sloping tower looks like the Eiffel Tower, which is why the adverse effect of changes of wind speed in the vertical direction on a sloping tower is called the Eiffel effect. This effect can be calculated with either the shear force ratio method or the reduction coefficient method.

A UHV tower generally has a greater height and a larger size, and the main leg has a larger slope. For wide cup towers and cat-head towers with a large root span, the slope changes sharply at the waist, and thus the Eiffel effect is prominent. For high double-circuit or multi-circuit towers, it is mainly the joint between the tower head and body and parts of the tower body where the slope changes are subject to the Eiffel effect. Figure 8.10 illustrates the Eiffel effect on a UHV tower.

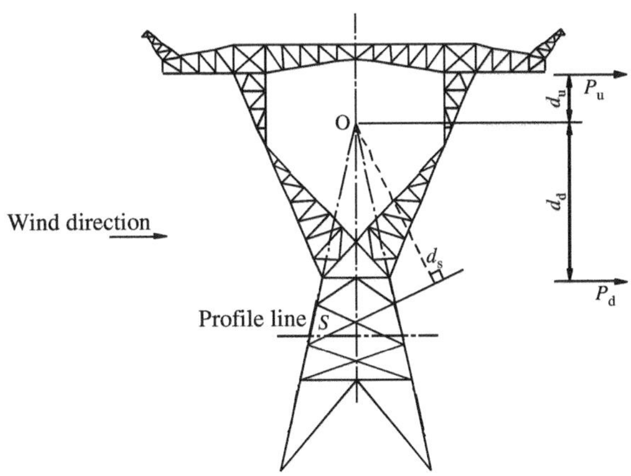

Figure 8.10
Schematic diagram of Eiffel effect on a UHV tower.

The internal force of a single diagonal bracing, as shown in Figure 8.10, is given by

$$S = \frac{P_u d_u - P_d d_d}{2d_S} \tag{8.1}$$

where

S—internal force of diagonal bracing

P_u, P_d—the resultant force of loads applied above point O and that of loads applied below point O and above the profile line, respectively

d_u, d_d, d_S—the arm of P_u, P_d, and internal force of the diagonal bracing, respectively.

The internal force S is discussed in the following three situations:

a. If $P_u d_u > P_d d_d$, then S is positive, indicating that the external loads generate a clockwise moment on point O.
b. If $P_u d_u = P_d d_d$, then S is 0, indicating that the external loads generate a zero moment on point O.
c. If $P_u d_u < P_d d_d$, then S is negative, indicating that the external loads generate an anti-clockwise moment on point O.

This analysis reveals that due to varying slopes of the main leg, the internal force of diagonal bracing S is not necessarily proportional to the total external loads, but rather depends on the difference between $P_u d_u$ and $P_d d_d$. When $P_u d_u > P_d d_d$, increasing P_d will reduce the internal force, and vice versa. This leads to the question of how to address the Eiffel effect in tower design. This is particularly an issue for wind loads because they are difficult to predict. For the distribution of an assumed external load, if the point O is located at a position where the internal force of diagonal members is nearly zero, then it would be very dangerous because the change of wind load will result in a substantial change to the internal force, even causing accidents under severe conditions.

6. **Bending moment at rod end**

The main leg of a UHV steel tube tower is generally connected with flanges, which increases the constraint stiffness at rod ends. Full-scale tests indicate that the bending moment at the point where the tower body slope changes, at ends of tower legs and cross-arms, is relatively large, and the effects of the bending moment on load-carrying capacity of the main leg should be considered.

The bending moment is calculated with the beam-bar hybrid element model [4], with the beam element representing the main leg and the bar element representing diagonal bracings and secondary members. The formulas for calculating the strength and load-carrying capacity (bending and compression) of a structural member given in GB 50017—2003 are applicable to commonly used members with an I-shape, H-shape, or box-shape cross-section. Corrections should be made as appropriate when they are used to calculate the strength and load-carrying capacity of members with a circular cross-section, because the geometrical features of a circular cross-section differ substantially from those of aforementioned cross-sections.

a. *Calculation of strength*: Because a circular cross-section has no edge points (or corner points), the stress could not be calculated for a given point on the cross-section. The vector sum should be used instead of scalar sum (i.e., the maximum vector sum of bending moments at both ends of the member should be used to calculate the strength), as given by:

$$\frac{|N|}{A} + \frac{\max\left(\sqrt{M_{yI}^2 + M_{zI}^2}, \sqrt{M_{yJ}^2 + M_{zJ}^2}\right)}{\gamma_y W_y} \le f \tag{8.2}$$

where

$|N|$—absolute value of the axial compression

A—gross cross-sectional area; downsizing of the cross-section is disregarded because the steel tube is connected with flanges and inserting plates

M_{yI}, M_{zI}—bending moment components in the two orthogonal directions at point I, respectively

M_{yJ}, M_{zJ}—bending moment components in the two orthogonal directions at point J, respectively

γ_y—plastic adaption coefficient of the cross-section

W_y—bending modulus of the cross-section

f—design value of steel strength.

b. *Stability calculation of a member subject to compression and bending*: A steel member with a circular cross-section and a common slenderness ratio will not become distorted and lose stability. In the calculation of overall stability of a steel tube subject to compression and bending at both ends, the vectors of equivalent bending moments in the plane of the two bending moments are added as the overall equivalent bending moment of the member. In case of different bending moments at the two ends, an equivalent bending moment coefficient will be used to convert the non-uniformly distributed bending moment to an equivalent uniformly distributed bending moment. Here, "equivalent" means the same effects of the bending moment appear when the member loses stability. Then, the formula for calculating the overall stability of a member subject to compression and bending at both ends can be expressed as:

$$\frac{N}{\varphi A} + \frac{\sqrt{(\beta_{my} M_y)^2 + (\beta_{mz} M_z)^2}}{\gamma_y W_y (1 - 0.8(N/N'_{Ey}))} \le f \tag{8.3}$$

where

$$N'_{Ey} = \pi^2 EA/(1.1\lambda_y^2)$$

where

N—axial compression

β_{my}, β_{mz}—equivalent bending moment coefficient of the bending moments in the two orthogonal directions, respectively

M_y, M_z—maximum bending moments at ends of the member in the two orthogonal directions, respectively; $M_y = \max(M_{yI}, M_{yJ})$, $M_z = \max(M_{zI}, M_{zJ})$

φ—stability factor of the axially compressed member in the bending moment plane

N'_{Ey}—Euler critical load accounting for a partial safety factor for resistance

E—elastic modulus

λ_y—equivalent slenderness ratio.

7. **Full-scale tests**
   Many countries set-up special bases for full-scale tests of transmission tower structures to ensure safe operation of transmission lines. A full-scale test is a means to verify the reliability of the tower structure to optimize the design and improve the design level.
   DL/T 5154—2002 *Technical Regulation of Design for Pole and Tower Structures of Overhead Transmission Line* stipulates that tests should be performed to examine the new theories, materials, and structural types used in design of poles and towers if no experience is available. Because UHV towers are subject to higher loads than EHV towers, their bodies are much larger than the latter, so new materials and structural types have to be introduced to improve their structures. As such, full-scale tests are required to verify these structures. All new tower types used in China's UHV AC and DC transmission projects are verified, thereby guaranteeing the safety of the projects.

8.1.3 Foundations

The foundation of a transmission tower is a structural body that anchors the tower to the soil or rock and transfers loads to the subsoil or rock through the connecting pieces at the bottom of the tower. The foundation type should be determined considering such factors as the tower type, local landform, geographical conditions, and construction and transport conditions [5].

8.1.3.1 Foundation by excavation and backfill

1. **Rigid concrete foundation**
   Rigid concrete foundations are suitable for various geographical conditions and tower types. They are formed by conducting heavy excavation, placing concrete with a formwork, and filling earth after molding. The soil and concrete combine to provide uplift resistance and the rigid slab provides resistance to compression.
2. **Flexible concrete foundation**
   Flexible concrete foundations are the most commonly used type of excavated foundation in UHV transmission, consisting of vertical column type and inclined column type. They are characterized by large slabs, shallow buried depth, and two-way

reinforced concrete sustaining the bending moment and shear force induced by uplift, compression, and lateral stresses of the tower. Vertical column foundations are applicable for various geographical conditions and tower types. As for an inclined column foundation, the gradient of the column is the same as that of the tower leg, and construction is complicated. It is generally used in plains or river areas with a simple landform and convenient transport access.

8.1.3.2 Undisturbed soil foundation

An undisturbed soil foundation, as shown in Figure 8.11, can take full advantage of the bearing capacity of the subsoil and does not require formwork and backfill, which shortens the construction period and reduces project cost. Such a foundation is widely used in UHV transmission.

8.1.3.3 Expanded pile foundation

To improve the bearing capacity, expanded pile foundation, as shown in Figure 8.12, is commonly used in areas with complicated geographical conditions. To address the problem that the surface soil beneath the tower has a low lateral load capacity, and to take full advantage of the structural features of inclined column foundation, a pile foundation with an inclined column on the top and an under-reamed vertical column at the bottom is often used.

8.1.3.4 Rock foundation

1. **Rock anchor foundation**
   A rock anchor foundation, as shown in Figure 8.13, can make full use of the strength of the rock. It is formed by directly drilling a hole in the rock, inserting the anchor into the rock, and then injecting grouting to fill voids between the anchor and rock for close binding.

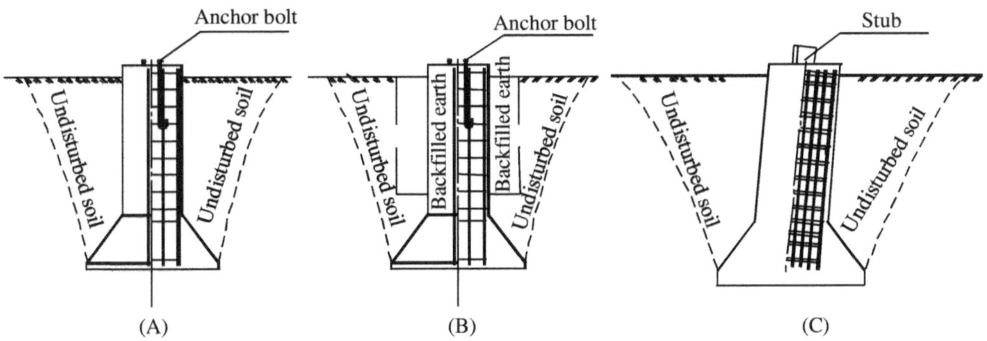

Figure 8.11: Schematic diagram of undisturbed soil foundations.
(A) Fully excavated foundation with a vertical column; (B) half-excavated foundation with a vertical column; and (C) fully excavated foundation with an inclined column.

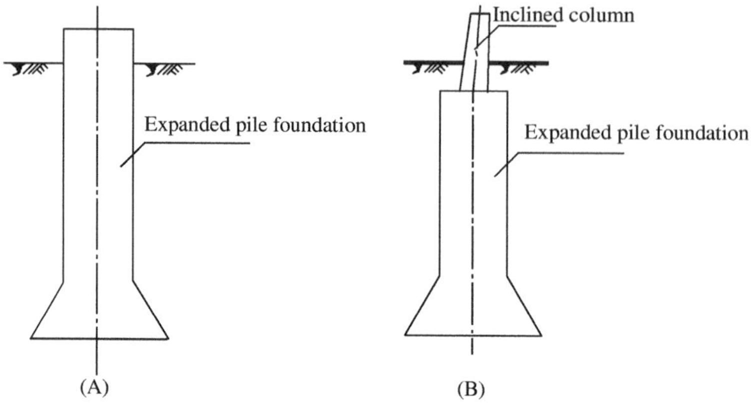

Figure 8.12: Expanded pile foundation.
(A) Vertical column type and (B) inclined column type.

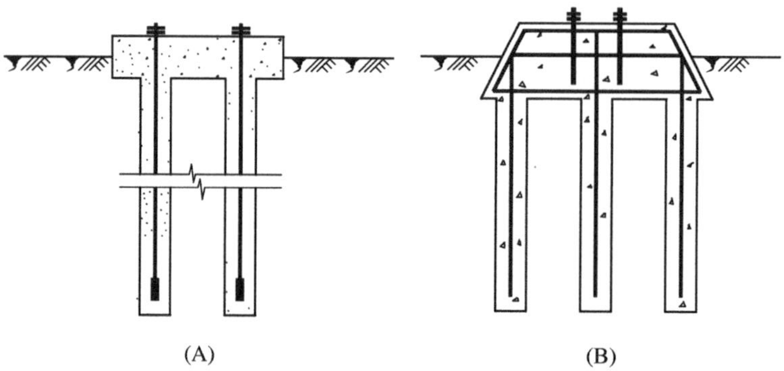

Figure 8.13: Rock anchor foundation.
(A) Vertical anchor type and (B) pile capping type.

It significantly reduces consumption of concrete and steel and is applicable for moderately or highly weathered rocks.

2. **Rock-embedded foundation**

   A rock-embedded foundation, as shown in Figure 8.14, is suitable for highly weathered rock subgrades with a shallow overburden or without overburden. It is characterized by a fully excavated foundation pit and no reinforcement provided for the base slab, and it does not require formwork during construction. This foundation type is widely used in UHV transmission because of its following advantages:

   a. Full use of the shear strength of the rock
   b. High uplift resistance
   c. Less consumption of concrete and reinforcing bars
   d. Small quantity of excavation for the pit
   e. High cost-effectiveness.

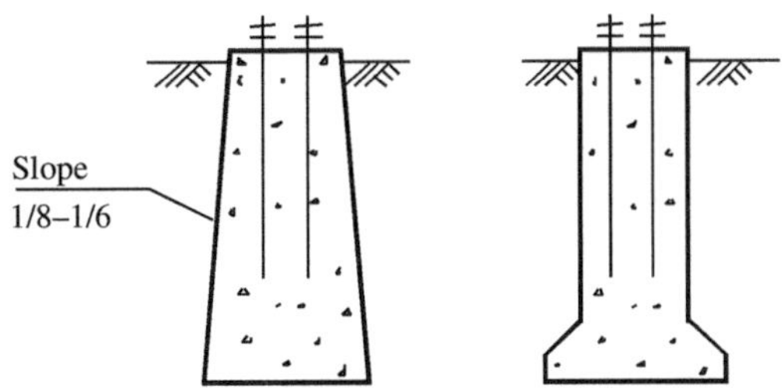

Figure 8.14: Schematic diagram of rock-embedded foundation.
(A) Pier type and (B) excavated type.

8.1.3.5 Combined foundation

In UHV transmission, if the subgrade is an interlayer between soil and bedrock, then the anchor foundation could be combined with another type of foundation, as shown in Figure 8.15. This can take full advantage of the bearing capacity of soil and bedrock, thereby improving the overall bearing capacity of the foundation, significantly reducing the buried depth of the foundation and the consumption of materials.

8.2 Conductors and Ground Wires

8.2.1 Types

As a major component, the conductor accounts for a large proportion of the cost of building an overhead line. The type of a conductor bundle should be determined based on the transmission capacity, the electromagnetic environment, and mechanical safety requirements, and by considering the capital and operating expenditures. Studies show that the effects of radio interference and audible noise arising from corona on the environment have become a decisive factor in selecting the conductor type.

The ground wire, erected above the conductor, is intended to protect the line from lightning strikes. To this end, it has to have high mechanical strength, proper conductivity, and heat capacity. Because the UHV line operates at a higher voltage class and is more important, the ground wire is expected to meet even higher requirements for lightning protection.

8.2.1.1 Types of conductors

The following three types of conductors are being used in UHV lines in China:

1. *Aluminum conductor steel-reinforced (ACSR)*: Features include very good electrical performance, high cost-efficiency, and abundant operating experience; it is widely used.
2. *Aluminum alloy conductor steel-reinforced (AACSR)*: Features include good electrical performance, higher draw-to-weight ratio than ACSR, and some operating experience; it is widely used in areas exposed to heavy icing.

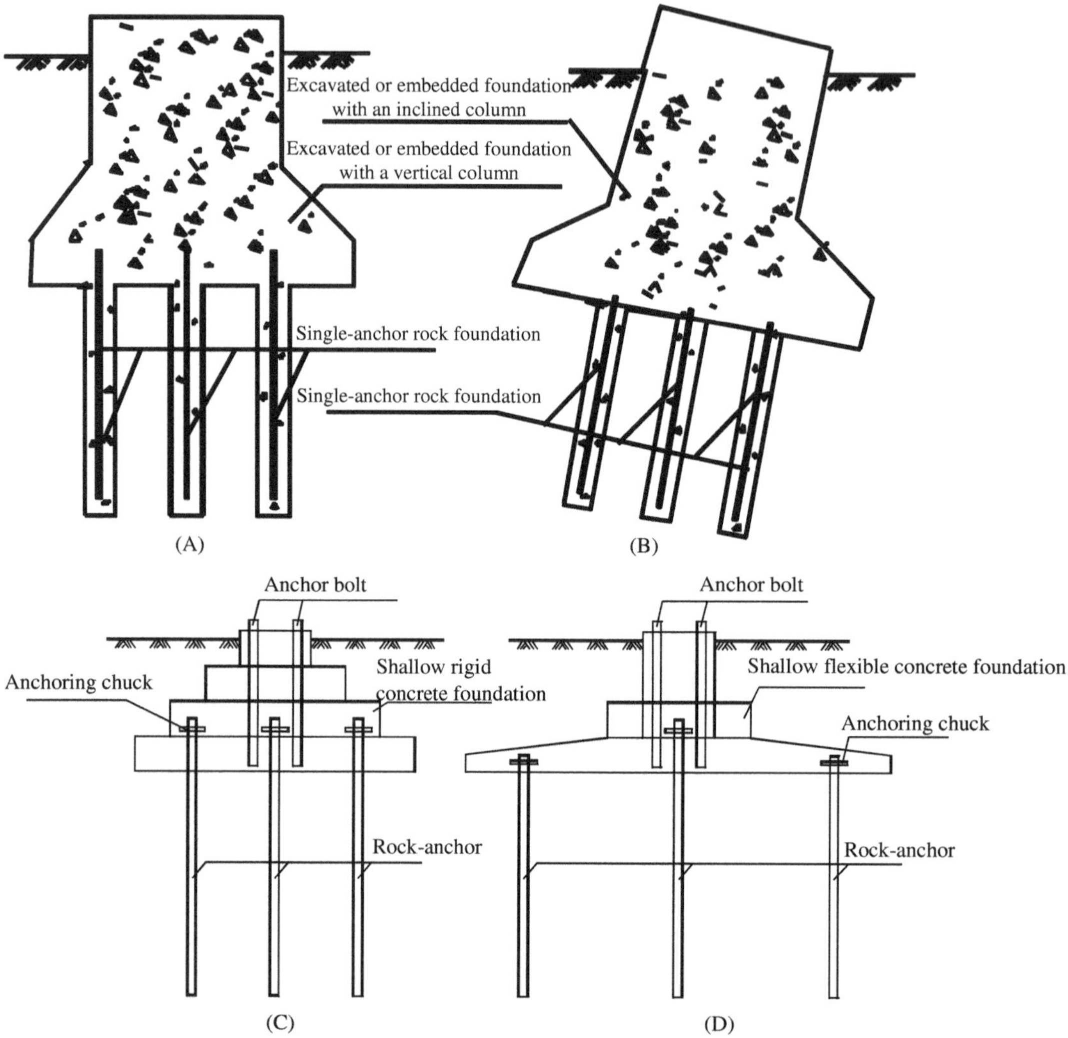

Figure 8.15: Schematic diagram of combined foundation.
(A) Combination of rock anchor foundation with excavated or embedded foundation with a vertical column; (B) combination of rock anchor foundation with excavated or embedded foundation with an inclined column; (C) combination of rock anchor and shallow rigid concrete foundation; and (D) combination of rock anchor and shallow flexible concrete foundation.

3. *Aluminum conductor extra-high-strength steel-reinforced*: Features include good electrical performance, high draw-to-weight ratio, and some operating experience; it is widely used at large crossings.

Table 8.1 summarizes the types and parameters of conductors for ordinary lines in existing UHV AC and DC projects in China. It can be seen that compared with a UHV DC

Table 8.1: Types and Parameters of Conductors in UHV Lines in China.

Parameter	UHV AC Pilot and Demonstration Project	Huainan–Shanghai Transmission Project	Yunnan–Guangzhou Project	Xiangjiaba–Shanghai Project	Jinping–Sunan Project	Haminan–Zhengzhou Project	Xiluodu–Zhejiang Project
Line voltage (kV)	1000	1000	±800	±800	±800	±800	±800
Number of circuits	1	2	1	1	1	1	1
Cross-section of aluminum wires (mm^2)	500	630	630	720	900	1000	900
Number of sub-conductors	8	8	6	6	6	6	6
Outer diameter of sub-conductor (mm)	30.00	33.60	33.60	36.23	39.90–40.59	42.08–42.82	39.90–40.59
Diameter of bundle (mm)	1045	1045	900	900	900	1000	900
Spacing between sub-conductors in a bundle	400	400	450	450	450	500	450

conductor, a UHV AC conductor has more subconductors but a smaller cross-sectional area. This is because a UHV DC line transmits more power and has more availability hours, and thus the resistance has a greater effect on the economy of the line. Large cross-section conductors can reduce the operating expenditures.

Table 8.2 outlines the types and current densities of conductors used in some recent UHV DC projects in China. In these applications, the current density ranges from 0.67 to 0.91 A/mm^2, and a lower value is used for transmission of electricity from thermal power plants. With the development of social economy, the current density shows a significant tendency to decrease.

8.2.1.2 Types of ground wires

Historically, galvanized steel strands were commonly used as the ground wire at the beginning and were later substituted by aluminum-clad steel strands in power grids rated above 220 kV in developed areas (Shanghai, for example) as a result of increased pollution. In China, the environment has been deteriorating for a long time, and dustfall and acid rain have become the predominant causes of corrosion of ground wires. As such, aluminum-clad steel strands are used in all UHV transmission projects.

Most UHV lines built in China are equipped with two ground wires, mostly an aluminum-clad steel-reinforced conductor and an OPGW.

OPGWs are available as three types: stranded stainless steel tube; central aluminum tube; and aluminum frame. The first type is widely used because of its advantages such as smaller cross-section, smaller temperature increase, and greater excess fiber length, despite low resistance to corrosion that can be compensated by filling cable compounds. Table 8.3 shows the configuration of ground wires in UHV projects.

Figure 8.16 is a schematic diagram of OPGW-240 used in the Huainan–Shanghai transmission project and a schematic diagram of OPGW-340 used for the large cross-section in the Haminan–Zhengzhou project.

8.2.1.3 New types of conductors for UHV transmission in China

1. **ACSR with a large cross-section and high conductivity**
   Large cross-section conductors are mostly ACSRs, with aluminum wires that are either round wires or formed wires. China has made great progress in developing and manufacturing duralumin in recent years. Based on the technical capability and need, duralumin of different conductivities is classified into four levels, namely L1, L2, L3, and L4, with respective conductivities of 61.5% IACS,[1] 62% IACS, 62.5% IACS, and 63% IACS.

[1] IACS, international annealed copper standard (%), is the ratio of the conductivity of a conductor to that of pure copper. Here, the conductivity of pure copper is assumed to be 100%.

Table 8.2: Types and Current Densities of Conductors in UHV DC Projects in China.

Project	Yunnan–Guangzhou Project	Xiangjiaba–Shanghai Project	Jinping–Sunan Project	Xiluodu–Zhejiang Project	Haminan–Zhengzhou Project	Lingzhou–Shaoxing Project
Power source	Hydro power	Hydro power	Hydro power	Hydro power	Thermal power/wind power	Thermal power/wind power
Configuration of conductor	6 × LGJ-630/45	6 × ACSR-720/50	6 × JL/G3A-900/40	6 × JL/G3A-900/40	6 × JL/G3A-1000/45	6 × JL/G2A-1250/70
Cross-sectional area of conductor (mm^2)	3780	4352	5400	5400	6000	7500
Rated voltage (kV)	±800	±800	±800	±800	±800	±800
Rated current (A)	3125	4000	4500	4750	5000	5000
Current density (A/mm^2)	0.827	0.912	0.833	0.879	0.833	0.666

Table 8.3: Configuration of Ground Wires in UHV Projects in China.

Parameter	UHV AC Pilot Project	Huainan–Shanghai Transmission Project	Yunnan–Guangzhou Project	Xiangjiaba–Shanghai Project	Jinping–Sunan Project	Haminan–Zhengzhou Project	Xiluodu–Zhejiang Project
Line voltage (kV)	1000	1000	±800	±800	±800	±800	±800
Configuration	One ground wire, one OPGW	One ground wire, one OPGW	One ground wire, one OPGW	One ground wire, one OPGW	One ground wire, one OPGW (N1–N46); two common ground wires for other sections	One ground wire, one OPGW	Two common ground wires
Specs of ground wire	LBGJ-170-20AC	LBGJ-240-20AC	LBGJ-180-20AC	LBGJ-180-20AC	LBGJ-180-20AC	LBGJ-150-20AC	JLB20A-150-19; JLB20A-240-19
Outer diameter of ground wire (mm)	17.00	20.00	17.50	17.50	17.50	15.75	15.75; 20.00
Specs of OPGW	OPGW-175	OPGW-240	OPGW-180	OPGW-180	OPGW-180	OPGW-150	–
Outer diameter of OPGW (mm)	17.50	20.30	17.40	17.40	17.40	16.60	–

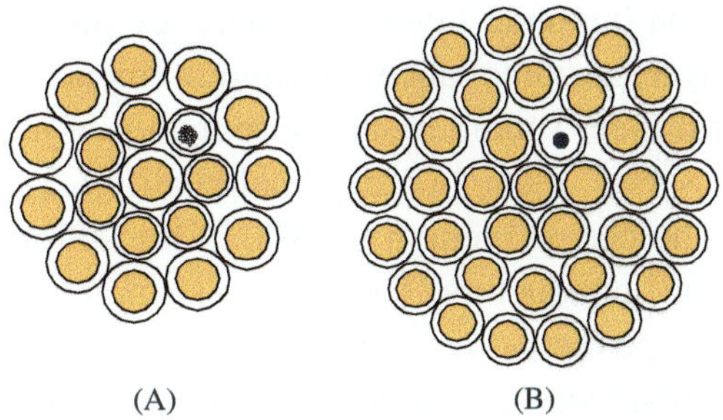

Figure 8.16: Schematic diagram of stranded stainless steel tube OPGWs.
(A) OPGW-240 and (B) OPGW-340.

Figure 8.17: High-conductivity 900-mm^2 ACSRs.
(A) JL1/G3A-900/40-72/7 and (B) JL1/G2A-900/75-84/7.

Figure 8.17 shows pictures of high-conductivity 900-mm^2 ACSRs. High-conductivity 900- or 1000-mm^2 ACSRs for use in UHV DC transmission have the following properties [6]:

1. The resistivity of aluminum strands is no higher than 0.028034 Ω mm^2/m (61.5% IACS), and below the required limit of 0.028264 Ω mm^2/m (61%IACS) in GB/T 1179—2008 *Round Wire Concentric Lay Overhead Electrical Stranded Conductors*.
2. Each conductor consists of four layers of aluminum strands, whereas conventional ACSR used for ordinary lines consists of only three layers or less.
3. For JL1/G3A-900/40-72/7 and JL1/G3A-1000/45-72/7, the ratio of the cross-section of aluminum to that of steel is 23; for ACSRs used for trunk transmission lines in China before 2010, the ratio was <20.
4. The range of tensile strength (difference between the maximum and minimum tensile strengths of wires of the same type) of aluminum wires is required to be no more than 25 MPa, and that of steel wires is required to be no more than 150 MPa; GB/T 1179—2008 has no requirements regarding this.

5. There are no joints on the outer layer, and there are no more than four joints per reel on the inner layer, in compliance with requirements in GB/T 1179—2008, which stipulates that there should be no more than five joints.
6. PL/4 2600 × 1500 × 1900 reels, in a detachable all-steel corrugated structure, are used for packing. The per-reel length is 2900 m for 900-mm^2 conductors and 2500 m for 1000-mm^2 conductors. This makes stringing faster and avoids waste because of using composite packing reels.

Table 8.4 gives a comparison of conductor JL1/G3A-1000/45-72/7 produced in China and the ASTM standard.

SGCC released the technical specifications for five types of conductors with a cross-section of 1250 mm^2 in September 2013, namely, ACSRs JL1/G3A-1250/70-76/7 and JL1/G2A-1250/100-84/19 (as shown in Figure 8.18), steel-reinforced formed aluminum conductors JL1X1/G3A-1250/70-431 and JL1X1/G2A-1250/100-437 (as shown in Figure 8.19), and aluminum alloy-reinforced formed aluminum conductor JL1X1/LHA1-800/550-452 (as shown in Figure 8.20).

Table 8.4: Comparison Between Conductor JL1/G3A-1000/45-72/7 Produced in China and ASTM Standard.

Index	Chinese Product	ASTM Standard	Engineering Purpose
Conductivity of aluminum wire	≥61.5%IACS	≥61%IACS	To reduce the DC resistance
Range of tensile strength of aluminum wires	≤25 MPa	Not required	To ensure aluminum wires are uniformly loaded
Range of tensile strength of steel wires	≤150 MPa	Not required	To ensure steel wires are uniformly loaded

Figure 8.18
JL1/G2A-1250/100-84/19.

Figure 8.19
JL1X1/G2A-1250/100-437.

Tables 8.5 and 8.6 summarize the parameters of the five types of conductors.

To increase the transmission capacity of a ±800-kV UHV DC line to 9600 MW or more, conductors with a cross-section up to 1520 mm^2 will be developed. Techno-economic comparison shows that JL1X/LHA1-1040/550 is superior to other conductors in a variety of scenarios.

2. **High-conductivity aluminum conductor aluminum alloy-reinforced (ACAR)**
Internationally, an ACAR comprises a core of aluminum alloy (designation: 6201) with hard-drawn aluminum wires (designation: 1350) stranded around the core. By increasing the conductivity of hard-drawn aluminum, China successfully develops high-conductivity ACAR.

Figure 8.21 is a schematic diagram of a high-conductivity ACAR JL1/LHA1-745/335-42/37, approximately 2000 tons of which are used in the Haminan–Zhengzhou Project. Table 8.7 compares the major parameters between JL1/LHA1-745/335-42/37 and JL1/G2A-1000/80-84/19.

It can be seen from the table that JL1/LHA1-745/335-42/37 and JL1/G2A-1000/80-84/19 have approximately the same diameter; therefore, the indicators of electromagnetic environments and corona losses of six conductors of both types are close at the same height. With smaller DC resistance, the former can reduce power losses by 9 kW/km, contributing to lower operating expenditure despite higher capital expenditure.

3. **Expanded conductors**
The corona onset voltage of conductors decreases with altitude. To prevent corona, the conductor diameter should be increased as appropriate. Also, in some cases, to reduce the field strength of the conductor and, hence, the noise resulting from the corona, the outer diameter of the conductor needs to be increased without increasing the cross-sectional area. To meet such requirements, an expanded conductor is developed.

Figure 8.20
JL1X1/LHA1-800/550-452.

Table 8.5: Major Parameters of ACSRs with a Cross-Section of 1250 mm^2.

Item			Unit	Parameter	
Specs			–	JL1/G3A-1250/70-76/7	JL1/G2A-1250/100-84/19
Schematic diagram			–	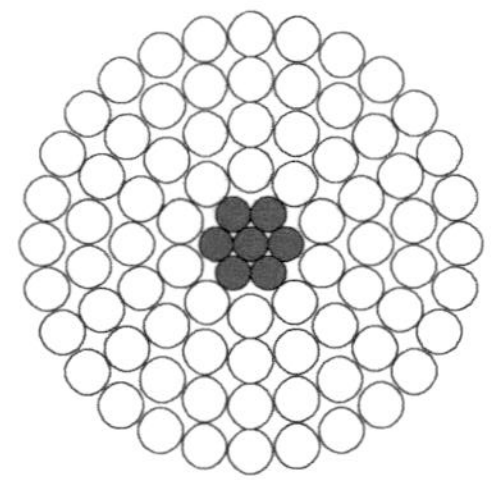	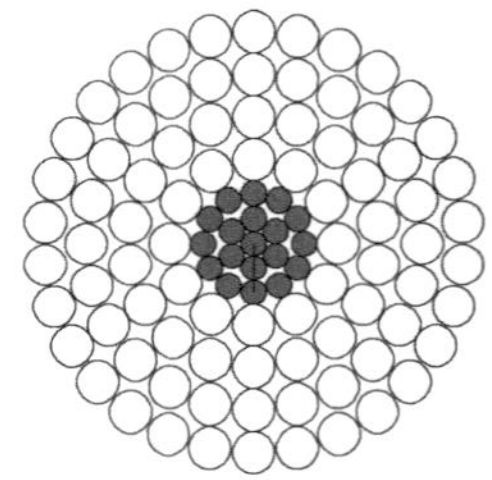
Configuration (number of wires/diameter)	Aluminum	Outermost layer	Piece/mm	28/4.58	30/4.35
		Second outermost layer	Piece/mm	22/4.58	24/4.35
		Second innermost layer	Piece/mm	16/4.58	18/4.35
		Innermost layer	Piece/mm	10/4.58	12/4.35
	Steel	Layer with 12 wires	Piece/mm	–	12/2.61
		Layer with six wires	Piece/mm	6/3.57	6/2.61
		Central layer	Piece/mm	1/3.57	1/2.61
Calculated cross-sectional area		Total	mm^2	1322.16	1350.03
		Aluminum	mm^2	1252.09	1248.38
		Steel	mm^2	70.07	101.65
Outer diameter			mm	$47.35_{0}^{+1\%}$	$47.85_{0}^{+1\%}$
Mass per unit length			kg/km	$4011.1_{0}^{+2\%}$	$4252.3_{0}^{+2\%}$
DC resistance at 20°C			Ω/km	≤ 0.02291	≤ 0.02300
Rated tensile strength			kN	294.23	329.85
Elastic modulus			GPa	62.2 ± 3	65.2 ± 3
Coefficient of linear expansion			1/°C	21.1×10^{-6}	20.5×10^{-6}

Table 8.6: Major Parameters of 1250 mm² Steel-Reinforced Formed Aluminum Conductors and 1250 mm² Aluminum Alloy-Reinforced Formed Aluminum Conductors.

Item			Unit	Parameter		
Specs			—	JL1X1/G3A-1250/70-431	JL1X1/G2A-1250/100-437	JL1X1/LHA1-800/550-452
Schematic diagram			—	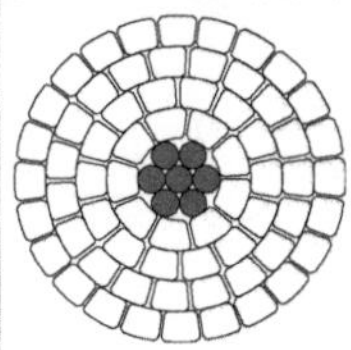	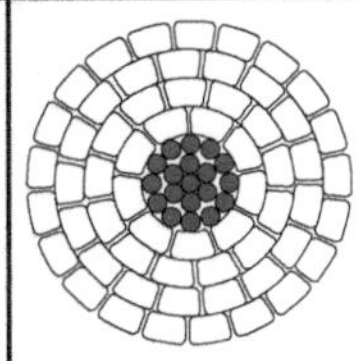	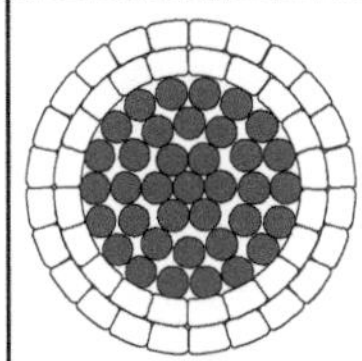
Configuration (number of wires/equivalent diameter)	Aluminum	Outermost layer	Piece/mm	24/4.93	21/5.16	24/4.82
		Second outermost layer	Piece/mm	19/4.93	17/5.16	—
		Second innermost layer	Piece/mm	14/4.93	13/5.16	—
		Innermost layer	Piece/mm	9/4.93	9/5.16	20/4.82
	Steel/aluminum alloy	Layer with 18 wires	Piece/mm	—	—	18/4.35
		Layer with 12 wires	Piece/mm	—	—	12/4.35
		Layer with six wires	Piece/mm	6/3.57	12/2.61	6/4.35
		Central layer	Piece/mm	1/3.57	6/2.61	1/4.35
Calculated cross-sectional area		Total	mm^2	1329.95	1356.35	1352.74
		Aluminum	mm^2	1259.88	1254.70	802.86
		Steel	mm^2	70.07	101.65	549.88
Outer diameter			mm	43.11 ± 0.4	43.67 ± 0.4	45.15 ± 0.4
Mass per unit length			kg/km	4055.1 ± 81	4290.1 ± 86	3737.6 ± 74
DC resistance at 20°C			Ω/km	≤ 0.02292	≤ 0.02301	≤ 0.02253
Rated tensile strength			kN	289.18	324.59	289.00
Elastic modulus			GPa	62.1 ± 3	65.1 ± 3	55 ± 3
Coefficient of linear expansion			1/°C	21.1×10^{-6}	20.5×10^{-6}	23×10^{-6}

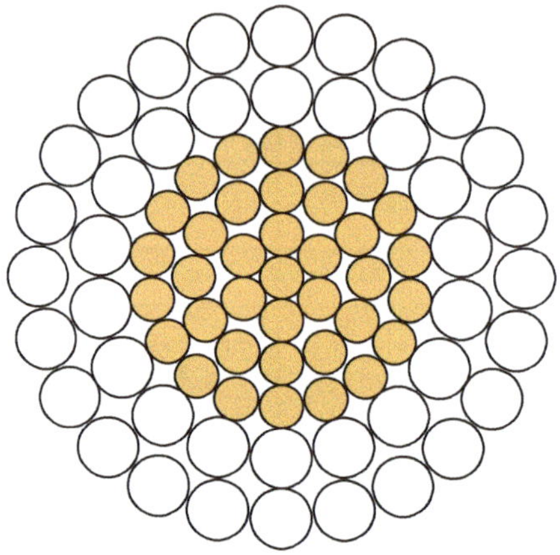

Figure 8.21
Schematic diagram of high-conductivity ACAR JL1/LHA1-745/335-42/37.

Table 8.7: Comparison Between JL1/G2A-1000/80-84/19 and JL1/LHA1-745/335-42/37.

Parameter		JL1/G2A-1000/80-84/19	JL1/LHA1-745/335-42/37
Configuration (number of wires and diameter, mm)	Aluminum	84 × Ø3.89	42 × Ø4.76
	Aluminum alloy	–	37 × Ø3.40
	Steel	19 × Ø2.34	–
Cross-sectional area (mm^2)	Aluminum	998.32	747.40
	Aluminum alloy	–	335.93
	Steel	81.71	–
	Total	1080.0	1083.33
Outer diameter (mm)		42.82 (100%)	42.84
Calculated weight (kg/km)		3403.8 (100%)	2995.9 (88%)
Rated tensile strength (kN)		264.32 (100%)	223.3 (84%)
Elastic modulus (MPa)		65,200	55,000
Coefficient of linear expansion (×10^{-6} 1/°C)		20.5	23
DC resistance at 20°C (Ω/km)		0.02876 (100%)	0.02774 (96.5%)
Power loss (kW/km)		251.8	242.8

Expanded conductors are a collective term referring to conductors with a diameter larger than that of a compact conductor of the same cross-section. Depending on applications, they may be divided into expanded bus, expanded conductor, or expanded jumper. The first is mainly used in substations, and the last two are used on transmission lines.

Expanded conductors are of various types, with highly varying structures. In UHV transmission in China, expanded conductors with sparse strands are preferred because of their low cost. In past years, expanded conductors with sparse round strands were used. However, many tests and studies have shown that such a conductor with a large cross-section has poor structure stability and easily suffers strand jump when going through a pulley. This is because, in this case, the evenly spaced aluminum wires move toward each other and, in addition, the wires of neighboring layers are in direct contact.

As such, China focuses on the analysis of cross-sectional stability by finite element modeling at the design stage with reference to related study results. Tests are also performed on sample conductors to verify their stability. In addition, the tensile force at strand jump, an index characterizing the cross-sectional stability, must be large enough (possibly 25% of the rated tensile strength) to meet the requirements for tension stringing of expanded conductors.

According to study results and experience available, expanded conductors with sparse strands should be designed following the principles given here:

a. No aluminum wires on the outermost layer are sparsely stranded, those on the second outermost layer and the second innermost layer are sparsely stranded, and few or none on the innermost layer are sparsely stranded.
b. Positive tolerance is used for the diameter of wires on the inner layer, whereas negative tolerance is used for that of wires on the outer layer.
c. The minimum tensile strength of aluminum wires is 170 MPa or higher.

Sparse stranding is not suitable for expanded conductors with a large expansion ratio, because they would have a poor cross-sectional stability. In this case, diameter may be expanded by filling other materials and using formed wires. Considering the technical maturity and economy, supporting expanded conductors filled with high-density polyethylene (HDPE) around the steel core are used. CEPRI led the development of expanded conductors filled with HDPE (JLXK/G2A-630(900)/50 and JLXK/G2A-720 (900)/50).

Tables 8.8 and 8.9 summarize the major parameters of various types of expanded conductors.

4. **Expanded jumpers**

In the double-circuit Huainan–Shanghai transmission line, the high electromagnetic environment requirements call for a minimum outer diameter far larger than the cross-sectional area required by the transmission capacity. Therefore, expanded jumpers are used.

Table 8.8: Major Parameters of Expanded Conductors with Sparse Strands.

Item			Unit	Parameter
Specs			—	JLK/G2A-630(720)/45
Schematic diagram			—	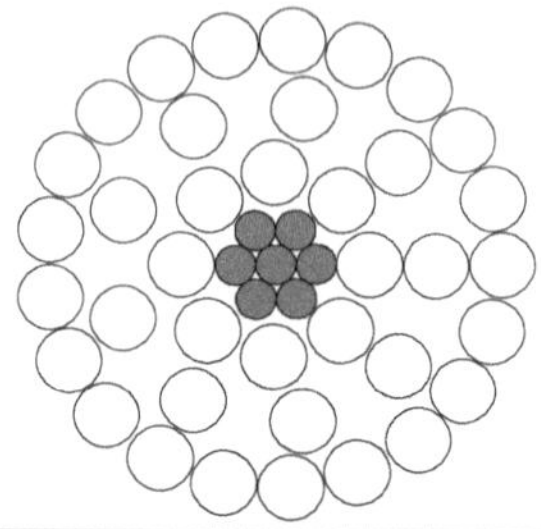
Configuration (number of wires/diameter)	Aluminum	Outer layer	Piece/mm	21/4.53
		Intermediate layer	Piece/mm	9/4.71
		Inner layer	Piece/mm	8/4.71
	Steel core	Layer with six wires	Piece/mm	6/2.80
		Central layer	Piece/mm	1/2.80
Calculated cross-sectional area		Total	mm^2	677.76
		Aluminum	mm^2	634.66
		Steel	mm^2	43.1
Outer diameter			mm	36.30 ± 1%
Mass per unit length			kg/km	2090 ± 2%
DC resistance at 20°C			Ω/km	≤0.04542
Rated tensile strength			kN	159.9
Elastic modulus			GPa	63.6 ± 3
Coefficient of linear expansion			1/°C	20.8×10^{-6}

According to studies of the project, if JL/G1A-630/45, an ACSR of the same model as the conductor, is used as the jumper, then for a 60° tension tower using aluminum tube rigid jumpers the field strength around the surface of the jumper may reach 2762 V/mm, and that of a single subconductor on the exterior angle side of a squirrel-cage rigid jumper may reach 2863 V/mm, exceeding the maximum allowed field strength of 2500 V/mm. Calculations show that using eight-conductor bundles with a diameter of 39.9 mm as the jumper for a tension tower can control the field strength within the allowable range.

Analysis and simulation calculations of cross-sectional stability of the conductors, with their cross-sections expanded from 630 and 725 mm^2 to 900 mm^2, show that

Table 8.9: Major Parameters of Expanded Conductors Filled with HDPE.

Item			Unit	Parameter	
Specs			–	JLXK/G2A-630(900)/50	JLXK/G2A-720(900)/50
Schematic diagram			–	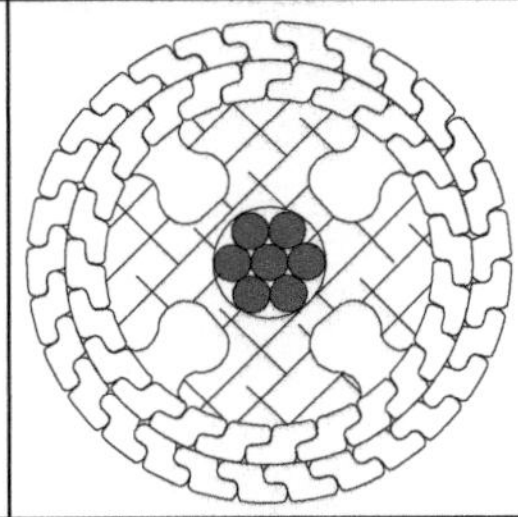	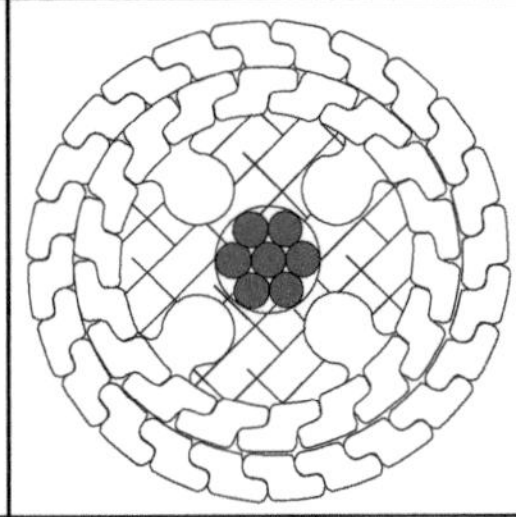
Configuration	Aluminum (number of wires/cross-sectional area of single wire)	Outer layer	Piece/mm^2	30/11.45	24/17
		Inner layer	Piece/mm^2	24/12.02	18/18
	HDPE (diameter)		mm	27.2	24.4
	Steel core (number of wires/diameter)	Layer with six wires	Piece/mm	6/3.02	6/3.02
		Central layer	Piece/mm	1/3.02	1/3.02
Calculated cross-sectional area		Total	mm^2	930	974
		Aluminum	mm^2	630	720
		Steel	mm^2	50	50
		HDPE	mm^2	250	204
Outer diameter			mm	40.0 ± 1%	40.0 ± 1%
Mass per unit length			kg/km	2366 ± 2%	2570 ± 2%
DC resistance at 20°C			Ω/km	≤ 0.04,576	≤ 0.04004
Rated tensile strength			kN	155	168
Elastic modulus			GPa	56.5 ± 3	57.5 ± 3
Coefficient of linear expansion			1/°C	20.5×10^{-6}	20.8×10^{-6}

JLK/G1A-725(900)/40 (as shown in Figure 8.22) should be used, because the tensile strength at strand jump, an index characterizing the cross-sectional stability, is high enough (possibly up to 15% the rated tensile strength) to meet the requirements for non-tension stringing.

Expanded jumpers should be made of non-stressed conductors. Corona test results show that the jumper fully meets the engineering requirements.

Figure 8.22
Expanded jumper JLK/G1A-725(900)/40.

8.2.2 Vibration of UHV Overhead Lines

Motions of conductors and ground wires of a UHV overhead line mainly include aeolian vibration, galloping, and subspan oscillation. Aeolian vibration may cause fatigue and breaking of conductors and wear of conductors and fittings; galloping may cause wear of conductors and fittings, or even breaking and tower collapse; subspan oscillation may easily cause damage to spacer clamps and wear of conductors in contact with clamps. These motions are likely to cause electrical faults, thereby reducing the long-term operational reliability of the line.

8.2.2.1 Aeolian vibration

Aeolian vibration is caused by the alternate shedding of vortex (Karman vortex) on the leeward side of the conductor, has high frequency (3–120 Hz), and has amplitude no greater than the diameter of the conductor. It occurs almost all the time because it can be initiated once the wind speed is above 0.5 m/s. Despite its low amplitude and minor impacts on the static strength of the line, long-duration vibration may cause fatigue and damage of wires and related fittings, thereby jeopardizing the operational safety of the line.

The frequency of aeolian vibration is related to the diameter of subconductors and wind speed:

$$f = s\frac{v}{D} \tag{8.4}$$

where

f—frequency of aeolian vibration, Hz
s—Strouhal number, ranging from 0.185 to 0.210, and taken as 0.200 in China
v—wind speed perpendicular to the overhead line, m/s
D—diameter of subconductors, m.

1. **Aeolian vibration of conductors in ordinary lines**
Aeolian vibration in EHV lines is a serious problem, often causing wire breaking. To suppress aeolian vibration, both spacer dampers and vibration dampers are required. Compared with EHV lines, the attachment points of UHV lines are higher; hence, the wind energy and frequency of aeolian vibration are also greater. If large cross-section conductors are used, then the minimum frequency of aeolian vibration will be reduced because of a larger diameter, making vibration control more difficult because the self-damping of the conductor is small for low-frequency vibrations. However, compared with EHV lines, UHV line conductors consist of more subconductors and the spacers can more effectively control the movement of subconductors, thus reducing aeolian vibration.
China has performed many studies of suppressing aeolian vibration of UHV conductors. In ordinary lines of the UHV AC pilot and demonstration project, the eight-conductor bundle is used, the aeolian vibration of which is much lower than a single conductor under the same conditions and it is easy to control. Furthermore, the eight-conductor bundle spacer dampers are used and arranged in a proper span, effectively suppressing the aeolian vibration. In ±800-kV UHV DC lines, large cross-section conductors are used, which perform better in absorbing wind energy and thus reduce the frequency of aeolian vibration. Special studies of anti-vibration of large cross-section conductors are performed, and vibration dampers suitable for conductors of 900 or 1000 mm^2 are designed. The configuration rules of such vibration dampers are finalized through theoretical calculations and simulation tests, addressing the problem of vibration of large cross-section conductors.
2. **Aeolian vibration of ground wires**
In contrast to EHV lines, UHV lines have higher attachment points of ground wires and greater spans, making aeolian vibration more prominent and control of vibration more difficult.
The use of optical fiber composite overhead ground wire (OPGW) makes control of vibration even more complicated. Through studies and experiments, it has been found that vibration of OPGW can be solved by using vibration dampers of preformed clamps and vibration dampers of bolt clamp.
In areas exposed to strong winds, aeolian vibration occurs more frequently and lasts longer. For example, in northwest China, aeolian vibration often causes breaking of ground wires. Therefore, special attention should be given to anti-vibration of ground wires in such areas. The Haminan–Zhengzhou line runs through Xinjiang, Gansu, Ningxia, Shaanxi, Shanxi, and Henan. Among them, Xinjiang, Gansu, and parts of

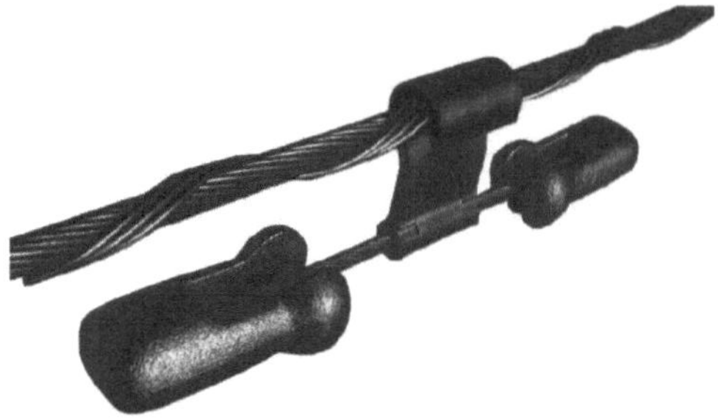

Figure 8.23
Vibration dampers of preformed clamp with four resonance frequencies.

Ningxia are areas exposed to long-term, steady, strong winds, and thus aeolian vibration in these areas is long-lasting and has high amplitude and a wide range of frequencies. To ensure safe and stable operation of the line, the anti-vibration measures must cover all vibration frequencies. To this end, vibration dampers of a preformed clamp with four resonance frequencies, as shown in Figure 8.23, are designed. As the name implies, the vibration damper has four resonance frequencies, with the highest being up to approximately 110 Hz, and can cover all aeolian vibration frequencies in windy areas, thereby effectively mitigating vibrations of ground wires.

3. **Aeolian vibration of large crossing lines**
   China has a vast territory and complicated geographical conditions, with many criss-crossed, large rivers. Therefore, it is inevitable for a UHV line to cross large spans; some of them are even more than 2000 m. Such a large crossing generally suffers more severe aeolian vibration due to higher attachment point, larger span, and greater wind speed, as well as the geographical conditions conducive to the formation of regular laminar flow. Also, any accident occurring on a large crossing line will result in heavy losses and repair will be difficult. Therefore, attention must be given to the anti-vibration of large crossing lines.
   a. *Anti-vibration scheme for conductors and ground wires.* Bate damping wire and vibration damper are generally used together. Considering the frequency response ranges of damping wire laces of different lengths, damping wire laces of various lengths are used in combination to ensure effective energy dissipation and vibration suppression within the entire frequency range of aeolian vibration and to provide focused protection against vibrations within the major frequency range. The smaller lace side (the right side in Figure 8.24) can be stripped as required, which can reduce the weight of the lace and the dynamic bending strain of the conductor in contact with the clamp for the lace, and it change the frequency response range of the lace and improve the frequency response characteristics of the anti-vibration scheme. In addition, vibration dampers are installed on the

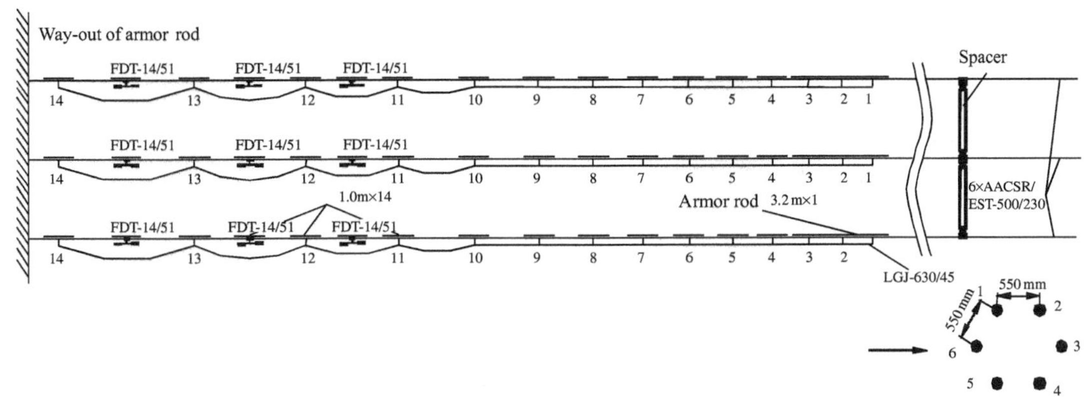

Figure 8.24
Schematic diagram of the aeolian vibration prevention scheme for a UHV large crossing line.

larger lace side (the left side in Figure 8.24) to better prevent low-frequency vibrations and achieve an optimum anti-vibration scheme by combining vibration dampers and damping wires. The anti-vibration scheme for a UHV large crossing line must be determined by combining theoretical calculation and laboratory simulation tests.

b. *Tests and studies of anti-vibration of conductors and ground wires.* The design and tests of an anti-vibration scheme are based on energy balance (i.e., the wind energy absorbed by the conductor is equal to the sum of the energy dissipated by the conductor due to self-damping and that dissipated by the anti-vibration devices). The purpose of tests is to optimize the anti-vibration scheme by simulating aeolian vibration to ensure the scheme meets related technical requirements and is fully functional in practical use. First, the self-damping of the conductor should be tested to understand its self-damping characteristics and calculate its vibration level in the absence of anti-vibration measures. Only on this basis can an appropriate anti-vibration scheme be developed. The anti-vibration scheme is expected to cover aeolian vibrations at all frequencies, particularly vibrations within the major frequency range. Then, the scheme should be tested with a single conductor to understand the vibration levels of the conductor at various resonance frequencies. The scheme is then optimized based on the test results to ensure that it meets related technical requirements. Because a spacer damper can suppress the vibration of subconductors to a certain extent, the aforementioned anti-vibration scheme should also be tested with a conductor bundle to verify its effects for the spacer-bundle conductor system. If the scheme meets related technical requirements, then it could be recommended for engineering use. The design, test procedure, and principles of the anti-vibration scheme for ground wires and OPGWs are basically the same as those for conductors.

Table 8.10: UHV Large-Crossing Lines for Which Anti-Vibration Design Has Been Completed.

No.	Project Name	Crossing Span	Configuration of Conductor
1	Large crossings over the Hanjiang River in Yanshantou and the Yellow River in Xihuagong of the UHV AC pilot and demonstration project	706 m–1650 m–600 m 450 m–1220 m–995 m–986 m	6 × AACSR/ EST-500/230
2	Large crossings over the Huai River and Yangtze River of the Huainan–Shanghai transmission project	520 m–1300 m–620 m 710 m–1817 m–653 m	6 × AACSR/ EST-640/290
3	Large crossings over the Yangtze River respectively in Yangjiachang, Hujiatan, Zhayinggang, and Xinjiyang of the Xiangjiaba–Shanghai project	500 m–1580 m–500 m 840 m–1705 m–856 m 509 m–1733 m–476 m 518 m–2052 m–415 m	4 × AACSR/ EST-640/290
4	Large crossings over the Yangtze River respectively in Yangjiachang, Hujiatan, Zhayinggang, and Xinjiyang of the Jinping–Sunan project	500 m–1610 m–500 m 632 m–1719 m–773 m 496 m–1639 m–635 m 454 m–1931 m–540 m	4 × AACSR/ EST-720/320
5	Large crossing over the Yellow River of the Haminan–Zhengzhou project	450 m–1200 m–1350 m–900 m	4 × AACSR/ EST-900/240

CEPRI's aeolian vibration test laboratory is equipped with a test system with a load-bearing capacity of 1280 kN and 192 test channels, sufficient for tests and studies of anti-vibration of an eight-conductor bundle. Table 8.10 outlines the UHV AC/DC large crossing lines for which anti-vibration design and tests have been completed in the laboratory.

8.2.2.2 Conductor galloping

1. **Anti-galloping studies around the world**
   Studies of conductor galloping date back to the 1930s. The most outstanding results are the vertical galloping mechanism, developed by Den Hartog in the United States, and torsional galloping mechanism, developed by O. Nigel in Canada [7,8].
   In 2007, the CIGRE issued a study report titled *Current Status of Studies on Conductor Galloping*, detailing the worldwide accomplishments made in the field over the past 20 years. According to the report, although great accomplishments have been made in anti-galloping theories, tests, and applications through years of efforts, the galloping mechanism of iced conductors remains unsolved. As such, further studies of the galloping inducing the mechanism and related parameters are required.
2. **Advancements of anti-galloping technologies in China**
   Systematic studies of conductor galloping started in the late 1980s in China. By comprehensively considering the coupling of the horizontal, vertical, and torsional components of winds, China established a dynamic stability mechanism of conductor galloping and set-up rational criteria for assessing the stability of conductor bundles

Table 8.11: Principle for Determining Galloping Grade.

Galloping Grade	Number of Days with Galloping in 10 Years
Grade III (strong)	≥140
Grade II (moderate)	(90, 140)
Grade I (weak)	(50, 90)
Grade 0 (no galloping)	(0, 50)

with anti-galloping devices under a given icing and wind condition. This mechanism was developed based on the vertical galloping mechanism by Den Hartog and torsional galloping mechanism by O. Nigel. Based on these mechanisms, a new double-pendulum anti-galloping device was invented that can improve the dynamic stability of conductors and thus suppress conductor galloping. This device has been proven successful in controlling conductor galloping.

Since the twenty-first century, conductor galloping has become increasingly severe in China. The CEPRI performed studies of anti-galloping technologies in collaboration with power companies and universities, and made great achievements. However, the non-linear bifurcation characteristics of conductor galloping were obtained, and the idea that galloping is a result of non-linear instability under combined effects of multiple parameters was introduced. However, an aerodynamic database of eccentrically iced conductors was established, and lines for full-scale galloping tests were set-up. Table 8.11 shows the principle for determining the galloping class based on the meteorological conditions. Specifically, the meteorological data of 10 consecutive years are analyzed to calculate the number of days when meteorology-caused conductor galloping occur, and then the calculation results should be corrected against the geographical conditions and operation experience to determine the galloping grade and draw the galloping grade distribution map.

For a common 330 ~ 750 kV single-circuit line, loose spacer damper, double-pendulum anti-galloping device, or corresponding combined anti-galloping devices are recommended; alternatively, detuning pendulum and eccentric weight may be used. For common lines with double or multiple circuits erected on the same tower, loose spacer damper, inter-phase spacers, or corresponding combined anti-galloping devices are recommended; alternatively, a double-pendulum anti-galloping device, detuning pendulum, and eccentric weight may be used. Different circuits may be provided with different anti-galloping devices. Inter-phase spacers are recommended for compact transmission lines.

When the phase conductors of a 110(66) ~ 220 kV line are in vertical or triangular configuration, inter-phase spacers are recommended, or a loose spacer damper, double-pendulum anti-galloping device, corresponding combined anti-galloping devices, detuning pendulum, and eccentric weight may be used. If the phase conductors are in a horizontal configuration, then loose spacer damper, double-pendulum anti-galloping device, corresponding combined anti-galloping devices, detuning pendulum, and eccentric weight may also be used.

3. **Anti-galloping technologies for UHV lines**

In the presence of an ice coating, a conductor bundle is more likely to experience galloping than a single conductor. And the larger the span and the higher the attachment point, the more severe the galloping will be. With many subconductors, large cross-section, high-attachment point, and large span, a UHV line will experience galloping under given conditions, sometimes even more severely than an EHV line.

The characteristics and triggering conditions of galloping of a UHV line are substantially the same as those of transmission lines at other voltage classes [9].

a. *Characteristics*: The envelope is in an oval form, and the highest amplitude occurs in the vertical plain.
b. *Triggering conditions*: Wind speed of 6–25 m/s, wind direction is substantially perpendicular to the line; icing thickness of 3–15 mm, temperature of 0–6°C; open and plains lands or rivers.

Conductor galloping occurs frequently in China. Figure 8.25 shows the geographical distribution pattern of galloping. It can be seen that galloping mainly occurs in mid and

Figure 8.25
Provinces with frequent galloping.

east China, particularly Hubei, Henan, and Liaoning provinces. Attention should be given to anti-galloping of new UHV lines running through these areas, and anti-galloping measures must be in place if necessary.

Considering the structural and electrical characteristics and anti-galloping requirements of UHV lines, and based on the theory that galloping can be suppressed by changing the icing shape, loose spacer dampers (as shown in Figure 8.26) have been invented. Because some clamps of such a spacer can rotate freely or within a given angle, the icing shape can be changed, thus preventing the occurrence of galloping. Application proves that it functions well, and it has become an important means for anti-galloping of UHV lines. In areas where transmission lines suffer severe galloping, two anti-galloping devices of different mechanisms, for example, a loose spacer damper and double-pendulum anti-galloping device (as shown in Figure 8.27), can be used together to achieve better anti-galloping effects.

8.2.2.3 Subspan oscillation

1. **Principles**
   Subspan oscillation of an overhead line is the oscillation of subconductors on the leeward side induced by the wake flow of subconductors on the windward side. It is peculiar to conductor bundles. The frequency generally ranges from 1 to 3 Hz, and the amplitude may be several times larger than the conductor diameter, greater than that of aeolian vibration, and smaller than that of galloping. Subspan oscillation may cause mechanical damage to spacers, conductors, and suspension fittings, thereby reducing operational safety of the line.

   The amplitude of subspan oscillation is mainly determined by the ratio of the spacing between subconductors S to the diameter of subconductor d, and the azimuth of

Figure 8.26
Loose spacer damper.

Figure 8.27
A combination of a loose spacer damper and double-pendulum anti-galloping device.

subconductors in relation to the wind direction, as shown in Figure 8.28. Table 8.12 gives the values of S/d for UHV lines recommended in various countries. It is generally believed that $S/d > 16$ for a two-conductor or three-conductor bundle and $S/d > 20$ or an azimuth $> 15°$ for a four-conductor bundle will reduce the probability of occurrence of subspan oscillation. However, this is impractical for UHV lines, because the conductor is generally composed of six to eight subconductors, the electrical requirements are higher, and S/d generally ranges from 10 to 17. Therefore, subspan oscillation of UHV lines deserves particular attention.

2. **Related studies**

   With the development of UHV transmission in China, there are more and more lines running through areas exposed to strong winds. To understand the subspan oscillation of conductor bundles, tunnel tests are performed, and the aerodynamic parameters of the subconductors on the leeward side under the effects of wake flow are obtained and spacings between subconductors are determined based on practical experience. Also, before construction of a UHV project, the effects of conductor type, span, microtopography, and micrometeorology on subspan oscillation are fully considered. Spacer dampers are widely used, and their arrangement is optimized, particularly in

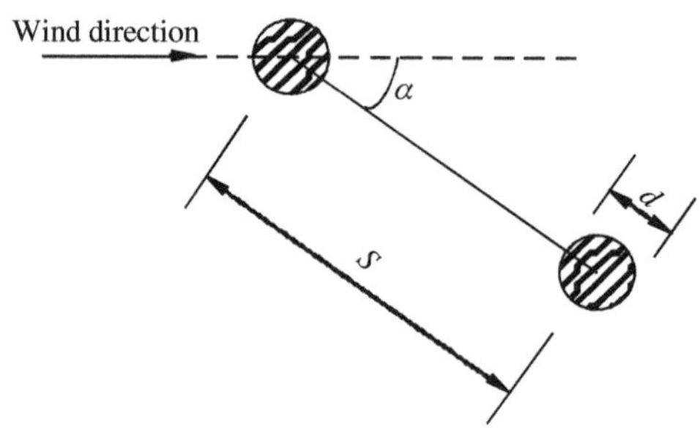

Figure 8.28
Schematic diagram of subconductor configuration.

Table 8.12: Values of *S/d* for UHV Lines Recommended in Various Countries.

Country	The Soviet Union	Japan	The United States	Italy	The United Kingdom	China	
Recommended by	The Ministry of Power and Electrification	Tokyo Electric Power Company	BPA	ENEL	CEGB	SGCC	
Line voltage (kV)	1150	1000	1100	1000	1300	±800	1000
Number of subconductors	8/10	8	8	8	8	6	8
S/d	16.7/15.9	10.4/11.7	10	14.3	12	11.1–11.7	16.7

Note: Two values of S/d are recommended for some lines considering the difference in the number of subconductors and outer diameter of conductor.

areas where subspan oscillation occurs more frequently, to mitigate subspan oscillation. The principles governing arrangement of spacers are summarized as follows:

a. *Maximum subspan.* This is mainly dictated by the maximum allowable amplitude of the subconductors.

   Under the effects of wind and electromagnetic field, subconductors will approach each other, and when the current is increased to a certain value, they will be suddenly absorbed by each other and cannot detach. For this reason, spacers must be arranged with a span that can limit the spacing between subconductors within an allowable range. The maximum subspan should be determined to ensure that the design current is smaller than the extreme current for the subconductors to detach.

   Figure 8.29 shows a schematic diagram of subspan oscillation. The amplitude of subspan oscillation varies as a function of wind speed, subspan, tension of the subconductors, mass per unit length of subconductors, spacing between subconductors, and diameter of subconductor. Therefore, the amplitude can be reduced by properly determining the subspan. In addition, the capability of the conductor bundles in recovering from torsion under heavy wind and icing conditions should also be considered in determining the maximum subspan.

b. *End subspan.* This is determined mainly by considering the capability of the conductor bundles in recovering from torsion and with reference to the empirical formula.

c. *Average subspan.* The spacers should function well to suppress the subspan oscillation, and they should be arranged far away from wave nodes and at an average spacing close to the half-wave length.

d. *Middle subspan.* The middle subspan should meet the aforementioned requirements and prevent propagation of oscillation to neighboring spans via spacers. The capability of the conductor bundles in recovering from torsion should be considered as well. The middle subspan should be determined following the principles given here:

   i. The length of neighboring subspans should be different from each other.
   ii. The overall amplitude of oscillations should be minimized under various resonance conditions after the end subspan is determined.
   iii. The ratio of a subspan to its neighboring subspan should be maximized to obtain a maximum tuning factor, provided that the aforementioned requirements 1 and 2 are satisfied.

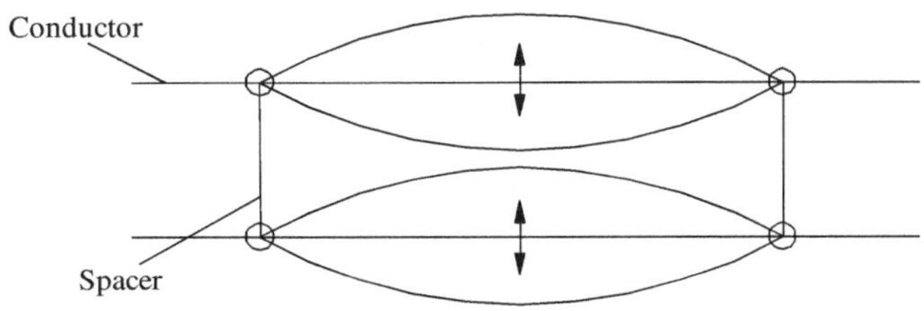

Figure 8.29
Schematic diagram of subspan oscillation.

Table 8.13: Recommended Values of Subspans for Typical Transmission Lines Subject to Strong Winds in China.

Item	±660 kV	±800 kV	
	4×JL/G3A-1000/45-72/7	6×JL/G3A-1000/45-72/7	6×JL/G2A-1000/80-84/19[a]
Spacing between subconductors	500	500	500
Maximum subspan (m)	72	55	60
Average subspan (m)	62	50	55
End subspan (m)	29–35	25–30	27–33

[a]Arrangement of subspans in Xinjiang. For arrangement in other areas, refer to the common configuration rules.

Table 8.13 gives the recommended values of subspans for typical transmission lines subject to strong winds in China.

8.3 Insulators

8.3.1 Insulators for UHV AC Overhead Transmission Lines

8.3.1.1 Porcelain and glass cap and pin insulators

1. **Technical data**
   In UHV AC transmission lines, 300, 420, and 550 kN porcelain and glass cap and pin insulators are used. For their technical data, see Chapter 7.
2. **Design and manufacture**
   Abundant experience has been accumulated with 300- and 420-kN common porcelain and glass cap and pin insulators, whereas less experience is available for insulators with a large disc diameter and creepage distance, and even less for 550-kN insulators. It is essential to ensure the electrical and mechanical stability of insulators in bulk production. The following summarizes the key technologies in design and manufacture of insulators:
   i. *Design of thickness of the head.* The internal insulation performance of an insulator mainly depends on its head structure. The breakdown withstand capability of the head is related not only to the formula, manufacture process, and techniques of the insulating part but also to the thickness of the head vulnerable to breakdown and optimization of electric field distribution over the head. As stipulated in GB/T 8411 *Ceramic and Glass Insulating Materials*, the electrical strength of porcelain should be no lower than 20 kV/mm. The electrical strength of high-aluminum porcelain insulating parts of a UHV insulator should be no less than 30 kV/mm. To improve the internal insulation strength of insulators used in 1000-kV UHV projects, the thickness of the head is increased.

Table 8.14: Technical Data of 1000-kV Composite Insulators.

No.	Item	Technical Data
1	Rated operating voltage (kV)	1000
2	Rated 1 min withstand tensile load (kN)	210, 300, 420, and 550
3	1 min wet power frequency withstand voltage (rms, kV)	≥990
4	Dry lightning impulse withstand voltage (peak value, kV)	≥3200
5	Wet switching impulse withstand voltage (peak value, kV)	≥1675
6	Visual corona voltage (kV)	≥700
7	Height (mm)	9750/10,530
8	Dry arcing distance (mm)	>9000
9	Creepage distance (mm)	>32,000
10	Markings of fittings for assembly	20, 24, 28, 32

Note: These parameters apply only for areas at altitudes below 1000 m and with a pollution class of B, C, D, or E. For altitudes above 1000 m, corrections should be made according to GB 311.1—2012 *Insulation Co-ordination—Part 1: Definitions, Principles and Rules.*

ii. *Control of shed shape.* Porcelain insulators used in UHV transmission projects are either double-shed or tri-shed. To achieve high anti-pollution performance and self-cleaning capability, characteristic parameters such as the parallelism between sheds, inclination of the upper shed, and smoothness of shed surface must be controlled.

iii. *Fine processing of raw materials.* Fine processing and strict control of raw materials is a prerequisite for appearance and electrical and mechanical performance. Incoming materials are strictly controlled regarding their chemical composition, plasticity number, and hydration rate of plastic clay, hydration rate of porcelain clay, residue on sieve, and true specific gravity of flux material.

iv. *Routine test.* For porcelain insulators, high-frequency and power-frequency discharge tests should be performed, whereas for toughened glass insulators, cold shock, heat shock, and uniformity treatment are conducted three times throughout the manufacturing process to remove insulators with internal insulation defects such as improper toughening or impurities in the head. The defective percentage during the routine test is no higher than 0.5%.

v. *Steep-front impulse voltage test.* The test voltage amplitude is $2.9^{0.07}_{-0.017}$ times the 50% negative lightning impulse discharge voltage $-U_{50\%}$ at maximum and 2.85 times $-U_{50\%}$ at minimum.

8.3.1.2 Rod suspension composite insulators

1. **Technical data**

   Composite insulators of 210–550 kN are used in UHV AC lines. Table 8.14 summarizes their technical data, and Figure 8.30 gives a picture of composite insulators used in the UHV AC pilot and demonstration project in China.

Figure 8.30
Composite insulators used in China's UHV AC pilot and demonstration project.

2. **Design and manufacture**
The following summarizes the key technologies involved in design and manufacture of composite insulators intended for use in UHV AC lines to ensure operational safety and reliability of UHV transmission projects:
   a. *Eccentricity during injection.* A 1000-kV composite insulator is generally 9750 mm or longer, much longer than a 500-kV insulator. This will easily result in eccentricity during injection and, thus, non-uniform thickness of the housing for the core. This problem can be solved by the following means:
      i. Design of an automatic eccentricity location device
      ii. Proper setting of process parameters, including the injection pressure, rate, and volume
      iii. Increase in the design thickness of the housing; the minimum thickness should be no less than 6 mm
      iv. Eccentricity check with the acoustic emission thickness gauge immediately after injection of each half-finished section
   b. *Shed mounting.* Compared with a 500-kV composite insulator, a 1000-kV AC composite insulator has a longer core, a larger shed diameter, and more sheds.

      By improving the manufacturing equipment and processes, insulator manufacturers have successfully developed a semi-automatic shed-mounting device, which can be used for manufacturing 12-m-long composite insulators. In addition, a new high-temperature vulcanization process is used, during which the extruded housing for the core is first vulcanized at high temperatures instantaneously for setting and then vulcanized at high temperatures, with more supports provided in the vulcanization can.

Table 8.15: Technical Data of Silicon Rubber Insulation Materials.

Item	Unit	Specified Value in DL/T 376—2010	Required Value for UHV Insulator
Tear strength (right-angle method)	kN/m^2	≥10	≥12
Ultimate tensile strength	MPa	≥4	≥4.0
Elongation at break	%	≥150	≥200

c. *Design optimization of grading devices.* With the increase in voltage class, distortion of electric field distribution of the insulator on the conductor side and earthing side becomes more severe, especially for UHV insulators. To solve this problem, the three-dimension simulation calculation technology is used for optimizing the dimensions of the grading device used for composite insulators, such as the structural type, installation location and diameter, pipe diameter of grading ring, radio interference tests, and corona tests, and potential distribution tests are performed, verifying that the grading devices intended for use for various tower types and various types of insulator strings comply with the design requirements of UHV projects.

d. *Technical data of silicon rubber.* Compared with EHV or lower-voltage insulators, UHV AC insulators place a higher requirement on silicon rubber. Table 8.15 gives the technical data of silicon rubber. It can be seen from the table that the mechanical data of the UHV insulator are higher than required in DL/T 376—2010 *General Technical Requirements of Silicon Rubber Insulation Materials for Composite Insulators.*

e. *120% rated mechanical loads 24-h withstand test.* This test is added for 1000-kV composite insulators. During the test, 1.2-times rated tensile loads are imposed on regular products. They are considered qualified if not damaged within 24 h. The purpose of this test is to assure a reliable mechanical strength of insulators for a long time, that is, the tensile failing load in the service life of 40 years is no less than the rated load.

f. *Water diffusion test on the rod.* The test is intended to verify the compactness and hydroscopicity of the rod made of epoxy and glass fiber and to check whether the interface between the housing and rod is free from defects. During the test, the leakage current for a 1000-kV composite insulator should be no more than 0.1 mA.

8.3.2 Insulators Used for UHV DC Overhead Transmission Lines

The design of a DC insulator is different from an AC insulator in two aspects: (i) under the same pollution condition, the pollution flashover voltage of the former is 15–30% lower than that of the latter; and (ii) due to effects of the unidirectional electric field, the dust accumulation on the surface of a DC insulator is, on average, twice that on the surface of an AC insulator. This places even higher requirements on the configuration of insulators in DC lines.

8.3.2.1 Porcelain and glass cap and pin insulators

1. **Technical data**
   Figure 8.31 shows a picture of porcelain insulators used in UHV DC lines. Table 8.16 shows the technical data of cap and pin insulators used in UHV DC lines. At present, the maximum mechanical strength of widely used insulators reaches 550 kN. In 2012, China developed heavy porcelain and glass cap and pin insulators with a mechanical strength of 760 kN, which have been certified and used in trials of ±800-kV UHV DC projects. Using heavy insulators in tension strings can significantly reduce the number of insulators, simplify the types of fittings and attachment points on towers, improve reliability of the transmission line, and effectively reduce the construction costs of the line with the wide use of such insulators.
2. **Design and manufacture**
   The design and manufacture of a DC cap and pin insulator differ greatly from that of an AC insulator. Under the effects of the DC electric field, ions always move in a specific direction on the head of the insulator, which will reduce the performance and service life of the insulator. As such, the contents of kalium ions and sodium ions in the porcelain and glass are reduced, whereas the contents of calcium ions and magnesium ions are increased based on the neutralization and suppression principles to achieve a volume resistivity of the order of 10^{10} Ω, suppress the movement of ions under the DC electric field, and improve the anti-aging performance of the insulator.
   For cap and pin insulators used in UHV DC transmission, the cap is provided with a zinc collar and the pin is provided with a zinc coating to form a sacrificial anode to protect

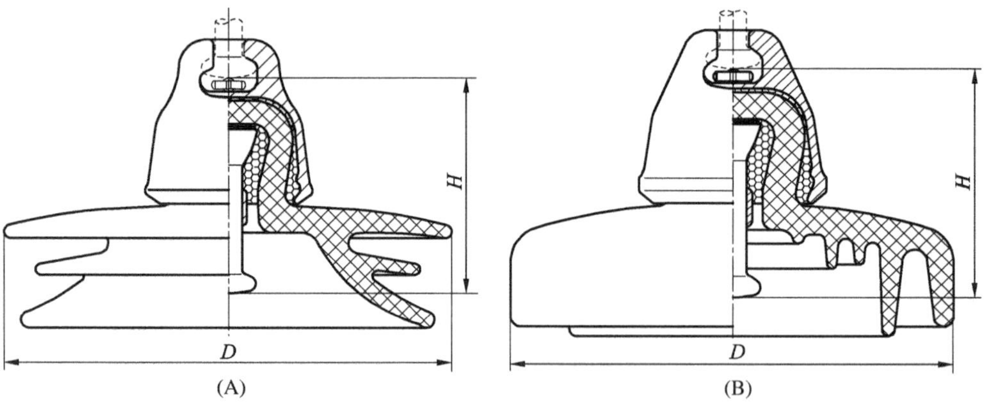

Figure 8.31
Schematic diagram of porcelain insulators used in a UHV DC line.
(A) DC porcelain tri-shed insulator and (B) DC porcelain bell insulator.

Table 8.16: Technical Data of Cap and Pin Insulators Used in a UHV DC Line.

Type	Model	Rated Electromechanical or Mechanical Loads (kN)	Height (mm)	Disc Diameter (mm)	Creepage Distance (mm)
Bell type	U210B/170H	210	170	320	560
	U300B/195H	300	195	400	635
	U420B/205H	420	205	340	560
	U550B/240H	550	240	380	635
Tri-shed type	U210B/170T	210	170	340	560
	U300B/195T	300	195	400	635
	U420B/205T	420	205	400	635
	U550B/240T	550	240	400	635

the fitting from electrolytic corrosion caused by DC leakage current. Both the coating and collar are made of zinc with a purity no less than 99.95%. No gap or loosening between the zinc coating and steel pin and between the zinc collar and cap is allowed. The quality and exposed dimensions of the zinc coating and zinc collar should meet environmental requirements. For a UHV DC project, the exposed part of the zinc coating for the steel pin in contact with cement should have a minimum mass of 10 g (for insulators of 300-kN or less) or 20 g (for insulators of 400 kN or more), and a length of 50% of the total. The fused area of the steel pin and zinc coating should be no less than 80% of the contact surface.

8.3.2.2 Rod suspension composite insulators

1. **Technical data**
   China is leading worldwide in design and manufacture of composite insulators, which exhibit an excellent anti-pollution performance in several regional and large-scale pollution flashover incidents. By now, composite insulators with high-temperature vulcanized (HTV) silicon rubber have been widely used in suspension strings of new UHV DC lines. Table 8.17 gives the technical data of composite insulators used in ±800 kV UHV DC lines.
2. **Design and manufacture**
   a. *Anti-drop design of V-string composite insulators.* For a V-string, the coupling of composite insulators is different from that of cap and pin insulators. Specifically, because a composite insulator has no hitch point its rigidness is very likely to show, resulting in compression on the R-pin by the ball where the ball and socket are jointed. In this case, the R-pin will be deformed and can no longer prevent fall-off of the ball and the insulators.

      Providing a hold hoop outside the socket can prevent the fall-off but cannot prevent compression on and deformation of the R-pin. The risks of excessive non-axial stress and local stress still exist, which may lead to breakage of the ball due to fatigue. But this measure is feasible as a remedy for existing lines.

Table 8.17: Technical Data of Composite Insulators Used in ±800-kV UHV DC Lines.

No.	Item	Technical Data
1	Rated operating voltage (kV)	±800
2	Rated mechanical failing load (kN)	160/240/300/420/550
3	1 min wet power frequency withstand voltage (kV)	≥900
4	Full-wave lightning impulse withstand voltage (kV)	≥3600
5	Wet switching impulse withstand voltage (kV)	≥1950
6	Visible corona voltage (kV)	No visible corona under normal operation
7	Height (mm)	10,600/11,500/12,000
8	Insulation distance (mm)	10,165/11,065/11,565
9	Creepage distance (mm)	36,300/39,600/41,400
10	Markings of fittings for assembly	20/20/24/28/32

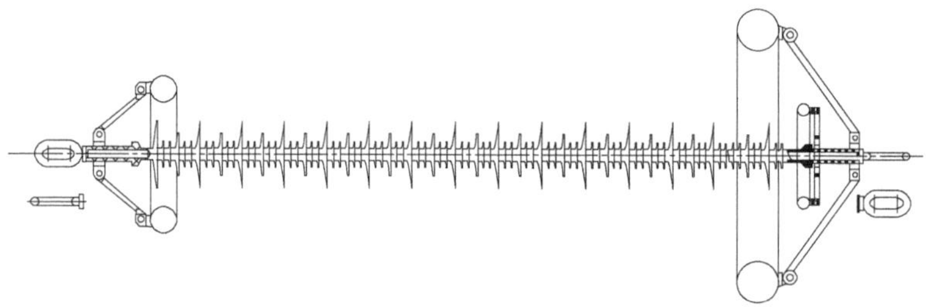

Figure 8.32
Composite insulators in ring–ring connection.

In the Xiangjiaba–Shanghai transmission project, composite insulators are in a ring–ring connection (as shown in Figure 8.32) instead of a ball and socket connection for the first time. This completely solves the problem of fall-off of V-string composite insulators from conductors. This connection scheme does not require any additional connection sections, so the leg length and tower head size remain unchanged.

b. *Configuration of zinc collar for fittings for anti-corrosion purpose*. Fittings on both ends of a composite insulator should be configured with zinc collars. The zinc collars must be in full contact with fittings, should not be jointed with bolts without gaps or bubbles, and should not be completely isolated from the air. The zinc collars should be made of zinc with a purity of 99.95% and cast onto the fittings. They should have a minimum thickness of 5 mm and a minimum length of 5 mm along the axis of the insulator. The fused area of the zinc collar should be >80% of the contact surface and is dim and non-reflective. If the fittings are high-temperature sealed, then the exposed part should be at least 3 mm long.

Figure 8.33
Use of tension composite insulators in the Jinping–Sunan transmission project.

3. **Application of heavy composite insulators in tension strings**
The reliability of mechanical performance and safety factors, other than the electrical performance, are key parameters of composite insulators for use in tension strings in transmission lines. To ensure reliability of rod suspension composite insulators when used in a UHV tension string, mechanical vibration tests should be performed by simulating its operation conditions by applying uniform tensile loads on all subconductors along their respective axis. According to the test requirements, the insulators undergoing tests must have all fittings required in place, including the grading ring and yoke plate, and the test span should be >40 m. The vibration angle, frequency, and time should be 30′, 20–40 Hz, and no less than 3×10^7, respectively. After the vibration test, water boiling test, appearance check, steep-front impulse voltage test, and dry power frequency voltage test should be performed in sequence on 50% of the insulators (if the insulators are in an odd number, round down to the nearest integer), and 120% rated mechanical load 24-h withstand test (or mechanical failing load test) should be performed on the other insulators.
Figure 8.33 shows a picture of tension composite insulators used in the Jinping–Sunan transmission project.
4. **UHV segmented composite insulators**
UHV DC lines operate at a high voltage class and cover a long distance, and thus a large number of composite insulators are required, which often have a length of more than 10 m. Transport of such a long insulator is difficult in areas with complicated geographical conditions. Moreover, the long core is likely to deform, which makes it inconvenient for transfer of such a composite insulator on the production line and may result in defects or eccentricity during extrusion of the housing and injection, thus

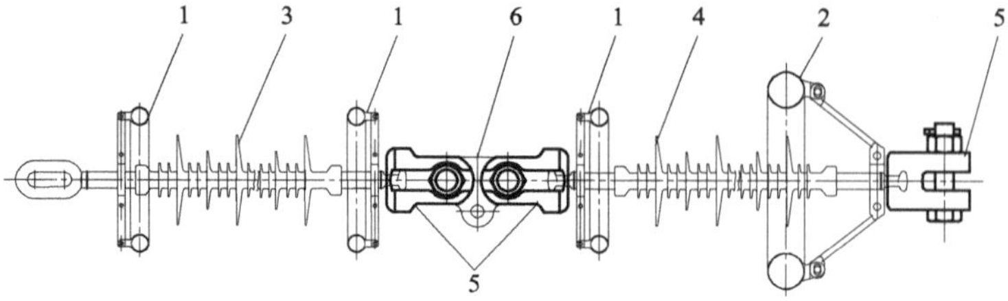

Figure 8.34
Schematic diagram of coupling of segmented composite insulators and accessory fittings (the left side is the HV end).
1—Small grading ring; 2—grading ring on the earthing side; 3—lower part of the composite insulator; 4—upper part of composite insulator; 5—enclosed socket; and 6—eccentric plate.

causing a high failure rate. As such, segmented composite insulators are used in ±800-kV UHV DC projects in mountainous areas with poor access. Segmented composite insulators have a symmetrical structure and are connected by enclosed sockets and U-ring at the HV end, whereas the total length of the segmented composite insulators remains unchanged. Figure 8.34 shows the connection of a segmented composite insulator and associated fittings.

China's mature crimping technology makes it possible to control the dispersion of crimp strength at ends of composite insulators. Similar to common composite insulators, end fittings of segmented composite insulators are connected by means of crimping. When sufficient crimp strength is ensured at the ends, only the mechanical characteristics of intermediate fittings, including the enclosed ball socket and eccentric plate, need to be verified. Through mechanical strength tests, it is found that all enclosed ball sockets of the rated mechanical strength remain in good condition when subjected to 1.2-times rated mechanical loads and should comply with requirements set forth in GB/T 2314—2008 *General Technical Requirements for Electric Power Fittings* and GB/T2317.1—2008 *Test Method for Electric Power Fittings—Part 1: Mechanical Tests.*

8.4 Fittings

8.4.1 Spacer

Spacer dampers are commonly used in UHV lines and are intended to keep a proper spacing between subconductors in a bundle and to mitigate damage to conductors due to aeolian vibrations and subspan oscillations by means of their damping characteristics.

In the 1000-kV UHV AC pilot and demonstration project [10] and the Huainan–Shanghai transmission project, both of which are in operation, spacer dampers for eight-conductor bundles are used in ordinary lines to keep a spacing of 400 mm between subconductors.

(A) (B) (C)

Figure 8.35
Spacer dampers used in UHV transmission lines in China.
(A) Spacer for eight-conductor bundles in a 1000-kV ordinary line; (B) spacer for six-conductor bundles in a 1000-kV large crossing line; and (C) spacer for six-conductor bundles in a ±800-kV ordinary line.

In the existing Xiangjiaba–Shanghai project and Jinping–Sunan project, and in the Haminan–Zhengzhou and Xiluodu–Zhejiang project under construction, spacer dampers for six-conductor bundles are used in ordinary lines to keep a spacing of 450 or 500 mm between subconductors in a bundle. The body of spacer dampers used in UHV AC/DC lines is in a dual-plate form. In areas where galloping occurs frequently, anti-galloping loose spacer dampers are used to change icing shapes to suppress galloping. Figure 8.35 shows a picture of spacer dampers used in UHV lines in China.

8.4.2 Suspension Fittings

Suspension fittings, mainly suspension clamps, serve to attach conductors to insulator strings on a suspension tower. A suspension clamp provides support to a conductor and sustains all loads applied by it. As a result, selecting a proper type of suspension clamp is crucial to operational safety of the line. In the case of a UHV line, both the cross-section of the conductor and the span are larger, and thus the suspension clamp is subject to higher loads. This makes selection and design of suspension clamps much more difficult.

Suspension strings in UHV AC lines in China are configured in a V or an I form, as shown in Figure 8.36. Suspension yoke plates of common type and of cross type are available, as shown in Figure 8.37. Suspension strings in UHV DC lines are mostly configured in a V form, as shown in Figure 8.38.

Anti-corona suspension clamps are used in UHV AC and in UHV DC lines in China, which can effectively prevent corona. In this case, no shielding ring is required when a suspension

Figure 8.36
Suspension strings in a UHV AC line.

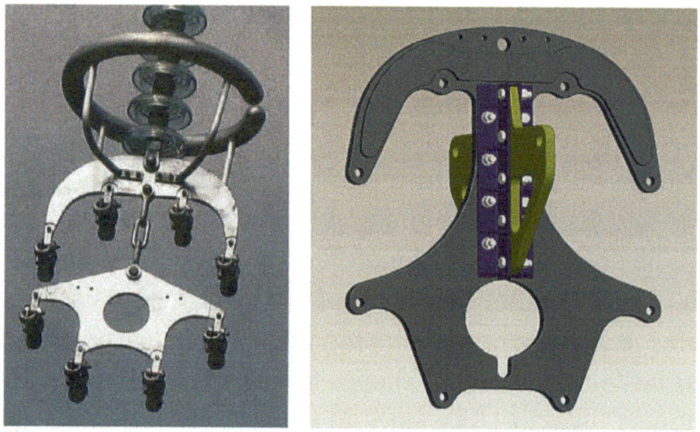

(A) (B)

Figure 8.37
Suspension yoke plates for suspension strings in a UHV AC line.
(A) Common suspension yoke plate and (B) cross-shaped suspension yoke plate.

Figure 8.38
Suspension strings in a UHV DC line.

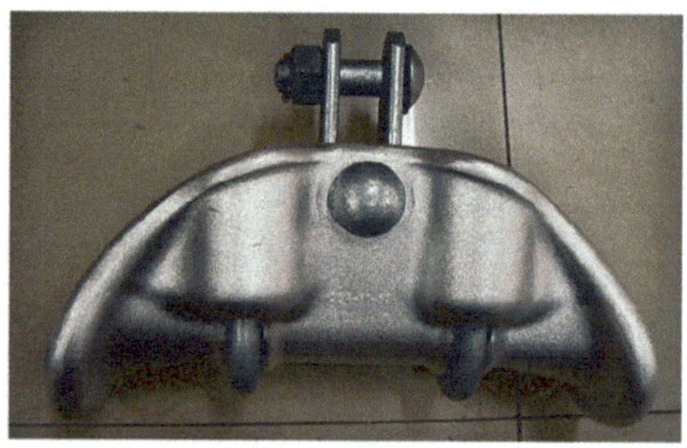

Figure 8.39
Handbag-type suspension clamp intended for UHV AC line conductors.

yoke plate is installed. Most suspension clamps are in a handbag form, as shown in Figure 8.39. In some cases, armor grip suspension (AGS) is applied as well.

8.4.3 Tension Fittings

Tension fittings serve to attach conductors to tension insulator strings on a tension tower. Installed at ends of a conductor in tension, tension fittings are subject to all tensile loads applied by the conductor.

In UHV AC and DC transmission lines in China, a tension set consists of two, three, or four strings. Figure 8.40 shows a double-string tension set of a UHV AC line and Figure 8.41 shows a three-string tension set of a UHV DC line.

8.4.4 Shielding Ring and Grading Ring

For a UHV line, the noises, radio interference, and voltage distribution non-uniformity across an insulator string resulting from corona of fittings are more severe due to a higher voltage class. Special attention should be given to these problems in the design of UHV lines.

In UHV transmission lines in China, the corona is suppressed mainly by installing shielding rings and making the surface of fittings arc-shaped [11,12]. For tension strings, both side shielding ring and circular shielding ring are available, and because a circular shielding ring has an opening, it can be installed last. For suspension strings, no shielding rings are provided yet. Both suspension strings and tension strings are provided with grading rings. Grading rings for I-shape suspension strings all have an opening, whereas those for V-shape suspension strings are either circular type or runway type. For tension strings, either a runway-type grading ring with an opening or two side rings are used. Figure 8.42 shows a

Figure 8.40
Double-string tension set of a UHV AC line.

Figure 8.41
Three-string tension set of a UHV DC line.

Figure 8.42
Grading ring and shielding ring for a double-string tension set in a UHV AC line.

Figure 8.43
Grading ring for a double I-string suspension set in a UHV AC line.

Figure 8.44
Grading ring and shielding ring for three-string tension set in a UHV DC line.

Figure 8.45
Grading ring for a double V-string suspension set in a UHV DC line.

picture of the grading ring and shielding ring for a double-string tension set, and Figure 8.43 shows a picture of the grading ring for a double I-string suspension set in a UHV AC line. Figures 8.44 and 8.45, respectively, show a picture of the grading ring and shielding ring for a three-string tension set and the grading ring for a double V-string suspension set in a UHV DC line.

8.4.5 Jumper Fittings

Jumpers are erected on a tension tower to connect the two conductors on two sides of the tower to allow electricity continuity.

Figure 8.46
Aluminum tube rigid jumper used in UHV transmission lines.

Figure 8.47
Squirrel-cage rigid jumper used in UHV transmission lines.

In China, two types of jumpers are available for UHV lines, namely, aluminum tube rigid jumper and squirrel-cage rigid jumper [13], as shown in Figures 8.46 and 8.47, respectively. The rigid part of the former comprises two parallel aluminum pipes, each of them containing two sections of aluminum pipes of different lengths. Aluminum pipe ends are connected with soft conductors through a 2-to-6 or a 2-to-8 clamp, and the shielding ring (a side ring) is installed at the joint of the two sections of aluminum pipes. The rigid part of the latter is a single steel pipe that is configured with a supporting spacer and counterweight, and the soft jumper is fixed directly by the spacer. Compared with the former, the latter eliminates the joints and shielding rings on both ends of the rigid part.

References

[1] S. Zhusen, C. Yongfeng, Z. Qiang, et al., Promotion and application of steel pipe towers in transmission lines, Power Syst. Technol. 34(6) (2010) 186–192.

[2] W. Jing, Z. Dachang, Y. Jingbo, et al., Test study on high-neck forged flanges of UHV steel tube transmission towers, J. Build. Struct. 31(1) (2010) 153–158.

[3] W. Jing, Parameter study on tube-inserting plate connections of transmission towers, Power Grids Clean Energy 24(4) (2008) 31–35.

[4] Y. Jingbo, H. Junke, L. Maohua, et al., Selection of calculation model for steel tubular tower of UHV power transmission line, Power Syst. Technol. 34(1) (2010) 1–5.

[5] L. Xianlong, C. Yongfeng, Review and new development on transmission lines tower foundation in China. 2008 CIGRE Session Technical Programme, B2-215, Paris; 2008.

[6] W. Jiancheng, Y. Jun, X. Kai, et al., Application of 900 mm^2 large cross-section conductor in UHV DC project, Power Syst. Technol. 33(15) (2009) 60–65.

[7] G. Yinglong, L. Guoxing, Y. Chuanyong, Conductor Galloping in Transmission Lines, China Electric Power Press, Beijing, 2002.

[8] O. Nigol, P.G. Buchan, Conductor galloping. Den Hartog mechanism, IEEE Trans. Power Apparatus Syst. 100(2) (1981) 699–707.

[9] Z. Kuanjun, L. Bin, L. Chaoqun, et al., Research on anti-galloping of UHV transmission line, Proc. CSEE 28(34) (2008) 12–20.

[10] L. Yongwei, Y. Jun, Z. Quanjiang, et al., Design of the first 1000 kV AC single-circuit overhead transmission line in China, Proc. CSEE 30(1) (2010) 117–126.

[11] W. Jingchao, F. Baozhen, Z. Lixian, et al. Study on structure design of grading and shielding rings of UHV AC transmission lines. Proceedings of International Conference on UHV Transmission Technology; 2009.

[12] D. Tao, S. Zhiyi, F. Jianbin, et al., Design optimization of grading ring of composite insulator for ±800 kV UHV DC transmission lines, Power Syst. Technol. 34(8) (2010) 31–35.

[13] X. Liang, G. Lili, Z. Huaiqing, et al., Corona test for UHV AC rigid jumper, High Voltage Eng. 35(3) (2009) 470–474.

[14] Z. Zifu, Y. Jingbo, L. Qinghua, et al., Full-scale test for Cup-type Steel Tubular Tower of UHV Transmission line, J. Build. Struct. 42 (2012) 403–407.

[15] G. Zhicheng, Insulators and External Insulation of Transmission and Transformation Equipment, Tsinghua University Press, Beijing, 2006.

[16] D. Jie, Manuals for Electric Power Fittings (the third version), China Electric Power Press, Beijing, 2010.

[17] Department of DC Construction of SGCC, Standard Drawings of Fittings of ±800 kV UHV DC Transmission Lines—Types and Standard Drawings of Fittings of Common Lines, China Electric Power Press, Beijing, 2012.

CHAPTER 9

UHV Substation and UHV AC Electrical Equipment

Chapter Outline

Ultra-high Voltage AC/DC Grids.
DOI: http://dx.doi.org/10.1016/B978-0-12-802161-3.00009-3

9.1 UHV Substation

9.1.1 Main Electrical Connection Scheme

The main electrical connection scheme of a substation is intended to connect the main electrical equipment in the substation according to certain regulations to realize power transmission and distribution as required and ensure compliance with the requirements regarding reliability, flexibility, and economy of operation.

The main electrical connection scheme of a substation should be determined through techno-economic comparison comprehensively, taking into account the role of the substation in the power grid, planned capacity, nature of loads, total number of connected components such as lines and transformers, and characteristics of equipment. It should be designed in such a manner that ensures reliable power supply, flexible operation, convenient operation and maintenance, easy transition or expansion, reasonable investment, and reduced land occupation.

Generally, 330- to 750-kV EHV switchgear arrangement may use double-bus, one-and-a-half breaker, transformer-bus, angle-type, and single-bus connection schemes. In China, 500- to 750-kV EHV substations usually use a one-and-a-half breaker scheme, and 330-kV substations use a one-and-a-half breaker or double-bus scheme.

With higher voltage level and larger transmission capacity, 1000-kV systems place higher requirements on the reliability of the main electrical connection scheme of UHV substations. The design institute has analyzed and researched the reliability of some typical connection schemes, such as two-breaker scheme, one-and-a-half breaker scheme, 4/3 breaker scheme, sectionalized double-bus scheme, and sectionalized double-bus with auxiliary bus scheme for a typical 1000-kV substation with four transformers and eight circuits of 1000-kV outgoing feeders.

The analysis results show that, of these five schemes, the former three schemes have comparable reliability and are significantly superior to the latter two schemes.

The manufacturing capability and operation experience of equipment also need to be considered during selection of the main electrical connection schemes. In a UHV project, if a 4/3 breaker scheme is used, then the current-carrying capability of the switchgear on the bus side cannot meet the requirements, whereas if a two-breaker scheme is used, then the investment for the equipment is high and the area occupied is large.

According to the comprehensive analysis described, in conjunction with the existing manufacturing capability of equipment, the one-and-a-half breaker scheme is selected for 1000-kV substations.

9.1.2 Electrical Equipment

The main equipment of a 1000-kV substation mainly include transformers, shunt reactors, switchgear (gas-insulated switchgear (GIS)/hybrid gas-insulated switchgear (HGIS)), metal oxide surge arresters (MOAs), capacitive voltage transformers, earthing switches for maintenance purpose, post insulators, and low-voltage reactive power compensation devices.

The main equipment should be selected to accommodate the requirements of normal operation, maintenance, short-circuit, and overvoltage conditions while allowing for long-term development, and they should be checked against the local environmental conditions. According to the engineering experience around the world, to reduce the failure rate of the equipment at new voltage levels during the early stage of the project operation, the following aspects should be focused on when selecting 1000-kV electrical equipment: (i) difference in system conditions; (ii) environmental constraints; (iii) increased reliability through the use of the state-of-the-art and proven technology, allowing for adequate safety margin for key equipment or components, such as transformers, switchgear, and bushings; (iv) reduction of equipment manufacturing difficulties; (v) giving full consideration to the equipment transportation restrictions; and (vi) placing emphasis on the economy of the UHV projects on the premise of ensuring reliability.

1. **Transformer**

   UHV transformers are single-phase, with high- (HV), medium- (MV), and low-voltage (LV) grades. The transformer impedance has a significant impact on the system stability, power distribution of HV and LV looped power network, system reactive power compensation capacity, and system short-circuit capacity. Also, it is directly related to the manufacturing difficulty of the equipment. Studies of 1000-kV transformers show that the short-circuit impedance $U_{\text{H-M}}$ (high–medium), $U_{\text{H-L}}$ (high–low), and $U_{\text{M-L}}$ (medium–low) should be considered 18%, 62%, and 40%, respectively.

   For a 1000-kV transformer, low-voltage reactive power compensation devices are usually installed on the tertiary winding side. Therefore, the rated capacity of the tertiary winding should be no less than the installation capacity of low-voltage reactive power compensation devices. For instance, the Jindongnan–Nanyang–Jingmen 1000-kV UHV AC transmission pilot and demonstration project requires a total of 840-Mvar capacitive and 960-Mvar inductive reactive power; therefore, the rated capacity of the tertiary winding of the 1000-kV transformer was designed to be 1000 MVA (334 MVA for single-phase three-winding transformers).

   The rated voltage of the tertiary winding is designed primarily based on the reactive power compensation requirement. In the existing 500-kV substation in China, the rated voltage of the tertiary winding of transformers is 35 or 66 kV; in the existing 750-kV substations, the rated voltage is 66 kV. Due to restrictions of short-circuit capacity of the equipment and rated current of the switching device, the rated voltage of the tertiary winding of 1000-kV/1000-MVA transformers has to be increased. Studies were performed on 110-, 132-, and 145-kV voltage levels, with the corresponding current ratings being 5249, 4374, and 3982 A, to decide which of them is more suitable. It was found that all the short-circuit currents of tertiary winding did not exceed 40 kA under those rated voltages. And the variation in the bus voltage on the transformer MV side when one reactive compensation bank is switched could be controlled within 2.5% by means of properly designing the capacity of reactive compensation banks. However, because voltages of 132 and 145 kV are not included in the list of China's standard voltage levels, 110 kV is chosen for the 1000-kV/1000-MVA transformers.

   According to the system requirements, the voltage of transformers is regulated through non-excitation tap changer. The non-excitation tap changer may be installed either at the MV-side line terminal or at the neutral. Because the MV-side voltage of a 1000-kV transformer is 500 kV, voltage regulation at the MV-side line terminal poses great difficulty in terms of ensuring insulation reliability and selecting circuit breaker types. Therefore, voltage regulation at the neutral is used.

However, in case of auto-transformers, voltage regulation at the neutral would result in voltage deviation at the tertiary winding. When HV-side voltage is increased, the number of turns of the common winding is required to be reduced to maintain constant secondary voltage. This will cause the electromotive force of each turn to increase, and the LV-side voltage will increase as the core becomes more saturated. As a result, when shunt compensation devices are installed at the tertiary winding, the magnitude of voltage variation would be increased, thereby deteriorating the operation characteristics. To solve this problem, additional compensation devices are installed at the tertiary winding of the transformer to control the voltage variations of the tertiary winding within 0.5%.

Figure 9.1 shows two 1000-kV transformers.

2. **Shunt reactor**

The capacity of shunt reactors should be determined to ensure not only that the requirements for limiting overvoltage and controlling voltage under normal light load operation can be met but also that the reactive power compensation and voltage control can be performed easily during heavy load operation. To protect the line from resonance during open-phase operation, the compensation of 1000 kV shunt reactors should not approach 100%.

(A) (B)

Figure 9.1: The 1000-kV transformers.

(A) Main transformer in Changzhi Substation and (B) main transformer in Jingmen Substation.

Moreover, the compensation capacity of reactors should be determined according to the line length. Given that the line sections of varying lengths are provided with shunt reactors of different capacities, through comprehensively analyzing the power frequency overvoltage and switching overvoltage under various operation modes as well as resonance overvoltage during open-phase operation, reactive power balance, voltage control, voltage profile along the line, availability strategy of spare parts, and difficulties in equipment development and manufacturing, the capacity series of 1000-kV shunt reactors is determined to be 320, 280, 240, and 200 Mvar in increments of 40 Mvar. In Changzhi, Nanyang, and Jingmen Substations, the rated capacity of the 1000-kV shunt reactor for each phase was determined to be 320, 240, and 200 Mvar, respectively, which can provide a compensation of 85–88%. Figure 9.2 shows an example of a 1000-kV shunt reactor.

3. **Switchgear**

After comprehensively considering a number of factors such as the insulation level of 1000-kV equipment and intensive use of land, 1000-kV GIS and HGIS are used.

Changzhi substation uses 1000-kV GIS manufactured by Henan Pinggao Electric Co., Ltd. This type of GIS is provided with series double breaks, hydraulic operating mechanism, and 600-Ω opening/closing resistor. The circuit breaker and its opening/closing resistor are housed in the same enclosure. The disconnector is provided with an

Figure 9.2
The 1000-kV shunt reactor.

electric spring operating mechanism and 500-Ω damping resistor. Surge arresters on incoming/outgoing feeders, voltage transformers, and 1000-kV shunt reactor circuits are open-type.

Nanyang Substation uses 1000-kV HGIS manufactured by New Northeast Electric Group High Voltage Switchgear Co., Ltd. This type of HGIS is provided with series double breaks, hydraulic operating mechanism, and 580-Ω closing resistor but no opening resistor. The circuit breaker and its closing resistor are housed in the same enclosure. The disconnector is provided with an electric spring operating mechanism and 500-Ω damping resistor. The 1000-kV bus is open-type. The equipment associated with the 1000-kV bus, incoming/outgoing feeders, and 1000-kV shunt reactor circuit are open-type as well.

Jingmen Substation uses 1000-kV HGIS manufactured by Zhejiang Xikai Electrical Co., Ltd. This type of HGIS is provided with serial-connected four breaks, hydraulic operating mechanism, and 560-Ω closing resistor but no opening resistor. The circuit breaker and its closing resistor are housed in separate enclosures. The disconnector is provided with an electric spring operating mechanism and no damping resistor. The 1000-kV bus is open-type. The equipment associated with the 1000-kV bus, incoming/outgoing feeder, and 1000-kV shunt reactor circuit are all open-type as well.

For grids that are 500 kV or less, the time constant of attenuation of the DC component is no more than 45 ms and the rated short-circuit making current of the circuit breaker is 2.5-times its rated short-circuit breaking current under the rated voltage. In the UHV AC pilot and demonstration project, the time constant of attenuation of the DC component is 120 ms according to study results, because the UHV AC grid has higher inductive impedance. Therefore, the rated short-circuit making current of the circuit breaker is selected to be 2.7-times its rated short-circuit breaking current. That is, when the rated short-circuit breaking current of the 1000-kV circuit breaker is 50 kA, its rated short-circuit making current is 135 kA, correspondingly.

According to the study results in conjunction with the operating experiences with 500-kV systems, in this project secondary arc current is compensated using neutral grounding reactors of 1000-kV line shunt reactors, rather than being forcibly extinguished by a high-speed grounding switch (HSGS). Therefore, no HSGS is required on the outgoing feeder side.

4. **Series compensation device**

UHV series compensation means that capacitors are connected in series with the line to compensate the inductive impedance of the UHV line to increase the transmission capacity and save the investment and land for transmission corridors. In the UHV grids planned by the SGCC, series compensation devices will be widely used. The worldwide first UHV series compensation devices, as shown in Figure 9.3, were completed and put into operation by the SGCC in 2011 to fulfill the target of ensuring that the UHV AC

Figure 9.3
Series compensation devices for the expansion of UHV AC pilot and demonstration project.

transmission pilot and demonstration project is able to deliver 5-GW power stably after expansion.

Based on the studies of system load flow and stability, installing fixed series compensation devices capable of providing a compensation of 40% on both Changzhi–Nanyang line I and Nanyang–Jingmen line I is planned for the expansion project. According to the requirements of the control of the voltage profile along the line, for the former line, series compensation devices are installed on both sides of the line, whereas for the latter line they are installed only on Nanyang side and arranged in two identical sections to reduce the difficulties in manufacturing. All series compensation devices will be installed on the line side of the HV shunt reactors. The rated current of UHV series compensation devices should be determined such that the system can transmit the maximum normal power in different target years with a certain margin. In addition, the overloading capability of capacitors should be utilized reasonably to accommodate the requirements for power transmission during contingency. Considering this, the rated current of the series compensation capacitor banks for both lines is determined to be 5.08 kA, which is approximately twice that of typical EHV series compensation devices and their overloading capability is 6.1 kA with a duration of 2 h. Therefore, the capacity of the series compensation devices for Changzhi–Nanyang line I is determined to be 1500 Mvar × 2, and that for Nanyang–Jingmen line I is determined to be 1144 Mvar × 2.

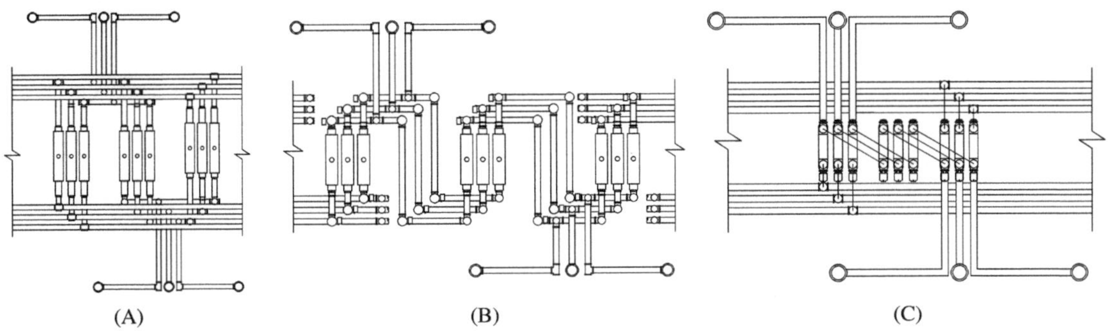

Figure 9.4: Schematic of single-row layout of one-and-a-half breaker scheme.
(A) Typical arrangement; (B) conventional diagonal arrangement; and (C) elevated diagonal arrangement.

9.1.3 Overall Layout

The design of the switchgear arrangement focuses on determining the bay width, heights of equipment supports, bus framework, incoming/outgoing feeders, longitudinal dimensions, as well as passage sizes through calculation and verification. It is also necessary to consider the equipment selection and study results on overvoltage and insulation coordination, which are the basis of recommending the switchgear layout that is safe and reliable, economically viable, and in compliance with the relevant technical requirements. On this basis, the reasonable electrical general layout can be determined. The layout of the substations in the UHV AC pilot and demonstration project is described.

1. **GIS layout**

   With GIS, the 1000-kV electrical connection scheme at Changzhi Substation uses a transitional scheme with dual-element and dual-circuit-breaker in this phase and one-and-a-half breaker scheme in the long-term.

   The GIS layout as determined should: (i) be as clear and simple as possible and minimize the land occupation; (ii) ensure that the width of switchgear bay accommodates the equipment size; (iii) facilitate the transportation and lifting of equipment to provide adequate room for installing equipment (de-energized) and maintaining equipment (with nearby equipment energized); and (iv) minimize and even avoid power outage during maintenance and expansion construction.

   Typical one-and-a-half breaker scheme for GIS may be classified into single-row layout and inline layout. The single-row layout may be further classified into typical arrangement, conventional diagonal arrangement, and elevated diagonal arrangement, as shown in Figure 9.4. From the perspective of bus arrangement, the inline layout may be further classified into phase-segregated bus arrangement, internal bus arrangement, and external bus arrangement, as shown in Figure 9.5.

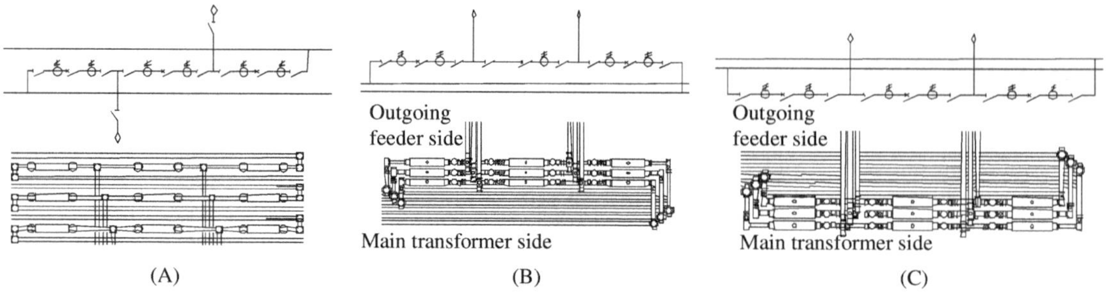

Figure 9.5: Schematic of inline layout of one-and-a-half breaker scheme.
(A) Phase-segregated bus arrangement; (B) internal bus arrangement; and (C) external bus arrangement.

In the single-row layout, circuit breakers are arranged perpendicular to the main bus: (i) the typical arrangement cannot achieve easy crossing connection but is suitable to the situation where the number of outgoing/incoming feeders on both sides is comparable and the two elements in the same string can only have outgoing feeders connected in different directions; (ii) the conventional diagonal arrangement, which is evolved from the typical arrangement, allows easy crossing connection and flexible arrangement of incoming/outgoing feeders; however, as compared with the typical arrangement, it has larger lateral dimension, longer bus branches, and requires higher manufacturing costs of equipment, and the main bus of GIS is close to the foundation surface and, therefore, a work platform straddling the main bus is required; (iii) the elevated diagonal arrangement, which is evolved from the conventional diagonal arrangement, has basically the same characteristics as the conventional diagonal arrangement with the exception that it has smaller lateral dimensions because the diagonal bus branches are elevated and therefore do not use the land laterally. In addition, some of the connecting bus in the string needs to be removed for maintenance and repair of a circuit breaker, especially when the intermediate circuit breaker is repaired, and power outage of the complete string may result. Furthermore, the supporting structure of elevated bus branches in the string should be reinforced.

In the inline layout, circuit breakers are arranged in parallel with the main bus. This layout allows flexible directions of incoming/outgoing feeders and meets the requirements of the Changzhi Substation regarding the main electrical connection scheme and outgoing feeder direction. It requires the smallest longitudinal dimensions, and its lateral dimension fits the width of the 1000-kV switchgear bay, making it possible to make full use of the occupied land area. In phase-segregated bus arrangement the buses are elevated, which makes it possible for convenient patrol inspection. However, this would cause drawbacks such as compromised equipment seismic performance and inconvenient equipment installation and maintenance. In both internal bus and external bus arrangements, the switchgear is placed at a low position

Table 9.1: Main Characteristics of GIS Layout.

Arrangement	Advantage	Disadvantage
Single-row layout, typical arrangement	Equipment in the string connected by short bus branches	Difficult to achieve crossing connection
Single-row layout, conventional diagonal arrangement	Allow easy crossing connection and flexible direction of incoming/ outgoing feeders	Long bus branches in the string, increasing the manufacturing costs
Single-row layout, elevated diagonal arrangement	Allow easy crossing connection and flexible direction of incoming/ outgoing feeders	Require long bus branches in the string; elevated GIS equipment causes inconvenient maintenance and repair
Inline layout, phase-segregated bus arrangement	Small longitudinal dimensions and short bus branches in the string; allow crossing connection and flexible arrangement direction of incoming/outgoing lines	Heavy-duty cranes required to lift the circuit breaker across main bus for installation; elevated main bus causes inconvenient installation, maintenance, and repair and compromises the seismic performance
Inline layout, internal bus arrangement	Small longitudinal dimensions, allow crossing connection and flexible direction of incoming/ outgoing line; outgoing bus branches and the main bus not intersected	Not convenient for equipment lifting & installation and operators' patrol inspection
Inline layout, external bus arrangement	Small longitudinal dimensions, allow crossing connection and flexible direction of incoming/ outgoing lines; convenient for equipment lifting & installation, operation, and patrol inspection	Long bus branches, and outgoing bus branches and the main bus crossed

and can be installed and repaired conveniently. Moreover, the switchgear has clear functional zones and is aesthetically pleasing. In the internal bus arrangement, the main bus is centrally arranged on the GIS transformer side and the outgoing bus branches are short. However, the installation and maintenance of circuit breaker units require a heavy-duty crane, and it is not convenient for the operators' patrol inspection. In case of the external bus arrangement, the main bus is centrally arranged on the GIS outgoing feeder side, and the lifting, installation, and transportation of the equipment can be through the transportation access of the transformers. It involves the use of a light-duty crane because of the smaller lifting radius and allows convenient operation patrol inspection, but the outgoing bus branches are longer compared with the internal bus arrangement.

Table 9.1 shows the main characteristics of the GIS layouts.

Through comprehensive techno-economic analysis, the GIS in the Changzhi Substation uses in-line layout with external bus, as shown in Figure 9.6.

Figure 9.6
GIS at Changzhi Substation.

In the in-line layout and the lateral dimension of the GIS depend on the internal lateral dimension of equipment units in each string and the number of strings, whereas its longitudinal dimension depends primarily on the longitudinal dimension of the equipment body and requirements of the transportation and lifting and installation of the GIS. The longitudinal dimension of the GIS dictates the length of the incoming/outgoing bus branches, and thus has a significant impact on the manufacturing costs of equipment. The longitudinal dimension of the equipment body is determined by the structural design determined by the manufacturer. Therefore, the optimization of longitudinal dimension of the GIS should focus on two aspects: (i) the longitudinal dimension on the incoming feeder side (i.e., the distance between bushing of the incoming feeder and the equipment body), which should be determined based on the guaranteed minimum safety clearance from the crane to the grading ring of the HV bushing during lifting the circuit breaker under live conditions; (ii) the longitudinal dimension on the outgoing feeder side (i.e., the distance between the lead-out bushing and the equipment body), which should be determined based on the guaranteed minimum safety clearance from the crane to the grading ring of the HV bushing during lifting the bus or the bus equipment under live conditions.

2. **HGIS layout**

   With HGIS, 1000-kV electrical connection at Nanyang Substation and Jingmen Substation uses a transitional dual-element and double-circuit-breaker scheme in this phase and a one-and-a-half breaker scheme in the long-term.

Table 9.2: Comparison Between Perpendicular and Parallel Arrangements.

Items	Arrangement	
	Perpendicular Arrangement	**Parallel Arrangement**
Land area occupied	Compact and occupying less land area	Not compact and occupying larger land area
Arrangement of overhead incoming feeder of transformer	Low-suspending crossing mode and additional incoming feeder gantry required	Inline arrangement, no additional incoming feeder gantry required
Coordination with the overall layout of the substation	Good	Difficult to coordinate with the layout of the switchgear at next voltage level
Incoming feeders of two different transformers connected to different buses	Satisfied	Satisfied but difficult, additional bays required leading to increased land occupation

For the HGIS, the typical one-and-a-half breaker scheme may be classified into three-row layout (parallel and perpendicular outgoing line type), single-row layout (diagonal and in-line type), and ring layout. The three-row layout is more adaptable to the main connection, with the outgoing feeders arranged on both sides (preferably). It occupies less land area, facilitates operation, and is therefore widely used. The single-row layout occupies a larger area and is mainly used for power plant step-up substations. It is in harmony with the main power building layout and facilitates routing of overhead outgoing feeders of transformers. The ring layout occupies a land area larger than the three-row layout and smaller than the single-row layout, and it is mainly used in confined areas requiring compact longitudinal dimension of the switchgear. As such, wherever no limitation of geographical conditions exists, a three-row layout is preferred for a 1000-kV switchgear arrangement using a one-and-a-half breaker scheme.

In terms of the incoming feeder direction of the transformer, the three-row layout may be classified into perpendicular arrangement and parallel arrangement. In the former arrangement, the incoming feeders are perpendicular to the outgoing feeders, whereas in the latter arrangement the incoming feeders are parallel with the outgoing feeders. Table 9.2 presents the comparison between perpendicular and parallel arrangements.

According to the actual conditions, the perpendicular arrangement as shown in Figure 9.7 was selected for Nangyang Substation.

The perpendicular arrangement may be further classified into a low-suspending crossing scheme and a high-suspending crossing scheme. Because the feeder attachment point on the bus framework of the 1000-kV switchgear is 38 m high, and because that on incoming/outgoing feeder framework is 55 m high, if a

Figure 9.7
Photograph of the HGIS in Nanyang Substation (perpendicular arrangement).

Table 9.3: HGIS Arrangement Feature.

No.	Item	Feature
1	Basic features	Access roads between phases provided Phase-segregated; outgoing feeders of transformers arranged in two directions and using high-suspending scheme and incoming feeders arranged using low-suspending crossing scheme Quad bundle flexible bus and jumper wires required; tubular bus used for connection between equipment
2	Bay width	54 m
3	Gantry	Bus and outgoing lines arranged on the same gantry, reducing the occupied land area
4	Technology	Proven and state-of-art technology, evolved from GIS
5	Maintenance repair	Same as the requirements of GIS with the exception of bus and outgoing line equipment which are of conventional type

high-suspending crossing scheme is used, then the height of the gantry will reach up to more than 72 m; therefore, it is not justifiable economically and difficult to implement during the project construction. Hence, in the HGIS, a low-suspending crossing scheme is used.

Table 9.3 presents the features of the HGIS. Figure 9.8 shows the photograph of the HGIS at Jingmen Substation.

Figure 9.8
Photograph of the HGIS at Jingmen Substation.

9.2 UHV Transformer and Shunt Reactor

9.2.1 UHV Transformer

As compared with the 500- and 750-kV transformers, UHV transformers have a higher voltage level. Currently, 1000 kV is the highest operating voltage level of AC transmission equipment around the world, which requires a high insulation level, thus causing great difficulties in the insulation design. They also have large single-unit/single-phase capacity. To ensure the economy of UHV transmission, the single-phase capacity of a UHV transformer is three- to four-times that of a conventional 500-kV single-phase transformer, making it difficult to control leakage flux and prevent local overheating. They also have rigorous requirements regarding transportation conditions. Due to HV level and large capacity of the equipment, they become larger and heavier and are limited by the transportation conditions. They have high reliability requirements. As the transmission power is greatly increased, higher reliability is required. No relevant standards and codes or mature experience is available for UHV transformers. Also, they have higher requirements regarding test technology and equipment capability. Particularly, the power frequency test of UHV transformers is performed under the ultimate limit conditions [1].

9.2.1.1 UHV transformer structure

A UHV transformer [2] is a single-phase, oil-immersed, off-circuit voltage regulation auto-transformer consisting of a main transformer and a regulating compensation transformer. The main transformer is a single-phase, oil-immersed auto-transformer that has a single-phase four-limb or five-limb core and is structured such that the HV, MV, and LV windings fitted on different limbs are connected in parallel. The regulating compensation transformer consists of a MV neutral-point off-circuit voltage regulation transformer and a LV compensation transformer sharing one tank in common. China has successfully developed a single-phase five-limb core UHV transformer that is structured such that HV, MV, and LV windings fitted on three main limbs are connected in parallel, and a single-phase four-limb core UHV transformer that is structured such that HV, MV, and LV windings fitted on two main limbs are connected in parallel. This section mainly deals with the structure of the latter transformer. Figure 9.9 shows the 1000-MVA UHV transformer (two limbs).

In the main transformer, HV, MV, and LV windings fitted on the main two limbs are connected in parallel [3]. The regulating compensation transformer houses a regulating transformer and an LV voltage compensation transformer. The compensation windings in the compensation transformer can effectively ensure stable LV winding voltages at different tap positions in variable flux voltage variation (VFVV) mode. The off-circuit tap changer is installed in the oil tank of the regulating transformer. The main transformer and regulating compensation transformer are externally connected through oil-air bushing.

Figure 9.9
The 1000-MVA UHV transformer (two limbs).

The LV winding is provided with 100% voltage compensation, which can effectively guarantee stable voltage at different tap positions. The HV, MV, and LV windings fitted on two main limbs are connected in parallel. HV winding is in a configuration with the wiring terminal at the intermediate location, whereas the MV and LV windings are in a configuration with the wiring terminal at an end location. Figure 9.10 shows the connection diagram of the UHV transformer.

The insulation of the transformer body is structured to comprise thin paper cylinders in combination with small oil gaps and non-directed circulation oil circuit. The transformer body is assembled in such a manner that individual windings are assembled first and then fitted into the overall body. To improve the reliability of the insulation, equipotential planar insulating barriers are provided on the outer surface of the HV winding shielding to divide the oil gap into small ones.

The HV windings fitted on two main limbs are connected in parallel via an inter-limb connector and connected to the bushing via HV outgoing lead. The inter-limb connections are all made of copper grading tubes to improve the electric field, and multiple insulating paper cylinders are provided outside the grading tubes to divide the oil gap into small ones.

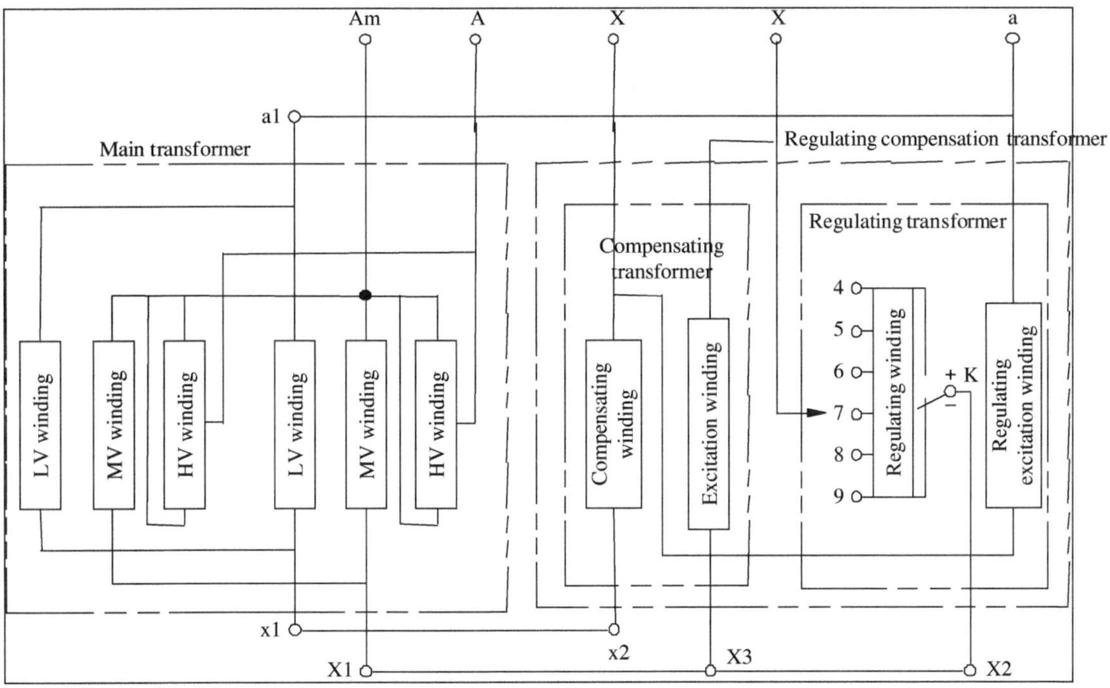

Figure 9.10
Connection diagram of the UHV transformer.

The 500-kV MV leads from the HV terminal of MV winding and from the LV terminal of HV winding are put through the grading tubes and then connected together to the bushing.

The regulating compensation transformer comprises two transformer bodies, each having a separate core. One is called the regulating transformer, which comprises a regulating winding and a regulating excitation winding, and the other is the compensating transformer, which comprises a compensating winding and a compensating excitation winding.

The regulating compensation transformer has three LV windings and three neutral point bushings. The outgoing line of the compensating excitation winding is separately led out such that the regulating transformer and the compensating transformer can be easily isolated during the test.

9.2.1.2 Key manufacturing technologies

1. **Window current**
   Because oil–air bushings are used in the UHV transformer, the insulation of HV and MV bushings at the air side is required to be spaced apart sufficiently, which would impose restrictions on the arrangement of bushings. To allow simple and reliable connections inside the transformer, the taps of HV and MV windings may be led out from both sides of the core. This configuration would lead to ampere-turn imbalance between the HV and MV sides, resulting in an unbalanced load current, which is also known as window current. Because the load current of the UHV transformer is large, it would excite the core, increasing the total loss of the transformer. Therefore, proper measures should be taken to eliminate the adverse effects resulting from the window current.
2. **Insulation design**
   a. *Frequency-to-time conversion.* In the long-duration-induced AC voltage test (ACLD) included in the type and routine test of the transformer, if the test frequency exceeds twice the rated frequency, then the test time t is given according to GB 1094.3 *Power transformers—Part 3: Insulation Levels, Dielectric Tests, and External Clearances in Air*:

$$t = 120 \times \frac{f_0}{f_t}\,(\text{s}) \tag{9.1}$$

where
t—test time, no less than 15 s
f_0—rated frequency
f_t—test frequency.

For UHV transformers, in contrast to the stipulations in technical code for transformers 750 kV or less, no 1-min short-duration power frequency voltage withstand test is required. To strictly assess a UHV transformer and verify its insulation margin, the test voltage U_1 is taken as U_m (the maximum permissible voltage) for type test and it is applied for 5 min in the ACLD with partial discharge measurement. No frequency–time conversion is required for the test voltage U_1 in this test. In the design phase, the impact of no frequency–time conversion on the transformer insulation must be taken into consideration. In routine testing, one may refer to the relevant chapters for frequency–time conversion.

b. *Permissible electric field strength.* In the design of a UHV transformer, full-field analysis should be performed at various positions inside the transformer. As the design criteria, the selected permissible field strength directly dictates the insulation margin of the objects under analysis and the reliability of the transformer. If the selected permissible field strength is too low, then the insulation design will be unable to meet the transformer sizing requirements placed by the transportation limitation. However, if it is too high, then it will be impossible to effectively control and prevent the transformer partial discharge, or even lead to insulation breakdown.

The power frequency voltage withstand test is performed to examine the insulation strength and whether partial discharge occurs. The probability of short-time and long-time partial discharge in a transformer can be calculated using the Weibull distribution equation:

$$P_1(U_t, t_t) = 1 - \exp(-kU_t^m t_t^m)$$

which is approximately 1–2%. According to the experience, the probability of partial discharge in the test is much lower than 2%. For important equipment, it should be controlled to the level of 0.1–0.2% and used as the basis to determine the test voltage U_t and test time t_t. For UHV transformers, the permissible field strength is determined from the existing UHV voltage level transformers while using partial discharge probability of 0.1% as the judgment criteria.

c. *High field strength at ground/low electric potential.* For transformers at conventional voltage classes, high field strength usually occurs at the structure of high potential. Therefore, the field strength in the structure of ground/low potential is not stressed in the insulation analysis and tends to be ignored in the design and design evaluation. However, as the transformer voltage class increases, the field strength at ground/low potential increases accordingly, and detailed analysis in the insulation design of UHV transformers is required. It follows from the calculation and analysis of the field strength of UHV transformers that the field strength at high potential and that at ground/low potential are comparable, with both at a high level. Therefore, the key structures such as the core and tank wall must be subject to full-field analysis and the insulation structures should be improved to ensure the required insulation margin and to prevent partial discharge.

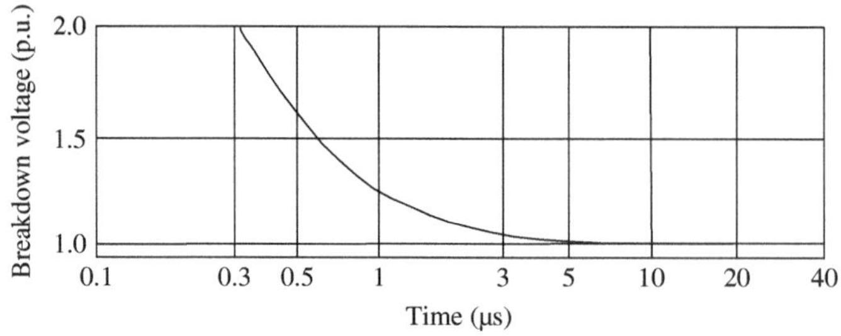

Figure 9.11
Volt-time characteristic of insulation breakdown of oil gap.

In the areas of ground/low potential with high field strength, the use of long threaded rods to fasten laminated paper board should be avoided as practically as possible, and the threaded rods should be placed in a staggered manner. In the areas of low field strength, the points on insulating barriers where threaded rods are used for fastening should be subject to detailed electric field calculations to avoid local concentrated electric field, and the length, materials, and quality of the threaded rods should be carefully controlled.

d. *Impacts of full-wave lightning impulse voltage with wave front time that is too long on insulation design.* The wave front steepness of lightning impulse voltage mainly affects the longitudinal insulation strength of the windings. The greater the wave front steepness, the greater the impact on the inter-turn and inter-segment insulation will be. Therefore, the longer front time may lead to the looser evaluation of some longitudinal insulation and to strict evaluation of the main insulation according to volt–time characteristics of combined insulation. Therefore, in the process of UHV transformer's design, manufacturing, and test, great importance should be attached to the impact of the longer wave front time on the insulation performance. The impact on the longitudinal insulation strength must be considered in the design when the wave front time T_1 is 2.5 μs or more. Figure 9.11 shows the volt–time characteristics of insulation breakdown of the oil gap.

   In addition, considering the actual arrangement of arresters in the substation, the wave crests of lightning-intruding overvoltage are suppressed using arresters, resulting in a larger potential difference between the series windings and shunt windings of the transformer than when the wave crests are not suppressed. This must be taken into account in the design to prevent the chopped wave caused by the operation of arresters from affecting the insulation of the transformer.

9.2.1.3 Key tests

1. **Type test of HV bushing**
   HV bushing is a key component of UHV transformers and its performance has a direct impact on the reliability of long-term operation of the transformers. Here, the

1200-kV, 5-min power frequency voltage withstand test of HV bushings is described. HV bushings are required to withstand the voltage of U_{m} for 5 min during the ACLD of the transformer. Therefore, they should be able to withstand the corresponding test.

Partial discharge measurement should be performed separately in the routine test of bushings. The test voltage should be applied in the following manner: turn on the power supply at a voltage not higher than $U_2/3$; increase the voltage to U_3 and hold it for 5 min; then, increase the voltage to U_2 and hold it for 5 min; further increase the voltage to U_1 and hold it for 5 min; then, immediately reduce the voltage to U_2 and hold it for 60 min and measure the partial discharge at the same time; then, continue to reduce the voltage to U_3 and hold it for 5 min; the power supply may be turned-off only when the voltage is reduced to less than $U_2/3$. In this description, $U_1 = U_m = 1100$ kV, $U_2 = 1.5 \times U_m/\sqrt{3}$, and $U_3 = 1.1 \times U_m/\sqrt{3}$. The apparent discharge should be no more than 10 pC under voltage U_2 and no more than 5 pC under the voltage U_3.

This test may also be performed during a short-duration power frequency voltage withstand test. In this case, U_1 is 1200 kV.

As the test voltage reaches 1200 kV, to monitor the partial discharge monitoring of bushings, the following anti-interference measures are taken to ensure satisfactory routine test.

a. Improve the environmental conditions at the test site to reduce corona discharge in the areas with high field strength.
b. Install double grading rings with a diameter larger than 2.6 m and use 50-mm metallic bellows as HV conductors. Properly polish the surface of all the grading rings on the HV side to remove burrs and other defects that may cause distortion of local field strength, thereby preventing corona from occurring.
c. Properly arrange the test objects at the test site to ensure that the HV side of the test circuit is more than 12 m from the walls and other unused site equipment to prevent electric discharge to the equipment nearby. Ensure that other unused site equipment are reliably grounded to prevent floating discharge.

Moreover, in the HV bushing sealing test, the pressure applied on the bushing should be 0.2 ± 0.01 MPa when it is placed vertically, which is lower than the value of 0.3 ± 0.01 MPa when placed horizontally. This is because the porcelain part of the bushing is long, and when it is placed vertically the effects of the pressure resulting from its self-weight should be taken into account in the test.

2. **Long-duration AC-induced voltage test with partial discharge measurement (ACLD)** Because the test voltage is high and the background noise is required to be low during partial discharge measurements, proper anti-interference measures should be taken to ensure that the test can be performed successfully.
   a. *Grading ring design.* Because many test items of UHV transformers involve partial discharge measurement and the test voltage is 1100 kV, it is necessary to take shielding measures on the air side of the bushing to ensure no corona occurs.

b. *Measures for preventing discharge on oil tank of ground potential.* During partial discharge test, the entire oil tank and all of its accessories are exposed to the electric field. Therefore, discharge on the tank of ground potential will create difficulties in the partial discharge test. This interference is more prominent in the test of UHV transformers. In addition, the UHV partial discharge test is performed synchronously with the HV terminal voltage withstand test and cannot be repeated. As such, greater caution must be exercised. Considering this, equipotential bonding and shielding measures should be taken before the test to deal with the problem associated with oil tank of ground potential and high field strength.
c. *Control of partial discharge of test power supply and intermediate transformer.* Reasonable compensation capacity should be determined and the compensation current should be accurately measured. The partial discharge of the test power supply and intermediate transformer should be controlled strictly to reduce the interference from the test system itself. The ACLD test at the substation should be conducted based on the parameters and experience obtained from the ACLD test in the factory.
d. *Controlling of environmental conditions.* Properly arrange the test objects. Ensure other unused test equipment is reliably grounded and that the test objects are sufficiently apart from the walls to meet the test requirements.
e. *Corona prevention.* Clean all grading rings on the HV side. Ensure that the grading shields and conductors of all terminals are reliably connected. The HV connector from the compensation reactors, test transformer, and test objects should have proper length and a diameter no less than 60 mm to prevent corona generation. Ensure that the connecting pipes within the HV zone at the top of the transformer oil tank are reliably grounded and all sharp corners are properly treated to prevent corona.
f. *Partial discharge measurement.* Select proper measuring impedance and measuring frequency range, improve the measuring sensitivity, and reduce the background interference.

Moreover, the clearance of the test hall must be taken into account.

The long-duration-induced voltage test for UHV AC transformers is somewhat different from that for the transformers of lower voltage classes. Specifically, in this test, the pre-applied voltage is equal to the withstand voltage on the HV feeder terminal. The voltage withstand test on the HV terminal is conducted synchronously with the partial discharge measurement, lasting 5 min under full voltage. In a type test, no frequency–time conversion is required. In a routine test, the test time is derived from frequency–time conversion as shown in the following equation:

$$t = 600 \times \frac{f_0}{f_t}\,(\mathrm{s}) \tag{9.2}$$

Each UHV transformer should be subject to an ACLD test as the routine test. The test procedure is as shown in Figure 9.12.

The test voltage to ground should be applied in the following manner:

- Turn on the power supply at a voltage no higher than $U_2/3$
- Increase the voltage to U_3 and hold there for 5 min
- Increase the voltage to U_2 and hold there for 5 min
- Further increase the voltage to U_1 and hold there for 5 min
- Immediately after the test time, reduce without interrupting the voltage to U_2 and hold there for 60 min; measure the partial discharge at the same time
- Reduce the voltage to U_3 and hold there for 5 min
- Reduce the voltage to a value less than $U_2/3$ before switching off.

(*Note*: This sequence is taken as a substitute for the short-duration power frequency voltage withstand test on the HV side. In the type test, no frequency–time conversion is required, whereas in the routine test such conversion is required.)

In this description, $U_1 = U_m$, $U_2 = 1.5 \times U_m/\sqrt{3}$, and $U_3 = 1.1 \times U_m/\sqrt{3}$.

Partial discharge should be monitored during the whole period of test voltage application.

Under voltage U_2, the partial discharge of HV, MV, and LV windings should not exceed 100, 200, and 300 pC, respectively.

Under the voltage U_3, the apparent discharge of all terminals should not exceed 100 pC.

Test results showed that the partial discharge of all windings of the UHV transformer can meet the design limits. Under test conditions, the partial discharge of the transformer can be satisfactorily controlled to the design intents.

3. **Short-duration AC induced voltage withstand test (ACSD)**
Generally, the switching impulse withstand test voltage is transmitted among the three windings according to the transformation ratio. Similar to the power frequency voltage withstand test, this test mainly aims to evaluate the main insulation of the transformer. Because of different insulation coordination principles for the three windings of a UHV

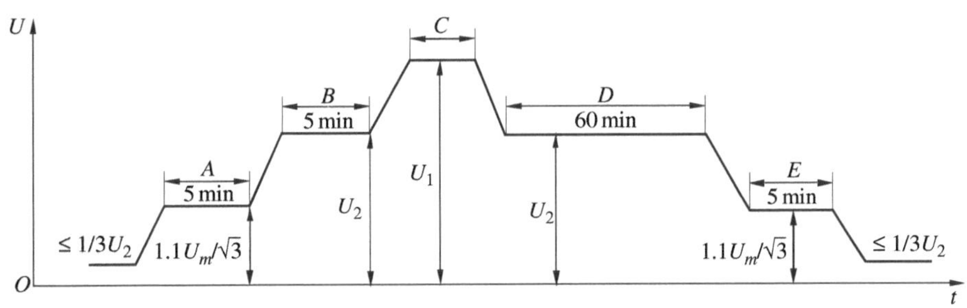

Figure 9.12
ACLD test sequence of UHV AC transformer.

transformer, the switching impulse insulation level on the HV and MV sides are designed to be 1800 and 1175 kV, respectively. When the switching impulse withstand test voltage on the 1000-kV HV side reaches the required 1800 kV, the test voltage on the 500-kV MV side will be just 900 kV, which is insufficient to evaluate the insulation level on the MV side. In comparison, if the switching impulse withstand test voltage on the MV side reaches the required 1175 kV, then the test voltage on the HV side will reach up to 2350 kV, which is far higher than the required insulation level (1800 kV). Therefore, the design level of switching impulse insulation on the MV side cannot be assessed by this test. Also, in the induced voltage withstand test where the voltage is applied in a conventional way (conducted synchronously with the voltage withstand test for the HV line terminal), the test voltage on the MV side still cannot reach the specified level. To deal with this problem, neutral point potential support is used during the 1-min short-duration power frequency voltage withstand test on the MV line terminal to achieve the required test voltage (630 kV). During the test, partial discharge should be monitored.

The ACSD test sequence is as shown in Figure 9.13.

During the ACSD test on the 500-kV MV side, the neutral terminal potential should be increased by connection to an auxiliary booster transformer such that the 500-kV line terminal of the UHV AC transformer can achieve the required test voltage (630 kV) under a lower induction ratio. However, in this case, the voltage of the UHV windings is lower than its power frequency insulation level. To address this, in the UHV AC pilot and demonstration project, the potential on the LV side (point X of 110-kV winding; see Figure 9.10) of the transformer is utilized to increase the neutral terminal potential, thereby increasing the potential on the MV line terminal. In this way, the ground insulation of the MV line terminal can be assessed. When the test voltage on the MV side reaches the insulation level 630 kV by using this method, the

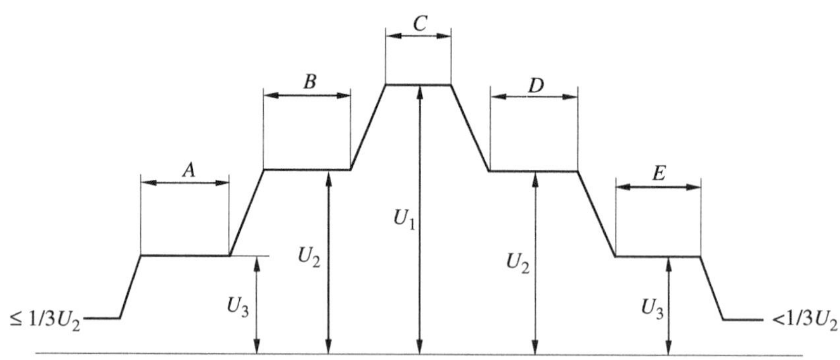

Figure 9.13: ACSD test sequence.

Note: $A = 5$ min, $B = 5$ min, and $C = 1$ min (conversion of frequency to time is required), $D \geq 5$ min, and $E = 5$ min.

test voltage on the 110-kV LV side is required to reach 167.8 kV. Therefore, it is necessary to increase the insulation level of the neutral terminal to the required level. Although, as described, the test voltage on the MV line terminal can reach up to 630 kV by elevating the neutral terminal potential, the test voltage of the longitudinal insulation (inter-turn insulation) on the MV side cannot be increased yet.

This test is successful if: (i) no audible discharge noise in the transformer under test occurs; (ii) no puncturing occurs inside the transformer under testing occurs; (iii) during the second 5 min when the voltage U_2 is applied, the apparent discharge does not exceed 100, 200, and 300 pC on the HV, MV, and LV line terminals, respectively; and (iv) during the second 5 min when the voltage U_3 is applied, the apparent discharge on all line terminals does not exceed 100 pC.

Test results showed that the insulation on the MV side can meet the test requirements, thus achieving the design objective.

9.2.2 UHV Shunt Reactor

9.2.2.1 UHV shunt reactor structure

UHV shunt reactors are core-type reactors. The core comprises core limbs and yokes, which are the main components of the magnetic circuit. It can be classified into single-core limb with two return yokes, two-core limbs with two return yokes, and two-core and winding assemblies, as shown in Figure 9.14. Table 9.4 lists the comparison results between the three core types.

Shunt reactors used in UHV application have a large capacity. If the type with a single-core and two return yokes is used, despite low loss and costs, then the single-phase capacity is larger, causing increased difficulties in flux leakage control and higher risk of local overheating as compared with 500- and 750-kV applications. In contrast, if two-core limbs with two return yokes and two-core and winding assembly types are used, then the

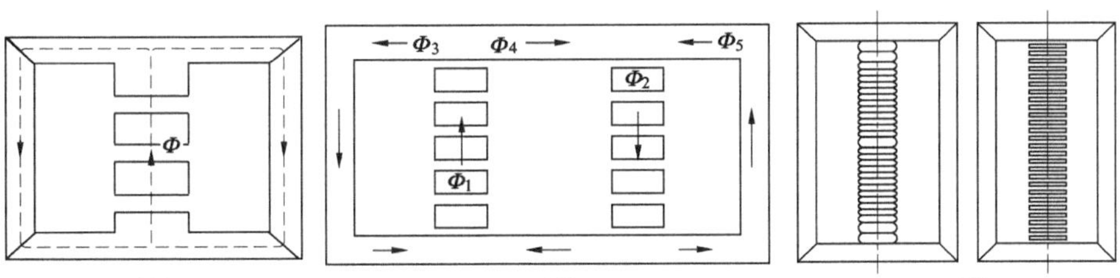

(A) (B) (C)

Figure 9.14: UHV shunt reactor core structure.
(A) Single-core limb with two return yokes; (B) two-core limbs with two return yokes; and (C) two-core and winding assemblies.

Table 9.4: Comparison Between Core Types.

Item	Type 1	Type 2	Type 3
Core type construction	Single-core limb with two return yokes	Two-core limbs with two return yokes	Two-core and winding assemblies
Capacity per phase	Total capacity	1/2 of total capacity	1/2 of total capacity
Advantages	1. Low loss 2. Low cost 3. Light weight 4. Simple core and winding assembly insulation and lead structure	1. Low capacity per phase, lower flux leakage, easy flux leakage control 2. With the same transportation height, larger insulation dimensions allowed; with increased height of winding reactance, insulation margin and reliability improved	1. Low capacity per phase, lower flux leakage, easy flux leakage control 2. With the same transportation height, larger insulation dimensions allowed; with increased height of winding reactance, insulation margin and reliability improved
Disadvantage	1. Higher axial working field strength per phase 2. High capacity per phase, difficulty in flux leakage control 3. Larger windings height	1. High loss 2. High cost 3. Increased weight 4. Complex lead structure 5. Difficult noise control	1. High loss 2. High cost 3. Increased weight 4. Complex lead structure

single-phase capacity is reduced by half and it is easier to control leakage flux and local overheating. Considering this, the UHV AC pilot and demonstration project selected these two types of shunt reactors. This not only can control the risks in equipment development and manufacturing but also can make technical preparations for the future use of single-core limb with two return yokes.

9.2.2.2 Key manufacturing technologies

1. **Connection of windings**
   For two-core limbs with two return yokes and for two-core with a winding assembly, their windings can be connected in two schemes, as shown in Figure 9.15. In the first scheme, the windings of the two columns are series-connected first and then parallel-connected. In this case, the insulation levels at both ends of the columns and the leads are in the 500-kV class and the insulation levels at the middle part of the two columns are changed from 1000 to 110 kV. With such a high voltage difference between the two columns, the insulation configuration becomes complex. In the second scheme, the windings of the two columns are parallel-connected first and then series-connected. In this case, the winding of one column is connected to the middle part of the other winding via a 500-kV lead and only one column has a voltage class of 500 kV at one

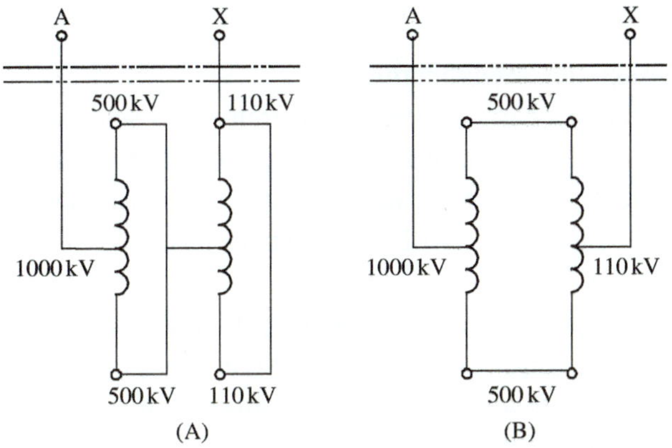

Figure 9.15: Connection diagram of windings of two columns.
(A) Parallel connection after series connection and (B) series connection after parallel connection.

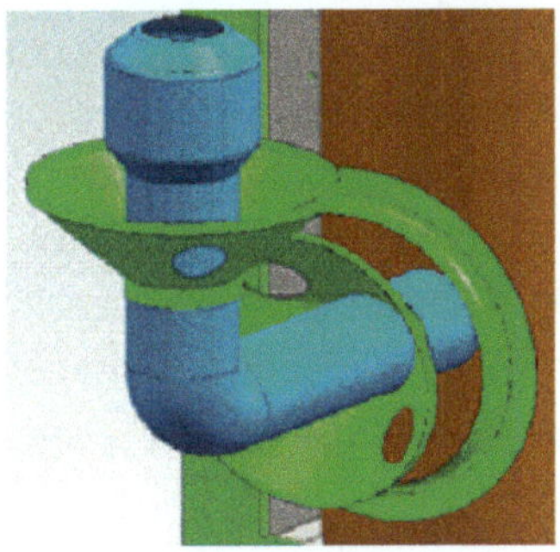

Figure 9.16
Profile of UHV exit lead.

end, which results in reduced high-potential area. The voltage level at the middle parts of the two columns is changed from 1000 to 550 kV. With lower voltage differences between the two columns, the inter-column electrical field and the structure of windings can be improved and the insulation configuration will be simpler. Through this comprehensive analysis, the second connection scheme is selected.

2. **UHV lead connection**

When designing the lead structure of UHV shunt reactors, the UHV outgoing lines and connections between columns should be considered.

For 500-kV shunt reactors, which have lower capacity and voltage levels, direct HV outgoing lines are used, that is, HV bushings are inserted directly into reactor oil through the tank cover. For UHV shunt reactors, taking into account the limitations on the transportation dimensions of the oil tank and the reliability of the insulation, the lead is connected outside from a UHV exit lead installed on the wall of oil tank. Figure 9.16 shows the profile of the UHV exit lead.

9.2.2.3 Key tests

1. **Excitation characteristics measurement**
   The key to the excitation characteristics measurement is selection of proper test equipment and optimum compensation capacity. For UHV shunt reactors, the test voltage reaches up to 1.4-times the rated voltage. To provide the required test voltage, a HV (1700 kV) test transformer with large capacity is used and proper connection scheme is used on the LV side.
   For such testing, large compensation capacity is required. For instance, the test on 320-Mvar shunt reactors requires at least 640-Mvar compensation capacity provided by capacitor towers. Therefore, an accurate calculation of the required compensation capacity is required before testing.
   In the test, a voltage divider is formed by an external capacitor connected at the HV bushing tap measurement terminal of the test transformer and the main capacitor of the bushing, and the voltage ratio is adjusted at a lower voltage. This enables the test voltage to be measured and thereby allows the test voltage to reach 1.4-times the rated voltage.
2. **Measurement of losses and impedance**
   As the capacity of the UHV shunt reactors increases, its power factor is reduced. For instance, the power factor of a 320-Mvar shunt reactor is only approximately 0.0018. With such a low power factor, accurate measurement of the losses becomes more difficult. To deal with this problem, capacitive voltage divider and current transformer with the required precision are indispensable and a special electrical bridge must be used.
   For UHV shunt reactors, their impedance should be measured both at the rated frequency and rated voltage and at relevant LV levels to provide reference data for comparative analysis of the on-site test results.

9.2.2.4 UHV stepped controllable shunt reactor

1. **Main configuration**
   HV controllable shunt reactors can be classified into magnetically controlled and high-impedance transformer types according to their configuration and principle. For the magnetically controlled type, the excitation system regulates the capacity of the reactor within the permissible range by adjusting the saturation degree of the core, thereby achieving continuous and smooth control of reactive power output. For the high-impedance transformer type, the reactor is in the form of a high-impedance transformer with a design leakage reactance equal to or approaching 100%, and a thyristor or circuit breaker is connected on the LV side of the transformer to regulate the capacity. Depending on the regulation mode, the high-impedance transformer type shunt reactor can be further classified into the stepped controllable type and thyristor controllable transformer (TCT) type.

Figure 9.17 shows the schematic diagram of the stepped controllable shunt reactor. The stepped controllable shunt reactor is designed to have a leakage resistance equal to or approaching 100% to make full use of the impedance of the high-impedance transformer. On the secondary side of the shunt reactor, several banks of auxiliary reactor, thyristor valves, and energy extraction reactor are installed. On the HV side, the star-connected three-phase windings are connected directly to the HV grid, and the winding neutral is grounded via a neutral grounding reactor. On the LV side, HV terminals of the three-phase windings are led out through bushings and connected, respectively, to the thyristor valve, auxiliary reactor, and the reactor in series with the circuit breaker, and their LV terminals are connected directly to the ground.

The capacity of a stepped controllable shunt reactor is controlled by using thyristors to switch on/off external reactors, thus meeting the primary system's need for reactive power compensation. In the event of a fault, it is possible to rapidly regulate the reactor to the maximum capacity to suppress overvoltage and secondary arc current. It has the advantages of rapid response speed, reliable operation, simple maintenance, and freedom from harmonic pollution.

2. **Key manufacturing technologies**
   a. *System demand and overall arrangement plan of UHV controllable shunt reactors.* By installing controllable shunt reactors in the UHV transmission system, it is possible to regulate the compensation capacity automatically when the lines operate

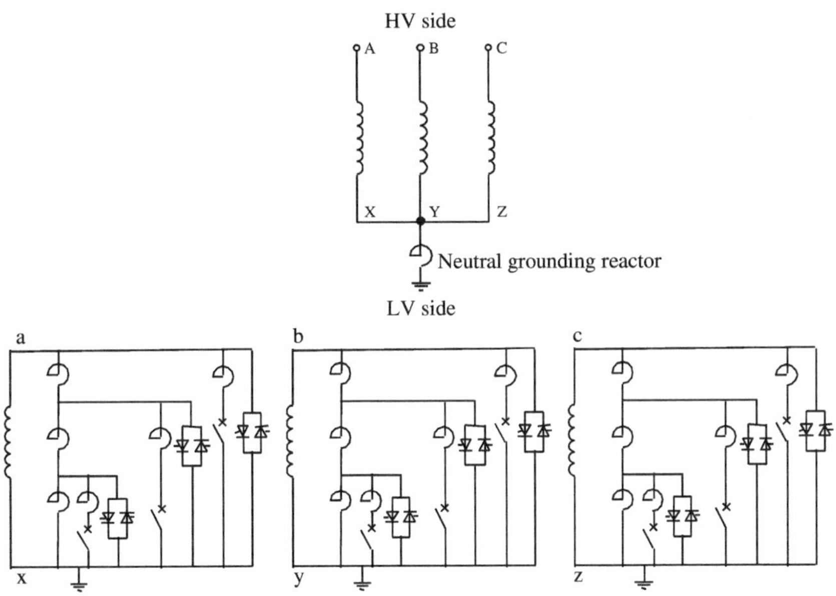

Figure 9.17
Schematic connection diagram of three phases of stepped controllable shunt reactor.

under both heavy load and light load, thus effectively addressing the conflict between limiting overvoltage and reactive power regulation. As shown by the analysis results on reactive power balance of the planned UHV grid, controllable shunt reactors are mainly required by the long-distance transmission lines with or without intermediate drop points and the intermediate substations with dense incoming/outgoing feeders.

For the typical long-distance transmission lines, the rated capacity of the required controllable shunt reactors should be properly determined based on both the long-term and short-term demand. In the event of a system fault, the compensation capacity should be rapidly increased to the maximum value to suppress power frequency overvoltage, resonance overvoltage, and secondary arc current, thereby ensuring the safe, stable, and reliable operation of the UHV transmission system.

By comprehensively considering the voltage fluctuation resulting from stepped switching of the controllable shunt reactors as well as the investment economy and operation reliability, the controllable shunt reactors in three-step configuration are selected, with a total installed capacity of 600 Mvar (200 Mvar for each step) and an adjustable capacity of 400 Mvar. Therefore, the voltage fluctuation of the UHV bus due to stepped switching at Xuzhou Substation is calculated to be 0.59% based on the short-circuit capacity, which is lower than the specified limit (2.5%) of the voltage fluctuation of the UHV bus.

Figure 9.18 shows the schematic diagram of the single-phase main circuit of the UHV stepped controllable shunt reactor, which is arranged in a compact layout to reduce the land occupation of UHV substations in south China. The shunt reactor comprises the following components:

i. Thyristor valve groups TK1 and TK2, which have a capacity of 100% and 67%, respectively, and can only become conductive during capacity switching

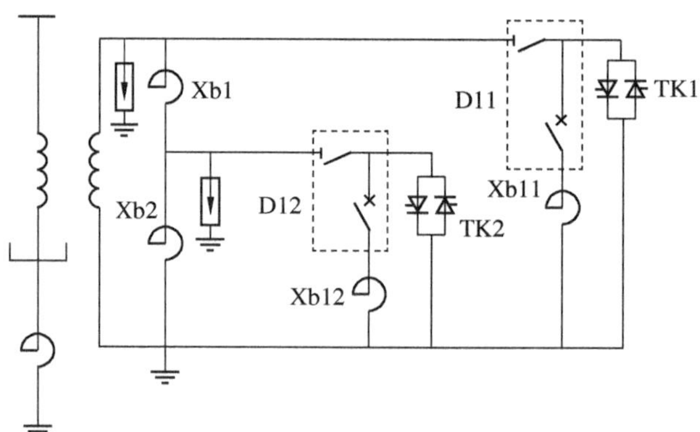

Figure 9.18
Schematic diagram of single-phase main circuit of UHV stepped controllable shunt reactor.

ii. Auxiliary reactors Xb1 and Xb2, which have an oil-immersed core and share the same enclosure, reducing the land occupation
iii. Indoor energy extraction reactors Xb11 and Xb12 of dry-type cores, which extract energy and provide the turn-on voltage for the thyristor valves of corresponding capacity levels
iv. D11 and D12 of 100% and 67% capacity GIS circuit breakers, which are designed according to the closing-priority principle to improve the operation reliability of controllable shunt reactors during transients.

b. *Fast capacity regulation using high-speed hybrid switch.* In contrast to ordinary shunt reactors, the output capacity of UHV controllable shunt reactors may be regulated automatically and rapidly according to the system's needs. This function is achieved by controlling a high-speed hybrid switch that comprises a high-speed bypass circuit breaker, thyristor valve, and energy extraction reactor, as shown in Figure 9.19. The operation of the high-speed hybrid switch involves the following two processes:
   i. During the stepped capacity regulation, the voltage division implemented by the energy extraction reactors can extract energy and provide turn-on voltage required by the thyristor valve. The thyristor valve can be turned-on and turned-off at the current zero-crossing point to allow rapid capacity regulation. Therefore, the bypass circuit breaker is not required to switch on/off large current.
   ii. After the thyristor valve becomes conductive, the bypass circuit breaker closes and carries the operating current for a long period, whereas the thyristor valve only works for a short period and is cooled naturally with air.

   With the high-speed hybrid switch technology, the capacity of the stepped controllable shunt reactor can be rapidly regulated to meet the requirements for frequent regulation of the controllable shunt reactors in the UHV transmission system. In addition, because the bypass circuit breaker does not need to switch on/off large current, its working conditions are improved and, hence, the service life is prolonged, thereby reducing the costs and maintenance efforts.

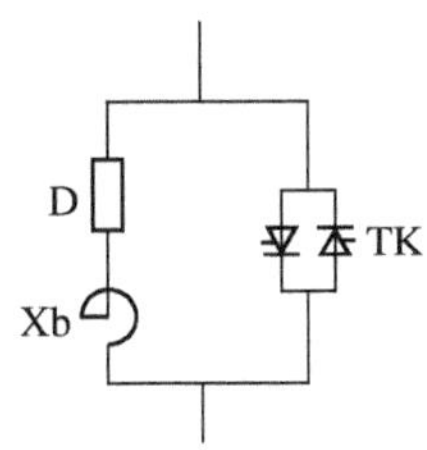

Figure 9.19
Schematic diagram of high-speed hybrid switch.

c. *UHV controllable shunt reactor*. By incorporating a high-impedance transformer configuration, the UHV controllable shunt reactor has the highest voltage level and unit capacity around the world and sets the strictest requirements on various technical performance indicators, including low loss, low noise, low partial discharge, low vibration, and high reliability. Because it has a larger unit capacity and leakage flux, the structure of its core and winding is more complex than the 500/750-kV stepped controllable shunt reactor and UHV transformer. The technical difficulties in the design and manufacturing lie in the control of magnetic leakage field, main and longitudinal insulation structure, HV lead connection, loss control, reduction of vibration and noise, mechanical strength and transportation of the oil tank, as well as cooling system design.

   Special studies have been performed for these technical difficulties and the following key technical measures have been taken:

   i. Perform simulation analysis to obtain the main technical parameters of the reactor.
   ii. Control the leakage flux of the reactor by using thick yoke and installing magnetic shield at both ends of the active part, and eliminate the local overheating problem with the magnetic shield.
   iii. Work out the main and longitudinal insulation design of the reactor and properly determine the gap of the main and longitudinal insulation of the winding and insulation sizes of the leads. Take measures to strictly control the electric field strength at various locations of the windings and lead.
   iv. Use the body with double-core limbs and two return yokes. Calculate and verify the magnetic leakage field, vortex field, and temperature field at metallic parts, such as the core, clamps, and oil tank, to ensure the reliability.
   v. Calculate and analyze the proportions of the winding losses, stray losses, and core losses in total losses of the reactor. Reduce the winding losses effectively by refining the winding coiling.
   vi. Provide a magnetic shield to effectively eliminate local overheating problems and reduce stray losses.

   The detailed design scheme of the reactor has been obtained based on the studies of the key technologies of the reactor. The prototype 1000-kV stepped controllable single-phase reactor has been developed and manufactured, as shown in Figure 9.20. It has a rated capacity of 200 Mvar and has passed all routine tests and type tests in one attempt.

d. *Large-capacity thyristor valve group*. Large-capacity thyristor valve groups are key to regulating the capacity of the UHV controllable shunt reactor. During operation of the reactor, the valve will become conductive for a short time to let through a large current that is far larger than that with 500- and 750-kV stepped controllable shunt reactors. To improve the operational reliability of the UHV controllable shunt

Figure 9.20
Prototype UHV stepped controllable single-phase reactor.

reactor, the design of the thyristor valve control system is improved and the valve group protection system is refined. The details are described here.

i. Insulation coordination for the thyristor valve group and the valve-end arresters is improved.
ii. The thyristor valve is cooled by a natural air cooling system instead of a water cooling system, reducing the maintenance efforts.
iii. The thyristor valve group is allowed to extract energy from both the voltage and current circuit, which improves the operational reliability.
iv. Multiple protections against BOD overvoltage, insufficient valve margin, triggering failure, and unintentional trigger are provided for thyristor valve groups.

The developed large-capacity thyristor valve equipment with a rated voltage of 66 kV and rated current of 3480 A are shown in Figure 9.21. These equipment have passed all type tests and operation tests.

3. **Key tests**
   a. *Capacity regulation test for valve control system of UHV controllable shunt reactor.* To better simulate the rapid switching between the stepped capacity of the UHV controllable shunt reactors, a capacity regulation test has been performed at the high-power switchgear test station in China Electric Power Research Institute.

Figure 9.21
Large-capacity thyristor valve of UHV controllable shunt reactor.

Before this test, an equivalent circuit of the valve control system has been developed to incorporate the thyristor valve, energy extraction reactor, GIS, and control and protection system such that the test is performed based on the actual operation conditions. This test mainly aims to verify:

i. The operation logic of the thyristor valve and bypass circuit breaker during stepped switching
ii. The process of current transfer between thyristor valve and circuit breaker during capacity regulation
iii. Electrical strength of the thyristor valve and circuit breaker during switching, switching time, and operating conditions of the energy extraction circuit of thyristor valve.

Figure 9.22 shows the wave shapes of the currents through bypass circuit breaker and thyristor valve and the voltage across thyristor valve measured in the test circuit.

Test results show that during the capacity regulation of UHV stepped controllable shunt reactor, the thyristor valve can be turned-on or turned-off at current zero-crossing point, no overvoltage occurs when the thyristor valve is turned-off, and no DC bias current flows through the bypass circuit breaker. This can effectively improve the operation conditions and prolongs the service life of the bypass circuit breaker.

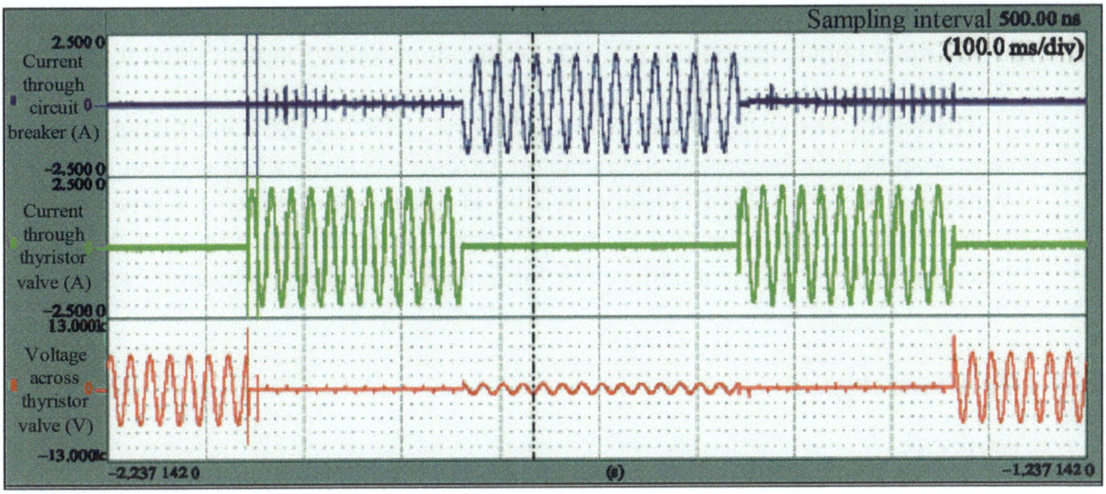

Figure 9.22
Wave shapes of valve control system of UHV controllable shunt reactor measured in capacity regulation test circuit.

Figure 9.23
Test site of the UHV stepped controllable shunt reactor together with auxiliary reactor.

b. *Test of UHV controllable shunt reactor together with auxiliary reactor.* To investigate the excitation characteristics and noise level of the UHV stepped controllable shunt reactor operating at a low capacity rather than full capacity, an auxiliary reactor should also be included in the test.

Before the test, the auxiliary reactor should be properly connected to the UHV shunt reactor, as shown in Figure 9.23. The excitation characteristic curve and noise level of the reactor are measured when the test voltage applied on the HV side is increased from zero to 1.4-times the rated voltage by using a step-up transformer connected on the HV side of the shunt reactor operating at a small capacity.

The results of this test performed for the first UHV stepped controllable shunt reactor show that the excitation characteristic curve is always linear when the test voltage on the HV side of the shunt reactor is increased from 0 to 1.4-times the rated voltage in the range of the 33–67% capacity and the noise level of the shunt reactor together with the auxiliary reactor can meet the design requirements.

9.3 UHV Switchgear

9.3.1 UHV GIS

UHV gas-insulated metal-enclosed switchgear (GIS) can only function when all its internal components are properly assembled and the metal enclosure is filled with sulfur hexafluoride (SF_6) gas at the required density. The density of filled SF_6 gas depends on the required internal arc extinction and insulation performance. Structurally, GIS mainly comprises the current-carrying part (or internal conductor), insulation structure, enclosure, operating mechanism, gas system, grounding system, auxiliary circuit, and components.

In GIS, some components have their own separate gas chambers and some components are connected together and share the same gas chamber. Different chambers may be filled with gas of different densities. The conductive parts inside the metal enclosure are supported on the latter using cast epoxy resin insulators and different chambers are connected electrically with metal fittings. The connecting flanges between the enclosures are precisely machined and high-pressure gas is thus tightly sealed within the enclosures with the aid of corrosion-resistant rubber O-rings. In addition, the enclosures are generally provided with relief valves or rupture discs.

GIS can be classified into two types (i.e., three-phase non-segregated type and phase-segregated type). For three-phase non-segregated GIS, the three phases are housed within the same enclosure and spaced apart using insulating supports or plates. This structure can reduce both the metal enclosure material consumption and land occupation. Moreover, when the three-phase currents flow through the bus concurrently, the resulting magnetic field lines will cancel-out each other in the enclosure, thereby reducing eddy-current losses. Figure 9.24 shows the schematic diagram of the internal structure of three-phase non-segregated GIS.

For phase-segregated GIS, the apparatus for individual phases is installed in separate metal enclosures so that the main circuit of each phase has its separate enclosure. This will create a coaxial cylinder electrode system. In this case, the electric field is evenly distributed, the GIS structure is simple, and the insulation problem can be easily addressed. As a result, no

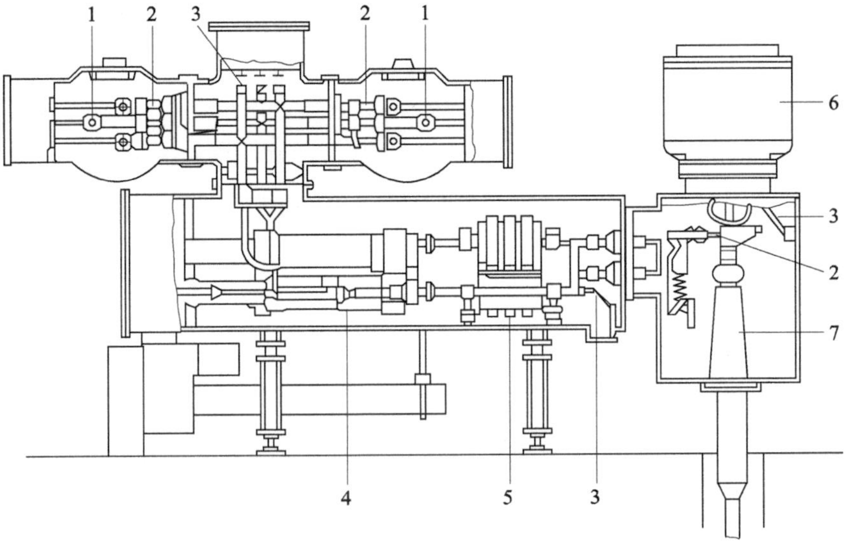

Figure 9.24: Schematic diagram of internal structure of three-phase non-segregated GIS.
1—bus; 2—disconnector; 3—earthing switch; 4—circuit breaker; 5—current transformer; 6—voltage transformer; and 7—cable terminal.

phase-to-phase short-circuit fault will occur. Therefore, GIS that are 330 kV or higher usually use this structure.

Compared with 750-kV switchgear, 1000-kV switchgear requires a higher insulation level. This causes increased equipment sizes and places higher requirements on the mechanical strength of the equipment and makes it more difficult to manufacture, transport, and install the equipment. Figure 9.25 shows the dimensions of 750-kV GIS and 1000-kV GIS for comparison.

GIS has the following advantages over air-insulated switchgear (AIS): (i) its land occupation is only approximately 40% that of the AIS, contributing to a large amount of saved land; (ii) all live parts are enclosed within a metal enclosure and the environmental electromagnetic pollution due to HV can thus be avoided; (iii) personal injuries and deaths due to electric shock can be avoided; (iv) the equipment overhaul interval can be prolonged and the equipment is generally not required to be disassembled and repaired for 10–20 years; and (v) the equipment insulation performance is not affected by atmospheric conditions, the seismic resistance performance is improved, and the operational reliability is enhanced.

GIS is more costly than conventional electrical equipment. However, when allowing for the land occupation and construction costs, the investment ratio of the GIS to conventional electrical equipment is reduced with increase of the voltage level, which justifies the use of GIS in the UHV transmission system.

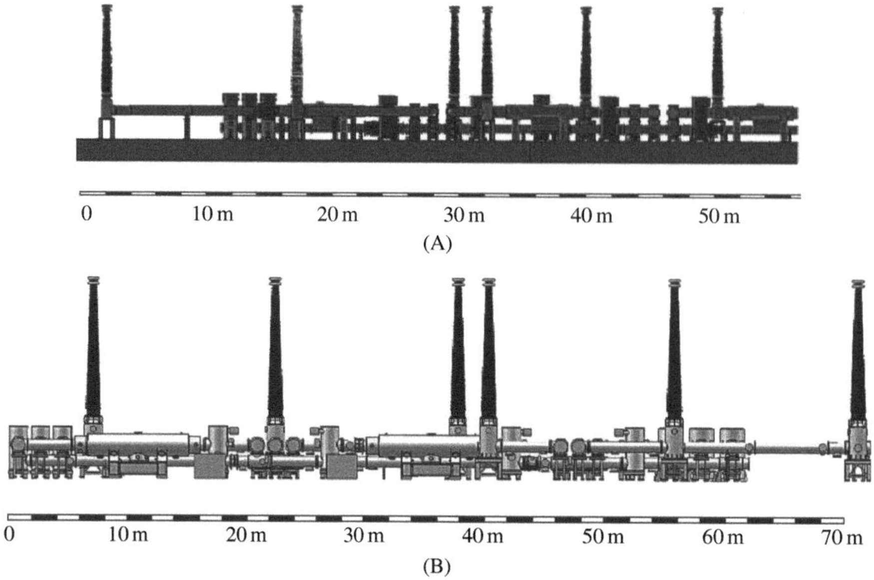

Figure 9.25: Dimensions of 750- and 1000-kV GISs.
(A) 750-kV GIS and (B) 1000-kV GIS.

In UHV substations, HGIS is also widely used. HGIS is comparable with GIS in terms of security and reliability. Because HGIS uses open-type bus, it can be expanded easily and save money. It occupies a land area larger than that of GIS but less than that of AIS, at approximately 70% of AIS. In the UHV AC pilot and demonstration project, Changzhi Substation uses GIS and Nanyang Switching Station and Jingmen substations use HGIS.

In Japan, most of the substations rated 275 kV or higher use GIS, and the development, manufacture, and test of 1100-kV GIS were started earlier. China has successfully developed and manufactured UHV GIS, which was designed jointly by China and foreign countries and was manufactured in China. China's UHV GIS has the following features:

1. Changzhi Substation uses the GIS in an in-line arrangement, which can reduce the longitudinal dimensions and allow crossed connection and flexible arrangement of incoming/outgoing feeders, as shown in Figure 9.6. Nanyang Switching Station uses the HGIS in an M-shaped arrangement, which can further reduce the total length of the equipment. Jingmen Substation uses HGIS in a horizontal arrangement, with the equipment center being only 1.3 m above the ground. Therefore, no working platform is required.
2. Circuit breakers are provided with 560- to 600-Ω shunt resistors. For instance, the disconnectors in Changzhi Substation and Nanyang Substation are provided with 500-Ω shunt reactors, which reduce the switching overvoltage.
3. Each chamber of the GIS is provided with an SF_6 gas density relay that serves to send alarm and blocking signals for monitoring by the site operators. Moreover, these signals

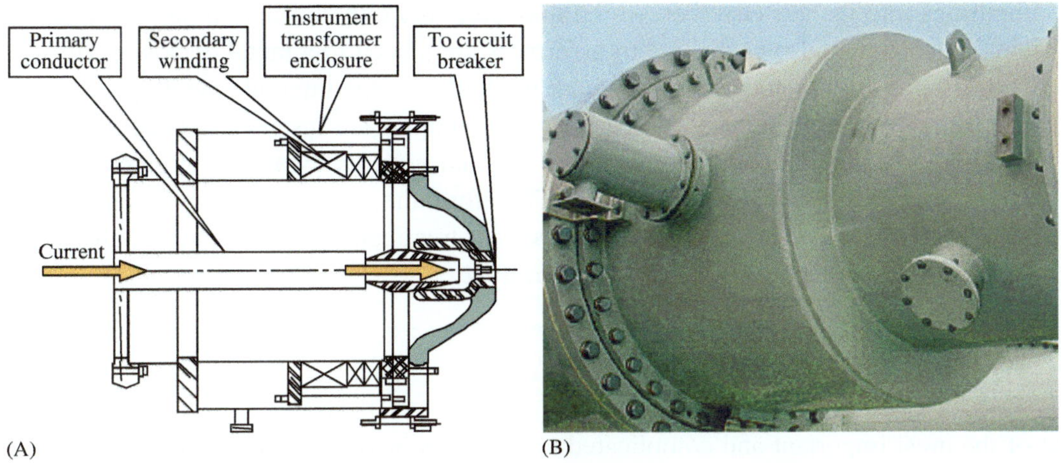

Figure 9.26: Internal-type current transformer.
(A) Structure and (B) picture.

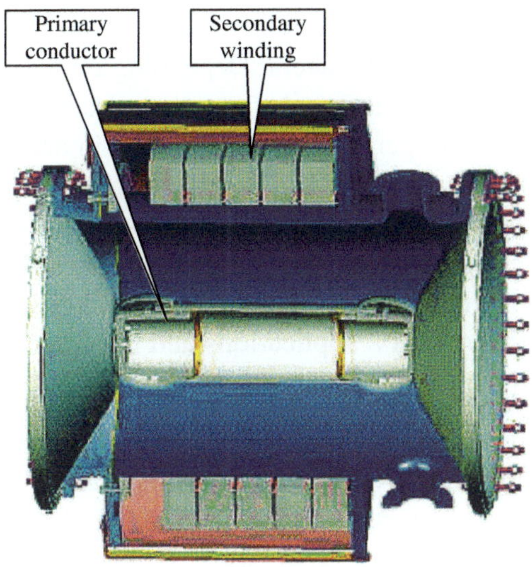

Figure 9.27
External current transformer.

can also be transmitted to the control room through cables for online monitoring of SF_6 gas density. In addition, GIS also has a partial discharge online monitoring system and sensors for measuring very fast transient overvoltage (VFTO).

4. There are two types of current transformer (i.e., internal type and external type), as shown in Figures 9.26 and 9.27. For the internal type, after single windings are coiled, they are usually cast integrally with epoxy resin to prevent entry of moisture and other

impurities into the gas chamber, which may deteriorate the insulation performance. The windings of internal-type current transformers are placed in a metal enclosure filled with SF_6 gas at the rated pressure, which is fixed onto the internal shielding cylinder. For the external-type current transformer, the windings are installed outside the enclosure of GIS/HGIS. In contrast to the internal type, the windings of external-type transformers are directly exposed to air. Therefore, the surface of the windings should be subjected to corrosion and moisture-resistant treatment and the windings are not required to be cast with epoxy resin integrally.

9.3.2 UHV Circuit Breaker

One of the most important and complicated apparatuses in UHV switchgear, the circuit breaker, can carry and cut-off both the normal current and specified overload current (such as short-circuit current) of an operating circuit. It is widely used in power plants, substations, switching stations, and transmission lines, assuming the dual tasks of control and protection.

There is a wide variety of HV circuit breakers. Depending on the arc extinction medium, they are classified into oil circuit breakers, air (vacuum) circuit breakers, and SF_6 circuit breakers. Because of the excellent performance of SF_6 gas in arc extinction and insulation, SF_6 circuit breakers have many advantages: (i) high break voltage and large breaking capability; (ii) can break short-circuit current many times in succession; (iii) allow frequent operations; (iv) can break capacitive current without restrike; and (v) can break inductive current without current chopping. In recent years, SF_6 circuit breakers have taken the place of air circuit breakers in HV and EHV applications and have become the dominant type in UHV applications.

SF_6 circuit breakers have strict requirements for the materials, manufacturing, and assembling processes, and particularly for the gas tightness; specifically, the annual leakage rate should be less than 1%. Therefore, the manufacturer of SF_6 circuit breakers should be equipped with special purified assembly workshop, gas recovery processing device, SF_6 gas leakage detector, micro-water detector, absorbents for low-fluorine compounds and moisture, and drying equipment.

In terms of the structural forms, SF_6 circuit breakers are classified into two types, dead tank and live tank, as shown in Figure 9.28.

9.3.2.1 Structural characteristics

In addition to performing the tasks of ordinary HV circuit breakers, UHV circuit breakers should be able to reduce the switching overvoltage as low as possible by means of special

(A) (B)

Figure 9.28: SF_6 circuit breakers.
(A) Live tank and (B) dead tank.

measures (such as opening and closing resistors), thus reducing the design insulation level and manufacturing costs of transmission lines and substation equipment.

Among the countries around the world, the former Soviet Union used compressed air circuit breakers instead of SF_6 circuit breakers in the UHV transmission systems because of cold weather; the SF_6 gas might be liquefied and therefore cannot work effectively. BHB-1150-4000-40 air circuit breaker was manufactured in accordance with the 110- to 750-kV BHB series and built with dual-break arc extinction units, with each break rated 250 kV. The 1150-kV circuit breakers are live tanks and three out of the five arc extinction units per pole are provided with closing resistors of 390 Ω $[6 \times (63 + 2)]$ to reduce the switching overvoltage on the line resulting from closing of the circuit breaker. In other counties, dead tank circuit breakers are used for the UHV test projects.

The arc extinguishing chamber of a circuit breaker consists of a number of breaks, each of which can withstand a certain voltage. For instance, if a single break can withstand 250 kV, then two breaks are required for a 500-kV circuit breaker and four are required for a 1000-kV circuit breaker (in this case, shunt capacitors are required between breaks for voltage grading). If a single break can withstand 500 kV, then only two breaks are required for a 1000-kV circuit breaker. In practice, 1000-kV circuit breakers may be classified into two types, dual-break and four-break.

As mentioned, to reduce the switching overvoltage, the UHV circuit breaker should be provided with opening and closing resistors. For its operating principle, see Figure 9.29. During the opening operation, the main contact of the circuit breaker is opened first and the opening resistor is still connected in the circuit through the auxiliary contact. After a lapse of approximately 30 ms, the auxiliary contact connected in series with the opening resistor will be opened. During the closing operation, the circuit breaker operates in the reverse order. Specifically, the auxiliary contact in series with the closing resistor is closed first; after a lapse of approximately 10 ms, the main contact will be closed. This is equivalent to closing a resistive circuit in parallel and, hence, no high closing overvoltage will be generated.

The parameters of opening and closing resistors as well as the time of delayed opening and early closing of the auxiliary contact should be determined through calculations based on the actual conditions of the system and line [4]. Usually, to reduce the switching overvoltage, the optimum values of closing resistance and opening resistance are different. The closing resistance is required to be of lower value and the opening resistance is required to be of larger value. To simplify the structure of the circuit breaker, only one resistor and one auxiliary contact may be used to serve as both closing and opening functions. In this case, proper resistance should be determined according to requirements for limiting overvoltage. However, because only one resistor is used, it should have a larger heating capacity to deal with the opening operation. Through analysis and comparison, usually only a closing resistor is used and the switching overvoltage resulting from the opening operation will be limited by surge arresters.

9.3.2.2 UHV circuit breakers in China

There are two types of UHV circuit breakers, series dual-break and series four-break. The dual-break circuit breaker has a simpler structure and less breaks. However, each break is required to withstand higher voltage, placing higher requirements on its insulation

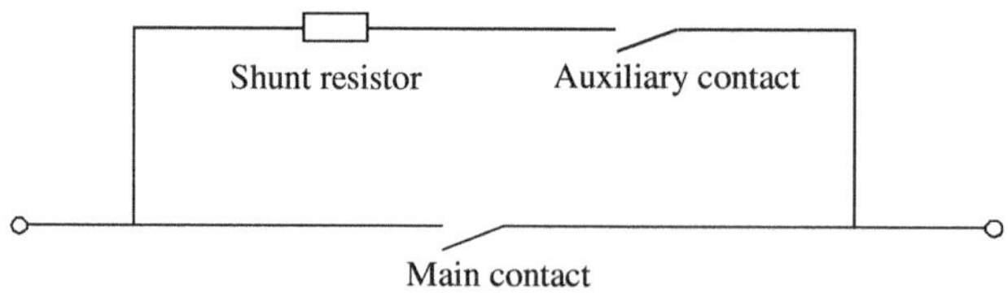

Figure 9.29: Schematic diagram illustrating the operations of the circuit breaker with opening and closing resistors.

Note: The auxiliary contact of the resistor is opened after a delay of approximately 30 ms during opening operation and is closed approximately 10 ms before delay during closing operation.

performance and requiring a very high power operating mechanism. In contrast, the four-break circuit breaker has standard and interchangeable parts and requires a lower power operating mechanism. However, it consists of a large number of parts and components and the structure of its closing resistor is relatively complex.

The UHV circuit breaker manufactured by Pinggao Electric has a shunt resistor that serves as both an opening and a closing resistor (only installing a closing resistor is also allowed). This shunt resistor and arc extinguishing chamber are housed in the same tank. The associated hydraulic operating mechanism is provided with two operating cylinders that cooperate with each other and serve to control the main break and resistor break, respectively, thus ensuring that the resistor break is closed before the main break during closing operation and is opened after the main break during opening operation. The arc extinguishing chamber is configured to contain two breaks connected in series and has a double-acting hybrid puffer structure. At the time of opening, arc is blown out from both directions. Each break is provided with a shunt capacitor, as shown in Figure 9.30. For the specific product parameters, see Appendix A.

As shown in Figure 9.31, the UHV dual-break circuit breaker manufactured by New Northeast Electric only has a closing resistor. The closing resistor and the arc extinguishing chamber are housed in the same tank. The circuit breaker is provided with a hydraulic operating mechanism, and the main break and auxiliary break can operate synchronously. Such mechanical structure ensures that the resistor break is closed and opened before the main break. For the specific product parameters, see Appendix A.

The UHV four-break circuit breaker manufactured by Xi'an XD Switchgear Electric has a shunt closing resistor. The closing resistor and arc extinguishing chamber are housed in separate two tanks, avoiding the interaction between them. This allows the total height of the product, the housing sizes, and consumption of SF_6 gas to be reduced. The operating mechanism and arc extinguishing chamber are directly connected. A disc cam drive

Figure 9.30
Dual-break circuit breaker with closing and opening resistor.

Figure 9.31
Dual-break circuit breaker with closing resistor.

Figure 9.32
Four-break circuit breaker with closing resistor.

mechanism is used to drive the closing resistor break to close before the main break. Such an operating principle has enhanced the reliability of the resistor. The arc extinguishing chamber contains four breaks in series, each of which is provided with a shunt capacitor, as shown in Figure 9.30. For the specific product parameters, see Appendix A (Figure 9.32).

China's UHV circuit breakers have been successfully subjected to all type tests at home. Also, UHV 63-kA circuit breakers have been tested for the first time around the world to break the full capacity of an entire pole and to break the peak transient recovery voltage (TRV), taking into account the effect of series compensation [3,4]. Figure 9.33 shows the real capacity test arrangements.

9.3.3 UHV Disconnector

Disconnector is a type of switching device that, once opened, can provide the specified insulation distance between contacts and visible break and, once closed, can carry normal

(A) (B) (C)

Figure 9.33: The 1000-kV 63-kA circuit breaker capacity test arrangements.
(A) Dual-break circuit breaker in Changzhi Substation; (B) dual-break circuit breaker in Nanyang Substation; and (C) four-break circuit breaker in Jingmen Substation.

operating current and short-circuit current. However, a disconnector cannot interrupt short-circuit current and large operating current. A disconnector primarily consists of porcelain insulating posts (post insulator), movable conducting arm, and operating mechanism, and it has no arc extinguishing chamber. Therefore, it has a structure much simpler than a circuit breaker and, hence, only the heating resulting from operating current and the dynamic and thermal stability under short-circuit current need to be considered.

In terms of the number of post insulators and operating behavior of the movable conducting arm, disconnectors can be classified as one of the following types:

1. *Single-post vertically retractable type*: This comprises a foldable conducting rod in the upper part and a vertical post insulator in the lower part. The conducting rod is movable (up and down) and may be arranged directly under the bus. Once opened, it can create a vertical air gap.
2. *Double-post horizontally rotatable type*: This comprises two post insulators that are spaced apart and vertically arranged. Once the disconnector is opened, the movable rods at the top can be rotated by 90° horizontally, creating a horizontal air gap. Another form is characterized by two post insulators that are arranged in a V-shape, known as the V-type disconnector.
3. *Double-post horizontally retractable type*: This comprises two post insulators, one of which has static contacts and the other has a foldable conducting rod. The conducting rod is horizontally retractable and can create a horizontal air gap during the opening position.

Figure 9.34: The 500-kV disconnector.
(A) Vertically retractable type and (B) horizontally retractable type.

4. *Triple-post horizontally rotatable type*: This comprises three post insulators that are spaced apart and vertically arranged. The two side post insulators are stationary and the middle one is provided with a horizontal movable conducting rod at the top. Once the disconnector is opened, the conducting rod is rotated by 60° horizontally, creating two horizontal air gaps.

The type selection and arrangement of the disconnector depend on the type of the switchgear and directly affect the land occupation and structure of the switchgear. For instance, the single-post disconnector used for buses may be arranged directly under the buses, thus reducing the longitudinal dimensions and land occupation of the switchgear. In contrast, if double-post horizontally rotatable disconnectors are used for buses, then a larger space is required because of increased bay width. Also, certain longitudinal dimensions are required for bus connection. All these will result in a larger land area being occupied by the switchgear. Figure 9.34 shows the vertically retractable and horizontally retractable 500-kV disconnectors.

Figure 9.35 shows the 1000-kV outdoor open-type disconnector used in UHV series compensation devices. It is in a triple-post horizontally rotatable (spinning) configuration that is characterized by a conducting rod that first rotates horizontally and, once the moving

(A) (B)

Figure 9.35: The 1000-kV outdoor open-type disconnector.
(A) Disconnector in Nanyang Substation and (B) disconnector in Changzhi Substation.

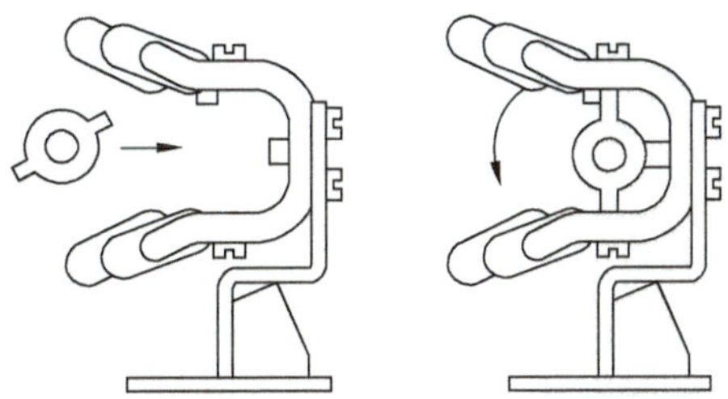

(A) (B)

Figure 9.36: Acting sequence of spinning contacts of 1000-kV outdoor open-type disconnector.
(A) Before closing and (B) after closing.

contact enters into and is blocked by the static contact holder, a flipping mechanism that drives the conducting rod to spin, thereby causing the moving contact to come into contact with the static contact finger. In contrast to the traditional triple-post with a middle post insulator that carries a large torque and is therefore susceptible to damage or breakage, if operated improperly, this triple-post has an advantage in that because only a little operating force is required, the post insulator is subject to a small lateral impact and that self-cleaning of the contact can be achieved. Figure 9.36 shows the acting sequence of contacts of this type of disconnector.

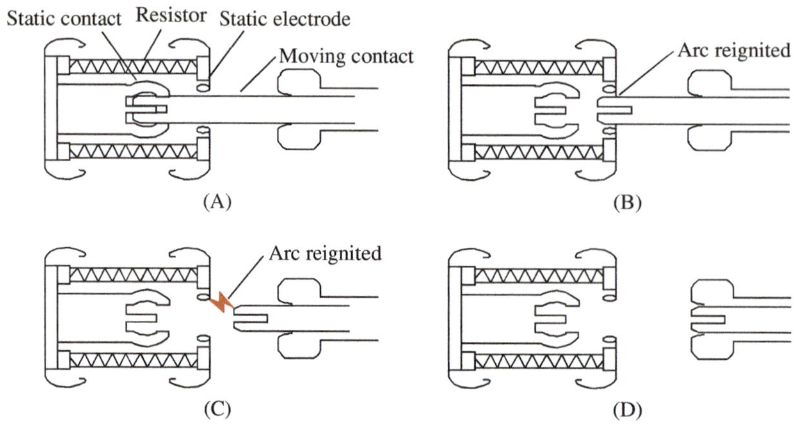

Figure 9.37: Schematic diagram of switching operation of disconnector where a resistor is series-connected.
(A) Closed state; (B) state 1 during opening operation; (C) state 2 during opening operation; and (D) opened state.

Because the disconnector in GIS opens and closes at a low speed, restrikes often occur in the SF_6 gas, leading to the VFTO, which is also known as steep-front overvoltage for its high oscillating frequency and steep front. Despite low amplitude (usually no higher than 2.0 p.u., but sometimes reaching up to 2.5 p.u.), the high frequency and steeply rising front of the VFTO make it difficult to protect the winding equipment (such as transformers) connected to the GIS bus even using gapless MOA, resulting in damage of the inter-turn insulation. Moreover, very fast transients outside the GIS may result, such as overvoltage on the enclosure of the GIS. This will cause a transient ground potential increase and consequent electromagnetic interference to the control and protection and other secondary equipment of the substation. This steep-front overvoltage will also expose the areas without sufficient machining accuracy and cleanliness inside the GIS to the risks of insulation breakdown. Numerous such accidents have been reported around the world. To mitigate this problem, in China, UHV GIS disconnectors are provided with a 500-Ω resistor, which is series-connected during switching operations, as shown in Figure 9.37. Test results show that this method can reduce the overvoltage less than 1.2 p.u.

Figure 9.37 shows the operating principle. With a resistor installed on the shield of the static contact of the disconnector, when the disconnectors are opened, an arc occurs between the moving contact and the shield of the static contact, and the resistor on the shield of the static contact will become connected in series in the circuit. Figure 9.37A–(D) illustrates different states during the opening operation. The closing operation is in reverse order, in which case, the resistor is also connected before the contacts are completely closed, greatly reducing the VFTO caused by restrikes. Figure 9.38 shows the disconnector in a 1000-kV GIS. For the specific product parameters, see Appendix A.

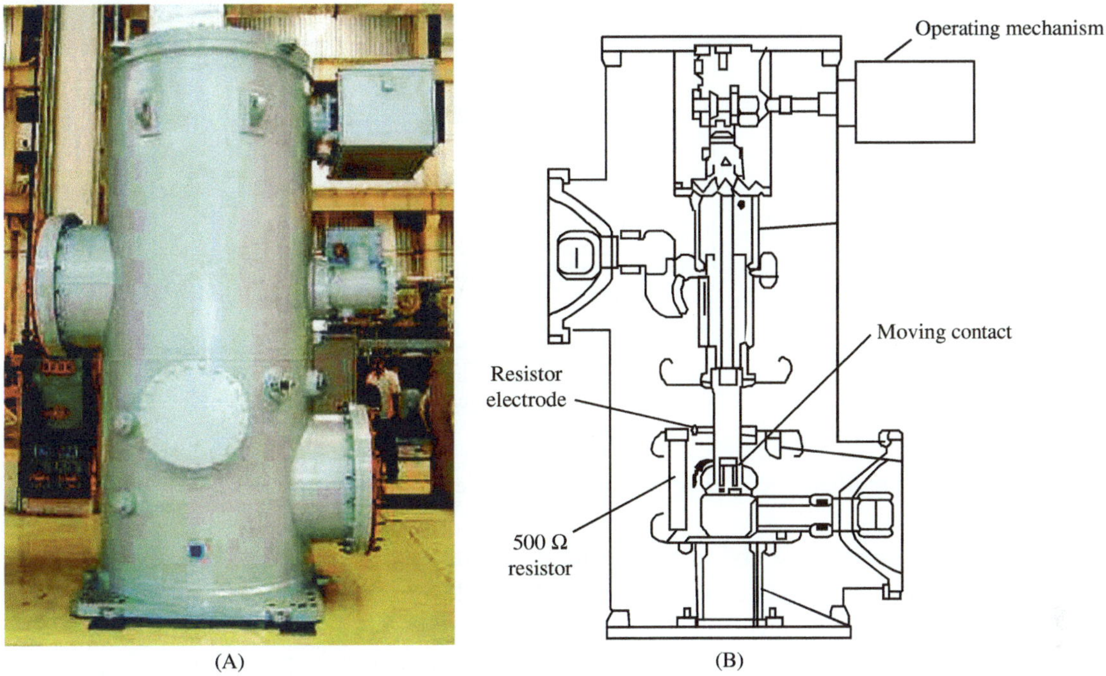

Figure 9.38: Disconnector in 1000-kV GIS.
(A) Picture and (B) internal structure diagram.

9.4 UHV Series Compensation Devices

9.4.1 Configuration

Figure 9.39 shows the simplified circuit of a UHV series compensation device that mainly consists of the capacitor bank, metal oxide varistor (MOV), trigger gap (gap for short), bypass circuit breaker, current-limiting damping equipment, and disconnectors. During system faults, a short-circuit current may flow through the series compensation device, causing terminal-to-terminal overvoltage on the series compensation capacitor bank. To ensure the system security and stability, it is generally required that bypassing the series compensation capacitor bank is not allowed in the case of an external fault (i.e., line fault occurring outside the protected line section containing the series capacitor bank), but it is only allowed in case of an internal fault. Therefore, reasonable and proper overvoltage protection measures must be taken. In the series compensation configuration shown in Figure 9.39, the MOV acts as the main overvoltage protection. If necessary, the gap may be bypassed to prevent the MOV's energy consumption from increasing beyond the withstand capability. Moreover, the bypass circuit breaker makes it possible to extinct arc across the gap, and it cooperates with the disconnectors to enable switching in/out of the series compensation capacitor bank. The current-limiting damping devices function to limit the

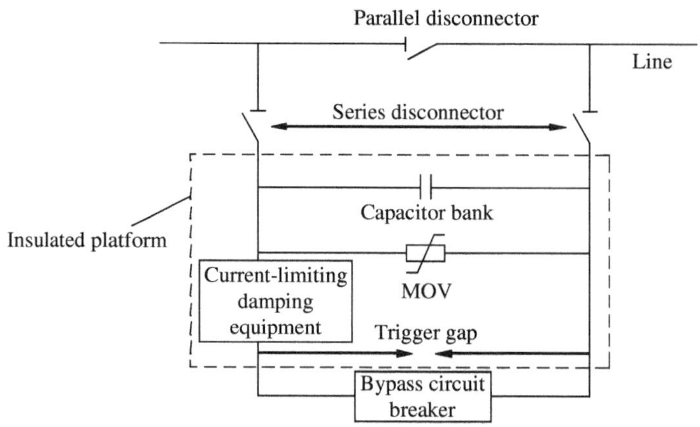

Figure 9.39
Simplified circuit diagram of a UHV series compensation device.

discharge current magnitude and to accelerate its attenuation. Through studies of overvoltage protections, it is determined in the basic design that 2.3 p.u. should be taken as the overvoltage protective level for the expansion of UHV AC pilot and demonstration project. For Changzhi–Nanyang line I and Nanyang–Jingmen line I, the MOV energy absorption capabilities are, respectively, 92 and 70 MJ, and the equipment, such as the gap and bypass circuit breaker, have a peak withstand current of 170 kA and short-time withstand current (r.m.s value) of 63 kA.

The series compensation device is a closely linked, complicated system comprising the capacitor, protection circuit, switchgear, and insulated platform, which rely on each other to fulfill the function and constrain each other to meet the performance requirements. To allow the series compensation device to be functional, proper coordination between the primary electrical equipment, and between the primary electrical equipment and the control and protection system, is required. As compared with series compensation in EHV applications, series compensation devices in UHV applications are featured with larger capacity and higher current and voltage ratings. They are closely related to the system's security and stability and, therefore, place higher requirements on the reliability, balance between electrical and mechanical performance, and synergism between the primary and secondary electrical equipment.

9.4.2 Key Technical Requirements

9.4.2.1 Selection of ratings

Ratings of the series compensation device mainly include the degree of series compensation, current, voltage, and capacity. Series compensation devices used in different systems vary in functions and operating conditions. To obtain the best technical and economic efficiency, these ratings should be determined through special studies while following the guidelines and allowing for the involved system's specific conditions and future developments.

1. **Degree of series compensation, grouping, and segmentation**
   The degree of series compensation should be selected based on the system voltage and reactive power, voltage profile along the line, system stability, distribution of power flow between parallel lines, and risk of sub-synchronous resonance in the vicinity of steam turbine generators. A higher degree of series compensation can improve the system operating performance but will lead to an increased cost of series compensation devices. Moreover, when series compensation devices are provided on the output channel of steam turbine generators, mitigating sub-synchronous resonance will lead to additional risks. For more than one line running in parallel, the degree of series compensation should be selected to prevent uneven distribution of power flow.
   The rated capacitive reactance will be determined according to the known degree of series compensation.

   If the line has a greater length and higher degree of series compensation and rated current, then series compensation devices may be either divided into two groups that are arranged at both ends of the line or concentrated at the middle without grouping to improve the voltage profile along the line. In the latter scheme, when the short-circuit current is smaller, it is possible that a less rated capacity of MOV is required, in which case, however, a series compensation station should be added. If the line has a shorter length and lower degree of series compensation or rated current, the series compensation devices should not be divided into groups.

   Usually, series compensation devices should not be segmented so that good economy can be ensured. However, if the rated voltage is too high, then it is necessary to segment the series compensation devices to decrease the rated voltage of each segment, thus reducing the requirements on main components of the series compensation devices, but the segmentation will incur additional equipment cost. For series compensation devices consisting of two segments, if any segment is bypassed, then the remaining one can still provide the required compensation. Therefore, the series compensation devices may be segmented to improve the power flow control.

   Using the expansion of the UHV AC pilot and demonstration project as an example, it is determined based on the studies on the system power flow stability that the fixed series compensation device with a compensation degree of 40% should be installed for both the Changzhi–Nanyang line I and Nanyang–Jingmen line I. Moreover, based on the studies of the voltage profile lines, arranging series compensation devices at both ends of the Changzhi–Nanyang line I but only at the Nanyang end of the Nanyang–Jingmen line I has been proposed. Considering the difficulties in equipment manufacturing, the series compensation devices centralized at the Nanyang end of the Nanyang–Jingmen line I are divided into two identical segments.

2. **Rated current**

To determine the rated current of the series compensation device, it is necessary to know the continuous operating current, fault current, and swing current beforehand when the series capacitor bank is switched-on and bypassed. Figure 9.40 shows a typical example of the swing current versus time curve during system fault. The series compensation device should accommodate these currents.

In China, the rated current of UHV series compensation devices is usually selected based on the line's thermal-stability current. Because the line's thermal-stability current may be as high as 6–7 kA, the determined rated current will result in excessive capacity, and excessive capacitive elements will jeopardize the device's reliability. In light of this, the following principles are indicated for determining the rated current and overload capability of the capacitor bank.

a. The system power flow and stability should be calculated according to the short-term and long-term grid planning and DL 755—2001 *Guide on Security and Stability for Power System* to determine the series compensation device's transmitted power in normal conditions and emergency transmitted power in the event that faults and abnormal conditions occur in the power system.
b. The capacitor bank should have a rated current larger than the current corresponding to the maximum normal transmitted power. On this basis, the overload capability of the capacitor bank should be reasonably exploited to accommodate the need for emergency power transmission in the event that faults and abnormal conditions occur in the power system.

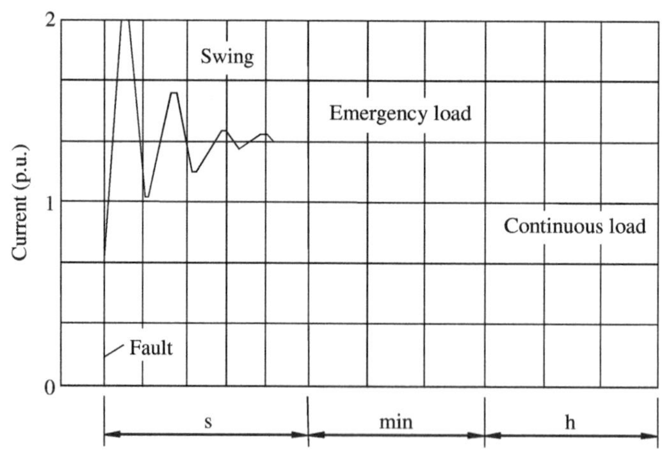

Figure 9.40
Typical swing current versus time curve during system fault (fault current not shown).

Generally, the swing current is determined using the transient stability program. If it is <1.7 p.u. of the rated current of the capacitor bank, then it has less effect on the capacitor bank design. If it is higher than 1.7 p.u., then it will largely affect the capacitor bank design.

Using the expansion of the UHV AC pilot and demonstration project as an example, based on these principles, the rated current of series compensation capacitor bank is determined to be 5.08 kA, which is approximately twice that of the EHV series compensation device. Also, an overload current of 6.1 kA (2 h of continuous operation) is proposed to accommodate the emergency power transmission during the system fault.

Table 9.5 gives the ratings of the series compensation device's capacitor bank determined based on the principles stated. These ratings all correspond to the ambient temperature of 20°C.

9.4.2.2 Overvoltage suppression and basic design principle

As the line current fluctuates, the series capacitor banks suffer much more voltage variation than shunt capacitor banks. If no measures are in place, then the voltage across the series capacitor bank would be very high when short-circuit faults occur, and designing the capacitor banks based on this voltage will not be justified economically. Therefore, in most cases, a shunt overvoltage protection device is provided to suppress capacitor overvoltage. As stated, the series compensation device for the expansion of the UHV AC pilot and demonstration project has a rated current of 5.08 kA. However, from a long-term view, the short-circuit current may reach 60 kA, which is almost 12-times the capacitor bank's rated current and is far beyond its withstand capability. Therefore, safe and reliable overvoltage suppression measures are a key to ensure the safety of series compensation devices.

1. **Overvoltage protection**
   When developing the specifications for series compensation technology, it is necessary to analyze and study the power system's sub-synchronous resonance or the TRV of the line circuit breakers to determine the overvoltage protective level.

Table 9.5: Ratings of the Series Compensation Device's Capacitor Bank.

Item	Series Compensation Device at Changzhi End of Changzhi–Nanyang Line I	Series Compensation Device at Nanyang End of Changzhi–Nanyang Line I	Series Compensation Device at Nanyang End of Nanyang–Jingmen Line I
Degree of series compensation	20%	20%	20% × 2
Rated capacitive reactance (Ω)	19.38	19.38	14.77 × 2
Rated current (A, r.m.s)	5080	5080	5080
Rated voltage (kV, r.m.s)	98.4	98.4	75.0 × 2
Rated three-phase capacity (Mvar)	1500	1500	1144 × 2

Today, the dominant overvoltage protection measure for EHV series compensation devices is using MOV in combination with gap, bypass circuit breakers, and current-limiting damping devices. See Figure 9.39 for the circuit diagram. The MOV acts as a passive protection and is directly parallel-connected across the capacitor bank. The gap acts as an active protection and is triggered by the control and protection system of the series compensation device. Figure 9.41 shows its typical protection logic, which usually works based on the MOV current and energy consumption. Specifically, on detecting either of the two parameters going beyond the protection setting, the control and protection system will send orders to trigger the gap and close the bypass circuit breaker, thus bypassing the series capacitor bank. The gap is generally turned-on before the bypass circuit breaker is closed. It is because once the bypass circuit breaker is closed, the voltage across the gap cannot support the arc discharge; therefore, the arc will soon become extinct. After a certain time lapse, the gap insulation will be recovered, making it possible for the series compensation device to be reinserted.

With such overvoltage protection measures, the protective level is typically in the range of 2.0–2.5 p.u., which can adapt well to the UHV series compensation device. Hence, it is also used in the expansion of the UHV AC pilot and demonstration project.

2. **Basic design**

The basic design mainly aims to establish an electromagnetic transient model of the series compensation device and the power system to calculate the operating conditions of the main equipment of the series compensation device under various operating and fault conditions and to determine the main wiring configuration of the series compensation device, key parameters of the main equipment such as the capacitor bank, MOV, gap, bypass circuit breaker, current-limiting damping device, and the series

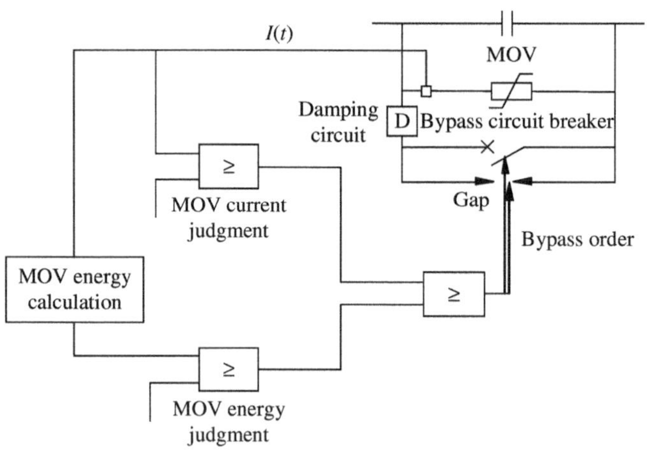

Figure 9.41
Schematic diagram of the overvoltage protection logic of series compensation device.

compensation device's insulation level. All of these will result in a basic design report or equipment specifications of the series compensation device.

Table 9.6 gives the series compensation voltage U_{SC}, peak current through MOV I_{MOV}, and the maximum energy consumption of MOV per phase E_{MOV} of the expansion of the UHV AC pilot and demonstration project in the event of various external faults in substations at both ends of the Changzhi–Nanyang line I and Nanyang–Jingmen line I under different operating modes.

According to the basic control and protection strategy of series compensation, the MOV's threshold energy absorption and current should be larger than the maximum energy consumption and current of the MOV during various external faults with certain margins to ensure that the series compensation device (gap and bypass circuit breaker) will not operate after these external faults.

In the event of an internal fault, if the MOV's current or energy consumption reaches the pre-determined setting, then the control and protection system will operate according to the logic as shown in Figure 9.42 (i.e., bypass the series compensation capacitor bank). The simulations based on this provide the operating conditions of main components of the series compensation devices in Changzhi and Nanyang substations of Changzhi–Nanyang line I and in Nanyang substation of Nanyang–Jingmen line I, as shown in Table 9.7. Figure 9.42 shows the typical waveform of relevant electrical parameters of the series compensation device during internal faults.

The simulations mentioned provide the technical basis for determining the technical requirements of the series compensation device's main components such as the capacitor bank, MOV, gap, and damping circuit.

9.4.2.3 Major concerns for development of series compensation device and its main components

The major concerns for development of the series compensation device and its main components are described here.

1. As compared with 500-kV series compensation devices, the UHV series compensation device's platform has a larger area (almost four-times larger), height of 12 m (more than two-times higher), and total load of 160 t (almost four-times heavier). It is designed with 8-degree seismic fortification intensity and places higher requirements on its coordination with the foundation, stability, and bearing capability of the supporting

Table 9.6: Operating Conditions of the Main Series Compensation Devices Following External Fault on the Protected UHV Line.

Line	Series Compensation Device	U_{SC} (p.u.)	I_{MOV} (kA)	E_{MOV} (MJ)
Changzhi–Nanyang line I	Changzhi substation	2.09	16.6	27.9
	Nanyang substation	2.09	16.5	24.9
Nanyang–Jingmen line I	Nanyang substation	2.10	18.9	26.0

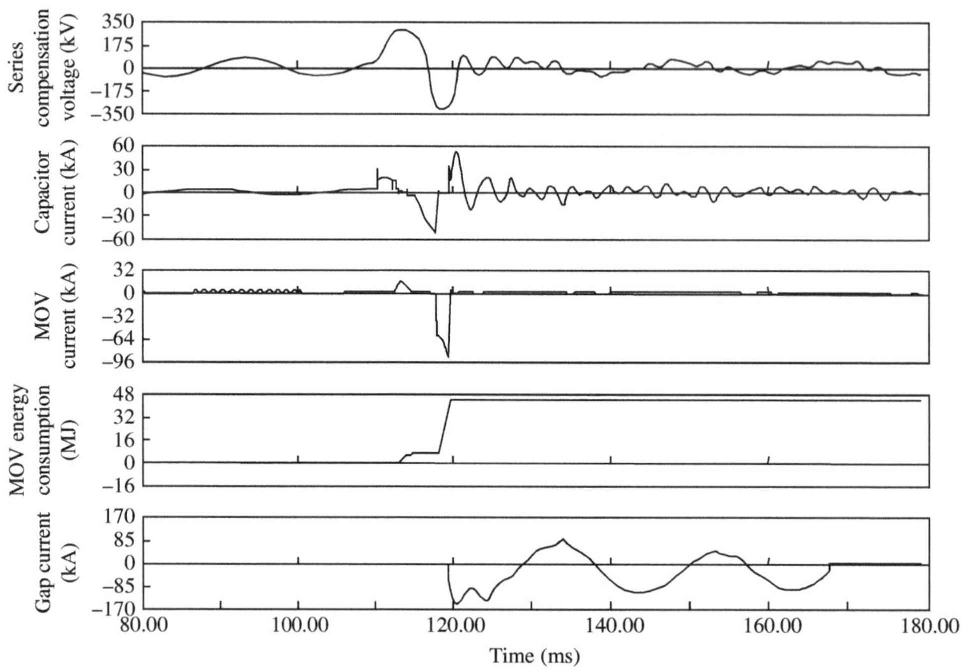

Figure 9.42
Typical waveform of relevant electrical parameters of the series compensation device during internal faults.

Table 9.7: Operating Conditions of Main Components of the Series Compensation Device During Internal Fault on the Protected UHV Line.

Relevant Parameters		Changzhi–Nanyang Line I		Nanyang–Jingmen Line I
		Changzhi Substation	Nanyang Substation	Nanyang Substation
Capacitor current (kA)		70	78	74
Series compensation voltage (p.u.)		2.3	2.3	2.3
MOV	Energy consumption (MJ)	57	52	45
	Current (kA)	90	93	81
Gap current (kA)		153	153	150

system. In light of this, special studies were performed regarding the aseismic performance of the platform to meet the engineering requirements of the expansion of the UHV AC pilot and demonstration project.

2. The UHV series compensation device has a higher voltage level and comprises a great number of components, the majority of which have irregular shapes. The field strength distribution around the series compensation device is complicated. Hence, to accurately

simulate the field strength distribution and optimize the design to satisfy the field strength, limits play an important role in optimizing the arrangement of series compensation device and reducing the land occupation. Therefore, special studies were performed regarding the space field strength around the platform for expansion of the UHV AC pilot and demonstration project, and the study results have been used in the engineering design.

3. Although the UHV series compensation device has phase-to-ground insulation levels and phase-to-phase insulation levels comparable with those of the conventional UHV equipment, no direct test results are available for the insulation characteristics of the series compensation device's platform, which is an electrode with a special shape. Therefore, CEPRI under SGCC conducted insulation withstand tests on a full-scale test platform and the test results directly support the platform design.
4. Considering that each UHV series compensation device contains as many as 2500 capacitor units (three-times the number of a 500-kV one), it is necessary to greatly increase the stability of each unit to ensure the device's reliability. Specifically, the annual capacitor failure rate is limited to one-third that required for an EHV series compensation capacitor.
5. In UHV applications, the short-circuit fault will result in a great amount of energy being injected into the MOV, which requires the MOV to have a capacity as high as 100 MJ. Because each MOV unit contains multiple varistors connected in parallel, high requirements are placed on the current sharing performance. Also, the unit is required to have reliable pressure relief capability.
6. Because the UHV grids have a large short-circuit capacity from a long-term view, it sets high requirements for the gap's current-conducting capability and post-conduction insulation recovery capability. Moreover, the gap should have high trigger reliability and interference immunity because the UHV system is required to be highly reliable.
7. As compared with those for 500-kV series compensation devices, the bypass circuit breakers used in UHV series compensation devices have considerably increased rated voltages, nearly doubling the break-to-break distance and requiring higher closing speed. Also, the breaker's phase-to-ground insulation level is required to be much higher, necessitating increasing the height of the porcelain post to 10 m and, consequently, higher overall mechanical strength and stability. Moreover, because the UHV system has a larger transmission capacity, higher requirements are placed on the bypass circuit breakers for series compensation devices in terms of series compensation reinsertion capability after current interrupting.
8. The measuring, control, and protection system of the series compensation device should be designed as a dual redundant system. It should be possible to monitor and control the status of main equipment, including the bypass circuit breaker, disconnector, and earthing switch from the series compensation station, and to monitor the status of the main equipment via the telecontrol equipment from the dispatching center. In addition,

because the UHV system's series compensation device is more closely related to system security and stability, the control and protection system is required to be more reliable and stable. The rigorous electromagnetic environment suffered by the UHV series compensation devices requires the control and protection system (especially its measuring subsystem) to be more resistant to electromagnetic interference.

Today, studies of the basic and key technologies of the controllable UHV series compensation devices have been completed. These studies have defined the application scenarios of such devices and key operating conditions of their main equipment, proposed a new controllable series compensation configuration in which a thyristor-controlled reactor (TCR) is connected in parallel with the capacitor and the associated control and protection strategy, developed and tested the LV model of such new controllable series compensation devices and the associated control and protection prototype, and developed prototypes of capacitance voltage measuring devices and insulated water pipes (both of which have passed the type test).

9.5 UHV Surge Arrester

9.5.1 Main Roles of UHV Surge Arrester

UHV surge arresters are usually installed in the vicinity of the main electrical equipment in substations to limit lightning and switching overvoltage, thereby protecting the electrical equipment in UHV substations. The protective characteristics of surge arresters directly dictate the design of the impulse insulation level of the equipment and air gap size of the substations. Meanwhile, the surge arresters work together with high-voltage shunt reactors and closing resistors of circuit breakers to limit overvoltage on transmission lines. Therefore, their design directly affects the air gap sizes of line tower head and provides a basis for the insulation coordination of transmission lines.

Because UHV equipment tends to have large sizes and causes transportation difficulties, improving the characteristics of the UHV surge arresters (i.e., reducing the protective level) can further reduce the impulse insulation level of the equipment, thereby reducing the equipment size and project costs. As such, UHV surge arresters play a more important role in the design of the insulation level of the substation equipment and transmission line than surge arresters of other voltage classes. Figure 9.43 shows the profiles of 1000-kV surge arresters.

9.5.2 Main Parameters of UHV Surge Arrester

The performance parameters of UHV surge arresters, such as the rated voltage, protective characteristics, energy absorption, and power frequency withstand characteristic, depend on the system conditions of UHV projects and manufacturing technologies of surge arresters.

Figure 9.43: Pictures of 1000-kV surge arresters.
(A) Porcelain UHV surge arrester and (B) GIS UHV surge arrester.

9.5.2.1 Rated voltage

As an important parameter characterizing the performance of surge arresters, the rated voltage should be verified through operating duty test. Generally, the rated voltage should be taken as the power frequency transient overvoltage at the places where surge arresters are installed, which are determined to be 1.4 p.u. on the line side and 1.3 p.u. on the bus side for 500- and 750-kV substations, respectively. In a 1000-kV UHV project, to further reduce the overvoltage level, circuit breaker–operated linkage protection should be used to reduce the duration of the line-side power frequency transient overvoltage to 0.2 s (allowing for the situation when the circuit breaker fails to operate and the standby protection trips). Because UHV surge arresters have better power frequency withstand characteristics, the rated voltage of the surge arresters on the line side is determined to be the same as that on the bus side (i.e., 828 kV), no longer following the practices of the past.

9.5.2.2 Protective characteristics

The protective characteristics are a key to suppress the overvoltage level. The lower the protective level of the arrester, the better its overvoltage limiting performance will be. However, given the specified reference voltage, the protective level should not be too low. Therefore, it is necessary to further reduce the voltage ratio of the varistors (i.e., the ratio of the residual voltage under nominal discharge current to the reference voltage). In view of this, 1000-kV UHV surge arresters should be in a four-parallel-column configuration with the diameter of varistors being 100 mm or larger, which reduces the voltage ratio by approximately 9% as compared with the single-column configuration.

9.5.2.3 Energy absorption

As an overvoltage suppressor, surge arresters can absorb tremendous energy without damage under lightning and switching overvoltage. The energy that a surge arrester is able to absorb is a function of the system parameters as well as the type, amplitude, waveshape, and duration of overvoltage and should be determined through simulation calculation.

Using the UHV AC pilot and demonstration project as example, the simulation results in terms of switching overvoltage shows that, in most cases, the surge arrester absorbs only a small amount of energy, and only when an oscillation overvoltage due to system splitting occurs will the surge arrester absorb a large amount of energy, which is sometimes as high as 27 MJ.

9.5.2.4 Power frequency withstand characteristics

To further reduce the insulation level of UHV transmission line, surge arresters should be able to withstand the power frequency transient overvoltage of the specified amplitude for the specified time period. This requires the rated voltage of an arrester on the line side of the substation to be designed the same as that on the bus side (i.e., 828 kV (r.m.s)). Therefore, the surge arresters are required to be able to withstand the power frequency transient overvoltage at least equal to 1.1-times the rated voltage of an arrester for at least 1 s.

Main performance parameters of a 1000-kV arrester are listed in Table 9.8.

Table 9.8: Main Performance Parameters of 1000-kV Surge Arrester.

Item	Parameter
Rated voltage (kV)	828
Power frequency reference voltage (kV, r.m.s)	≥828
Reference voltage at 8 mA DC (kV)	≥1114
Continuous operating voltage (kV, r.m.s)	638
Nominal discharge current (kA)	20
Steep impulse residual voltage (1/10 μs, 20 kA) (kV)	≤1782
Lightning impulse residual voltage (8/20 μs, 20 kA) (kV)	≤1620
Switching impulse residual voltage (30/60 μs, 2 kA) (kV)	≤1460
Total energy absorbed during two switching operations (MJ)	≥40

9.5.3 Structural Design of UHV Surge Arrester

9.5.3.1 Porcelain-housed surge arrester

As with the conventional EHV AC surge arresters, UHV porcelain-housed surge arresters have a column structure [5]. Taking into account the various factors such as the manufacturing, fabrication, installation, and transportation, EHV/UHV surge arresters usually comprise several units connected in series and grading ring(s). Each unit mainly comprises the ZnO varistor core and porcelain housing.

In the UHV AC pilot and demonstration project, surge arresters are all more than 12 m high and weigh more than 10 tons, with the diameter of the largest shed reaching 750–890 mm, and they are installed on supports. Figure 9.43. A shows the picture of such surge arresters. These features have caused prominent concerns regarding voltage grading and mechanical strength.

For 1000-kV UHV surge arresters, the relevant codes and standards require the uneven coefficient of potential distribution to be <1.15. The test results available from the manufacturers show that simply using grading rings will lead to a value of 1.15–1.17; if additional grading capacitors are installed, then the value is approximately 1.0–1.1.

9.5.3.2 Gas-insulated metal-enclosed surge arrester

In contrast to porcelain-housed surge arresters, gas-insulated metal-enclosed surge arresters (GIS-arrester) have a ZnO varistor core enclosed in an air-tight steel tank [6,7] and are fabricated in a monoblock structure, as shown in Figure 9.44. With the confined interior volume of GIS-arresters, high-gradient varistors should be used as practically as possible to reduce the size of the core, and each column of the varistors is in triple-helix configuration, as shown in Figure 9.45. To this end, insulating pad plates with excellent electrical (particularly dielectric strength) and mechanical performance is required.

Moreover, tank surge arresters are as high as 4.6 m, making it impossible to transport them in typical vertical orientation because of the limitations on transportation height on roads [8]. Instead, they should be transported in horizontal orientation, necessitating the use of a configuration with the varistor columns placed into the insulating tubes. The sufficiently strong insulated tubes can reliably support the columns, thereby protecting the insulating bars for fixing the varistors from being subjected to excessive loads because of their self-weight and vibration. Moreover, these insulating tubes make other components inside the arresters capable of resisting higher vibration and impact, thus ensuring that the internal structure of the arresters can remain stable without degraded performance when the arresters are transported in horizontal orientation.

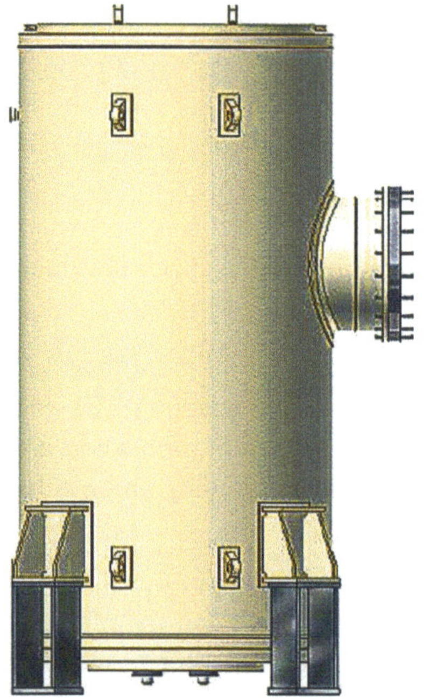

Figure 9.44
Profile of 1000-kV gas-insulated metal-enclosed surge arrester.

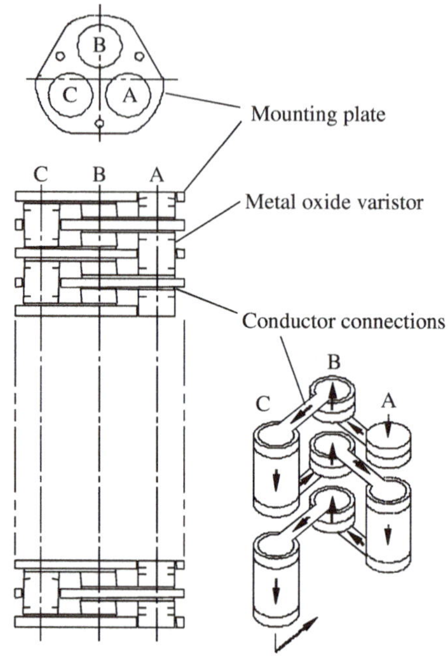

Figure 9.45
Schematic of triple-helix arrangement.

9.6 UHV Post Insulators and Bushings

9.6.1 UHV Post Insulators

Several post insulators are required in a UHV substation, especially in open-type switchgear. As compared with an EHV line, a UHV line places higher requirements on post insulators in terms of bending failing load and shock resistance, which require a higher manufacture level. In the UHV AC pilot and demonstration project, post insulators also function to provide insulation and support for the series compensation device's platform. These solid-core post insulators have the largest diameter and weight in the world,; therefore, they are extremely difficult to manufacture. In UHV transmission projects, porcelain post insulators are mainly used [9]. For their parameters, see Table 9.9.

9.6.1.1 Tolerance of form and position

The tolerance of form and position for a post insulator should be as specified in Table 9.10.

The post insulators used in UHV applications are required to have a highly reliable electrical and mechanical performance. These performance, especially the mechanical

Table 9.9: Dimensions and Technical Parameters of Post Insulators Used in a UHV Line.

No.	Technical Parameter	Guaranteed Value
1	Dry lightning impulse withstand voltage (peak value, kV)	≥ 2400
2	Wet switching impulse withstand voltage (peak value, kV)	≥ 1800
3	Wet power frequency withstand voltage (r.m.s, kV)	1100
4	Height (mm)	10,000
5	Creepage distance (mm)	30,250
6	Average diameter (mm)	320
7	Rated bending failing load (kN)	16
8	Rated torsion failing load (kN m)	10
9	Radio interference voltage at a voltage of $1.1U_m/\sqrt{3}$ kV (μV)	≤ 500
10	Pollution class	d
11	Specific creepage distance (mm/kV)	≥ 25
12	Dry arcing distance (mm)	≥ 8100

Table 9.10: Tolerance of Form and Position of a Post Insulator.

Height (mm)	Straightness of Axis (mm)	Parallelism Between the Upper and Lower Surfaces of Accessory	Maximum Deviation Between Central Axes of the Upper and Lower Mounting Holes
$0.5 + 0.001h$	$1.5 + 0.006h$	$0.001h$	$2 \times (1 + 0.001)h$

Note: *h* refers to the installation height of the post insulator above ground.

performance, of a post insulator is related to its tolerance of form and position. This means that UHV projects place higher requirements on the tolerance of form and position of post insulators [10].

9.6.1.2 Electric performance

1. **Pollution withstand voltage**
   The external insulation design of a post insulator should meet the requirements of external insulation in areas with a pollution class of D. Specifically, the post insulator should withstand an SDD of 0.08 mg/cm^2 and non-soluble deposit density (NSDD) of 0.5 mg/cm^2 under a power frequency voltage of 698 kV.
2. **Radio interference voltage and visual corona voltage**
   The 1000-kV post insulators must be provided with grading devices of proper design and configuration [11] to ensure that no visual corona occurs on clear nights under the voltage of $1.1 \times 1100/\sqrt{3}$ kV, and to ensure that the radio interference voltage is no higher than 500 μV under a test voltage of 700 kV and a measurement frequency of 1 MHz.

9.6.1.3 Mechanical performance

Post insulators should have a minimum bending failing load of 16 kN and a minimum torsion failing load of 10 kN m.

Post insulators should be designed according to 8-degree seismic fortification intensity. Specifically, it should be able to withstand 3-m/s^2 horizontal acceleration and 1.5-m/s^2 vertical acceleration at ground level (sine-shape seismic wave lasting for three cycles and a safety factor of 1.67).

9.6.2 UHV Bushings

Bushings can be classified as:

1. Porcelain bushings and composite bushings by the main insulating material
2. Capacitance-graded bushings and non-capacitance graded bushings by the structure
3. Bushings for transformers, reactors, GIS, circuit breakers, cable terminals, transformers, lightning arresters, and other electrical equipment (classified by to the intended use)
4. Indoor and outdoor bushings (classified by installation position and operating status)
5. Vertical, inclined, and horizontal bushings (classified by the installation orientations). In 1000-kV UHV AC projects in China, both porcelain and composite insulators are used mainly for outdoor transformers, reactors, GIS, circuit breakers, transformers, and lightning arresters. They are installed either vertically or with an inclination.

In China, most bushings previously used for electrical equipment are made of porcelain. They are manufactured as one piece or using bonding processes with organic or inorganic agents. The porcelain insulators used for the electrical equipment in the UHV AC pilot project have a height and diameter more than twice those of a 500-kV bushing. These ultra-large hollow insulators are manufactured using bonding processes with inorganic agent. These place a higher requirement on the manufacture processes of the bushings. In the Jingmen Substation of the UHV AC pilot and demonstration project, three 1000-kV composite insulators developed by Jiangsu Shenma Electric Co., Ltd. are used for the external insulation of GIS. Figure 9.46 shows the 1000-kV bushings.

9.6.2.1 Electric performance

The porcelain bushings used in the UHV AC pilot and demonstration project can withstand Class D pollution conditions, can have a specific creepage distance of no less than 25 mm/kV, and can have minimum spacing between sheds of 80 mm. The bushings used for reactors, transformers, and switchgears have rated currents of 800, 2500, and 4000 A, respectively.

Figure 9.46: The 1000-kV bushings.
(A) Porcelain bushing for GIS; (B) composite bushing for GIS; and (C) oil paper–insulated bushing for transformer.

9.6.2.2 Mechanical performance

A UHV bushing is more than 10 m long, and even a maximum of 15 m. Its mechanical performance should not only meet the normal environmental requirements but also accommodate special operating conditions and the requirements for transport and lifting. Hollow insulators should pass internal pressure test and bending test one-by-one. To be specific, during the internal pressure test, 1.6 MPa air fills the insulator and is maintained for 5 min; no pressure drop should occur and the insulator should be free of damage, During the bending test, 18-kN tensile force is applied at the top of the insulator and the residual offset should be $<5\%$ of the maximum offset.

For the major performance parameters of bushings used in the UHV AC pilot and demonstration project, see Appendix A.6.

9.7 UHV Voltage Transformer and Current Transformer

9.7.1 UHV Voltage Transformer

In China's UHV AC pilot demonstration project, column-type capacitive voltage transformers (CVTs) are adopted, In the expansion project, tank-type SF_6 gas-insulated metal-enclosed voltage transformers (tank-type voltage transformers for short) are also used.

9.7.1.1 Types and operating principle

In China's UHV AC transmission projects, column-type CVTs and tank-type electromagnetic VTs are used. China has developed and manufactured a new type of CVT (tank CVT) that has been put into operation in the Huainan–Shanghai UHV AC transmission project.

A column-type CVT mainly comprises a capacitor voltage divider, MV transformer, compensation reactor, and damper, the latter three of which are collectively known as an electromagnetic unit. The main insulation of a column-type CVT consists of a coupling capacitor voltage divider that comprises multiple series-connected capacitive elements with uniform voltage distribution. From the viewpoint of security and reliability of the project, column-type CVTs are desirable for open-type substations. Figure 9.47 shows a column-type CVT used in China's UHV projects.

The tank-type CVT is a wire-wound transformer, which transforms a primary voltage to a secondary voltage at a certain transformation ratio based on the electromagnetic principle and uses polyester film and SF_6 gas as the insulation media. It is mainly used in GIS substations. Figure 9.48 shows a UHV tank-type CVT used in China's UHV projects.

Figure 9.47
Column-type CVT in China's UHV projects.

Figure 9.48
Tank-type CVT in China's UHV projects.

As shown in Figure 9.49, the tank-type CVT is a new type of voltage transformer developed by China Electric Power Research Institute. It works using the same operating principle as the conventional column-type CVT. The main difference between them lies in the structure of the capacitor voltage divider. For the tank-type CVT, the capacitor voltage divider uses SF_6 gas with recoverable insulation performance as the main insulation and its high-voltage arm capacitor C1 has a coaxial cylindrical structure.

Technical parameters of various types of UHV voltage transformers are listed in Table 9.11.

Figure 9.49
Tank-type CVT prototype in China's UHV projects.

Table 9.11: Technical Parameters of Various Types of UHV Voltage Transformers.

Types	Application	Rated Voltage Ratio	Accuracy Class	Secondary Load	Rated Capacity
UHV column type CVT	Already used in UHV projects	$1000/\sqrt{3}/0.1/\sqrt{3}/0.1/\sqrt{3}/0.1/\sqrt{3}/0.1$ kV	0.2/0.5/0.5 (3P)/3P	15/15/ 15/15VA, $\cos\varphi = 1$	5000 pF
UHV tank type CVT	Already used in UHV projects			10/10/ 10/10VA, $\cos\varphi = 1$	500 pF
UHV tank type TV	Already used in UHV projects	$1000/\sqrt{3}/0.1/\sqrt{3}/0.1/\sqrt{3}/0.1/\sqrt{3}$ kV	0.2/0.5 (3P)/3P	30/30/30 VA, $\cos\varphi = 0.8$	–

9.7.1.2 Accuracy test of UHV voltage transformer

China has developed a 1600-kV standard capacitor, a 1000-kV standard voltage transformer for traceability, and a 1000-kV standard voltage transformer for field tests, all of which incorporate the proprietary intellectual property rights. Figure 9.50 shows these transformers.

Because the 1000-kV standard voltage transformer for field testing is designed to measure a voltage in the range of 20–120% of the rated primary voltage U_{pr}, the error is directly measured using such 1000-kV standard voltage transformers for 120% rated voltage and lower. However, for 150% of the rated primary voltage, the error should

Figure 9.50: Accuracy test of UHV voltage transformer.
(A) 1600-kV standard capacitor; (B) 1000-kV standard voltage transformer for traceability; and (C) 1000-kV standard voltage transformer for field test.

be derived from the measurements by the 1600-kV standard capacitor using the voltage factor method.

9.7.1.3 UHV electronic voltage transformer

The electronic voltage transformer (EVT) has the following advantages over the conventional voltage transformer: (i) it has a simple insulation structure and small size; (ii) no magnetic saturation occurs; and (iii) its output signals can be directly transmitted to the computer-based measuring and protection equipment. In UHV projects, capacitor voltage divider–based EVTs as shown in Figure 9.51 are preferred. In this case, MV signals are isolated by Pockels cells. In China, the development of a UHV EVT prototype is underway.

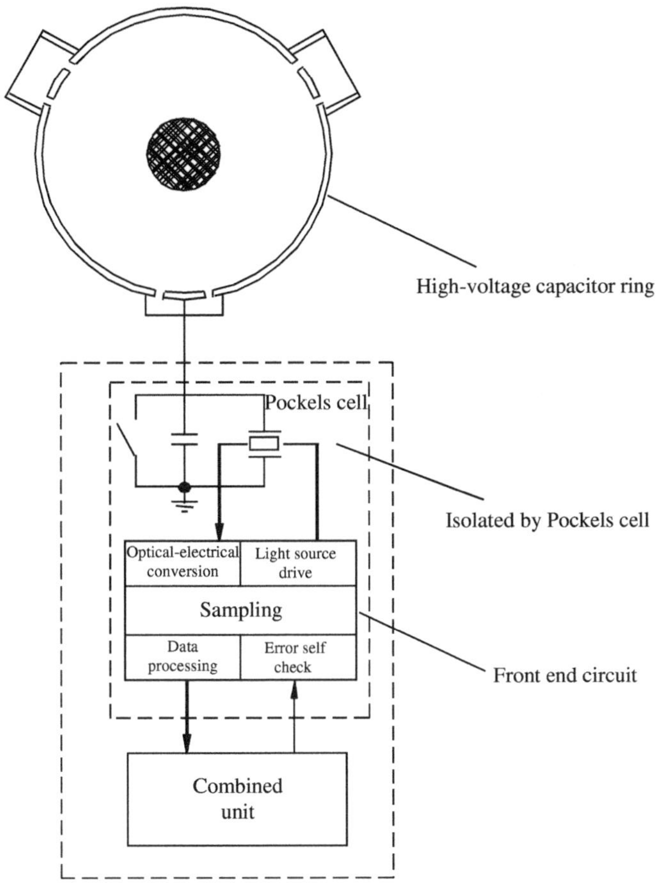

Figure 9.51
Schematic diagram of UHV EVT.

9.7.2 UHV Current Transformer

9.7.2.1 Technical parameters

The rated primary current is the design basis for a current transformer (TA). From the viewpoint of system planning, if a one-and-a-half breaker connection scheme used, then the natural power passing through each 1000-kV circuit breaker is assumed to be natural transmission power of each circuit of a 1000-kV line (5000 MV). When a circuit breaker is repaired, the power passing through one 1000-kV circuit breaker will be two-times the natural transmission power. In this case, the rated current of the circuit breaker, disconnector, and current transformer will reach up to 6300 A.

In the UHV AC pilot and demonstration project, the transformation ratio of the measuring TA and protection TA on the primary side is 6000/1 A, and their tap positions are at 3000/1 A, 1500/1 A, and 3000/1 A. If TAs are used for the GIS equipment, then there is a MV side and a low-voltage side in addition to the primary side. The transformation ratio may be determined based on the capacity ratio and the need for protection (such as differential protection).

9.7.2.2 Structure

In China's UHV AC transmission projects, all TAs are fit over the bushings of GIS/HGIS and can be classified into two types (i.e., internal type and external type). The TA is a window-type and comprises a single-turn primary winding and several secondary windings, with each having a number of tap positions according to the engineering needs.

For the internal-type TA, after a single winding is coiled, it is usually cast integrally with epoxy resin to prevent entry of moisture and other impurities into the gas chamber, which may deteriorate the insulation performance. The windings of internal-type TA are placed in a metal enclosure, which is fixed onto the internal shielding cylinder. Because the UHV GIS/HGIS has a higher insulation level and, hence, a larger-diameter shielding cylinder. The diameter and weight of the windings (especially TPY class) are far larger than the conventional windings, requiring the internal shielding cylinder to have an adequate bearing capacity.

For the external-type TA, the windings are installed outside the enclosure of GIS/HGIS. In contrast to the internal type, the windings of external-type transformers are directly exposed to air. Therefore, the surface of the windings should be subjected to corrosion and moisture-resistant treatment, and the windings are not required to be cast with epoxy resin integrally. In addition, neither special shielding cylinder nor special insulation design is required. Consistency with the insulation structure of buses ensures compliance with the relevant requirements.

9.7.2.3 UHV electronic current transformer

All TAs used in UHV AC projects are electromagnetic, are installed on a GIS with a large volume, and weigh more than 1 ton. In contrast, the electronic current transformer (ECT) incorporates a novel type of sensor, making it compact in size and lightweight. It not only can be used together with GIS but also can be used independently. The ECT used in the UHV AC projects may have a complete fiber optic structure.

In addition, the development of combined transformer comprising a current transformer and a voltage transformer housed in a common enclosure and optical current transformer is underway.

9.8 Seismic Resistance of Major Electrical Equipment in UHV Substation

9.8.1 Structural Characteristics of UHV Electrical Equipment

The seismic resistance of UHV electrical equipment, an important part of a substation, has a significant impact on the operational security and stability of the entire substation, or even the entire power grid. The seismic resistance is closely related to the structural characteristics of such equipment. Electrical equipment can be classified into the following three types according to the structural characteristics:

1. Equipment with a small mass, a low center of gravity, and high overall performance, such as GIS and some switchgear. Such equipment has a large overall stiffness and its natural vibration frequency is clear of the predominant frequencies of earthquakes, subjecting it to a small earthquake force. Therefore, such equipment are more capable of withstanding earthquake shock, placing lower requirements on seismic resistance design.
2. Equipment with large mass and low center of gravity, but attached to important accessories, such as transformers. The transformer body has a natural frequency of 10 Hz or higher, large stiffness, and high seismic resistance. However, the outgoing bushings are commonly made of brittle porcelain materials and are prone to damage from earthquake shock. Moreover, any earthquake shock applied to the outgoing bushings is transferred from and magnified by the transformer body, making them more susceptible to damage from the earthquake shock. As such, more stringent requirements are placed on seismic design of the outgoing bushings.
3. Open-type equipment in the form of a post, such as surge arrester, transformer, disconnector, and earthing switch. Generally, such equipment have a porcelain cantilever structure and are slim and flexible, yet heavy. The natural frequency ranges from 1 to 10 Hz, close to the predominant frequencies of earthquakes; therefore, resonance or pseudo-resonance is very likely to occur, causing the equipment to be susceptible to damage from the earthquake shock. As such, more stringent requirements are placed on the seismic design of such equipment.

9.8.2 Studies on Seismic Resistance

To date, fewer studies of seismic resistance of UHV equipment have been performed around the world. Relying on the UHV projects built in China, CEPRI and other research institutes have conducted in-depth studies of the seismic resistance of electrical equipment.

1. *Theoretical studies*. The seismic fortification criteria, standard earthquake response spectrum for seismic design, methods for anti-seismic evaluation, and methods for seismic absorption were established for planning UHV projects and with reference to available codes and the latest study results, thus providing a basis for design of seismic resistance and absorption of UHV electrical equipment.
2. *Numerical simulation*. Analysis and calculation of seismic resistance were conducted for the UHV high-impedance circuits, UHV series compensation circuits, UHV bypass circuit breakers, and UHV electrical equipment using finite element models, providing a reference for optimizing the seismic design of electrical equipment of UHV projects.
3. *Tests*. The equivalent flexural stiffness of porcelain connections of UHV equipment was tested with the static loading method; the overall performance of a variety of UHV equipment regarding seismic resistance and mitigation was tested by simulating earthquake shock on a vibrating table. These tests have resulted in ample test data used for theoretical analysis and numerical simulation. Figure 9.52 shows the test site.

Figure 9.52: Seismic simulation tests for typical UHV equipment with the vibrating table. (A) UHV voltage transformer and (B) UHV surge arrester.

9.8.3 Seismic Design

1. *Seismic fortification standards.* Seismic fortification standards for UHV electrical equipment are established after considering factors such as the social economic level, technical level, and importance.
   a. *Objectives.* The equipment is free from any damage in the event of earthquakes of seismic fortification intensity or less.
   b. *Criteria.* Fortification intensity with a 2% exceedance probability within 50 years is recommended. Table 9.12 lists the corresponding design basic acceleration of ground motion.
2. *Coefficient of earthquake effect.* Figure 9.53 shows a coefficient curve of an earthquake effect as a function of time, which differs from that contained in GB 50011—2010 *Code for Seismic Design of Buildings* in that a critical point of 0.03 s (33 Hz) is included, the characteristic period of the plateau is 0.9 s, and the magnification factor is 2.5 (based on 5% damping ratio). For more details on parameters presented in Figure 9.54, see GB 50011—2010.

Table 9.12: Design Basic Earthquake Acceleration for UHV Electrical Equipment.

Seismic fortification intensity	6°	7°	8°
2% Exceedance probability within 50 years	0.10g	0.20g (0.30g)	0.40g (0.50g)
Maximum coefficient of earthquake effect α_{max}	0.25	0.50 (0.75)	1.00 (1.25)

Note: A seismic fortification intensity of 9° and above is rarely used UHV projects, and is not included in this table. If required, related studies will be carried out. The figures inside and outside the parentheses should be referred to for areas with a seismic fortification intensity with a 10% exceedance probability within 50 years as specified in *Seismic Ground Motion Parameter Zonation Map of China*.

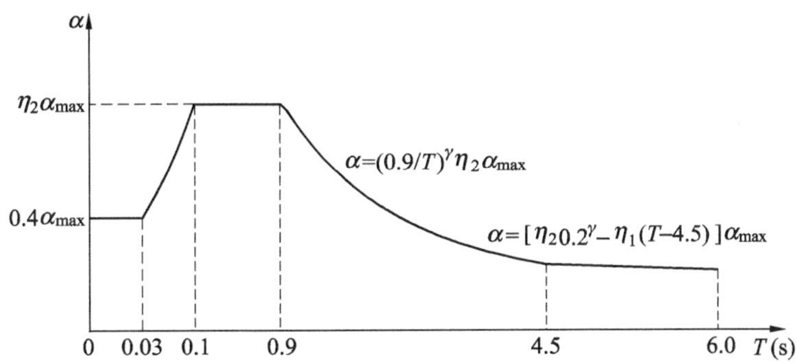

Figure 9.53: Coefficient of earthquake effect versus time curve for UHV electrical equipment. $\sqrt{3}$—coefficient of earthquake effect; $\sqrt{3}$—maximum coefficient of earthquake effect; T—natural period of vibration; γ—exponent of attenuation for the declining curve; η_1—regulation coefficient of the linear declining slope, taken as 0 if <0; and η_2—regulation coefficient of damping, taken as 0.55 if <0.55.

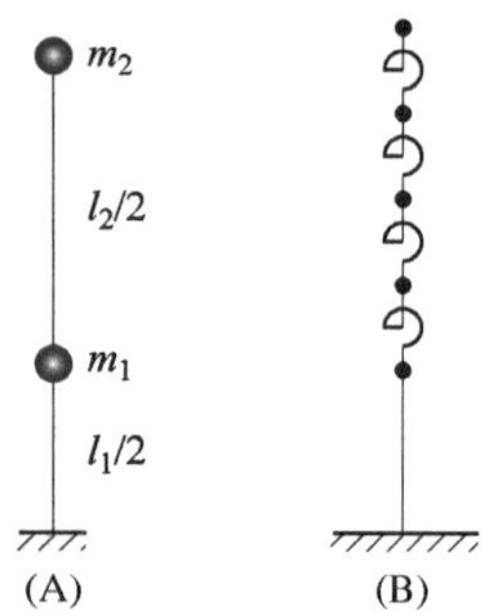

Figure 9.54: Simplified mechanical models.
(A) Multi-particle model and (B) spring-mass model.

Figure 9.55
Finite element model.

3. Seismic resistance evaluation and calculation
   a. *Calculation model.* The calculation model should be determined based on the structural characteristics, precision requirements, and calculation methods.
      For evaluating seismic resistance, the simplified multi-mass model and spring-mass model as shown in Figure 9.54 can be used. If high calculation precision is required, then a finite element model as shown in Figure 9.55 needs to be established.
   b. *Damping ratio.* The seismic effect on UHV electrical equipment should be calculated with the actual damping ratio of the equipment structure. In the case of electrical equipment made of electrical porcelain material, if the actual damping ratio is unknown, then a value no more than 2% is recommended.
   c. *Calculation method.* The following methods may be used as appropriate:
      i. The static method can be used for equipment with a fundamental frequency higher than 33 Hz.
      ii. The equivalent base shear method can be used for equipment that is mainly subjected to shear deformation or that can be equivalent to a single-mass system.
      iii. The mode analysis response spectrum method is recommended for equipment other than that mentioned previously.
      iv. For irregular equipment or equipment placing special requirements, the time history method should be used for supplementary calculation.
4. *Load combination.* The combination of earthquake action and other loads, such as the self-weight of equipment, wind load, actual tension of the conductor, internal pressure of equipment, and weight of fittings.
5. *Seismic mitigation and isolation design.* The seismic mitigation and isolation design of UHV electrical equipment should be determined based on structural characteristics following the principle of using simple and convenient processes while improving the anti-seismic performance and safety margin of the equipment with the minimum costs.
6. *Criteria for determining the seismic resistance.* After the seismic resistance is assessed, the stress developing at the root or other critical cross-sections of the UHV equipment

should be smaller than the allowable stress of the equipment or material. When the fracture stress σ is used as a criterion, the ratio of σ_v to the calculated stress or the actually measured stress σ_{tot} should be no less than 1.67 (i.e., $\sigma_v/\sigma_{tot} \geq 1.67$).

7. *Seismic requirements for installation of electrical equipment.* Electrical equipment and its leads should be joined through flexible connections with a certain length margin. When a rigid tubular bus is used, a flexible connection, such as expansion clamps, expansion joints, or sliding fittings, should be in place for transition. The strength of installation bolts or welds on the electrical equipment must comply with the seismic requirements. In addition, the electrical equipment should incorporate seismic mitigation devices such that they have sufficient stiffness and therefore allow the equipment to function normally. The supports of such seismic mitigation devices must be flat enough to ensure uniform load carrying.
8. *Seismic performance tests.* UHV electrical equipment are characterized as slim and flexible, and as having complicated connections between nodes as well as prominent non-linearity. The seismic simulation test with the vibrating table provides the best method for examining the seismic resistance of UHV electrical equipment. It not only can accurately simulate the loading state of the tested equipment but also can reflect its vibration intuitively.

References

[1] L. Guangfan, et al., Insulation level and test technology of 1000 kV power transformers, Power Syst. Technol. 32(3) (2008) 1–6.
[2] L. Guangfan, L. Bo, L. Jingzhong, Z. Zhigang, et al., Research and development of UHV AC transformer and shunt reactor, Eur. Trans. Elect. Power (2012) 49–59.
[3] L. Jiming, G. Nihong, W. Xiaogang, et al., A study on transient recovery voltage of UHV circuit breakers, Power Syst. Technol. 31(1) (2007) 1–5.
[4] G. Yuanyuan, C. Boyuan, W. Chengyu, et al., Development of gas-insulated metal-enclosed switchgear with a breaking capacity of 63 kA for 1000 kV power transmission project, Power Syst. Technol. 12(35) (2011) 20–25.
[5] Z. Qin, W. Shaowu, W. Guangya, et al., The manufacturing technology of post porcelain insulator for 1000 kV AC UHV system, High Voltage Eng. 36(6) (2010) [China].
[6] C. Wenjun, S. Jijun, Z. Xiaoxing, et al., R&D of 1000 kV UHV AC GIS and porcelain type metal oxide arrester, Elect. Equip. 4 (2007) 14–18.
[7] L. Minggang, S. Jijun, H. Shumo, et al., Development and application of gas-insulated metal enclosed surge arrester for 1000 kV system in China, High Voltage Eng. 35(11) (2009) 2618–2623.
[8] W. Baoshan, X. Yi, T. Lin, et al., Summary and conclusion of chinese ultra high voltage AC porcelain-housed metal oxide surge arrester's development and type test, High Voltage Eng. 36(4) (2010) 891–899.
[9] W. Xiaoqi, Development and application of 1000 kV standard voltage transformer for field test, High Voltage Eng. 35(6) (2009) 1254–1259.
[10] F. Ye, Application of method of increasing burden on error calibration of 1000 kV GIS current transformer, High Voltage Eng. 34(9) (2008) 1862–1868.
[11] M. Graovac, R. Iravani, S. Member, X. Wang, R.D. McTaggart, Fast ferroresonance suppression of coupling capacitor voltage transformers, IEEE Trans. Power Delivery 17(4) (2002) 158–163.

CHAPTER 10

UHV Converter Station and UHV DC Electrical Equipment

Chapter Outline

Ultra-high Voltage AC/DC Grids.
DOI: http://dx.doi.org/10.1016/B978-0-12-802161-3.00010-X

10.1 UHV Converter Station

10.1.1 DC Main Electrical Connection Scheme

UHV converter stations refer to those with DC $\pm$800 kV or higher. With the development of UHV DC transmission technology, the design and construction of $\pm$800-kV converter stations have become mature, and the technical achievements may be used to guide the design and construction of converter stations at higher voltage levels.

A converter station can be divided into several functional areas including the valve hall and control building, converter transformer area, DC switchyard, and AC switchyard. The valve hall and control building are of an integral structure. The valve hall mainly contains the converter valves, associated switchgears, overvoltage protection, and fire alarm devices. The control building is equipped with communication equipment, control and protection equipment, DC power supply equipment, and valve cooling equipment. The converter transformer area is mainly installed with converter transformers and associated fire protection equipment. The DC switchyard consists of smoothing reactors, DC filters, DC surge arresters, DC measuring devices, and DC switchgears required for operation mode transferring and fault clearing. The AC switchyard mainly consists of switchgears, AC filters and reactive power compensation devices, AC surge arresters, and AC measuring devices [1,2].

The main electrical connection scheme of a converter station should meet the operation requirements of the DC transmission system, ensure operational safety, reliability,

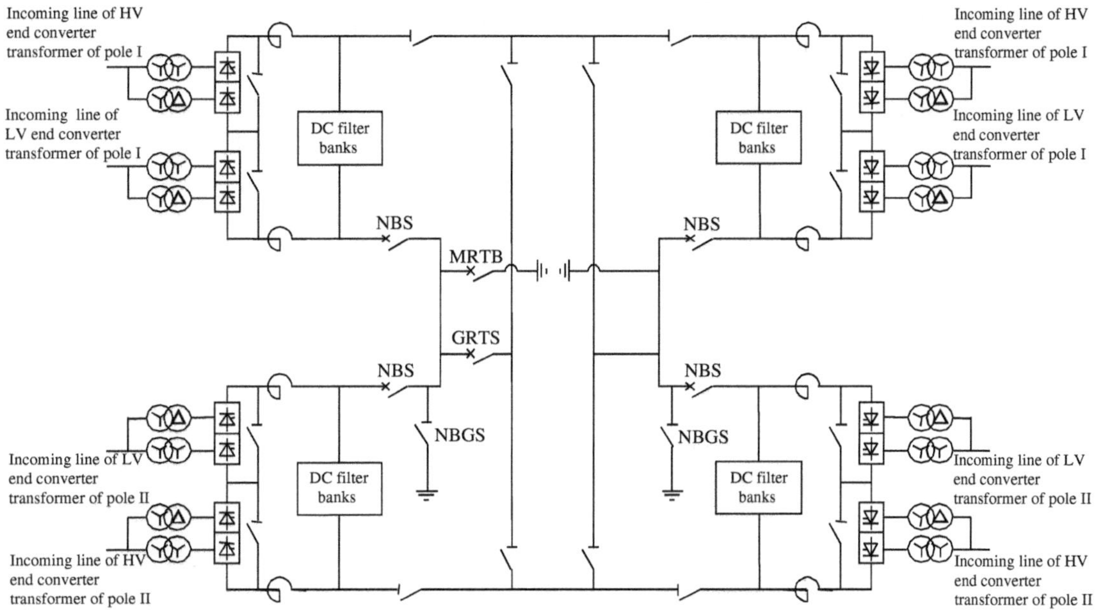

Figure 10.1
DC main electrical connection diagram of a UHV converter station.

flexibility, and economy, and allow easy operation and maintenance. The ±400- to ±660-kV HVDC projects use the connection scheme in which each pole has one 12-pulse converter unit, whereas in typical UHV converter stations in China each pole has two series-connected 12-pulse converter units.

Figure 10.1 shows the typical DC main electrical connection scheme of a UHV converter station.

In China, the ±800-kV UHV converter station uses a bipolar configuration in which each pole has two 12-pulse converter units (400 kV + 400 kV) connected in series. Such a converter station has four 12-pulse converter units in total, of which the two near the pole bus are called the HV end converter unit and the two near the neutral bus are called the LV end converter unit. Each converter unit is connected in parallel with one group of bypass circuit breakers. In this way, if any 12-pulse converter at any end of a pole is out of service, the rest converter can still operate as an incomplete pole.

A UHV converter station has a total of 24 + 4 converter transformers that are usually of the single-phase two-winding type. Each pole has 12 converter transformers, and the six transformers belonging to each converter unit are respectively connected in Y/Y and Y/Δ configurations. Both the HV and LV end transformers of either winding configuration are provided with standby transformers. Therefore, there are four standby transformers in total per station.

A DC switchyard contains pole bus, neutral bus, smoothing reactors, DC filters, DC switchgears, DC measuring devices, surge arresters, and other associated equipment. Generally, multiple dry-type smoothing reactors are used and are divided into two groups that are connected in series with the pole and neutral buses, respectively. Generally, two double-tuned DC filter banks or one triple-tuned DC filter bank are installed between the neutral bus and pole bus of each pole, and they mainly serve to eliminate the DC-side characteristic harmonics and harmonics of high orders impairing communication and to suppress resonance on the DC side. When two DC filter banks are used, one group of disconnectors and earthing switches for common use are installed on the HV and LV sides of the DC filters, respectively. Surge capacitors are installed on the neutral bus side and cooperate with the DC filters to eliminate harmonics. In addition, the DC switchyard is equipped with DC switchgears to transfer operating modes and clear faults. In UHV DC projects with de-icing demand, it is possible to switch over the system to the de-icing configuration, where the two 12-pulse converter valve groups are connected in parallel.

10.1.2 AC Main Electrical Connection Scheme

The AC (line) side of UHV converter stations usually adopts a one-breaker-and-a-half configuration, connecting the station to grids of 500 kV or higher. For UHV DC projects connected to 500-kV AC grids, the AC switchyard mostly uses the GIS. The incoming line of the converter transformer and the AC incoming line are directly connected to the one-breaker-and-a-half bus string. The AC filter banks serve to eliminate the harmonics generated by converters and compensate the reactive power consumed by converter stations. According to the performance and compensation capabilities of AC filters and the voltage fluctuations of AC grid, the reactive power compensation devices, such as the AC filters and compensating capacitors, are usually divided into 16–20 sub-banks that will then form 4–5 banks. Each bank is connected as an electrical component to the one-breaker-and-a-half bus string through an individual AC bus. The sub-banks are connected to the AC bus through circuit breakers and disconnectors. These circuit breakers are generally the outdoor open type and are capable of switching relevant capacitive current and withstanding high power frequency recovery overvoltage. Figure 10.2 shows the typical AC main electrical connection scheme of a UHV converter station.

10.1.3 General Layout

As stated, a UHV converter station can be divided into several functional areas including valve hall, converter transformer area, DC switchyard, AC switchyard, and AC filter area. Figure 10.3 shows the typical layout of a UHV converter station.

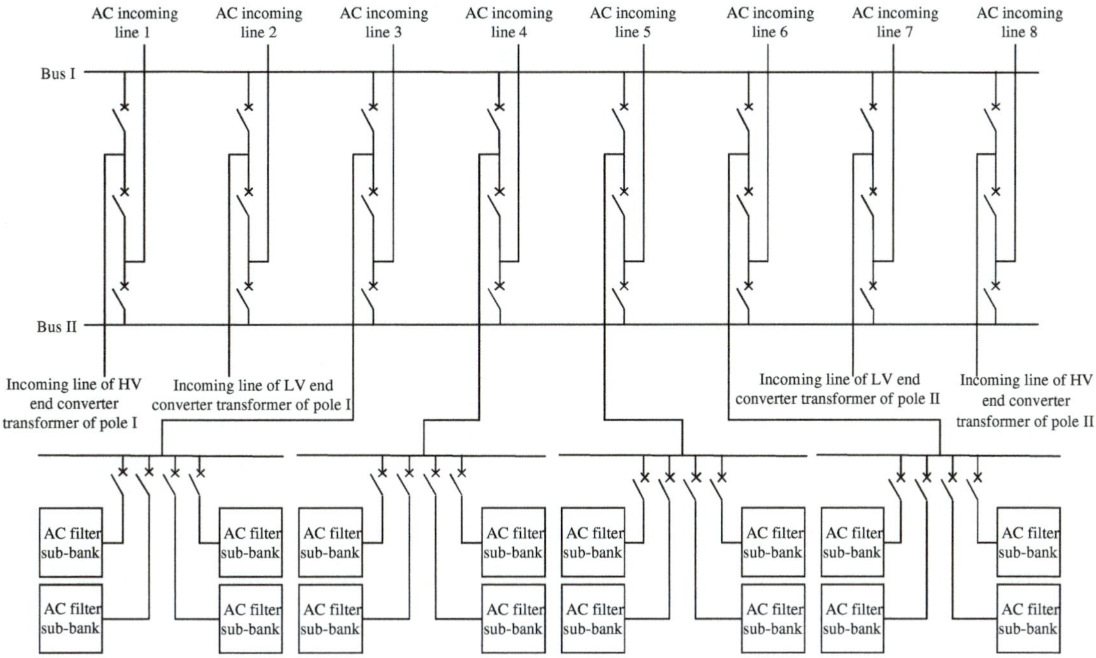

Figure 10.2
Typical AC main electrical connection diagram of a UHV converter station.

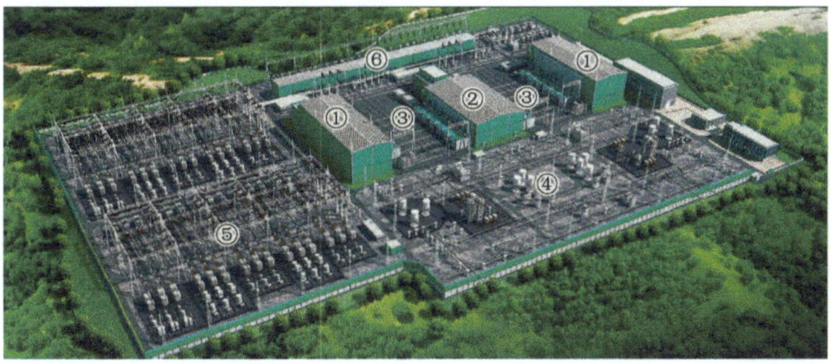

Figure 10.3: Typical layout of a UHV converter station.
①—HV valve hall; ②—LV valve hall; ③—converter transformer area; ④—DC switchyard; ⑤—AC filter area; ⑥—indoor GIS.

Figure 10.3 shows that a UHV converter station is configured as follows:

1. *Valve hall.* Each 12-pulse converter valve group comprises two series-connected six-pulse converter valves, each of which has two series-connected valve arms per phase. The two arms are closely connected to a valve tower to form a double valve. Each 12-pulse converter valve is arranged in one individual valve hall. Hence, there are four

valve halls in total at one station. Each pole is provided with two valve halls, one HV and one LV, arranged face-to-face. The LV valve halls of two poles are arranged back-to-back.

2. *Converter transformer*. Converter transformers are often placed adjacent to the valve hall with the valve side bushings directly protruding into the valve hall and being connected inside the valve hall. The six single-phase two-winding converter transformers belonging to each valve hall are arranged in line outside the valve hall and spaced apart from each other through firewalls. The HV and LV end converter transformers of each pole are arranged back-to-back to effectively reduce the noise.
3. *DC switchyard*. The outdoor DC switchyard has a symmetrical configuration of poles with the DC neutral earthing equipment at the center of the yard. The DC pole bus equipment is arranged on both sides of the DC switchgears and the two DC filter banks of each pole are arranged between the neutral earthing equipment and the pole bus equipment.
4. *AC switchyard and AC filters*. The 500-kV switchgears have an indoor GIS arrangement. The four banks of 500-kV AC filters and capacitors are arranged in "田" formation.

10.2 UHV Converter Valve and Valve Control System

10.2.1 UHV Converter Valve

The converter valves are the core equipment of a UHV DC system and mainly serve to convert alternating current to direct current, or to convert direct current to alternating current. Thyristor valves are the most widely used type in DC transmission projects and are classified as ETT and LTT. Most DC projects use ETTs. In China's UHV DC projects such as Xiangjiaba–Shanghai and Jinping–Sunan links, the current rating is increased to higher than 4 kA and all thyristors used are 6-in ETTs [3].

10.2.1.1 Valve structure

In China, a typical UHV DC link has two series-connected 12-pulse bridges per pole at each end, and the converter valves in use are air-insulated, water-cooled, and suspended. Each 12-pulse bridge has a rated voltage of 400 kV and comprises two six-pulse bridges connected in series. Every six-pulse bridge valve has a double-valve configuration suspended from the ceiling of the valve hall. Each UHV converter station requires a total of 24 double valves, 12 arranged in the two HV valve halls and 12 in the two LV valve halls. Figure 10.4 shows a ±800-kV, 4000-A double-valve tower.

Each arm of a double valve comprises several valve sections. Each section further comprises multiple thyristors and their thyristor control units (TCUs), voltage grading

Figure 10.4
The ±800-kV, 4000-A double-valve tower.

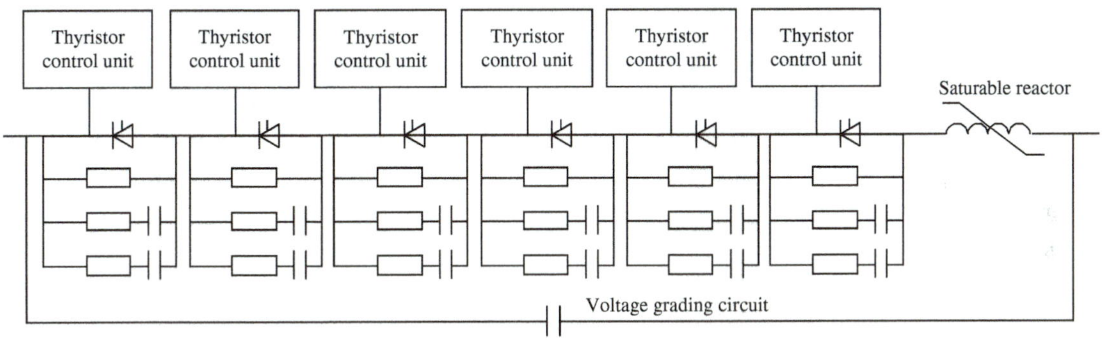

Figure 10.5
Structure schematic diagram of a valve section.

circuit, and saturable reactor. Figure 10.5 shows the schematic diagram of the structure of a valve section.

The most important part of a valve section is the thyristor module, which combines together with the TCUs and snubber circuit to form a thyristor level. A thyristor level is the most fundamental functional unit of a converter valve. Figure 10.6 shows the schematic diagram of a typical thyristor level.

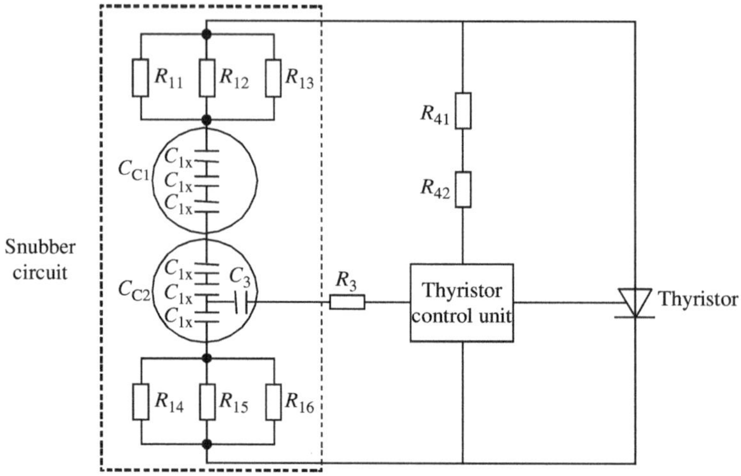

Figure 10.6
Schematic diagram of a thyristor level.

10.2.1.2 Electrical performance

1. **Current withstand capability**
   The converter valve should be capable of withstanding the rated current, overload current, and various transient impulse currents. Generally, parallel connection of thyristors is not considered in design of the valve electrical circuit.
   In terms of transient overcurrents resulting from faults on the DC link, the converter valves should be able to withstand the short-circuit current under all possible conditions irrespective of whether follow-up blocking and load rejection occur. The converter valves used in UHV DC projects are usually supposed to withstand short-circuit currents up to 50 kA.
2. **Voltage withstand capability**
   In electrical design of converter valves, due consideration should be given to the uneven voltage distribution along the series-connected thyristor circuit under various impulse conditions.
   The electrical design of converter valves should also give consideration to the dispersion of voltage protection levels and the impacts of other nonlinear factors in valves on voltage withstand capability of valves. When the converter valves operate at the maximum design junction temperature and all the redundant thyristor levels are damaged, the insulation levels of both the single and multiple valves under switching, lightning impulse, and steep-front overvoltages should have a safety factor of no less than 10–15%.
   At the maximum design junction temperature, the thyristors of inverter valves should be able to withstand a forward transient voltage peak equivalent to the protective firing voltage during the recovery period after commutation.

Surge arresters are the main equipment used to protect the converter valves from both forward and reverse overvoltages, and each single valve is protected by a gap-less zinc-oxide arrester.

In the Xiangjiaba–Shanghai UHV DC link, each individual converter valve has a rated DC voltage of 200 kV, no-load DC voltage U_{dioN} of 230 kV, switching impulse withstand voltage of 450 kV, lightning impulse withstand voltage of 425 kV, and an upper 12-pulse bridge DC bus-to-ground insulation level of 1600 kV (for switching impulse)/1800 kV (for lightning impulse).

3. **Operating capability under AC system faults**
   The thyristor valves should be able to operate under various AC system faults, including 30% of the rated voltage and system dynamic overvoltages.
   The thyristor valves should be capable of withstanding various AC system faults, such as the grounding fault on AC and DC sides and short circuit of thyristor valve.

10.2.2 UHV Converter Valve Control System

The valve control system is an essential part of converter valves. Its main functions are to execute the firing pulse orders from the DC control and protection system, monitor the operating conditions of converter valves, and ensure the safety of the valve operation.

1. **Valve base electronics**
   The valve base electronics (VBEs) are the intermediate equipment to connect the control and protection system and the converter valves, and they can be regarded as the former's fast remote I/O terminals. These electronics trigger the converter valves following orders from the control and protection system and protect the converter valves according to the operating status information of the converter valves monitored. Figure 10.7 presents the schematic diagram of VBEs.

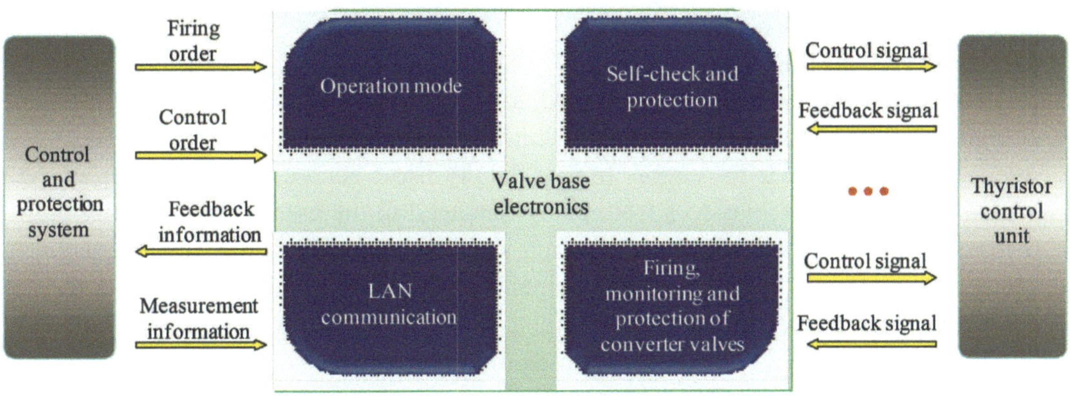

Figure 10.7
Schematic diagram of VBEs.

When the DC system is put into operation normally or as required by a system test, the VBEs will deblock the converter valves, receive, decode, and recode the orders issued by the control and protection system, and then will send the recoded orders to the thyristor firing and monitoring units at the converter valves. When the DC system is shut down normally or due to faults, these electronics will block the converter valves and switch in the bypass pair according to orders from the control and protection system. After blocking the converter valves, the electronics will stop sending firing pulses to the converter valves. When switching in the bypass pair, the electronics will only send firing pulses to single valves selected by the control and protection system.

When the converter transformer is energized and its valve side voltage can support the energy demand of the thyristor firing and monitoring unit, the VBEs will monitor in real time the operating condition of every thyristor level. On detecting any abnormal condition, the electronics will take proper protective measures. Apart from the operating conditions of thyristors, these electronics can also monitor the operation of surge arresters and detect water leakage of valve towers. These auxiliary functions are sometimes not included in the monitoring scope of the VBEs.

The VBEs have a dual redundant configuration and are provided in two completely independent sets, one in operation and the other in hot standby. They should have sound self-check and protection functions. Usually, these electronics can operate in multiple modes such as pre-cycling check before energization, single thyristor level test, voltage boosting in case of undervoltage, and normal operation.

The thyristors currently in use include LTT and ETT, which correspond to two types of VBEs. The major difference between the two types of electronics lies in the transmitted optical power. The VBEs for controlling LTTs serve to transmit optical signals with higher powers that, after being split by a beam splitter, will directly trigger the thyristor. In contrast, the VBEs for controlling ETTs serve to transmit optical signals with lower powers that, after being received by the thyristor firing and monitoring unit, will first undergo photovoltaic conversion and amplification before triggering the thyristors.

2. **Thyristor control unit**

In a UHV DC link, the TCU is positioned at each thyristor level of the converter valves at high potential and serves several functions: (i) provide pulses with sufficient steepness and amplitude to reliably trigger the thyristors according to orders from the VBEs; (ii) protect the thyristors from being damaged in the event of abnormally high forward voltages across them by triggering; and (iii) transmit the real-time thyristor status information and protective firing signals to the VBEs, which will protect the converter valves based on the signals received. The information transmission between the TCU and the VBEs is achieved through the optical fiber to ensure a satisfactory dielectric strength between the high potential and the ground potential. Figure 10.8 shows the schematic diagram of the TCU of a converter valve.

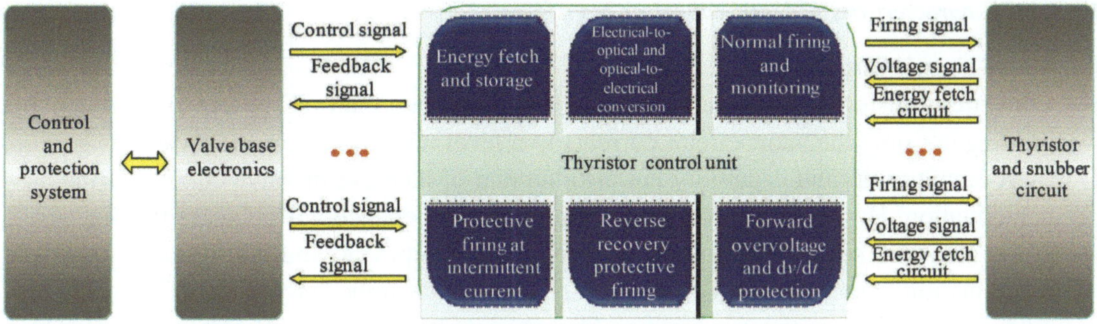

Figure 10.8
Schematic diagram of TCU of a converter valve.

A TCU has the following functions: energy fetch and storage; electrical-to-optical and optical-to-electrical conversion; monitoring on normal firing of thyristors; protection against intermittent current; reverse recovery protection; and overvoltage protection.

Along with the rapid development of electronic information science and technology, the TCU has evolved from a digital–analog hybrid circuit to the present large-scale integrated circuit, which offers a higher integration level as well as higher reliability. It is evolving toward an intelligent solution that has the following merits: (i) it can predict faults on any module of a thyristor level to extend the scheduled maintenance interval of converter valves, simplify the maintenance process, and reduce maintenance times; (ii) with such an intelligent unit, it is possible to conduct unscheduled maintenance early; (iii) it reduces the comprehensive manufacturing cost; and (iv) it can increase the reliability of converter valves.

10.3 UHV Converter Transformer and Smoothing Reactor

10.3.1 UHV Converter Transformer

The converter transformer serves several functions: (i) isolate the AC from DC systems to prevent the DC potential entering the AC system; (ii) supply controllable AC voltage needed by the converter valves, i.e., supply AC voltages in two separate circuits with a relative phase shift of 30 electrical degrees for the two series-connected six-pulse bridges of a 12-pulse converter valve to reduce the low-order characteristic harmonics, especially the fifth and seventh harmonics; (iii) reduce short-circuit currents under faults and control the rate of increase in valve current during commutation.

10.3.1.1 Characteristics of UHV converter transformer

1. **Short-circuit impedance**
   Converter transformers generally have higher impedance than AC power transformers for limiting not only the short-circuit currents to a level within the converter valve's short-circuit withstand capability but also the rate of increase in valve current during commutation. However, an excessive short-circuit impedance will increase the reactive power loss and, hence, the required compensation equipment, and will result in excessive commutation voltage drops. Generally, the short-circuit impedance ranges from 15% to 18%. For instance, in the Xiangjiaba–Shanghai project, the short-circuit impedances at Fulong and Fengxian converter stations are 18% and 16.7%, respectively.
   As the DC transmission voltage increases, the capacity of a single converter transformer is further increased. Because of the manufacturing difficulties and limitation of transportation of oversized equipment, the short-circuit impedance may potentially increase up to 23%. Moreover, the impedance difference between individual phases of converter transformers must be maintained to the minimum, which is generally not higher than 2%. Otherwise, the noncharacteristic harmonic components in the converter transformer current will increase.
2. **AC and DC voltage withstand capability**
   The valve side windings of a converter transformer are subject to the combined load stress of AC and DC voltage. Added to this is the polarity reversal. These all make its insulating structure more complicated than that of an AC power transformer.
   The main insulation of an AC power transformer is designed based on the theory of thin paper cylinders and small oil gaps; in other words, in the AC electric field, the paperboard mainly serves to separate the oil gaps and is not supposed to take up any voltage. Hence, the paper cylinder is made thin and light. However, for a converter transformer, due to the presence of DC electric fields, the main insulation requires more insulation papers to withstand the majority of the field strength. The valve side windings need to be wrapped with multilayered paper cylinders and horn rings. As a result, both the thickness and consumption of paper cylinders greatly exceed those of an AC power transformer. Also, more insulators are required at the valve side leads and the exit lead should take into account factors such as the AC, DC, and polarity reversal fields and is therefore more complicated than that of an AC power transformer.
   In addition, because the valve side windings are subject to DC voltage during operation for a long time, the valve side bushings selected should have a satisfactory creepage distance under DC voltage to not affect the operation of converter transformer. Therefore, the valve side bushings are much longer than the AC side bushings.
3. **Harmonic current withstand capability**
   A converter transformer operates with a great amount of characteristic and noncharacteristic harmonic currents, causing increased stray loss and local overheating

of some metallic components and oil tank. The magnetostrictive noise due to large harmonic flux is in the frequency band sensitive to the human auditory system and, consequently, more effective noise reduction measures such as the Box-in must be taken.

4. **DC magnetic bias current withstand capability**
   The DC bias current present in the converter transformer will lead to increase of the loss, temperature increase, and noise. Generally, the DC bias current of a UHV converter transformer can be designed as 10 A.
5. **Short-circuit current withstand capability**
   Due to the presence of DC component in a fault current, it takes a long time for the maximum unbalanced short-circuit current through the converter transformer to attenuate and it remains at a high level until the protection operates. The electrodynamic force is directly proportional to the square of short-circuit current amplitude, and it is applied on the windings and outgoing line supports. In light of this, the converter transformer should be able to withstand a larger short-circuit stress. Additionally, commutation failures of converter valves may also expose the converter transformer to larger electrodynamic force.
6. **On-load tapping range**
   The converter transformer usually has a large on-load tapping range to ensure that variations in voltage and firing angles are controlled within a proper range. When the DC system operates at a reduced voltage, the tap changer has a maximum of more than 20 tapping positions. For instance, the tapping range of converter transformers used in the Xiangjiaba–Shanghai project is $+23/-5 \times 1.25\%$.

10.3.1.2 Structure of UHV converter transformer

Because the UHV converter transformer has a large capacity and a high valve-side insulation level, it is usually of a single-phase two-winding design with a single-phase four-limb core configuration where the central two (or three) limbs are fitted with windings and two side limbs free of windings act as the magnetic flux loop. The windings on the two main limbs are electrically connected in parallel. Figure 10.9 shows the winding arrangement of a converter transformer with a single-phase four-limb core.

Figure 10.10 shows the schematic diagram of converter transformer windings. It shows the windings comprising core, regulating winding, line side winding, and valve side winding sequentially. The two valve side windings are connected in parallel through the valve-side exit lead. The two line side windings have their HV terminals connected together through a lead, and their LV terminals are first connected to the regulating windings and then connected to the on-load tap changers before being led out through terminals of the on-load tap changers and connected to the neutral point. Such a configuration eases the arrangement of the outgoing lines of valve side windings and can effectively optimize the design

Figure 10.9
Windings of a converter transformer.

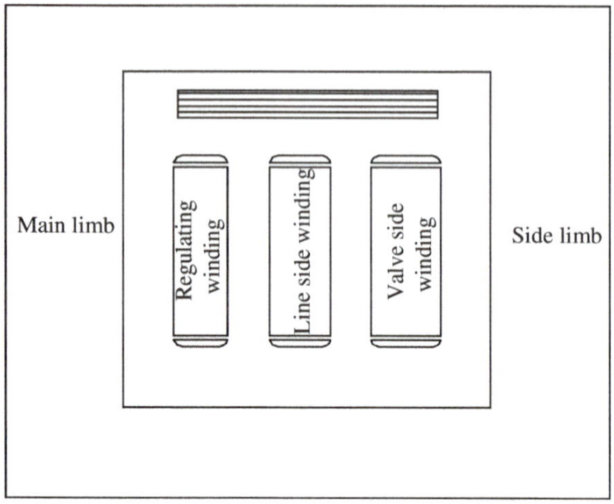

Figure 10.10
Schematic diagram of the configuration of converter transformer windings.

structure of the HV end converter transformers. The LV end converter transformers also use such a winding configuration so that the impedance of all four converter transformers on the same phase of each pole fluctuates consistently.

Figure 10.11 shows the structure of a single-phase two-limb core with side yokes. The core assembly is held together by clamping plates. The upper and lower yokes are of an elliptical cross-section instead of a "D" cross-section, and they are equipped with a magnetic shielding facility to absorb leakage flux escaping from the winding ends. The core limbs and side yokes are all tied with special heat-shrink high-strength tie bands, and the upper and lower yokes are clamped by steel clamps. The core clamps and laminations are grounded via a special grounding conductor at the oil tank, and the insulation resistance can be easily measured from outside of the transformer.

Considering the constraints on transportation of oversized equipment, the UHV converter transformer may also use a single-phase five-limb design in which the central three limbs are fitted with windings and the two side limbs are free of windings. Windings on these main limbs are electrically connected in parallel. Converter transformers of such configuration are used in Jinping converter station. Figures 10.12 and 10.13, respectively, show the converter transformers used in Fengxian and Fulong converter stations.

The oil tank is cylindrical-shaped with a flat cover, which can withstand the mechanical strength tests under 13.3 Pa vacuum pressure and 98 kPa positive pressure. On its inner wall, aluminum shielding (Fulong converter station) or cooper shielding (Fengxian converter station) is applied.

Figure 10.11
Structure of a converter transformer core.

Figure 10.12
Converter transformer in the Fengxian converter station.

Figure 10.13
Converter transformer in the Fulong converter station.

The line side bushings are protruded out from the oil tank top, whereas the two valve-side DC bushings are penetrated into the valve hall on a slant at one end of the transformer. The Box-in is installed outside the converter transformer, with its muffling boards being securely mounted onto the transformer body. The transformer cooling system comprises independent cooler banks that are external to the Box-in for better heat dissipation.

Line side bushings use porcelain bushings with oil-impregnated paper insulation and oil level gauge. Valve side bushings use dry-type or SF_6 filled bushings and are provided with a pressure gauge. The taps and bushing tap are led out through small bushings.

10.3.1.3 Main technical parameters of UHV converter transformer

The main technical parameters of converter transformers used in Fulong converter station of the Xiangjiaba–Shanghai project are shown here:

Rated capacity: 321.1 MVA.
Voltage ratio:
$\frac{550}{\sqrt{3}} / \frac{170.3}{\sqrt{3}}$ kV Y/Y (800 and 400 kV end)
$\frac{550}{\sqrt{3}}$ /170.3 kV Y/Δ (600 and 200 kV end).
Tapping range of on-load tap changer: $+23/-5 \times 1.25\%$.
Short-circuit impedance: 18%.
No-load loss: 188 kW.
Load loss: 646 kW.
Noise limit: 78 dB(A).

Table 10.1 provides the temperature increase limits at rated capacity, and Table 10.2 provides this information for the test voltages.

10.3.2 UHV Smoothing Reactor

The smoothing reactor serves several primary functions: (i) it limits abrupt changes in DC current to reduce the probability of commutation failures; (ii) in case of faults on the DC line, it cooperates with the rectifier side regulators to limit the short-circuit current peak, and it can also limit the currents resulting from discharges of lines and shunt capacitors installed at the line ends through the inverter; (iii) it combines with DC filters to form a DC line harmonics filtering circuit to reduce the harmonic content in the DC line voltage and current; (iv) it prevents the steep-front impulses generated by the DC switchyard or DC lines from intruding into the valve hall, thus protecting converter valves from being exposed to overvoltages; (v) it smoothes the ripples in the DC current to avoid current discontinuity during low DC power transmission; and (vi) it avoids resonance on the DC side.

10.3.2.1 Structure and characteristics

In UHV DC projects, naturally cooled, outdoor, dry, hollow, smooth reactors are used. The windings of a single reactor have a rated inductance of 75 mH in the Xiangjiaba–Shanghai

Table 10.1: Temperature Rise Limits at Rated Capacity of Converter Transformer.

Top Oil Temperature Rise (K)	Average Winding Temperature Rise (K)	Hot-Spot Temperature Rise of Winding (K)	Temperature Rise of Oil Tank, Core and Structural Part (K)	Hot-Spot Temperature of Winding Under Short-Time Overload (°C)
50	55	68	75	120

Table 10.2: Converter Transformer Test Voltages (kV).

Item		Line Side Winding	Valve Side Winding			
			Y1 (HV End)	Δ1 (HV End)	Y2 (LV End)	Δ2 (LV End)
Full-wave lightning impulse	Terminal 1	1550	1800	1550	1300	1175
	Terminal 2	200	1800	1550	1300	1175
Chopped-wave lighting impulse (type test)	Terminal 1	1705	1980	1705	1430	1293
	Terminal 2	—	1980	1705	1430	1293
Switching impulse	Terminal 1	1175	—	—	—	—
	Terminal 2	—	—	—	—	—
	Terminals 1 and 2	—	1600	1300	1175	1050
Short-time AC applied voltage (neutral point)	Terminals 1 and 2	95	—	—	—	—
Short-time AC induced voltage	Terminal 1	680	—	—	—	—
Long-time AC induced voltage and partial discharge	Terminal 1 (U1)	550	—	—	—	—
	Terminal 1 (U2)	476	—	—	—	—
Long-time AC applied voltage and partial discharge	Terminals 1 and 2	—	912	695	479	262
Long-time DC applied voltage and partial discharge	Terminals 1 and 2	—	1258	950	646	341
DC polarity reversal and partial discharge	Terminals 1 and 2	—	970	715	460	205

project and 60 mH in the Jinping–Sunan project. A total of four reactors are installed per pole at each station, with every two being connected in series and then connected to the line pole circuit and neutral line circuit, respectively. Hence, the overall inductance per pole is 300 mH (in Xiangjiaba–Shanghai project) or 240 mH (in Jinping–Sunan project). Figures 10.14 and 10.15 show the ±800-kV line pole and neutral bus smoothing reactors in the Fulong converter station.

The HV side (line pole side) smoothing reactors may be either vertically supported using porcelain insulators or supported using composite insulators at an angle of inclination of 10°.

Figure 10.16 shows a smoothing reactor vertically supported using porcelain insulators. The supporting structure comprises twelve 12-m-high porcelain insulator posts, each consisting of six 2-m-high insulators. Two insulators are defined as one unit, and a rigid metal platform is provided between every two successive units to secure all insulator posts together. In addition, twenty-four 800-mm-high stainless steel supports are fitted between the top rigid platform and the reactor to provide a transitional supporting. The smoothing

Figure 10.14
The ±800-kV line pole smoothing reactor in the Fulong converter station.

Figure 10.15
Neutral bus smoothing reactor in the Fulong converter station.

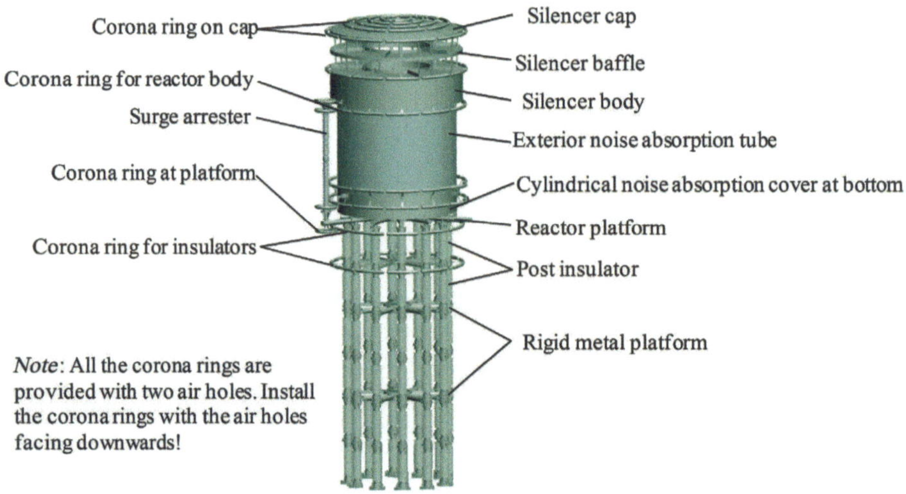

Figure 10.16
Installation diagram of an HV side smoothing reactor vertically supported using porcelain insulators.

reactor body is mounted on these supports. Noise reduction devices are disposed at the top, middle, bottom, as well as inside of the smoothing reactor body to completely encapsulate it, thereby reducing noises and protecting the reactor from rain. The HV side reactor winding is also provided with brackets at both terminals to mount arresters. Moreover, because such smoothing reactors are used in the UHV DC line, corona rings as shown in

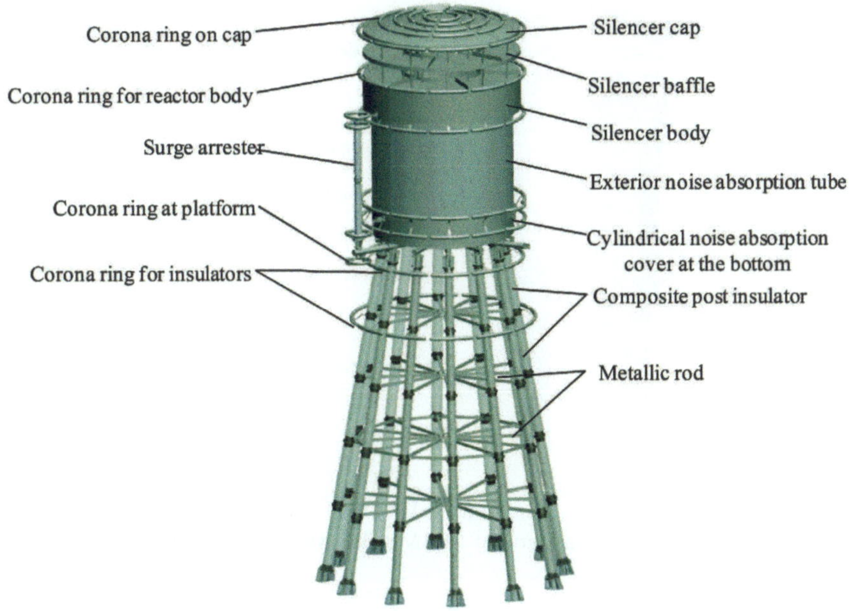

Figure 10.17
Installation diagram of an HV side smoothing reactor supported on a slant using composite insulators.

Figure 10.16 are provided at any point of the smoothing reactor where high-strength fields occur to avoid coronas.

Figure 10.17 shows a smoothing reactor supported on a slant using composite insulators. The supporting structure comprises twelve 12.27-m-high solid composite insulator posts inclined at an angle of 10°, each consisting of five insulators with slightly varying lengths. Metallic rods are attached between individual insulators to hold the insulator posts together.

Figure 10.18 shows the installation diagram of an LV (neutral line) side smoothing reactor irrespective of its being supported using porcelain or composite insulators.

As compared with an ordinary LV dry hollow reactor, a UHV smoothing reactor has higher voltage and capacity. Instead of single-wire round aluminum conductors, Class H uniformly insulated aluminum transposed conductors are used to reduce the harmonic loss and to increase the insulation level and reliability. The HV and LV end windings are of the same configuration and hence are interchangeable.

10.3.2.2 Technical requirements

In the Xiangjiaba–Shanghai project, the smoothing reactors have a rated DC voltage of 800 kV, rated DC current of 4000 A, maximum continuous overload DC current of 4497 A,

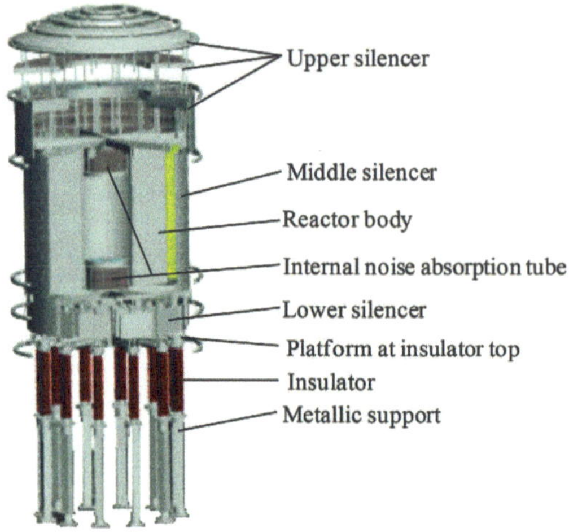

Figure 10.18
Installation diagram of an LV side smoothing reactor.

and peak transient fault current of 40 kA. Also, they are allowed to operate at 4731 A for 2 h and 5909 A for 3 s.

The steady-state temperature under the maximum continuous current is limited as follows. For outdoor applications, the hot spot temperature increase induced by the maximum continuous current I_{mcc} plus the equivalent harmonic current is up to 100 K and the average temperature increase is up to 80 K; for indoor applications, the absolute average temperature and hot spot temperature should be consistent with those for outdoor applications, based on the average and hot spot temperature increases.

Apart from sunlight, ground, and building reflection, the local ambient temperature increase due to noise reduction measures should also be taken into consideration.

Noise level limits: The noise level (sound pressure level) measured at 2 m above the ground and 5 m beyond the vertical projection of an operating smoothing reactor should not exceed 70 dB(A). To meet this requirement, noise reduction devices are installed at the top, middle, and bottom of the reactor body, thus decreasing the sound level (sound pressure level) from 81.7 to 65.7 dB(A).

The nominal full-wave lightning and switching impulse withstand voltages between terminals are 1175 and 950 kV, respectively; the lightning and switching impulse withstand voltages between the HV terminal and ground are 1950 and 1600 kV, respectively; the lightning and switching impulse withstand voltages between the LV terminal and ground are both 550 kV.

Requirements on thermal class of insulation: It should be at least Class H for interstrand insulation and Class F for inter-turn insulation.

10.4 Filters in UHV Converter Station

10.4.1 UHV DC Filter

An operating HVDC converter will generate harmonic voltage and current on the DC side of the DC system, thus causing noise on telephone lines adjacent to the DC line. By installing appropriate DC filters, smoothing reactors and neutral earthing capacitors in the DC system, it is possible to limit those harmonics to an acceptable level. Moreover, in a UHV DC project, the single phase-to-ground or phase-to-phase short-circuit faults on the AC side may, due to the propagation effects of the converters, result in low-frequency second harmonic resonance on the DC side, which may give rise to the need for installation of filtering branches to suppress such resonance [4].

10.4.1.1 DC filter configuration

Generally, two DC filter configurations are adopted for DC transmission projects: (i) a neutral earthing surge capacitor is connected between each pole's neutral bus and the ground to offer a low impedance path through which the third-order harmonic current flows from the stray capacitance of the converter transformer's windings to ground, thereby suppressing such noncharacteristic harmonics; and (ii) two banks of double-tuned or one bank of triple-tuned passive DC filters are connected between each pole's DC bus and neutral bus in parallel such that when any DC filter bank at the system's two converter stations is taken out of service due to faults, the remaining filter can still operate satisfactorily.

The configuration of UHV DC filters should consider the amplitude of harmonics of individual orders and the proportion of them in the equivalent disturbing current. Theoretically, a 12-pulse converter only generates characteristic harmonics on the DC side, which account for the majority. Characteristic harmonic orders below the 50th include 12th, 24th, 36th, and 48th. Figure 10.19 shows the schematic diagram of a typical configuration of DC filters.

However, noncharacteristic harmonics of orders such as 3rd, 6th, 9th, and 39th also occur due to various reasons, such as the harmonic contents in the AC bus voltage, differences in leakage reactance and voltage ratio of converter transformers, leakage reactance unbalance between three phases of converter transformers, inconsistency between any operating parameter of the two poles' converters, and the ground stray capacitance of converter transformers.

Based on the analysis results of DC harmonics, the DC filters in a UHV DC project are arranged either in the 2/12/39 configuration or in a combination of 12/24 and 2/39. Figure 10.20 shows the DC filter configuration used by the Xiangjiaba–Shanghai project.

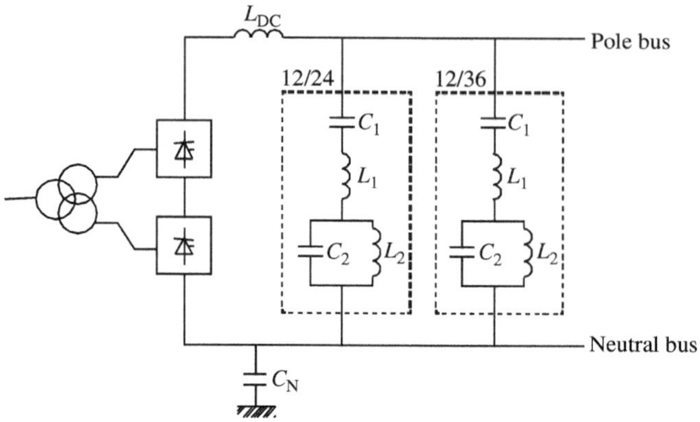

Figure 10.19
Schematic diagram of a typical configuration of DC filters.

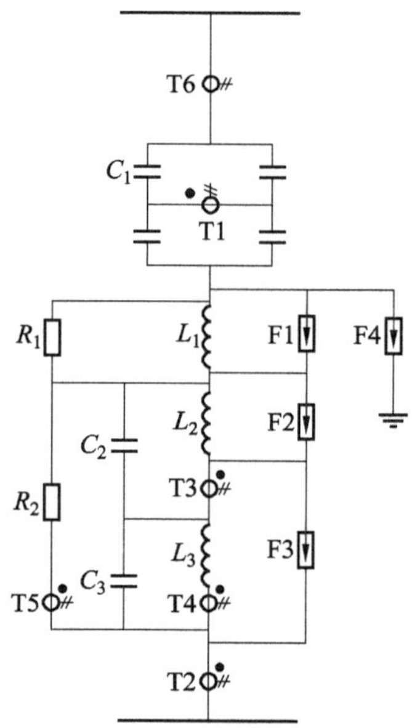

Figure 10.20
Schematic diagram of an HP2/12/39 triple-tuned DC filter bank.

At one converter station, each pole is equipped with one triple-tuned DC filter bank with tuning frequencies of 100, 600, and 1950 Hz and an HV capacitance of 1.05 μF.

The use of the triple-tuned DC filter reduces the manufacturing costs of HV capacitors but may result in excessive harmonics if the filter is taken out of service after faults. Considering this, the Jinping–Sunan project used a combination of 12/24 and 2/39 double-tuned filters, as shown in Figures 10.21 and 10.22, respectively.

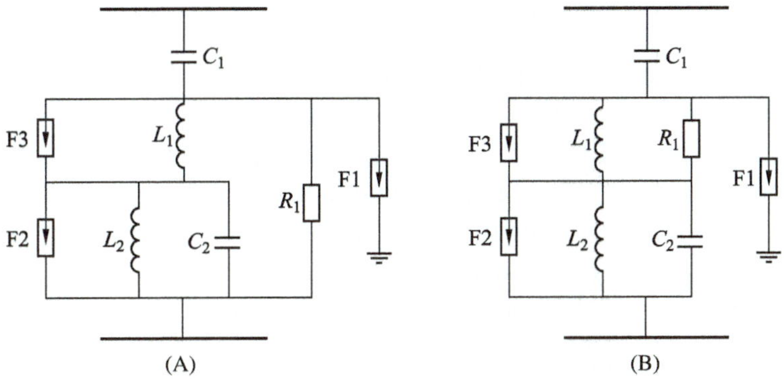

Figure 10.21: Schematic diagram of double-tuned DC filters.
(A) HP12/24 and (B) HP2/39.

Figure 10.22
DC filter yard in the Jinping (Yulong) converter station.

The HV capacitor tower of DC filters has a three-tower supporting structure, each having a height of 21 m. As the voltage classes increase, the capacitor tower will become even higher, thus making its design a new concern in the R&D of DC filters.

10.4.1.2 DC filter performance requirements

The filtering performance of DC filters is evaluated based on the equivalent disturbing current calculations. The calculation formula is given by:

$$I_{eq}(x) = \sqrt{I_e(x)_J^2 + I_e(x)_H^2} \text{ (mA)}$$

where

$I_{eq}(x)$—noise-weighted equivalent disturbing current (up to 800 Hz) at any point along the line corridor, mA
$I_e(x)_J$—magnitude of the RSS equivalent disturbing current component induced by harmonic voltage sources at the rectifier station, mA
$I_e(x)_H$—magnitude of the RSS equivalent disturbing current component induced by the harmonic voltage sources at the inverter station, mA
x—relative location along the line corridor.

For a ±800-kV UHV DC transmission project, the maximum equivalent disturbing current is set at 3000 mA under the bipolar mode and at 6000 mA under the monopolar ground return or monopolar metallic return mode. Given that the telephone links nearly all use optical fiber cables instead of open-wire lines today, relevant tests and studies were conducted at the R&D stage of ±800-kV UHV DC transmission, which indicates that the equivalent disturbing current is greatly increased, thereby greatly reducing the costs of DC filters.

10.4.1.3 Parameters of HV capacitors

The DC filter comprises HV and LV capacitors, reactors, and resistors, among which the HV capacitors are the main components. The HV capacitor mainly serves to withstand the DC operating voltage and harmonic voltage. The former is distributed among the series-connected capacitor units according to their discharge resistance. The average operating field strength of HV capacitors ranges from 90 to 110 kV/mm, and for a UHV DC project it is 90 kV/mm. The distribution of DC voltage should take into account the unevenness of leakage resistance distribution among the porcelain bushings of capacitor units, especially the effects of insulator pollution.

Most capacitor units are protected using internal fuses, with a lower voltage limit of no higher than $0.7U_{DC}/S$ and the upper voltage limit of no less than U_{SIPL}/S (where $U_{SIPL} = U_{SIWL}/1.2$). Here, S is the number of series-connected elements in the capacitor unit; U_{DC}, U_{SIPL}, and U_{SIWL} are the DC operating voltage, switching impulse protection level, and withstand level, respectively, of each HV capacitor unit.

The HV capacitor bank has a rated voltage of approximately 1400 kV, rated current of approximately 200 A, HV end-to-ground lightning impulse withstand voltage of 1950 kV, switching impulse withstand voltage of 1600 kV, and noise limit no larger than 70 dB.

10.4.2 UHV AC Filter

As a nonlinear component of the power system, the HVDC converter will generate various harmonics during its operation. It acts as a harmonic current source seen from the AC system. Specifically, the converter produces harmonic currents that are injected into the AC system from the line side of the converter transformer and result in distortion of the AC bus voltage at the converter station.

The inrush of a large amount of harmonic currents into the AC side will inevitably cause serious impact on the stable operation of both the AC and DC systems. The AC filters can limit the distortion rate of the converter bus voltage to guarantee satisfactory voltage quality and provide the reactive power required for valve commutation. By reasonably configuring AC filters, it is possible to meet the requirements of harmonic control and reactive power balance.

10.4.2.1 AC filter configuration

The characteristic harmonic currents produced by a converter depend on its pulse number. In a ±800-kV UHV DC project, each pole has two 12-pulse converter units that produce characteristic harmonics of the following orders on the AC side:

$$n = 12k \pm 1 \quad (k = 1, 2, \ldots, \infty)$$

Only those below the 50th are considered, including 11th, 13th, 23rd, 25th, 35th, 37th, 47th, and 49th.

The AC filters should be properly configured according to the harmonic currents, and the number of types of the filters should not be too many, generally two or three. The number of filter banks should be as few as possible provided that the requirements of filtering performance and reactive power balance of converter station are met. Specifically, when all filters are put into operation, they should meet the filtering requirements of operation at continuous overload and reduced voltage; if any filter bank is out of service, then the rest can still meet the filtering requirement under the rated operating conditions. When the system operates at a small load (10% of the rating), the total capacity of filters in service should be kept at the minimum.

Following these principles, the Xiangjiaba–Shanghai project uses the following configuration of AC filters (connected to 500-kV AC bus): the Fulong converter station is provided with four banks of double-tuned high-pass filters HP24/36 to remove characteristic harmonics of the 23rd, 25th, 35th, and 37th and other high-order harmonics; four banks of double-tuned filters HP11/13 to remove harmonics of the 11th and 13th; one HP3 filter bank to remove the third and other low-order harmonics; and five shunt capacitor banks to compensate reactive power. In sum, there are 14 sub-banks, each having a capacity of 220 Mvar. They comprise four banks, each consisting of three or four sub-banks.

The Fengxian converter station is provided with eight banks of double-tuned high-pass filters HP12/24 to remove characteristic harmonics of the 11th, 13th, 23rd, and 25th and other high-order harmonics. It has seven shunt capacitor banks to compensate reactive power. In sum, there are a total number of 15 sub-banks, each having a capacity of 260 Mvar. They comprise four banks, each consisting of three or four sub-banks.

Figure 10.23 shows the circuit diagram and impedance curve of a typical single-tuned filter and Figure 10.24 shows those of a double-tuned high-pass filter.

As compared with the conventional HVDC project, the ±800-kV UHV DC project has a much larger filter capacity. With the increase of system transmission capacity and voltage

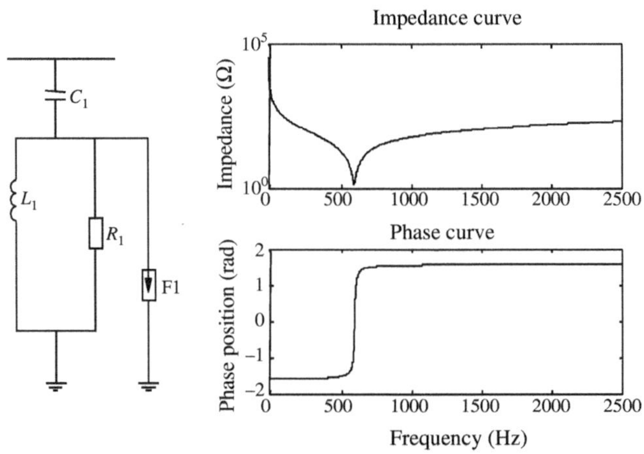

Figure 10.23
Circuit diagram and impedance curve of a typical single-tuned filter.

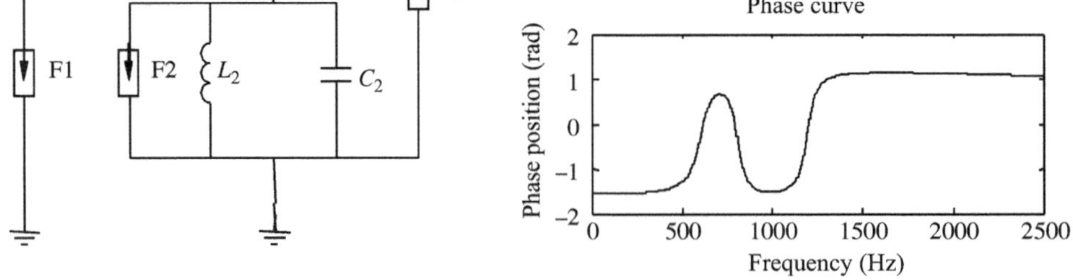

Figure 10.24
Circuit diagram and impedance curve of a typical double-tuned high-pass filter.

level, if a double-tuned filter is used to filter-out the 11th and 13th characteristic harmonics, its parallel resonant circuit will be subject to great equipment stress. Using two parallel-connected single-tuned filters instead can avoid excessive resonance stress on the parallel equipment. The two single-tuned filters are connected in parallel to form a sub-bank, and they share one set of switches.

The HV shunt capacitors are a supplement to the required number of filter banks and are intended for reactive power compensation. An HV shunt capacitor usually comprises one HV capacitor and one small damping reactor. The latter is mainly used to suppress the inrush currents caused by the switching of two identical capacitors on the bus.

10.4.2.2 AC filter performance requirements

The AC filters configured according to the characteristics of converter station harmonics should be such that the harmonic interference indexes of the AC buses at the converter station meet the following standards.

1. **Single harmonic distortion D_n**
   During calculation, when selecting the impedance points within the defined harmonic impedance envelope diagram that result in the highest distortion of individual orders of harmonics, it should be ensured that the third and fifth harmonic distortions of the AC system do not exceed 1.25%, other odd-order harmonic distortion D_n does not exceed 1.0%, and the even-order harmonic distortion D_n does not exceed 0.5%.
2. **Total harmonic distortion D_{eff}**
   The harmonic impedance of the AC system should be taken from the values within the given harmonic impedance envelope diagram for the two harmonics that result in the highest total harmonic distortion. For all other harmonics, the AC system harmonic impedances should be considered to be an open circuit, and the calculated D_{eff} should not exceed 1.75%.
3. **Telephone harmonic form factor (THFF)**
   The harmonic impedance of the AC system should be taken from the values within the given harmonic impedance envelope diagram for the two harmonics that result in the highest THFF. For all other harmonics, the AC system harmonic impedances should be considered to be an open circuit and the calculated THFF should not exceed 1.0%.

10.4.2.3 Parameters of HV capacitor

An AC filter comprises HV and LV capacitors, reactors, and resistors. The cost of HV capacitors constitutes a large portion of the entire filter. It sets high requirements for the design and manufacturing technologies and affects the performance and reliability of the filter.

In China's DC transmission projects, the average operating field strength of HV capacitors of the AC filters ranges from 50 to 60 kV/mm. For UHV DC projects, it is approximately 57 kV/mm. The HV capacitors are internally fused.

The HV capacitor banks connected to a 500-kV AC bus have a maximum continuous operating voltage of approximately 460 kV (RMS value), maximum continuous operating current of approximately 340 A (RMS value), HV end switching impulse withstand voltage of 1175 kV, and HV end-to-ground lightning impulse withstand voltage of 1550 kV. All AC filters have a two-tower structure with a height $<$13 m, and the noise is controlled to less than 65 dB.

10.5 Surge Arresters in UHV Converter Station

10.5.1 Types and Characteristics of Arresters

As compared with those in a substation, surge arresters in a converter station have two major characteristics: (i) the required number is larger and most of them are installed immediately close to the protected equipment; and (ii) they absorb much more energy because numerous reactive power compensation equipment (e.g., filters) are provided in the converter station. Because the UHV DC transmission systems operate at high voltage levels, the rated voltage and protection level of arresters are inevitably required to be higher. The use of arresters with superior protection performance and an optimized configuration can effectively decrease the requirements of the equipment's insulation level, thus facilitating the R&D and manufacturing of UHV equipment at a lower cost [5].

Both DC and AC arresters are installed in a UHV DC converter station. All these arresters may be classified into the following three categories in terms of their characteristics.

1. **DC arresters with significant continuous operating voltage**
   This category includes the DC pole bus arresters (DB, DL, and CB2), ±400-kV bus arresters between the two 12-pulse converter units (CB1A), arrester of six-pulse bridge of the upper 12-pulse converter unit (M2), arrester of six-pulse bridge of the lower 12-pulse converter unit (M1), and the valve arresters (V1, V2, and V3) (see Figure 4.32).
   The valve arrester is subject to valve voltage only when the valve is off in an AC cycle, and the heat generated by the leakage current under the valve voltage in a cycle is very low, on average. Hence, a higher applied voltage ratio ranging from 0.95 to 1 may be selected.
   The arresters C2, M1, M2, and CB1A are subject to DC voltage plus harmonic voltage. Some of the current resulting from the harmonic voltage is discharged through the stray capacitance of the arresters and, especially, the commutation overshoot generates less heat on the varistors than the DC voltage component. In addition,

because these arresters are installed indoors, the effects of pollution and ambient temperature may be disregarded. Hence, an applied voltage ratio of approximately 0.9 may be selected.

The arresters DB and DL are subject to a very high, pure DC voltage. If they are installed outdoors, then pollution may lead to nonuniform potential distribution over the ceramic or silicon rubber housing, causing local overheating on the varistors. In addition, the ambient temperature has a great impact on heat dissipation and voltage–current characteristics of the arresters. Consequently, a lower applied voltage ratio of approximately 0.8 seems more reasonable. However, the performance of arrester DB plays an essential role in decreasing the insulation level of UHV DC equipment. By optimization studies, using varistors with excellent performance can increase its applied voltage ratio to 0.85.

2. **Arresters without significant continuous operating voltage**
This category mainly includes those on the neutral bus (E1, E1H, E2H, EL, and EM), which have lower continuous operating DC voltage. Arresters in both the AC and DC filters have low continuous operating voltages, which, however, contain high harmonic content.

Although the neutral bus arresters have low continuous operating voltages, the neutral bus may be subject to switching overvoltage that will cause a large amount of stress on its arresters under faults like short-circuit between the DC pole and the ground, as well as between the outlet of the converter transformer's valve side bushing and the ground, asymmetrical fault at the AC side, and others. Figure 10.25 shows the maximum energy demand of a neutral bus arrester as a function of its rated voltage. The rated voltage of the arrester is determined to be higher than its continuous operating voltage to control the maximum energy demand at a low level while meeting the requirements of continuous operating voltage.

Arresters in filters have a low operating voltage, whereas the capacitor connected in series with them may charge up to several hundred kilovolts. In the case of an external short circuit, the charged capacitor will be connected in parallel with the LV reactor, giving rise to high overvoltage. In light of this, the rated voltage of an arrester should

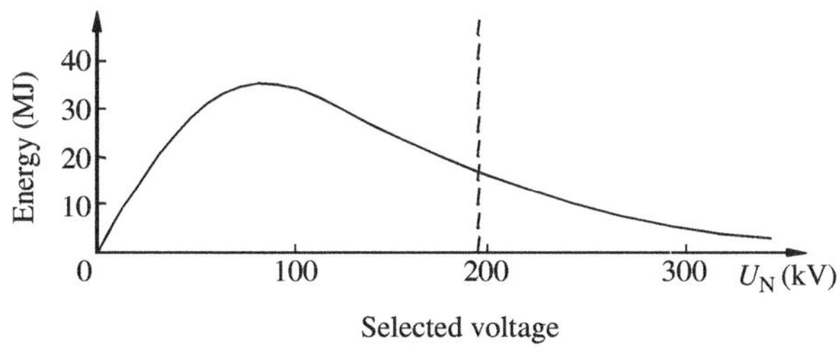

Figure 10.25
Maximum energy demand versus rated voltage of neutral bus arresters.

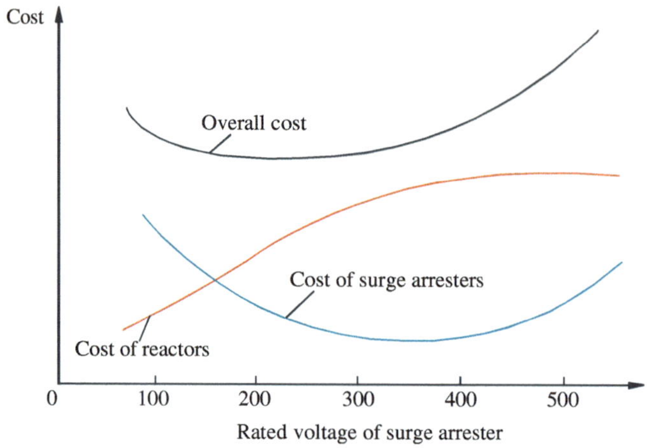

Figure 10.26
Overall equipment cost versus rated voltage of arresters.

be determined following the same principle as the neutral bus arrester, i.e., according to its relationship with the overall equipment cost (including reactors and arresters), as shown in Figure 10.26, rather than according to the arrester's maximum continuous operating voltage. Generally, given a lower overall cost, it is not sensitive to the equipment's rated voltage. To ensure the equipment safety, the rated voltage should be as high as possible provided that the overall cost is not considerably increased.

Generally, such arresters take no account of the applied voltage ratio, which is generally low.

This category of surge arresters consumes more energy in UHV DC applications than in conventional HVDC applications.

3. **AC bus arresters**
The AC bus arresters in a converter station need to absorb much more energy than those in an ordinary substation. This is because the converter station is equipped with a large number of reactive power compensation equipment (e.g., filters) that will be subject to long-duration and large-amplitude temporary overvoltage after an AC system fault. If such overvoltage exceeds 1.3 p.u., then the reactive power compensation equipment (e.g., filters) should be quickly switched off. It is generally designed so that when the temporary overvoltage increases to higher than 1.3 p.u. but lower than 1.5 p.u., the compensation equipment should be switched off after 100 ms; when the temporary overvoltage exceeds 1.5 p.u., the compensation equipment should be immediately switched off. Therefore, the AC bus arresters should be able to absorb a large amount of energy.

Moreover, the circuit breakers used in these cases must be capable of switching off capacitors under high recovery voltages, which is obviously a harsh requirement on the circuit breakers. To properly reduce the recovery voltage across breaks of the circuit breaker, it is necessary to control the arrester's protection level at a low value, which further increases the energy absorption requirements of the AC bus arresters. This is particularly important when the UHV DC system is connected to 750 or 1000-kV AC systems.

10.5.2 Structure of UHV DC Pole Bus Arrester

The essential structural difference between a UHV DC pole bus arrester and others lies in the external insulation. Under DC voltage, the arrester housing is severely polluted, which therefore requires a larger specific creepage distance and consequently affects the electrical performance of arresters greatly. When the DC system operates at a very high voltage, especially in the UHV applications, the external insulation of porcelain-housed arresters is not satisfactory and, therefore, the UHV DC arresters are usually polymer-housed or porcelain-housed with RTV coating.

The UHV DC pole bus arrester DB is generally suspended, as shown in Figure 10.27, but it may also be supported. Because the arrester DB has a high continuous operating voltage and protection level, and because it plays an important role in decreasing the insulation level of equipment in UHV DC projects, several arresters are connected in parallel to decrease the protection level. In contrast, if the arrester has a structure with varistors connected in parallel, then its total weight will be increased. Once the weight exceeds the bearing capacity of the suspending structure, the supporting-type arrester has to be used. Using varistors with good performance and optimized structure is a key to arrester DB.

See Table C.8 in Appendix C for the main performance parameters of UHV DC arresters.

Figure 10.27
The ±800-kV UHV DC pole bus arrester.

10.6 UHV DC Post Insulators and Bushings

10.6.1 Pollution Characteristics of DC Insulators

Because the DC insulators have lower pollution flashover withstand voltage than AC insulators, and because the electrostatic adhesion of DC voltage makes the DC insulators suffer more severe pollution, the insulators for DC applications should have much larger specific creepage distances than AC insulators.

For instance, in regions of Class D site pollution severity (SPS), the equivalent salt deposit densities (ESDD) of post insulators in the AC and DC switchyards are approximately 0.05 and 0.063 mg/cm^2, respectively. Therefore, the DC insulators are required to have a much larger specific creepage distance than AC insulators. For example, if the specific creepage distance of porcelain insulators in AC application is required to be 25 mm/kV, then it would be 54 mm/kV for porcelain insulators in DC applications.

Actually, even at sites with the same SPS class, for example, class D, the DC insulators have different ESDD values depending on the local wind direction as well as the characteristics and geographic location of pollution sources. The ratio between the ESSDs of DC and AC insulators should be determined according to the wind speed and the diameter of pollution particles, as shown in Figure 10.28.

Because the UHV DC project operates at an ultrahigh voltage, achieving the required specific creepage distance will inevitably cause the insulator height to be significantly increased, making it difficult to meet the requirements of mechanical strength. Hence, the

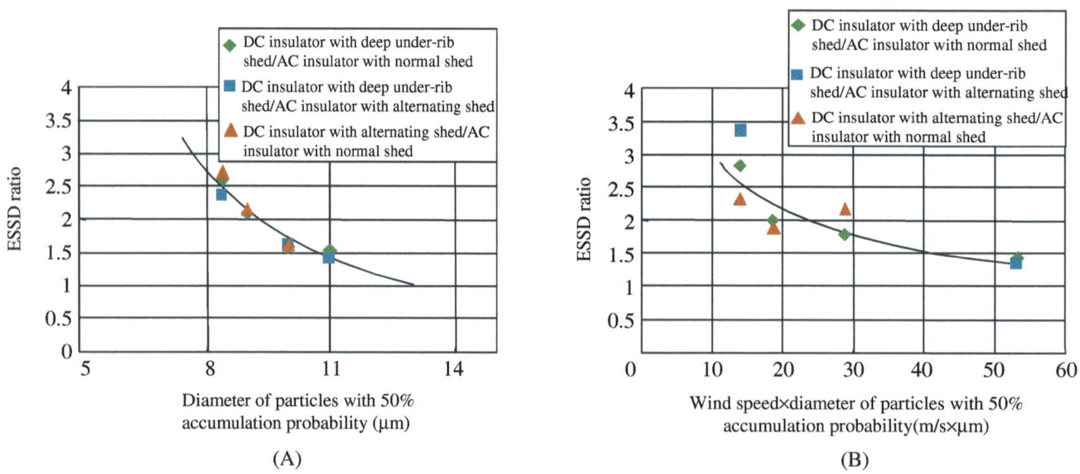

Figure 10.28: Ratio between ESDD of DC and AC insulators.
(A) Wind speed less than 1.5 m/s and (B) wind speed more than 1.5 m/s.

selection of insulators in the DC switchyard follows the principle that equipment insulators are composite external insulation and that the ±800-kV post insulators are porcelain with RTV coating or composite insulation.

Considering factors such as flashover due to uneven wetting, the wall bushings used are composite insulation.

10.6.2 UHV DC Post Insulators

The UHV DC converter station requires a great number of DC post insulators with high mechanical bending strength and aseismic ability that place high requirements on their manufacturing process. Two types of post insulators (i.e., the porcelain type with RTV coating and composite insulation type) are used in the UHV DC applications.

10.6.2.1 Structure of post insulator

The porcelain post insulator and vertical bushing are both the deep under-rib shed-type, with the minimum spacing of the shed being no less than 95 mm, ratio of the shed spacing to the shed overhang being no less than 1.0, and specific creepage distance being no less than 54 mm/kV.

The composite post insulator and vertical bushing are both the alternating shed-type with a ratio of the shed spacing to the shed overhang being no less than 0.9 and specific creepage distance being no less than 48 mm/kV. Figure 10.29 shows the schematic diagram of a vertically mounted post insulator. Its key profile parameters are given below.

1. Spacing of large shed: $S \geq 65$ mm.
2. The difference between the overhangs of the large and small sheds P-$P1$ should be larger than that of the AC porcelain insulator and should be as large as possible (e.g., 20 mm).

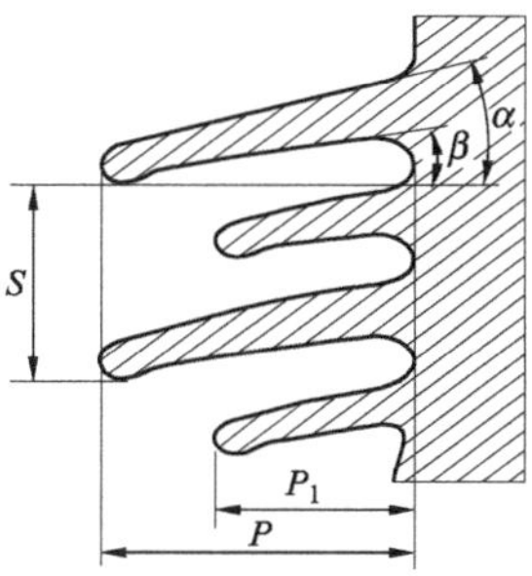

Figure 10.29
Schematic diagram of an insulator with alternating shed.

3. Inclination of shed's upper surface: $\alpha > 10°$; inclination of shed's lower surface: $\beta > 3°$. Too small or big of an inclination angle of the shed's lower surface should be avoided, because an angle that is too small can lead to backflow of rain water and an angle that is too large can lead to pollution deposit under the shed.

10.6.2.2 Main technical parameters

1. **Porcelain post insulator**
   The most important parameter characterizing the DC post insulators is the bending strength, which can reach up to 10 kN for ±800-kV porcelain post insulators. Table 10.3 lists the main technical parameters of the porcelain post insulators used in the Xiangjiaba–Shanghai project.
2. **Composite post insulator**
   The composite post insulators are made of high-temperature vulcanized silicone rubber formed using a vacuum integral injection molding process. There are two types of insulator cores. One is made of glass fiber-reinforced plastics and is formed and cured using continuous filament wet winding process (also called solid-core composite insulator type I). The other is formed by wrapping an insulating tube with glass fiber-reinforced plastics filled with insulating media (also called solid-core composite insulator type II). Table 10.4 lists the main technical parameters of the UHV DC post insulators used in the Jinping–Sunan project.

10.6.3 UHV DC Wall Bushing

The DC wall bushings are used to connect the inside and outside of the valve hall. Those for the UHV DC applications may be classified into ±800- and ±400-kV line pole wall

Table 10.3: Main Technical Parameters of Porcelain Post Insulators with RTV Coating Used in Xiangjiaba–Shanghai Project.

No.	Item	Unit	Magnitude
1	Dry lightning impulse withstand voltage (peak)	kV	1950
2	Wet switching impulse withstand voltage (peak)	kV	1600
3	DC withstand voltage (wet)	kV	1236
4	Structure height	mm	10,970
5	Number of sections	Section	6
6	Section height	mm	1825
7	Maximum trunk diameter	mm	280
8	Maximum shed diameter	mm	472
9	Outer diameter of grading ring	mm	1800
10	Bending strength	kN	10

Table 10.4: Main Technical Parameters of UHV DC Composite Post Insulators in Jinping–Sunan Project.

Item	Unit	Main Technical Parameters	
		Solid-Core Composite Insulator I—Supports of Smoothing Reactor	Solid-Core Composite Insulator II—Supports of Bus
Rated DC voltage	kV	±800	±800
Lightning impulse withstand voltage (peak)	kV	2750	1800
Wet switching impulse withstand voltage (peak)	kV	1800	1600
DC withstand voltage (wet, 60 min)	kV	1236	1236
Structure height	mm	12,270	12,000
Number of sections	Section	5	5
Section height	mm	2454	2400
Minimum creepage distance	mm	40,800	39,168
Bending strength	kN	16	12.5

Figure 10.30
The ±800- and ±400-kV line pole wall bushings.

bushings (Figure 10.30) and ±200-kV neutral bus wall bushing according to their installation positions. Bushings installed at different positions are provided with different insulation structure.

10.6.3.1 Structure and characteristics

The ±800- and ±400-kV line pole wall bushings have a high insulation level and use composite insulation structure. Next to the current-carrying conductor is a tapered solid insulation structure that mainly comprises a condenser core of epoxy resin-impregnated paper. The condenser core is made of insulation papers and aluminum foils that are wrapped on the current-carrying conductor and then subject to vacuum drying before being impregnated with resin. The condenser core uses capacitance screens to achieve even distribution of both axial and radial electric fields. Outside, the tapered structure is filled with insulating gas, which mainly serves to withstand the longitudinal electric fields of the bushing. In addition, an internal shield is usually provided at the bushing end to ensure even distribution of axial electric fields in the bushing. The outermost layer is the epoxy glass cylinder and silicone rubber housing.

The external insulation of neutral bus wall bushings is implemented using an organic composite housing. The space between the external insulation and main insulation is filled with a single insulating medium that may be epoxy resin-impregnated insulation paper, insulating silica gel, or insulating gas. Bushings using the first two insulating media have a cylindrical capacitor structure. For bushings using the third insulating medium, internal shield electrodes are provided only at the bushing end and flanges to ensure even electric fields.

A DC wall bushing comprises current-carrying conductor, insulation, and metal flange. The current-carrying conductor runs through the cylindrical insulating body along its axis and should be able to withstand both the DC and harmonic currents. The insulation consists of the internal and external insulation. Generally, the DC wall bushing is the dry type, whose external insulation is usually implemented using a silicone rubber housing and internal insulation using resin-impregnated paper or insulating gas. As with other bushings, the ring metallic flange of the DC wall bushing is fitted on its insulation housing for grounding purposes. However, a DC wall bushing is subjected to more complicated electric fields, including DC, AC, and polarity reversal electric fields. Because bushing generally involves multiple insulating materials, and because the distribution of electric fields under DC voltage varies from that under AC voltage, it is necessary to take measures for DC wall bushings, including voltage grading and shielding, which all make the structure of a DC wall bushing more complicated.

10.6.3.2 Main technical parameters

In the Xiangjiaba–Shanghai project, the ±800-kV line pole wall bushings have a total length of 18,804 mm. The indoor part (inside the valve hall) is 8327 mm long and has a flashover distance of 6720 mm and a creepage distance of 23,200 mm; the outdoor part (outside the valve hall) is 10,477 mm long and has a flashover distance of 8870 mm and a creepage distance of 39,800 mm; the SF_6 pressure is 0.57 MPa and the bushing weighs 4051 kg. The service temperature in the valve hall must not exceed 60°C. The main technical parameters of the wall bushings used in the Xiangjiaba–Shanghai project are given in Table C.5 in Appendix C.

10.7 DC Switchgears

10.7.1 DC Transfer Switches

1. **Types and Duties**

   The DC transfer switches are mainly used for shifting between different operating modes of DC transmission systems, such as shifting between the current return paths. Figure 10.1 shows the configuration of DC transfer switches in a UHV DC project.

   a. *Metallic Return Transfer Breaker (MRTB)*. It is installed in the earth electrode line circuit and used to transfer DC current from an earth return path to a metallic return path without interrupting the DC power flow.

   b. *Ground Return Transfer Switch (GRTS)*. It is installed between the earth electrode line and the line pole, and is used to transfer DC current from a metallic return path to an earth return path. It must be used in combination with MRTB.

   c. *Neutral Bus Switch (NBS)*. It is installed on the neutral buses in both the rectifier and inverter stations. At a planned outage of one pole, the converter unit is blocked without switching on the bypass pair. The converter will decrease the DC current to zero, at which the NBS will open. This is also a normal procedure of isolating converter faults (other than ground faults). During the normal bipolar operation, in the event of an earth fault on one pole, that faulty pole will be blocked after the bypass pair is switched-on. In such a circumstance, the NBS serves to divert the DC current injected by the healthy pole into the fault point to the earth electrode line.

   d. *Neutral Bus Ground Switch (NBGS)*. It is installed between the neutral bus and the station earth in both the rectifier and inverter stations. When the earth electrode line is open during the bipolar earth return operation, the NBGS can temporarily connect the neutral bus to the converter station earth. After the earth electrode line restores, the NBGS must be able to divert the current flowing through it from the converter station earth to the earth electrode line.

   Before the MRTB operates, the GRTS closes first to establish two parallel circuits, thus dividing the DC current. After the system reaches a steady state, the MRTB operates to divert the current, after which the disconnector in series with the MRTB will be opened to prevent the MRTB from being subjected to continuous voltage.

   Before the GRTS operates, the MRTB closes first to establish two parallel circuits, namely the earth and metallic return paths, thus dividing the DC current. After the system reaches a steady state, the GRTS operates to divert the current, after which the disconnector in series with the GRTS will be opened to prevent the GRTS from being subjected to continuous voltage.

2. **Basic composition and working principle**

   Because the DC current contains no zero-crossing point, arc extinction at zero-crossing points as in the case of AC current is impossible. Therefore, as evolved from ordinary

AC circuit breakers, DC transfer switches incorporate an additional auxiliary circuit to forcibly create zero-crossing points in the DC current, thereby allowing shifting between different operating modes and fault clearing.

The DC transfer switches are classified into active and passive types. The active DC transfer switch requires a complicated operation because it is necessary to pre-charge capacitors before switching. Moreover, it has an additional power supply as compared with a passive one, which increases the land occupation and maintenance efforts. Therefore, passive DC transfer switches are used in UHV DC projects. Figures 10.31 and 10.32, respectively, show the schematic diagram and configuration.

As shown in Figure 10.31, *B* represents the transfer switch modified from an AC SF_6 circuit breaker, *L* represents the commutation inductor, *C* represents the commutation capacitor, and *R* represents the nonlinear varistor, which is usually a surge arrester. The inductor, capacitor, and nonlinear varistor, after being assembled, are mounted on the same insulated platform.

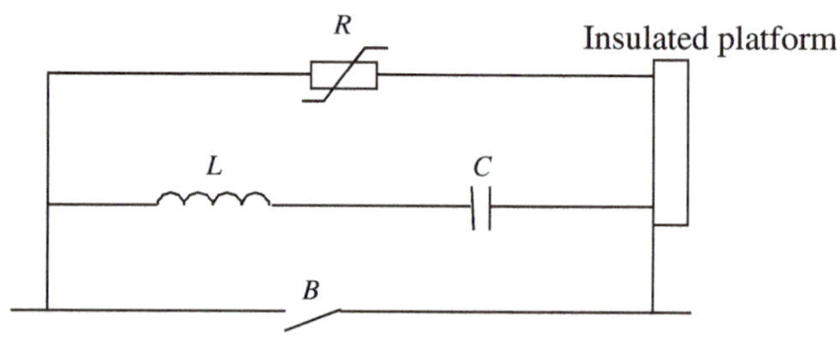

Figure 10.31
Schematic diagram of a passive DC transfer switch.

Figure 10.32
Configuration of a passive DC transfer switch.

Before arc extinction: After the contacts of SF_6 circuit breaker *B* are opened, the arc voltage will cause oscillatory current in the loop formed by the SF_6 circuit breaker and the L–C branch. When the reverse crest of the oscillatory current is exactly equal to the DC current, zero-crossing points occur in the current through the SF_6 circuit breaker and the arc at the breaks will become extinct.

After arc extinction: The DC current still exists but is diverted to the L–C branch. The current flowing through the capacitor will charge it to a certain voltage, which is defined as the transfer voltage. Its magnitude is dependent on the surge arrester (*R*) connected in parallel with the circuit breaker. When the transfer voltage increases to the arrester's operation voltage, it operates and the current through the L–C branch is diverted into the arrester. Then, this current will gradually diminish to zero.

In this way, the DC current flowing through this DC transfer switch is gradually diverted to other circuits connected in parallel with it, during which process the arrester will absorb a large amount of energy.

Because the UHV DC transfer switch is intended to divert a large current, 252 kV or higher two-pole AC SF_6 circuit breakers with higher arc extinction capability are selected as the switching elements.

3. **Main technical parameters**
Generally, a DC transfer switch is designed to perform no more than 20–30 normal transfers every year. The MRTB and GRTS are only allowed to perform two successive transfers without cooling each year. The permissible transfer current depends on the rated continuous overload conditions of the DC system. In conventional HVDC applications, the passive DC transfer switch is designed to transfer a maximum current of 2.5 kA. For a UHV DC transfer switch, the maximum current is increased up to 5.1 kA and the energy absorption capability of the arrester is increased from 40 to 53 MJ. For instance, in the Xiangjiaba–Shanghai project, the MRTB and GRTS, respectively, have a maximum transfer current of 4.32kA and 1.2 kA.

Table C.9 in Appendix C provides the main parameters of DC transfer switches used in the Xiangjiaba–Shanghai project.

10.7.2 DC Disconnector

The DC disconnectors and earthing switches serve the functions of isolation and earthing during equipment maintenance, and they cooperate with the DC circuit breaker to transfer operating modes.

In UHV DC projects, the outdoor DC disconnectors installed at the DC switchyard may be classified into three categories by voltage class: the line pole disconnector; disconnector at midpoint of the two 12-pulse converter bridges; and neutral bus disconnector. To enhance the operational safety and increase the energy availability of the DC system, in a UHV DC project where two series-connected 12-pulse converter units are provided for each pole at each end, bypass circuit breakers are provided for the 12-pulse converter bridges and valve-side disconnectors are also added to cooperate with the bypass circuit breakers. To reduce the equipment types, the smoothing reactor valve-side disconnectors and the line-side disconnectors have the same design.

The DC disconnectors can be modified from ordinary single-phase AC disconnectors. The major difference between the two is that in the UHV DC project, the DC disconnector is designed to withstand a large DC current of approximately 4500 A (40°C) for a long time. To address the disconnector heating problem, special attention should be given to the design of their contacts and flexible connection.

In addition, the line pole disconnector, as shown in Figure 10.33, has a very high external insulation level, i.e., a switching impulse withstand voltage of 1600 kV, a lightning impulse withstand voltage of 1950 kV, and a creepage distance of 39,168 mm (composite insulation). The disconnector body has a large size (approximately 15 m high) and requires high mechanical stability. Therefore, it is supported by three parallel insulators. Because the DC disconnector of line pole filter is required to switch the DC filters while energized, the disconnector should be capable of interrupting harmonic current, which is usually 200 A at 60 kV.

Figure 10.33
UHV DC line pole disconnector.

10.7.3 Bypass Circuit Breaker

The bypass circuit breaker of the 12-pulse valve group, as shown in Figure 10.34, is specific to the ±800-kV UHV DC system and is installed on the DC side of each 12-pulse converter bridge, increasing the system reliability and isolating the converter unit, which becomes faulty or requires maintenance without affecting the normal operation of the other converter unit.

For example, the Xiangjiaba–Shanghai UHV DC transmission system has a rated transmission capacity of 6400 MW, and the outage of one pole due to valve fault will cause a power loss of 3200 MW, which is even larger than the transmitted power of a ±500-kV DC bipolar system, and will greatly affect the AC systems at both ends. The use of bypass circuit breakers allows the system to operate in the unbalanced monopolar and bipolar operating modes, thereby reducing the power loss and its effects on AC systems at both ends and increasing the energy availability.

Closing sequence of bypass circuit breaker: Although the bypass circuit breaker is in open status, the valve group's bypass pair is switched-on and a closing order is sent to the bypass circuit breaker. On receiving the order, the bypass circuit breaker will be closed and will

Figure 10.34
UHV DC 12-pulse valve group bypass circuit breaker.

allow the rated current through. The closing time should not exceed 100 ms to avoid overstress on the bypass pair.

Opening sequence of bypass circuit breaker: Deblock the 12-pulse converter bridge connected in parallel with the bypass circuit breaker, and when the firing angle declines to approximately 90°, the rated DC current will flow through the converter bridge and the 12th harmonic dominated ripple current (300–400 A) will flow through the bypass circuit breaker. At this moment, an opening order will be sent to the bypass circuit breaker, which is then opened.

Despite the high insulation level between the bypass circuit breaker and the ground and between its breaks, the bypass circuit breaker is not subject to high breaking current and recovery voltage during the switching process because the bypass pair is switched on and the valve group is deblocked at 90°. Therefore, the UHV DC bypass circuit breaker may be an ordinary AC circuit breaker that can meet the current and voltage requirements.

However, the bypass circuit breaker is quite different from an AC circuit breaker in external insulation and voltage grading between breaks of the arc-extinguishing chamber. Specifically, to meet the external insulation requirements under pollution conditions of the DC switchyard, the arc-extinguishing chamber bushings and post insulators of the bypass circuit breaker are all of composite insulation type; the bypass circuit breaker uses voltage grading resistors instead of grading capacitors, as used in normal AC circuit breakers, to enable it to withstand DC voltage for a long period.

In the Xiangjiaba–Shanghai project, the bypass circuit breaker connected to the HV end converter unit is 12.781 m high and the associated post insulators are 8.7 m high; the bypass circuit breaker connected to the LV end converter unit is 8.069 m high and the associated post insulators are 4.33 m high. These breakers are all 5.857 m wide and have 2.124-m-long arc-extinguishing chambers. Table 10.5 gives the main technical parameters of the UHV DC bypass circuit breaker.

Table 10.5: Main Technical Parameters of UHV DC Bypass Circuit Breaker.

Technical Parameters		Unit	Bypass Circuit Breaker of the HV End 12-Pulse Converter Unit	Bypass Circuit Breaker of the LV End 12-Pulse Converter Unit
Rated breaker-to-ground voltage		kV	800	400
Maximum continuous voltage between breaks		kV	408	408
Lightning impulse withstand voltage (peak)	Breaks	kV	916	929
	Breaker-to-ground	kV	1800	903
Switching impulse withstand voltage (peak)	Breaks	kV	927	941
	Breaker-to-ground	kV	1600	825
DC withstand voltage	Terminal 1	kV	1224	612
	Terminal 2 (of the same pole)	kV	612	–

10.8 UHV DC Measuring Devices

10.8.1 DC Current Measuring Devices

The DC current measuring devices are usually installed at the high-voltage DC line end and the neutral bus and earth electrode line in the converter station. They output signals for control and protection of the DC system. In terms of their technical performance, the DC current measuring devices are required to ensure sufficient dielectric insulation strength between the output circuit and the measured main circuit. They have high anti-electromagnetic interference capability, high measuring accuracy, and quick response. For DC current measuring devices intended for control purposes, where the measured current falls between the minimum guaranteed value and the 2-h overload current, the measuring error should not exceed ±0.2% of the rated current, and the instantaneous output signal during the transient state may be as high as 600% of the rated current.

Generally, the UHV DC current measuring devices are classified into three types: inductive, hybrid-optical, and optical.

10.8.1.1 Inductive DC current transformer

The inductive DC current measuring device used in DC transmission applications is generally the zero-flux type, which mainly comprises the saturable reactor, auxiliary AC power supply, rectifier circuit, and burden resistor. When the DC current through the primary circuit varies, a secondary current in proportion to the primary current will flow through the burden resistor. Main performance parameters of an inductive DC current transducer are measuring accuracy, generally 0.2–1.5%; response time, 50–100 μs; incorrect response when the primary current is smaller than 10% of the current rating, 0.5–3%.

The zero-flux current transducer can be used to measure the DC current and AC current. A zero-flux DC current transducer comprises a flux integrator and a magnetic modulator. Figure 10.35 shows the schematic diagram of an inductive DC current transducer. It comprises three cores, T1, T2, and T3, with each having an auxiliary winding, N1, N2, and N3. A compensating winding, N4, and a correction winding, N5, are wound around all three cores. N4 and N5 have the same number of turns and are connected in parallel with the three auxiliary windings. N5 is only put in use when correcting N4, and it is connected in parallel with N4 during its normal operation. This figure shows that the flux integrator comprises N3, N4, N5, a power amplifier, and a burden resistor; the magnetic modulator comprises N1, N2, a peak detector, and an oscillator.

The flux integrator works in a way such that it eliminates the induced electromotive force generated on winding N3 using the power amplifier to achieve a complete balance between

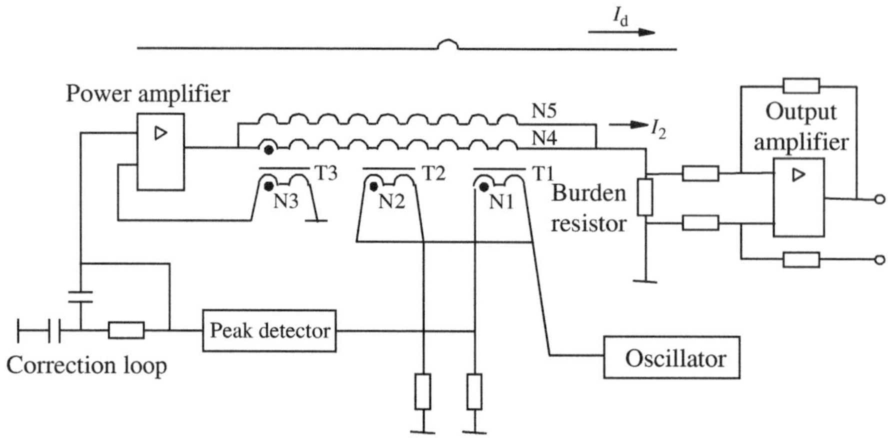

Figure 10.35
Schematic diagram of an inductive DC current transducer.

the magnetomotive forces induced by the primary winding and the secondary winding N4 (i.e., ampere-turns balance between the primary and secondary windings).

The magnetic modulator works in a way such that the oscillator causes cores T1 and T2 to reach a saturation state. At that instant, the current flowing through N1 will increase sharply, which will be detected by the peak detector. Winding N2 serves to balance the magnetic flux generated by N1. If pure DC magnetic flux passes through the core, then the peak detector will sense the positive and negative peaks and send a correction signal to the power amplifier.

When the measured DC current is $I_d \neq 0$, the peak detector outputs the correction voltage U_r, which will control the flux integrator's power amplifier to output a secondary compensating current I_2, thus achieving a complete ampere-turns balance between the cores. The larger the I_d, the larger is the compensating current I_2 produced by the power amplifier. Actually, because of the limited gain of the power amplifier and the drift of magnetic flux, it is impossible to keep a complete balance between the magnetomotive forces of the primary and secondary windings. Therefore, to recover the ampere-turns balance, a system capable of negative feedback is required, and the flux integrator is used exactly for this purpose. Any flux variation will result in an induced voltage in tertiary winding N3. The induced voltage is applied to the inverting input of the flux integrator and therefore changes the compensating current I_2 output by the power amplifier, thus achieving the required balance. By measuring the DC voltage across the burden resistor produced by the compensating current, the magnitude of the primary current can be obtained, allowing the DC current to be measured.

10.8.1.2 Hybrid-optical DC current transducer

Depending on the applications, the hybrid-optical DC current transducers are classified into two types: one for measuring the DC current and the other for measuring the harmonic components of DC current. A hybrid-optical DC current transducer mainly consists of high-accuracy shunt (shunt resistor or Rogovski coil), electro-optical module, signal transmission optical fiber, and optical interface. It works in such a way that the current signals are converted to voltage signals via the sampling coil, and then to digital signals via the multichannel analog-to-digital sampling system. Subsequently, such digital signals together with the clock signal are transmitted through the light-emitting diode (LED) to the signal-receiving module on the low potential side over the optical fiber.

The hybrid-optical current transducer has the following advantages over a conventional current transformer: (i) the HV side is completely electrically isolated from the LV side and the insulation structure is simplified, but with excellent insulating performance and safety; (ii) the use of optical transmission allows immunity from interference; (iii) the core is eliminated, avoiding problems such as magnetic saturation and ferro-resonance; (iv) it offers a full spectrum of functions and high measuring accuracy; (v) it is environmentally friendly because it causes no noise and less pollution; and (vi) it is compact and lightweight.

10.8.1.3 Optical DC current transducer

An optical DC current transducer is substantially the same as a hybrid-optical one in signal transmission and processing circuit. Specifically, both of them have the measuring signals transmitted to the control and protection room through optical fiber, where they are converted to digital signals and are then transferred to the control and protection system. The difference lies in that the optical DC current transducer utilizes the fiber optic measurement technique to measure the velocity difference of high-frequency light waves in electric fields based on the photoelectric effect, which can reflect the variations in DC analog quantities.

10.8.2 DC Voltage Measuring Devices

Most of the DC voltage measuring devices are resistive-capacitive voltage dividers plus DC amplifiers and, therefore, are usually called the DC voltage divider.

The DC voltage divider substantially consists of resistive-capacitive voltage division elements and incorporates voltage grading measures. Figure 10.36 shows the schematic diagram of a DC voltage divider. The core component of its voltage sensor is the DC resistance voltage divider (including the high-voltage arm resistor R_{13} and low-voltage arm resistor R_{24}, which are connected in series). Because the current through R_{13} is equal to that through R_{24}, it is possible to calculate the voltage of the high-voltage terminal by measuring the voltage across

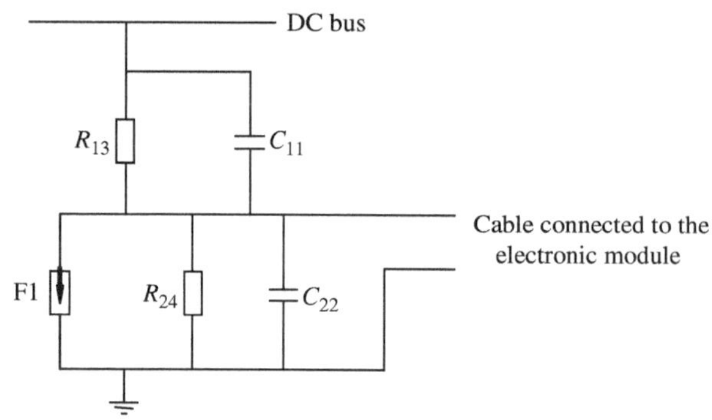

Figure 10.36
Schematic diagram of a DC voltage divider.

the low-voltage arm resistor R_{24}. To ensure the measuring accuracy under different ambient temperatures and voltages, the resistors constituting the arms must have a small temperature coefficient and inductance and can remain stable when voltage is applied. Because the value of a high-voltage resistor is as high as hundreds of megohms, the resistive current flowing through the voltage divider is usually of the order of milliamperes.

For a purely resistive voltage divider, although the voltage is uniformly distributed when a DC current flows through it, the voltage distribution under lightning impulses will be extremely nonuniform because of the different ground stray capacitances of resistor elements at varying heights. Specifically, a resistor element at the high-voltage side is subject to a much higher voltage than that applied on the middle and bottom resistor elements; therefore, it is susceptible to breakdowns and unable to meet the requirements of test and operation. In light of this, two compensation capacitors, C_{11} and C_{22}, are connected in parallel across the resistor to improve the distribution of electric fields. Their compensating capacity should match with the resistor's resistance. Moreover, a purely resistive voltage divider usually has unsatisfactory frequency response characteristics, and the compensating capacitors can improve such characteristics, ensuring compliance with the relevant requirements of monitoring, control, and protection requirements of HVDC transmission lines.

The major difference between a UHV DC voltage divider and a $\pm$500-kV one lies in the internal and external insulation. The UHV DC voltage divider operates at a higher voltage class and requires an increased insulation level. Therefore, it is necessary to re-design the terminal insulation and compensation devices to avoid coronas and improve the distribution of impulse voltages. During the design and manufacturing, consideration should be given to the temperature increase control of resistor elements to ensure a desirable measuring accuracy. Also, the effects of wet pollution on the distribution of internal and external voltages should be taken into account. To ensure system safety and reduce interference, an

Figure 10.37
DC voltage divider for the ±800-kV pole bus (outdoor).

electrostatic shield should be provided between the primary circuit and the secondary output circuit of the UHV DC voltage divider.

The UHV DC voltage measuring device is connected to the signal input terminal of the control and protection system and its measuring error is controlled within ±0.2% of the rated voltage, for a voltage range from zero to the maximum steady-state DC voltage. Moreover, its measuring accuracy should meet the requirements for measuring 1.5 p.u. of the DC voltage.

The DC voltage measuring device should have a desirable transient response and frequency response to ensure that, at the maximum tolerances, the measured value is within the accuracy requirements for UHV DC system control and protection purposes.

On the DC side of a ±800-kV converter station, voltage dividers are mainly installed on the pole bus (in the outdoor DC switchyard as shown in Figure 10.37), neutral bus (in valve hall), and bypass circuit breaker of each 12-pulse converter unit (in valve hall, no voltage divider is installed at this location in SGCC's UHV projects).

10.9 UHV DC Control and Protection Equipment

10.9.1 Characteristics

By virtue of novel hardware and software platforms, the host-based UHV DC control and protection system incorporates a real-time operating system and therefore has improved system performance. The communications between control and protection equipment are realized using the high-speed real-time communication techniques such as electrical time division multiplexing (eTDM) to accommodate the requirements of UHV control and protection with respect to the processing and communication capability of the host computer. The host computer uses the general-purpose commercial processor featuring high computing capability, less heat generation, and easy upgrade; the use of a passive cooling method can prevent adverse effects on the computer board caused by inhaled dust in the case of an active cooling method.

The UHV DC transmission project usually uses a configuration in which each pole has two series-connected 12-pulse converter units, which allows more flexible and reliable system operation and offers a total number of 45 operating modes. The control and protection system should meet the control requirements of all the 45 modes as well as ensure fast and stable switching between individual modes. The controls and protections are configured on a single 12-pulse converter unit basis. The implementation of control functions and protection configuration for each converter unit is kept as independent as possible, thus enabling a single converter unit to be switched-out without causing any impact on the normal operation of other equipment. Regarding the DC protection configuration, additional protections for the disconnectors and bypass circuit breakers of the 12-pulse converter units and for the connections between the two 12-pulse converter units are required when compared with conventional HVDC projects.

10.9.2 Hierarchical Structure

The UHV DC transmission control system is implemented in a hierarchical structure where all the control functions of the converter station are divided into several levels, as shown in Figure 10.38. Such a structure can enhance the system operational reliability such that the impact and risks caused by faults with any control link can be minimized and the convenience and flexibility of the operation and maintenance can be improved. Generally, the UHV converter station's hierarchical control system consists of remote dispatch control level, converter station control level, local control, and equipment control level.

1. **Remote dispatching control level**
   Through the national Electric Power System Dispatching Data Network or other dedicated channels, the remote dispatching center (such as the National Electric Power Dispatching and Communication Center and backup dispatching centers) remotely controls all the equipment in converter stations via the telecontrol workstation. Also, the telecontrol workstation will upload information regarding the operating status and alarming message to the remote dispatching center. The involved equipment mainly includes the telecontrol workstation, scheduling workstation, maintenance workstation, and remote local area network (LAN).

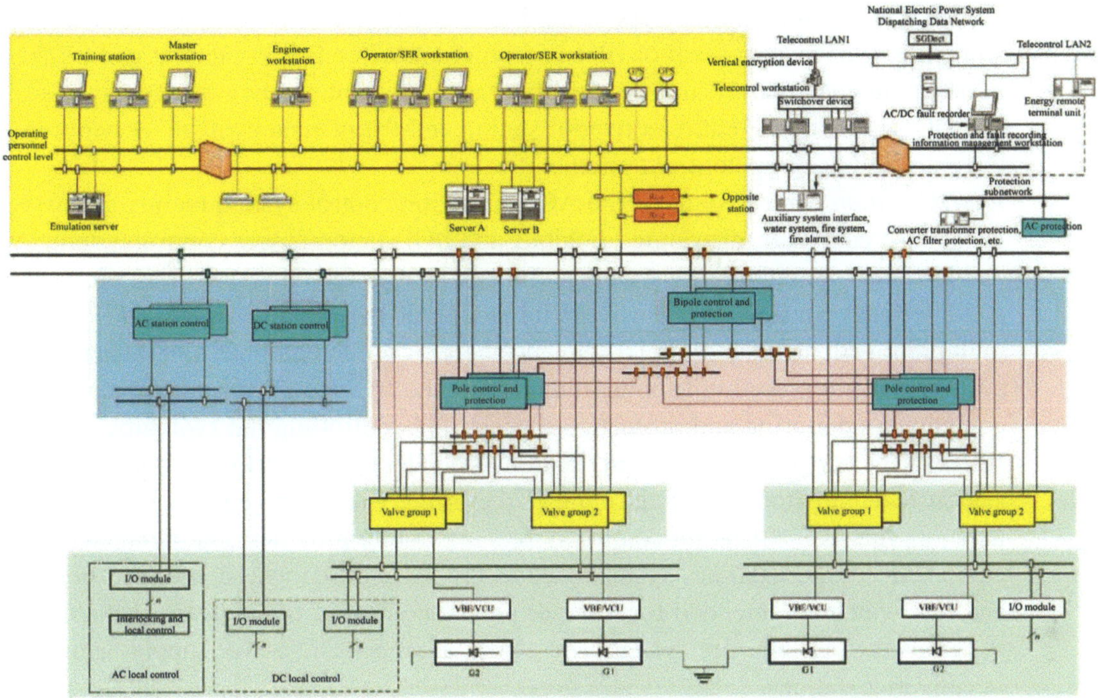

Figure 10.38
Hierarchical structure diagram of the converter station control and protection system in the Xiangjiaba–Shanghai project.

2. **Converter station control level**

   The control-level equipment includes the station operator control system, AC station control system (including the station power control and auxiliary system interfaces), DC station control system, DC bipole/pole/converter control system, and DC protection system. This level serves to control and protect both the AC and DC equipment in the converter station.

   a. *Operator control system.* It mainly consists of the LAN, operator workstation, remote monitoring workstation, centralized control workstation, valve cooling control workstation, engineer workstation, master workstation, SER (sequence of events recorder) station, server, and MIS interface workstation. This subsystem is the main human–machine interface and is an important component of the station monitoring data acquisition system during normal operation of the converter station. It serves the following functions: (i) receiving normal operation orders sent from operators or remote dispatching centers to the converter stations via the operator workstation and telecontrol workstation; (ii) monitoring and troubleshooting of faults or abnormal operating conditions; (iii) station-wide sequence of events recording and event alarming; (iv) network and time synchronization of the control system; (v) adjusting parameters of the DC control system; (vi) historical data archiving; (vii) management of files and programs of converter station; and (viii) basic training. Moreover, it should enable the manual emergency outage of the AC and DC system of the entire converter station and the manual switchover between the main and back-up control systems.

b. *AC/DC station control system.* The AC/DC station control system mainly consists of the host computer, distributed fieldbus and I/O, station power control system, and auxiliary system interface workstation, as well as the duplicated host computers and interfaces configured exclusively for station power monitoring. It serves multiple functions, including: (i) sequence control of the equipment in AC and DC switchyard; (ii) executing tripping orders from the AC and DC protection or the equipment monitoring and protection devices and controlling the switching of filters in the converter station; (iii) station-wide electrical interlocking; and (iv) station-wide monitoring and alarming of the primary/auxiliary equipment. In addition, the AC/DC station control system separately provides dual redundant equipment for the auxiliary system such as the monitoring host, distributed fieldbus, and I/O, which are intended for monitoring and control of the industrial monitoring system, water system, fire system, fire alarm, auxiliary DC power supply, and UPS in converter stations.

c. *DC bipole/pole/converter control system.* This subsystem mainly consists of the bipole control host, pole control host, converter unit control host, distributed fieldbus, and I/O. It mainly serves the following functions: (i) generate control pulses for converter valves and maintain synchronization with the commutating voltage and equal spacing; (ii) deblock/block converter valves and ensure steady start-up and outage of DC system; (iii) adjust firing angles of converter valves within the permissible range through calculations and analysis to achieve the required transmitted power/current and reactive power control logic; (iv) adjust the tap changers of the converter transformer according to operating conditions of the AC and DC systems; and (v) perform power flow reversal and sequence control of the switching in/out of converters.

d. *DC protection system.* It mainly consists of the DC converter unit protection, pole protection, bipole protection, DC filter protection, converter transformer protection, AC filter protection, distributed fieldbus, and I/O, and it protects the DC switchyard, valve hall, converter transformers and associated leads, and AC filters and associated leads.

3. **Local control and equipment control level**

On receiving orders from other control levels, the local control and equipment control levels will operate and control the associated equipment, including the circuit breakers, disconnectors, converter transformers, station power supply, and other equipment in the converter station.

References

[1] Z. Wanjun, HVDC Power Transmission Engineering Technology, second ed., China Electric Power Press, Beijing, 2011.
[2] L. Zhenya, Material Procurement Standards of SGCC, China Electric Power Press, Beijing, 2010.
[3] Y. Qingyun, HVDC Converter Valves and Firing and On-line Monitoring System, China Electric Power Press, Beijing, 1999.
[4] SGCC DC Construction Branch, Standardized System Design of HVDC Transmission System, China Electric Power Press, Beijing, 2012.
[5] Zhejiang University HVDC Research Group, Direct Current Transmission, Electric Power Industry Press, China, 1982.

CHAPTER 11

Construction of UHV Power Grids in China

Chapter Outline

11.1 Forecast of Power Demands

11.1.1 Development Trend of National Economy

During the past 30 years, China's economy has seen tremendous success. Socioeconomics and culture have been greatly enriched and improved. In 2012, China's economic aggregate amounted to RMB51.9 trillion, ranking second in the world.

Ultra-high Voltage AC/DC Grids.
DOI: http://dx.doi.org/10.1016/B978-0-12-802161-3.00011-1

In parallel with the economic development, energy and electricity demands have been increasing, giving strong support to economic and social development. However, the energy mix dominated by fossil energy has caused increasing concerns about a number of issues, such as energy resources and climatic environment, which poses a bottleneck to the sustainable development of the economy and society. As such, it is imperative to follow the path of green, low-carbon, harmonious, and sustainable development to build an energy-conserving and environment-friendly society. As pointed out in *Outline of the 12th Five-Year Plan for National Economic and Social Development of the People's Republic of China* (hereinafter referred to as the 12th Five-Year Plan), during the 12th Five-Year Plan period, China will implement a scientific outlook on development, speed up the transformation of the economic growth model, and deepen its endeavor of reform to ensure and improve well-being and promote steady and rapid economic development and social harmony in the long term. Furthermore, the 12th Five-Year Plan also calls for a higher priority for energy conservation and emission reduction, including implementation of binding index and reasonable control of the total energy consumption.

The Chinese Government has worked out a comprehensive plan regarding China's future development, according to which China will build a moderately prosperous society and double its GDP and per capita income by 2020 from 2010 levels. With the rapid development of industrialization and urbanization, the future energy and power demands will keep growing rapidly, the power consumption structure will be further optimized, and allocation of energy sources among various regions will become more reasonable.

To achieve these objectives, the average annual economic growth rate would have to reach a minimum of 7.2% from 2011 to 2020. Judging by the past experiences of previous five-year plans and their actual achievement, the growth rate for the 12th Five-Year Plan period is likely to reach 7.5–8.5%. If so, then China's economy will maintain a steady and rapid growth during the 12th and 13th Five-Year Plan periods. According to studies performed by the Development Research Center of the State Council and Chinese Academy of Social Sciences, China's economy is expected to see an average annual growth of 7.2–7.8% from 2011 to 2020, with growth slowing to 6.8–7.6% in the latter 8 years (from 2013 to 2020). Based on the fact that the electric power consumption grew slightly faster than the economy in the past decade, the power consumption is estimated to maintain a growth rate of 7% in the next 10 years. Moreover, the plan of doubling per capita income will favor the power consumption in the tertiary industry and households. It is estimated that the power consumption by the tertiary industry and households will account for 32% of the total consumption in 2020, an increase of approximately 10% from 22.8% in 2010. Furthermore, the growth in peak demand load is expected to be more than 1% higher than that of energy consumption.

China's urbanization level reached 49.95% and 51.3% in 2010 and 2011, respectively. The Chinese Government aims to speed the construction of a new type of urbanization to significantly improve the urbanization quality by 2020. The plan for a new type of urbanization currently under preparation deals with the construction of more than 20 city clusters, more than 180 cities at and above the prefecture level, and more than 10,000 towns. As predicted in the 12th Five-Year Plan and by the United Nations, the urbanization level in China will reach 53–55% by 2015 and 58–60% by 2020. As a result, it is likely that more people will be moving to the eastern and central regions and settling in towns, which will cause the power demand to increase dramatically.

In 2010, China's industrialization index reached 66.6, indicating that China was entering into the middle to later stage of the industrialization process. At this stage, industrial growth will shift from quantitative expansion to quality improvement and the industrialization process will be accelerated. It is expected that China will be fully industrialized by 2018–2020, which is consistent with the Chinese Government's goal to achieve basic industrialization by 2020. Given the uneven level of industrialization across different regions of China, the industrialization process is likely to progress from east to west. It follows from the development experience of other countries at a similar industrialization stage that the energy-intensive industries such as heavy industry and chemical industry have rigid energy demands and will contribute to a rapid growth of the economy and electricity demand.

The Chinese Government gives high priority to promoting ecological progress and building a beautiful country to achieve sustainable development of the Chinese nation. This will vigorously boost the development of clean energy, improve energy utilization efficiency, and speed the shift of energy consumption to electricity from coal and oil. In recent years, the severe environmental pollution symbolized by frequent haze has attracted more and more attention from all walks of life. To address this issue and thereby improve the environment, it is essential to constantly optimize the energy mix and increase the share of electricity in the end-user energy consumption.

11.1.2 Power Demand and Its Distribution

With the development of economy and improvement in people's lives, China's total power consumption has seen a rapid increase. In 2012, it reached up to 4.96×10^{12} kWh, 18.9-times more than that in 1978, representing an average annual growth rate of 9.2%.

By region, east China accounts for the largest proportion, although central and west China have seen their shares increasing in recent years. In 2012, the power consumption in east China[1] accounted for 48.4% of the national total, 9.4% higher than that in 1978. With the increase in labor and land costs in the eastern region, energy-intensive industries have been

[1] The 10 provinces (municipalities) in east China: Beijing, Tianjin, Hebei, Shandong, Shanghai, Jiangsu, Zhejiang, Fujian, Guangdong, and Hainan.

gradually moved inland. As a result, the proportion of power consumption by the central[2] and western regions[3] began to increase from 2007 and 2004, respectively, and they have reached 19.3% and 25.4%, respectively, in 2012. Mid-east China (consisting of 12 provinces and municipalities),[4] one of the most economically developed regions across the country, is now at the middle to later stage of industrialization and has a large economic aggregate and, consequently, large power consumption. In 2012, the power consumption by these 12 provinces and municipalities accounted for 51.7% of the nationwide power consumption, 2.3% more than that in 1978, and 64.8% of the total power consumption across the service area of SGCC.

Over the recent decade, the Chinese Government has promoted and urged industry restructuring and transformation; however, because of differences in many factors such as investment environment, technology, and human resources, industry transformation is a progressive process. The power demand from mid-east China has been increasing relatively quickly. During the 10th and 11th Five-Year Plan periods, the national share of power consumption by this region increased. The rates of increase were 1.2% and 0.7%, respectively, showing a moderate slowdown. During the 12th and 13th Five-Year Plan periods, the national power demand will maintain rapid increase. However, as the industrial transformation process progresses, more traditional energy-intensive industries will move from eastern and central regions to the western region, resulting in a further decrease in the power consumption by the eastern and central regions. But the magnitude and rate of decrease will not be significant because some new strategic industries are emerging in these regions and shift of traditional industries is happening progressively.

Tables 11.1 and 11.2 show the forecast power demand in China according to the economic development and power demand forecast models developed by China's Power Supply and Demand Research Laboratory. According to the forecast, by 2015, the national power consumption will reach 6.15×10^{12} kWh and the maximum load will reach 1004 GW, representing annual growth rates of 7.9% and 8.7% during the 12th Five-Year Plan period, respectively; by 2020, the total power consumption will reach 8.42×10^{12} kWh and the maximum load will reach 1409 GW, representing annual growth rates of 6.5% and 7.0%, respectively, during the 13th Five-Year Plan period.

As can be seen from these tables, the power demand from the 12 provinces (municipalities) in mid-east China will still account for more than half of the nationwide demand, and more than 60% of the demand of the service area of SGCC; therefore, they remain the load

[2] The six provinces in central China: Shanxi, Henan, Hubei, Hunan, Anhui, and Jiangxi.

[3] The 12 provinces (municipalities) in west China: Shaanxi, Inner Mongolia, Gansu, Ningxia, Qinghai, Xinjiang, Sichuan, Chongqing, Guizhou, Yunnan, Tibet, and Guangxi.

[4] The 12 provinces (municipalities) in mid-east China: Beijing, Tianjin, Hebei, Shandong, Shanghai, Jiangsu, Zhejiang, Fujian, Henan, Hubei, Hunan, and Jiangxi.

Table 11.1: Forecast of National Total Power Consumption (10^8 kWh).

Scope	Year 2010 (Published Data)	Year 2015	Year 2020
National	41,999	61,552	84,196
Service area of SGCC	33,689	49,460	67,749
North China	8906	12,696	17,242
East China	10,375	14,592	19,127
Central China	7732	11,564	16,584
Northeast China	3318	4659	6443
Northwest China	3357	5949	8353
The 12 provinces (municipalities) in mid-east China	22,299	31,548	42,885

Table 11.2: Forecast of National Peak Demand (GW).

Scope	Year 2010 (Published Data)	Year 2015	Year 2020
National	660.70	1004.33	1409.02
Service area of SGCC	529.26	809.00	1138.86
North China	141.75	206.73	287.81
East China	174.16	261.52	359.15
Central China	129.10	202.11	299.12
Northeast China	48.33	71.92	100.84
Northwest China	50.50	91.09	128.46
The 12 provinces (municipalities) in mid-east China	353.75	529.71	742.24

center of the country and SGCC. It is expected that by 2015 and 2020, their power consumption will reach 3.15×10^{12} and 4.29×10^{12} kWh, respectively, with an average annual growth rate of 6.8% from 2011 to 2020. Their maximum load will reach 530 and 742 GW in 2015 and 2020, respectively, with an average annual growth rate of 7.7% from 2011 to 2020. By 2015 and 2020, the national share of their power consumption, although still accounting for more than half, will decrease to 51.3% and 50.9%, respectively. Moreover, by 2015 and 2020, the proportion of their power consumption to the total consumption by the service area of SGCC will be 63.8% and 63.3%, respectively.

11.1.3 Power Source Structure and Layout

As of the end of 2012, China's total installed generation capacity reached 1145 GW, including:

819.17 GW of thermal power, accounting for 71.5% (a year-on-year decrease of 1.0%);
248.9 GW of hydropower, accounting for 21.7% (a year-on-year decrease of 0.1%);
60.83 GW of wind power, accounting for 5.3%;
12.57 GW of nuclear power, accounting for 1.1%;
3.44 GW of power from other resources (see Figure 11.1).

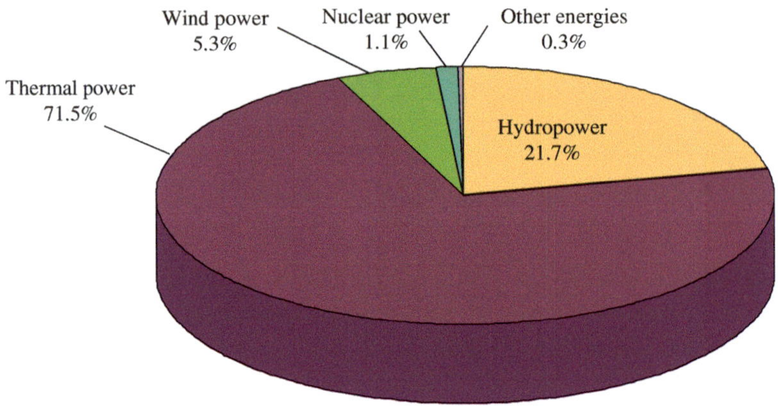

Figure 11.1
China's power source structure by the end of 2012.

Of regional power grids, the north China grid and the east China grid are dominated by thermal units, where the installed thermal power capacity accounts for up to 86.4% and 84.6%, respectively; whereas the central China grid, south China grid, and northwest China grid have a relatively large proportion of hydropower, with thermal power accounting for 53.8%, 55.3%, and 65.9%, respectively, lower than that in the north China grid and that in east China grid.

As for the power sources with good load following performance, namely, hydro, pumped storage, and gas power, most hydropower is located in the central China grid and south China grid, accounting for 41.5% and 31.7% of the national total, respectively. Most pumped storage is situated in the east China grid, north China grid, and south China grid, accounting for 27.3%, 23%, and 25.6% of the national total, respectively. Most gas-fired power is in the east China grid, south China grid, north China grid, and central China grid, accounting for 59.7%, 21.9%, 8.6%, and 6.7% of the national total, respectively.

Primary energy resources are mostly found in the southwest, northwest, and north, which are mostly sparsely populated with low energy demands and a less developed economy. More than 80% of coal reserves are located in the northwest, Inner Mongolia, and Shanxi. Taking into consideration the local water demand, water resources that are available for power generation in the aforementioned energy bases could support potential development of up to 638.80 GW in coal-fired generation. The water used for power generation only accounts for approximately 5% of the total water consumption. Table 11.3 shows the exploitable potential of coal bases across the country.

In Sichuan, Yunnan, and Tibet, the exploitable hydropower potential that has yet to be realized is as high as 308 GW, accounting for 82% of the national total. So far, only 18.3% of hydropower has been exploited. As such, the future hydropower development will

Table 11.3: Exploitable Power Potential of Coal Power Bases Across the Country.

Base	Remaining Coal Reserves (10^8 tons)	Water Consumption for Power Generation/ Total Water Resources (10^8 m^3)	Developable Power Potential (GW)
Shanxi	2663	7.10/123.8	100.00
North Shaanxi	1291	1.48/48.4	43.80
Ningdong	309	1.67/3.2	48.80
Junggar	256	1.78/3.6	60.00
Ordos	560	1.81/25.8	60.00
Xilin Gol League	484	1.52/26.1	50.00
Hulunbeir League	338	1.24/127.4	37.00
Huolin River	118	0.45/2.4	14.20
Baoqing	52	1.50/34.6	12.00
Hami	373	0.62/5.7	25.00
Zhundong	789	0.84/13.9	35.40
Ili	129	3.00/170	87.00
Binchang	88	0.42/15.1	14.00
Longdong	142	0.79/12.5	26.60
Huainan	139	3.6/58	25.00
Total	7731	27.8/670.5	638.80

Table 11.4: Development Potential of the Major Hydropower Bases in Southwest China by 2015 and 2020 (MW).

River	Year 2015		Year 2020	
	Potential	Capacity for External Export	Potential	Capacity for External Export
Mainstream of Chin-sha River	32,160	23,700	46,330	39,700
Mainstream of Yalong River	14,250	7200	21,700	18,200
Mainstream of Dadu River	13,490	4000	24,640	8000
Mainstream of Lancang River	20,060	11,000	25,965	18,000
Medium- and small-size hydropower projects	59,140	11,000	72,980	5800
Total	139,100	56,900	191,615	89,700

mainly take place in these regions. Table 11.4 shows the hydropower capacity that will be built on the four large rivers in Sichuan and Yunnan during the 12th and 13th Five-Year Plan periods.

Most of China's wind resources are located in the nine large wind power bases, namely, Xinjiang, Gansu, west Inner Mongolia, Hebei, east Inner Mongolia, Heilongjiang, Jilin, Shandong, and Jiangsu. It is expected that by 2020, the total installed wind power capacity will reach up to 200 GW, with 85% coming from these nine bases, as shown in Figure 11.2.

Figure 11.2
Distribution of large wind power bases and development capacity by 2020 (GW).

By 2020, taking into account local consumption in these nine bases, there would be approximately 98 GW of wind power available for exporting to areas outside these bases. This accounts for 49% of the total installed wind power capacity nationwide and 58% in the northeast, north, and northwest China (the "Three North" region).

The *12th Five-Year Plan for Renewable Energy Development* points out that diversified utilization of solar energy shall be further promoted following the principle of combining concentrated development with distributed utilization. The Qinghai–Tibet Plateau, Gansu, north Ningxia, south Xinjiang, and west Inner Mongolia are rich in solar energy, with more than 3000 sunshine hours annually, creating a favorable condition for concentrated development of solar power. In contrast, east and central China are good candidates for developing distributed solar PV power. Table 11.5 shows the installed solar power capacity as planned for 2015 and 2020.

Table 11.5: Planned Solar Power Capacity for 2015 and 2020 (GW).

Type of Power Generation	Year 2015	Year 2020
1. Solar power stations	11	23
Solar PV power stations	10	20
Solar thermal power stations	1	3
2. Distributed PV generation systems	10	27
Total	21	50

Table 11.6: Total Installed and Planned Capacity Nationwide by Fuel Type (GW).

Type of Power Source	Year 2010	Year 2015	Year 2020
Hydropower	199.24	290.14	346.65
Pumped storage hydropower	16.94	25.68	50.04
Coal power	681.01	961.53	1201.14
Gas power	26.41	63.71	80.21
Nuclear power	10.81	40.08	61.60
Wind power	29.58	100.00	200.00
Biomass power	2.13	13.00	30.00
Solar power	0.29	21.00	50.00
Total	966.41	1515.14	2019.64

According to *the 12th Five-Year Plan and the 2020 Vision of Nuclear Safety and Radioactive Pollution Prevention*, China will resume approval of new nuclear power projects following the principle of "developing efficient nuclear power with guaranteed safety, especially in eastern coast regions." It is expected that by 2020, the national nuclear power capacity will reach up to 61.60 GW, most of which will be located in the coastal regions including Zhejiang, Jiangsu, Fujian, Shandong, and Guangdong, and 50.80 GW will be built during the period from 2011 to 2020.

Based on the balance of electric power and energy, it is estimated that by 2015, China's total generation capacity will reach up to 1515 GW, with approximately 550 GW built during the 12th Five-Year Plan period, representing an average annual increase of 110 GW; by 2020, this will further increase to 2020 GW, with approximately 500 GW built during the 13th Five-Year Plan period, representing an average annual increase of 100 GW. Table 11.6 shows the national generation capacity by fuel type. Figures 11.3 and 11.4 show the power source composition by 2015 and 2020, respectively.

11.1.4 Power Flow Patterns [1]

1. *Current power flow pattern.* Given the characteristics of energy resource distribution, China has developed and implemented the west-to-east and

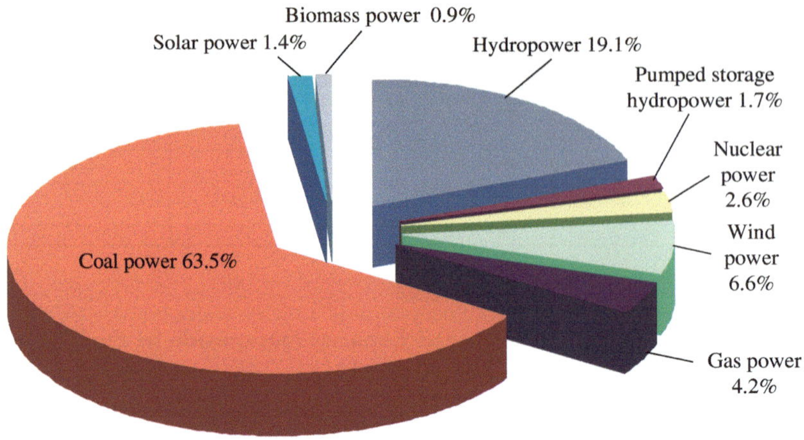

Figure 11.3
Composition of planned generation capacity by fuel type in 2015.

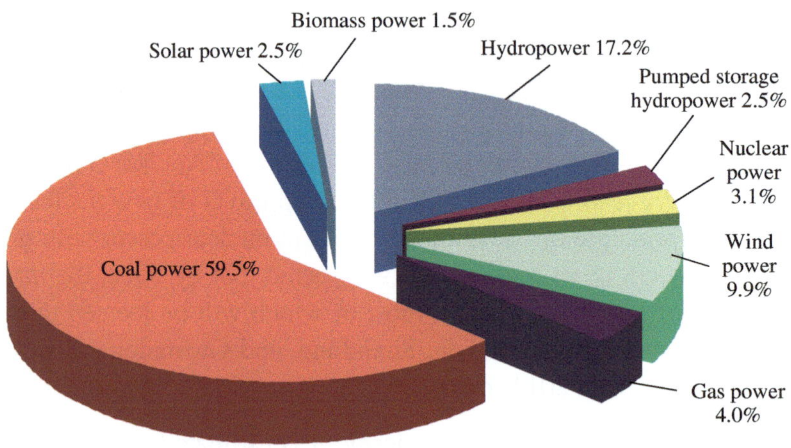

Figure 11.4
Composition of planned generation capacity by fuel type in 2020.

north-to-south power transmission strategy to achieve sustainable development of energy and electric power, address geographical mismatch between energy resources and productive forces, convert resource advantages of the western and northern regions into economic advantages, and promote mutual and balanced economic development among various regions. In 2012 alone, there was a total of approximately 80 GW of power flowing into the service areas of SGCC in east and

Table 11.7: Summary of Cross-Region Power Flows into East and Central China in 2012.

Item	Receiving End	Sending End	Capacity (GW)
Thermal power inflow	North China	From Shanxi to the Beijing–Tianjin–Hebei region	11.02
		From west Inner Mongolia to the Beijing–Tianjin–Hebei region	12.30
		From Shaanxi to Hebei	3.60
		From Ningxia to Shandong	4.00
	East China	From Anhui to east China	7.20
		Yangcheng	3.30
	Central China	Through UHV AC transmission	5.00
	Total		46.42
Hydropower inflow	East China	The Three Gorges and Sichuan	22.20
	Central China	The Three Gorges and Sichuan	11.20
	Total		33.40
Total power flows			79.82

central China, including 46.42 GW of thermal power and 33.40 GW of hydropower, as shown in Table 11.7. As the power grids become more able to optimize allocation of resources across large areas, cross-region power exchange will occur in increasingly large quantities.

2. *Future power flow patterns*. The forecast of China's power demand indicates that by 2020, 12 provinces (municipalities) in mid-east China will account for up to 63.3% of the total power consumption within the service area of SGCC. This situation will persist for a long term. Mid-east China has little primary energy resources; furthermore, limited water and coal resources that are available in this region have already been exploited. As a result, almost all energies required for economic development have to be imported. The extensive economic development mode of the past few decades and coal utilization model of meeting demand from power supply generated locally can no longer be sustained because of depleted environmental resources, increased power supply costs, and insufficient coal transportation capacity. At present, only 69 GW of power is transmitted to mid-east China from west Sichuan, The Three Gorges, Shanxi, Shaanxi, and west Inner Mongolia, which is insufficient to satisfy the energy demands for sustaining economical and social development. Hence, there is an urgent need to vigorously implement the strategy of "replacing coal and oil with electricity generated from remote sites" and speed the development of power transmission links. In view of

the geographical distribution of primary energy resources in China and the optimal siting and development of power bases, it is clear that all large hydro, coal, wind, and solar power bases will be situated in the southwest, northwest, Inner Mongolia, and other remote regions, which are far away from load center in eastern and central regions. The *12th Five-Year Plan for National Energy Development* explicitly pointed out that construction of coal power stations in the eastern and central regions should be strictly controlled, more efforts should be made to promote the development of large energy bases in the western and northern regions and greatly increase the scale of power transfer from west to east and north to south, fully utilizing UHV transmission technology. It is apparent that the future power flow pattern in China is that of west-to-east and north-to-south bulk power transfer. The hydropower in the southwest and coal power and wind power in the west and north will be delivered through cross-region power grids to the load centers in the north China grid, central China grid, east China grid, and south China grid. Such a power flow pattern will require higher transmission capacity in China's power grids. Hence, the cross-region interconnection capacity needs to be strengthened or built to enable interconnection of power grids and optimal allocation of resources over a wider area.

3. *Optimization of power flows in 2020.*
   a. Investigation of import headroom in the receiving end market. As shown in Figure 11.5, demand in the receiving-end region will first be met by the existing hydro, gas, nuclear, and renewable power, and also by coal power generation under

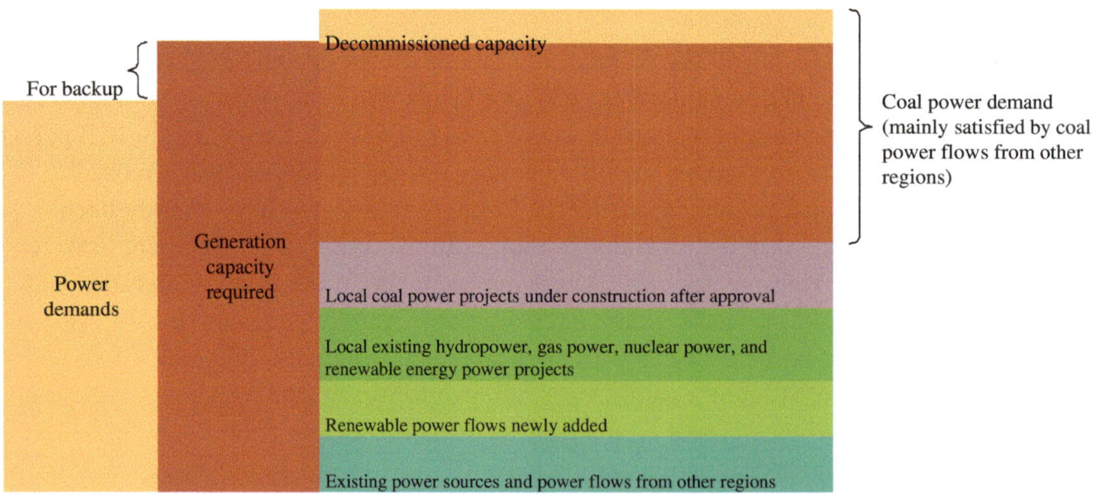

Figure 11.5
Import headroom in the receiving-end market.

construction as well as renewable power to be added within the region. The shortfall will be covered by power import from other regions and power plants to be built locally. The exact proportion of import and new power stations will be determined through economic comparison.

b. Determination of export requirements for clean energy and coal-fired power capacity from the sending-end region. Priority should be given to exporting power from hydro, wind, solar, and other clean power ahead of coal. The development of large coal power bases should take into account the available water resources, and they will act as an alternative source while satisfying the local power demands.
c. Determination of coal power flows through matching multiple sending ends with multiple receiving ends. Determine the reasonable transmission directions of various power bases according to the geographical locations of the sending ends and receiving ends. Rank all the "one-to-one" transmission options in terms of their cost-effectiveness by considering the factors such as power demand, load characteristics, existing and identified cross-province and cross-region transmission scales and directions, identified power sources, as well as the techno-economic indices of various power sources. The most cost-effective power transmission scheme will be given the top priority. Repeat this process until the power demands from the 12 provinces (municipalities) in mid-east China are fully satisfied or the capacity of all the coal power bases is fully utilized. In this way, the coal power flows can be determined.
d. Determination of the optimal power transmission scheme to achieve efficient utilization of clean energies and minimum power supply costs. Determine the optimal accommodation scheme of renewable energy (such as wind power) by simulating hour-by-hour operation of the power system incorporating wind power generation all the year. Through adjusting and comparing the cross-region power transmission options repeatedly, considering the transmission corridors and routes from an overall view, and optimizing the scale and direction of power transmission from hydropower, wind power, and coal power bases to the eastern and central regions, the optimal transmission scheme with the minimum costs can be finalized.

Based on the process described, it is expected that by 2020, the 12 provinces (municipalities) in mid-east China will import up to 350 GW of power from other regions, including 288 GW new plants during the period from 2011 to 2020. From the perspective of exporting regions, power export consists of output from coal-fired and renewable power and hydropower primarily from those energy bases identified in the *Outline of the 12th Five-Year Plan*. Specifically, 256 GW of the export is from coal-fired and renewable generation, of which 98 GW of wind power from the "Three North" regions, 74 GW from the hydropower bases, and 16 GW is imported from abroad (Table 11.8).

Table 11.8: Composition of Export Power Flows from the Sending Ends by 2020.

No.	Energy Base	Evacuated Capacity (GW)
I	Hydropower bases	74.00
1	The Three Gorges	19.40
2	Chin-sha River and Yalong River	32.40
3	Sichuan-to-east China power transmission	22.20
II	Coal power and wind power bases	256.12
1	Shanxi	42.20
2	North Shaanxi	24.20
3	West Inner Mongolia	48.20
4	Xilin Gol League	30.12
5	Ningdong	18.40
6	Hami	8.00
7	Zhundong	20.00
8	Hulunbeir League	18.00
9	Jiuquan	8.00
10	Longbin	15.00
11	Shanghaimiao	8.00
12	Jarud	8.00
13	Baoqing	8.00
III	Power transmission from abroad	16.00
	Total	346.12

11.2 Options of Transmitting Power from Large Power Bases

11.2.1 Overview of Large Power Bases

China's energy resources are suitable for concentrated development of large hydro, coal, and wind power bases.

1. *Large hydropower bases.* Water resources are geographically uneven in China, featuring more resources in the west and few in the east. Most water resources are located in the southwest, with Yunnan, Sichuan, Tibet, Guizhou, and Chongqing accounting for approximately 71% of the national total. Sichuan, Yunnan, and Tibet are China's key hydropower bases, with a technically exploitable capacity of 120, 102, and 140 GW, respectively. In view of this geographical feature, 13 large hydropower bases will be constructed, respectively, in the upper reaches of Yangtze River, Chin-sha River, Dadu River, Yalong River, Wu River, Nanpan and Hongshui River, Lancang River, upper reach of Yellow River, north tributary of Yellow River, northeast China, west Hunan, Fujian–Zhejiang–Jiangxi region, and Nu River. Figure 11.6 shows the geographical distribution of these bases.
2. *Large coal power bases.* In view of the distribution of coal resources and energy flow pattern, China will construct large coal power bases in regions rich in coal resources, including Shanxi, Shaanxi, Inner Mongolia, Xinjiang, and Ningxia, from where both

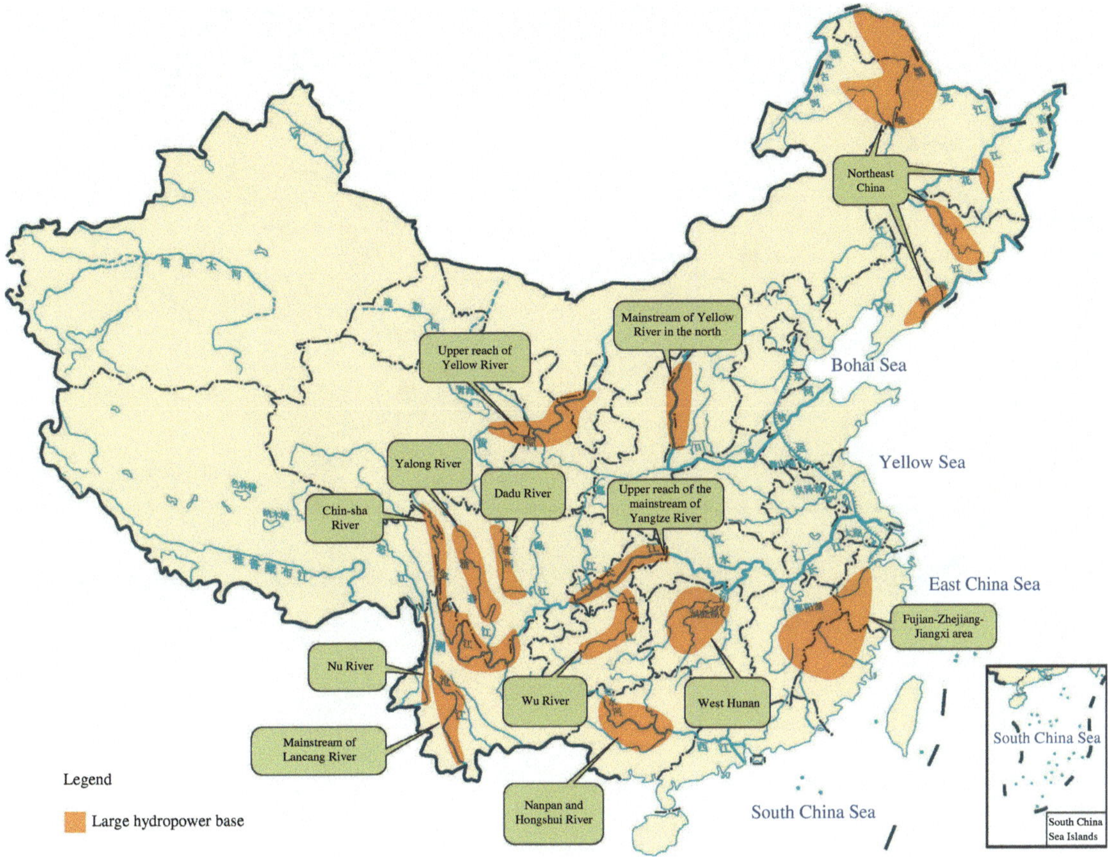

Figure 11.6
Geographical distribution of major hydropower bases across China.

power and coal are to be transmitted to other regions. At present, 18 large coal bases have conditions conducive for coal-fired power development, namely, north Shanxi, central Shanxi, southeast Shanxi, north Shaanxi, Ningdong, Junggar, Ordos, Xilin Gol League, Hulunbeir League, Huolin River, Baoqing, Hami, Zhundong, Ili, Binchang, Longdong, Huainan, and Guizhou. Their total coal-fired power development potential is more than 600 GW, and the power will mostly be delivered to the load centers in north China, central China, and east China. Figure 11.7 shows the geographical distribution of these bases.

3. *Large renewable energy bases*. Here, the renewable energy mainly refers to wind and solar energy. China is endowed with relatively abundant wind resources that are geographically located in remote areas, away from the load centers. This requires China to adopt an approach of "building large renewable bases connected to large grids," focusing on developing the "Three North" regions, eastern coastal areas, and off-shore areas where wind resources are abundant. Nine large wind power bases will be

Figure 11.7
Geographical distribution of large coal power bases across China.

constructed, respectively, in Hebei, east Inner Mongolia, west Inner Mongolia, Jilin, Gansu, Shandong, Jiangsu, Xinjiang, and Heilongjiang, as shown in Figure 11.2. By 2020, there will be approximately 98 GW of wind power available for cross-region delivery. China is rich in solar energy resources with large geographical areas and buildings suitable for solar power development. Some areas are also suitable for large, concentrated development. A number of PV farms will be constructed, respectively, in Qinghai, Gansu, Xinjiang, Inner Mongolia, Tibet, Ningxia, Shaanxi, Yunnan, and north and northeast China with suitable conditions. By 2020, the total installed capacity of PV power stations is expected to reach up to 20 GW.

11.2.2 Power Transmission Modes of Large Power Bases

As stated, the overall power flow pattern of west-to-east and north-to-south bulk power transfer will remain for quite a long time. In other words, the power from the large hydro, thermal, and wind power bases to be developed in the north and west will be mostly transmitted to the load centers in the eastern and central regions. Over time, China would establish large power bases in the west and north as exporting regions. The Beijing–Tianjin–Hebei–Shandong region, the four provinces in east central China (namely, Hubei, Hunan, Henan, and Jiangxi), and east China would be importing regions. The progress of UHV AC and UHV DC transmission technology would be essential to transmitting power from exporting to importing regions in large quantities over a long distance.

Considering the features of UHV DC and UHV AC transmission, the former is intended for long-distance and large-capacity transmission, such as power export from large energy bases including the hydropower bases in the southwest, coal power bases and wind power bases in the northwest and Xinjiang, as well as cross-border exchange. The latter is targeted to provide fortitude and cross-region power transmission facilitating connection of large coal, hydro, wind, and PV power generation. A UHV AC grid not only allows for energy delivery but also contributes to a robust UHV synchronous grid. It can provide strong support to a large-scale multi-DC-infeed transmission system, thereby achieving direct transmission of electricity to load centers through UHV DC lines.

It is essential to construct hybrid power grids consisting of UHV AC and UHV DC systems to take full advantages of both technologies to accommodate the demand for exporting electricity from large power bases in the west and north and to ensure the safety and economy of power grids.

11.2.2.1 Principles of selecting transmission mode

The following principles are to be followed in determining the power transmission mode for large power bases:

1. Adhere to government energy development strategies. Construct UHV power grids to facilitate bulk, long-distance, and efficient delivery of power; boost intensive development of large coal, hydro, nuclear, and renewable energy bases; ensure stable energy supply in the long term; optimize the power structure; reduce the costs; and achieve energy conservation and emission reduction.
2. Determine appropriate mode and voltage class of power transmission from power bases taking into account factors such as geographical location, export capacity, distance to load centers, as well as exporting power grid conditions with the aim of minimizing the transmission corridor and maximizing the transmission capacity.
3. Follow the principle of economics and rationality. Minimize total costs while meeting the general principles governing the market economy and power system characteristics. Ensure the price of the power delivered to the load centers is not higher than the local feed-in tariff for coal-fired generation to make bulk power transfer commercially competitive.

11.2.2.2 Mode of power transmission for large power bases

1. **Large hydropower bases**
   a. *Chin-sha River*. Four cascade power stations, namely, Xiangjiaba, Xiluodu, Wudongde, and Baihetan, will be built in the lower reach areas of Chin-sha River between Sichuan and Yunnan. So far, the Xiangjiaba–Shanghai ±800-kV UHV DC demonstration project, with a transmission capacity of 6.4 GW, has been put into operation; the left river bank at Xiluodu–Jinhua ±800-kV UHV DC transmission project, with a transmission capacity of 8 GW, is under construction.

      Considering that the distance from Wudongde Power Station to Wenzhou, Zhejiang, is approximately 2100 km, it is justifiable to use ±800-kV UHV DC transmission. By 2020, the power from Wudongde and Baihetan power stations will be transmitted through the Wudongde–Zhejiang ±800-kV DC transmission system and Baihetan–Hubei ±800-kV DC transmission system (each 8 GW).

      The hydropower stations in the upper reaches of Chin-sha River are expected to be operational in 2020–2030. As stated in the *Planning of Power Transmission for the Cascade Hydropower Stations along the Mainstream and Tributary of Chin-sha River in the Upper Reaches*, the timing and order of these hydro station developments has a major impact on the chosen power transmission mode. Specifically, if the construction of these stations keeps in step with other hydropower projects in Sichuan, then bulk power will be delivered over a long distance to the four provinces in east central China and to east China through UHV DC lines. Conversely, the power will be delivered subsequently through UHV AC lines.

   b. *Yalong River*. After meeting the growing local demand in Sichuan, the surplus power from this base can be delivered to Chongqing and east China. So far, the Jinping–Sunan ±800-kV UHV DC transmission project has been completed, which has a transmission capacity of 7.2 GW and is intended to deliver power to the east China grid.

      The cascade power stations in the middle reaches of Yalong River are approximately 400–500, 700–800, and 1400–1500 km away from the load centers in Chengdu and south Sichuan, Chongqing, and east China, respectively. It is planned that by 2020, the total generation capacity of these power stations will reach 11.56 GW. All electricity generated during the dry season and most of that during the wet season should be first delivered to the consumers and load centers in the Sichuan grid. The surplus generation in wet seasons should be transmitted through 1000-kV UHV AC lines to the four provinces in east central China. This can save precious rights-of-way in west Sichuan and leave room for future export of hydropower from west Sichuan.

   c. *Dadu River*. As of 2020, almost all hydropower resources in the Dadu River will have been developed. After meeting the growing local demand in Sichuan, the surplus power will be delivered through 500-kV EHV AC lines and UHV AC lines to Chongqing and the four provinces in the east central China.

2. **Large coal power bases**
   a. *Shanxi*. The coal power bases in southeast Shanxi, central Shanxi, and north Shanxi are 500–1000 km away from the load centers in the Beijing–Tianjin–Tangshan region, Shandong, the four provinces in east central China, and east China, and the application of 1000-kV UHV AC transmission is justifiable.
   b. *Xilin Gol League*. This base is 300–1600 km away from load centers in the Three-Hua region. It is not only a major coal power base but also has abundant wind resources. According to the projected power flows, by 2020, there will be 30.12 GW of power (both wind and thermal power) available for export from this base. Bearing in mind the magnitude and scale of export from thermal and wind power generation export, it is proposed that at the early stage, the power be delivered to the Three-Hua grid through UHV AC lines to take full advantage of AC grid and meet load demands of multiple receiving-end regions; as the development at the base accelerates, an increasing amount of power will have to be exported externally, and this necessitates construction of ±800-kV UHV DC transmission lines. In the long term, the power generated at the base will be transmitted through an AC/DC hybrid grid.
   c. *Hami*. This base is part of northwest China grid that operates at 750 kV and is asynchronously connected to the Three-Hua grid. Hence, it is desirable to deliver power from the base to the Three-Hua grid through UHV DC lines. This base has favorable conditions for large-scale development of both coal and wind power. According to the projected power flows, by 2020, 16 GW of power (both wind and thermal power) will be delivered to Henan and Chongqing. In view of the power capacity available for export, a double circuit ±800-kV UHV DC system may be constructed to transmit both wind and thermal power from the base.
   d. *Zhundong and Ili*. These two bases are located in northwest China, and HVDC transmission is also proposed for transmitting electricity from these bases to the Three-Hua grid. The distance from the bases to load centers in the eastern and central regions is 2700–3700 km, which is economically justifiable for ±1100-kV UHV DC transmission. In the coming decade, another 40 GW will be available for export from these bases and ±1100-kV UHV DC lines should be built to transmit power to the four provinces in east central China, Sichuan, and Chongqing.
   e. *Junggar*. This base, located in Junggar, Inner Mongolia, is 600–1600 km away from the load centers in the Beijing–Tianjin–Tangshan region, Shandong, the four provinces in east central China, and east China. According to the projected power flows, there will be 15 and 35.1 GW of power available for export from this base by 2015 and 2020, respectively. Given the power capacity available for export and transmission distance, hybrid transmission consisting of UHV AC and UHV DC is

preferred. Depending on the power transmission demand, some power may be first connected to the UHV AC substation nearby and exported to the UHV AC grid to directly load to centers through ±800-kV UHV DC lines.

f. *Ningdong*. This base is located in the north of Ningxia and belongs to the northwest China grid, which is asynchronously connected with the Three-Hua grid. At present, the Ningdong–Shandong ±660-kV HVDC transmission demonstration project has been completed and delivers electricity to Shandong with a capacity of 4 GW over a distance of approximately 1335 km. According to the projected power flows, during the 12th Five-Year Plan period, there will be 8 GW of power exported from the base to Zhejiang over a distance of approximately 1700 km, which is economically justifiable for ±800-kV UHV DC transmission. Therefore, a single ± 800-kV UHV DC transmission line may be constructed. Moreover, as UHV AC grids become increasingly robust and extend to these energy bases, these bases can be directly connected to the UHV AC substations nearby and export to the Three-Hua grid through the UHV AC grid.

g. *North Shaanxi*. This base is located in the north of Shaanxi Province and belongs to the northwest China grid. It is 600–1400 km away from the load centers in the Beijing–Tianjin–Tangshan region, Shandong, the four provinces in east central China, and east China. According to the planned power flows, there will be 3.6 and 11.6 GW of power available for export by 2015 and 2020, respectively. At the early stage, the power can be directly delivered to south Hebei grid using 500-kV AC lines. As the capacity available for export increases, the power from this base may be combined with that from Junggar base, Ningdong base, and other bases in the vicinity for export through a UHV AC and DC hybrid grid. In view of the transmission distance, ±800-kV UHV DC transmission lines may be justifiable, and the power sources can be connected to the UHV AC substation or DC converter station in the immediate vicinity.

h. *Longdong and Binchang*. Longdong and Binchang are China's key coal power bases. They are adjacent and quite close to load centers in the Three-Hua region. Their generation capacity is expected to exceed 30 GW. Longdong base is further away from load centers in Gansu; therefore, its power will be mostly exported. Binchang base is close to load centers in the Guanzhong region, so its power will first be used to meet local demand and the surplus can be exported. In line with the planned UHV AC grid development, at the early stage, UHV AC lines will be constructed to deliver the power from these two bases to load centers in east and central China; as more power generation is developed, UHV DC lines will be constructed. According to the projected power flows, during the 13th Five-Year Plan period, 8 GW of power will be delivered from Longdong base to Jiangxi over a distance of approximately 1400 km, which is justifiable for ±800-kV UHV DC transmission.

3. **Large renewable energy bases**
Given the distribution of renewable energy bases, excluding output consumed by local demands, all or part of the power output from the nine wind power bases will be delivered to the Three-Hua grid. Furthermore, solar power generation should be built in coordination with wind and hydropower development to take advantage of complementary features between wind and solar and between hydro and solar.
   a. *Wind power bases in Hebei.* Wind energy in Hebei is mainly found in Zhangjiakou and Chengde. At the early stage, the power is mostly consumed in north China. As the wind power capacity increases, the power needs to be delivered to the Three-Hua grid through AC grids.
   b. *Wind power bases in east Inner Mongolia.* Wind power bases in east Inner Mongolia are mainly located in Chifeng city, Tongliao city, Hinggan League, as well as Hulunbeir League. Most of the wind power is exported to the Three-Hua grid, with only a small portion consumed locally. The east Inner Mongolia grid is asynchronously connected to the Three-Hua grid; therefore, HVDC transmission is proposed. According to the projected power flows, the wind power will be delivered to Shandong and Anhui. For this purpose, one ±800- or ±1100-kV UHV DC circuit may be constructed to transmit both wind and thermal power from Hulunbeir League.
   c. *Wind power bases and solar power stations in Gansu.* The wind farms in Gansu are mainly situated in Guazhou (a county of Jiuquan), Yumen, Subei, and Minqin. Gansu is also rich in solar energy. Hence, solar power stations will be constructed along the Hexi Corridor to complement wind power generation. By 2020, there will be 4.75 GW of wind power available for exporting along with thermal power through the Jiuquan–Hunan UHV DC link, with the remainder consumed locally and in the northwestern region.
   d. *Wind power bases and solar power stations in Xinjiang.* Wind power bases in Xinjiang are mainly located in Hami and Dabancheng (in Urumqi). By 2020, the wind power developed in the northeastern area will be delivered together with thermal power to the Three-Hua grid through double circuit ±800-kV DC from Hami base, with the remainder consumed in the Xinjiang, or delivered to the main grid of northwest China through two circuits of 750-kV EHV AC lines that connect the Xinjiang grid to the main grid of northwest China. Solar farms will be constructed in Bazhou, Hotan, Turpan, and Hami.
   e. *Wind power bases in west Inner Mongolia.* The wind power bases in west Inner Mongolia are mainly located in Ulanqab, Xilin Gol League, and Baotou. Considering meeting local demand, the surplus will be delivered to the Three-Hua grid. The wind power from Xilin Gol League is mostly transmitted together with thermal power through the UHV AC/DC hybrid grid, and a portion of wind power

from Ulanqab is first collected in the nearby 500-kV substations and then sent out through UHV AC grids.

f. *Wind power bases in Shandong and Jiangsu.* The wind power is delivered through 500-kV AC or less to load centers in the respective provinces and to Three-Hua grid.

11.2.3 Relationship Between UHV AC/DC Grid and Large Power Bases

11.2.3.1 Relationship between UHV AC/DC grid and power plants

Connection of a power plant to the grid should be considered from many aspects, including planned station capacity, unit size, transmission distance, its role in the power system, flexibility in operation and dispatch, and emergency management, as well as impact on grid stability. A power plant or generating unit with a certain capacity should be connected at an appropriate voltage class. Given that the power plants at large energy bases are far from load centers and have a larger capacity, they are usually connected to a UHV grid.

With the development of major energy bases moving toward the west and north, large power stations will be constructed mainly in the western and northern regions in the future. For example, large pithead power plants will be constructed at coal power bases. Each power plant is expected to have a total capacity more than 4000 MW (or 6000 MW in some cases) with unit capacity of 600 MW or larger; therefore, their preferred connection to the grid is through a UHV grid. In general, a power plant is connected to the grid at a voltage class of either 500 kV or 1000 kV, which can be achieved by: (i) increasing the voltage of the generating units to 500 kV and then connecting them to a UHV substation or (ii) increasing the voltage of the generating units to 1000 kV directly and then connecting them to a UHV grid.

The construction of UHV AC and UHV DC grids creates conditions for the construction and connection of large energy bases to the grid. At the early stage of a UHV AC and UHV DC grid construction, its capability may not be fully utilized due to the weak grid conditions and mismatch between the capacity of power source and transmission lines. A large number of generating units have to be equipped with inter-trip devices to manage the situation arising from a fault occurring to the AC/DC transmission lines. In addition, some generating units need to be increased to 1000 kV directly, which places a high requirement on the step-up transformers. However, as the UHV grid develops, it will become more and more robust with high levels of safety and stability and hence able to transmit power from the power plants in a safe and reliable manner.

11.2.3.2 Relationship between UHV AC/DC grid and safety of power plants

1. **Tripping of generating units following a fault on AC/DC lines**
   If some or all of the power station connection circuits are lost after a fault, it is necessary to trip some or all of the units to ensure the safety of the units and system. For

an AC-connected power station, it is usually assumed that it can withstand a permanent single-phase-to-earth fault. However, against three-phase fault (i.e., $N-1$ security criteria), control measures are required to trip some of the units for safety reasons.

At the early stage of UHV grid construction, the grid structure has not been fully formed, stability control measures are usually required to trip some or all of units at the sending end when one of "point-to-grid" connection lines is lost after a permanent three-phase fault, or all the connection circuits are lost ($N-2$ fault). As the UHV power grid develops, some existing transmission lines would be incorporated into the backbone grid, increasing the connection strength between the power stations and the grid. The number of circuits and stations requiring measures for unit tripping is reduced. Except for the line section between the power plant and the first substation, the system can remain stable when an $N-1$ fault occurs on any other circuits. Even when an $N-2$ fault occurs on the majority of transmission lines, the system can remain stable. In other words, an $N-2$ fault occurring on just a small number of circuits requires measures for tripping some units. The studies show that the power stations connected to the UHV transmission are usually large, concentrated, and can easily implement unit inter-tripping schemes. The safety and stability control systems based on distributed and regional decision-making principles are also applicable to a UHV grid, and the existing safety and stability control measures designed for the 500-kV grid are also suitable for the UHV system. At the early stage of construction of a UHV grid, the safety and stability control measures of the UHV grid interact with those of the 500-kV grid, requiring coordination. After the UHV grid is fully constructed, their interaction will be gradually reduced and, finally, they will be decoupled.

The power transmitted through UHV DC is more than twice that of HVDC. As such, the impact of UHV DC monopolar blocking is higher than HVDC. At the early stage of construction of a UHV grid, the requirement for tripping units after monopolar blocking mainly depends on the sending-end grid's ability to withstand the surge of surplus power. During the middle and late stages, some existing cross-region UHV DC channels will be incorporated into the same synchronous grid to form a parallel AC and DC transmission system. In most cases, the AC system in parallel with the DC system can withstand the surge of the power transferred to it after DC monopolar blocking and, hence, no unit tripping is required. However, the power surge resulting from a bipolar blocking is more than twice the UHV DC monopolar blocking, thus the unit tripping measures will still be required.

2. **SSR/SSO**

Sub-synchronous resonance (SSR) refers to a phenomenon in which AC transmission system with series compensation capacitors experiences LC resonance after a disturbance and the sum of the resonance frequency and that of turbo-generator's shaft torsional vibration mode is close to or equal to the synchronous frequency of the system. Sub-synchronous oscillation (SSO) refers to torsional vibration of the shaft of a fossil-fired generator unit connected to the load center through a DC link, and its mechanism is different from that of shaft torsional vibration of the turbo-generator unit induced with series capacitor compensation. Theoretical analysis and practical experience indicate that SSR and SSO only occur with large-capacity turbo-generator units and are inherent in their shaft system structure, whereas for hydro turbine generating units the shaft torsional vibration is not a concern.

Considering the mechanisms of SSR and SSO and grid operation experience, there is a high risk that SSR will occur when a large fossil-fired power plant is connected to an AC grid with series compensation capacity; there is also a high risk that SSO will occur when the power is transmitted through a DC link. The phenomena of SSR and SSO, which are caused by interaction between the electrical system and mechanical system and characterized by long duration and high amplitude, could shorten the mechanical life of the shaft system or even damage it in a short time if the oscillation amplitude increases with time. This is also the case for both EHV/UHV AC and DC transmission. For a specific project, whether there is a risk of SSR or SSO occurring, analysis is required according to the actual conditions. The likelihood that the extension project of UHV AC pilot demonstration project (with 40% series capacitor compensation), Xilin Gol League UHV AC transmission link (with 40% series capacitor compensation), and Haminan–Zhengzhou UHV DC transmission link are exposed to the risk of SSR or SSO has been analyzed. Results are as follows:

a. *Risk of SSR arising from the expansion project of UHV AC pilot demonstration project*. With series capacitors installed for the expansion project, the power is normally transmitted from one grid to another, and SSR is unlikely to occur. Although in some special transmission modes (point-to-grid transmission, for example), the frequency scanning results show that the impedance drops significantly or series resonance occurs at some points, and SSR is likely to occur. In these situations, SSR monitoring and protective devices may be installed on units susceptible to the influence of SSR.
b. *Risk of SSR caused by Xilin Gol League UHV AC transmission link*. The analysis using planning data suggests that if the power generated from the three power plants at Xilin Gol League Coal Power Base is transmitted over a long distance in a point-to-grid connection through a UHV AC transmission system with series capacitor compensation, then SSR will occur under conditions when only a small number of units in a power plant are put into service. In this case, static Var compensator (SVC) with accessory SSR suppression function may be installed to mitigate the risks of SSR.
c. *Likelihood of SSO on Haminan–Zhengzhou UHV DC transmission link*. The analysis using planning data indicates that SSO will not occur when the Haminan Converter Station is connected to the 750-kV main grid in the northwestern region. However, under DC islanded transmission mode and certain operating conditions,

the torsional vibration of the generator shaft system may spread, causing a high risk of SSO. By selecting proper control parameters for the sub-synchronous damping controller of the DC control system, the SSO condition can be effectively suppressed.

3. **Transfer of power flow following a DC fault**
The transfer of power flow following a DC fault is closely related to the sending-end and receiving-end grids connected through a DC link. It is considered in the following two scenarios:
   a. The sending-end grid and receiving-end grid are asynchronously connected. Once a DC blocking occurs, surplus power appears and is transferred to other AC circuits within the sending end. If AC circuits cannot withstand the power transferred, then instability may result, requiring units to be tripped. In contrast, power deficit will occur at the receiving end and will be provided by the units on the receiving-end grid, causing redistribution of flows on the AC circuits. If AC circuits cannot withstand the power transferred, there is a potential that the local points on the receiving-end grid will experience undervoltage and that the line thermal stability limit will be exceeded. In this case, load rejection may be required.
   b. The sending-end and receiving-end grids are synchronously connected, representing a parallel AC and DC transmission system. DC blocking will not result in power imbalance of the entire system, but will result in all the power on the DC lines being transferred to the AC interconnection circuits. If the AC interconnection circuits cannot withstand the power transferred, then the system instability may result, requiring units to be tripped at the sending end and loads to be rejected at the receiving end. Moreover, the bulk power transferred to the parallel AC interconnection circuits will lead to increase in the power flow on the related circuits and consequent voltage drop on some buses. In this case, it is essential to trip units to effectively mitigate voltage drop along the transmission line.

Because the UHV DC system transfers more power than EHV DC system, DC blocking would result in more power flow being transferred. Any mismatch between the power transfer capability of the AC grid and the transmission capacity of the DC system would require tripping units and rejecting loads of larger capacity. In the early stage of a UHV grid development, the task of power transfer is still assumed by the 500-kV grid. However, because its power transfer capability is insufficient, to avoid cascading events it is necessary to trip units with a large capacity at the sending end and reject a certain amount of loads at the receiving end accordingly. A UHV grid has a power transfer capability much larger than a 500-kV grid. As the UHV grid develops, the power transfer after DC blocking will be mainly assumed by the UHV grid. As a result, the capacity of units to be tripped will be reduced significantly and load rejection is barely required. In addition, the voltage drop resulting from bulk power transfer can be controlled within a reasonable range.

4. **DC islanded power transmission**
   Of the regions surrounding China, the Russian Far East is rich in water resources and Mongolia is rich in coal resources. These resources can be converted into electricity to meet China's growing energy demands through international cooperation. The cross-border transmission links to be developed in the future feature large capacity, long distance, and greatly varying grid development levels between the both ends, making UHV DC transmission an inevitable choice. Additionally, the power grids in the Russian Far East and Mongolia are less robust and the capacity of DC links is larger than that of the AC grid. To protect the AC grid from severe impacts after a DC fault and to ensure safety of the local power grid, the converter station and power sources at the sending end will not be connected to the local power grid and, hence, the islanded transmission mode will be used.

   In DC islanded transmission mode, the problems such as DC start-up, SSO, overvoltage, and DC islanded operation mode and system stability control draw more concerns: (i) because the sending-end system is far away from the main system and has a weak grid structure, any fault disturbance in the DC transmission system or the AC system connected thereto will possibly result in fluctuation in the transmitted power, causing the power angle and frequency of the units at the sending end to vary considerably, consequently compromising the safety and stability of the system; (ii) the DC transmission system has a minimum operating power as dictated by the minimum current; however, because no support from the grid is available for the sending-end system operating under islanded mode, the sending-end generators cannot accommodate the requirements of the DC system, and this gives rise to a problem concerning the matching between the DC system and the sending-end generators during start-up; (iii) because the power source has a simple configuration and is closely connected to the DC system, there is a risk of SSO, especially when fossil-fired units are involved; therefore, provisions should be made to suppress the SSO; and (iv) the overvoltage problem and associated suppressing measures are also of great concern. So far, less experience has been gained for DC islanded transmission links around the world. Therefore, it is essential to perform an in-depth study of this transmission mode and related operating and maintenance technologies to accumulate experience.

11.3 Development Pattern of Power Grids in China [2]

11.3.1 Features of Future Power Grids

To meet the power demand of load centers in east and central China, the future backbone grids will have three features: (i) large-capacity and long-distance transmission, (ii) large-scale parallel AC and DC transmission, and (iii) multiple DC infeeds at the receiving end.

These grids will serve to transmit power from the northwest, northeast, west Inner Mongolia, west Sichuan, and foreign countries to load centers in east and central China over a distance of 1000–3000 km. To meet power delivery demands of these bases, the transmission capacity of both AC corridor (two circuits on the same tower) and DC corridor is expected to reach 8–10 GW. In the future, large-capacity and long-distance transmission will be a typical feature of power grids.

Currently in China, parallel AC and DC transmission connecting the same grids at both ends is only used in the northeast China grid and the south China grid. In the northeast China grid, the Hulunbeir–Liaoning ±500-kV DC line with a capacity of 3 GW runs in parallel with the 500-kV AC tie lines; in the south China grid, four DC circuits with a total capacity of 12.8 GW and a maximum single circuit capacity of 5 GW run in parallel with the 500-kV west-to-east AC transmission lines. Sichuan has abundant hydropower resources. The power from the large hydropower stations constructed on the Chin-sha River, Yalong River, and Dadu River will be exported through UHV DC transmission. Xiangjiaba–Shanghai and Jinping–Sunan UHV DC transmission lines have been completed; in the future, more DC transmission links will be constructed to transmit power generated at Wudongde Power Station and Baihetan Power Station. Moreover, there are several medium and small hydropower stations in west Sichuan, and their combined capacity is large. Therefore, it is appropriate to have a system collecting power generated from these stations and deliver it to east China and east central China through UHV AC circuits. Correspondingly, a typical parallel UHV AC/DC transmission system will be constructed to export power from west Sichuan, which is expected to reach up to 29.6 GW. Compared with existing parallel AC/DC transmission systems, a UHV AC/DC hybrid grid has a higher voltage class and a larger transmission capacity. An AC/DC hybrid grid will also be constructed to evacuate the power from Xilin Gol League. The major disadvantage of parallel AC/DC transmission is that the surge on the AC grid caused by a DC fault may lead to loss of system stability.

UHV DC transmission is an effective "point-to-point" transmission mode. The converter station at the receiving end is usually directly connected to the load center. In the future, the sending-end grids in China will evolve into a pattern where multiple DC links are densely fed into the same regional grid. For example, by 2020, more than 13 DC links will be fed into east China grid, among which more than nine will be fed into the Yangtze River Delta region. However, the multiple DC infeeds have the disadvantage that when the system is subjected to a disturbance, multiple DC systems may attempt to recover simultaneously, thereby requiring greatly increased reactive power support from the AC grid. If the AC grid cannot provide sufficient voltage support, then multiple DC systems may fail to restore normal operation, thus leading to system collapse.

11.3.2 Selection of Grid Development Plans

The Chinese Government aims to develop a moderately prosperous society in all respects, double its GDP and per capita income both in rural and urban areas, and make progress in overall planning regarding industrialization, urbanization, agricultural modernization, and ecology by 2020. To achieve these goals, China's future power grids should meet the following requirements:

1. *Higher transmission capacity*. China's primary energy resources are mainly distributed in remote southwestern, northwestern, and northern regions characterized with low population density, low energy demand, underdeveloped economy, and a large environmental footprint, which are 1000–3000 km away from the eastern and central regions with high population density, high energy demand, advanced economy, and heavy pollution. The future overall power flow pattern in China is still featured by the west-to-east and north-to-south bulk power transmissions. It is estimated that by 2020, the 12 provinces (municipalities) in mid-east China within the service area of SGCC will need to take 350 GW of power from other regions. How to safely and efficiently transmit such a large capacity of power to the receiving-end grid over a long distance is of great concern.
2. *Higher security*. The primary concern is how to ensure the planned main grid at the receiving end can accommodate 350 GW or more power transmitted from other regions.
3. *Higher capability in accommodating variable clean energy*. As at the end of 2012, the installed wind power capacity nationwide had exceeded 60 GW. It is estimated that by 2020, the installed capacity of wind power and solar power will reach at least 200 and 50 GW, respectively. As a result, it is urgent to plan and build the power grids that can accommodate the development of clean energy.
4. *Higher requirements on grid's functions*. As the use of UHV transmission and other advanced technologies becomes wider, the grid is not only a carrier of electricity in the traditional sense but also an essential means to attain energy conservation and emission reduction goals and achieve coordinated and sustainable development of the eastern and western regions. It is also a prerequisite for diversifying China's energy mix and ensuring energy security, a platform for optimizing resource allocation nationwide and a major component of China's energy system.

At the end of 2012, China's power grids comprised six regional grids and cross-region (cross-border) AC/DC transmission systems that supply power to 31 provinces, autonomous regions, or municipalities covering a territory area of 9.4×10^6 km^2. Among them, cross-region AC/DC transmission systems are all point-to-grid transmission links consisting of two UHV DC links, one UHV AC link, nine ±500-kV DC links, one ±660-kV DC link, one ±400-kV DC link, and four 500-kV AC links. Except that the north China grid and the central China grid are synchronously connected to each other through a UHV AC link, all other regional grids and connection to the Russian grid are asynchronously connected to each other through DC links. Of the six regional grids, except for the northwest China grid operating at a voltage class of 750 kV, northeast China grid, north China grid, central China grid, east China grid, and south China grid all operate at 500 kV.

In view of the future power flow patterns across China, the south China grid will mainly rely on the west-to-east power transmission within its service area, with the shortfall supplied from other regions through DC channels. It will remain asynchronously connected to other regional grids. From the viewpoint of grid power flow patterns in the regions served by the SGCC, the northwest China grid and northeast China grid are both sending-end grids that are more than 1000 km away from load centers in the Three-Hua region. Making these two grids robust is a prerequisite for bulk power transmission, because a UHV DC system requires strong support from the AC grids at sending and receiving ends. In the coming decade, in parallel with the construction of UHV DC projects, the 750-kV main grid connecting all load centers and energy bases in the northwestern region will be optimized and a 330-kV grid will be operated in some regions. The 500-kV main grid in the northeastern region will be optimized and a 220-kV grid will be operated in some regions and, if necessary, the 1000-kV AC main grid will be planned and constructed in due time. The north China grid, central China grid, and east China grid are all receiving-end grids and neighbor each other. Their interconnection can complement each other in terms of hydro, thermal, and wind power output. They allow optimal power allocation between dry seasons and wet seasons, thereby reducing investments and contributing to a higher grid security. The voltage class and interconnection mode of those three grids dictate the development pattern of power grids in China. Based on the voltage class and interconnection mode, three candidate plans are screened out, namely, the Three-Hua 500 kV asynchronous grid, the Three-Hua UHV asynchronous grid, and the Three-Hua UHV synchronous grid, which are detailed as follows:

1. *Three-Hua 500-kV asynchronous grid.* The Three-Hua 500-kV main grid will be expanded, with all the power required from other regions transmitted to it through UHV DC lines. By 2020, the capacity of transmission lines and transformation of the 500-kV grid will be 1.7-times and 2.1-times, respectively, those at the end of 2012. The north China grid and central China grid will remain interconnected synchronously and weakly. UHV AC transmission link between Anhui and east China will be completed and commissioned, and the east China grid will remain asynchronously interconnected with the central China grid and the north China grid. The Three-Hua 500 kV grid will be asynchronously interconnected through DC links with the northwest China grid and the northeast China grid as the sending end, as shown in Figure 11.8.

   According to this plan, there will be five 500-kV synchronous grids formed across China, of which four cover the service area of SGCC.

Figure 11.8
Schematic of Three-Hua 500-kV asynchronous grid.

2. *Three-Hua UHV asynchronous grid.* The north China–central China 1000-kV AC grid and the east China 1000-kV AC grid will be constructed separately. The grids are constructed centered on the load centers in the Beijing–Tianjin–Hebei–Shandong region in north China and in the four provinces in east central China. To accommodate the multiple infeeds of UHV DC links from southwestern and northwestern regions, and to connect the large capacity of nuclear power units located in the coastal areas, the east China 1000-kV AC grid at the receiving end will be constructed to form two UHV double-circuit loop networks (i.e., north Jiangsu network and Yangtze River Delta network) and extend up to Fuzhou to the south. The aforementioned two 1000-kV AC grids are asynchronously interconnected through a DC link, and the Three-Hua grid is then asynchronously interconnected with the northwest China grid and the northeast China grid through DC links. The power is transmitted to the east China grid solely through UHV DC links and to the north China–central China grid through both UHV DC and UHV AC links, as shown in Figure 11.9.

Figure 11.9
Schematic diagram of Three-Hua 1000-kV UHV asynchronous grid.

According to this plan, there will be five synchronous grids formed across China, of which two 1000-kV synchronous grids, the 750-kV northwest China synchronous grid, and the 500/1000-kV northeast China synchronous grid cover the service area of SGCC.

3. *Three-Hua UHV synchronous grid.* The Three-Hua 1000-kV AC grid will be constructed and asynchronously interconnected with the northwest China grid and the northeast China grid through DC links. The bulk coal power and wind power from Shanxi, west inner Mongolia, north Shaanxi, Ningxia, and Xilin Gol League, hydropower from Sichuan, as well as nuclear power from coastal areas will be transmitted to the grid through UHV AC circuits, whereas the bulk coal power and wind power from the northwestern and northeastern regions and hydropower from the southwestern region will be transmitted to the Three-Hua UHV AC grid through UHV DC lines, as shown in Figure 11.10.

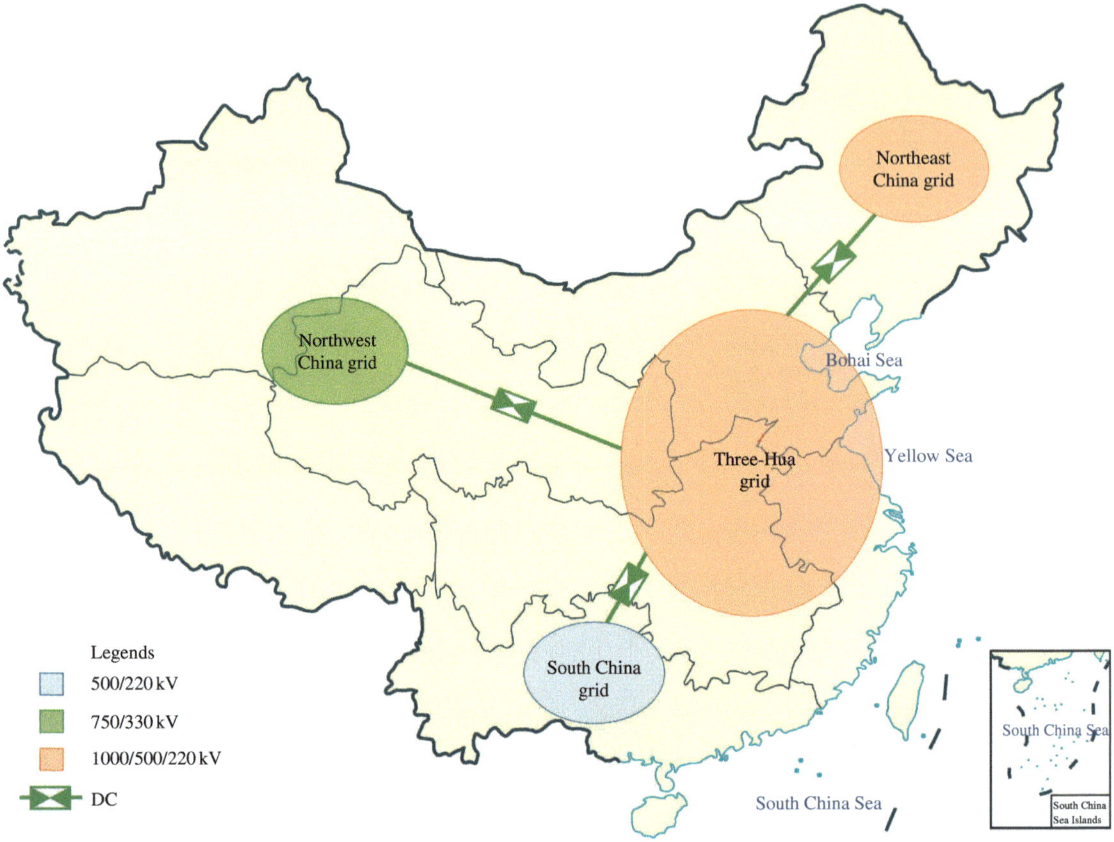

Figure 11.10
Schematic diagram of Three-Hua UHV synchronous grid.

According to this plan, there will be four synchronous grids formed across China, among which the Three-Hua 1000-kV UHV synchronous grid, 750-kV northwest China grid, and 500/1000-kV northeast China grid cover the service area of SGCC.

Table 11.9 summarizes the basic information of these three plans. Whether the three plans can meet the basic requirements of power grids to achieve sound and sustainable development of the economy and society of China is the major consideration when selecting the development pattern of power grids in China.

11.3.3 Security Analysis on Grid Development Plans

1. **Standards, contents, and methodology**
   Based on the medium- and long-term power balance forecast for the eastern and central regions and the pattern of energy flow of China, and considering the volume and path

Table 11.9: Basic Information of the Three Plans.

Basic Information of Grids	Three-Hua 500-kV Asynchronous Grid	Three-Hua UHV Asynchronous Grid	Three-Hua UHV Synchronous Grid
Number of synchronous grids within the service area of SGCC	4	4	3
Voltage class of backbone grid (kV)	500	1000	1000
Number of DC infeeds to the 12 provinces (municipalities) in Three-Hua region	39	34	27

of power transmission from energy bases, planned installed capacity, and market demands at receiving ends, the power balance calculation shows that by 2020, the Beijing–Tianjin–Hebei–Shandong region in north China, Hubei–Hunan–Henan–Jiangxi region in central China, and Shanghai–Zhejiang–Jiangsu–Fujian region in east China will receive 112, 130, and 104 GW of power, respectively. Of the five comprehensive energy bases planned in the *Outline of the 12th Five-Year Plan*, the hydropower bases in the southwest will deliver 74 GW of power and other energy bases (such as those in Shanxi, Shaanxi, Inner Mongolia, Ningxia, and Xinjiang) will deliver 260 GW of power. These figures will become even bigger after 2020. In view of such a power flow pattern, a comparison is made among the three candidate grid development plans in terms of security and economy to determine which one can fulfill the future power transmission requirements and guarantee sound and sustainable development of China's economy and society.

See Figure 11.11 for the grid security calculation process.

Of all the planning data input, the models and parameters used in calculation of the dispatching operation mode are the same as those used in the existing grid operation. For the transmission lines, transformers, and existing generating units of 300 MW or more, the measured parameters are used. For planned grid equipment and generating units, typical parameters are used following the principle that the same models and parameters are used for equipment and units of same type and capacity. The power flow calculation, static security analysis, and security and stability analysis are performed in two operating modes, the large mode in wet season and in dry season. The fault type, clearing time, and stability control measures are selected in accordance with DL 755—2001 *Guide on Security and Stability for Power System* and GB/T 15544—1995 *Short-Circuit Current Calculation in Three-Phase AC Systems.*

The $N-1$ transient security analysis shown in Figure 11.11 aims to check whether the grid meets the first-level security and stability standards specified in DL 755—2001, that is, when faults such as permanent three-phase short circuit on an AC line or

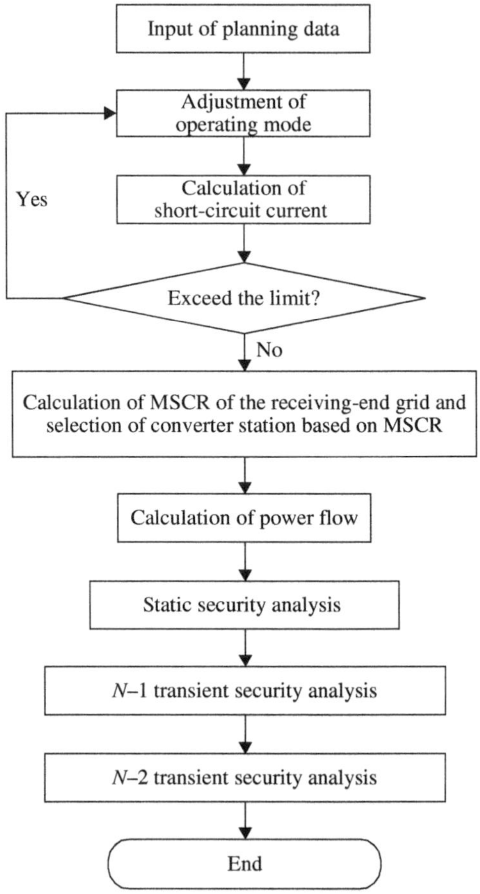

Figure 11.11
Grid security calculation process.

blocking of a single DC pole occur, the grid can remain stable without tripping generators or rejecting loads. The $N-2$ transient security analysis is to check whether the grid meets the second-level security and stability standards specified in DL 755—2001, that is, when faults such as tripping of two circuits of the AC line and DC bipolar blocking occur, the grid can remain stable under the condition that partial load loss is allowable and steps are taken to trip generators and reject loads.

Because the power system is one of the most complicated human-made systems and permits no interruption, simulating calculation must be conducted to understand the characteristics of modern grids, thereby determining the grid development plan and operating modes and increasing the grid control capability.

Currently, the power system simulation, planning, and operation control are calculated and analyzed using nearly the same simulation technology and software

Table 11.10: Short-Circuit Current Control Measures for the Three-Hua 500-kV Asynchronous Grid.

Item	Number of Measures			
	East China Grid	North China Grid	Central China Grid	Subtotal
Switching lines to hot standby	6	1	2	9
Disconnecting lines from the string	8	8	4	20
Bus splitting in substations	1	1	0	2
Adding series reactors	1	2	1	4
Total	16	12	7	35

around the world. The calculation results have been proven to be effective and reliable by China's major projects (such as the UHV AC pilot demonstration projects, Three Gorges Power Transmission Project, West-to-East Electricity Transmission Project by China Southern Power Grid, interconnection of Xinjiang and northwest grids, etc.) in the past and the grid's daily dispatch and operation.

To fully appreciate the characteristics of modern grids and ensure the operational security and stability, SGCC has established a data platform at its simulation center. This platform incorporates the data of more than 11,000 nodes and 35,000 branches involving various large hydro, thermal, nuclear, and wind power generating units that are connected to grids 220 kV and higher. With this platform, more than 100,000 fault simulation calculations have been conducted according to the standards, including DL 755—2001; numerous statistics, analyses, and conclusions that have been made were based on these standards.

2. **Three-Hua 500-kV asynchronous grid**
   a. *Short-circuit current overstress.* In the Three-Hua 500-kV asynchronous grid plan, the short-circuit current overstress problem is most severe in central China grid, with approximately 19% of its substations overstressed, exceeding 63 kA, mainly in the Yangtze River Delta region. Currently, no circuit breakers with a breaking capacity higher than 63 kA are available.

      The short-circuit current overstress problem can be solved using 35 measures such as replacing the circuit breakers, disconnecting the circuit breaker from the substation line, and switching the lines to hot standby. See Table 11.10 for details. Replacing the circuit breakers will involve the replacement of the primary circuit equipment and modification to grounded grid; therefore, it is a difficult task during operation.
   b. *Multi-DC-infeed short-circuit ratio (MSCR).* Preliminarily screen the inverter stations based on the ranking in terms of the value of MSCR. The MSCR calculation results show that seven inverter stations in the east China grid and the central China grid have an MSCR less than 2.5, of which six are located in the east China grid. This indicates that the east China grid is exposed to a high risk of voltage instability. See Figure 11.12 for its MSCRs.

Figure 11.12
MSCRs of east China grid.

c. N − *1 transient security analysis*. As analyzed, an $N-1$ short-circuit fault on more than 140 AC lines in the Three-Hua grid will result in voltage instability and large-scale blackouts. For example, when a fault occurs on a 500-kV tie line in south Jiangsu, it will cause voltage instability and the generating units in the receiving-end grid will lose synchronism. Areas affected by this instability include south Jiangsu, Zhejiang, Shanghai, and Fujian.

Analysis of system instability shows that the process after the fault occurring involves three stages: (i) during the fault, the DC system close to the fault point will experience a short commutation failure; (ii) after the fault is cleared, the DC system attempts to recover and absorbs reactive power from the AC system; and (iii) the DC system fails to restore to normal operation and the consequent severe reactive power imbalance causes voltage collapse. As a consequence, the active power imbalance will be exacerbated, leading to large-scale blackouts.

Analysis on the grid voltage instability mechanism with multiple DC infeeds. The active power fed into the AC grid by the inverter station, the reactive power consumed, and the extinction angle of converter valves are calculated using the following equations:

$$P_2 = U_{d2} I_2 \tag{11.1}$$

$$Q_2 = I_d\sqrt{U_{d02}^2 - U_{d2}^2} \tag{11.2}$$

$$\cos\gamma = K\frac{I_d}{U_2} + \frac{U_{d2}}{U_{d02}} \tag{11.3}$$

where

P_2, Q_2—the active power fed into the AC grid by inverter station and the reactive power consumed by it

U_{d02}, U_{d2}, U_2—no-load voltage and operating voltage of the DC system on the valve side and voltage of the AC bus

I_d—DC current

γ—extinction angle of converter valves in inverter station

K—constant.

Equations (11.1)–(11.3) show that as the voltage drops at the instant when a short-circuit fault occurs in the grid with multiple DC infeeds, the extinction angle will instantaneously fall to zero, resulting in commutation failures in inverter stations. It is equivalent to the situation where another single-phase short-circuit fault occurs on the valve side of the converter transformer before the AC system short-circuit fault is cleared, thus leading to further drop in voltage and considerable increase in current of both the AC and DC systems and also dramatic increase in the reactive power consumed by the inverter station. Because the reactive power supplied by the reactive power compensation devices installed in the inverter station is significantly reduced because of the voltage drop, the inverter station has to absorb reactive power from the AC grid, contributing to a further decrease in AC bus voltage. When multiple inverter stations suffer commutation failures at the same time as increased demand of reactive power, the AC system voltage will continue to drop until the receiving-end grid collapses. To improve the voltage stability of the receiving ends of the Three-Hua grid, simulation analysis studies were performed for new inverter stations with SVCs or static synchronous compensators (STATCOMs) installed. Results show that the voltage stability is notably enhanced, although the voltage instability problem cannot be completely eradicated.

d. *Analysis of DC system faults and simultaneous faults in two UHV DC links sharing the same right of way*. The purpose of the DC system fault analysis is mainly to assess the AC grid's capability of withstanding power flow redistribution arising from such a fault. DC system faults are classified into two categories: (i) fault in the parallel AC and DC transmission system, which will cause the power transmitted by the DC system to be directly transferred to the parallel AC interconnection lines and (ii) fault with the DC equipment at the receiving end, which will cause power loss and, therefore, the voltage supports from AC system will give rise to a power surge. If the delivery capacity of the AC grid does not match that of the DC system, then the system stability may be compromised or cascading events may occur, resulting in major blackouts.

When an AC system runs in parallel with a DC system, in the event of a fault in the DC system, the power originally delivered by the DC system will be transferred to the AC system. If the AC system cannot accommodate such power, then an instability problem may occur. For instance, if the Sichuan–Chongqing link and Chongqing–Hubei link use their original transmission circuits (both have two 500-kV double circuits), then a monopole blocking in the Baihetan–east Hubei DC link will cause unattenuated low-frequency oscillation, which can only be eliminated when the Sichuan–Chongqing and Chongqing–Hubei links are strengthened.

During normal operation of the east China grid, the monopolar or bipolar blocking of a DC link will cause overload on AC lines. For example, a bipolar blocking in the Jinping–Suzhou ± 800-kV UHV DC link will cause the relevant 500-kV lines to be severely overloaded, in which case the active power through a single line will be up to 3.62 GW. This may result in cascading tripping and propagation of events.

When large units in southeast Zhejiang are out of service, a bipolar blocking in the Wudongde–Wenzhou ±800-kV UHV DC link will cause instability of the Zhejiang and Fujian grids. During normal operation of the central China grid, a bipolar blocking in any ±800-kV UHV DC link or monopolar blocking in any ±1100-kV UHV DC link may result in grid instability. Such faults necessitate a large amount of loads being rejected at the receiving end, posing a high risk of large-scale blackouts. For instance, a bipolar blocking in the Hami–Zhengzhou ±800-kV UHV DC link will cause the Jindongnan–Nanyang 1000-kV UHV AC tie line to operate beyond its steady-state stability limit and is consequently disconnected, thus leading to instability of the central China grid. In this case, it is necessary to reject more than 3.5 GW of loads in Zhengzhou, Wuhan, and other locations to maintain the grid stability. According to the *Regulations on Emergency Response, Investigation and Handling of Electricity Safety Accidents* (No. 599 Decree of the State Council of the People's Republic of China, hereinafter referred to as the *Regulations*) issued in July 2011, such a situation is considered to be an accident. See Table 11.11 for the classification of possible consequences resulting from DC system faults in the central China grid.

When a bipolar blocking fault occurs simultaneously in two UHV DC links of any grid, regardless of the north China, central China, or east China grid, the gird will suffer instability. Therefore, up to 9.5 GW of load must be rejected to maintain the grid stability. When a monopolar or bipolar blocking occurs at the sending-end grid, the relevant generating units must be tripped over a short period of time, and up to a maximum of 6 GW could be tripped in the case of a bipolar blocking.

Table 11.11: Classification of Consequences Resulting from Blocking Faults in UHV DC Links.

Fault Type	Voltage Instability of System	Classification of Consequences Resulting from Stability Control Measures			
		Below the Accident Criteria	Minor Accident	Moderate Accident	Major Accident
Monopolar blocking	3	2	1	0	0
Bipolar blocking	10	0	3	2	5

These calculations and analyses show that the Three-Hua 500-kV asynchronous grid plan has the following disadvantages: (i) the short-circuit current of a number of substations exceeds the maximum breaking capability of circuit breakers, which cannot be addressed effectively; (ii) as an enormous amount of power is delivered to load centers in the eastern and central regions through DC systems, the receiving-end AC systems are too weak compared with the DC system, and any fault occurring in either the AC or DC systems will cause cascading events and voltage collapse, leading to large-scale blackouts; (iii) even equipped with SVCs or STATCOMs, the grid still cannot completely meet the criteria of the first-level security and stability specified in DL 755—2001; (iv) in this plan, the grid has a low capability to withstand accidents and is exposed to a high risk of large-scale blackouts, thus failing to meet the stipulations of the *Regulations* or the requirements of sustainable development of China's economy and society.

3. **Three-Hua UHV asynchronous grid**
   a. *Short-circuit current calculation.* First, split the looped 500-kV network where it forms a 500/1000-kV electromagnetic loop (i.e., 500-kV interconnection operates in parallel with 1000-kV system, coupled via 500/1000-kV transformers). See Figure 11.13 for the schematic diagram of splitting the inter-provincial electromagnetic loop within the east China grid. Then, calculate the short-circuit current. After measures like disconnection of a few lines are taken, the short-circuit current of each substation falls below 63 kA, satisfying the requirements of the breaking capacity of circuit breakers.
   b. *MSCR calculation.* The MSCR values of the north China–central China UHV synchronous grid are considerably increased, with the lowest MSCR being 2.94, therefore meeting the security requirements. However, six substations in the east China UHV synchronous grid have an MSCR value less than 2.5, posing a high risk of voltage instability. See Figure 11.14 for MSCRs of the east China grid.
   c. $N-1$ *transient security calculation and analysis.* The north China–central China UHV synchronous grid is not subject to instability in the case of an $N-1$ fault. In the event of an $N-2$ fault, it is only necessary to trip the sending-end units rather than reject loads. In summary, the grid can remain stable under faults.

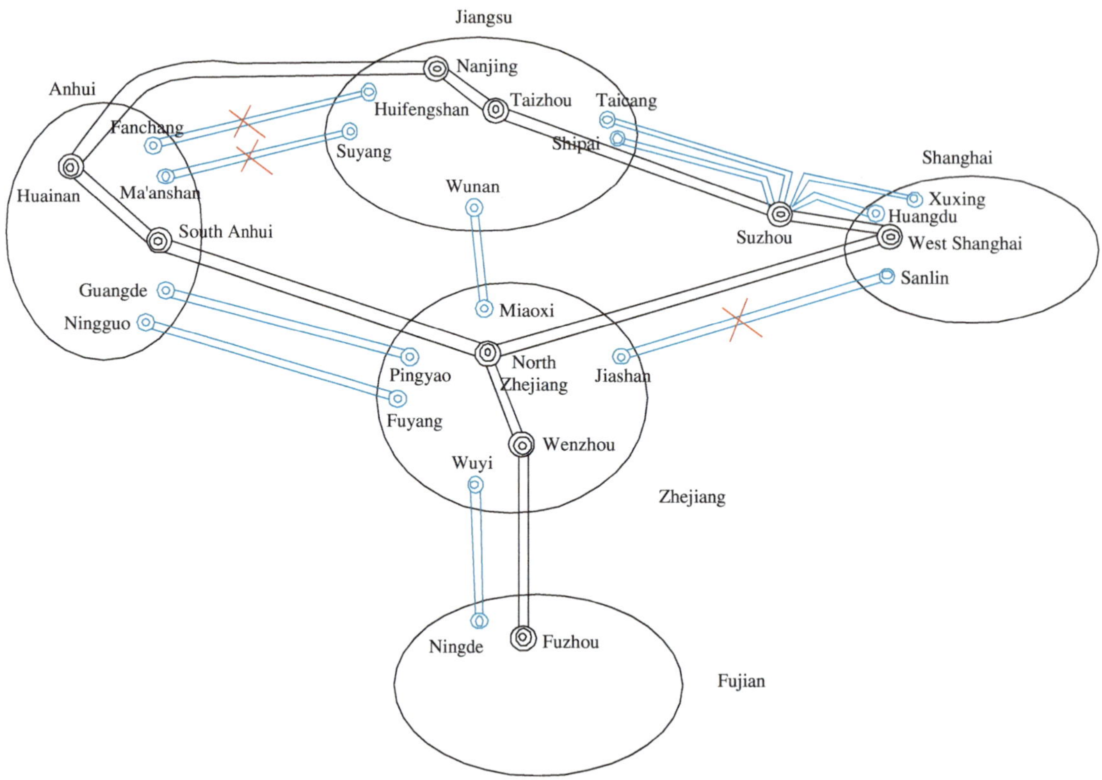

Figure 11.13
Schematic diagram of opening the inter-provincial electromagnetic loop within the east China Grid.

However, the east China grid will still suffer voltage instability when an $N-1$ fault occurs on any of more than 70 lines following the same mechanism as analyzed before. Installing SVCs and STATCOMs also cannot completely resolve the instability problem. The main reason is that the DC power fed into the east China grid accounts for more than 60% of its total generating capacity, which significantly exceeds the withstand capability of the AC system according to the relevant CIGRE study reports. If it is required for the receiving-end AC grid to receive more electricity, then more efforts should be made to build AC grids to offer more AC transmission channels.

d. Analysis of AC network security against a monopolar or bipolar blocking or concurrent blocking of two UHV DC links sharing the same right of way. For both north China–central China and east China UHV synchronous grids, a monopolar or bipolar blocking of a DC link will not cause overload on AC lines and the grids can maintain stability. When a bipolar blocking occurs concurrently in two UHV DC links supplying Shanghai and Suzhou, the grid can be maintained stable only by

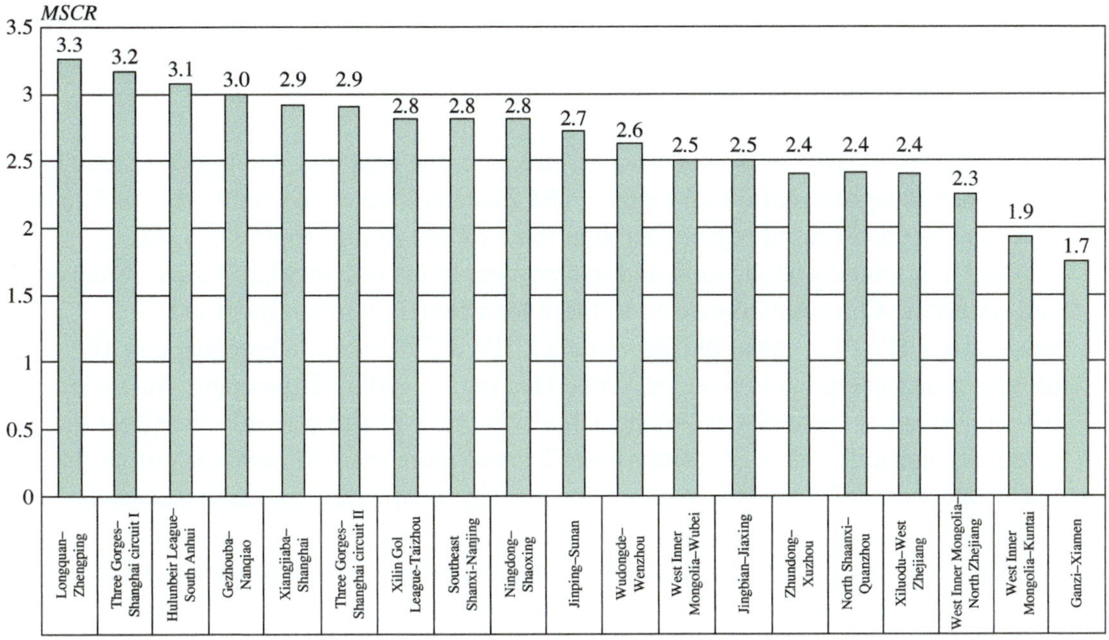

Figure 11.14
MSCRs of east China grid.

rejecting 3.6 GW of loads in these two cities. For concurrent bipolar blocking in any two of the other DC links, the grids can remain stable without load rejection. Compared with the Three-Hua 500-kV asynchronous grid plan, the Three-Hua UHV asynchronous grid plan has much higher capabilities of withstanding complicated incidents.

These calculations and analyses demonstrate that the Three-Hua UHV asynchronous grid plan has two advantages: (i) it can solve the widespread high short-circuit current problem faced by the Three-Hua grids, especially the east China grid, and (ii) it can solve the problem regarding the AC systems not being able to accommodate the enormous amount of power transferred from the DC systems after a monopolar or bipolar blocking fault.

The capability of the north China–central China UHV synchronous grid to withstand various incidents is significantly improved and meets the security and stability requirements specified in DL 755—2001. For the east China grid, the amount of DC infeeds exceeds the withstand capability of the AC grids and, therefore, the grid still suffers serious voltage instability in the case of $N-1$ short-circuit faults on AC lines. Even installing a number of SVCs/STATCOMs cannot allow the east China grid to meet the relevant "$N-1$" requirements specified in DL 755—2001. Moreover, the grid has a low capability to withstand incidents and is still exposed to a high risk of large-scale blackouts. In conclusion, the Three-Hua UHV synchronous grid plan cannot ensure a secure and reliable power supply of the east China grid and fulfill the requirements of economic and social development.

4. **Three-Hua UHV synchronous grid**

Under the Three-Hua UHV synchronous grid plan, the coal and wind power bases in west Inner Mongolia, Xilin Gol League, and Zhangbei are connected to the load centers in the eastern and central regions through three to five north–south UHV AC transmission routes, and the coal and hydropower bases in Ningxia, Shaanxi, Shanxi, and southwest China are connected to these load centers through three to five west–east UHV AC routes, thus forming a UHV AC network interconnecting 16 provinces and municipalities in the Three-Hua region and all large energy bases located within 800 km of these places. With such a network it is also possible to deliver the electricity generated at large energy bases located more than 800 km away from the load centers in the Three-Hua region. The security verification and analysis results are as follows:

a. The Three-Hua UHV synchronous grid provides a solution to short-circuit current overstress problem. The short-circuit current of all 1000/500-kV substations is contained to less than 60 kA by operating the 500-kV grids in a radial manner.
b. All MSCRs are more than 3.0; therefore, the risk of voltage instability is dramatically decreased. See Figure 11.15 for MSCRs of the east China grid.
c. The transient security analysis results show that in the case of an $N-1$ or $N-2$ fault on any lines, the Three-Hua UHV synchronous grid can remain stable without load rejection. An $N-2$ fault occurring on certain receiving-end lines requires the tripping of relevant generating units.
d. When monopolar or bipolar blocking occurs in any individual UHV DC link or a bipolar blocking simultaneously occurs in two UHV DC links, the grid can maintain stability by tripping some generating units at the sending end rather than rejecting loads.

In conclusion, the Three-Hua UHV synchronous grid plan can solve the problem of short-circuit current overstress, which restricts the grid development as well as causes the severe voltage instability at load centers in the eastern and central regions, fully meeting the requirements of DL 755—2001 and the *Regulations*. Moreover, the grid has much higher capability to withstand complicated faults and to receive electricity from regions other than the Three-Hua UHV asynchronous grid plan; therefore, it can guarantee a secure and reliable power supply to support the long-term sustainable development of China's economy and society.

5. **Comparisons and recommended plan**

Based on these analyses, a comparison was made among the three grid development plans in terms of the short-circuit current, transfer of power flow under faults, voltage

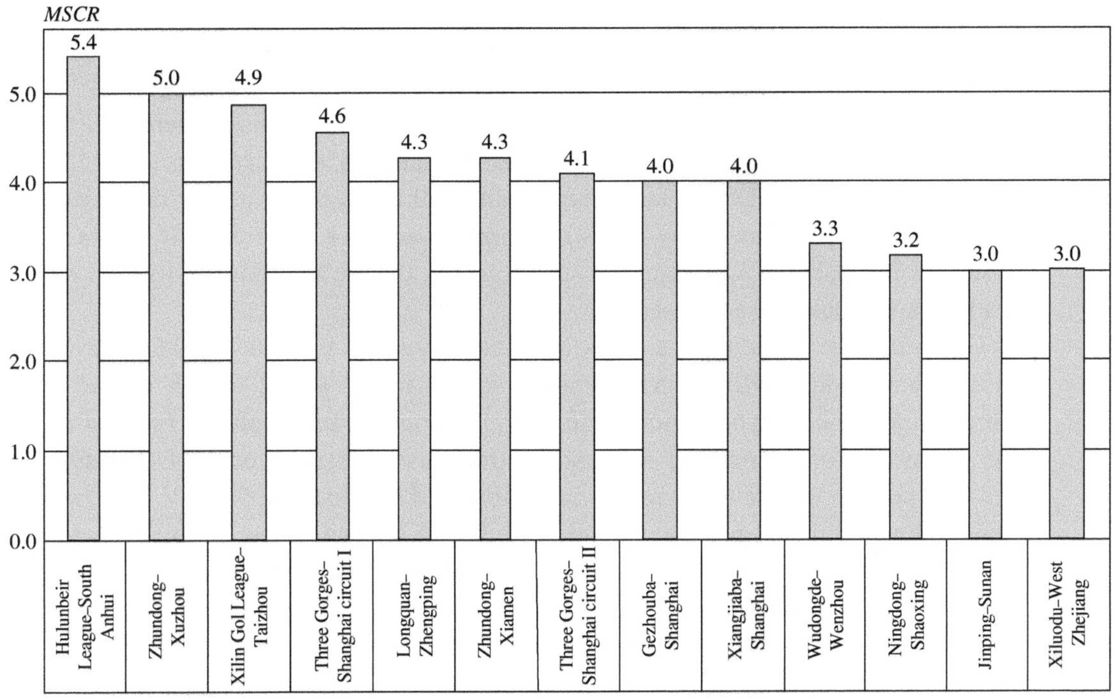

Figure 11.15
MSCRs of east China grid.

stability of the receiving-end AC grid with multiple DC infeeds, and the risk of large-scale blackouts.

a. *Short-circuit current.* At present, the Shanghai and Sunan grids suffer serious excess short-circuit currents, leading to the frequent use of measures like disconnecting and switching out lines. This has severely affected the operational reliability and flexibility of the grids, resulting in power imbalance and low transmission capacity, and hindering the maintenance and accident management. If China continues to develop the 500-kV grids, then by 2020 the short-circuit current will inevitably increase as more and more power sources at the receiving end are connected and new 500-kV substations are put into operation. In the Three-Hua 500-kV asynchronous grid plan, it is unavoidable to take some measures such as replacing a great number of circuit breakers, reconfiguring substations, and switching lines to hot standby. These measures are difficult to implement; however, even if they can be implemented in some places, the grid structure will possibly be reconfigured suboptimally, exposing the grid to potential security and stability risks. In both Three-Hua UHV asynchronous grid and synchronous grid plans, UHV AC grids are built at the load centers, and the short-circuit current of the 500-kV grids can be effectively controlled through opening the 1000/500-kV electromagnetic loops and optimizing the structure of some 500-kV grids.

b. *Post fault power transfers*. Any faults in both the AC and DC systems would involve power transfer. By 2020, the AC grids will become denser; therefore, the power flow transfer following faults on the AC channels will have less consequence. As a UHV DC system transmits bulk power over a long distance, the power flow transfer caused by faults is enormous across a large area. If the AC system is incapable of withstanding such transfer, then system instability will result, leading to serious consequences.

   Given that the capacity of a UHV DC link may reach up to 6.4–10 GW, even monopolar blocking will trigger a power transfer as high as 3.2–5 GW. In the Three-Hua 500-kV asynchronous grid plan, the power transfer after a fault on either the DC system of the parallel AC and DC transmission system or the DC system in the load centers will cause dynamic instability and overload on 500-kV lines, which may go beyond the steady-state stability limit of a section. These problems mainly result from the mismatch between the power rating of the 500-kV grid and the UHV DC system. In contrast, both the Three-Hua UHV asynchronous and synchronous grid plans incorporate the construction of UHV grids that can transmit an enormous amount of power and match the power rating of a UHV DC system. Therefore, the UHV grids can withstand the power transfer after a fault on the DC system, thus avoiding system instability.

c. *Voltage stability of the receiving-end AC grid with multiple DC infeeds*. Studies performed in China and abroad show that the more DC infeeds at the receiving-end AC grid, the more concentrated terminations and larger amounts of DC power transmission would lead to the more serious system voltage instability problem. Considering that there will be more than 13 DC infeeds to east China in the future, the voltage stability will become a major concern for grid security and stability.

   In the Three-Hua 500-kV asynchronous grid plan, the short-circuit current control measures will require network reconfiguration, which will compromise the capability of AC grids to provide voltage support to the DC system. Therefore, any fault in the AC grid may render multiple DC systems unable to restore normal operation, leading to widespread blackouts. As compared with the Three-Hua 500-kV asynchronous grid plan, the Three-Hua UHV asynchronous grid plan feeds a comparable amount of DC power into the east China grid. Although the MSCR is somewhat larger, the DC power to be fed into the AC grids still exceeds their withstand capability. Therefore, there is still a high risk that an $N-1$ fault on many AC lines may cause system instability. In the Three-Hua UHV synchronous grid plan, the east China grid and north China–central China grid are interconnected through several UHV AC routes, which not only presents a robust network but also makes it possible for the thermal power in the north and hydropower in west

Sichuan to be delivered to east China. As a result, the amount of DC power fed into east China is greatly reduced to such a level that the AC grids are robust. In addition, the higher MSCRs enable the grid to maintain stability in the event of $N-1$ faults on AC lines. In conclusion, from the perspective of the voltage stability at receiving ends, only the Three-Hua UHV synchronous grid plan is satisfactory.

d. *Risk of large-scale blackouts.* In the future, a complex pattern would emerge within the service area of SGCC in that UHV AC and DC links share the same sending and receiving ends and run in parallel in the same right of way. It will expose the UHV AC and DC systems sharing the same right of way to the risk of simultaneous faults. In the Three-Hua 500-kV grid plan, the 500-kV grid itself has low capability of withstanding the transfer of power flow. Hence, simultaneous faults with two DC circuits will cause instability of the receiving-end grid and consequent large-scale blackouts. If the Three-Hua UHV asynchronous grid is adopted instead, then the system stability will be greatly improved under similar fault conditions; however, in some cases when simultaneous bipolar blocking of two DC circuits results in loss of some loads, the system can remain stable. By contrast, the Three-Hua UHV synchronous grid can withstand the surge resulting from the simultaneous bipolar blocking of any two DC circuits.

In conclusion, the Three-Hua UHV synchronous grid plan can accommodate the requirements of China's national condition of uneven distribution of energy and loads and the national strategic development of energy, and can ensure secure electricity supply for the long-term sustainable development of China's economy and society. Therefore, it is an inevitable choice for China's future grid development.

11.3.4 Assessment on Economy of Three-Hua UHV Synchronous Grid

A UHV synchronous grid is characterized by large-scale high investments and notable social effects. Its economy dictates the planning, decision-making, and development pattern of the grid structure, which are prerequisites for minimizing investment waste in electric power construction.

This section deals with a comprehensive assessment of the economy of the recommended plan, the Three-Hua UHV synchronous grid, mainly in terms of its financial feasibility, competitiveness of electricity price, and contribution to the national economy.

11.3.4.1 Approaches and methodologies

1. **Analysis approaches**
   a. Perform financial analysis of the Three-Hua UHV synchronous grid as a whole and calculate its payback period, internal rate of return (IRR), rate of return on common

stock holders' equity, and net present value (FNPV) to assess the return on investment (ROI) and profitability of the plan.

b. Analyze the price of power transmitted through the Three-Hua UHV synchronous grid to examine whether the price is competitive at the receiving end. This can be performed by simplifying the UHV AC main grid as multiple transmission links that deliver power from large energy bases to load centers. By assuming the transmission links to be all AC links, all DC links, or a combination of AC and DC links, analyze the competitiveness of the electricity prices and assess the affordability. It should be noted that the UHV AC grid is merely simplified as transmission links here without regard to its networking function. If the calculation results show the price is still competitive, then it is demonstrated that the UHV AC grid has better economy.

c. Assess the contribution of the Three-Hua UHV synchronous grid to national economy. Evaluate the various benefits of the Three-Hua UHV synchronous grid quantitatively to assess its contribution to the national economy and social benefits using the input–output analysis method.

2. **Methodologies**

a. *Financial assessment.* Financial assessment of an electric power project is based on the current national finance and tax system and pricing system to analyze and estimate the benefits and costs of the project, prepare a financial statement, and analyze the ROI, profitability, and financial feasibility of the project. The related financial metrics mainly include IRR and payback period.

i. *Payback period (P_t)*: This refers to the time required to recoup the funds expended on a project, usually expressed in years. It is a major static financial metric used to assess the ROI of a project. It can be calculated based on the cumulative net cash flows (the total investment) using:

$$P_t = n_\mathrm{p} - 1 + \frac{|\mathrm{CI} - \mathrm{CO}|_{n_\mathrm{p}-1}}{(\mathrm{CI}-\mathrm{CO})_{n_\mathrm{p}}} \tag{11.4}$$

where

n_p—the year when the cumulative net cash flow becomes positive

$|\mathrm{CI} - \mathrm{CO}|_{n_\mathrm{p}-1}$—absolute value of cumulative net cash flow of the previous year

$(\mathrm{CI}-\mathrm{CO})_{n_\mathrm{p}}$—net cash flow of the current year.

Criterion: If P_t is equal to or less than P_0 (reference payback period in the electric power industry), then the project is considered financially feasible.

ii. *IRR*: The internal rate of return on an investment or project is the "annualized rate of return" at which the sum of NPV of cash flows during its construction and operating life is equal to zero. It is a major dynamic financial metric that reflects the profitability of an investment, which is given by:

$$NPV = \sum_{t=1}^{n} (CI - CO)_t (1 + IRR)^{-t} = 0 \qquad (11.5)$$

where

CI, CO—cash inflow and cash outflow of each year, respectively

NPV—net present value.

Criterion: If IRR is greater than the reference IRR in the electric power industry, then the project is financially feasible. A greater IRR means a higher profitability.

b. *Competitiveness of electricity price*. The competitiveness of electricity price is assessed based on the "additive" principle. Specifically, the price of electricity delivered to the receiving-end grid is calculated as the sum of three components: (i) on-grid tariff at the sending end, (ii) transmission price, and (iii) line loss price. If the price is lower than the benchmark on-grid tariff (as per data published by the National Development and Reform Commission) at the receiving end, then the electricity price is deemed competitive.

The construction of UHV grids allows higher transmission capacity across regions, thus optimizing allocation of resources over a larger area and making it possible for power transmission from western and northern regions to load centers in eastern and central regions. The price competitiveness of power in the receiving-end region is an important basis for determining the economy of a UHV grid. For a UHV grid, the competitiveness of electricity price can be summarized as follows:

i. Calculate the transmission price of a UHV AC grid. The UHV AC grid can be simplified as a number of decoupled equivalent networks linked by transmission lines that transmit power from large energy bases to load centers and calculate the transmission price of individual links (wheeling costs) within the operation period using the expected IRR.
ii. Calculate the transmission price of a single UHV DC link.
iii. Calculate the weighted average transmission price using the total amount of energy transmitted through UHV AC and UHV DC links.
iv. Calculate the price of electricity transmitted to the eastern and central regions based on the on-grid tariff at coal power bases and hydropower bases, and compare such price with the benchmark on-grid tariff in the receiving-end region. If the difference is greater than zero, then the electricity delivered through UHV AC is competitive at the receiving end. And the greater the difference, the more cost-effective the transmitted power is. Their relation is given by:

$$\text{Price at the receiving-end grid} = \text{on-grid tariff at the sending end} + \text{transmission cost} + \text{line loss cost}$$

Competitiveness of electricity price =
benchmark on-grid tariff at the receiving end −
price of electricity transmitted to the receiving-end grid ≥ 0

c. *Contribution to national economy*. Calculate the costs of a project and its net contribution to the national economy from the viewpoint of comprehensive factors to make a general assessment of the economic feasibility of the project. As for the Three-Hua UHV synchronous grid, first estimate the social resources required for construction of the grid and the economic, social, and ecological benefits that the completed grid will bring, prepare a cash flow statement, and calculate the economic IRR (EIRR) and economic net present value (ENPV) (as illustrated) to analyze its contribution to the national economy.

i. *ENPV*: The economic net present value is defined as the sum of NPVs of annual economic cash flows during the period of analysis using the social discount rate. It can be calculated by Eq. (11.6):

$$\text{ENPV} = \sum_{t=1}^{n} (B-C)_t (1+i_s)^{-t} \tag{11.6}$$

where
B—cash inflow
C—cash outflow
$(B-C)_t$—net cash flow within the *period t*
i_s—social discount rate
n—total number of years.

Criterion: If the ENPV of a prospective project is positive, then it should be considered as acceptable because it meets the profitability requirement at the social discount rate.

ii. *EIRR*: This is the "rate of return" at which the sum of NPV of annual cash flows over the period of analysis is equal to zero. It can be calculated by using Eq. (11.7):

$$\text{ENPV} = \sum_{t=1}^{n} (B-C)_t (1+\text{EIRR})^{-t} = 0 \tag{11.7}$$

Criterion: If EIRR is not smaller than the social discount rate, then the project is considered acceptable in terms of economic resources allocation.

11.3.4.2 Financial analysis

In accordance with the *Methods and Parameters for Economic Evaluation of Construction Projects* (the third edition) jointly issued by the National Development and Reform

Commission and the Ministry of Housing and Urban–Rural Development of the People's Republic of China, the profitability of the Three-Hua UHV synchronous grid is assessed by calculating the IRR, payback period, and other related financial metrics, assuming that the investment on this project will be recouped by charging the power transmitted through the grid (based on the energy price).

Basic parameters for calculation: The reference IRR is taken as 7%. Loan interest is calculated on a quarterly basis with the repayment period of 15 years. Added-value tax, urban construction, and maintenance tax, educational surcharge, and income tax are taken as 17%, 7%, 3%, and 25%, respectively. The ratio of statutory reserve funds and that of the public welfare fund are taken as 10% and 5%, respectively. As per national applicable policies, the ratio of capital to total investment is considered to be 20% (the other 80% is from bank loans).

It is estimated that the NPV of total investments on the Three-Hua UHV synchronous grid is worth RMB154.8 billion, the payback period is 11.6 years (excluding the construction period), and the IRR is 9%, indicating that this project has good financial viability.

11.3.4.3 Analysis of competitiveness of electricity prices

Estimate the competitiveness of electricity price for energy delivered through the Three-Hua UHV synchronous grid using basic parameters such as on-grid tariff at the sending end and average on-grid tariff at the receiving end.

On-grid tariff at the sending end: RMB0.20–0.28/kWh for coal bases in Xilin Gol League, Ningdong, and Hami based on the feasibility study reports of power plants at these bases, allowing for reasonable profit of the generating units; RMB0.28/kWh tentatively for coal power bases in central Shanxi, Shaanxi, Inner Mongolia, and Ningdong; RMB0.25/kWh in Xinjiang (the benchmark on-grid tariff); and RMB0.3/kWh for hydropower bases in west Sichuan taking into consideration the on-grid tariff during averaging over wet, normal, and dry seasons as well as the early study results for hydropower projects.

For the average on-grid tariff at the receiving end, calculate the weighted average on-grid tariff in the Three-Hua region using the estimated power consumption of each individual province in north China, central China, and east China in 2020 based on the latest electricity price stipulated by the National Development and Reform Commission in November 2011.

It is estimated that by 2020, the prices of energy delivered to the load centers in north, central, and east China regional grids for coal-fired power from the northern region and hydropower from west Sichuan are, respectively, RMB0.1033/kWh, RMB0.0909/kWh, and RMB0.0590/kWh and are lower than the average on-grid tariff of these three grids. As such, the Three-Hua UHV synchronous grid is highly competitive, as shown in Table 11.12.

Table 11.12: Competitiveness of the Three-Hua UHV Grid in Electricity Price by 2020 (RMB/kWh).

Item	Receiving Area	Price of Power Delivered to the Receiving-End Grid	Average Price of Power Delivered	Average On-Grid Tariff at the Receiving End	Competitiveness of Delivered Power
Coal power from the northern region through AC links	North China	0.3075	0.3297	0.4329	0.1033
Through DC links		0.3476			
Coal power from the northern region through AC links	Central China	0.3992	0.3767	0.4675	0.0909
Hydropower from west Sichuan through AC links		0.4092			
Through DC links		0.3651			
Coal power from the northern region through AC links	East China	0.4008	0.4053	0.4644	0.0590
Hydropower from west Sichuan through AC links		0.4280			
Though DC links		0.4062			

11.3.4.4 Contribution to national economy

The construction of the Three-Hua UHV synchronous grid will further expand the size of interconnected areas and allow optimization of energy resources across a larger region. This will likely reduce coal-related costs, save installed generation capacity, enhance utilization of hydropower and wind power, save land resources, and gain environmental benefits. An assessment is made regarding the contribution of the Three-Hua UHV synchronous grid to the national economy in terms of reasonable resource allocation using grid investments as inputs, benefits gained as outputs, and NPV as evaluation criterion.

The assessment should be performed based on the shadow price system. On the basis of financial analysis, the national economy cash flow statement is prepared after adjusting the resource price and deducting the transfer payment, including taxes, interest, and subsidies.

The analysis on inputs and outputs for construction of the Three-Hua UHV synchronous grid is as follows:

1. *Inputs*. Mainly consisting of the construction investment of RMB980 billion, in addition to operating costs.

2. *Outputs*. The benefits of the grid are calculated using "with-and-without" comparison and analysis methods. For details, see Section 11.3.5.

Assuming a calculated period of 25 years and a base discount rate of 8%, it is estimated that the national economy IRR of the Three-Hua UHV synchronous grid is 21% (greater than the base discount rate), and the NPV is as high as RMB482 billion. As such, the Three-Hua UHV synchronous grid has a higher ROI and input–output ratio.

11.3.5 Social Benefits of Three-Hua UHV Synchronous Grid

With the Three-Hua UHV synchronous grid, the power from coal power bases can be directly delivered to load centers, thereby reducing coal-related costs. The maximum load level can be reduced and, hence, less installed capacity is required. The hydropower from the southwest can be delivered to load centers in the Three-Hua regions, thereby achieving mutual supplement of hydropower and thermal power and reducing hydropower loss resulting from abandoned water. Peak-shaving demands can be reduced, and more wind power can be accommodated while maintaining the peak-shaving capability of the existing units. Precious land resources in the eastern and central regions can be saved, thereby gaining benefits from price difference in land. It is possible to optimize and coordinate the environmental resources across the eastern, central, and western regions, thus minimizing the environmental losses.

The Three-Hua UHV synchronous grid can bring about the following social benefits:

1. Clean energy will be utilized efficiently, contributing to significant effects of energy conservation and emission reduction. By 2020, the grid will be able to accommodate all the hydropower produced in the southwest in wet seasons, reducing water abandonment by 3.43×10^{10} kWh, which is equivalent to saving 1.2×10^7 tons of standard coals, reduction in carbon dioxide emission by 3×10^7 tons, sulfur dioxide by 1.37×10^4 tons, nitrogen oxide by 1.51×10^4 tons, and smoke dust by 3.4×10^3 tons (combustion of 1 kg of standard coals will emit 2.493 kg of carbon dioxide, and generation of 1 kWh of power will emit 0.4 g of sulfur dioxide, 0.44 g of nitrogen oxide, and 0.1 g of smoke).

   After this grid is constructed, the planned wind power from a larger area can be accommodated efficiently and loss of wind power due to wind curtailment can be reduced by 6×10^{10} kWh (including 1.5×10^9 kWh from Hebei, 1.95×10^{10} kWh from west Inner Mongolia, 9×10^9 kWh from Jilin, 8.5×10^9 kWh from Heilongjiang, 1.55×10^{10} kWh from east Inner Mongolia, 2.5×10^9 kWh from Gansu, and 3.5×10^9 kWh from Xinjiang, equivalent to combustion of 2.04×10^7 tons of standard coals), which is equivalent to reduction in carbon dioxide emissions by 5.085×10^7 tons, sulfur dioxide by 2.4×10^4 tons, nitrogen oxide by 2.64×10^4 tons, and smoke dust by 6×10^3 tons.

The reserve capacity of the system can be reduced through mutual supplement of hydropower and thermal power and the measures of peak load shifting. By 2020, the total loads of the Three-Hua grid are expected to reach 956.69 GW (with the maximum loads occurring in August), and 795.98 GW of thermal power will be required. Compared with three independent grids, the Three-Hua synchronous grid can contribute to a reduction of approximately 31.76 GW of installed capacity.

2. Allocation of environmental resources nationwide can be optimized, thereby minimizing environmental losses in the eastern and central regions. The expansion of grid interconnection scale will take full advantage of large grids and boost intensive development of large energy bases and development of the western region to convert the resource advantages in the western region into economic advantages. By constructing power plants in the western region and transmitting power from there to the eastern and central regions, it is possible to save 5000 hm^2 of highly valued land in the eastern and central regions. In this way, the coordinated economic and social development between the eastern and western regions can be achieved.

   Taking into consideration the population density and level of economic development in the sending-end area and receiving-end area, and assuming that an additional 230 GW of coal power will be transmitted to the eastern and central regions by 2020, the Three-Hua UHV synchronous grid could reduce emissions of sulfur dioxide by 5.5×10^5 tons, nitrogen oxide by 5.6×10^5 tons, and smoke dust by 1.2×10^5 tons annually, and reduce environmental losses worth approximately RMB4.5 billion annually in the eastern and central regions.

3. Optimization of energy mix and energy transportation system. The inter-regional power transmission capacity will exceed 450 GW, enabling 350 GW of power to be transmitted to the eastern and central regions, helping optimization of energy mix and meeting energy demand for sustainable economic and social development in these regions. The share of electric power transmission in energy transportation will increase notably. From 2011 to 2020, the ratio of energy transported in the form of electricity and coal from the four large coal bases in Shanxi, Ordos, east Inner Mongolia, and Xinjiang will decline from 14:1 to 3.4:1, and the amount of coal transported by road networks will decline from 4.0×10^8 tons to 1.0×10^8 tons. This will significantly improve the energy security and energy transportation efficiency, thereby reducing the effective cost of coal-fired generation and power supply. Under the Three-Hua UHV synchronous grid plan, an additional 230 GW of coal power would be transported through the grid by 2020. It is estimated that if these plant were built at the receiving end, then the generation cost would be approximately RMB322 billion, and if they were built at the sending end, then the generation cost would decrease to approximately RMB172.5 billion, resulting in a savings of RMB149.5 billion.

   Taking into account the transmission costs, the tariff of electricity delivered to the eastern and central regions from major energy bases is RMB0.06–0.11/kWh lower than the average feed-in tariff for local coal power, effectively helping suppress the increase of coal price and resulting in an increase in electricity price at the receiving-end areas. It is estimated that approximately RMB70 billion of power supply cost can be saved annually across the eastern and central regions.

4. Promoting economic development in the west and facilitating coordinated development in the eastern and western regions. Construction of the Three-Hua synchronous grid will boost intensive development of coal bases in that coal resources are converted into electricity and then transmitted externally through the UHV network. This would extend the value chain of coal development and utilization, and contribute to economic and social development in the western region. The ratio of contribution to local GDP growth from transmitting electricity (generated from coal locally) and transporting coal directly is 6:1, and that to the creation of job opportunities is 2:1. By 2020, the western regions will see an increase in cumulative investment of RMB1.45 trillion in total and tax revenue of RMB68 billion annually. The plan will help stimulate domestic consumption, drive economic growth, and promote development in high-tech industries, such as new energy and new materials, and in emerging industries, such as The Internet of Things and electric automobiles. During the period from 2013 to 2020, investments in smart grids will increase by RMB50 billion annually, contributing to the gross social production value by RMB194 billion, creating 140,000 jobs and increasing people's income by RMB3.5 billion annually.

References

[1] L. Zhenya, Electric Power and Energy in China, China Electric Power Press, Beijing, 2012.
[2] L. Zhenya, Z. Qiping, Studies on state grid development mode, Proc. CSEE 33(7) (2013) 1–10.

CHAPTER 12

UHV Engineering Practices in China

Chapter Outline

Ultra-high Voltage AC/DC Grids.
DOI: http://dx.doi.org/10.1016/B978-0-12-802161-3.00012-3

12.1 UHV AC Transmission Projects

12.1.1 1000-kV Jindongnan–Nanyang–Jingmen UHV AC Pilot and Demonstration Project

12.1.1.1 Project overview

UHV AC power transmission is a complex and systematic project, the technology and equipment of which cannot be fully examined and verified by laboratory tests or short experimental lines. It is therefore necessary to build a pilot and demonstration project to prove the UHV AC technologies so that they can be promoted and to avoid severe loss due to potential risks in system design and equipment manufacturing.

The purpose of this project is to improve capabilities in system design and equipment manufacturing and to assess the equipment performance and verify system design parameters through tests under rated voltage, power, and parameters. The project will begin commercial operation after it is demonstrated to be a success, and more UHV AC projects will be built on this basis.

Based on China's power development plan, it is justifiable to choose the Jindongnan–Jingmen project as the pilot and demonstration project. This project is consistent with the mainstream energy delivery direction in China. After completion of the project, its system and equipment are evaluated for a certain period and improved. It is then turned into commercial operation as an important energy transmission channel in China. This project links the north China and central China grids, both of which are robust networks. All power sources in the whole power network could be conveniently organized and the power flow in the transmission line could be flexibly controlled to conduct tests of the pilot and demonstration project under rated voltage and power to verify the capability of equipment, such as transformer and circuit breaker, to withstand large current. However, it does not impede the export of existing power sources or affect the normal operation of the power grid, thus showing strong adaptability.

On August 9, 2006, the project was commenced after approval; the schematic transmission line route, shown in Figure 12.1, consists of Changzhi and Jingmen substations, Nanyang Switching Station, and Changzhi–Nanyang–Jingmen single-

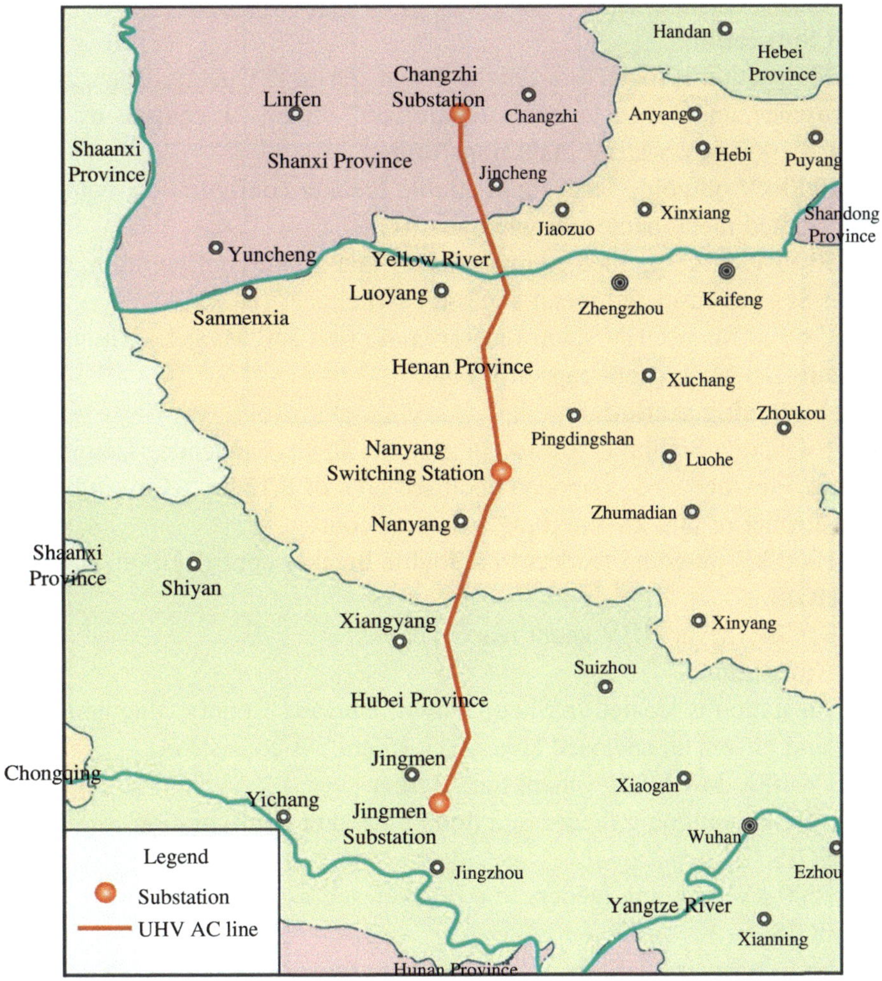

Figure 12.1
Schematic diagram of the UHV AC pilot and demonstration project.

circuit transmission line. The two 1000-kV substations are equipped with a main transformer rated 3000/3000/1000 MVA and 1050/525/110 kV. The Changzhi–Nanyang line segment is approximately 360 km long, and the Nanyang–Jingmen is approximately 280 km. Changzhi–Nanyang Circuit I is equipped with 960 and 720 Mvar HV shunt reactors, respectively, on the Changzhi and Nanyang sides. Nanyang–Jingmen Circuit I is equipped with 720 and 600 Mvar HV shunt reactors, respectively, on the Nanyang and Jingmen sides. Both substations are equipped with four LV shunt capacitor banks (210 Mvar per bank) and two LV shunt reactor banks (240 Mvar per bank) [1].

12.1.1.2 Substation and switching station

1. **Changzhi Substation**
   Changzhi Substation is located in Shizhe Town, Zhangzi County, Changzhi City, Shanxi Province, and covers an enclosed area of 7.78 hm^2. It consists of:
   - One 3 × 1000-MVA UHV main transformer
   - One 1000-kV outgoing feeder in a double breaker configuration, equipped with SF_6 gas-insulated metal-enclosed switchgear (GIS)
   - Five 500-kV outgoing feeders in a breaker-and-a-half configuration, equipped with hybrid SF_6 gas-insulated metal-enclosed switchgear (HGIS).
   - One 3 × 320-Mvar UHV shunt reactor bank, two 240-Mvar LV shunt reactor banks, and four 210-Mvar shunt capacitor banks
2. **Nanyang Switching Station**
   Nanyang Switching Station is located in Zhaohe Town, Fangcheng County, Nanyang City, Henan Province, and covers an enclosed area of 8.15 hm^2. Currently, with no main transformer or 500-kV outgoing feeder, it consists of:
   - Two 1000-kV outgoing feeders in a double breaker configuration, equipped with HGIS.
   - Two 3 × 240-Mvar UHV shunt reactor banks
3. **Jingmen Substation**
   Jingmen Substation is located in Shenji Town, Shayang County, Jingmen City, Hubei Province, and covers an enclosed area of 11.45 hm^2. It consists of:
   - One 3 × 1000-MVA UHV main transformer
   - One 1000-kV outgoing feeder in a double breaker configuration, equipped with HGIS
   - Three 500-kV outgoing feeders in a breaker-and-a-half configuration, equipped with HGIS
   - One 3 × 200-Mvar UHV shunt reactor bank, two 240-Mvar LV shunt reactor banks, and four 210-Mvar shunt capacitor banks

12.1.1.3 Transmission line

The transmission line is totally 640 km long, crossing 22 administrative regions of three provinces, namely, Shanxi, Henan, and Hubei. The Jindongnan–Nanyang section, which is 360 km long, starts from Changzhi Substation in Shanxi, passes through Zhangzi County, Qinshui County, and Zezhou City in Shanxi, Qinyang City, Bo'ai County, Wen County, Mengzhou City, Gongyi City, Yanshi City, Yichuan City, Ruzhou City, Baofeng County, Lushan County, Nanzhao County, and Fangcheng County in Henan, and ends at the Nanyang Switching Station. The Nanyang–Jingmen section, which is 280 km long, starts from Nanyang Switching Station, passes through Sheqi County, Tanghe County, Zaoyang City, Xiangfan City, Yicheng City, Zhongxiang City, and Shayang County in Hubei, and ends at the Jingmen Substation. There are 1284 towers along the line, including 1275 regular towers, five towers of large-crossing line over the Yellow River, and four towers of the large-crossing line over the Han River. For a regular line, there are 49 subcategories under four categories, namely, cup type tower, cat-head type tower, "干" type tower, and portal tower, averaging 77.2 m and 70.5 tons. The 8 × LGJ-500/35 ACSR conductors are used for a regular line, with 8 × LGJ-630/45 (3.946 km) for Henan Taihangshan Macaca mulatto chinensis National Nature Reserve. One JLB20A-170 aluminum-clad steel conductor and one OPGW-175 fiber optic cable serve as the ground wire.

The large-crossing section over the Yellow River is located in Mengzhou City, Henan Province, at the north segment. It is in a "tension tower–suspension tower–suspension tower–suspension tower–tension tower" arrangement with a 1.22-km main span and a 3.651-km tension segment. There are two "干"-type angle steel tension towers (the height from lowest cross-arm to ground: 38 m; total height: 68 m) and three cup-type steel tubular crossover suspension towers (the height from lowest cross-arm to ground: 112 m, total height: 122.8 m). The large-crossing section over the Han River is located in Zhongxiang City, Hubei Province, at the south segment. It is in a "tension tower–suspension tower–suspension tower–tension tower" arrangement with a 1.65-km main span and a 2.956-km tension segment. There are two angle tension towers (the height from lowest cross-arm to ground: 40 m, total height: 72 m) and two crossover suspension towers (the height from lowest cross-arm to ground: 168 m, total height: 181.8 m). The 6 × AACSR/EST-500/230 extra-high-strength aluminum alloy conductor steel reinforced (AACSR) is used for this large crossing, with one JBL20-240 aluminum-clad steel conductor and one OPGW-24B1-254 fiber optic cable as the ground wire.

12.1.1.4 Commissioning and operation

System commissioning is the last checkpoint before the UHV AC project is put into operation. At this stage, the performance of the primary and secondary equipment and the operational control characteristics of the system are fully examined and tested. Also, this provides the required technical experience for the acceptance and operation of the project. Through analyzing the characteristics of the project and its requirements for test and assessment, a feasible commissioning plan is developed and, according to this plan, the system is rigorously, fully commissioned. The commissioning results are analyzed and evaluated to provide a reliable basis for putting the project into operation. The commissioning was conducted from December 2008 to January 2009. All 15 test items are completed and are summarized as follows:

1. Test for raising generator voltage/current from zero to energize the following apparatuses: the main transformers at both Jingmen and Changzhi substations; Changzhi–Nanyang Circuit I and associated HV shunt reactors on both sides; and Nanyang–Jingmen Circuit I and associated HV shunt reactors on both sides. All passed the test of raising the voltage to the rating.
2. The UHV transformers, circuit breakers, UHV line, HV shunt reactors, and LV compensation equipment were subjected to and passed multiple switching impulse overvoltage tests.
3. The secondary system was subjected to a comprehensive check and immunity test, the results of which show that the system and auxiliary equipment are operating normally.
4. The UHV system withstood the tests of interconnection line power control, dynamic system disturbance, heavy load, and artificial single-phase instantaneous ground fault.
5. The results of the electromagnetic environment test met the requirements of the State Environmental Protection Administration (now the Ministry of Environmental Protection).
6. On December 29, 2008, the active power on the Changzhi side was 2828 MW, and that on the Jingmen side was 2780 MW, representing a line loss rate of approximately 1.7% over the UHV line (including the UHV shunt reactors). To transfer 1000-MW power, the line loss rate of a 500-kV line having the same length is approximately 5%.

The commissioning results show that the primary and secondary equipment have good performance and meet the requirements of technical specifications. The voltage and power control of the UHV interconnection function well and withstand tests such as interconnection line power control, dynamic system disturbance, heavy load, and single-phase ground fault. The project is electromagnetically friendly and has a much smaller line loss compared with a 500-kV transmission line. The whole project and the UHV interconnection are sound and ready for operation.

The project has been operating well since being put into operation and has become an important north–south energy transmission channel in China. From 2009 to 2011, it transferred 277.1×10^8 kWh of electricity, with 88.6×10^8, 117.9×10^8, and 70.6×10^8 kWh in 2009, 2010, and 2011, respectively. During this period, it delivered 185×10^8 kWh thermal power from north China to central China in lieu of approximately

6×10^6 tons of coal, mitigating the power shortage in central China. In addition, it transferred 92×10^8 kWh hydropower from central China to north China, equivalent to a reduction of 55×10^8 kWh energy loss due to abandoned water, a savings of more than 3×10^6 tons coal, and a reduction of more than 9×10^6 tons of CO_2 emissions.

12.1.2 1000-kV Jindongnan–Nanyang–Jingmen UHV AC Expansion Project

12.1.2.1 Project overview

The 1000-kV Jindongnan–Nanyang–Jingmen UHV AC Pilot Project has a weak electrical connection due to its single-circuit configuration. Its transmission capability is not fully exploited. By expanding the UHV substations (switching stations) in Changzhi, Nanyang, and Jingmen, and by installing series compensation devices, the capacity constraint of the main transformers is eliminated, the voltage and reactive power support are strengthened, and the transmission capability is increased to more than 5000 MW.

After completion of the expansion, Changzhi, Nanyang, and Jingmen UHV substations are each equipped with two main transformers, further improving the security and reliability. Through 500 kV system in Nanyang UHV sustation, Henan grid is directly connected with UHV AC Pilot Project. This project provides an additional UHV AC channel to the existing four 500-kV interconnection lines between Hubei and Henan grids, enabling a closer grid connection, more flexible operation, and higher security and reliability of central China grid.

The four provinces in eastern Central China (Henan, Hubei, Hunan, and Jiangxi) have favorable geological locations and huge economic growth potential, but they lack primary energy resources. With a coal reserve accounting for only 4.5% of the nation's total reserve, they are heavily dependent on coal from other regions. In recent years, the power demand in central China has increased rapidly. However, due to the limited transport capacity of coal, the power shortage during dry seasons has become increasingly prominent. Expanding the pilot and demonstration project can fully exploit the power transmission capability, effectively ease the power shortage in central China, and meet the continuously and rapidly increasing power demand.

On December 29, 2010, the 1000-kV Jindongnan–Nanyang–Jingmen UHV AC Expansion Project (the expansion project) was commenced after approval.

The expansion project mainly involves:

- One fixed series compensation bank with a compensation degree of 20% on both sides of Changzhi–Nanyang Circuit I
- One fixed series compensation bank with a compensation degree of 40% on the Nanyang side of Nanyang–Jingmen Circuit I

- One transformer at Changzhi Substation, equipped with four LV shunt capacitor banks and two LV shunt reactor banks
- Two transformers at Nanyang Switching Station that are expanded to a substation, each equipped with four LV shunt capacitor banks and two LV shunt reactor banks on the LV side
- One transformer at Jingmen Substation, equipped with four LV shunt capacitor banks and two LV shunt reactor banks on the LV side [1]. The transmission line is not expanded.

12.1.2.2 Substation

1. **Changzhi Substation**
   At the Changzhi Substation, one 1000-kV main transformer of 3000 MVA with two 500-kV outgoing bays are added. On the Jindongnan side of the Jindongnan–Nanyang line segment, series compensation devices with a compensation degree of 20% and a capacity of 1382 Mvar are installed, 0.76 km line is retrofitted or removed, and 0.5 km is built with a conductor cross-sectional area of 8×500 mm^2.
2. **Nanyang Substation**
   At the Nanyang Substation, two 1000-kV main transformers of 2×3000 MVA, one standby single-phase main transformer of 1000 MVA, and four 500-kV outgoing bays (two to Baihe Substation and two to Pingdingshan Substation) have been built. On the Nanyang side of Jindongnan–Nanyang 1000-kV line segment, series compensation devices with a compensation degree of 20% and a capacity of 1382 Mvar are installed, 0.55 km of line is retrofitted or removed, and 0.31 km is built with a conductor cross-sectional area of 8×500 mm^2. On the Nanyang side of Nanyang–Jingmen 1000 kV line segment, series compensation devices with a compensation degree of 40% and a capacity of 2160 Mvar are installed, 0.91 km of line is retrofitted or removed and 0.55 km built with a conductor cross-sectional area of 8×500 mm^2.
3. **Jingmen Substation**
   At the Jingmen Substation, one 3000-MVA main transformer is installed, equipped with two LV shunt reactor banks (240 Mvar per bank) and four LV shunt capacitors (210 Mvar per bank) on the LV side.

12.1.2.3 Commissioning and operation

Commissioning of the expansion project was conducted from November 7 to December 9, 2011. All 39 test items were completed and are summarized as follows:

1. The UHV transformers, circuit breakers, line, UHV shunt reactors, and LV compensation devices were subjected to and passed multiple switching impulse tests.
2. The first UHV series compensation device in the world, independently developed by China, passed the platform energization test, proper energization test, interlocking test

between the bypass circuit breaker and line protection, on-load test, artificial single-phase instantaneous short-circuit test, and 5000-MW heavy load test, the results of which show that the device is functional and meets the requirements of the project.

3. All substation control and protection systems, including system splitting devices, steady-state overvoltage control system, and other safe controls, are subjected to comprehensive check and immunity test, the results of which show that the system and auxiliary equipment are operating normally.
4. The UHV interconnected transmission system passes tests under various operating modes and conditions, such as 1000-kV connection/disconnection, 1000/500-kV loop closing/opening, and heavy load test, and is subjected to an artificial instantaneous single-phase-to-ground short-circuit test, during which the system is operating stably, indicating that the UHV grid has good dynamic operating characteristics and strong anti-disturbing capability.
5. During the 5000 MW normal heavy load test, the maximum power transferred by Changzhi–Nanyang Circuit I reaches approximately 5726 MW (measured at the Changzhi Substation). During the 5000-MW special heavy load test, the maximum power transferred by Changzhi–Nanyang Circuit I reaches approximately 5474 MW (measured at Changzhi Substation) and that transferred by the Nanyang–Jingmen Circuit I reaches approximately 5391 MW (measured at Nanyang Substation).
6. All electromagnetic environmental factors meet the requirements given in the approval of the Ministry of Environmental Protection on the environmental impact assessment.

The commissioning results show that the primary and secondary equipment have good performance and meet the requirements of technical specifications. The voltage and power control of the UHV interconnection function well and withstand tests such as loop closing/opening, connection/disconnection, artificial single-phase ground fault, and 5000-MW heavy load test. The project is electromagnetically friendly and meets the environmental protection requirements. The whole project and the UHV interconnection are sound and ready for a test run.

The expansion project was completed and passed the acceptance test on October 31, 2011. The system commissioning was performed in early December, leading to the operation stage. It has been operating well since then. By the end of 2012, it had transferred 409.01×108 kWh of electricity.

12.1.3 1000-kV Huainan–Shanghai UHV AC Demonstration Project

12.1.3.1 Project overview

With abundant coal resources, Huaibei and Huainan are energy bases for east China. With rapid economic growth, Yangtze River Delta is the load center of the east China grid. Because of the severe shortage of primary energy, it has to turn to power from outside the region. Building coal-fired power plants in Huaibei and Huainan to supply the Yangtze River Delta conforms to China's overall strategic energy plan. It can meet the increasing power demand and promote the economic development in the delta region.

By the end of the 11th Five-Year Plan period, there was 8.55 GW of operating power sources and more than 6 GW under early construction in the Huainan coal power base. Anhui's 500-kV grid had established three river-crossing channels for "north-to-south power transmission" and four exporting lines, with each having a transmission capability of approximately 7 GW. These lines can basically meet the need of the generating units in Anhui to export power to east China. However, with further development of power sources, more transmission lines are required to deliver additional power from Huaibei and Huainan.

Considering the rapid load growth in the east China grid, even with all the approved or permitted power source projects being operational, the power shortage in Shanghai, Jiangsu, and Zhejiang still cannot be bridged during the 12th Five-Year Plan period. To enable exporting of new generating units in Huaibei and Huainan, and to enhance the capacity and reliability of the Yangtze River Delta grid in receiving power, it is necessary to build a bulk AC transmission link from Huainan to Shanghai as soon as possible. As planned, during the 12th Five-Year Plan period, the east China grid will gradually form a UHV loop in the delta and connect with central China and north China grids through UHV AC links. Huainan–Shanghai UHV AC Transmission Project is an important part of the UHV loop network in the Yangtze River Delta region. It helps further improve the capability of the regional grid to accommodate imported power and to withstand severe faults on the Xiangjiaba–Shanghai ±800-kV UHV DC Demonstration Project and other DC projects, optimize allocation of resources, and minimize line corridors.

Compared with the UHV AC pilot and demonstration project, this project has the following two features:

1. Point-to-grid connection scheme. Different from the pilot and demonstration project, this project has a number of 500-kV tie-lines at cross-provincial interfaces that enable a close connection between province networks. Therefore, the power fluctuation on the UHV line is not a key consideration when reviewing the project's transmission capability and configuring its security controls.

   The point-to-grid connection scheme simplifies the security control strategy such that most faults on the UHV line or main transformer can be cleared by disconnecting the sending-end thermal power units connected to the Huainan Substation.
2. Double-circuit arrangement. The double-circuit arrangement greatly increases the transmission capability of the UHV project and reduces the risk of widespread transfer of power flow caused by the complete disconnection of the UHV line after fault. If a fault occurs on a single UHV circuit, then 80–90% of the power flow will be

transferred to the healthy circuit. However, according to DL 755—2001 *Guide on Security and Stability for Power System* and the estimated actual risks, it is necessary to consider security controls for an *N*-2 fault on the UHV line.

On September 27, 2011, the project was commenced after approval. This project will form a UHV transmission channel in the east China grid, which, in addition to the 500-kV exporting lines, will enable a delivery capability of up to 14 GW by the end of 2014. The transmission line, crossing Anhui, Zhejiang, and Shanghai, is 2 × 647.4 km long and includes the Huainan–Wuhu segment, which is 2 × 333.6 km, the Wuhu–Anji segment, which is 2 × 151.9 km, and the Anji–Shanghai segment, which is 2 × 161.9 km, as shown in Figure 12.2.

The Huainan–Shanghai project has four UHV substations, namely, Huainan, Wuhu, Anji, and Liantang. Wuhu Substation has a capacity of 1 × 3000 MVA, whereas the other three each have a capacity of 2 × 3000 MVA. The 500-kV outgoing feeders of Huainan Substation are connected to Yuanzhuang and Pingwei Power Plants. Power generated by two 1000-MW units of Pingwei Power Plant Phase III is directly transferred to the 1000-kV transmission line. Two 500-kV circuits from Wuhu Substation are connected to Chucheng

Figure 12.2

Schematic diagram of the Huainan–Shanghai power transmission project.

Substation as a support for the channel. Four large-section conductors of outgoing feeders from Anji Substation transfers power to the Zhebei main grid. Liantang Substation's main transformers are directly connected to the 500-kV bus and transfer power to the Shanghai grid through four 500-kV large-section conductors of lines.

12.1.3.2 Substation

1. **Huainan Substation**
   Huainan UHV Substation is located in Pingwei Town, Panji District, 15 km northwest of Huainan City, Anhui Province, covering an enclosed area of 9.56 hm^2. Currently, it consists of:
   - Two 3000-MVA main transformers (three transformers total)
   - Two 1000-kV outgoing feeders (both to Wannan Substation; eight circuits total) in a breaker-and-a-half configuration, forming one complete string and two incomplete strings, equipped with GIS with seven circuit breakers
   - Four 500-kV outgoing feeders (two to Tianji and Pingwei power plants, respectively; eight circuits total) in a breaker-and-a-half configuration, forming two complete strings and one incomplete string, equipped with GIS with four circuit breakers
   - Two 720-Mvar fixed UHV shunt reactor banks connected to the outgoing feeders to Huainan Substations
   - 110-kV single bus connection on the LV side of transformer
   - One 240-Mvar LV reactor bank on the 110-kV side of each main transformer
2. **Wuhu Substation**
   Wuhu UHV Substation is located in Hongyang Town, Wuhu County, 35 km south of Wuhu City, Anhui Province, covering an enclosed area of 14.43 hm^2. Currently, it consists of:
   - One 3000-MVA main transformer (three banks total, with space for future expansion)
   - Four 1000-kV outgoing feeders (two to Huainan and Zhebei substations, respectively; eight circuits total) in a breaker-and-a-half configuration, forming one complete string and three incomplete strings, equipped with GIS with nine circuit breakers
   - Two 500-kV outgoing feeders (to Wuhu Substation; eight circuits total) in a breaker-and-a-half configuration, equipped with GIS with two circuit breakers
   - Two 720-Mvar HV shunt reactor banks and one 600-Mvar HV shunt reactor bank connected to the outgoing feeders to Huainan and Zhebei substations, respectively
   - 110-kV single bus connection on the LV side of the transformer
   - Two 240-Mvar LV reactor banks and four 210-Mvar LV capacitor banks on the 110-kV side of the main transformer

3. **Anji Substation**
   Anji UHV Substation is located in Meixi Town, Anji County, 52 km southwest of Huzhou City, Zhejiang Province, covering an enclosed area of 14.98 hm^2. Currently, it consists of:
   - Two 3000-MVA main transformers (four banks total).
   - Four 1000-kV outgoing feeders (two to Wannan and Huxi substations, respectively; eight circuits total) in a breaker-and-a-half configuration, forming two complete strings and two incomplete strings, equipped with GIS with ten circuit breakers.
   - Four 500-kV outgoing feeders (two to Hangbei and Miaoxi substations, respectively; 10 circuits total) in a breaker-and-a-half configuration, equipped with GIS with four circuit breakers
   - One 720-Mvar UHV shunt reactor bank connected to the outgoing feeders to Wannan and Huxi substations, respectively
   - 110-kV single bus connection on the LV side of the transformer
   - One 240-Mvar LV reactor bank and two 210-Mvar LV capacitor banks on the 110-kV side of the main transformer
4. **Liantang Substation**
   Liantang UHV Substation is located in Liantang Town, Qingpu District, 45 km southwest of Shanghai, covering an enclosed area of 8.98 hm^2. Currently, it consists of:
   - Two 3000-MVA main transformers (four banks total)
   - Two 1000-kV outgoing feeders (to Zhebei Substation; four circuits total) in a breaker-and-a-half configuration, forming one complete string and two incomplete strings, equipped with GIS with seven circuit breakers
   - 500-kV outgoing feeders in a breaker-and-a-half configuration jointly built with Liantang 500-kV Substation, 500-kV side of 1000-kV main transformer directly connected to the fourth and ninth strings of the 500-kV switchgear, equipped with HGIS with two circuit breakers
   - One 720-Mvar HV shunt reactor bank connected to the outgoing feeder to Zhebei Substation
   - 110-kV single bus connection on the LV side of transformer
   - One 240-Mvar LV reactor bank and two 210-Mvar LV capacitor banks on the 110-kV side of each main transformer

12.1.3.3 Transmission line

The project starts from Huainan Substation, passes through Huainan City, Changfeng County, Hefei City, Feixi County, Lujiang County, Chaohu City, Wuwei County, Fanchang County, Tongling County, Nanling County, Wuhu County, Xuanzhou District, Langxi County, and Guangde County in Anhui, Anji County, Changxing County, Huzhou urban area, and Jiaxing City in Zhejiang, Wujiang City in Jiangsu, and Qingpu and Songjiang Districts in Shanghai, and it ends at Huxi Substation. The transmission line is

648.7 km long with 1421 towers. The Huainan–Wannan segment is 334.4 km long, the Wannan–Zhebei segment is 151.7 km, and the Zhebei–Shanghai segment is 162.6 km. The line has a double-circuit arrangement supported by steel tubular towers averages 108 m and 185 tons. The conductors used are 8 × LGJ-630/45 ACSR conductors. One LBGJ-240-20AC aluminum-clad steel conductor and one OPGW-240 fiber optic cable serve as the ground wire.

The large crossing section over the Huai River is located 600 m north of Yingou Gate on the left bank of the Huai River in the northeast of Huainan. It is in a "tension tower–suspension tower–suspension tower–tension tower" arrangement with a span of 520, 1300, and 600 m, respectively, and a 2420-m tension segment. There are four tension towers (the height from lowest cross-arm to ground: 40 m, total height: 75 m) and two crossover suspension towers (the height from lowest cross-arm to ground: 131 m, total height: 197.5 m).The large crossing section over the Yangtze River is located approximately 500 m upstream of the Digang large crossing section along the Three Gorges Right Bank–Shanghai ±500-kV DC Transmission Line. It is in a "tension tower–suspension tower–suspension tower–tension tower" arrangement with a span of 710, 1817, and 653 m, respectively, and a 3180-m tension segment. There are four tension towers (the height from lowest cross-arm to ground: 39 m, total height: 74 m) and two crossover suspension towers (the height from lowest cross-arm to ground: 206 m, total height: 277.5 m). The 6 × AACSR/EST-640/290 extra-high-strength AACSR conductors are used for the large crossing section, with two OPGW-350 fiber optic cables as the ground wire.

12.1.3.4 Commissioning and operation

In a comprehensive and rigorous manner, 38 system commissioning items under six categories were conducted, including switching on/off no-load transformer on the 500-kV side/1000-kV side, switching on/off no-load line, loop closing/opening, artificial short-circuit to ground (under three operating conditions, namely, "two circuits in operation," "one in operation and one under maintenance," and "one in operation and one on hot standby," among which the third is the most severe and used for the first time), and loop closing test, as well as 13 types of tests such as on-load voltage regulation test, ferroresonance test, very fast transient overvoltage (VFTO) test, transient voltage and current test, and electromagnetic environment test. During the commissioning and test run, all functional indices meet the design requirements, and the system and equipment are operating normally. On September 25, 2013, the project was put into operation.

Using the world's leading comprehensive simulation analysis system, the power flow, steady-state security, large-disturbance transient stability, small-disturbance stability, severe fault withstand capability, security and stability measures, AC/DC interaction, and multiterminal DC feed system security of the UHV grid are fully analyzed, and corresponding measures and control requirements for safe and stable operation are proposed. After the project begins operation, the security and stability of east China grid is further improved to fully meet the requirements of guidelines on security and stability.

12.2 UHV DC Transmission Projects

12.2.1 Xiangjiaba–Shanghai ±800-kV UHV DC Demonstration Project

12.2.1.1 Project overview

To export electricity from Xiangjiaba, Xiluodu, and other hydropower stations in the lower reaches of Jinsha River according to the west-to-east power transmission strategy and improve China's innovative and manufacturing capabilities, the voltage class of ±800 kV is introduced over ±500 kV. On April 26, 2007, Xiangjiaba–Shanghai ±800-kV UHV DC Demonstration Project (Xiangjiaba–Shanghai project) was commenced after approval.

The project is rated ±800 kV, 4000 A, and 6400 MW. Fulong Converter Station (the sending end) is located in Jinsha River area, Sichuan, and has nine 500-kV AC outgoing feeders: two to Xiangjiaba Left Bank Power Plant, two to Right Bank, two to Xiluodu Left Bank Converter Station, and three to Luzhou Substation. Figure 12.3 shows a photo of the Fulong Converter Station.

The Fengxian Converter Station (the receiving end) is located in Pudong New Area, Shanghai, with four planned 500-kV AC outgoing feeders: two to Nanhui Substation and

Figure 12.3
Fulong converter station.

Figure 12.4
Fengxian converter station.

two to Sanlin Substation. The current phase consists of three 500-kV AC outgoing feeders to Nanhui Substation. Figure 12.4 shows a photo of the Fengxian Converter Station.

This project has two complete monopoles, each consisting of two 400-kV 12-pulse converters connected in series. This way, any 12-pulse converter quitting service does not refrain the remaining converters from operating as an incomplete monopolar system. To this end, each 12-pulse converter is connected in parallel with one bypass circuit breaker.

The smoothing reactors on the DC side are dry-type. Four 75-mH coils are arranged for each pole at each station, two on the pole bus and two on the neutral bus. To reduce radio interference produced by the DC system, RI trap inductors are provided at the outlet of the HV end on the DC side of the converter and RI filter capacitors on the AC side of the converter transformer. A triple-tuned DC filter bank is provided at each end of each pole. For easier repair and maintenance, DC filters are equipped with a disconnector for live switching. A power frequency blocking filter is provided on the neutral bus of each pole at the sending end.

As the first large-capacity UHV DC project, the Xiangjiaba–Shanghai project involves a lot of groundbreaking work. It introduces the scheme of connecting two 12-pulse converters in series and splitting smoothing reactors equally on the pole bus and the neutral bus. This scheme is applied in subsequent projects as a typical $\pm$800-kV design. The introduction of 6-inch thyristor converter valves lays a foundation for enhanced transmission capacity of subsequent UHV DC projects. The 50- and 100-Hz resonance suppression measures on the

DC side are studied and a solution is proposed: installing 50-Hz DC blocking filters on the neutral bus and 100-Hz shunt filters integrated with the DC filter on the DC side of the converter station. Based on the connection characteristics of the project, a deicing plan for DC line is proposed, offering a deicing DC current of up to 7000 A, making the design and manufacturing of towers and conductors for heavy icing areas less difficult. The DC control system is divided into bipolar control, pole control, and valve group control, in duplicated configuration. A real-time operating system is used to improve the data processing speed. The DC protection is in triple configuration and a two-out-of-three voting logic is used to prevent inadvertent operation caused by fault. In addition, a 3D design is used for modeling, arrangement, and air gap check for the equipment in the valve hall, which helps solve the major technical difficulties in the design of valve hall for the project.

In engineering design, the reactive power support from the AC system is fully used to reduce the reactive power compensation capacity of the converter station by 1000 Mvar. GIS is used as the AC 500-kV switchgear at the converter station. AC filter sub-banks are in a grid-shaped arrangement to reduce the floor area occupied by the converter station. The HV and LV valve halls are arranged face-to-face. The distance between the valve halls is determined to translate a converter transformer with the others remaining in operation. To reduce noise produced by transformers, box-in sound insulation devices are used, AC and DC filter reactors and smoothing reactors are designed to be low-noise, and the enclosure is partially provided with sound barriers. Thus, the noise is reduced to a level conforming to applicable national standards.

12.2.1.2 Transmission line

The transmission line starts from Fulong Converter Station; passes through Sichuan, Chongqing, Hubei, Hunan, Anhui, Jiangsu, Zhejiang, and Shanghai; crosses the Yangtze River four times; and ends at Fengxian Converter Station, as shown in Figure 12.5. The transmission line is 1891 km long, with DC conductors of ACSR-6 $\times$ 720 mm^2 and a nonlinear factor of 1.103. Over the length of the line, high mountains account for 16.26%, common mountains account for 34.41%, hilly lands account for 20.76%, plains areas account for 16.06%, and river networks account for 12.51%. The line is routed as follows:

- 185.977 km in Sichuan, crossing six counties (cities, districts), namely, Yibin, Gao County, Jiang'an, Changning, Luzhou, and Hejiang
- 288.665 km in Chongqing, crossing seven counties (districts), namely, Jiangjin, Qijiang, Nanchuan, Wulong, Pengshui, Qianjiang, and Chaoyang
- 490.684 km in Hubei, crossing 14 counties (cities), namely, Xianfeng, Laifeng, Songzi, Gong'an, Jiangling, Qianjiang, Xiantao, Honghu, Jiayu, Chibi, Xianning, Yangxin, Daye, and Qichun
- 255.64 km in Hunan, crossing five counties (cities), namely, Longshan, Sangzhi, Cili, Shimen, and Li County

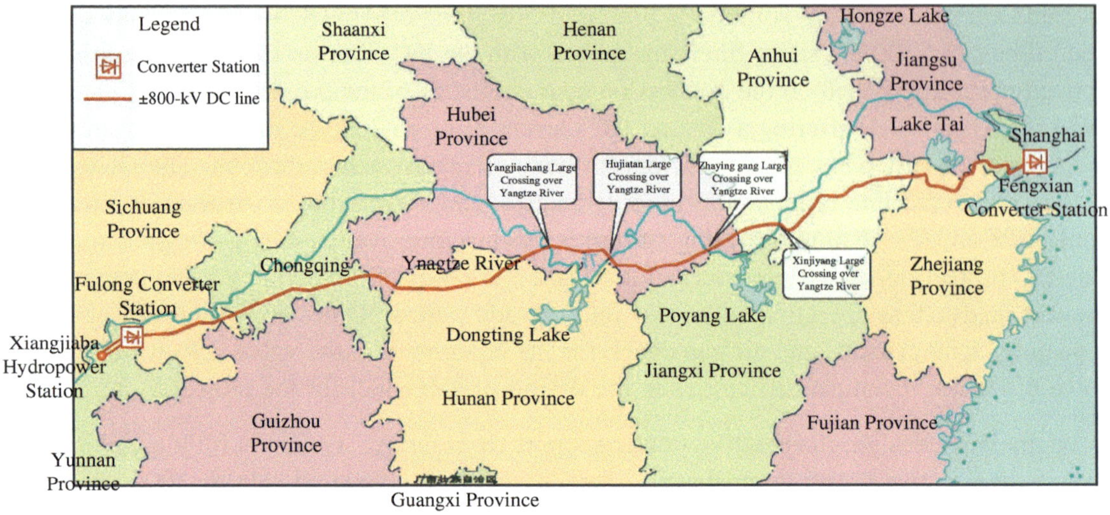

Figure 12.5
Schematic diagram of the Xiangjiaba–Shanghai project.

- 385.404 km in Anhui, crossing 12 counties (cities, districts), namely, Susong, Taihu, Wangjiang, Zongyang, Dongzhi, Chizhou, Qingyang, Nanling, Jing County, Xuanzhou, Langxi, and Guangde
- 160.656 km in Zhejiang, crossing seven counties (cities), namely, Anji, Changxing, Huzhou, Deqing, Tongxiang, Jiaxing, and Jiashan
- 17.4 km in Jiangsu, crossing Wujiang
- 106.574 km in Shanghai, crossing four districts, namely, Qingpu, Songjiang, Jinshan, and Fengxian.

In double-pole line arrangement, the line is placed on angle steel towers and steel tubular towers (for sites with an average wind speed of 35 m/s). The average height of towers is 65 m and the weight is 62 tons. The total tower weight is approximately 2.4×10^5 tons. The conductors used are 6 × ACSR-720/50 and 6 × AACSR-720/50 conductors. One ground wire is made of LBGJ-180-20AC and LBGJ-240-20AC aluminum-clad steel conductor and the other is made of OPGW-180 and OPGW-250 composite fiber optic cable.

In line design, a large amount of pioneering work is performed. The electric field and sharing of the line corridor by UHV AC and UHV DC lines are studied. F-shape towers with conductors arranged vertically are used in developed areas, greatly reducing the amount of house-relocation, and project costs. The pole line and earth electrode line share a tower, which greatly reduces the land required for the line corridor. The Helava technique is used to prevent routes from going through areas subject to heavy ice, galloping, goaf, and adverse geological conditions. Icing observations are used to determine the length in heavy icing areas where split tension towers are used to reduce wire breakage and torsion under non-uniform icing. In addition, the tower weight is reduced by 11–22%. According to the characteristics of ±800-kV UHV DC transmission lines, anti-corona and grading schemes for fittings are optimized with satisfactory results. Rigid jumpers are used to reduce jumper sag, the size of jumper cross-arm, and the height of tension tower. To solve the "dropout" of ball and socket previously found on suspension V-strings, ring–ring connection configuration is adopted for composite insulators of suspension V-strings and fittings. Composite insulators for suspension strings on suspension towers and jumper insulator strings on tension towers are used in lightly and moderately icing areas. A new type of anchor foundation suitable for complex terrains is studied and introduced, saving a large amount of concrete and earthworks. An unequal foundation with unequal tower leg length is used to maintain landform, leave the topography undisturbed, protect vegetation, and reduce soil erosion as much as possible.

12.2.1.3 Technical Data

1. **Converter transformer**
   The converter transformers used at Fulong Converter Station have the following parameters:
   - Type: single-phase two-winding converter transformer
   - Rated capacity: 321.1 MVA
   - Number of tap positions: +23/−5
   - Step size: 1.25%
   - Short-circuit impedance at rated tap position: 18%

   The converter transformers used at Fengxian Converter Station have the following parameters:
   - Type: single-phase two-winding converter transformer
   - Rated capacity: 297.1 MVA
   - Number of tap positions: +22/−6
   - Step size: 1.25%
   - Short-circuit impedance at rated tap position: 16.7%

   Oil-directed, air-forced (ODAF) and oil-forced, air-forced (OFAF) are used for cooling converter transformers at Fulong Converter Station and Fengxian Converter Station, respectively.
2. **Converter valve**
   The UHV DC thyristor converter valves used for the Xiangjiaba–Shanghai project have the following parameters:
   - Type: 6-inch thyristor converter valve
   - Valve structure: double-valve configuration

- triggering mode: electrically triggered
- Rated through-current capability: 4000 A
- Maximum short-circuit withstand current: 50 kA
- Switching impulse withstand voltage: 927 kV
- Lightning impulse withstand voltage: 916 kV

3. **Smoothing reactor**
   The Xiangjiaba–Shanghai project uses 4000-A, 75-mH dry-type smoothing reactors and is the first 4000-A reactor of this type in the world used in a ±800-kV system.
4. **AC filter**
   Fulong Converter Station has four banks of reactive power compensation equipment with a total capacity of 3080 Mvar, configured in 14 220 Mvar sub-banks including four HP11/13 filter sub-banks, four HP24/36 filter sub-banks, one HP3 filter sub-bank, and five shunt reactor sub-banks. The AC filter banks are provided with a separate bus. Fengxian Converter Station has four banks of reactive power compensation equipment with a total capacity of 3746 Mvar, configured in 15 sub-banks including eight 260 Mvar AC filter sub-banks and seven 238 Mvar shunt capacitor sub-banks.
5. **DC filter**
   The resonance is found in the DC system near 50 and 100 Hz; therefore, a 50-Hz DC blocking filter is installed on the neutral bus of each pole on the DC side of Fulong Converter Station. The blocking filter consists of a 75-mH smoothing reactor and a parallel-connected capacitor that form a parallel 50-Hz resonant circuit. In the design of DC filter, suppressing the system resonance at 100 Hz on the DC side is also considered. Ultimately, a 2/12/36 triple-tuned DC filter bank is installed for each pole in the DC switchyard at converter stations at both terminals.

   According to studies, the DC filter design standard is specified such that the equivalent disturbing current does not exceed 3000 mA under bipolar operation and 6000 mA under monopolar operation. Through this optimized design, the number and type of equipment, and thus the project cost, are reduced.

12.2.1.4 Commissioning and operation

The Xiangjiaba–Shanghai project is subjected to 2200 joint commissioning items in the laboratory, more than 20,000 subsystem commissioning items, 147 station system commissioning items, and 591 system commissioning items under 12 categories in the field. Since the beginning of bipolar operation on July 8, 2010, it passed thunderstorm, high temperature, heavy load, and other severe tests, playing an important role in exporting hydropower from Sichuan and meeting the peak demand of the east China grid. By the end of June 19, 2013, it had exported 3.37×1010 kWh of electricity. On October 9, 2013, the project passed the final acceptance test.

12.2.2 Jinping–Sunan ±800-kV UHV DC Transmission Project

12.2.2.1 Project overview

To export hydropower from Yalong River and Jinping according to the west-to-east power transmission strategy, in November 2008, the Jinping–Sunan ±800-kV UHV DC Transmission Project (Jinping–Sunan project) was commenced after approval.

The project is rated ±800 kV, 4500 A, and 7200 MW. The Jinping (Yulong) Converter Station (the sending end, see Figure 12.6) is located in Xichang, Sichuan, with seven 500-kV AC outgoing feeders: three to the Jinping—I Hydropower Station; two to the Jinping—II Hydropower Station; and two to the Yuecheng (Xichang) Substation.

The Suzhou (Tongli) Converter Station (the receiving end, see Figure 12.7) is located in Wujiang, Jiangsu, with six 500-kV outgoing feeders: two to the Suzhouxi Substation, one to the Chefang Substation, and three to the Wujiang Substation. These circuits and the substation form a 500-kV loop network.

This project has two complete monopoles, each consisting of two 12-pulse converters connected in series. Each converter is connected in parallel with one bypass switch. In this way, any pair of 12-pulse converters quitting service does not stop the remaining converters from operating as an incomplete monopolar system. Each pole is equipped with four 60-mH smoothing reactors, two on the pole bus and two on the neutral bus, one triple-tuned DC

Figure 12.6
Jinping (Yulong) converter station.

Figure 12.7
Suzhou (Tongli) converter station.

filter bank, and one 100-Hz second harmonic filter bank. To effectively eliminate the radio interference produced by the DC system, RI trap inductors are provided at the outlet of the HV end on the DC side of the converter and RI filter capacitors on the incoming circuit of AC-side converter transformer. To suppress the 50-Hz power frequency resonance, one 50-Hz blocking filter bank is installed on the neutral bus of each pole at Yulong Converter Station. The blocking filter uses a 60-mH coil to form a resonant circuit.

As the third ±800-kV UHV DC transmission project independently designed by China, the Jinping–Sunan project is characterized by larger capacity and more complex combination of main equipment. At each converter station, one 12-pulse converter valve unit has independent intellectual property rights among all 12-pulse converter valve units. In addition, the difficulties in interfacing and control coordination among valve units manufactured by different manufactures are solved, further demonstrating the technical feasibility of the independently developed valve units. The AC switchyard control system at the sending end uses an independently developed platform and the interfacing difficulties for communication with DC control and protection systems are solved. Due to transportation constraints caused by the Chengdu–Kunming Railway, three-leg HV converter transformers with external tap changers are used at the sending end. Based on the connection characteristics of Yulong Converter Station, the reactive power support from the AC system is fully used to reduce the reactive power compensation capacity by 1600 Mvar, allowing the unified coordination and control over the reactive power equipment. In addition, frequency controllers are independently designed to achieve islanded operation of hydropower generators at the sending end.

The Jinping–Sunan project marks an increase of the realized UHV DC transmission capacity from 6400 to 7200 MW. Through properly arranging arresters and optimizing parameters, the insulation levels at key nodes of the DC system are consistent with that of the Xiangjiaba–Shanghai project, making the equipment manufacturing less difficult. In addition, considering the high altitude and seismic intensity of the Yulong Converter Station, various altitude correction methods are evaluated and a practical method is identified, which is applied in subsequent UHV DC projects.

By optimizing arrangement, the number of earthing switches in the valve hall is reduced from 32 to 16, which means a smaller investment. A full 3D digital design method is used to optimize the arrangement of electrical equipment and reduce the dimensions of the valve hall. The design of fittings is improved so they produce less heat and corona compared with the Xiangjiaba–Shanghai project.

12.2.2.2 Transmission line

The transmission line is 2059 km long (including four large crossing sections over the Yangtze River, totaling 11.43 km), with DC conductors of ACSR-6 × 900 mm^2, a linear distance of approximately 1840 km, and a nonlinear factor of 1.136. Over the length of the line, steep mountains account for 1.17%, high mountains account for 19.35%, common mountains account for 40.20%, hilly lands account for 21.95%, plains areas account for 9.40%, and river networks account for 7.93%. It passes through Yunnan, Sichuan, Chongqing, Hunan, Hubei, Anhui, Zhejiang, and Jiangsu, as shown in Figure 12.8. The line is routed as follows:

- 420.616 km in Sichuan, crossing 13 counties (cities, districts), namely, Xichang City, Puge County, Zhaojue County, Meigu County, Leibo County, Pingshan County, Yibin City, Yibin County, Gao County, Jiang'an, Changning, Luzhou, and Hejiang
- 62.602 km in Yunnan, crossing two counties, namely, Suijiang County and Shuifu County
- 288.887 km in Chongqing, crossing seven counties (districts), namely, Jiangjin, Qijiang, Nanchuan, Wulong, Pengshui, Qianjiang, and Chaoyang
- 489.719 km in Hubei, crossing 14 counties (cities), namely, Xianfeng, Laifeng, Songzi, Gong'an, Jiangling, Qianjiang, Xiantao, Honghu, Jiayu, Chibi, Xianning, Yangxin, Daye, and Qichun
- 255.288 km in Hunan, crossing five counties (cities), namely, Longshan, Sangzhi, Cili, Shimen, and Li County
- 385.567 km in Anhui, crossing 12 counties (cities, districts), namely, Susong, Taihu, Wangjiang, Zongyang, Dongzhi, Chizhou, Qingyang, Nanling, Jing County, Xuanzhou, Ningguo, and Guangde
- 115.065 km in Zhejiang, crossing five counties (cities), namely, Anji, Changxing, Huzhou, Deqing, and Tongxiang
- 40.832 km in Jiangsu, crossing Wujiang

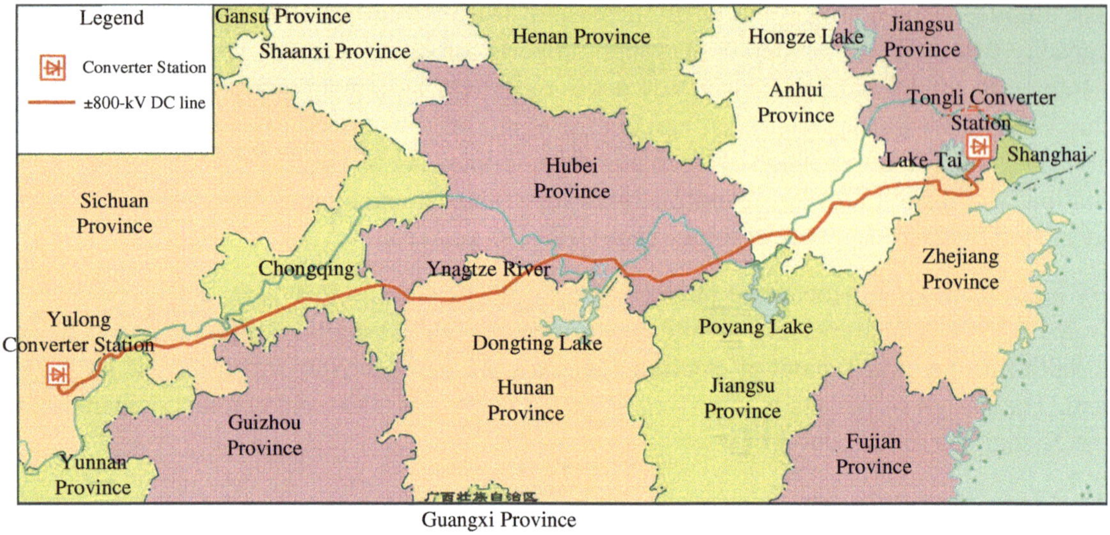

Figure 12.8
Schematic diagram of the Jinping–Sunan project.

In double-pole line arrangement, the line is erected on angle steel towers with an average height of 65 m and weight of 73 tons. The total tower weight is approximatey 3.1×10^5 tons. The conductors used are 6 × JL/G3A-900/40 and 6 × JL/G2A-900/75 ACSR conductors and JLHA1/G1A-800/55 and JLHA1/G2A-800/70 AACSR conductors. LBGJ-180-20AC, LBGJ-240-20AC, and LBGJ-300-20AC aluminum-clad steel conductors are used for ground wire.

The line crosses the Yantze River at four locations, namely, Yangjiachang, Hujiatan, and Zhayinggang in Hubei and Xinjiyang in Anhui. These crossings are in a "tension tower–suspension tower–suspension tower–tension tower" arrangement. Steel tubular towers and 4 × AACSR/EST-720/320 AACSR conductors are used for these large crossing sections, with JLB14-340 aluminum-clad steel conductor used for ground wire.

With a maximum altitude of 3650 m, the Jinping–Sunan project is the world's highest UHV DC project. Overall, the line lengths in high mountains and common mountains account for one-fifth and two-fifths, respectively Unequal foundations with unequal tower legs are used for these terrains, which greatly reduce the volume of earthworks. To facilitate transportation, composite insulators of segmental design are introduced in regions with difficult access. Based on engineering needs, Q420 high-strength steel and grade 8.8 connection bolts are used to reduce the area of inserting plate; consequently, the tower weight is reduced by 6–8%. Rock bolt foundations are used to reduce the volume of earthworks and avoid blasting and artificial excavation of deep foundation pits, making the foundation construction safer and more environmentally friendly. In addition, 760-kN insulators, composite tension insulators, and glass-disk composite insulators are tested in the project to gather experience for subsequent projects.

12.2.2.3 Technical data

1. **Converter transformer**

   The converter transformers used at Yulong Converter Station have the following parameters:
   - Type: single-phase two-winding converter transformer
   - Rated capacity: 363.4 MVA
   - Number of tap positions: +22/−6
   - Step size: 1.25%
   - Short-circuit impedance at rated tap position: 19%

   The converter transformers used at Suzhou (Tongli) Converter Station have the following parameters:
   - Type: single-phase two-winding converter transformer
   - Rated capacity: 340.8 MVA
   - Number of tap positions: +24/−4
   - Step size: 1.25%
   - Short-circuit impedance at rated tap position: 19%

   OFAF is used for cooling the converter transformers at both ends. To reduce noise produced by the transformers, box-in sound insulation devices are used.
2. **Converter valve**

   The thyristor converter valves used for Jinping–Sunan project have the following parameters:
   - Type: 6-inch thyristor converter valve
   - Valve structure: double-valve configuration
   - Triggering mode: electrically triggered
   - Rated voltage of thyristor element: 8.5 kV
   - Rated current of thyristor element: 4500 A
   - Maximum short-circuit withstand current: 50 kA
   - Switching impulse withstand voltage: 927 kV
   - Lightning impulse withstand voltage: 916 kV
3. **Smoothing reactor**

   The Jinping–Sunan project uses 4500-A, 60-mH dry-type air-core smoothing reactors.
4. **AC filter**

   The Yulong Converter Station has four reactive power compensation banks with a total capacity of 3010 Mvar, configured in 14 215-Mvar sub-banks including four BP11/BP13 filter sub-banks, four HP24/36 filter sub-banks, one HP3 filter sub-bank, and five shunt reactor sub-banks. The AC filter banks are provided with a separate bus. The Tongli Converter Station has four reactive power compensation banks with a total capacity of 4320 Mvar, configured in 16 270-Mvar sub-banks including four HP12 filter sub-banks, four HP24/36 filter sub-banks, and eight shunt capacitor sub-banks. The AC filter banks are provided with a separate bus.

5. **DC filter**
   Two 12/24 filter banks with 0.35-μF HV capacitors and two 2/39 filter banks with 0.8-μF HV capacitors are provided for each pole at each station. Like the Xiangjiaba–Shanghai project, the DC filter is designed such that the equivalent disturbing current does not exceed 3000 mA under bipolar operation and 6000 mA under monopolar operation.

12.2.2.4 Commissioning and operation

The Jinping–Sunan project began bipolar operation with the LV end on June 27, 2012, and all 568 system commissioning tests were completed on November 28, 2012. It began its test run on November 29, 2012, and bipolar commercial operation was performed on December 12, 2012. The commissioning of this project helps export the hydropower from the Guandi and Jinping Hydropower Stations and the surplus power in Sichuan during the flood season, and helps mitigate the power shortage in east China during the peak demand days in summer. The Jinping–Sunan project can export approximately 3.6×1010 kWh of electricity to east China each year. It is of great significance in solving the contradiction between "surplus power during the wet season and deficit power during the dry season" in Sichuan, meeting the power demand of sustainable development of economy and society in eastern regions, and mitigating the increasingly severe environmental problems.

12.2.3 Haminan–Zhengzhou ±800-kV UHV DC Transmission Project

12.2.3.1 Project overview

To export thermal power and wind power from Hami, address the power shortage in Henan, promote Xinjiang's economic development, and implement the West-to-East power transmission strategy, in May 2012, the Haminan–Zhengzhou ±800-kV UHV DC Transmission Project (Haminan–Zhengzhou project) was commenced after approval.

The project is rated ±800 kV, 5000 A, and 8000 MW. The Haminan Converter Station (the sending end, see Figure 12.9) is located in Hami, Xinjiang, with six 500-kV AC outgoing feeders connected to the 750-kV grid through two 750/500/66-kV, 2100-MVA tie transformers, four 750-kV AC outgoing feeders, and six planned 750-kV AC outgoing feeders.

The Zhengzhou Converter Station (the receiving end, see Figure 12.10) is located in Zhengzhou, Henan, with six 500-kV AC outgoing feeders: two to Zhengbei, Guandu, and Bianxi substations, respectively.

This project has two complete monopoles, each consisting of two 12-pulse converters connected in series. Each converter is connected in parallel with one bank of bypass circuit

Figure 12.9
Aerial view of Haminan converter station.

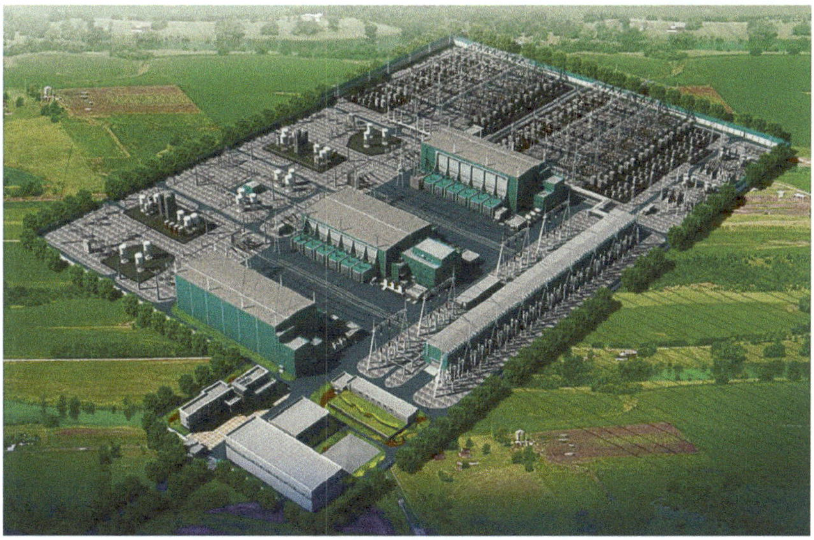

Figure 12.10
Aerial view of Zhengzhou converter station.

breakers. This way, any pair of 12-pulse converters quitting service does not stop the remaining converters from operating as an incomplete monopolar system. Each pole is equipped with six 50-mH smoothing reactors, three on the pole bus and three on the neutral bus, two double-tuned DC filter banks, and one 100-Hz second harmonic filter bank. To reduce the radio interference produced by the DC system, RI trap inductors are provided at the outlet of the HV end on the DC side of the converter and RI filter capacitors are on the incoming circuit of AC-side converter transformer. To suppress the 50-Hz power frequency resonance, one 50-Hz blocking filter bank is installed on the neutral bus of each pole at Haminan Converter Station. The blocking filter uses a 50-mH coil to form a resonant circuit.

The current-conducting capability of 6-inch thyristor valves is fully used in the project, increasing the transmission capacity to 8000 MW. Considering the low temperature and strong sandstorm in Hami, thermal insulation, heating measures, and sealed design are adopted to ensure the safe and stable operation of the project.

12.2.3.2 Transmission line

The transmission line starts from Haminan Converter Station in Hami, Xinjiang, and ends at Zhengzhou Converter Station in Zhongmou County, Henan. It is 2191.525 km long, with a nonlinear factor of 1.11, among which the large crossing over the Yellow River is 3.9 km and other common segments are 2187.625 km. Figure 12.11 shows the route of the project.

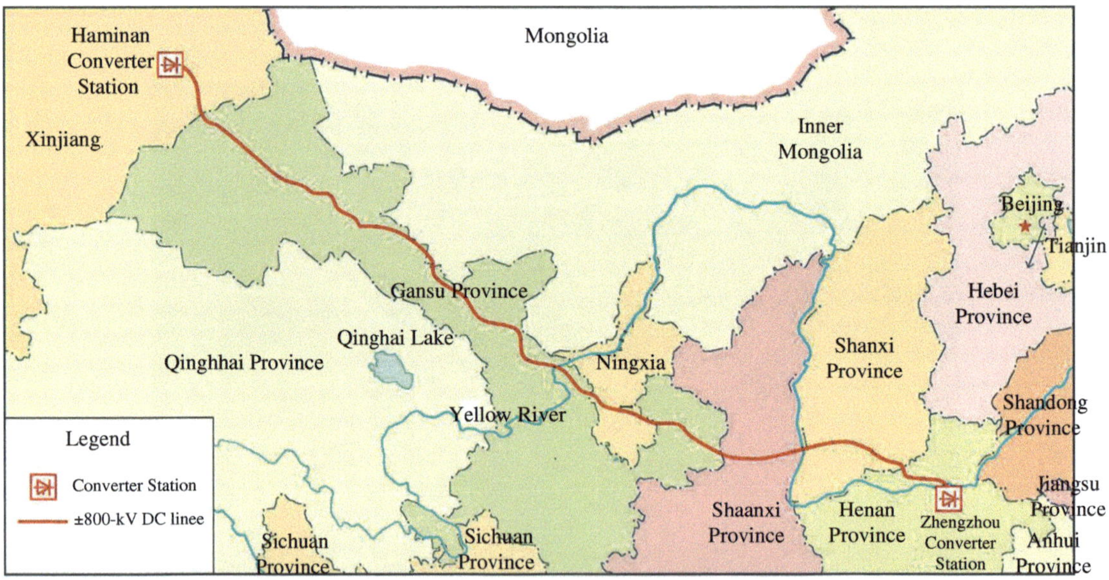

Figure 12.11
Schematic diagram of the Haminan–Zhengzhou project.

Over the length of the line, the maximum altitude is 2300 m, high mountains account for 5.6%, common mountains account for 29.8%, hilly lands account for 27.1%, plains areas account for 39.3%, bog and river networks account for 1.2%, and dunes and deserts account for 2.0%. It passes through 42 county-level administrative regions under 16 prefecture-level administrative regions in six provinces, namely, Xinjiang, Gansu, Ningxia, Shaanxi, Shanxi, and Henan.

In double-circuit arrangement, the line is placed on angle steel towers with an average of 57.01 m (the height from lowest cross-arm to ground) and 62.05 tons. The total tower weight is approximately 2.6×10^5 tons. The conductors used are 6 × JL/G3A-1000/45 and 6 × JL/G2A-1000/80 ACSR conductors. LBGJ-150-20AC aluminum-clad steel conductor and 24-core OPGW-150 optical fiber cables are used as the ground wire.

The large crossing over the Yellow River is located in Henan. It is in a "tension tower–suspension tower–suspension tower–suspension tower–tension tower" arrangement. Steel tubular towers and 4 × JLHA1/EST-900/240 extra-high-strength AACSR conductors are used at this crossing, with 36-core OPGW-300 overhead composite optical fiber cable as the ground wire.

The Haminan–Zhengzhou project is the longest UHV DC transmission project in the world. The use of $6 \times 1000\ \text{mm}^2$ large-section conductors greatly reduces the line loss. The conductor spacing is reduced to 18–20 m, which reduces the width of the line corridor. The line runs through a variety of terrains, including deserts, Gobis, high mountains, common mountains, river networks, and bogs. Unequal foundations with unequal tower leg lengths suitable for complex terrains are introduced to reduce the volume of earthworks. To reduce the conductor spacing, and thus the width of line corridor, V-strings are introduced for crowded segments. Based on engineering needs, large angle steel is widely used in the design, which balances the forces on the two legs and reduces the tower weight by 3–5%. In cold areas, Q420C high-strength steel is used, reducing the tower weight by 7%. In saline soil areas with serious liquefaction, enclosed cast-in-place concrete piles are used to solve corrosion and reduce cost.

12.2.3.3 Technical data

1. **Converter transformer**

   The converter transformers used at Haminan Converter Station have the following parameters:
   - Type: single-phase two-winding converter transformer
   - Rated capacity: 405.2 MVA
   - Number of tap positions: +23/−5

- Step size: 1.25%
- Short-circuit impedance at rated tap position: 20%

The converter transformers used at the Zhengzhou Converter Station have the following parameters:

- Type: single-phase two-winding converter transformer
- Rated capacity: 376.6 MVA
- Number of tap positions: +25/−5
- Step size: 1.25%
- Short-circuit impedance at rated tap position: 19%

OFAF is used for cooling converter transformers at both ends. To reduce noise produced by transformers, box-in sound insulation devices are used.

2. **Converter valve**

The thyristor converter valves used for the project have the following parameters:

- Type: 6-inch thyristor converter valve
- Valve structure: double-valve configuration
- Triggering mode: electrically triggered
- Rated voltage of thyristor element: 8.5 kV
- Rated current of thyristor element: 5000 A
- Maximum short-circuit withstand current: 50 kA
- Switching impulse withstand voltage: 927 kV
- Lightning impulse withstand voltage: 916 kV

3. **Smoothing reactor**

The project uses 5000-A, 50-mH high-voltage dry-type air-core smoothing reactors.

4. **AC filter**

Haminan Converter Station has four reactive power compensation banks with a total capacity of 3880 Mvar, configured in five 270-Mvar sub-banks and 11 230-Mvar sub-banks, including four BP11/BP13 filter sub-banks, four HP24/36 filter sub-banks, three HP3 filter sub-banks, and five shunt capacitor sub-banks. The AC filter banks are provided with a separate bus.

The Zhengzhou Converter Station has four reactive power compensation with a total capacity of 4940 Mvar, configured in 19 260-Mvar sub-banks including eight HP12/24 filter sub-banks, two HP3 filter sub-banks, and nine shunt capacitor sub-banks. The AC filter banks are provided with a separate bus.

5. **DC filter**

One HP12/24 DC filter bank with 0.35-μF HV capacitors and one HP2/39 filter bank with 0.8-μF HV capacitors are provided for each pole at each station of Haminan−Zhengzhou project. Like the Xiangjiaba−Shanghai and Jinping−Sunan projects, the DC filter is designed such that the equivalent disturbing current does not exceed 3000 mA under bipolar operation and 6000 mA under monopolar operation.

12.2.3.4 Commissioning and operation

The Haminan–Zhengzhou project began bipolar operation with the LV end on October 17, 2013, and its bipolar commissioning was completed in December 2013. It began bipolar operation in January 2014. It is the first UHV channel for exporting electricity from Xinjiang. It transmits power produced from the abundant resources of coal and wind to central China. This helps improve Xinjiang's self-development capability, solve the contradiction among coal, electricity, and transportation in central China, and mitigate the power shortage in Henan and other regions.

12.3 UHV Test Facilities

12.3.1 UHV AC Test Base

1. **Overview**

   On October 10, 2006, a UHV AC test base began construction in the south of Fenghuang Mountain in Jiangxia District, Wuhan, Hubei. On December 26, 2008, it passed the acceptance test. The base is located at an altitude of 36 m, covering an area of 24 hm^2. As an important part of the UHV AC pilot and demonstration project, a key project during the 11th Five-Year Plan period, the UHV AC test base is composed of 1-km-long single-/double-circuit UHV experimental lines, an electromagnetic environment measurement laboratory, a UHV AC corona cage, an artificial climate laboratory, a 7500-kV outdoor impulse test field, a UHV equipment long-term energization test field, and other test facilities. It could be used for studies of UHV electromagnetic environmental effects, external insulation characteristics under special natural conditions such as high altitude, heavy pollution, and severe icing, long-term full-voltage energization test on electrical equipment, as well as the operation and maintenance of transmission equipment. Also, it provides strong technical support for construction and safe, stable, and economic operation of the UHV AC pilot and demonstration project and subsequent projects. In 2008, the UHV AC test base was determined to be a national engineering laboratory for UHV engineering technology. In April 2013, it was awarded the title of a "National Science Popularization Education Base."

   As an important part of the pilot and demonstration project, construction of the test base could be deemed as a rehearsal of the project. A lot of useful experience is obtained from constructing the test base in UHV electromagnetic environment, engineering design, technical requirements for equipment, equipment manufacturing, project construction, test, commissioning, and operation. Key design parameters of the project are verified and then used as a guide for selection of technologies and development of UHV equipment. UHV construction technologies are rehearsed on the test base, which contribute to speeding the construction of the pilot project.

Since energization of the single UHV experimental line in the test base, a number of tests and studies have been conducted, including 12 pioneering tests. For example, during the experimental study of VFTO of UHV GIS, the world's first VFTO prototype test platform was built. A 7500-kV long wave front impulse voltage generator was developed, with the highest voltage level and output efficiency and the longest wave front in the world. The world's most versatile climate and environment laboratory was studied and built. A solution was proposed to address the interference between UHV lines and radio stations. Also, a study of live work under equipotential on a 1000-kV line was conducted. Throughout the construction and operation of the test base, 53 patents, including 37 innovation patents and 16 utility model patents, were awarded, among which eight were core patents. In addition, more than 10 national, industrial, and enterprise-level standards were established.

2. **Main functions and technical data**
   a. UHV AC experimental line. In designing the experimental lines, complex circumstances that may be encountered in UHV power transmission project are assumed. The conductor height in relation to ground and the phase-to-phase spacing are made adjustable to suit the experimental study for optimized design of UHV lines. This experimental line can be used for studying the electromagnetic characteristics of various types of towers and conductors.

      The single-circuit and double-circuit experimental lines are approximately 1000 m long, with a center-to-center distance of approximately 135 m. The towers of both lines are in a "tension tower–suspension tower–suspension tower–tension tower" arrangement, as shown in Figure 12.12.

Figure 12.12
UHV AC experimental lines.

The span lengths of the single-circuit experimental line are 68, 333, 344, and 64 m. The towers are 65, 70, 73, and 65 m high, respectively. The outside phase conductors each have five suspension points spaced 1 m apart on a suspension tower. The middle suspension is a normal suspension point. The two phase conductors on both sides of the normal suspension point are 33 m apart.

The span lengths of the double-circuit experimental line are 74, 226, 398, and 294 m. The towers are 92.5, 101, 98, and 92.5 m high, respectively. The outside phase conductors each have five suspension points spaced 1 m apart on a suspension tower. The middle suspension point among five suspension points is a normal suspension point. At this point, the upper and lower phase conductors of the same circuit are approximately 21 m apart, and two phase conductors of different circuits at the same level are 31 m (upper), 37 m (middle), and 33 m (lower) apart, respectively.

b. Electromagnetic environment measurement laboratory. This laboratory may be used for studies of important subjects such as the electromagnetic environment test field for UHV lines, protective distance between UHV lines and radio stations, and electromagnetic compatibility of UHV equipment. Through these studies, the laboratory provides basic data and measures to meet the environmental protection requirements and guide for the engineering design.

It is composed of a 200-m^2 measuring room (including a 60-m^2 shielded enclosure), a 120-m × 10-m test field (as shown in Figure 12.13), and an

Figure 12.13
Electromagnetic environment test field.

observation room. The test field is located below the midpoint of the UHV experimental lines. Equipped with automatic detection systems for radio interference, audible noise, and power frequency electric field, and automatic measuring points for ground electric field strength and house models, it can be used to measure radio interference, audible noise, and power frequency electric field of the experimental line under all weather conditions.

c. UHV AC corona cage. The UHV AC corona cage can be used for experimental studies of various corona characteristics of different types of bundle conductors (12-bundle or less) under different weather conditions, including studies of corona inception voltage, characteristics of audible noise and radio interference induced by corona, and corona loss characteristics. These studies provide information for design and selection of conductors for UHV AC transmission lines.

   The UHV AC corona cage has a symmetrical arrangement, is 35 m long, with a cross-sectional area of 8 m × 8 m, and has a power source rated 400 kVA, 800 kV, and 0.5 A, as shown in Figure 12.14.

d. Artificial climate laboratory. This laboratory can be used for experimental studies of external insulation characteristics of equipment such as UHV insulator strings under various special weather conditions like rain, fog, pollution, and high altitude, as shown in Figure 12.15.

   It is composed of an artificial low-pressure tank, HV testing equipment, security systems, and auxiliary equipment. The tank shell is ϕ 20 m × 25 m and can be used to simulate severe natural conditions such as high altitude, heavy pollution, and icing under low temperatures. The HV testing equipment can provide a power source of 1000 kV/6 A at power frequency and a DC power source of 1200 kV/2 A. The auxiliary systems include vacuum pumping system, refrigeration system, icing spray system, artificial fog system, CCTV system, and measurement and control system. With these, the conditions at an altitude of 5500 m under −19°C could be simulated in the tank.

Figure 12.14
UHV AC corona cage.

Figure 12.15
Artificial climate laboratory.

Figure 12.16
7500-kV outdoor impulse test field.

e. The 7500-kV outdoor impulse test field. The 7500-kV outdoor impulse test field is composed of a 7500-kV impulse voltage generator, a voltage divider, a measuring system, and a gantry structure. It could be used for studies of characteristics of external insulation for EHV AC/UHV AC transmission lines under long wave front switching impulse, enables more reasonable external insulation design for UHV lines, and is a cost-effective and reasonable design for UHV tower windows, as shown in Figure 12.16.

Figure 12.17
Test field for UHV equipment under long-term energization.

The 7500-kV impulse voltage generator, with the highest voltage class in the world and a highest test voltage (nominal voltage) of 7500 kV, can generate standard switching impulse voltage with a front time of 250 μs and switching impulse voltage with a long front time of 1000–3000 μs.

f. Test field for UHV equipment under long-term energization. After passing the type test, the UHV equipment has to endure a long-term full-voltage test to ensure safe operation. The test field is provided with foundations for mounting CVT, TA, surge arresters, bushings, disconnectors, post insulators, GIS bays, and other equipment, which can be T-connected to the UHV equipment through the lower bus in the test field for the long-term energization test.

This test field (216 m × 58 m) is located below the outgoing feeder frame of the single-circuit and double-circuit experimental lines, as shown in Figure 12.17. Three voltage classes, 500, 750, and 1000 kV, are available from the UHV test power source.

g. UHV VFTO test circuit. This test circuit is used for experimental studies on VFTO in UHV GIS under full voltage, including various measuring methods, VFTO levels, and probability characteristics of VFTO. Typical VFTO wave shapes for UHV GIS are obtained through tests. The test results provide a basis for safe switching of GIS disconnectors and suppression of VFTO. This is also the first full-scale test of VFTO in the world.

This test circuit, as shown in Figure 12.18, is equipped with various types of UHV GIS disconnectors, for which the connection scheme could be

Figure 12.18
UHV VFTO test field.

Figure 12.19
Long-term observation base for ecologic impacts of extremely low-frequency electromagnetic fields.

changed. In addition, UHV tank-type arresters and magnetic rings can also be connected.

h. Long-term observation base for ecologic impacts of extremely low-frequency electromagnetic fields (Figure 12.19). This base is located below the 1000-kV UHV

AC experimental lines, covering an area of 10 μm. It is divided into the observation area for plants exposed to power frequency electric fields, the observation area for animals exposed to power frequency electromagnetic fields, the observation room for animals exposed to power frequency magnetic field tests, and the power frequency magnetic field observation area. This base is qualified for a national Class II biological laboratory. The attainable electric field strength is 0–7 kV/m and magnetic field strength is 0–500 μT. The base can be used for long-term exposure tests using animals and plants.

12.3.2 UHV DC Test Base

1. **Overview**
   To perform studies of UHV transmission projects, on February 23, 2007, a UHV DC test base began construction in Changping Park East, Zhongguancun Science and Technology Park, Beijing. On February 22, 2009, it passed the acceptance test. Covering an area of 26 hm^2, the base is composed of an outdoor test field, a test hall, a pollution and environment laboratory, an insulator laboratory, an arrester laboratory, a UHV DC experimental line, a corona cage, and an electromagnetic environment simulation test field. With complete and practical functions, it provides strong technical supports for the construction and operation of the ±800-kV UHV DC demonstration project.
   This base is capable of conducting experimental studies of UHV electromagnetic environments, external insulation, operational security of systems, test technology, and operating characteristics of equipment. It is a world leader in comprehensive testing capabilities, with 15 equipment parameters and performance indices ranking first. As an open platform for fundamental and leading-edge studies, the base has become a leading UHV DC research center in the world. In 2008, it was recognized as a national laboratory for UHV engineering technologies.
2. **Main functions and technical data**
   a. Outdoor test field. The outdoor test filed is 180 m long and 90 m wide, as shown in Figure 12.20. It contains a 7200-kV/480-kJ outdoor impulse voltage generator, a 27-m-high single-column impulse voltage divider, a 50-m (W) × 60-m (H) gantry tower, and two tension anchor towers. This field can be used for lightning/switching impulse tests on DC tower insulators and various air gaps, live working tests, and lightning/switching tests on large equipment.
   b. Test hall. The test hall (86 m × 60 m × 50 m) provides an electromagnetic shielding effectiveness of 70 dB. It contains a 1500-kV power frequency test transformer, a ±1800-kV DC voltage generator (Figure 12.21), and a 6-MV impulse voltage generator. This test hall can be used for electrical performance tests on DC transmission equipment rated from ±800 to ±1100 kV, such as lightning/

Figure 12.20
Outdoor impulse test field.

Figure 12.21
±1800-kV DC voltage generator in the test hall.

switching impulse voltage withstand tests, AC/DC voltage withstand tests, partial discharge tests, visible corona tests, and radio interference tests.

c. Pollution and environment laboratory. The pollution and environment laboratory provides an artificial climate tank with a net test space of ϕ 20 m × 25 m and 800 kV/6 A AC/ ±1000 kV/2 A DC power sources for AC/DC pollution tests.

It can simulate external environmental conditions at altitudes ranging from 0 to 5500 m and low temperatures ranging from $-19°C$ to 0°C. It could be used for pollution, icing, rain, and high altitude tests for AC/DC long suspension insulator strings, post insulators, and large bushings for EHV/UHV applications, as shown in Figure 12.22.

d. Insulator laboratory. The insulator laboratory is used for electromechanical performance tests on insulators and bushings of various materials and types used for UHV DC transmission projects, including power frequency, DC, lightning and steep-front impulse voltage tests, tensile load tests, flexural–torsional buckling tests, and thermal mechanical tests. It contains a 1200 kV impulse voltage generator, a 500 kV/1 A power frequency test transformer, a 600 kN/ $-60 \pm 80°C$ thermal mechanical test facility, a 2000 kN/15 m horizontal tensile tester, a 700 kN · m/14 m flexural torsional buckling tester, and temperature cycling and physical and chemical test facilities, as shown in Figure 12.23.

e. Arrester laboratory. The arrester laboratory contains an 8/20 μs/20 kA residual voltage and current dividing test facility (maximum residual voltage: 600 kV), an accelerated aging test facility, a power frequency voltage withstand test facility, and a square wave impulse current generator, as shown in Figure 12.24. It can be used for performance and test studies of AC/DC arresters and resistors, including type tests, sampling tests, quality supervision and inspection, and technical performance tests.

Figure 12.22
Pollution and environment laboratory.

Figure 12.23
Power source field in the insulator laboratory.

Figure 12.24
Arrester laboratory.

f. UHV DC experimental line. The experimental line is 1080-m-long in double-circuit ±1200-kV design, as shown in Figure 12.25. Supplemented by tests in the corona cage, it is mainly used to examine the electromagnetic environment of UHV DC transmission lines and determine forecasting methods suitable for China, is supplemented by tests in the simulation test field, and is used to study the electromagnetic environment of lines in different configurations.

g. Corona cage. The corona cage has a two-compartment structure, with a cross-sectional area of 22 m × 10 m and dimension of 70 m × 22 m × 13 m (L × W × H), as shown in Figure 12.26. It can be used for corona characteristic tests of various types of conductor bundles for ±800 kV lines, including monopolar or bipolar conductors. By grasping the corona characteristics of bundle conductors in various configurations and analyzing the test results obtained from the experimental line, a corona effect formula suitable for the conditions in China is derived for use in engineering design.

h. Electromagnetic environment simulation test field. Covering an area of 1 hm^2, this test field contains an 800-kV DC voltage generator and a 500-kV AC voltage generator. It has two pairs of gantry towers used for electromagnetic environment tests on multi-circuit DC lines sharing a corridor, AC and DC lines sharing a corridor, and AC and DC lines on the same tower, as shown in Figure 12.27.

Figure 12.25
UHV DC experimental line.

Figure 12.26
Corona cage.

Figure 12.27
Electromagnetic environment simulation test field.

12.3.3 UHV Tower Test Base

1. **Overview**
   The UHV tower test base is located in Jingang Industrial Park (Yangfengang Town), Bazhou, Hebei, covering an area of 20 hm^2, as shown in Figure 12.28. It began construction on October 18, 2008, passed the full-scale test on the first tower on March 28, 2009, and was completed at the end of 2009.

Figure 12.28
Aerial view of the UHV tower test base.

To satisfy both the current and long-term needs, the base is constructed by holistic planning and step-by-step implementation. Its designed capabilities include full-scale tests on towers for the 1000-kV double-circuit Huainan–Shanghai UHV transmission line, towers of all possible dimensions, and design loads for 1000-kV double-circuit, ±1100-kV single-circuit, and ±800-kV double-circuit lines during the development of UHV grids, towers of emerging structures, and existing and upcoming towers for 500-kV multi-circuit lines.

2. **Main functions and technical data**

The UHV tower test base could be used for full-scale tests, bearing capacity tests of members and components, optimized design tests on node structures, force transfer tests on members, strength tests under cyclic loads, and dynamic stability tests of structures as required. The full-scale test capabilities cover most conventional tower types, including:

a. Suspension towers and 60° angle towers for 1000-kV double-circuit lines and 500-kV four-circuit lines in light icing areas
b. Suspension towers and 40° angle towers for 1000-kV double-circuit lines in moderate icing areas
c. Suspension towers and 40° angle towers for UHV DC lines in heavy icing areas

The main capability parameters of the UHV tower test base are given in Table 12.1.

By function, the test facilities in the UHV tower test base can be classified into the following categories:

i. Load bearing system. This mainly includes a universal testing foundation, a transverse loading tower, a longitudinal loading tower, a longitudinal reverse loading tower, various steering ground anchors, hydraulic cylinder

Table 12.1: Main Capability Parameters of the UHV Tower Test Base.

Main Technical Parameters of Test Object		Performance Parameters of Test Equipment	
Height of testable tower	≤140 m	Height of loading tower	Transverse and longitudinal towers: 153 m, longitudinal reverse loading tower: 133 m
Cross-arm width of testable tower	≤66 m	Width of loading tower	70 m
Root span of testable tower	≤40 m	Size of universal testing foundation	50 m × 50 m
Uplift force of foundation	≤2000 tons/leg	Anti-uplift force of universal testing foundation	2000 tons/leg
Number of loading channels	≤188	Number of loading channels	188
Strain measurement	192 strain gauges	Number of channels for strain measurement	192
Displacement measurement	3D deformation measurement, rounded up to mm	Total station	3

foundations, associated test girders, and other anchoring and force transferring facilities and equipment.

This system occupies an area of approximately 3650 m^2. The transverse and longitudinal loading towers are each 153 m high, and the longitudinal reverse loading tower is 133 m high.

ii. Hydraulic loading system. This mainly including four independent hydraulic loading systems, namely, two transverse loading rooms, a longitudinal loading room, and a longitudinal reverse loading room, with 188 hydraulic cylinders (total of 206 hydraulic cylinders by reserving 18 positions).

iii. Measurement and control system. The VXI fieldbus technology enables the coordinated, synchronous multi-point loading of 206 hydraulic cylinders in four loading rooms. The system has a data polling cycle of less than 10 ms and loading control accuracy higher than 1%. It is also provided with load and protection control.

iv. Data acquisition and management system. This includes video surveillance, displacement and stain data acquisition, and a test database. The data acquisition system is provided with reserve capacity for expansion.

v. Electrical and protection systems. These including an external power source, an internal distribution system, and an internal lightning protection and grounding system. The power source is led from a 10-kV distribution grid and supplies through two auxiliary transformers.

vi. Auxiliary facilities. These include a stationary tower crane (lifting height: 150 m), a track travelling crane, a tester, and riggings.
vii. Buildings. These include an office building, an observation building, four loading rooms, a component laboratory, and a distribution room, covering an area of approximately 8635 m^2.

12.3.4 Tibet High-Altitude Test Base [2]

1. **Overview**

The Tibet high-altitude test base, with an altitude of 4300 m, is located in Yangbajing Town, Dangxiong County, Lhasa, Tibet, covering an area of approximately 6 hm^2. It began construction on June 15, 2008, and it was completed on October 31, 2008. The test base is composed of an outdoor test field, a pollution test laboratory, and an experimental line, as shown in Figure 12.29.

This base, commanding the highest altitude in the world, can be used for studies of insulation and electromagnetic environment characteristics of transmission lines and equipment at an altitude of 4000 m and higher. It was independently developed, designed, and built by China with full intellectual property rights. It involves a number of major technological breakthroughs in functional design, equipment development, control and testing technology, as well as engineering application.

Figure 12.29
Aerial view of Tibet high-altitude test base.

2. **Main functions and technical data**
   a. Outdoor test field. The outdoor test field (120 m × 100 m) is composed of a 4200-kV impulse voltage generator, a 660-kV power frequency cascade test transformer (which would ultimately be upgraded to 1000 kV), +1000- and −1500-kV DC voltage generators, and a 50-m × 50-m gantry tower.

      The impulse voltage generator is capable of tests of discharge characteristics of various air gaps for DC transmission lines up to ±1100 kV and air gaps between typical electrodes at an altitude of 4300 m, as shown in Figure 12.30. The DC voltage generator can produce a voltage up to −1500 and +1000 kV and is capable of testing of electromagnetic environment indices, testing DC discharge characteristics of air gaps, corona inception tests on fittings, and DC withstand tests of power equipment at an altitude of 4300 m. The power frequency test transformer (currently, two cascade transformers connected in series) can produce 660 kV of power frequency test voltage. It can be used for power frequency discharge voltage tests on various air gaps, power frequency voltage withstand tests of power equipment, and live working tests of transmission equipment at high altitudes.
   b. Pollution laboratory. The pollution laboratory contains a 9-m × 9-m × 11-m (high) fog room, a ±250-kV/2-A DC power supply system, a 1000-kVA/200-kV AC test power source, a water treatment system, and a steam boiler (maximum continuous

Figure 12.30
Impulse discharge test on large air gaps at an altitude higher than 4000 m.

Figure 12.31
Experimental line.

rating: 0.3 tons/h of saturated steam). It can be used for artificial pollution tests on various station and line insulators at high altitudes.

c. Experimental line. The experimental line, which is 500 m long, consists of a current guiding segment and a test segment, as shown in Figure 12.31. It includes three towers designed at ±500 kV, each with nine horizontal suspension points. By hanging the conductors at different points, the conductor spacing of the DC experimental line can be changed. The pole conductors are in a single-circuit bipolar arrangement with a normal test voltage of ±500 kV and maximum operating voltage of ±800 kV. Below the midpoint of the test segment is a measuring field. The experimental line shares the DC power source and power frequency power source with the outdoor test field. It can be used to study the electromagnetic environment parameters of lines at an altitude of 4300 m, such as ground total electric field, audible noise, and radio interference.

12.3.5 High-Power Laboratory

1. **Overview**
   The High-Power Laboratory, located in Xi'an, Shaanxi, is attached to the China National High Voltage Apparatus Quality Supervision and Inspection Test Center. In the laboratory, the performance of transmission equipment could be evaluated by

simulating various faults in the power system. The laboratory is composed of two stations, the old station and the new station. The former consists of two 2500-MVA short-circuit generators, a 50-MVA short-circuit generator and associated supporting facilities, and a 363-kV 7-MJ oscillation circuit system. The latter consists of a 6500-MVA short-circuit generator, three 300-MVA short-circuit transformers, a 260-MVA short-circuit transformer, and associated supporting facilities. Two independent voltage circuits rated 1000 kV and 7.5 MJ serve as the oscillating circuit. The test circuit for converter valve operation is rated ±500 kV/3600 A. This laboratory is capable of tests of UHV 63 kA circuit breakers.

2. **Main functions and technical data**

From January 2006 to December 2008, type tests of approximately 1500 pieces (sets) of equipment were performed in the laboratory. Breaking tests on generator circuit breakers, 252-kV/40-kA three-phase sudden short-circuit tests, short-time withstand current tests on 800-kV/50-kA disconnectors, full-pole tests on 550-kV/63-kA circuit breakers, and breaking tests on 800-kV/50-kA circuit breakers and on 1100-kV/50-kA circuit breakers were conducted for the first time in China. The current test capability of this laboratory is given in Table 12.2.

Main testing equipment in the laboratory includes:

a. Short-circuit generator (Figure 12.32). As the power source for high-power tests, the short-circuit generator is a core piece of equipment developed and manufactured by Harbin Electric Machinery Company Limited. It is one of the short-circuit generators with the largest capacity in the world. Its main parameters are as follows:

   Terminal short-circuit capacity: 6500 MVA
   Rate frequency: 50/60 Hz
   Rated speed: 3000/3600 r/min

b. Drive motor. The 4000-kW drive motor is a wound-rotor asynchronous motor produced by Harbin Electric Machinery Company Limited. Rotor-resistance started,

Table 12.2: Testing Capabilities of the High-Power Laboratory.

Classification of Tests	Testing Capability	
Synthetic test	Three-phase test	126–363 kV, 63 kA
	Single-phase test	126–1100 kV
	Breaking and making tests on circuit breakers	63 kA, 1100 kV
Direct test	Three-phase test	14 kV, 100 kA
		24 kV, 40 kA
	Single-phase test	31.5 kV, 31.5 kA
		40.5 kV, 31.5 kA
	Thermodynamic stability current test	Peak: 320 kA, r.m.s.: 63 kA/3 s
	Sudden short-circuit test on transformers	220 kV and below
	Short-circuit test on HV converter valve assemblies	±800 kV

Figure 12.32
Short-circuit generator.

the motor can drive the short-circuit generator to the rated speed. The drive motor is rated 4000 kW, 10 kV, and 50/60 Hz.

c. Short-circuit transformer. The short-circuit transformer, a part of the generator system, is mainly used for high-power short-circuit tests at 15 kV and higher and synthetic short-circuit tests. It is rated 50/60 Hz and 1500 MVA.
d. Synthetic circuit. The high-power laboratory has two independent UHV voltage power circuits. The main circuit is rated ±1000 kV, and the operating voltage of the regulating circuit when simulating the system transient recovery voltage is 1690 kV. The capacity of the voltage source is 13.6 MJ. In addition, a ±1000-kV high-accuracy, high-voltage firing device with a control accuracy of ±5 μs is available in the laboratory.

12.3.6 SGCC Simulation Center

The SGCC Simulation Center (Figure 12.33) is located in the China Electric Power Research Institute (CEPRI) in Beijing. To facilitate planning, design, construction, and operation of large hybrid UHV AC and UHV DC grids in China, test devices for digital simulation, dynamic simulation, digital/analog hybrid simulation, operation, and safety supervision of the power system are established. These devices are combined to form an integrated, fully functional, and advanced power system simulation system. This allows for the multilevel, multiperspective, all-around simulation of the power system and provides strong technical support and quality technical services for developing and operating grids.

Figure 12.33
SGCC simulation center.

The digital/analog hybrid simulation technology, an advanced real-time simulation technology combining digital and analog simulators, can make full use of the advantages of both simulators (i.e., large-scale simulation and high accuracy). This way, it can accurately represent the physical phenomena of an AC/DC system. Currently, a hybrid simulation platform for large AC and DC grids containing multiple HVDC simulators has been built, which is leading in the world.

For dynamic simulation, based on the characteristics of UHV AC transmission systems, analog simulators for various key equipment were developed and the first Chinese UHV AC dynamic simulation system was built with the world's leading simulation technologies, reflecting the true characteristics of UHV equipment and systems. The huge value of this simulation system has been proven in the UHV AC pilot and demonstration project.

The simulation center developed the "PSD power system analysis software" with the most state-of-the-art technology in the world. This achievement enabled multiple time-scale unified simulation for large power systems. It can simulate the whole dynamic process of large power systems with more than 50,000 nodes on a time scale of a few hours. This software package is widely used in the power industry as a routine calculation and analysis tool and for simulation analysis of major transmission project planning, system commissioning, and grid fault in China.

The simulation center developed an all-digital real-time simulation system (ADPSS) with full intellectual property rights, enabling real-time simulation of power systems with 10,000 nodes for the first time anywhere. With a maximum simulation capacity of up to 20,000 nodes and 3000 generators, this system is a worldwide leader in both technology and economy. This achievement was awarded the first prize of 2009 National Science and

Technology Progress Award. It has been used by more than 20 power companies and research and education institutions as a powerful tool for simulation analysis, experimental studies, and equipment performance testing for large power grids.

The SGCC Simulation Center has been recognized as a national engineering laboratory for power system simulation and has been named a key laboratory of grid security and energy savings. It is a worldwide leader in comprehensive testing capabilities.

12.3.7 R&D Center for Packaged Design of UHV DC Projects

1. **Overview**
   The R&D Center for Packaged Design of UHV DC Projects (Figure 12.34) is located in the State Power Economic Research Institute in Beijing. After four-phase construction of 2003, it now covers approximately 2000 m^2 and is capable of simultaneously performing secondary system testing and experimental studies of one UHV DC transmission project and three conventional HVDC transmission projects. This center is equipped with 13 real-time digital simulators (RTDSs) for power systems and three DC control and protection systems that have the same software/hardware platform as those used in the Three Gorges HVDC project, Gaoling back-to-back HVDC project, and Xiangjiaba–Shanghai UHV DC project.

Figure 12.34
R&D center for packaged design of UHV DC projects.

In 2009, it was awarded the "State Energy R&D (Test) Center for Packages Design of UHV DC Transmission Projects." SGCC accredits the "HVDC Real-Time Simulation Laboratory" as a key laboratory of the corporation.

2. **Main functions and technical data**

The role of this R&D center is to promote localization of HVDC equipment, improve reliability, reduce construction cost, and provide all-around services and support for HVDC transmission projects in China. It aims to provide technical support and acts as a resource-sharing and management platform for key technology research, system design, valve hall design, packaged equipment design, equipment procurement specifications, equipment manufacturing supervision, system commissioning, and field commissioning of EHV/UHV DC transmission projects.

This center is mainly composed of a system test support platform, a design-supporting system for key techniques of valve halls, and a special software system for integrated design of UHV DC projects, forming an integrated, comprehensive support system.

a. System test support platform. Based on the RTDSs, the system test support platform is built for joint commissioning and overall function and performance tests for DC control and protection systems (pre-commissioning of DC projects). This way, possible defects on secondary equipment and interfacing problems with primary equipment may be resolved before the equipment is delivered to the site, so that the DC project begins operation after a shortened field commissioning.

   The joint testing of a secondary system for DC projects covers the DC control and protection system, AC protection and telecontrol system, auxiliary control system, DC electronic measuring components, and VBE. In addition, this platform can also be used for studies of cutting-edge technologies such as the UHV DC multi-terminal system, interaction after interconnection, sub-synchronous resonance, islanded operation, and black start.

b. Design supporting system for key techniques of valve halls. This system involves determination of the air gaps in the valve hall and design of special fittings and shielding for the valve hall by setting-up proper calculation models and simulation methods. With this system, the valve hall design can be optimized while ensuring its operational security and greatly reducing its construction cost.

c. Special software system for integrated design of UHV DC projects. The software is composed of a special program, interfaces, a database, and an analysis system, mainly including the tool and software, various interface software, and DC system engineering database for integrated design of HVDC (UHV DC) systems. This system is developed by absorbing imported technologies and taking into account the characteristics and difficulties of UHV DC projects, the goal of independent design, and the actual need of DC projects, all specifications, standards, design software, and actual operating data related to the packaged design of DC projects are integrated to make this system the one that truly meets the needs of UHV and ±660-kV HVDC projects. It provides necessary technical support for the packaged design and technical breakthrough of DC projects, equipment manufacturing supervision, factory test, field commissioning, and data analysis.

12.4 Standardization of UHV Transmission Technologies

12.4.1 Standards System of UHV AC Transmission Technologies

When building UHV projects, China attaches great importance to the standardization of UHV technologies. Relying on the pilot and demonstration project, SGCC proposed to advance research, construction, and standardization at the same time. With the completion and commissioning of the project, major innovations were achieved in studies of standards for electric engineering and electric power.

1. **A strong organization and management system was set-up**
   With the support of Standardization Administration of the People's Republic of China and China Electricity Council, SGCC coordinated organizations involved in the construction and operation of the UHV AC pilot and demonstration project to establish a UHV AC transmission technology standardization committee under the project leadership team as a unified, efficient, and tiered management body for UHV AC standards.
2. **A UHV AC transmission technology standard system was proposed**
   Establishing a standards system provides technical basis for standardization. With proactive plans and thorough discussions, and by relying on study results and the pilot and demonstration project, the world's first comprehensive, systematic, coordinated, and reasonable standards system for UHV AC transmission technologies was proposed. This system covers 77 standards under seven categories, namely, basic and generic, environmental protection, planning and design, equipment and materials, construction, measurements and tests, and operation and maintenance, all of which are approved by competent standard authorities. These standards include 36 national standards and 41 industrial standards.

   All of these standards have been completed, among which 62 (28 national standards and 34 industrial standards) have been issued and applied in the UHV AC expansion project and the Huainan–Shanghai project. The standards system plays an important role in summarizing the experience of the UHV AC projects, promoting the application of universal design, universal equipment, universal cost, and standard process, and in improving the quality, efficiency, and benefit of UHV projects.
3. **Significant breakthroughs were made in promoting China's UHV AC standards to become international standards**
   China's great progress in UHV AC standardization attracts extensive attention from the international power industry. A number of China's technical achievements have been

recognized by the International Electrotechnical Commission (IEC), International Conference on Large High Voltage Electric Systems (CIGRE), and Institute of Electrical and Electronics Engineers, greatly strengthening China's voice and influence on international electric and electrotechnical standards. China's UHV AC standard voltage has been adopted by the IEC. To develop or revise related IEC standards to cover UHV technologies, China prompted CIGRE to initiate technical preparations and establish a number of working groups led by China, including:

WG A3.22—Technical requirements for substation equipment exceeding 800 kV

WG B3.22—Technical requirements for substation exceeding 800 kV

WG C4.306—Insulation Coordination of UHV AC Systems

WG A3.28—Switching phenomena and testing requirements for UHV and EHV equipment

WG B3.29—Field test technology on UHV substation construction and operation

A large number of China's study results are referenced in the technical reports of these working groups. In addition, IEEE established a Chinese-chaired Ultra-High-Voltage Working Group. A substantial portion of three active projects sponsored by China and managed by this working group, including IEEE P1862: Guide for Voltage Regulation and Reactive Power Compensation at 1000-kV AC and Above, IEEE P1861: Guide for On-site Acceptance Tests of Electrical Equipment and System Commissioning of 1000-kV AC and Above, and IEEE P1860: Recommended Practice for Overvoltage and Insulation Coordination of Transmission Systems at 1000-kV AC and Above, is also referenced in China's study results.

The Standardization Administration of the People's Republic of China gave the UHV AC pilot and demonstration project of China the "National Major Standardization Demonstration Project" award in 2009. In addition, SGCC won the first prize, the 2010 China Standards Innovation and Contribution Award, with 73 UHV AC corporate standards.

The rapid growth and extensive application of UHV AC transmission technologies in China will promote the coordinated development of construction, research, and standardization; will contribute to the use of the latest study results; will improve the standards system; and will make UHV AC transmission projects more standardized. The building of the standards system is a process of summarizing mature technologies and practices. It will greatly improve the security, cost-effectiveness, and construction of UHV projects, and it will provide technical support for building safe, reliable, innovative, environmentally friendly, and quality projects.

12.4.2 Standards System of UHV DC Transmission Technologies

China's ±800-kV UHV DC transmission technologies are original. Initially there was no standard available for this voltage class. A standards system must be well-established before actual construction of a UHV DC project. Through the Xiangjiaba–Shanghai project, the world's first series of ±800-kV UHV DC standards is established in a comprehensive, integrated, and coordinated way for maximum overall benefits. The series consists of 64 standards covering planning and design, equipment and materials, construction, measurements and tests, operation, and maintenance of UHV DC projects. Since it was commissioned in July 2010, the Xiangjiaba–Shanghai project has transferred more than 2.5 × 1010 kWh of clean hydropower. The annual power delivered can meet 24% of Shanghai's power demand, fully showcasing the results of resource optimization, energy conservation, and emission reduction.

These standards are developed based on a large amount of research of key technologies and independent innovations through 4-year efforts (2006–2009). Since 2005, SGCC has completed five studies under the National Key Technology R&D Program, 87 key technology studies, and 37 project-specific studies, independently built the UHV DC test base, state energy R&D (test) center for packaged design of UHV DC transmission projects, high-altitude test base, and UHV tower test base, forming a UHV test platform with the most complete functions, the strongest test capability, and the highest technical level in the world. These research results lay a solid foundation for the development of the standards series, providing strong support for construction of projects and test bases.

This standards series contributed to breakthroughs in R&D and in the application of UHV DC transmission technologies, greatly improved the capabilities in technology development and innovation, upgraded the electrotechnical equipment manufacturing industry, and significantly enhanced the core competitiveness of relevant industries. The Xiangjiaba–Shanghai project, completed under the support of these standards, features the highest voltage class, largest transmission capacity, longest transmission distance, and most advanced technologies in the world, signifying China's leading position in HVDC transmission. This standards series is widely used in a number of subsequent UHV DC projects and has been involved with historic breakthroughs in international projects, showing a perfect combination with engineering practices.

In June 2013, the Standardization Administration of the People's Republic of China approved the application of China Electricity Council for building an "international standardization innovation and demonstration base" for a 3-year term (2013–2015) in SGCC. SGCC won the first prize at the 2013 China Standards Innovation and Contribution Awards with 64 UHV DC standards such as *Guide for Overvoltage Protection and Insulation Coordination of ±800-kV UHV DC Converter Station.*

Based on the ±800-kV UHV DC standards series, SGCC has currently developed seven national standards and 18 industrial standards, prompted the IEC to establish a new HVDC technical committee (TC115) and set-up its secretariat in China, and led to the development of four IEC standards. In addition, an expert from SGCC was elected as the vice chairman of the IEC. All these characteristics greatly strengthen the voice and influence of China in the community of international standards.

The ±800-kV UHV DC transmission technologies are the world's leading DC transmission technologies and an important means to optimize the allocation of energy resources and guarantee the energy security in China. The standards series plays an extremely important role in guiding UHV DC progress, supporting project construction and enhancing standard internationalization.

12.5 Technological Innovation in UHV Engineering

12.5.1 Technological Innovation in UHV AC Engineering

12.5.1.1 Technological innovations already made

1. **Key technologies in UHV AC transmission are mastered**
During the construction of the UHV AC pilot and demonstration project, through system analysis and studies of key technologies such as external insulation characteristics, overvoltage and insulation coordination, electromagnetic environment, live working, and parameters of main equipment, major technical difficulties are solved for system security and stability control, external insulation characteristics under complicated environmental conditions, deep suppression of overvoltage and electromagnetic environment control, and key technical parameters, conditions are determined, and a series of standards, codes, and specifications are established. This way, UHV AC transmission technologies are fully mastered and used to guide grid development in China, with overall capabilities surpassing those of Europe, the United States, Japan, and other developed countries.
   a. System analysis: The AC transmission project with the world's highest voltage class is successfully planned, designed, and operated. Key technologies in planning, operation, security, and stability of UHV grids are mastered. The security and stability of UHV grids, reasonableness of grid structure, and power flow distribution, as well as overvoltage and electromagnetic transient problems of UHV transmission system are solved. This way, the safe and stable operation is ensured for the pilot and demonstration project and interconnection. Also, the capabilities of optimizing allocation of energy resources and mutually supporting the regional networks are improved.
   b. External insulation characteristics: The external insulation characteristics at 1000-kV AC under high altitude, heavy pollution, heavy icing, and other special environmental conditions are fully learned. China stretches across a vast area covering different natural conditions, including altitude, pollution, and icing, which have great impacts on the external insulation characteristics of UHV equipment.

Limited by technical and testing capabilities, no relevant studies have been conducted in the world before the pilot project. Therefore, it is necessary to devise a set of systematic methods and develop proper equipment for full-scale tests of UHV external insulation characteristics under different combinations of environmental conditions. Test results are scientifically analyzed to determine the external insulation design data for UHV AC projects.

c. Overvoltage and insulation coordination: HV shunt reactors, high-performance arresters, and circuit breakers with closing resistors are, for the first time, combined to substantially suppress possible overvoltage for lines and substations. Insulation coordination is optimized to ensure the cost-effectiveness, security, and reliability of UHV transmission lines. Different lightning protection for lines and optimized configuration of MOAs for substations are proposed with a lightning trip-out rate of approximately 0.1-times/(100 km · a) lower than that of UHV lines in the Soviet Union and Japan and 500-kV lines in China. Basic parameters such as insulation level for UHV equipment and air gaps for UHV lines and substations are proposed to determine the external insulation in heavily polluted areas. A neutral grounding reactor is installed on the neutral of HV shunt reactors to greatly suppress the secondary arc current, which allows successful single-phase reclosing within 1 s. It is concluded that HGIS disconnectors require no switching resistors, and the effects on VFTO, a key consideration for the reliability and cost-effectiveness of equipment, are verified for the first time. Key parameters such as the transient recovery voltage (TRV) for circuit breakers and time constant of the DC component of short-circuit current (120 ms) are determined, which indicates for the first time that UHV circuit breakers are incapable of reducing such time constants.
d. Electromagnetic environment: Bundle conductors and optimized design of fittings are used to make UHV projects as electromagnetically friendly as conventional transmission technologies. A number of studies and tests are conducted according to control standards for ecological, information and communication, and aerospace environments. This results in electromagnetic interference levels along UHV line corridors that are superior to international standards.
e. Live working: Difficulties in technologies and equipment for live working under equipotential on a 1000-kV line are solved. This makes it possible to perform maintenance activities on 1000-kV live lines without cutting the power. Studies of such maintenance are conducted, applicable technical standards are established, and a complete set of tools are developed and used.
f. Studies of specifications for main equipment and parameters and acceptance methods for key equipment: Based on the independently developed technical specifications (standards) for UHV AC equipment, a complete set of the world's leading UHV AC equipment is independently developed with a number of world records. These lead to enhanced capabilities in equipment manufacturing and international competitiveness.

Through studies, the parameters for 1000-kV main equipment are determined and a complete set of technical specifications are prepared.

2. **A full package of UHV AC equipment is independently developed**
The 1000-kV UHV AC pilot and demonstration project is the start point of UHV transmission and a vehicle for localization of UHV equipment. It is challenging to develop advanced and highly localized equipment suitable for commercial operation in the project within a short period of time. During this process, relying on independent innovation and with the purpose of boosting China's manufacturing industry, collaboration of enterprises, universities, and research institutions is promoted to fully utilize all resources. Through these efforts, the following innovations are achieved:
   a. The world's first 1000-kV, 1000-MVA UHV transformer: Key technologies for insulation design, leakage flux control, and local overheating prevention are mastered, and UHV transformers are developed, installed, tested, and operated on-site successfully, laying a foundation for safe operation of the pilot and demonstration project.
      Because no applicable standards or mature experience is available, it is difficult to design and manufacture such a UHV transformer with high-voltage and insulation levels. In the pilot and demonstration project, these difficulties were solved and the world's first 1000-kV, 1000-MVA transformer with a single-limb winding rated 1000 kV and 334 MVA was developed, creating a world record.
   b. UHV shunt reactor with the world's largest single-phase capacity: Difficulties in controlling the leakage flux, temperature increase, noise, and vibration under high-voltage and large-capacity conditions are solved. A large-capacity testing transformer is developed, and the voltage withstand and partial discharge test capabilities are improved.
      The UHV shunt reactor has similar difficulties in insulation design for the UHV transformer. A number of key technical difficulties have to be solved. The 320-Mvar UHV shunt reactor developed for the pilot and demonstration project used the single-limb design for the first time and is a leading product in the world.
   c. GIS/HGIS with the world's highest operating voltage: This ensures the reliability of UHV switchgears. The capabilities of domestic switchgear manufacturing industry in design, manufacturing, production management, test, and quality control are improved. Substantial experience is gained for installing and operating products at different voltage classes in subsequent projects.
      As an essential control and protection device for a UHV transmission system, switchgears are important in determining the operating scheme and safe operation of the system. With increasing breaking capacity of UHV circuit breakers, the requirements of the breaking performance are higher. The 1000-kV GIS developed for the pilot and demonstration project is a world leader with the highest operating voltage.

d. UHV arresters, capacitive voltage transformers, and UHV insulators for moderately and heavily polluted areas, as well as LV reactive power compensation equipment for UHV systems: Difficulties are solved for the development of UHV arresters, transformers, and insulators, which are all important parts of the pilot and demonstration project, and the manufacturing capabilities, testing equipment, and technologies are all improved, so that on-schedule operation of these equipment and safe and reliable operation of the project are ensured.

e. Secondary equipment for UHV system: New algorithms and principles are proposed to improve the software and hardware design, and then the overall performance of the secondary equipment. All equipment and systems have passed stringent testing, joint commissioning, and acceptance. By concentrating the best human knowledge, material, and management resources of related organizations, all uncertainties are solved before the equipment is delivered to the site, so that field commissioning is shortened and system reliability and stability are ensured.

   Secondary equipment is important for a substation. It includes relay protection and control, computer-based monitoring and control, metering, auxiliary DC systems, the performance and reliability of which are of vital importance to system security and stability. Because the UHV grid is of particular importance to system security, the secondary equipment has to be highly reliable. The monitoring and control system used in the pilot and demonstration project uses, for the first time, the panoramic data analysis-based post-disturbance review and 3D simulation training system.

3. **Full technologies for the design, construction, test, operation, and maintenance of UHV AC projects are independently developed**
   Currently, no mature standards or specifications are available for design of UHV AC projects. Through studies, a complete set of integrated technologies for planning, design, equipment integration, construction, installation, commissioning, test, operation, and maintenance are mastered, giving rise to a technically and economically viable, safe, reliable, advanced, and practical transmission.

12.5.1.2 Continuous technological innovation

1. **UHV step-up transformer**
   UHV step-up transformer has promising prospects in China. Connecting power plants to the UHV system through UHV step-up transformer could avoid short-circuit current overstress on the 500-kV side and help increase the transmission distance and suppress voltage fluctuation. UHV step-up transformers developed by China (Figure 12.35) involve the following four technological innovations:
   a. The main technical parameters of 400-MVA UHV step-up transformers were determined after demonstration. Technical specifications for prototype for research purposes were developed. Two feasible design schemes were completed, one with a

Figure 12.35
UHV step-up transformer.

single active part (four-legged single-phase structure) and the other with two active parts (a separate active part for regulating winding).

b. For each design scheme, in-depth research and analysis were conducted for the main and longitudinal insulation strength, short-circuit withstand capability, leakage flux, local overheating, and mechanical strength of oil tank. Main difficulties, such as insulation design, leakage flux control, temperature increase, and short-circuit withstand capability were solved. Three prototypes have been developed and all passed type tests with sound performance. Partial discharge was basically avoided in the design schemes.
c. The concept design of the ±800-kV UHV step-up transformer was proposed based on the studies conducted using 400-MVA transformers. Four structure schemes and main parameters such as dimensions and weight were obtained.
d. The cost-effectiveness was demonstrated by analyzing the cost of plant equipment under different connection schemes. A conclusion was reached that the use of UHV step-up transformers has certain economic advantages under certain conditions.

2. **UHV large-capacity transformer**

To fully exploit the transmission capability, maximize the cost-effectiveness, and meet the demand of all possible UHV AC transmission projects including double circuits sharing the same tower, it is necessary to develop UHV AC transformers with a capacity of 4500 MVA (UHV large-capacity transformers).

Increased capacity results in larger internal and external dimensions, which must be controlled due to transport constraints. Therefore, the dimensions, transportation, insulation design, leakage flux control, and temperature increase control must be considered in developing these transformers.

The single-phase 1500-MVA/1000-kV UHV large-capacity transformer developed by China has a construction similar to the 1000-MVA/1000-kV transformers at the Changzhi Substation. It is composed of an autotransformer and a regulating transformer. The former uses four-frame five-limb iron cores, namely, three main limbs and two side limbs. Windings on the three limbs are connected in parallel, and each limb has three windings, namely, LV, common, and series windings, respectively, from inside to outside. The regulating transformer is equipped with a regulator and a compensator that are integrated through a lead in the same tank. It is connected to the autotransformer via an overhead tubular bus to form a complete single-phase off-circuit-tap-changing autotransformer. Figure 12.36 shows the outline of an autotransformer.

Test results show that this transformer is free from partial discharge, capable of controlling local overheating with high-voltage class, and has large capacity, small loss, and low noise.

Figure 12.36
Outline drawing of autotransformer.

The successful development of the UHV large-capacity transformer helps fully exploit UHV AC transmission capability, ensure cost-effectiveness of UHV transmission, and enhance capabilities in transformer manufacturing.

3. **UHV oil-SF_6 bushings**
   In UHV substations, transformers and GIS are generally connected through transformer bushings, overhead lines, and GIS bushings. Connection method mentioned above has the following disadvantages: (i) large height; (ii) long production duration; (iii) large land occupation; (iv) vulnerability of exposed parts to environment; and (v) large maintenance workload. The use of oil-SF_6 bushings for direct connection between transformer and GIS has the strengths of small land occupation, high reliability, no corona noise pollution, and small maintenance workload. These bushings are so named because one end is immersed in transformer oil and the other is immersed in SF_6.

   The development of UHV oil-SF_6 bushings is groundbreaking in many aspects: (i) a platform for optimized design is built to achieve homogeneous radial, axial, and tail electric fields and reasonable structure, and to solve difficulties in designing and optimizing main insulation; (ii) the dielectric property of epoxy resin and permeability, bonding, and compatibility performance of crepe paper are learned to obtain the formula for epoxy resin for UHV dry-type oil-SF_6 bushings; (iii) crepe paper bushing technology, epoxy resin impregnation, and curing technologies, as well as the packaged manufacturing process of 1000-kV AC dry-type oil-SF_6 bushings, are learned to master the design and manufacturing capabilities and develop 1000-kV/3150 A bushings; (iv) packaged manufacturing process of 1000-kV AC-immersed oil-SF_6 bushings and 1000-kV/3150-A bushings are developed (as shown in Figure 12.37); and (v) packaged testing facilities are developed and demonstrated, and test technologies are mastered for assessing the performance of UHV AC oil-SF_6 bushings.

4. **UHV controllable shunt reactor**
   Like fixed HV reactors, controllable shunt reactors are directly parallel-connected to both ends of the line or the bus on the HV side of the system. Their reactance can be regulated continuously or in a stepped way. They can greatly mitigate the contradiction between reactive power regulation and overvoltage suppression with the following functions:
   a. Under steady-state conditions, the output can be flexibly regulated based on operating conditions of the system to dynamically compensate charging power of lines.
   b. When the line operates under a light load, the capacity is maximized to suppress the power frequency overvoltage; when the line operates under a heavy load, the capacity is minimized to maintain voltage stability, increase transmission capability, balance reactive power distribution, and reduce line loss. When the line is open-circuited, the capacity is rapidly maximized to limit switching and power frequency overvoltages. If a fault occurs on the line, then they can help extinguish the secondary arc and improve the success rate of single-phase reclosing and transient stability.

Figure 12.37
UHV oil-immersed oil-SF_6 bushings.

In addition, as a flexible means of reactive power regulation, controllable shunt reactors can improve the dynamic stability of power grid under weak damping.

The studies of key technologies of EHV controllable shunt reactors and product demonstration began in 2005. The world's first 500-kV stepped controllable shunt reactor and 500-kV magnetically controllable shunt reactor began operation in 2006 and 2007, respectively. Research and prototype development for UHV controllable shunt reactors began in 2008. This is the world's first effort to develop UHV controllable shunt reactors. Because there is no experience to reference, and because the technical requirements are high, the task is very difficult. Through 4-year efforts, a single-phase prototype has been developed and the technical scheme of packaged UHV controllable shunt reactors, design scheme, control and protection strategies, and prototype of high-power self-cooling thyristor valves were independently developed and have passed the type test. The prototype is rated 1000 kV/600 Mvar and is capable of three-step regulation with a transient response time of less than 50 ms. The schematic diagram and prototype are shown in Figures 12.10–12.18 and 12.10–12.23.

5. **UHV double circuits sharing the same tower**
With the development of industry and agriculture, the shortage in land resources is becoming worse. To minimize line corridor, reduce transmission cost, and increase transmission efficiency, it is an effective way to use double-circuit transmission.

With experience from the Huainan–Shanghai project, a comprehensive study was conducted on the double-circuit arrangement, the results of which are as follows:

a. A prototype experimental line was built and studies were performed regarding discharge characteristics of I-strings and V-strings under power frequency voltage, long wave front switching impulse voltage, and lightning impulse voltage to understand the insulating properties of tower clearances of different phases.
b. For interphase insulation safety between upper and lower conductors, studies of discharge characteristics under long wave front switching overvoltage were conducted to master the law of interphase insulation at different air gaps and proportion of switching impulse voltage of different polarities.
c. According to lightning statistics along the line, the protective angle afforded by a ground wire was determined to be smaller than 3° in plains and hilly areas and 5° in mountainous areas. The minimum air gap under lightning overvoltage is 6.7 m for lines at altitudes of 500 m or less.
d. Through optimizing design, shed-type towers were used, which greatly reduced the investment cost and improved the operational reliability.

6. **UHV compact transmission**
The compact transmission technology may be used to improve land efficiency, reduce line corridor, and increase transmission capacity. Having always attracted wide attention, it is one of the major trends of UHV transmission technologies. China has conducted in-depth studies of this technology, such as discharge characteristics of tower air gaps and interphase air gaps under long wave front impulse voltage, electromagnetic environment, system overvoltage and insulation coordination, live working, anti-galloping measures, and fitting design for UHV single-circuit compact lines. Configuration and arrangement of conductors and interphase spacers that could make the projects more electromagnetically friendly have been proposed. The overvoltage level and suppression measures were determined. In addition, key parameters were obtained for live working on UHV compact lines. These study results can be directly used for engineering design of UHV compact lines. This offers a solution to grid development in areas with limited line corridors or higher demand for transmission capacity [3].

7. **UHV expanded diameter conductor**
Transmission loss is a major problem that harms the cost-effectiveness of UHV transmission. It includes substation (or converter station) loss and line loss. The latter is mainly caused by resistance and corona.

A techno-economic analysis was conducted for operating UHV AC projects in China. The economic current density of conductors was determined by analyzing investment, operating loss, line loss hours, on-grid tariff, and line construction cost. A conclusion was reached that new conductors with a larger expanded diameter ratio should be developed to reduce investment and improve economic benefits [4].

Based on the economic current density obtained, and combined with analysis of mechanical properties, electromagnetic environment, and economic benefit, a design scheme for expanded diameter conductors with HDPE sustentation was proposed to improve existing composite expanded diameter conductors.

12.5.2 Technological Innovation in UHV DC Engineering

12.5.2.1 Technological innovations [5]

UHV DC transmission projects are diversified in configuration, operating mode, and other aspects. By performing groundbreaking work on the overvoltage, insulation coordination, external insulation, and performance of key equipment, China has made several innovations.

1. **Key technologies in UHV DC transmission**
   a. The configuration scheme of connecting two 12-pulse converters in series per pole with voltage being equally distributed is introduced. Schemes of each pole having one converter and two series-connected or parallel-connected converters were compared. After considering factors such as the difficulties in manufacturing and transportation of large equipment, system stability, and project economy, the series connection scheme was chosen for ±800-kV UHV DC systems to boost the system voltage.

      Detailed comparisons were made among three voltage distribution proposals (600 + 200) kV, (500 + 300) kV, and (400 + 400) kV, respectively. The results show that the (400 + 400) kV scheme has prominent advantages in technology, operational flexibility, and economy; therefore, it is selected as the converter connection scheme in ±800-kV UHV DC applications.

      The online switching in/out of converters achieved by a bypass path increases the system operational reliability as well as reduces the monopolar fault rate and the impact of converter faults on the AC system. Also, the scheme of connecting multiple smoothing reactors in series and splitting them equally on the line pole and neutral bus reduces the difficulties in equipment manufacturing and overvoltage level in line pole equipment.

      Based on the AC system conditions and system configuration studies, main circuit parameters of the UHV DC pilot project are determined, including the rated voltage, rated current, overload capacity, reversed power flow, range of reduced-voltage operation, and short-circuit impedance of converter transformer.

      Based on characteristics of the configuration scheme of two series-connected converters, a total of 45 operating modes are determined for ±800-kV UHV DC systems. Added to these is a deicing configuration in which the two HV-end converter valve groups are connected in parallel, whereby the line current is increased to twice that (8000 A) under normal operation and the DC line is capable of preventing and melting ice.

b. Studies have been performed regarding overvoltage and insulation coordination. The amplitude and distribution characteristic curve of overvoltages along ±800-kV DC lines are obtained by fully simulating the characteristics and levels of overvoltages in converter stations and on lines of a ±800-kV DC link under various switching operations and faults.

   According to characteristics of the ±800-kV UHV DC project configuration and overvoltage simulation results, a comprehensive technical scheme is proposed to control the insulation levels at various points of the HV-end converter by measures like properly arranging surge arresters, optimizing surge arrester ratings, and equally arranging smoothing reactors on line pole and neutral bus. Also, the arrangement plan and parameters of surge arresters are proposed.

   The insulation levels of ±800-kV electrical equipment are determined. Specifically, the lightning and switching impulse withstand voltages of the valve side equipment and bus of the HV-end converter transformer are 1800 and 1620 kV, respectively, and those of the DC bus are 1900 and 1600 kV, respectively. Considering factors such as the insulation coordination factor and safety margin as well as altitude correction of external insulation, the switching, lightning, and steep-front impulse withstand voltage margins are selected as 10%, 10%, and 15% for converter valves, respectively, and 15%, 20%, and 25% for other DC equipment, respectively. The lightning and switching impulse withstand voltages of bushings on the converter transformer are taken as 1.05-times those of the windings.

c. Studies are conducted on external insulation. Full-scale towers and electrodes are used based on which of the discharge characteristics of large air gap and full-scale electrode air gap and which of the relevant altitude correction factors are obtained; values of air gap for tower, air clearance for DC switchyard, and valve halls are recommended for ±800-kV projects. In addition, the pollution flashover characteristic curve of UHV insulator strings with a maximum length of 15.6 m under −920 kV was obtained and, relying on this curve, the pollution and ice flashover characteristics of UHV DC full-scale insulators and altitude correction schemes have been obtained for the first time. These results can be directly used to guide the external insulation design of UHV projects and provide reference for future projects.

d. Studies have been performed regarding the electromagnetic environment and noise control. Full-scale tests under rated voltage are conducted on ±800-kV DC lines, which, combined with theoretical studies, determine the conductor type, minimum conductor height, and minimum line corridor width for ±800-kV DC lines in China. Other results of these studies include the lateral profile and changing pattern of various electromagnetic environmental factors of lines with different conductor heights and pole spacing in different seasons and climate. The effects of ±800-kV DC lines on nearby trees have been studied, through which the minimum line-tree distance can be determined for eco-friendly consideration. Full-scale tests are also conducted to analyze the shielding effects of residences on the total electric field on ground, audible noise, and radio interference where the ±800-kV DC line approaches the residences as well as the effects of DC lines on radio and television reception. The electromagnetic environment limits have been proposed. Full-scale tests are conducted regarding the electromagnetic environment of a ±800-kV DC line with vertically arranged pole conductors, for which a recommended arrangement with the optimal electromagnetic environment is given. Based on the UHV DC corona cage tests, China derived its first audible noise prediction formula with IPR, which has a wider scope of application than any other peer international formulas. A UHV DC corona–loss measuring device has been developed and the usability of method for calculating corona loss on UHV DC line can be proven through full-scale tests. In addition, an optimized scheme of arranging the ±800-kV converter station face-to-face has been proposed. The use of noise reductions measures like Box-ins for converter transformers and integral noise reduction devices for smoothing reactors greatly mitigates the effects of noise from converter stations on surroundings, thus ensuring the noise generated by ±800-kV converter stations meets applicable environmental protection requirements.

For AC and DC lines sharing the same corridor, an electromagnetic coupling model is established. The minimum conductor height and corridor width are determined for the ±800-kV DC line and the 1000-kV AC line based on this model.

e. Studies have been conducted regarding specifications for equipment of ±800-kV UHV DC transmission systems. The type and main parameters of equipment used in ±800-kV UHV DC project are determined and a complete set of technical specifications for the equipment, such as 6-inch thyristor, converter valve, converter transformer, smoothing reactor, and DC switchyard equipment, are developed.

2. **Development of equipment for converter station**
   a. Development of 6-inch thyristors and converter valves. During the process, the following work was performed:
      i. According to the specifications for UHV converter valves and the state-of-the-art manufacturing technology of high-rating power electronics, main parameters of 6-inch thyristors are determined, setting a correct path for their development. The special pretreatment process of monocrystalline silicon solves the key difficulties caused by increasing the thyristor chip diameter to 6 inches, such as control of the uniform diffusion of impurities, uniform thickness and parameter of chip, and chip deformation, thus making it possible to fabricate thyristors with high current and blocking voltage ratings by 6-inch

monocrystalline silicon. The blocking voltage/wafer thickness ratio is introduced to determine the high-rating and high-compatibility thyristor structure, achieving balance between the high blocking voltage and low loss with the chip thickness and other parameters. A special combined process of pre-deposition photo etching selective erosion–oxidation and redistribution is developed, enabling accurate control of the P-type impurity's junction depth, selectivity, and impurity concentration gradient at the junction edge and solving difficulties in manufacturing large-section thyristors.

ii. Compatibility between the thyristor and converter valve is optimized. The converter valves and 6-inch thyristors are developed synchronously. Their compatibility is optimized at all valve development stages, which ensures the full performance of both.

iii. The UHV valve tower structure, shielding electrode, and air clearance are optimized and checked. Through studies and tests of the valve tower's interlayer and tower-to-ground shielding electrodes, the design of various electrodes and air clearances are optimized, enabling the UHV converter valves to have a higher insulation level and a more compact and lighter valve tower when the dynamic voltage is distributed more uniformly. Also, the requirements of air clearance in the valve hall are greatly reduced.

iv. A novel design of the valve tower cooling water loop has been developed. The use of a spiral water tube greatly reduces the number of joints and, consequently, the probability of leakage while fulfilling the cooling requirements of converter valves.

v. The conventional mechanical design platform for converter valves has been upgraded and improved. The new platform is capable of performing static structural analysis, transient dynamic analysis, and contact analysis of valve structural models, ensuring that the new valve tower design is reliable and reasonable.

b. Development of a converter transformer for ±800-kV UHV DC applications. In the Xiangjiaba–Shanghai project, the converter transformer has its highest voltage at 800 kV (DC) on its valve side and 500 kV (AC) on its line side. To withstand such high voltages, the converter transformers are subject to complicated electric, magnetic, thermal, and mechanical stresses, and their overall dimensions are constrained because of transportation conditions. There are also great difficulties in insulation design. These are great challenges to the electrotechnology, materials technology, and HV test technology. Through the development process, the following breakthroughs have been made.

i. Difficulties in insulation design are solved. The primary difficulty is the main and longitudinal insulation performance when the significantly increased insulation level (switching impulse withstand voltage of 1600 kV, lightning full-wave impulse withstand voltage of 1800 kV, applied long-duration AC voltage of 912 kV, DC partial voltage of 1258 kV, and polarity reversal voltage of 970 kV) with limited dimensions causes the valve side windings to suffer combined AC and DC voltage and polarity reversal voltage. Studies have been conducted regarding the effects of combined AC and DC voltage on insulation characteristics, effects of AC and DC voltage components on partial discharge characteristics of oil-paper insulation, effects of space charges on oil-paper insulation characteristics, dielectric properties of transformer oil, bushing end finished parts and porcelain parts, distribution of dynamic electric fields in oil-impregnated paper insulation structure at polarity reversal, and arrangement and optimization of grading/shielding devices, oil gap separation, and finished parts. The study results have created a theoretical foundation and basis for the converter transformer insulation design, material characteristics, effects of space charges, and optimization of insulation design. The insulation design has been optimized by analyzing the electric field distribution under different voltages and conducting model tests. These efforts ensure the proper distribution of internal fields, increase the dielectric strength, and solve the difficulties in insulation design. Also, through the process, China has mastered the core technology of designing internal insulation of UHV converter transformers and manufacturing technology of finished insulating parts.

ii. Valve side bushing and exit leads are developed. Both the withstand voltage and length of the converter transformer's valve side bushings are greatly increased with the increasing insulation levels, therefore making the internal insulation design and manufacturing much more difficult. Methods such as 3D holographic modeling of the distribution of AC, DC, and harmonic fields are used to achieve their uniform distribution in bushings approximately 14 m long. China has developed an insulating resin formula with the highest fluidity worldwide and a special vacuum impregnation process that can avoid tiny bubbles, solving the problems in uniform impregnation of multilayered insulation papers of long condenser cores in resin. An integral insulating device has been developed for seamless connection of the valve side exit lead and the bushing end and for prevention of nonuniform distribution of electric fields on different insulating materials. So far, this insulating device has the highest voltage withstand capability, most complex structure, and highest homogeneous field strength.

iii. The connection scheme of regulating windings are optimized. Alternate series connections of two limbs and a proper copper shielding structure are used to meet the capacitance requirement of the regulating lead, to resolve the local heating caused by the lead, and to reduce the number of leads connected to the switch by 50%.

iv. Difficulties in partial discharge control are solved. Electric fields of the main insulation, insulation of finished parts, and outgoing line insulation are analyzed and calculated using the finite element software, thus helping to understand the mechanism of partial discharges within the converter transformer under combined AC/DC electric fields. By controlling the maximum field strength at various positions, the partial discharge is effectively controlled and, hence, the equipment quality is greatly improved. Moreover, the use of an advanced manufacturing process and effective quality control of materials also contribute to the control of partial discharge due to fabrication tolerance and dispersity of materials.
v. Difficulties in leakage flux, harmonic, and temperature increase control are solved. By analyzing and calculating the distribution of static magnetic fields, leakage magnetic fields, and eddy current loss, reference criteria for determining partial discharges in converter transformers are provided; the internal cooling system and oil circulation are properly designed to effectively control the temperature increase at hot spots.
vi. Single-phase four-limb cores are used. The core is made of high-grade, low-loss cold-rolled grain-oriented silicon steel and is six step–laminated, which can greatly reduce the no-load loss and noise. The core limbs are tied with high-strength tie bands. The tie plates and lateral tie bands are all made of high-grade steel. The top and bottom clamps are welded with positioning devices that are secured to the oil tank's bottom and cover, respectively, during final assembly to prevent the active part from displacement during transportation. All tightening bolts and nuts on the core are provided with reliable locking measures to prevent them from becoming loose under violent vibration. The core clamps and laminations are grounded via special grounding conductors at the oil tank; therefore, the insulation resistance can be easily meggered from outside the transformer.
vii. The converter transformer's line side and valve side windings are all wound by special transposed conductors. According to the rated current, voltage, and insulation level, the winding may be of the interleaved type, capacitor shield type, or helical type. In terms of insulation performance, these windings types feature lower potential gradient under impulses, which makes it easier to control impulse fields. Field strength at all main positions are controlled through relevant insulation studies, thus ensuring the overall reliability of converter transformer.
viii. The active part is compact and the insulation structure is well-designed according to the distribution of AC and DC electric fields and the properties of insulating materials. The active part is provided with a pressure plate. Before press fitting, the active part must be thoroughly dried. A multiple-point

hydraulic cylinder will be used to press the plate synchronously to make the active part an integral structure. Such an integral structure not only protects the windings from displacement during transportation but also greatly increases the transformer's capability of preventing short-circuit. Moreover, the active part is provided with electrical shield-to-shield electric fields.

ix. The converter transformer leads are of a new design. The valve side leads are all arranged inside the oil tank and the two valve side bushings protrude out of the oil tank from its side wall. Based on the maximum width allowed for transportation, the internal leads are reasonably arranged, the on-site installation procedure is simplified, and the installation time is shortened.

x. The converter transformer uses a cylindrical oil tank with flat cover and flat side walls reinforced with the plate. The oil tank's walls, bottom, cover, and reinforcements are all made of high-strength structural steel and, particularly, the bottom and cover are made of one-piece steel plates that have all passed the ultrasonic testing before welding. All welds on the tank body and important reinforcing ribs are also required to pass the ultrasonic testing. The oil tank is checked for strength and is supposed to withstand a vacuum pressure of 13.3 Pa and positive pressure of 98 kPa.

c. Development of dry-type smoothing reactors.

i. Design program of smoothing reactors. A complete set of design and calculation programs and methods are developed to analyze the distribution pattern of electric, thermal, and force fields. The reactor design program "Reactor91" is used to calculate the temperature increase under combined AC and DC voltage; the calculation program "Mag" is used to analyze and calculate the magnetic fields; the calculation program "Spiderforce" is used to analyze forces applied on spiders; and the program "Coilforce" is used to analyze forces applied on coils.

ii. Class H transposed conductor. Ordinary low-voltage AC reactors use single-wire round aluminum conductors featuring large interlayer insulation stress, large harmonic loss, high temperature increase at hot spots during operation, and difficulty indentifying local defects during manufacturing. Based on the experience in manufacturing LV reactor-used conductors and configuration of transposed conductors used for transformers, the manufacturing technology of class H uniformly insulated aluminum-transposed conductors for reactors with high voltage and large capacity have been developed. Such conductors not only can reduce the eddy current loss and heating of smoothing reactors but also have advantages such as low additional loss, small interlayer insulation stress, and simple manufacturing process.

iii. Supporting system. Considering the high-voltage level, large capacity, and weight of dry-type smoothing reactors used in the Xiangjiaba–Shanghai project, and to ensure their operational stability and reliability, a finite element-based computational algorithm has been developed to perform anti-seismic calculations for the supporting system of smoothing reactors. On this basis, a supporting system with satisfactory seismic stability has been developed, and a method has been proposed for checking the seismic stability of 70-ton, 800-kV smoothing reactors with a high current rating.

iv. Noise reduction measures. To minimize the noise generated by smoothing reactors, a new noise reduction scheme has been proposed after studies and improvements. A soundproof enclosure that provides protection against rain, sunlight, and noise and that offers good ventilation has been developed for large-capacity smoothing reactors, which are completely soundproof enclosures for large smoothing reactors. With such an enclosure, the noise level generated by smoothing reactors can be reduced to 16 dB(A), thus greatly lowering the noise level in converter stations and mitigating the conflict between the ventilation and heat dissipation and the noise reduction.

v. Electric field shielding. To avoid partial discharges on the outgoing line, 800-kV smoothing reactors must be provided with integral shielding structures. The electrostatic fields on and around smoothing reactors are modeled and analyzed using the numerical analysis method. The shielding scheme is selected based on the analysis results of electric field distribution for both the HV and LV smoothing reactors at different distances from nearby walls. In addition, multiple grading rings are added to the top cover according to the UHV corona tests so discharges will not occur on the soundproof enclosure.

vi. Overvoltage distribution between series smoothing reactors. The characteristics of transient overvoltage distribution between two ±800-kV smoothing reactors in series have been proposed.

d. Development of equipment in DC and AC switchyards.

   i. UHV DC wall bushings. Because 800-kV wall bushings are required to have much higher internal and external insulation levels and current-conducting capabilities, their length and conductor dimensions are considerably increased. Also, they are supported only at one point, which makes it more difficult to achieve coordination among the internal and external insulation, ventilation and heat dissipation, and electrical and mechanical reliabilities. The innovative manufacturing process of bushings solves the difficulties in manufacturing large-size, high-strength composite insulating housing and develops the world's state-of-the-art DC wall bushings. Such bushings are 18.8 m in length and 4.2 tons in weight. They have a lightning and switching impulse withstand voltage of 1950 and 1600 kV, respectively, 60 minDC voltage withstand capability of 1020 kV, and maximum long-time current-conducting capability of 4970 A.

ii. UHV DC post insulators. The increase in the DC voltage makes the coordination between the external insulation strength and mechanical reliability of post insulators (including equipment supports) more difficult. If the 800-kV post insulators were designed according to the same specific creepage distance of 54–60 mm/kV for site pollution severity class 2–3 as that used in ±500-kV HVDC applications, then the insulators would be higher than 14 m, in which case the required mechanical performance will be far beyond the mechanical performance limits of existing materials. After considering the mechanical reliability and external insulation requirements, the RTV-coated porcelain insulators with a specific creepage distance of 48 mm/kV and structure height of 10.5 m are selected for UHV DC projects, solving the problem of pollution flashover and mechanical reliability of insulators intended for supporting equipment such as the dry-type smoothing reactor.

iii. UHV DC surge arresters. Surge arresters used in ±800-kV UHV DC transmission projects have high-rated voltage, energy release capability, and insulation structure, and involve numerous types. In addition, several arresters may be connected in series or parallel, if necessary. Through in-depth studies of many aspects such as the aging, discharge energy withstand capability, residual voltage, operation characteristics, transient overvoltage, current division characteristics of parallel-connected arresters, antipollution, and heat dissipation characteristics, a complete set of technical indexes and test methods have been proposed for surge arresters for ±800-kV UHV DC applications. By optimizing the performance of surge arrester's resistors and using an innovative surge arrester structure, the current is equally distributed among all columns and the voltage is equally distributed among all series-connected varistors.

iv. UHV DC voltage and current measuring devices. The use of new terminal insulation design and compensation devices improves the distribution characteristics of impulse voltages on the DC voltage divider to avoid coronas. The current measuring devices for ±800-kV UHV DC applications are all optical fiber-based current transformers, except for some neutral buses, in which zero-flux current transformers are used. A beam-powered remote module has been developed that can acquire signals locally and transmit them through optical fiber. In addition, composite insulators are provided to ensure good insulation and, hence, satisfactory measuring accuracy and reliability.

v. UHV DC disconnectors and earthing switches. The DC disconnectors have a double-column structure. The main knife has a single folding-arm design and supports have a triangle frame design to increase the rigidity. Ring-type contact fingers with multiple contact points are used to assure good contact and, hence, high current-conducting capability. The disconnector is closed by inserting the moving contact in the fixed contact. The two contacts are sealed during both closing and opening operations, weakening the adhesion effects of DC electric fields on the contacts. The disconnector is supported by three parallel-connected porcelain insulators with RTV coating, whereby the difficulties in achieving high insulation level, high mechanical strength, and accurate control of contact positions are solved. The newly developed current-conducting structure and contact system greatly increased the current-conducting and harmonic current breaking capabilities, which are 4500 and 150 A, respectively. The harmonic current breaking capability allows the DC filter's disconnector to be switched while energized if DC filters at one end quit, which makes the maintenance easier.

vi. UHV DC bypass circuit breakers and transfer switches. Bypass circuit breakers for ±800-kV UHV DC applications have been developed that can realize the online switching in/out of converters. A passive DC transfer switch with a maximum transfer current of 4320 A has been developed. The residual voltage of the nonlinear resistor and parameters of the transfer circuit are determined according to simulation results. The LC circuit directly uses the stray inductance in the transfer circuit and therefore needs no reactors. The arrangement of the capacitor bank and nonlinear resistor on the insulated platform is optimized. A spherical bearing unit is designed between the post insulators and supports and lacing bars are designed between the insulated platform and its supports, ensuring that forces are uniformly distributed on the platform. In addition, it dodges the peak frequency of earthquakes. Through calculations and analyses of electric fields, the distribution of electric field strength on switches and electrical stress on capacitors and nonlinear resistors during the DC current diversion are obtained. A proper insulation structure is provided to meet the insulation requirements of switches and breakers.

vii. Fittings in the valve hall and DC switchyard. According to the coordination between shapes of fittings in the valve hall and air gaps, and considering the requirements on corona noise control, current-conducting capability, and mechanical reliability, a complete set of fittings have been developed.

viii. GIS circuit breakers and open-type circuit breakers capable of breaking and making capacitive current. The world's first open-type circuit breaker for 500-kV AC filters, which is capable of breaking and making a capacitive current of 500 A and recovery voltage of 1470 kV and that meets the requirements of class C2, and the GIS-used circuit breaker capable of breaking a capacitive current of 1800 A have been developed.

e. Development of control and protection equipment. Based on the control and protection system in conventional HVDC applications and the distinct feature of UHV DC systems (i.e., each pole having two series-connected 12-pulse converters), full analyses and studies have been performed regarding the overall architecture, control strategy, hierarchy and redundancy, allocation of control functions, and arrangement of protections of the control and protection system. Finally, a real-time operating system-based distributed control and protection system that is adaptable to operation of more than one converter is developed and used. This system is designed in multiple redundancy and is capable of self-checking. Also, a complete UHV DC digital simulation system and a dynamic analog simulation system have been established.

3. **Development of UHV DC line equipment**

   a. Large-section conductors. China had made great achievements in studies of conductors that are 1000 mm^2 and higher. Before the 1000-mm^2 conductors were developed, ACSR-720/50 conductors had the largest cross-sectional area among all types of conductors used in China's transmission projects and had a maximum of three layers of aluminum strands. The 1000-mm^2 conductor has a complicated configuration, large aluminum/steel ratio, and four layers of aluminum strands. The proper strength range of the steel and aluminum wires are determined through studies. To increase the overall performance of conductors, the conductivity is increased, which can reduce the line loss. To verify the overall performance of conductors, tests are performed using conductor samples according to applicable national standards. Based on the test results, technical standards on a 1000-mm^2 conductor have been developed and issued. To ensure the quality of conductors, all four layers of aluminum strands of the 1000-mm^2 conductor are required to be stranded at the same time, for which the four-section frame stranders or even more advanced stranding machines are used, which improves China's conductor manufacturing capability.

      The Ningdong–Shangdong $\pm$660-kV DC demonstration project uses $4 \times$ 1000-mm^2 conductors and the South Kumul–Zhengzhou $\pm$800-kV UHV DC demonstration project uses $6 \times$ 1000-mm^2 conductors. Along with the continuous improvements in conductor manufacturing, even 1250- and 1520-mm^2 conductors will be used in the future.

   b. $\pm$800-kV UHV DC line insulators. To prevent corona discharges on the end surface of composite insulator's fittings and galvanic corrosion of insulator's silicone rubber housing, the composite insulators are equipped with large and small grading rings to improve the electric field distribution. The $\pm$800-kV DC long rod composite insulators and devices used to protect composite insulators from bird pecking have been developed. Large-tonnage composite insulators with rated bearing capacity of 550 kN have been developed to meet the requirements of line insulators on large load and mechanical strength.

   c. A complete set of fittings for $\pm$800-kV UHV DC lines. A large-section six-conductor bundle-spacer system subspan arrangement model is established for $\pm$800-kV UHV DC transmission lines, and relevant anti-corona suspension clamps and vibration dampers have been developed. A multiple degree-of-freedom conductor bundle torsion calculation model has been established based on the 3D curved beam elements, and a two-step geometrically nonlinear analysis method consisting of the initial static balance and torsional deformation has been proposed. A complete set of conductor galloping analysis methods for ±800-kV UHV DC transmission lines have been proposed. Three anti-galloping devices, including a double pendulum anti-galloping device, a spacer with pivot-type clamps, and a two-conductor spacer with detuning pendulum, as well as recyclable, detachable, all-steel corrugated reels have been designed.

12.5.2.2 Continuous technological innovation

Based on the remarkable achievements already made in UHV DC engineering, China is making more innovations in developing the ±1100-kV UHV DC transmission system and connecting UHV DC systems to UHV AC systems.

1. **Development of the ±1100-kV UHV DC transmission system**

   In China, the HVDC schemes have to meet diversified requirements in transmission capacity and distance. Projects delivering hydroelectric power from Tibet and thermal power from Zhundong and cross-border transmission projects all transmit more than 100 GW of power over more than 3000 km. If these links are designed at ±800 kV, then the line loss will exceed 10%. Therefore, UHV DC transmission at higher voltage classes should be developed to decrease the power loss while increasing the capacity of a single line.

   The step differences in the rated capacity of UHV DC links at ±1000, ±1100, and ±1200 kV are 2000, 3000, and 4000 MW with respect to a ±800-kV link. It seems justified to choose ±1100 kV as the one-step higher voltage class above ±800 kV. The ±1000- and ±1100-kV UHV DC links have a rated capacity of 9500 and 10,450 MW, respectively; obviously, the latter delivers 10% more than the former. The power loss of ±1000- and ±1100-kV links is 6.7–7.29% and 5.34–5.88%, respectively, and the energy loss calculated based on annual availability of 5500 h is 4.6–5% and 3.7–4.1%, respectively. These figures for a ±1100-kV link are 20.1% and 19.6% lower than those for a ±1000-kV link. With larger capacity and distance but lower power loss, ±1100-kV DC links can better suit China's grid developments.

   The ±1100-kV UHV DC technology can further increase the economic transmission distance and transmission capacity of UHV DC systems, and it makes full use of UHV grid's capability of optimizing allocation of energy resources by efficient bulk power delivery over larger areas, thus facilitating substantial breakthroughs in electrotechnology and innovations in equipment manufacturing technology.

   In 2009, China launched studies of key technologies of ±1000-kV and higher UHV DC transmission, covering the justification for different system schemes relying on the proposed Zhundong–Chengdu project, package design of the ±1100-kV UHV DC transmission system, and development of specifications for the main electrical equipment in ±1100-kV converter stations. Also, studies have been conducted regarding air gaps for towers and DC switchyards, on-site assembly schemes of converter transformers, and connection of the AC side to 750- or 1000-kV grids. According to these equipment specifications, manufacturers in various countries developed prototypes and models of the ±1100-kV DC equipment and completed relevant tests. All these study results provide a basis for the development of ±1100-kV UHV DC transmission technology. System simulation analysis and calculation show that the scheme of connecting the Zhundong–Chengdu ±1100-kV UHV DC system to AC grids is reasonable, which can ensure the stability of sending-end and receiving-end grids and meet the requirements of DL 755—2001 *Guide on Security and Stability for Power System.*

a. Studies of system configuration scheme and parameters of main equipment. Studies are conducted on schemes of each pole having two or three series-connected converters with voltage uniformly distributed. The study results show that it is possible to reduce the capacity, dimensions, and weight of a single converter transformer by increasing the number of converters in series and converter transformers, but it will lead to increased number and cost of equipment. Specifically, the three-converter scheme requires 14 more converter transformers and 12 more valve towers at each converter station as compared with the two-converter scheme.

   The scheme of each pole having two series-connected 12-pulse converters (550 kV + 550 kV) and connecting the rectifier and inverter stations to 750- and 500-kV grids, respectively, is recommended. The short-circuit impedance of ±1100-kV converter transformers is determined as 24%; the valve side no-load DC voltage of converter transformers U_{dioN} in rectifier and inverter stations is 326.6 and 317.67 kV, respectively; the voltage ratio, rated capacity, and on-load tapping range of the high-voltage end Y/Y converter transformers in rectifier and inverter stations are, respectively, $770/\sqrt{3}/241.84/\sqrt{3}$kV, 541.52 MVA, $+25/-3\times0.86\%$ and $550/\sqrt{3}/235.23/\sqrt{3}$kV, 526.72 MVA, $+25/-3\times1.25\%$.

b. Studies of overvoltage and insulation coordination. It is recommended that MOAs in ±1100-kV converter stations should be configured the same way as a ±800-kV converter station, except that an additional MOA A2 should be installed on leads of the valve side bushings of the Y/Y converter transformer.

   Through analyses and calculations, the switching and lightning impulse withstand voltages of equipment operating at locations with the highest voltage, such as the valve top in ±1100-kV converter stations, are recommended to be 2100 and 2300 kV, respectively. The highest overvoltage occurs at the line center at 1.7 p.u., which can be reduced to 1.55 p.u. by installing MOAs.

Figure 12.38
Converter transformer prototype 1.

c. ±1100-kV converter transformers. The converter transformers and wall bushings are directly exposed to the system-to-ground voltage of the DC system and are a key constraint on the selection of voltage class. Therefore, great efforts should be made in their development.

China has developed and tested several converter transformer prototypes. Prototype 1, as shown in Figure 12.38, has its valve side outgoing line led from the oil tank top and is designed based on the DC-side voltage of 1100 kV, AC-side voltage of 750 kV, and on-site assembly. The insulation conditions are determined according to the overall electrical design of the converter transformer. The insulation design is incorporated into the prototype and tested. During the development process, the major difficulties lie in the main insulation and bushing turret.

Test items and parameters of the valve side exit lead are as follows:

Switching impulse test	−2100 kV;
DC voltage withstand test	+1745 kV, 2 h;
Polarity reversal voltage withstand test	±1345 kV, (90 + 90 + 45) min;
AC voltage withstand test	1260 kV, 1 h.

Test items and parameters of the line side exit lead are as follows:

Switching impulse test	−1550 kV;
Lightning impulse test	−1745 kV;
Short-duration AC voltage withstand test	900 kV, 1 min;

Long-duration AC voltage withstand test: (800 kV, 60 s) and (693 kV, 1 h).

The converter transformer prototype 2, as shown in Figure 12.39, has its valve side outgoing line led from the oil tank's side walls. It is designed based on the DC-side voltage of 1100 kV and AC-side voltage of 750 kV. The major difficulties are related to the main insulation and bushing turret.

Test items and parameters of prototype 2 are as follows:

Switching impulse test	−2100 kV;
Lightning impulse test	−2300 kV;
DC voltage withstand test	+1754 kV, 2 h;
Polarity reversal voltage withstand test	±1745 kV, (90 + 90 + 45) min;
AC voltage withstand test	1270 kV, 1 h.

d. ±1100-kV converter valves. Converter valves have a series-connected structure and the voltage class can be boosted by increasing the number of valve sections. Therefore, the primary concern is the insulation of the HV-end valve tower, which

Figure 12.39
Converter transformer prototype 2.

is intended to withstand a switching impulse voltage of 2100 kV and lightning impulse voltage of 2300 kV.

Converter valve prototypes KGWF-5000 A/±1100 kV and HVTV2000–1100/5000E made of 8.5-kV/5000-A electrically triggered thyristors have passed all type tests. Also, type tests are performed on the insulation of ±1000-kV converter valves under equivalent load, solving difficulties in ensuring the equivalent effects of stray capacitance and inductance under ±1100 kV. The peak short-circuit current of all ±1100-kV converter valves tested is 50 kA, meeting the requirements of technical specifications.

e. ±1100-kV converter transformer valve side bushings and wall bushings. The ±1100-kV converter transformer valve side bushings and wall bushings have been developed and tested. Figures 12.40 and 12.41 show the test scenarios. Valve side bushings (GGF1100) for converter transformer prototype 1 are the ordinary oil/gas-insulated type, and are 15.6 m in length (the silicone rubber housing has a flashover distance of 11 m) and 7900 kg in mass. Wall bushings associated with prototype 1 (GGFL1100) are the gas-filled type, and are 25.5 m long (the silicone rubber housing has an indoor and outdoor flashover distance of 11 and 13 m, respectively) and 6600 kg in mass.

Figure 12.40
Converter transformer valve side bushing under AC voltage withstand test.

Figure 12.41
Wall bushing undergoing testing.

Valve side bushings (GSETF 2950/1100) for converter transformer prototype 2 are the gas-filled type with a condenser core made of resin-impregnated paper. They are 16 m (the silicone rubber housing has a flashover distance of 12 m), 7000 kg, and have a rated current of 5000 A. Wall bushings associated with prototype 2 have the same design as their valve side bushings. The indoor and outdoor flashover distances of their silicone rubber housing are estimated to be 12 and 14 m, respectively.

f. ±1100-kV DC filters. Two design schemes have been proposed for the HV capacitor tower of ±1100-kV DC filters (i.e., the capacitor banks are supported or suspended). In both schemes, the capacitor banks are arranged into four towers. In the suspension scheme, the portal frame is approximately 50 m high. In the supporting scheme, a single tower is approximately 27.4 m high. Mechanical force calculations are performed for towers in both schemes and the results show that the mechanical load on towers in both schemes can meet the engineering requirements. This proves that the four-tower design is feasible. With the proceeding of the DC filter design, anti-seismic characteristics of capacitor towers in the two design schemes can be analyzed using the finite element method to further compare the two schemes.
g. Equipment in DC switchyard.
   i. ±1100-kV bypass circuit breakers. The mechanical and electrical design, assembly of prototype (Figure 12.42), test at rated current of 5000 A, and insulation test are all completed.
   ii. ±1100-kV post-insulators and disconnectors. The long rod post-insulators used by ±1100-kV disconnectors have been developed. Such an insulator

Figure 12.42
Prototype of bypass circuit breaker.

Figure 12.43
Prototype of DC line pole disconnector.

consists of eight porcelain elements of the same length and has a total height of 14.7 and a bending strength over 8 kN. They have already been tested in ±1100-kV DC disconnectors. Design schemes for post-insulators intended for ±1100-kV bus and smoothing reactors have been developed.

A horizontally pivoting, three-column semi-pantograph disconnector prototype (Figure 12.43) for use on the ±1100-kV line pole has been developed. ZGW6-1120/J6300-25 HVDC disconnectors and ZJW1-1120/J25 HVDC earthing switches have passed all type tests.

iii. ±1100-kV DC surge arresters. Electric field simulation calculations have been performed and a prototype, as shown in Figure 12.44, has been designed and assembled.

h. Design of lines and towers. Considering factors such as the audible noise and radio interference, which should not exceed 45 dB(A) and 58 dB, respectively, 8×1000 or $8 \times 1250\ mm^2$ conductor bundles are selected for ±1100-kV lines. According to relevant test results, the insulator string is determined to be 15 m high, the tower air gap is taken as 10 m, and the minimum line conductor-to-ground height is taken as 27.5 m.

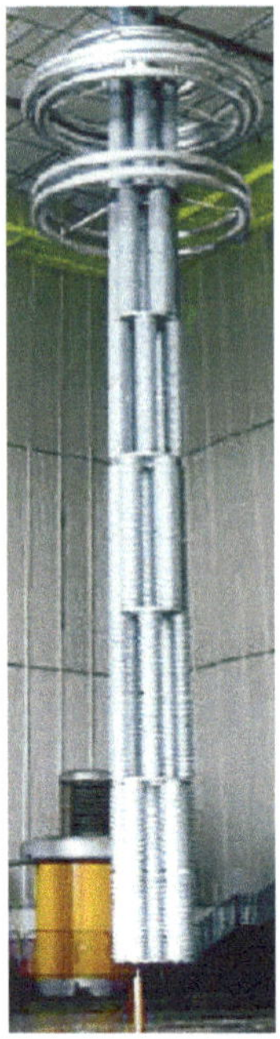

Figure 12.44
Prototype of surge arrester.

2. **Studies of schemes of connecting UHV DC systems to UHV AC grids**
Because a UHV DC transmission system delivers a huge amount of power, it has a great impact on the receiving-end AC grid. To better accommodate the delivered power and offer necessary support to converter stations, associated AC lines must be built. Moreover, 500-kV networks are becoming more and more mature, it is difficult to find sites for new converter stations, and the line corridor is limited. Along with the increasing transmission capacity of UHV DC systems, DC system faults cause more and more impacts on the AC grids and will greatly affect the operational stability and security of 500-kV AC grids. To maximize the economic efficiency of interconnection and guarantee grid safety, the following three schemes of connecting UHV DC systems to 1000-kV UHV AC grids have been proposed.
   a. Scheme of direct connection to 1000-kV AC grids. In this scheme, the HV-end converter transformers and AC filters are directly connected to the 1000-kV AC bus, as shown in Figure 12.45. If the $\pm$800-kV/$\pm$1100-kV UHV DC systems are directly connected to 1000-kV UHV AC grids, then many difficulties need to be solved, including the manufacturing of DC switchyard equipment, insulation design of converter transformers and AC switchyard equipment, arrangement of the 1000- and 1100-kV outgoing lines within converter transformers as well as the external bushings, and greatly increased dimensions and weight of converter transformers. In

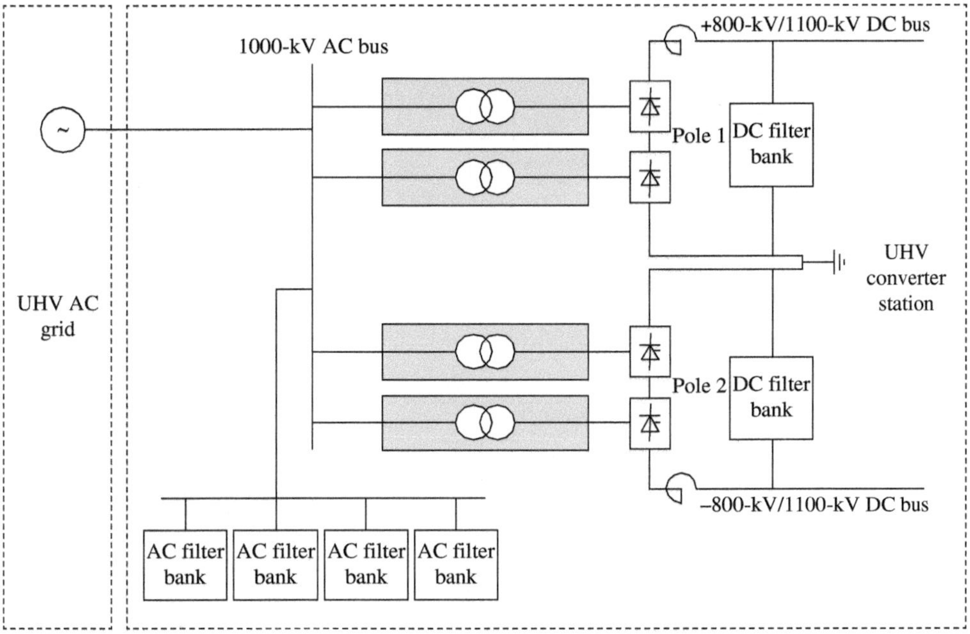

Figure 12.45
Scheme of direct connection to 1000-kV UHV AC grids.

addition, the converter transformers need to be assembled on-site and there are no readily available on-load tap changers. Higher technical requirements are placed on circuit breakers used for 1000-kV AC filters; switches in the AC filter yard are intended to withstand recovery voltages beyond capabilities of the existing switches.

b. Scheme with pad-mounted transformers. By developing new pad-mounted transformers, the following four connection schemes, as shown in Figure 12.46, are derived from the aforementioned scheme (1).

Scheme a: A new pad-mounted transformer consisting of one converter transformer unit and one AC step-up transformer unit interconnected through a 500-kV lead has been developed. A 500-kV bus is led from the transformer. The 1000-kV AC side of the converter transformer is connected to the 1000-kV AC bus in the converter station, and its DC side is connected to the valve hall. The 500-kV AC bus led from the transformer is connected to the reactive power compensation equipment such as AC filters.

Scheme b: A new pad-mounted transformer consisting of one converter transformer unit and one AC step-up transformer unit connected through a 500-kV lead has been developed. No outgoing lines are led from the transformer. The 1000-kV AC side of the converter transformer is connected to the 1000-kV AC bus in

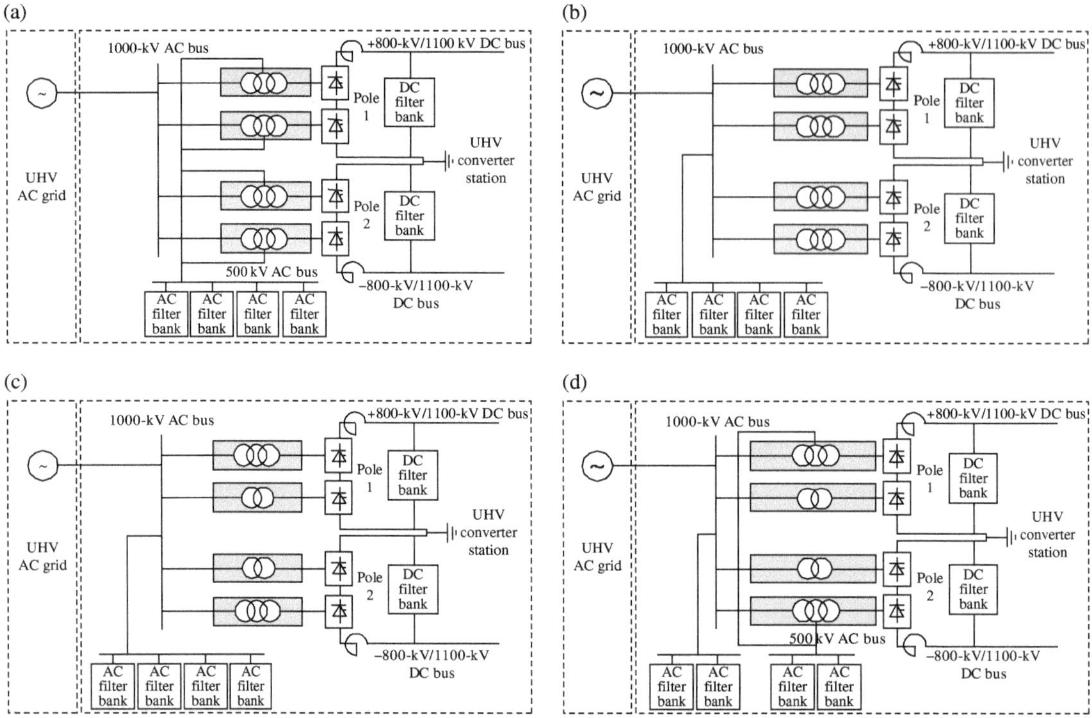

Figure 12.46
Scheme with pad-mounted transformers.
(A) Scheme a, (B) scheme b, (C) scheme c, and (D) scheme d.

converter station and its DC side is connected to the valve hall. The reactive power compensation equipment, such as AC filters, are connected to the 1000-kV AC bus in a converter station.

Scheme c: The HV ends of both poles are designed the same as scheme b, and the LV ends are directly connected to 1000-kV AC systems.

Scheme d: The HV ends of both poles are designed the same as scheme a, and the LV ends are directly connected to 1000-kV AC systems.

In these four schemes, differences exist in neither the DC side, which has a bipolar configuration with each pole having two series-connected converters, nor the DC-side switches and DC filters as well as their arrangement. The main difference lies in the configuration of the converter transformer unit and AC filters. The main difficulties are in the development and manufacturing of new pad-mounted transformers.

c. Hierarchical connection scheme. To reduce difficulties in the manufacturing of HV-end converter transformers as well as the overall cost of the project, the LV-end converter transformers can be directly connected to 1000-kV AC systems, whereas

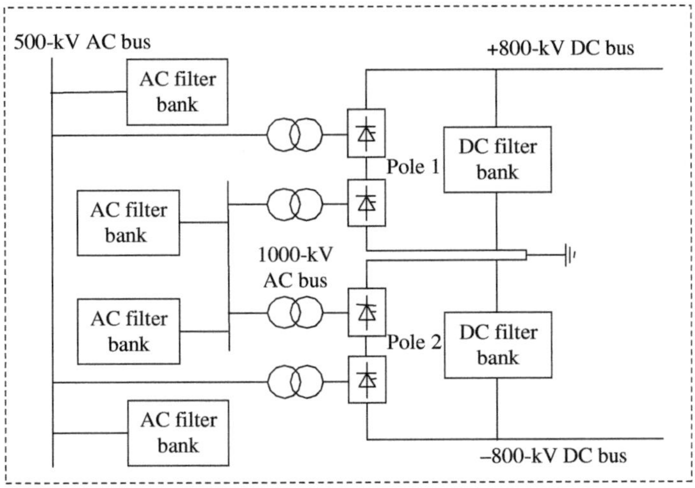

Figure 12.47
Hierarchical connection scheme.

the HV-end converter transformers are connected to 500-kV AC systems, as shown in Figure 12.47. To use this scheme, the following aspects should be considered:

i. DC voltage control is the core of the steady-state control strategy of the receiving-end AC system. In both the Xiangjiaba–Shanghai and Jinping–Sunan projects, the pole control level serves to send firing angles to the converter unit control level, and the tap position control at the pole control level cooperates with the extinction angle control to control the DC voltage. Under steady-state operation, the extinction angle equals the rated extinction angle and the DC voltage is controlled mainly by the tap changer. Such a hierarchical control system may result in different DC voltage outputs of the HV-end and LV-end converters, leading to unbalanced DC voltage. Although under the steady-state, the HV-end and LV-end converter units have the same extinction angle and tap position, the voltage ratio of the 500- and 1000-kV AC buses is uncertain because it depends on the turns ratio and impedance of the transformer interconnecting the two buses, as well as the distribution of system power flow. If it is confirmed through calculations that the voltage unbalance between the HV-end and LV-end converters caused by variations in the aforementioned voltage ratio are acceptable, then the hierarchical structure used in the Xiangjiaba–Shanghai and Jinping–Sunan projects can be used. If not, then other alternatives should be used.
ii. If the reactive power is controlled in a station-wide unified manner, attention should be given to the distribution of AC filters on the 500- and 1000-kV AC buses. When the reactive power control is realized by controlling the AC bus

voltage, there will be two targets under control. Even the repeated switching of 500- and 1000-kV filters will occur. When it is realized by controlling the reactive power exchanged with the AC system, there will be difficulties in selecting the 500- and 1000-kV filters.

If the HV-end and LV-end converters use independent reactive power control (i.e., a set of compensation equipment is provided for the HV-end and LV-end converters of both poles, respectively), then HV and LV ends are independent from each other in terms of the electrical equipment and the control system. Also, AC voltage control should not be used.

When comparing the three connection schemes, the hierarchical connection scheme has the advantages of less difficult technology and manufacturing of converter transformers and therefore has been selected by the Xilin Gol League–Taizhou and West Inner Mongolia–Wuhan UHV DC transmission projects.

12.6 Localization of UHV Equipment and Transport of Large Equipment

12.6.1 Manufacturing Capabilities of UHV AC Equipment

1. **UHV transformer and HV shunt reactor**
   In the UHV AC pilot and demonstration project, the UHV transformers are produced by TBEA Shenyang Transformer Co., Ltd. (TBEA Shenyang) and Baoding Tianwei Baobian Electric Co., Ltd. (Baobian), and the reactors are produced by Xian XD Transformer Co., Ltd. (XD Transformer) and TBEA Hengyang Transformer Co., Ltd. (TBEA Hengyang). For the expansion project and the Huainan–Shanghai project, UHV AC reactors are developed by Shandong Power Equipment Co., Ltd. (Shandong Power Equipment) and TBEA Shenyang and Baobian, and UHV AC transformers are developed by Changzhou XD Transformer Co., Ltd. (Changzhou XD Transformer), Tianwei Baobian (Qinhuangdao) Transformer Co., Ltd. (Tianwei Baobian (Qinhuangdao) Transformer), Shandong Power Equipment, and TBEA Hengyang. As such, TBEA Shenyang, TBEA Hengyang, Changzhou XD Transformer, Baobian, Tianwei Baobian (Qinhuangdao) Transformer, and Shandong Power Equipment are capable of producing 1000-kV/1000-MVA AC transformers, and TBEA Shenyang, TBEA Hengyang, XD Transformer, Baobian, and Shandong Power Equipment are capable of producing 1000-kV AC shunt reactors.
2. **UHV switchgear**
   Currently, China has the capability of manufacturing some components of UHV switchgear. Chinese suppliers work with foreign suppliers to jointly design and produce switchgears in China and seek to share property rights and master the core technologies.

They are working to localize some core components that still have to be imported, such as arc-extinguishing chambers, operating mechanisms, and porcelain bushings.

3. **UHV series compensation equipment**
All equipment for China's first UHV series compensation project were supplied by CEPRI. The series compensation equipment comprise a complicated system containing a control and protection system and a number of major components, such as series capacitor banks, spark gaps, metal oxide varistors (MOVs), damping devices, bypass circuit breakers, disconnectors, current transformers, fiber optic columns, post insulators, tension insulators, and the series compensation platform. Because of their high degree of specialization, these systems and components are supplied by different manufacturers.
For the first UHV series compensation project, the suppliers include:
- Series capacitor banks: ABB Xi'an Power Capacitor Co., Ltd., Guilin Power Capacitor Co., Ltd., and Shanghai Sieyuan Power Capacitor Co., Ltd.
- Bypass circuit breakers: Henan Pinggao Electric Co., Ltd., Xian XD Switchgear Electric Co., Ltd., and ABB corporation.
- Disconnectors: Henan Pinggao Electric Co., Ltd., Shandong Taikai Disconnectors Co., Ltd., and Hunan Changgao High Voltage Switchgear Group Co., Ltd.
- Post insulators: Xian XD High Voltage Porcelain Insulator Co., Ltd. and Sinoma Advanced Materials Co., Ltd.
- Tension insulators for series compensation platform: Shandong Taiguang Electrical Equipment Company Limited (Zibo Taiguang Electrical Equipment Factory)
- Other equipment such as spark gaps, MOVs, damping devices, current transformers, fiber optic columns, control and protection systems, and series compensation platforms: EPRI and China EPRI Science & Technology Co., Ltd.

With strong production capacity that could meet the needs of series compensation, suppliers in China are capable of manufacturing more than 10 sets of series compensation equipment each year.

4. **Other equipment**
A number of major manufacturing difficulties have been solved, including shaping, turning, sintering, and shed design for post insulators. Examples include a 2-m lower component with a maximum bending moment of 160 kN · m and a 10-m single-column porcelain post insulator composed of five single components with a structural height of 10 m, a specific creepage distance of 25 mm/kV, a rated bending failing load of 16 kN, and a rated torsion failing load of 10 kN · m. Tests and operation demonstrate that the mechanical strength, seismic withstand capability, pollution withstand voltage, radio interference, and visual corona characteristics of independently developed ZSW1-1100/16 post insulators fully meet the requirements of UHV AC projects. FXBW-1000/550-9750 rod suspension composite insulators with a rated mechanical strength of 550 kN were introduced in the Huainan–Shanghai project, representing breakthroughs in

manufacturing, testing, and installation techniques. Xian XD High Voltage Porcelain Insulator Co., Ltd., Fushun Electric Porcelain Manufacturing Co., Ltd., Dalian Insulator Group Co., Ltd., and Jiangsu Nanci Insulators Co., Ltd. are capable of producing UHV AC porcelain insulators. Jiangsu Shenma Power Co., Ltd., Xiangfan Guowang Composite Insulator Co., Ltd., Shandong Taiguang Electrical Equipment Company Limited, Jiangsu Xiangyuan Electric Equipment Co., Ltd., and Zhejiang Gaoneng Electric Installation Co., Ltd. are capable of producing UHV AC composite insulators.

China's top four arrester manufacturers, PG Toshiba (Langfang) Arrester Co., Ltd., Xian XD Arrester Co., Ltd., Nanyang Jinguan Electric Co., Ltd., and Fushun Electric Porcelain Manufacturing Co., Ltd., are all capable of producing UHV AC arresters, and the first two are also capable of producing tank-type arresters. Porcelain arresters supplied by these manufacturers and tank arresters from PG Toshiba (Langfang) Arrester Co., Ltd. are used at the Changzhi, Nanyang, and Jingmen Substations.

12.6.2 Manufacturing Capabilities of UHV DC Equipment

1. **Converter valve**
   Six-inch thyristor converter valves are usually built as double 12-pulse circuits. This is a serial connection of double-valve towers suspended from the ceiling. For the Xiangjiaba–Shanghai project, Xi'an Power Electronics Research Institute and CSR Zhuzhou Electric Locomotive Research Institute Co., Ltd. develop a series of 4000- to 5000-A 6-inch thyristors involving a number of technical achievements such as "voltage and resistivity optimization with large-diameter FZ Si single crystals," "very thin design for power semiconductor device," "cathode layout design and plate making technology," and "compatibility and optimized design for specially welded positive–negative bevel." The 6-inch thyristors independently produced by China have been operating safely for more than 2 years in the Xiangjiaba–Shanghai project. The Jinping–Sunan project, which began commercial operation at the end of 2012, is also a world leader in the failure rate of thyristors.

   All 500-kV or lower converter valves are supplied domestically. Before the Xiangjiaba–Shanghai project, ±800-kV UHV converter valves were jointly developed by Chinese and foreign suppliers and the major components were supplied by the latter. Since the project, Chinese suppliers began to master the design and manufacturing capabilities, with all the major components purchased on their own. A number of studies of voltage distribution parameters, valve tower's seismic withstand capability, and key technologies for valve control have been performed, a practical ±800-kV converter valve simulation application was developed, and methods and standards have been formulated for multiple valve operation tests on 4000-A and higher converter valves. XJ Group Corporation (XJ Group), C-EPRI Electric Power Engineering Co., Ltd. and Xian XD Power Systems Co., Ltd. are capable of independently designing, manufacturing, procuring, and testing converter valves and auxiliary equipment for ±800-kV and lower HVDC projects.

2. **Converter transformer**
With the commissioning of the Xiangjiaba–Shanghai project, China's design and manufacturing capabilities of UHV converter transformers have been greatly improved. The 7200-MW Jinping–Sunan project that began operation in 2012 further enhanced these capabilities. TBEA Shenyang, Baobian, and XD Transformer are capable of independently designing and manufacturing LV converter transformers. China has accumulated vast design experience with HV converter transformers, and has begun to design and manufacture them independently. Currently, China is capable of performing tests on ±1100-kV converter transformers. Most components such as windings, oil tanks, main insulating materials, insulating oil, and coolers are available in China. Only a few components have to be imported, including bushings and tap changers. Despite its late start in development, the 400-kV and lower bushings have already passed type tests and are ready for field use. Through continuous efforts, imported bushings will soon be eliminated. A prototype of an on-load tap changer has been developed.
3. **Smoothing reactor**
UHV smoothing reactors are the dry type and are independently developed by TBEA Shenyang, XD Transformer, Shandong Power Equipment, Xi'an Zhongyang Electric Corporation, and Beijing Power Equipment Group. The 4000-A/75-mH and 4750-A/60-mH smoothing reactors have been operating well in the Xiangjiaba–Shanghai and Jinping–Sunan projects. China has independent intellectual property rights over the insulation, mechanical, and seismic design of smoothing reactors, insulating platforms, and post insulators, and all components are domestically manufactured, indicating that China has become a world leader in designing, procuring, manufacturing, and testing UHV smoothing reactors. In addition, China is capable of producing dry-type smoothing reactors with a maximum current of 5000 A.
4. **DC filter**
Through the Baoji–Deyang ±500-kV HVDC project, Hulunbeier–Liaoning ±500-kV HVDC project, and Ningdong–Shandong ±660 kV HVDC demonstration project, Guilin Power Capacitor Co., Ltd., Xi'an XD Power Capacitor Co., Ltd., and Shanghai Sieyuan Power Capacitor Co., Ltd. have developed capabilities for designing and manufacturing DC filters for ±800-kV and lower projects. DC filters used in the Jinping–Sunan project have been operating well after being put into operation. All DC filters used in the Haminan–Zhengzhou and Xiluodu Left Bank–Zhejiang Jinhua projects are provided by Chinese suppliers. All components, including capacitors, resistors, reactors, arresters, and insulators, are manufactured domestically. Specifically, optical current transformers (OCTs) for unbalance protection with independent intellectual property rights will be applied in the Haminan–Zhengzhou and Xiluodu Left Bank–Zhejiang Jinhua projects. The reactors are leading the world in unique noise

reduction and anti-bird design. Through technical upgrades in recent years, a world-class capacitor production line has been built. Some parameters of the Chinese-made DC capacitors outperform peer products in the world.

5. **DC measuring device**
   In recent years, the world's leading power transformer suppliers (such as Alstom and Reitz) introduced their funds and technologies to China through joint ventures, which helps China improve its design and manufacturing capabilities of power transformers. The ±500-kV and lower current transformers and voltage dividers have been developed by TBEA Shenyang, NR Electric Co., Ltd. (NR Electric), and ABB Sifang Power System Co., Ltd. (ABB Sifang Power System) and have been put into field use. In the past, zero-flux current measuring devices were monopolized by foreign manufacturers (HITEC and Reitz). Now, a 1-V/3000-A prototype with an accuracy class of 0.2 developed by CEPRI has passed all type tests. Efforts also have been made to develop ±800-kV electronic current transformers and voltage dividers. The prototypes have passed type tests and have been awarded the contract for the Xiluodu Left Bank–Zhejiang Jinhua project.
6. **Protection and control equipment**
   NR Electric, ABB Sifang Power System, and XJ Group are capable of manufacturing various DC control and protection equipment. For conventional HVDC projects that are already in operation, such as the Hulunbeier–Liaoning, Baoji–Deyang, Gaoling back-to-back, Lingbao back-to-back, and Qinghai–Tibet projects, Chinese control and protection equipment are used. Long-term operation shows that the performance of the equipment meets applicable technical requirements. Based on the characteristics of two series-connected 12-pulse valve groups, suitable structure and functions are designed for ±800-kV UHV DC control and protection systems. The main computer runs a real-time operating system to improve reliability. High-speed, real-time communication technologies such as eTDM are used to control converter valves with higher integration and compatibility of primary equipment with the system. In July 2010, the Xiangjiaba–Shanghai project began operation. Three-year operation shows that the performance of the UHV control and protection system fully meets the technical requirements of the project. With manufacturing technologies becoming mature, Jingping–Sunan, Haminan–Zhengzhou and Xiluodu Left Bank–Zhejiang Jinhua UHV DC transmission projects use domestically supplied control and protection equipment.

12.6.3 Transport of Large Equipment

UHV AC and DC technologies mean higher voltage class, larger transmission capacity, and larger DC equipment capacity. These pose higher requirements for the transport of large equipment.

The large equipment are transported over a long distance, covering a wide area vulnerable to various difficulties. Large equipment are key to transmission projects, and production and transportation are costly and time-consuming. AC main transformers, DC converter transformers, soothing reactors, and HV shunt reactors in AC substations, DC converter stations, and switching stations usually have excessive dimensions and weights with sophisticated structures, which require high stability during transport. However, common cargo transport means are unable to satisfy such requirements. Therefore, it is necessary to investigate road conditions and use special vehicles and tools and work out a detailed transport scheme. Considering that converter transformers are larger and heavier than smoothing reactors, converter transformers are the main consideration in determining the transport scheme.

Large equipment are transported in China by road, rail, water, or a combination of these. Different transport forms vary greatly in construction standards and capacity and require special loading tools.

Roads in China are classified into highways, Class I, II, III, and IV, and substandard roads. The parameters of these road classifications are as follows:

Highways: Headroom 5 m, lane width 3.75 m
Class I roads: Headroom 5 m, lane width 3.5 m for design speed of 80 km/h and 3.5 m for design speed of 40 or 60 km/h
Class II roads: Headroom 5 m, lane width 3.5 m for design speed of 80 km/h and 3.5 m for design speed of 60 km/h
Class III roads: Headroom 4.5 m, lane width 3.5 m for design speed of 40 km/h and 3.25 m for design speed of 30 km/h
Class IV roads: Headroom 4.5 m, design speed 20 km/h, lane width 3 m

Railways are classified into three classes. Classes I and II are important players in the road network. The former has an ultimate passenger and freight capacity of more than 20×10^7 tons, and the latter is supplementary to the former as a means of linking. Class III railways provide regional transport services.

Limitations of railway transport include:

1. Equipment dimensions have to be compared with structure gauge. When transporting oversize equipment, gauges along the route must be verified through a variety of ways to determine whether the equipment meets the requirements of rail transport.
2. The equipment cannot be transported by rail if its internal or external width is larger than the structure gauge when traveling around a curve.
3. Bridge capacity: The equipment must be checked against the bridge capacity to determine whether it can be transported by rail according to the actual conditions of the bridge. Appropriate measures must be taken to reinforce the bridge if necessary.

4. Railway curve radius: Based on different curve radiuses, the dimension of the equipment must be checked against the structure gauge and the minimum curve radius specified in the instructions.
5. Height of overhead line: Because of terrain and tunnel constraints, overhead lines along some railroads are very low. It is a key consideration and whether measures such as covering insulation plates and powering-off are required should be determined as appropriate.

Navigable inland waterways are classified into seven classes by navigable tonnage, namely, 3000, 2000, 1000, 500, 300, 100, and 50 tons.

Transport by road, rail, and water is different in terms of vehicle type, deadweight, duration, adaptability, and constraints. Transport by road is characterized by low loads, high freight index, short duration, and relatively strong adaptability, but is restrained by headroom, bridge load, and multiple competent road authorities. Transport by rail is characterized by moderate loads, freight index, and duration and poor adaptability, but is less restrained by vehicle gauge, bridge load, and overhead lines. Transport by water is characterized by high loads, low freight index, long duration, and moderate adaptability, but is restrained by waterway class, climate, and water level changes. Combined rail–water–road transport is characterized by high loads, low freight index, moderate duration, and strong adaptability, but is restrained by intermediate procedures such as transfer, loading, and unloading. Therefore, when selecting transport mode, water transport should be fully used and combined transport is preferred to ensure safe and reliable transport while minimizing the cost of troubleshooting facilities.

Main transport vehicles for transformers include road, rail, and water transport. Flatbed trucks are the major tool for road transport. The maximum axle loads for domestic and imported hydraulic flatbed trucks are 20–25 and 30–32 tons (including 3–4 tons of axle weight), respectively. With a minimum operating height as low as 0.78 m, Nicholas hydraulic flatbed trucks are the lowest, whereas the other types have a minimum operating height of 0.8–0.9 m. Vehicles for rail transport include well cars and Schnabel cars. Currently, China mainly uses DK36 well cars and is researching D45 Schnabel cars (for transformers). In waterways offshore and in the lower reaches of the Yangtze River, 3000 tons self-propelled deep-hold ships may be used to transport large transformers. The transportable dimensions of transformers are given in Table 12.3.

To transport converter transformers, the following should be considered: (i) site conditions; (ii) transport conditions of large equipment; and (iii) dimensions of converter transformers. The equipment used in a converter station cannot be transported by only one means. If conditions permit, then combined water–road transport is a preferred means. A less preferred means is combined rail–road transport. In very few cases, the equipment is transported only by road. This is because manufacturers are often located far away from

Table 12.3: Transportable Transformer Dimensions by Transport Type.

No.	Transport Mode	Weight (tons)	Length (m)	Width (m)	Height (m)
1	Road (500 tons bridge carrier)	<500	13.5	5.2 (including the width of shoulder base)	5.0
2	Rail D45 Schnabel car (new)	<450 (including the weight of special casing)	13	3.84	4.9
	Rail DK36 (existing)	<360	13	3.5	4.85
3	Water	Unrestricted	Unrestricted	Unrestricted	Unrestricted

converter stations, making it difficult to transport large equipment by road over a long distance. In this case, water and rail transport are mainly used. The only time that long-distance road transport is used is when sending transformers from Xian XD Transformer Co., Ltd. to Jingzhou Converter Station.

For operating HVDC projects in China, converter transformers are transported by two or more modes. For example, combined rail–road transport and combined water–road transport are used, respectively, for transporting converter transformers to Yinchuandong and Jiaodong Converter Stations of Ningdong–Shandong ±660-kV HVDC demonstration project. Combined water–road transport is used for transporting converter transformers to both Fulong and Fengxian Converter Stations of the Xiangjiaba–Shanghai project. Combined road–water transport and combined road–water–rail transport are used, respectively, for transporting converter transformers to Tongli and Yulong Converter Stations of the Jinping–Sunan project.

The ±500-kV converter transformers weigh 250–280 tons and ±800-kV 300–350 tons. Given their weights and dimensions, mature vehicles are available for all three transport means. Dimensions and weights of converter transformers increase after the increase of voltage class and capacity of HVDC transmission. Existing vehicles are no longer able to meet the transport need. Therefore, it is necessary to develop new means for rail transport or disassemble converter transformers for transport and assemble them on the installation site.

References

[1] China Electric Power Research Institute, AC Transmission Section for UHV Transmission Technology, China Electric Power Press, Beijing, 2012.

[2] Z. Wenliang, Y. Yongqing, L. Guangfan, et al., Function and design of the high altitude Tibet test base, High Voltage Eng. 8(35) (2009) 1785–1790.

[3] H. Yi, W. Baoquan, H. Huiwen, Key technologies of 1000 kV AC compact transmission, High Voltage Eng. 8(31) (2011) 1825–1831.

[4] L. Hanqing, W. Xiong, L. Xuming, Operation losses and economic evaluation of UHV AC and HVDC transmission systems, High Voltage Eng. 39(3) (2013) 630–635.
[5] Z. Wenliang, Y. Yongqing, L. Guangfan, et al., Researches on UHV AC technology, Proc. CSEE 22(27) (2007) 1–7.
[6] L. Zhenya, Research Collection of UHV AC Power Transmission Technology (2005–2011), China Electric Power Press, Beijing, 2006–2012.

Technical Data of UHV AC Electrical Equipment

Table A.1: Main Technical Data of GIS.

Manufacturer		**Pinggao Electric**	**NHVS**	**XD Switchgear**
Rated voltage (kV)		1100	1100	1100
Rated current (A)	Bus Bushing Other equipment	8000 6300 6300	8000 6300 8000	8000 6300 6300
1 min power frequency withstand voltage (kV, rms)	Pole-to-earth Between breaks	1100 1100 (+635)	1100 1100 (+635)	1100 1100 (+635)
Lightning impulse withstand voltage (kV, peak)	Pole-to-earth Between breaks	2400 2400 (+900)	2400 2400 (+900)	2400 2400 (+900)
Switching impulse withstand voltage (kV, peak)	Pole-to-earth Between breaks	1800 1675 (+900)	1800 1675 (+900)	1800 1675 (+900)

Circuit breaker

	Pinggao Electric		**NHVS**		**XD Switchgear**
Manufacturer	**Cooperative Development Scheme**	**Independent Development Scheme**	**Cooperative Development Scheme**	**Independent Development Scheme**	**Independent Development Scheme**
Number of breaks of each phase	2	2	2	4	4
Rated current (A)	6300	6300	8000	6300	6300
Rated short-circuit breaking current (kA)	63	63	63	63	63
Rated short-time withstand current (kA, 2 s)	63	63	63	63	63

(*Continued*)

Table A.1: (Continued)

Manufacturer	Pinggao Electric: Cooperative Development Scheme	Pinggao Electric: Independent Development Scheme	NHVS: Cooperative Development Scheme	NHVS: Independent Development Scheme	XD Switchgear: Independent Development Scheme
Rated peak withstand current (kA)	171	171	171	171	171
First pole-to-clear factor	1.3	1.3	1.3	1.3	1.3
Time constant (ms)	120	120	120	120	120
Closing resistance (Ω)	600	600	580	600	560
Rated SF_6 gas pressure (MPa)	0.65	0.60	0.60	0.60	0.58
Type of operating mechanism	Hydraulic operating mechanism			Spring hydraulic operating mechanism	
Weight per pole (t)	30	22	19	8.5	10
Disconnector and grounding switch					
Rated current of disconnector (A)	8000		8000		6300
Rated short-time withstand current (kA)	63(2 s)		Disconnectors: 63 (3 s) Grounding switches: 63 (2 s)		63 (2 s)
Rated peak withstand current (kA)	171		171		171
Switching-induced capacitive current of disconnectors (A)	2		2		2
Switching-induced inductive current of disconnector (A)	1		1		1
Bus-transfer current/voltage of disconnectors (A/V)	1600/400		1600/400		1600/400
Operating mechanism of disconnectors	Stored-energy spring mechanism		Stored-energy spring mechanism		Motor operating mechanism
Mechanical life of disconnectors	2000		2000		2000
Opening/closing resistor of disconnectors (Ω)	500		500		N.A.

(Continued)

Table A.1: (Continued)

Manufacturer	Pinggao Electric		NHVS		XD Switchgear
	Cooperative Development Scheme	Independent Development Scheme	Cooperative Development Scheme	Independent Development Scheme	Independent Development Scheme
Operating mechanism of earthing switches	Motor operating mechanism		Motor operating mechanism		Motor operating mechanism
Rated SF_6 gas pressure (MPa)	0.45		0.5		0.58
Weight per pole (t)	7.6		7.0		1.1
Bus					
Rated SF6 gas pressure (MPa)	0.45		0.5		0.36

Table A.2: Main Technical Data of Transformer.

Manufacturer		XD Transformer	TBEA	Tianwei Baobian	Shandong Power Equipment
Maximum voltage U_m (kV, rms)		1100	1100	1100	1100
Rated capacity (MVA)		1000/1000/334	1000/1000/334	1000/1000/334	1000/1000/334
Rated voltage (kV)	HV	$1050/\sqrt{3}$	$1050/\sqrt{3}$	$1050/\sqrt{3}$	$1050/\sqrt{3}$
	MV	(1) $525/\sqrt{3}$ (2) $520/\sqrt{3}$ (3) $515/\sqrt{3}$	(1) $525/\sqrt{3}$ (2) $520/\sqrt{3}$	$525/\sqrt{3}$	(1) $525/\sqrt{3}$ (2) $515/\sqrt{3}$
	LV	110	110	110	110
Primary tapping		± 5% (9 steps)	± 5% (9 steps)	± 5% (9 steps)	± 5% (9 steps)
Cooling method		OFAF	OFAF	OFAF	OFAF
Leading-out scheme		Bushing	Bushing	Bushing	Bushing
Voltage regulation		Neutral point off-circuit/on-load tap changing	Neutral point off-circuit tap changing	Neutral point off-circuit tap changing	Neutral point off-circuit tap changing

(Continued)

Table A.2: (Continued)

Manufacturer			XD Transformer	TBEA	Tianwei Baobian	Shandong Power Equipment
Insulation level (kV)	HV	Full/chopped lightning impulse	2250/2400	2250/2400	2250/2400	2250/2400
		Switching impulse	1800	1800	1800	1800
		Power frequency (5 min)	1100	1100	1100	1100
	MV	Full/chopped lightning impulse	1550/1675	1550/1675	1550/1675	1550/1675
		Switching impulse	1175	1175	1175	1175
		Power frequency (1 min)	630	630	630	630
	LV	Full/chopped lightning impulse	650/750	650/750	650/750	650/750
		Power frequency (1 min)	275	275	275	275
	Neutral point	Full/chopped lightning impulse	325/–	325/–	325/–	325/–
		Power frequency (1 min)	140	140	140	140
No-load loss P_0 (kW)			215	198	176.8	200
Load loss P_k (kW)			(1) 1387 (2) 1414 (3) 1470	(1) 1462 (2) 1382	1422	1400
Short-circuit impedance U_k (%)			18	17.57	17.31	18
Noise (dB(A))			75	75	73	75

Table A.3: Main Technical Data of HV Shunt Reactors.

Manufacturer			XD Transformer	TBEA	Tianwei Baobian	Shandong Power Equipment
Rated capacity (MVA)			320/240/200	320/240/200	240	240
Rated voltage (kV)			$1100/\sqrt{3}$	$1100/\sqrt{3}$	$1100/\sqrt{3}$	$1100/\sqrt{3}$
Rated current (A)			504/378/315	504/378/315	378	378
Insulation level (kV)	Power frequency		1100 (5 min)	1100 (5 min)	1100 (5 min)	1100 (5 min)
	Lightning impulse	Full wave	2250	2250	2250	2250
		Chopped wave	2400	2400	2400	2400
	Switching impulse		1800	1800	1800	1800
	1 min power frequency voltage test on the neutral point		230/275	230/275	275	275
Loss (kW)			568.7/438.3/389.0	467.5/422.8/373.9	359.9	414.8

Table A.4: Main Technical Data of Arresters.

Manufacturer		XD Arrester	Fushun Electric Porcelain	Nanyang Jinguan	PG Toshiba
Nominal system voltage (kV, rms)		1000	1000	1000	1000
Maximum system operating voltage (kV, rms)		1100	1100	1100	1100
Rated voltage (kV, rms)		828	828	828	828
Continuous operating voltage (kV, rms)		638	638	638	638
Nominal discharge current (kA)		20	20	20	20
Power frequency reference voltage (kV, peak/$\sqrt{2}$)		$\geq$828/24 mA	$\geq$828/24 mA	$\geq$828/24 mA	$\geq$828/24 mA
Reference voltage at DC 8 mA (kV)		$\geq$1114	$\geq$1114	$\geq$1114	$\geq$1114
Residual voltage (kV, peak)	Lightning impulse residual voltage	$\leq$1616	$\leq$1606	$\leq$1606	$\leq$1584
	Switching impulse residual voltage	$\leq$1450	$\leq$1460	$\leq$1447	$\leq$1392
	Steep current impulse residual voltage	$\leq$1730	$\leq$1714	$\leq$1722	$\leq$1740
Total injected energy of two switching operations (MJ)		50	50	40	63

(*Continued*)

Table A.4: (Continued)

Manufacturer		XD Arrester	Fushun Electric Porcelain	Nanyang Jinguan	PG Toshiba
Insulation level (kV)	Wet power frequency withstand voltage	1200	1221	1221	1223
	Lightning impulse withstand voltage	2550	2664	2664	2667
	Wet switching impulse withstand voltage	1950	1998	1998	2000
Partial discharge (pC)		≤10	≤10	≤10	≤10
Radio interference level (μV)		≤500	≤500	≤500	≤500

Table A.5: Main Technical Data of Capacitive Voltage Transformers.

Manufacturer	XD Power Capacitor	Guilin Power Capacitor	Nissin Electric Wuxi	Shanghai MWB
Rated voltage (kV, rms)	$(1000/\sqrt{3})/(0.1/\sqrt{3})/(0.1/\sqrt{3})/(0.1/\sqrt{3})/0.1$			
Accuracy class	0.2/0.5(3P)/0.5(3P)/3P			
Rated output capacity (VA)	15/15/15/15	15/15/15/15	15/15/15/15	10/5/5/2
Rated capacitance (pF)	5000			
Power frequency/lightning/switching impulse insulation level (kV)	1300/2600/1860			
Number of sections	4	5	5	5
Anti-seismic measures	Porcelain housing			

Table A.6: Main Technical Data of Gas-Insulated Bushings.

Manufacturer	Jiangsu Shenma	XD	NGK
Type of hollow insulators	Composite insulators	Composite insulators	Porcelain insulators
Rated voltage (kV, rms)	1100	1100	1100
1 min wet power frequency withstand voltage (kV, rms)	1100	1100	1100
Lightning impulse withstand voltage (kV, peak)	2400	2400	2400
Wet switching impulse withstand voltage (kV, peak)	1800	1800	1800
Insulation distance (mm)	10,000	10,000	10,500

(Continued)

Table A.6: (Continued)

Manufacturer	Jiangsu Shenma	XD	NGK
Internal pressure for routine test (MPa)	2.4	2.0	2.4
Failing internal pressure (MPa)	4.0	3.4	3.4
Specified mechanical load (kN)	45	37.5	33.5
Maximum mechanical load (kN)	18	15	23.5

Table A.7: Main Technical Data of Oil Impregnated Paper Capacitor Bushing.

Manufacturer		XD Bushing	P&V	ABB
Rated voltage (kV, rms)		1100	1100	1100
Rated current (A)		2500	2500	2500
1 min wet power frequency withstand voltage (kV, rms)		1200 (5 min)	1200 (5 min)	1200 (5 min)
Lightning impulse withstand voltage (kV, peak value)		2400 (full-wave) 2760 (chopped-wave)	2400 (full-wave) 2760 (chopped-wave)	2400 (full-wave) 2760 (chopped-wave)
Switching impulse withstand voltage (kV, peak value)		1950	1950	1950
Total height (mm)		13,903	13,495	13,921
Insulation distance (mm)		11,568	11,340	11,465
Dielectric loss factor		≤0.40%	≤0.40%	≤0.40%
Partial discharge (pC)	$1.05U_r/\sqrt{3}$	≤5	≤5	≤5
	$1.5U_r/\sqrt{3}$	≤10	≤10	≤10

Table A.8: Main Technical Data of Series Capacitor Compensation.

Manufacturer	C-EPRI		
Line and Substation	Changzhi–Nanyang Circuit I		Nanyang–Jingmen Circuit I
	Changzhi Substation	Nanyang Substation	
Rated current (A)	5080	5080	5080
Rated impedance (Ω/phase)	19.38	19.38	14.77 + 14.77
Three-phase rated capacity (Mvar)	1500	1500	1144 + 1144
Rated voltage (kV, rms)	98.4	98.4	75.0 + 75.0
Degree of compensation	20%	20%	40%
Overvoltage protection level (p.u.)	2.3	2.3	2.3

Capacitor		
Manufacturer	Shanghai Sieyuan Power Capacitor/ABB Xi'an Power Capacitor/Guilin Power Capacitor	
Line	Changzhi–Nanyang Circuit I	Nanyang–Jingmen Circuit I
Capacitance (μF)	164.3	107.8
Three-phase rated capacity (Mvar)	1500	1144 + 1144
Rated voltage (kV, rms)	98.4	75.0 + 75.0
8 h overcurrent measured at an interval of 12 h (A)	5588	
2 h overcurrent measured at an interval of 8 h (A)	6096	
30-min overcurrent measured at an interval of 6 h (A)	6858	
10-min overcurrent measured at an interval of 2 h (A)	7620	
10 s overcurrent (A)	9144	

MOV		
Manufacturer	CEPRI	
Line	Changzhi–Nanyang Circuit I	Nanyang–Jingmen Circuit I
Allowable energy absorption (MJ per phase)	91.6	70
Rated voltage (kV, rms)	169.7	130

Clearance	
Manufacturer	CEPRI
Peak withstand current (kA)	≥170
Short-time withstand current (kA, 0.2 s)	≥63
Number of maintenance-free times under maximum fault current	25
Minimum breakdown voltage in the presence of trigger signals (p.u.)	1.8

Bypass circuit breaker	
Manufacturer	Pinggao Electric/ABB/XD Switchgear
Rated current (kA)	6.3
Peak withstand current (kA)	≥170
Short-time withstand current (kA, 2 s)	63
Closing time (ms)	≤35

Bypass disconnector		
Rated current (kA)	6.3	6.3
Short-time withstand current (kA, 2 s)	63	63
Recovery voltage (kV, rms)	≥4	≥6

Damping circuit (reactor + MOV)		
Manufacturer	**CEPRI**	
Line	**Changzhi–Nanyang Circuit I**	**Nanyang–Jingmen Circuit I**
Damping inductance (μH)	1683	1276
Peak discharge current (kA)	≥170	≥170
Design natural frequency (p.u.)	6.0	6.0

Technical Data of UHV AC Transmission Lines

Table B.1: Technical Data of UHV AC Transmission Lines.

Project	UHV AC Single-Circuit Line (UHV AC Pilot and Demonstration Project)	UHV AC Double-Circuit Line (Anhui–Shanghai Project)
Nominal voltage (kV)	1000	1000
Maximum operating voltage (kV)	1100	1100
Total length (km)	639.8	2 × 656 (including large crossing over the Huai River 2.42 and large crossing over the Yangtze River 3.18)
	Minimum air gap	
Required air gap under power frequency voltage (m)	2.7 for altitudes up to 500 m 2.9 for altitudes up to 1000 m 3.1 for altitudes up to 1500 m	2.7 for altitudes up to 500 m
Required air gap under switching overvoltage (m)	6.7/7.9[a] for middle-phase V-string and 5.6[b] for outside-phase I-string for altitudes up to 500 m 7.2/8.0[a] for middle-phase V-string and 6.0[b] for outside-phase I-string for altitudes up to 1000 m 7.7/8.1[a] for middle-phase V-string and 6.4 for outside-phase I-string for altitudes up to 1500 m	6.0 for altitudes up to 500 m
Required air gap under lightning overvoltage (m)	Not specified	6.7 for plain areas, 7.0 for mountainous areas for altitudes up to 500 m

(Continued)

Table B.1: (Continued)

Project		UHV AC Single-Circuit Line (UHV AC Pilot and Demonstration Project)	UHV AC Double-Circuit Line (Anhui–Shanghai Project)
		Insulator strings	
Type of insulator string		I-string on outside phases and V-string on the center phase	I-string
Type of insulators		Cap and pin type, and long rod type	Cap and pin type and long rod type
Insulating material		Porcelain, glass, composite material	Porcelain, glass, and composite material
Number of insulators in a suspension string		54 cap and pin insulators of double-shed type for a pollution class of II; composite insulators for a pollution class of III or IV	Composite insulators
Length of insulator string (m)		Cap and pin insulator string: 10.53; composite insulator string: 9.75	Composite insulator string: 9.0
		Tower grounding resistance and ground wire protective angle	
Design value of power frequency grounding resistance (Ω)	Soil resistivity ≤ 100 Ω m	10	10
	100 Ω m < soil resistivity ≤ 500 Ω m	15	15
	500 Ω m < soil resistivity ≤ 1000 Ω m	20	20
	1000 Ω m < soil resistivity ≤ 2000 Ω m	25	25
	Soil resistivity > 2000 Ω m	30	30
Protective angle for ground wire to phase conductor		No more than 5° for plain areas No more than 5° for mountainous areas Three ground wires used in the span where the incoming and outgoing feeders of substation are connected	No more than −3° for plain areas No more than −5° for mountainous areas No more than 0° for protective angle of ground wire to the jumper of a tension tower No more than −4° for the incoming and outgoing feeders within 2 km range in the substation

(Continued)

Table B.1: (Continued)

Project		UHV AC Single-Circuit Line (UHV AC Pilot and Demonstration Project)	UHV AC Double-Circuit Line (Anhui–Shanghai Project)
Ground wire			
Number of wires		2	2
Model		One aluminum-clad steel-reinforced conductor JLB20A-170 and one OPGW-175	One aluminum-clad steel-reinforced conductor LBGJ-240-20AC and one OPGW-240
Electromagnetic environment			
Limiting design value of power frequency electric field strength (kV/m)	Residential areas	4	4
	Artery traffic, such as roads and railways, that a line crosses	7	7
	Other areas, such as farmlands	10	10
Limiting design value of power frequency magnetic induction density (μT)		100	100
Limit value of radio interference (dB (μV/m))		No greater than 55 at 20 m from projection of outside phase conductor under a test frequency of 0.5 MHz on sunny days	No greater than 55 at 20 m from projection of outside phase conductor under a test frequency of 0.5 MHz on sunny days
Noise	Substation	The border of noise control area should satisfy the standard of GB 12348–2008 *Emission Standard for Industrial Enterprises Noise at Boundary*	The border of noise control area should satisfy the standard of GB 12348–2008 *Emission Standard for Industrial Enterprises Noise at Boundary*
Noise	Limiting design value for a line (dB(A))	No more than 55 at 20 m from the projection of wet outside phase conductor	No more than 55 at 20 m from the projection of wet outside phase conductor
Minimum height of conductor to ground			
Minimum height (m)	Residential areas	27	25
	Non-residential areas (farmlands)	22	
Tower and conductor			
Tower	Type	Mainly angle steel tower, steel tube tower at large crossings	Steel tube tower along the entire length of line
	Number	1284	1421
	Average height (m)	77.2	108
	Average weight (t)	70.5	180

(Continued)

Table B.1: (Continued)

Project		UHV AC Single-Circuit Line (UHV AC Pilot and Demonstration Project)	UHV AC Double-Circuit Line (Anhui–Shanghai Project)
Conductor	Model	8 × LGJ-500/35	8 × LGJ-630/45
	Spacing between subconductors	400	400
	Outer diameter (mm)	30.00	33.60
	Elastic modulus (N/mm^2)	63,000	63,000
Conductor	Calculated mass (kg/km)	1642	2060
	Tensile strength (kN)	119.5	148.7
	Number of sub-conductors in each bundle	8	8

[a]The figures following the slash refer to the minimum clearance to the upper cross-arm.
[b]The values used in engineering design differ slightly from those recommended in the text considering the width of tower body and correction of front time and based on full-scale test results.

Main Technical Data of UHV DC Electrical Equipment[1]

Table C.1: Converter Valves.

Main Technical Data	Unit	Value
1. DC current ratings		
Nominal DC current I_{dN}	A	4000
Minimum continuous operating DC current	A	367
Maximum continuous operating DC current (rated transmission power, maximum ambient temperature, 50°C valve hall temperature, without redundant cooling in service)	A	4037
Maximum continuous operating DC current (1.1 p.u. of rated transmission power, maximum ambient temperature, 50°C valve hall temperature, with redundant cooling in service)	A	4497
Maximum current when operated with 1.4 p.u. overload for 3 s (maximum ambient temperature, 50°C valve hall temperature)	A	–
2. DC voltage ratings		
Nominal pole-to-neutral DC voltage U_{dRN}	kV	800
Maximum continuous pole-to-neutral DC voltage under rated operation (including errors)	kV	816
Nominal no-load DC voltage U_{dioN}	kV	230.0
Maximum no-load DC voltage $U_{dioabsmax}$ (design value)	kV	237.6
Load rejection factor at temporary overvoltage (valve blocked)	p.u.	1.4
3. Transient current of thyristor valve		
Peak short-circuit current of a single valve ($\alpha_{min} = 5°$, $f = 49.8$ Hz, followed by blocking)	kA	43.9
Recovery time with follow-up blocking	ms	2.12
4. Insulation level		
Valve (*SIWL/LIWL*)	kV	450/426
Upper 12-pulse bridge's DC bus-to-ground insulation level (*SIWL/LIWL*)	kV	1600/1800

1 The technical data in this appendix are parameters of the main equipment in converter stations in the Xiangjiaba–Shanghai project.

Table C.2: Six-Inch Thyristors.

Main Technical Data	Symbols	Unit	Average *SCR*	Worst *SCR*	Rated *SCR*	Test Conditions
Continuous off-state DC voltage	VD	V	–	–	–	7020 (1 min)
Peak off-state voltage	VDM	V	4953	5585	–	6423
Off-state operating voltage (peak)	VDWM	V	4953	5585	–	6423
Repetitive peak off-state voltage	VDRM	V	6017	6019	7900	7675
Non-repetitive peak off-state voltage	VDSM	V	8257	8500	8500	8500
Forward breakover voltage (DC)	V(BO)	V	8257	8500	–	8500
Continuous on-state voltage (DC)	VT	V	1.81	1.94	1.97	–
Continuous reverse voltage (DC)	VR	V	–	–	–	7020 (1 min)
Peak reverse voltage (crest)	VRWM	V	4953	5585	–	6423
Repetitive peak reverse voltage	VRRM	V	6017	7001	7900	7675
Non-repetitive peak reverse voltage	VRSM	V	6750	7611	8500	8198
Continuous on-state current (DC)	IT	A	4500	4542	4200	4500
On-state overload current	I(OV)	A	–	5934	–	–
Surge (non-repetitive) on-state current	ITSM	A	–	45,800	46,000	46,000

Table C.3: Converter Transformer.

Main Technical Data	HV End Y/Y	LV End Y/Y	HV End Y/Δ	LV End Y/Δ
Type	Forced-oil and air cooled, oil-immersed type single-phase two-winding transformer with on-load tap changer			
Connection of transformer winding (after three phases are connected)	YNy		YNd11	
Rated voltage ratio	$(530/\sqrt{3})/(170.3/\sqrt{3})$		$(530/\sqrt{3})/170.3$	
Cooling method	ODAF			
Rated capacity (MVA)	321.1			
Average temperature rise under operation at rated capacity (K)	55			
Hot-spot temperature rise under operation at rated capacity (K)	68			
Top oil temperature rise (K)	50			
Short-circuit impedance of main tap (%)	18			

(*Continued*)

Table C.3: (Continued)

Main Technical Data			HV End Y/Y	LV End Y/Y	HV End Y/Δ	LV End Y/Δ
Insulation level (kV)	Lightning impulse	500 kV side	1550			
		Line side neutral	200			
		Valve side	1800	1300	1550	1175
	Chopped lightning impulse	500 kV side	1705			
		Valve side	1980	1430	1705	1293
	Switching impulse	500 kV side	1175			
		Valve side	1600	1175	1300	1050
Noise	Under rated power frequency voltage (dB(A))		75			
Tap changer	Rated current (A)		1391			
	Tapping range		−5 to +23			
	Tap step (%)		1.25			
	Tap transfer rate (steps/min)		12			
	Switching times without maintenance		1×10^5			
	Electrical life (switching times)		5×10^5			
	Mechanical life (switching times)		1×10^6			
	Full lightning impulse withstand voltage (kV)		325			
Permissible continuous DC current flowing through neutral of line side winding of each phase's converter transformer (A)			10			

Table C.4: Dry-Type Smoothing Reactor.

Main Technical Data		Unit	Value
Rated inductance		mH	75(1 + 5%)
Nominal rated DC current I_{dN}		A	4000
Maximum continuous operating DC current under 1.10 p.u. of the rated transmitting power		A	4497
Maximum DC current under 2 h overload operation		A	4731
Maximum current when operated at 1.4 p.u. overload for 3 s		A	5909
Transient fault current (peak)		kA	40
Nominal rated pole-to-neutral DC voltage		kV	800
Maximum continuous operating DC voltage under rated operation (including errors)		kV	816
Rated neutral bus voltage		kV	120
Maximum neutral bus voltage		kV	123
Winding temperature rise	Average temperature rise	K	80
	Hot-spot temperature rise	K	105
Insulation level and test voltage (applicable to two series-connected 75 mH devices)	Switching impulse withstand level (peak, between terminals)	kV	1675
	Switching impulse withstand level (peak, terminal-to-earth)	kV	1600
	Lighting impulse withstand level (peak, between terminals)	kV	2100
Insulation level and test voltage (applicable to two series-connected 75 mH smoothing reactors)	Lightning impulse withstand level (peak)	kV	1950
	Chopped lightning impulse withstand level (peak, between terminals)	kV	2310
	DC withstand voltage (kV, DC, 120 min, to ground)	kV	1236
Sound pressure level		dB	70

Table C.5: Wall Bushings (kV).

Main Technical Data	± 800 kV Line Pole Bushing	± 400 kV Line Pole Bushing
Rated operating voltage (DC)	800	400
Maximum continuous operating voltage (DC)	816	408
DC withstand voltage (DC)	1224	612
Power frequency withstand voltage	865	433
Lightning impulse withstand level	1800	1300
Switching impulse withstand level	1620	1050
Polarity reversal withstand voltage (DC)	1020	510

Table C.6: DC Voltage Measuring Device.

Main Technical Data	Unit	Line Pole (Fulong)	Neutral Bus (Fulong)
Nominal DC voltage U_{dN}	kV	8	10
Maximum continuous DC voltage U_{dmax}	kV	816	95
Measuring ranges	kV	+ 1200	± 120
Output bus voltage of isolated A/D system	p.u.	1.0	1.0
	kV	800	60
DC voltage withstand test (dry) (60 min), simultaneously with polarity reversal and partial discharge test	kV	1224	143
Lightning impulse tests	kV	1950	574

Table C.7: DC Current Measuring Devices.

Main Technical Data		Unit	Line Pole (in Valve Hall and Outdoor)	Neutral Bus (in Valve Hall and Outdoor)	Between Two 12-pulse Converters	
					Converter on HV Side	Converter on Neutral Bus Side
Rated DC current I_{dN}		A	4000	4000	4000	4000
Steady-state DC measuring limit		A	5000	5000	5000	5000
Transient measuring limit		kA	24	24	24	24
Lightning impulse withstand level		kV	2182	450	1800	1175
Switching impulse withstand level		kV	1792	325	1600	950
Measurement accuracy	(0−134%) I_{dN}	%	0.5	0.2	0.5	0.2
	(134−300%) I_{dN}	%	1.5	1.5	1.5	1.5
	(300−600%) I_{dN}	%	10	10	10	10

Table C.8: DC Arresters.

Main Technical Data		Unit	DC Arresters			Valve Arresters		
			DB	CBH	CBL	V11	V12	V2/V3
Continuous voltage (peak)		kV	824	878	404.5	248.8	248.8	248.8
Continuous voltage with commutation overshoot (peak)		kV	–	916	496.4	288.4	288.4	295.6
Reference voltage U_c (DC)		kV	969	1070	553	$204\sqrt{2}$	$204\sqrt{2}$	$210\sqrt{2}$
Maximum residual voltage under steep front current impulse	1000 A (peak)	kV	<1536	<1627	<842	<420	<420	<430
	3000 A (peak)	kV	1536	<1627	<842	<420	<420	<430
	10 kA (peak)	kV	1653	1753	907	432	454	467
	20 kA (peak)	kV	1751	1864	964	454	482	497
Maximum residual voltage under 8/20 μs current impulse	1000 A (peak)	kV	1381	1453	751	369	376	387
	3000 A (peak)	kV	1445	1518	785	380	393	405
	10 kA (peak)	kV	1547	1642	849	406	425	437
	20 kA (peak)	kV	1621	1731	896	425	448	461
Maximum residual voltage under 30–45/60–90 μs current impulse	300 A (peak)	kV	1316	1402	725	<360	363	363
	1000 A (peak)	kV	1371	1447	748	368	374	385
	3000 A (peak)	kV	1431	1510	781	379	391	402
Energy absorption rating	For a single impulse (4 ms)	kJ/kV	7.0	7.0	7.0	7.0	7.0	7.0
	For the entire surge arrester	kJ	13,500	19,000	10,100	10,300	5100	5300
Number of parallel columns			3	4	4	8	4	4

Table C.9: DC Transfer Switch.

Main Technical Data	Unit	MRTB	GRTS	NBS	NBGS
Rated current	A	4970	4970	4970	—
Rated voltage	kV	95	95	95	95
Maximum transfer current	A	4330	1200	4970	4970
Maximum transfer voltage (peak)	kV	120	54	55	45
Maximum arcing time if the switch operates successfully	ms	20	20	20	20
The integral of arcing current over time if the switch fails and recloses	As	1860	262	950	794
LIWL between breaks (dry)	kVc_{rest}	189	85	88	71
LIWL to ground (dry)	kVc_{rest}	574	574	574	574
SIWL between breaks (wet)	kVc_{rest}	138	62	64	52
SIWL to ground (wet)	kVc_{rest}	452	452	452	452
The (maximum) rate of rise of recovery voltage	kV/ms	280	280	280	280
Normal closing time	ms	50	50	50	50
Normal reclosing time (C—O—C) if the switch fails	ms	430	218	191	160
Operation cycle		O—C—O	O—C—O	C—O—C	C—O—C
Number of series poles		2	1	2	2
Rated capacitance	μF	60	22	60	60
Maximum energy absorbed by arrester	kJ	53,050	1270	5570	1700
Nominal inductance	μH		60		

Table C.10: Main Technical Data of Other UHV DC Projects.

Main Technical Data		Yunnan–Guangdong Project	Haminan–Zhengzhou Project	Xiluodu–Zhejiang Project
DC voltage (kV)		± 800	± 800	± 800
DC current (kA)		3125	5000	5000
DC power (MW)		5000	8000	8000
Transmission distance (km)		1373	2210	1669
Main connection scheme		Each pole having two series connected 12-pulse converter units	Each pole having two series connected 12-pulse converter units	Each pole having two series connected 12-pulse converter units
Converter transformers	Capacity (MVA)	250.21 (Sending end) 244 (Receiving end)	405.13 (Sending end) 376.61 (Receiving end)	404 (Sending end) 382 (Receiving end)
	Short-circuit impedance	18% (Sending end) 18.5% (Receiving end)	20% (Sending end) 19% (Receiving end)	19% (Sending end) 19% (Receiving end)
	Step size	1.25%	1.25%	1.25%
	Number of tap positions	+ 18/ − 6 (Sending end) + 16/ − 8 (Receiving end)	+ 23/ − 5 (Sending end) + 25/ − 5 (Receiving end)	+ 23/ − 5 (Sending end) + 23/ − 5 (Receiving end)
	Type	Single-phase two-winding	Single-phase two-winding	Single-phase two-winding
	Cooling method	ODAF (Sending end) ODAF (Receiving end)	OFAF (Sending end) ODAF (Receiving end)	OFAF (Sending end) ODAF (Receiving end)
Converter valves	Size (inch)	5	6	6
	Valve tower structure	Double valve unit	Double valve unit	Double valve unit
Smoothing reactors	Inductance (mH)	300	300	300
	Type	Dry	Dry	Dry
	Arrangement	Line pole: 2 × 75 mH Neutral bus: 2 × 75 mH	Line pole: 3 × 500 mH Neutral bus: 3 × 500 mH	Line pole: 3 × 500 mH Neutral bus: 3 × 500 mH
AC filters/shunt capacitors	Total compensation capacity (Mvar)	3366 (Sending end) 3010 (Receiving end)	3575 (Sending end) 4940 (Receiving end)	5200 (Sending end) 4879 (Receiving end)
DC filters	Type (for each pole of a converter station)	12/24 and 36/48	12/24 and 2/39	12/24 and 2/39
	Equivalent disturbing current (mA)	Monopolar: 2000 Bipolar: 2000	Monopolar: 6000 Bipolar: 3000	Monopolar: 6000 Bipolar: 3000

Technical Data of UHV DC Transmission Lines

Table D.1: Technical Data of UHV DC Transmission Lines.

Technical Data		Xiangjiaba–Shanghai ±800 kV UHV DC Transmission Line
Rated operating voltage (kV)		±800
Maximum operating voltage (kV)		±816
Transmission capacity (MW)		6400
Rated current (A)		4000
Minimum clearance	Under switching overvoltage for common lines	At an altitude of 500 m: 5.5 m At an altitude of 1000 m: 5.7 m At an altitude of 2000 m: 6.2 m
	Under operating voltage for common lines	At an altitude of 500 m: 2.1 m At an altitude of 1000 m: 2.3 m At an altitude of 2000 m: 2.5 m
	For large-crossing lines	6.1 m under switching overvoltage 2.55 m under operating voltage
Minimum height of conductor above ground (m, considering the ground total electric field)		18 (non-residential areas), 21 (residential areas)
Minimum height of conductor above ground at different approach distances from 1000 kV lines (m)	Approach distance 70 m	20 (non-residential areas), 24 (residential areas)
	Approach distance 80 m	18 (non-residential areas), 23 (residential areas)
Line corridors	Within 7 m from the projection of pole conductor	All occupied houses removed
	Beyond 7 m from the projection of pole conductor	Occupied houses not meeting the requirements of ground total electric field (maximum total electric field not larger than 25 kV/m, and 80% measured value not exceed 15 kV/m) relocated

(*Continued*)

Table D.1: (Continued)

Technical Data		Xiangjiaba–Shanghai ±800 kV UHV DC Transmission Line
		If the requirements of conductors on vertical distance and wind deflection distance are met, non-residential buildings with non-combustible roofs not relocated, and buildings with combustible roofs and maximum total electric field over 25 kV/m relocated
Width of line corridor at different approach distances from 1000 kV lines (m)	Approach distance 70 m Approach distance 80 m	146 (non-residential areas), 141 (residential areas) 156 (non-residential areas), 152 (residential areas)
Insulator strings	Light and medium icing areas	Suspension strings on common lines: 240, 300, 420, 550 kN single and double composite V-strings Tension strings: triple 420 kN disc insulator strings Jumper strings: 160 kN composite V-strings
Insulator strings	Heavy icing areas	Suspension strings on common lines: 210, 300, 420, 550 kN single and double disc V-strings Tension strings in 20 mm heavy icing areas: triple 420 kN disc insulator strings Tension strings in 30 mm heavy icing areas: triple 550 kN disc insulator strings Jumper strings: 210 kN disc V-strings
Lightning protection	Common lines	Protective angle afforded by ground wire to outside conductor smaller than −10° in mountainous areas Protective angle afforded by ground wire to outside conductor smaller than 10° in plain areas
	Large crossing lines	Horizontal displacement between ground wire and conductor no smaller than 3 m, protective angle afforded by ground wire to outside conductor is smaller than 0° Enlarging the distance between conductor in the middle of span and ground wire (>20 m) to enable a lightning withstand level of 200 kA Design required grounding resistance smaller than 5 Ω

(*Continued*)

Table D.1: (Continued)

Technical Data		Xiangjiaba–Shanghai ±800 kV UHV DC Transmission Line
Ground wires	Light, medium, and heavy icing (20 mm) areas Heavy icing (30 mm) areas	One LBGJ-180-20AC and one OPGW-250 One LBGJ-240-20AC and one OPGW-250
Electric field strength on the surface of conductor under maximum operating conditions (kV/cm)		22.2
Design limit of maximum total electric field strength	Limit of technical specification Measuring position Maximum total electric field (kV/m)	Industrial standards Ground Under the transmission line: 30 Close to houses: 25 (80% value not exceed 15 kV/m) In converter stations: 30
Design limit of maximum DC magnetic field	Limit of technical specification Maximum magnetic flux density (μT)	Industrial standards 10,000
Ion current density	Limit of technical specification Maximum magnetic induction density (nA/m^2)	Industrial standards Under the transmission line: 100
Radio interference	Limit of technical specification Design criteria Frequency (MHz) Technical specifications Weather conditions Radio interference level (dB) Statistical parameter Altitude correction	Industrial standards 20 m from the projection of positive conductor 0.5 CISPR All-weather conditions 58 80%//80% (CISPR) When the altitude exceeds 1000 m, the limit should be corrected by increasing 3 dB for every 1000 m increase in altitude
Noise	Limit of technical specification Design criteria Technical specifications for measurement	Industrial standards 20 m from the projection of positive conductor National standards

(Continued)

Table D.1: (Continued)

Technical Data		Xiangjiaba–Shanghai ± 800 kV UHV DC Transmission Line
Noise	Weather conditions Noise limit (dB) Statistical parameter Altitude correction	Fair weather conditions 45 L_{50} When the altitude exceeds 1000 m, the audible noise limit should be corrected by increasing 3 dB for every 1000 m increase in altitude
Conductor type		Light, medium, and heavy icing (20 mm) areas: ACSR-720/50; heavy icing (30 mm) areas: AACSR-720/50
Number of subconductors per phase		6
Arrangement of subconductors		Symmetrically arranged

Standards for UHV AC and DC Transmission Technologies

Table E.1: Standards for UHV AC Transmission Technologies.

No.	Standard No.	Designation of Standard
I National standards		
1	GB/T 156-2007	Standard Voltages
2	GB/T 24833–2009	Technical Specification for 1000 kV Substation Automation System
3	GB/T 24834–2009	Technical Specification for Fittings of 1000 kV AC Overhead Transmission Line
4	GB/Z 24835–2009	Regulation of Operation and Maintenance for 1000 kV Gas-Insulated Metal-Enclosed Switchgear
5	GB/Z 24836–2009	Specification for 1100 kV Gas-Insulated Metal-Enclosed Switchgear
6	GB/Z 24837–2009	Specification for 1100 kV Alternating-Current Disconnectors and Earthing Switches
7	GB/Z 24838–2009	Specification for 1100 kV Alternating-Current High-Voltage Circuit-Breakers
8	GB/Z 24839–2009	Technical Specification of Post Insulators for 1000 kV AC System
9	GB/Z 24840–2009	Technical Specification for Bushing of 1000 kV AC System
10	GB/Z 24841–2009	Technical Specification for Capacitor Voltage Transformers of 1000 kV AC System
11	GB/Z 24842–2009	Overvoltage and Insulation Coordination of 1000 kV UHV AC Transmission Project
12	GB/Z 24843–2009	Technical Specification of 1000 kV Single Phase Oil-Immersed Auto-Transformer
13	GB/Z 24844–2009	Technical Specification for Oil-Immersed Shunt Reactor of 1000 kV AC System
14	GB/Z 24845–2009	Specification of Metal-Oxide Surge Arresters Without Gaps for 1000 kV AC System
15	GB/Z 24846–2009	Preventive Test Standards of 1000 kV AC Electrical Equipments
16	GB/Z 24847–2009	Technical Guide on 1000 kV AC System Voltage and Reactive Power
17	GB/T 25726–2010	Screen Clothes for Live Working on 1000 kV AC
18	GB/T 25737–2010	Acceptance Norms for 1000 kV Substation Monitoring System
19	GB/Z 25841–2010	Guide for Protection Relaying for 1000 kV Power System

(Continued)

Table E.1: (Continued)

No.	Standard No.	Designation of Standard
20	GB/T 29322–2012	Specification for 1000 kV Power Transformer Protection Equipment
21	GB/T 29323–2012	Specification for 1000 kV Circuit Breaker Protection Equipment
22	GB/Z 29327–2012	Specification for 1000 kV Reactor Protection Equipment
23	GB 50665–2011	Code for Design of 1000 kV Overhead Transmission Line
24	GB 50697–2011	Code for Design of 1000 kV Substation
25	GB/T 50832–2013	Standard for Acceptance Test of Electric Equipment of 1000 kV System Electric Equipment Installation Engineering
26	GB 50834–2013	Code for Construction and Acceptance of 1000 kV Lattice Frame and Support
27	GB 50835–2013	Codes for Construction and Acceptance of 1000 kV Power Transformer, Oil-Immersed Reactor and Mutual Inductor
28	GB 50836–2013	Code for Construction and Acceptance of 1000 kV High-Voltage Electric Equipment (GIS, HGIS, Disconnector, and Surge Arrester)
II Industrial standards		
1	DL/T 306.1–2010	Code of Operation for AC 1000 kV Substation—Part 1: Equipment
2	DL/T 306.2–2010	Code of Operation for AC 1000 kV Substation—Part 2: Operating
3	DL/T 306.3–2010	Code of Operation for AC 1000 kV Substation—Part 3: Inspecting
4	DL/T 306.4–2010	Code of Operation for AC 1000 kV Substation—Part 4: Handling of Equipment Failure
5	DL/T 306.5–2010	Code of Operation for AC 1000 kV Substation—Part 5: Typical Operation
6	DL/T 306.6–2010	Code of Operation for AC 1000 kV Substation—Part 6: Diagram
7	DL/T 307–2010	Code of Operation for AC 1000 kV Overhead Transmission Line
8	DL/T 309–2010	Execution Guide for Field Test of 1000 kV AC System Power Equipment
9	DL/T 310–2010	Guide for AC 1000 kV Oil-Immersed Transformer and Shunt Reactor Maintenance
10	DL/T 311–2010	Guide for AC 1100 kV GIS/HGIS Maintenance
11	DL/T 312–2010	Guide for AC 1000 kV Capacitor Voltage Transformer Maintenance
12	DL/T 313–2010	Technical Specification for Field Inspection of 1000 kV Power Transformers
13	DL/T 392–2010	Technical Guide for Live Working on 1000 kV AC Transmission Line
14	DL/T 1176–2012	Regulation of Operation and Preventive Maintenance for AC 1000 kV Oil-Immersed Transformer and Shunt Reactor
15	DL/T 1177–2012	Guide for Technology Supervision of 1000 kV AC Transmission & Transformation Equipment
16	DL/T 1178–2012	Corona and RIV Test Methods for Fittings of 1000 kV AC Transmission Line
17	DL/T 1179–2012	Guide for Power Frequency Parameters Measurement of 1000 kV AC Overhead Transmission Lines
18	DL/T 1180–2012	Guide of Supervision for Manufacturing 1000 kV Electrical Equipment

(*Continued*)

Table E.1: (Continued)

No.	Standard No.	Designation of Standard
19	DL/T 1181–2012	Technical Specification of 1000 kV AC Composite Long Rod Insulators
20	DL/T 1182–2012	Technical Specification of 110 kV Shunt Capacitor Installations in 1000 kV Substations
21	DL/T 1183–2012	Non-Contact Voltage Detectors for 1000 kV UHV AC
22	DL/T 1184–2012	The Guide of Supervision for Manufacturing 1000 kV Transmission Line Tower, Conductors, Electric Power Fittings and Optical Fibre Composite Overhead Ground Wire
23	DL/T 1185–2012	Technical Regulations for Impact Assessment of Electromagnetic Environment Produced by 1000 kV Transmission and Transfer Power Engineering
24	DL/T 1186–2012	Technical Specification for Tank-Type Voltage Transformers of 1000 kV
25	DL/T 1187–2012	Control Levels for Electromagnetic Environment Produced by 1000 kV Overhead Transmission Line
26	DL/T 1188–2012	Control Levels of Electromagnetic Environment Generated for 1000 kV Substation
27	DL/T 1237–2013	Testing Regulations on AC 1000 kV Protection and Security Automatic Equipments
28	DL/T 1238–2013	Electrostatic Shielding Clothing for 1000 kV AC System
29	DL/T 1239–2013	Code for Operation and Management of AC 1000 kV Protection and Security Automatic Equipments
30	DL/T 1240–2013	Preventive Test Code of Tools, Devices, and Equipment for Live Working on 1000 kV AC
31	DL/T 5289–2013	Construction Technology Guide to the Assembly and Erection of the Steel Towers of 1000 kV Overhead Transmission Line
32	DL/T 5290–2013	Construction Technology Guide for Tension Stringing of 1000 kV Overhead Transmission Line
33	DL/T 5291–2013	Crimping Technology Guide for Conductor and Ground Wire of 1000 kV Power Transmission and Transformation Project
34	DL/T 5292–2013	Code of System Commissioning for 1000 kV AC Transmission and Transformation Project
III Corporate standards of SGCC		
1	Q/GDW 153–2006	Code for Construction and Acceptance of 1000 kV Overhead Transmission Line
2	Q/GDW 154–2006	Construction Technology Guide for Tension Stringing of 1000 kV Overhead Transmission Line
3	Q/GDW 155–2006	Construction Technology Guide to the Assembly and Erection of the Steel Towers of 1000 kV Overhead Transmission Line
4	Q/GDW 163–2007	Specification for Construction Quality Checkout and Evaluation of 1000 kV Overhead Transmission Line

(*Continued*)

Table E.1: (Continued)

No.	Standard No.	Designation of Standard
5	Q/GDW 164—2007	Code for Manufacture, Construction, and Acceptance of Gantry and Support for 1000 kV Substation
6	Q/GDW 165—2007	Code for Construction of Lattice Frame and Support for 1000 kV AC Substation
7	Q/GDW 178—2008	Provisional Technical Code for Designing 1000 kV Overhead Transmission Line
8	Q/GDW 183—2008	Code for Construction Quality Acceptance and Evaluation of 110—1000 kV Substation (Converter Station)
9	Q/GDW 189—2008	Specification for Construction Quality Checkout and Evaluation of Electric Equipment for 1000 kV Substation
10	Q/GDW 190—2008	Guide for Construction of Secondary Wiring for 1000 kV Substation
11	Q/GDW 191—2008	Guide for Construction of Earthing Device for 1000 kV Substation
12	Q/GDW 192—2008	Code for Construction and Acceptance of 1000 kV Power Transformer, Oil-Immersed Reactor, and Mutual Inductor
13	Q/GDW 193—2008	Guide for Construction of 1000 kV Power Transformer and Oil-Immersed Reactor
14	Q/GDW 194—2008	Guide for Construction of 1000 kV Capacitance Potential Transformer, Surge Arrester, and Post Insulator
15	Q/GDW 195—2008	Code for Construction and Acceptance of 1000 kV High-Voltage Electric Equipment (GIS, HGIS, Disconnecting Switch, and Surge Arrester)
16	Q/GDW 196—2008	Guide for Construction of 1000 kV Disconnecting Switch
17	Q/GDW 197—2008	Guide for Construction of 1000 kV Bus Device
18	Q/GDW 198—2008	Code for Construction and Acceptance of 1000 kV Bus Device
19	Q/GDW 199—2008	Guide for Construction of 1000 kV Gas-Insulated Metal-Enclosed Switchgear
20	Q/GDW 239—2009	Testing Regulations on AC 1000 kV Protection and Security Automatic Equipments
21	Q/GDW 278—2009	Technical Specifications for 1000 kV Substations Grounding
22	Q/GDW 284—2009	Commissioning Standards for 1000 kV AC Transmission and Distribution System
23	Q/GDW 285—2009	Specification for Start-Up and Completion Acceptance of 1000 kV AC Transmission and Distribution Projects
24	Q/GDW 286—2009	Technical Guide on 1000 kV System Voltage and Reactive Power
25	Q/GDW 290—2009	Technical Specification for Fittings of 1000 kV AC Overhead Transmission Line
26	Q/GDW 291—2009	Technical Specification of Fittings for 1000 kV Substation
27	Q/GDW 292—2009	Guide of Supervision for Manufacturing 1000 kV Transmission Line Tower, Conductors, Electric Power Fittings, and OPGW
28	Q/GDW 294—2009	Technical Code for Design of 1000 kV AC Substations (Electrical Section)

(*Continued*)

Table E.1: (Continued)

No.	Standard No.	Designation of Standard
29	Q/GDW 295—2009	Technical Specifications for 1000 kV AC Current Transformer
30	Q/GDW 297—2009	Technical Code for Tower Design of 1000 kV Overhead Transmission Line
31	Q/GDW 298—2009	Technical Regulation of Investigation and Surveying for 1000 kV AC Overhead Transmission Line
32	Q/GDW 303—2009	Technical Specification of Post Insulators for 1000 kV Substation
33	Q/GDW 304—2009	Technical Regulations for Impact Assessment of Electromagnetic Environment Produced by 1000 kV Transmission and Transfer Power Engineering
34	Q/GDW 305—2009	Limits for Electromagnetic Environment Generated by 1000 kV Overhead Power Transmission Lines
35	Q/GDW 306—2009	Technical Specification for Oil-Immersed Shunt Reactor of 1000 kV AC System
36	Q/GDW 307—2009	Specification of Metal-Oxide Surge Arresters without Gaps for 1000 kV AC System
37	Q/GDW 308—2009	Technical Specification for Bushing of 1000 kV AC System
38	Q/GDW 309—2009	Technical Specification for Capacitor Voltage Transformers of 1000 kV AC System
39	Q/GDW 310—2009	Standard for Acceptance Test of Electric Equipment of 1000 kV System Electric Equipment Installation Engineering
40	Q/GDW 311—2009	Corona and RIV Tests for 1000 kV AC Transmission Line Electric Power Fittings
41	Q/GDW 312—2009	Technical Specification of Oil-Immersed Transformer for 1000 kV System
42	Q/GDW 313—2009	Specification for High-Voltage Circuit Breaker in 1000 kV System
43	Q/GDW 314—2009	Specification for Alternating-Current Disconnectors and Earthing Switches in 1000 kV Power System
44	Q/GDW 315—2009	Specification for Gas-Insulated Metal-Enclosed Switchgear in 1000 kV System
45	Q/GDW 316—2009	Technical Code and Operation Technical Requirements for Optical Fiber Composite Overhead Ground Wire (OPGW) Under UHV
46	Q/GDW 317—2009	Technical Code for Construction and Final Acceptance of Optical Fiber Composite Overhead Ground Wire (OPGW) Engineering Under UHV
47	Q/GDW 318—2009	Overvoltage and Insulation Coordination for 1000 kV Transmission and Distribution Projects
48	Q/GDW 319—2009	Guide for Overvoltage and Insulation Coordination of 1000 kV UHV AC Transmission System
49	Q/GDW 320—2009	Guide of Supervision for Manufacturing 1000 kV AC Electrical Power Equipments
50	Q/GDW 321—2009	Technical Specification of Field Test Equipment for 1000 kV AC Electrical Equipments
51	Q/GDW 322—2009	Preventive Test Standards of 1000 kV AC Electrical Equipments

(Continued)

Table E.1: (Continued)

No.	Standard No.	Designation of Standard
52	Q/GDW 323–2009	Environmental Pollution Classification for 1000 kV System
53	Q/GDW 324–2009	Technical Guide for Live Working on 1000 kV AC Transmission Line
54	Q/GDW 325–2009	Specification for 1000 kV Power Transformer Protection Equipment
55	Q/GDW 326–2009	Specification for 1000 kV Reactor Protection Equipment
56	Q/GDW 327–2009	Specification for 1000 kV Transmission Line Protection Equipment
57	Q/GDW 328–2009	Specification for 1000 kV Busbar Protection Equipment
58	Q/GDW 329–2009	Specification for 1000 kV Circuit Breaker Protection Equipment
59	Q/GDW 330–2009	Inspection Specification for 1000 kV Power System Protection Relay and Automation Device
60	Q/GDW 331–2009	Code for Operation and Management of AC 1000 kV Protection and Security Automatic Equipments
61	Q/GDW/Z 207.1–2008	Guide for 1000 kV Transformation Equipment Maintenance–Part 1: Oil-Immersed Transformer and Shunt Reactor
62	Q/GDW/Z 207.2–2008	Guide for 1000 kV Transformation Equipment Maintenance–Part 2: 1000 kV GIS/HGIS
63	Q/GDW/Z 207.3–2008	Guide for 1000 kV Transformation Equipment Maintenance–Part 3: Surge Arrester
64	Q/GDW/Z 207.4–2008	Guide for 1000 kV Transformation Equipment Maintenance–Part 4: Capacitor Voltage Transformer
65	Q/GDW/Z 208–2008	Code for 1000 kV Substation Overhauling Administration
66	Q/GDW/Z 209–2008	Maintenance Code for 1000 kV AC Overhead Transmission Line
67	Q/GDW/Z 210–2008	Operating Code for AC 1000 kV Overhead Transmission Line
68	Q/GDW/Z 211.1–2008	Code of Operation for 1000 kV UHV Substation–Part 1: Equipment
69	Q/GDW/Z 211.2–2008	Code of Operation for 1000 kV UHV Substation–Part 2: Operating
70	Q/GDW/Z 211.3–2008	Code of Operation for 1000 kV UHV Substation–Part 3: Inspecting
71	Q/GDW/Z 211.4–2008	Code of Operation for 1000 kV UHV Substation–Part 4: Handling of Equipment Failure
72	Q/GDW/Z 211.5–2008	Code of Operation for 1000 kV UHV Substation–Part 5: Typical Operation
73	Q/GDW/Z 211.6–2008	Code of Operation for 1000 kV UHV Substation–Part 6: Diagram

Table E.2: Standards for UHV DC Transmission Technologies.

No.	Standard No.	Designation of Standard
I National standards		
1	GB/T 28813–2012	Operating Code for ±800 kV DC Overhead Transmission Line
2	GB/T 28814–2012	Guide for Composition of the ±800 kV Converter Station Operating Code
3	GB 50729–2012	Code for Acceptance of Construction Quality of ±800 kV and Under HVDC Converter Substation
4	GB 50774–2012	Code for Construction and Acceptance of Dry-Type Smoothing Reactors in Converter Stations at ±800 kV and Below
5	GB/T 50775–2012	Code for Construction and Acceptance of Converter Valve in Converter Station at ±800 kV and Below
6	GB 50776–2012	Code for Construction and Acceptance of Converter Transformer in Converter Station at ±800 kV and Below
7	GB 50777–2012	Code for Construction and Acceptance of Frame Works in Converter Station at ±800 kV and Below
II Industrial standards		
1	DL/T 251–2012	Maintenance Code for ±800 kV DC Overhead Transmission Line
2	DL/T 273–2012	Preventive Test Code for ±800 kV HVDC Electric Equipments
3	DL/T 274–2012	Hand-Over Test of ±800 kV HVDC Electric Equipments
4	DL/T 275–2012	Limits of Electromagnetic Environment at ±800 kV UHV DC Converter Station
5	DL/T 399–2010	Supervision Guide for Main Electric Equipment of ±800 kV and Under DC Transmission Projects
6	DL/T 1131–2009	System Test Standard for ±800 kV HVDC Project
7	DL/T 1242–2013	Technical Specification for Live Working on ±800 kV DC Transmission Line
8	DL/T 5231–2010	Specification for Construction and Acceptance of Earthing Pole in ±800 kV and Under DC Transmission
9	DL/T 5232–2010	Specification for Construction and Acceptance of Electric Equipments in ±800 kV and Under Converter Station
10	DL/T 5233–2010	Specification for Construction Quality Checkout and Acceptance of Electric Equipments in ±800 kV and Under DC Converter Station
11	DL/T 5234–2010	Specification for Starting and Acceptance of ±800 kV and Under DC Transmission Project
12	DL/T 5235–2010	Code for Construction and Acceptance of ±800 kV and Under DC Overhead Transmission Line
13	DL/T 5236–2010	Specification for Construction Quality Checkout and Evaluation of ±800 kV and Under DC Overhead Transmission Line
14	DL/T 5275–2012	Code for Construction Quality Checkout and Evaluation of Earth Electrode in DC Transmission System at ±800 kV and Below

(*Continued*)

Table E.2: (Continued)

No.	Standard No.	Designation of Standard
15	DL/T 5276–2012	Construction Technology Guide for the Assembling of Busbar and Jumper in Converter Station at ±800 kV and Below
16	DL/T 5286–2013	Construction Technology Guide for Tension Stringing of ±800 kV Overhead Transmission Line
17	DL/T 5287–2013	Construction Technology Guide for the Assembly and Erection of the Steel Towers of ±800 kV Overhead Transmission Line
18	DL/T 5426–2009	System Design Standard for ±800 kV HVDC System
III Corporate standards of SGCC		
1	Q/GDW 144–2006	Guide for Overvoltage Protection and Insulation Coordination of ±800 kV UHV DC Converter Station
2	Q/GDW 145–2006	Control Level of the Electromagnetic Environment of ±800 kV DC Overhead Transmission Line
3	Q/GDW 146–2006	Technical Guide for Reactive Power Compensation and Allocation of HVDC Converter Stations
4	Q/GDW 147–2006	General Specification for ±800 kV UHV DC Converter Transformers
5	Q/GDW 148–2006	General Specification for Smoothing Reactors of ±800 kV UHV DC Applications—Part 1: Oil-Immersed Smoothing Reactors
6	Q/GDW 149–2006	General Specification for ±800 kV UHV DC Smoothing Reactors—Part 2: Dry-Type Smoothing Reactors
7	Q/GDW 150-2006	General Specification for ±800 kV UHV DC Wall Bushings
8	Q/GDW 151–2006	Disc Suspension Composite Sheds Insulator String for HVDC Transmission Lines Applications
9	Q/GDW 217.1–2008	Code for Construction Quality Checkout and Evaluation of ±800 kV Converter Station—Part 1: General Rules
10	Q/GDW 217.2–2008	Code for Construction Quality Checkout and Evaluation of ±800 kV Converter Station—Part 2: Converter Valve
11	Q/GDW 217.3–2008	Code for Construction Quality Checkout and Evaluation of ±800 kV Converter Station—Part 3: Converter Transformer
12	Q/GDW 217.4–2008	Code for Construction Quality Checkout and Evaluation of ±800 kV Converter Station—Part 4: HVDC Equipment
13	Q/GDW 217.5–2008	Code for Construction Quality Checkout and Evaluation of ±800 kV Converter Station—Part 5: AC Harmonic Filter
14	Q/GDW 217.6–2008	Code for Construction Quality Checkout and Evaluation of ±800 kV Converter Station—Part 6: Busbar
15	Q/GDW 217.7–2008	Code for Construction Quality Checkout and Evaluation of ±800 kV Converter Station—Part 7: Electric Panel and Cabinet, Control Wiring Diagram
16	Q/GDW 217.8–2008	Code for Construction Quality Checkout and Evaluation of ±800 kV Converter Station—Part 8: Valve Hall

(Continued)

Table E.2: (Continued)

No.	Standard No.	Designation of Standard
17	Q/GDW 218–2008	Code for Construction and Acceptance of Valve Hall in ±800 kV Converter Station
18	Q/GDW 219–2008	Code for Construction and Acceptance of HVDC Equipment in ±800 kV Converter Station
19	Q/GDW 220–2008	Code for Construction and Acceptance of Converter Transformer in ±800 kV Converter Station
20	Q/GDW 221–2008	Code for Construction and Acceptance of Converter Valve in ±800 kV Converter Station
21	Q/GDW 222–2008	Code for Construction and Acceptance of AC Harmonic Filter in ±800 kV Converter Station
22	Q/GDW 223–2008	Code for Construction and Acceptance of Busbar in ±800 kV Converter Station
23	Q/GDW 224–2008	Code for Construction and Acceptance of Electric Panel and Cabinet, Control Wiring Diagram in ±800 kV Converter Station
24	Q/GDW 225–2008	Code for Construction and Acceptance of ±800 kV Overhead Transmission Line
25	Q/GDW 226–2008	Code for Construction Quality Checkout and Evaluation of ±800 kV Overhead Transmission Line
26	Q/GDW 227–2008	Code for Construction and Acceptance of Earth Electrode of ±800 kV DC Transmission System
27	Q/GDW 228–2008	Code for Construction Quality Checkout and Evaluation of Earth Electrode of ±800 kV DC Transmission System
28	Q/GDW 229–2008	Code for Construction and Acceptance of Overhead Earth Electrode Line of ±800 kV DC Transmission System
29	Q/GDW 230–2008	Code for Construction Quality Checkout and Evaluation of Overhead Earth Electrode Line of ±800 kV DC Transmission System
30	Q/GDW 253–2009	General Specification for ±800 kV UHV DC Transmission Project System Design
31	Q/GDW 254–2009	System Test Standard for ±800 kV UHV DC Project
32	Q/GDW 255–2009	Construction Technology Guide to the Main Equipment Installation of ±800 kV UHV DC Converter Station
33	Q/GDW 256–2009	Guide for Construction of Lattice Frame and Support for ±800 kV DC Substation
34	Q/GDW 257–2009	Guide for Construction of Busbar and Jumper Wiring of ±800 kV UHV DC Substation
35	Q/GDW 258–2009	Guide for from Starting to Accomplishment Checking of ±800 kV and Lower Level HVDC Transmission Project
36	Q/GDW 259–2009	General Specification for Sensors of ±800 kV UHV DC Applications

(*Continued*)

Table E.2: (Continued)

No.	Standard No.	Designation of Standard
37	Q/GDW 260–2009	Construction Technology Guide for Tension Stringing of ± 800 kV Overhead Transmission Line
38	Q/GDW 261–2009	General Specification for ± 800 kV UHV DC 6 Inch Thyristors
39	Q/GDW 262–2009	Construction Technology Guide to the Assembly and Erection of the Steel Towers of ± 800 kV Overhead Transmission Line
40	Q/GDW 263–2009	Guide of Supervision for ± 800 kV UHV DC Electrical Equipments
41	Q/GDW 264–2009	Standard for Hand-Over Tests of Secondary Circuit Equipment of Converter Stations of ± 800 kV UHV DC Transmission Project
42	Q/GDW 265–2009	Technical Specification for Control and Protection Equipment of ± 800 kV Ultra-High-Voltage Direct Current (UHV DC) Transmission System
43	Q/GDW 275–2009	Hand-Over Test of Electric Equipment for ± 800 kV HVDC System
44	Q/GDW 276–2009	Technical Specification of Metal Oxide Arrester for ± 800 kV UHV DC Converter Stations
45	Q/GDW 277–2009	Limits of Electromagnetic Environment at ± 800 kV UHV DC Converter Station
46	Q/GDW 279–2009	Technical Specification of ± 800 kV DC Post Insulators
47	Q/GDW 280–2009	Technical Specification of the Cap and Pin Insulators for ± 800 kV DC Systems
48	Q/GDW 281–2009	Technical Specification for DC Bushing of ± 800 kV Converter Station
49	Q/GDW 282–2009	Technical Specification for ± 800 kV DC Long Rod Composite Insulators
50	Q/GDW 283–2009	Measuring Method of Total Electric Field and Ion Current Density Under DC Lines and at Converter Stations
51	Q/GDW 287–2009	The Anti-Jamming Requirement for the Secondary Equipment in the ± 800 kV DC Converter Station
52	Q/GDW 288–2009	General Specification for ± 800 kV UHV DC Converter Valves
53	Q/GDW 289–2009	Technology Specification of Direct-Current Disconnectors and Earthing Switches for ± 800 kV Converter Station
54	Q/GDW 293–2009	Technical Rule for Designing ± 800 kV Converter Station
55	Q/GDW 296–2009	Design Code for 110–500 kV Overhead Transmission Lines
56	Q/GDW 299–2009	Preventive Test Code for ± 800 kV UHV DC Equipment
57	Q/GDW 300–2009	Measurement Method for Radio Interference of DC Equipments
58	Q/GDW 301–2009	Icing Flashover Test Methods on High-Voltage Insulators to be Used in DC System
59	Q/GDW 302–2009	Technical Guide for Live Working in ± 800 kV DC Transmission Lines
60	Q/GDW 332–2009	Code of Operation for ± 800 kV DC Overhead Transmission Line
61	Q/GDW 333–2009	Code of Operation for ± 800 kV UHV DC Substation
62	Q/GDW 334–2009	Maintenance Code for ± 800 kV DC Overhead Transmission Line

(*Continued*)

Table E.2: (Continued)

No.	Standard No.	Designation of Standard
63	Q/GDW 335–2009	Specification of Maintenance Management for ±800 kV UHV DC Converter Station
64	Q/GDW 336–2009	Specification of Equipment Maintenance for ±800 kV UHV DC Converter Station
65	Q/GDW 700.1–2012	Control and Protection Equipment of ±800 kV Ultra High-Voltage Direct Current Transmission—Part 1: General
66	Q/GDW 700.2–2012	Control and Protection Equipment of ±800 kV Ultra High-Voltage Direct Current Transmission—Part 2: Operator Control System
67	Q/GDW 700.3–2012	Control and Protection Equipment of ±800 kV Ultra High-Voltage Direct Current Transmission—Part 3: AC and DC Station Control Equipment
68	Q/GDW 700.4–2012	Control and Protection Equipment of ±800 kV Ultra High-Voltage Direct Current Transmission—Part 4: DC Pole Control Equipment
69	Q/GDW 700.5–2012	Control and Protection Equipment of ±800 kV Ultra High-Voltage Direct Current Transmission—Part 5: DC Protection Equipment
70	Q/GDW 700.6–2012	Control and Protection Equipment of ±800 kV Ultra High-Voltage Direct Current Transmission—Part 6: Auxiliary Secondary Equipment in Converter Station

Afterword

The living environment for humans is going through profound changes, including climate warming, glaciers melting, energy shortage, and ecological deterioration in the twenty-first century. In contrast, the human beings are pursuing higher and higher material standard of living. This conflict brings serious challenges to social development.

In China, the conflict between fast economic growth and depleting fossil energy as well as the deteriorating environment is even more serious. These facts push me to think about the transformation of energy mix and grid development in China. Despite all the differences in economic development, energy endowment, and grid development among countries all over the world, they have the same requirement: for the grid to be strong and smart. Building advanced large or nationwide power systems and planning the European super grid are aimed at building strong and smart grids that ensure a secure, quality, efficient, and clean power supply. To solve the energy shortage of load centers in east and central China, reduce pollution, develop new energy sources, and achieve energy mix transformation, it is necessary to establish a strong and smart grid with ultra-high-voltage (UHV) networks as the backbone. To accommodate hydro, wind, solar, nuclear, biomass, tidal, and other primary energy sources in a complementary, coordinated, and reasonable way, we should bridge the coal, wind, solar, hydro, nuclear, and other large energy bases to the load centers by UHV networks. In addition, UHV networks will also act as a powerful energy conversion and resource allocation platform. Only by transmitting bulk electricity efficiently through UHV projects over long distances and optimizing energy allocation on a larger scale can we ensure secure, efficient, and clean power supply to meet the diverse needs of customers and guarantee economic and social development.

With the world's increasingly acute supply–demand imbalance in energy and the profound change of energy mix, UHV grids will play an even more important role in constructing nationwide, cross-region, cross-border, or even cross-continent interconnections.

Constructing UHV AC and DC grids is a significant and innovative project. The success of China's UHV transmission projects is attributable to the strong government leadership, support of the society, and the hard work of scientific workers and SGCC employees. Looking back on the hardships in the past decade, my heart is full of appreciation. It is my obligation and aspiration to summarize the technical achievements and innovative practices in the construction of UHV grids. I hope this book benefits the development of China's UHV grids and all the readers. I am sure there are imperfections in this book that need improvement, and your criticism and corrections are welcome.

During the writing of this book, some research institutes provided copious data and some of my colleagues at SGCC were of great help in collecting and collating data. Also, many experts made valuable recommendations and comments. The exchanges with international peers regarding grid technology expanded our vision and insight about UHV grids. Heartfelt thanks are extended to all these contributors!

The Author

December 2013

Index

Note: Page number followed by "*f*" and "*t*" refer to figures and tables, respectively.

F

G

J

L

M

N

Kolibri 360 GmbH
Ferdinand-Jühlke-Straße 7
99095 Erfurt, Deutschland
produktsicherheit@kolibri360.de